CONFERENCE PROCEEDINGS NO. **235**

AMERICAN VACUUM SOCIETY SERIES 11

SERIES EDITOR: GERALD LUCOVSKY
NORTH CAROLINA STATE UNIVERSITY

PHYSICS AND CHEMISTRY OF MERCURY CADMIUM TELLURIDE AND NOVEL IR DETECTOR MATERIALS

SAN FRANCISCO, CA 1990

EDITOR:

DAVID G. SEILER
NATIONAL INSTITUTE OF
STANDARDS AND TECHNOLOGY

AIP
American Institute of Physics

New York

Authorization to photocopy items for internal or personal use, beyond the free copying permitted under the 1978 US Copyright Law (see statement below), is granted by the American Institute of Physics for users registered with the Copyright Clearance Center (CCC) Transactional Reporting Service, provided that the base fee of $2.00 per copy is paid directly to CCC, 27 Congress St., Salem, MA 01970. For those organizations that have been granted a photocopy license by CCC, a separate system of payment has been arranged. The fee code for users of the Transactional Reporting Service is: 0094-243X/87 $2.00.

© 1991 American Institute of Physics.

Individual readers of this volume and non-profit libraries, acting for them, are permitted to make fair use of the material in it, such as copying an article for use in teaching or research. Permission is granted to quote from this volume in scientific work with the customary acknowledgment of the source. To reprint a figure, table or other excerpt requires the consent of one of the original authors and notification to AIP. Republication or systematic or multiple reproduction of any material in this volume is permitted only under license from AIP. Address inquiries to Series Editor, AIP Conference Proceedings, AIP, 335 E. 45th St., New York, NY 10017.

L.C. Catalog Card No. 91-55493
ISBN 0-88318-931-3
DOE CONF 9010359

Printed in the United States of America.

Contents

Proceedings of the 1990 U.S. Workshop on the Physics and Chemistry of Mercury Cadmium Telluride and Novel Infrared Detector Materials

Preface ix

Program Committee x

Keynote Address

Mercury cadmium telluride and related compounds: The last ten years and the next ten years
    Charles F. Freeman 1613

Growth I

Diffusion mechanisms in mercury cadmium telluride
    D. A. Stevenson and M-F. S. Tang 1615

Selective annealing for the planar processing of HgCdTe devices
    K. K. Parat, H. Ehsani, I. B. Bhat, and S. K. Ghandhi 1625

The metalorganic chemical vapor deposition growth of HgCdTe on GaAs at 300 °C using diisopropyltelluride
    R. Korenstein, P. Hallock, and B. MacLeod 1630

The effect of growth orientation on the morphology, composition, and growth rate of mercury cadmium telluride layers grown by metalorganic vapor phase epitaxy
    G. Cinader, A. Raizman, and A. Sher 1634

Growth and carrier concentration control of $Hg_{1-x}Cd_xTe$ heterostructures using isothermal vapor phase epitaxy and vapor phase epitaxy techniques
    S. B. Lee, D. Kim, and D. A. Stevenson 1639

Growth II

Dislocation density reduction by thermal annealing of HgCdTe epilayers grown by molecular beam epitaxy on GaAs substrates
    J. M. Arias, M. Zandian, S. H. Shin, W. V. McLevige, J. G. Pasko, and R. E. DeWames 1646

Molecular-beam epitaxy of CdTe on large area Si(100)
    R. Sporken, M. D. Lange, J. P. Faurie, and J. Petruzzello 1651

Characterization of CdTe, HgTe, and $Hg_{1-x}Cd_xTe$ grown by chemical beam epitaxy
    B. K. Wagner, D. Rajavel, R. G. Benz, II, and C. J. Summers 1656

Low-temperature growth of midwavelength infrared liquid phase epitaxy HgCdTe on sapphire
    S. Johnston, E. R. Blazejewski, J. Bajaj, J. S. Chen, L. Bubulac, and G. Williams 1661

Doping I

A review of impurity behavior in bulk and epitaxial $Hg_{1-x}Cd_xTe$
    P. Capper 1667

Impurities and metalorganic chemical-vapor deposition growth of mercury cadmium telluride
    B. C. Easton, C. D. Maxey, P. A. C. Whiffin, J. A. Roberts, I. G. Gale, F. Grainger, and P. Capper 1682

The growth and properties of In-doped metalorganic vapor phase epitaxy interdiffused multilayer process (HgCd)Te
    J. S. Gough, M. R. Houlton, S. J. C. Irvine, N. Shaw, M. L. Young, and M. G. Astles 1687

Arsenic doping in metalorganic chemical vapor deposition $Hg_{1-x}Cd_xTe$ using tertiarybutylarsine and diethylarsine
    D. D. Edwall, J.-S. Chen, and L. O. Bubulac 1691

Doping II

Dynamics of arsenic diffusion in metalorganic chemical vapor deposition HgCdTe on GaAs/Si substrates
    L. O. Bubulac, D. D. Edwall, and C. R. Viswanathan 1695

Extrinsic p-doped HgCdTe grown by direct alloy growth organometallic epitaxy
    N. R. Taskar, I. B. Bhat, K. K. Parat, S. K. Ghandhi, and G. J. Scilla 1705

Determination of acceptor densities in p-type $Hg_{1-x}Cd_xTe$ by thermoelectric measurements
    J. Baars, D. Brink, and J. Ziegler 1709

Devices I

Mechanisms of incorporation of donor and acceptor dopants in (Hg,Cd)Te alloys
H. R. Vydyanath 1716

Growth and characterization of P-on-n HgCdTe liquid-phase epitaxy heterojunction material for 11-18 μm applications
G. N. Pultz, Peter W. Norton, E. Eric Krueger, and M. B. Reine 1724

Improved breakdown voltage in molecular beam epitaxy HgCdTe heterostructures
R. J. Koestner, M. W. Goodwin, and H. F. Schaake 1731

Review of the status of computational solid-state physics
A. Sher, M. van Schilfgaarde, and M. A. Berding 1738

Devices II

Mercury cadmium telluride junctions grown by liquid phase epitaxy
C. C. Wang 1740

HgZnTe for very long wavelength infrared applications
E. A. Patten, M. H. Kalisher, G. R. Chapman, J. M. Fulton, C. Y. Huang, P. R. Norton, M. Ray, and S. Sen 1746

A high quantum efficiency in situ doped mid-wavelength infrared p-on-n homojunction superlattice detector grown by photoassisted molecular-beam epitaxy
K. A. Harris, T. H. Myers, R. W. Yanka, L. M. Mohnkern, and N. Otsuka 1752

Subband spectroscopy and steady-state measurement of dark current in metal–insulator–semiconductor devices
R. A. Schiebel 1759

Dark current processes in thinned p-type HgCdTe
D. K. Blanks 1764

Properties of Schottky diodes on n-type $Hg_{1-x}Cd_xTe$
Patrick W. Leech and Martyn H. Kibel 1770

Uncooled 10.6 μm mercury manganese telluride photoelectromagnetic infrared detectors
P. Becla, N. Grudzien, and J. Piotrowski 1777

Devices III

Novel device concept for silicon based infrared detectors
G. Scott, D. E. Mercer, and C. R. Helms 1781

Characterization of PbTe/p-Si and SnTe/p-Si heterostructures
G. Scott and C. R. Helms 1785

Long-wavelength infrared detection in a photovoltaic-type superlattice structure
Byungsung O, J.-W. Choe, M. H. Francombe, K. M. S. V. Bandara, E. Sorar, D. D. Coon, Y. F. Lin, and W. J. Takei 1789

High-efficiency infrared light emitting diodes made in liquid phase epitaxy and molecular beam epitaxy HgCdTe layers
P. Bouchut, G. Destefanis, J. P. Chamonal, A. Million, B. Pelliciari, and J. Piaguet 1794

Superlattices

Electrical properties of modulation-doped HgTe–CdTe superlattices
S. Hwang, Y. Lansari, Z. Yang, J. W. Cook, Jr., and J. F. Schetzina 1799

Optical and magneto-optic properties of HgTe/CdTe superlattices in the inverted-band semiconducting regime
Z. Yang, Z. Yu, Y. Lansari, J. W. Cook, Jr., and J. F. Schetzina 1805

Magneto-optical transitions between subbands with different quantum numbers in narrow gap HgTe–CdTe superlattices
H. Luo, G. L. Yang, J. K. Furdyna, and L. R. Ram-Mohan 1809

Shubnikov–de Haas oscillations and quantum Hall effect in modulation-doped HgTe–CdTe superlattices
C. A. Hoffman, J. R. Meyer, D. J. Arnold, F. J. Bartoli, Y. Lansari, J. W. Cook, Jr., and J. F. Schetzina 1813

Theory for electron and hole transport in HgTe–CdTe superlattices
J. R. Meyer, D. J. Arnold, C. A. Hoffman, F. J. Bartoli, and L. R. Ram-Mohan 1818

Defects I

Minority carrier lifetimes of metalorganic chemical vapor deposition long-wavelength infrared HgCdTe on GaAs
R. Zucca, D. D. Edwall, J. S. Chen, S. L. Johnston, and C. R. Younger 1823

Trapping effects in HgCdTe
Y. Nemirovsky, R. Fastow, M. Meyassed, and A. Unikovsky 1829

Defects II

Correlation of HgCdTe epilayer defects with underlying substrate defects by synchrotron x-ray topography
B. E. Dean, C. J. Johnson, S. C. McDevitt, G. T. Neugebauer, J. L. Sepich, R. C. Dobbyn, M. Kuriyama, J. Ellsworth, H. R. Vydyanath, and J. J. Kennedy 1840

Photoexcited hot electron relaxation processes in n-HgCdTe through impact ionization into traps
D. G. Seiler, J. R. Lowney, C. L. Littler, I. T. Yoon, and M. R. Loloee 1847

Dislocation density variations in HgCdTe films grown by dipping liquid phase epitaxy: Effects on metal–insulator–semiconductor properties
D. Chandra, J. H. Tregilgas, and M. W. Goodwin 1852

Surfaces and Miscellaneous

Surface energies for molecular beam epitaxy growth of HgTe and CdTe
M. A. Berding, Srinivasan Krishnamurthy, and A. Sher 1858

Structural characterization of the (111) surfaces of CdZnTe and HgCdTe epilayers by x-ray photoelectron diffraction
M. Seelmann-Eggebert and H. J. Richter 1861

X-ray photoelectron diffraction from the HgCdTe(111) surface
G. S. Herman, D. J. Friedman, T. T. Tran, C. S. Fadley, G. Granozzi, G. A. Rizzi, J. Osterwalder, and S. Bernardi 1870

Electrochemical approaches to cleaning, reconstruction, passivation, and characterization of the HgCdTe surface
S. Menezes, W. V. McLevige, E. R. Blazejewski, W. E. Tennant, and J. P. Ziegler 1874

Composition, growth mechanism, and stability of anodic fluoride films on $Hg_{1-x}Cd_xTe$
Eliezer Weiss and C. R. Helms 1879

Study of temperature dependent structural changes in molecular-beam epitaxy grown $Hg_{1-x}Cd_xTe$ by x-ray lattice parameter measurements and extended x-ray absorption fine structure
D. Di Marzio, M. B. Lee, J. DeCarlo, A. Gibaud, and S. M. Heald 1886

Characterization of CdTe, (Cd,Zn)Te, and Cd(Te,Se) single crystals by transmission electron microscopy
R. S. Rai, S. Mahajan, S. McDevitt, and C. J. Johnson 1892

Optical techniques for composition measurement of bulk and thin-film $Cd_{1-y}Zn_yTe$
S. M. Johnson, S. Sen, W. H. Konkel, and M. H. Kalisher 1897

Author Index 1902

NOTE: These page numbers reflect the pagination of the proceedings as they appear in the special issue of *JVSTB* 9(3), May/Jun 1991.

Preface

The 1990 U.S. Workshop on the Physics and Chemistry of Mercury Cadmium Telluride and Novel Infrared Detector Materials was held in San Francisco, California during October 2–4, 1990. The scientific and technological interest in mercury cadmium telluride (MCT) materials and their related devices continues since the first workshop was initiated in 1981. The last few years have seen HgCdTe technology broaden and mature. New growth techniques, greater materials understanding, and increasingly diverse applications have marked this change.

The Workshop plays a vital role in this technological evolution. It provides the principal open forum for the exchange of information relative to theory and experiment, synthesis, and analysis. It brings together university, governmental, and industrial research in a highly interactive manner. The Workshop focuses on fundamental research on the major scientific problems in HgCdTe and related materials systems. Its primary goal is to promote an understanding of the relationship among the physical and chemical properties. Specific subject areas included: material growth, defects and impurities, mechanical and structural properties, electronic properties, surfaces and interfaces, and novel materials and structures.

This volume of papers is the proceedings of the ninth workshop. Over 200 scientists and engineers from universities, industry, and government agencies attended the meeting. Of the 57 invited and contributed papers presented during the three days of the workshop, 49 are in this volume! Participants for a panel discussion on "Growth-Related Defects in MCT" included: R. L. Aggarwal, MIT and E. R. Gertner, Rockwell (moderators); J. H. Tregilgas, Texas Instruments and N. Duy, SAT, France (Bulk MCT); B. F. Pelliciari, LETI, France and T. Tung, SBRC (LPE MCT); D. D. Edwall, Rockwell and B. Easton, Phillips, United Kingdom (MOCVD); and R. J. Koestner, Texas Instruments and T. Meyers, General Electric Company (MBE). The many informal discussions between participants helped to make the workshop successful.

Many individuals and institutions have made important contributions to the success of the Workshop. Special thanks are due to Dr. R. L. Aggarwal and Dr. E. R. Gertner, co-chairman of the Workshop. Significant contributions were made by the following institutions: National Institute of Standards and Technology, Rockwell International, Santa Barbara Research Center, and Texas Instruments. The following individuals made up the Program Committee or were Government Advisors: P. M. Amirtharaj, R. L. Denison, J. P. Faurie, J. K. Furdyna, S. K. Ghandhi, C. R. Helms, P. R. Norton, P. W. Norton, H. F. Schaake, J. F. Schetzina, D. G. Seiler, A. Sher, C. J. Summers, J. R. Waterman, and H. Wittmann. The work of Jay Morreale and Mark Goldfarb and their colleagues at Palisades Institute for Research Services is acknowledged for providing efficient and smooth coordination of the Workshop. The editor wishes to especially thank the Semiconductor Electronics Division of the National Institute of Standards and Technology for the resources provided during the editing of these Proceedings, and Ms. Jane Walters for her expert and tireless assistance with the manuscripts.

The Workshop was held under joint sponsorship of the CECOM Center for NV and EO, the Office of Naval Technology, the Air Force Office of Scientific Research, and the Electronic Materials and Processing Division of the American Vacuum Society.

The editor would also like to thank the numerous referees who diligently assisted in reviewing the manuscripts published here.

David G. Seiler
Proceedings Editor

PROGRAM CO-CHAIRMEN

R. L. Aggarwal
Massachusetts Institute of Technology

E. R. Gertner
Rockwell International

PROGRAM COMMITTEE

J. P. Faurie
University of Illinois

J. K. Furdyna
Notre Dame University

S. K. Ghandhi
Rensselaer Polytechnic Institute

C. R. Helms
Stanford University

P. R. Norton
Santa Barbara Research Center

P. W. Norton
Loral IR and Imaging Systems

H. F. Schaake
Texas Instruments

J. F. Schetzina
North Carolina State University

A. Sher
SRI International

C. J. Summers
Georgia Institute of Technology

GOVERNMENT ADVISORS

P. M. Amirtharaj
CECOM Center for Night Vision
and Electro-Optics

R. L. Denison
Air Force Wright Aeronautical
Laboratories

D. G. Seiler, Proceedings Editor
National Institute of Standards
and Technology

J. R. Waterman
Naval Research Laboratory

H. Wittmann
Air Force Office of
Scientific Research

WORKSHOP COORDINATOR

J. Morreale
Palisades Institute for
Research Services, Inc.

Mercury cadmium telluride and related compounds: The last ten years and the next ten years

Charles F. Freeman
CECOM Center for Night & Electro-Optics, Fort Belvoir, Virginia 22060

(Received 19 December 1990; accepted 22 January 1991)

The series of workshops on the Physics and Chemistry of Mercury Cadmium Telluride (MCT) and related compounds has been in existence for about ten years now. It is worthwhile to take a look at its origins; the progress in these past ten years and some projections for the next ten years as we prepare to enter the 21st century.

In the late 1950's there was a search for an infrared (IR) sensitive material that not only had a good response in the 8–12 μM spectral region but had the capability of photon shot noise limited behavior at 77 K operation. This translated into a material that had a direct band gap of about 0.1 eV and could operate as an intrinsic device. In 1959, Lawson, Nielson, Putley, and Young of the Royal Research Establishment in Malvern, U.K. published a paper on the alloy system $Hg_{1-x}Cd_xTe$[1] that showed continuously varying direct gap semiconducting characteristics through the full alloy range; E_g from <0 eV (semimetal) to 1.6 eV. As a direct gap intrinsic material it showed high quantum efficiency. Because the bandgap could be tailored, the highest operating temperature for any desired wavelength was possible.

Over the years since then, most of the effort on Mercury Cadmium Telluride resided with the government and with industry. Because of its properties, it had become the material of choice for IR sensing, particularly for the 8–12 μM spectral region. As a consequence of this, there was a desire to increase participation of the university community with the government and industrial communities that led to formulation of the Workshop on the Chemistry and Physics of Mercury Cadmium Telluride in 1980 (The program committee consisted of T. Casselman, W. Spicer, R. Aggarwal, D. Cheung, T. McGill, A. Nurmikko, P. Raccah, K. Riley and G. Roberts).

Ten years ago, the state-of-the-art for MCT IR sensors was based upon bulk growth of crystals of this alloy system. Photoconductivity was the baseline principle for these sensors.

Experiments in epitaxial growth on cadmium telluride substrates were in their infancy. In the case of equilibrium growth, both liquid phase epitaxy and close space vapor epitaxy were being pursued. The latter was dropped because of the difficulty in obtaining a sufficiently small x gradient throughout the grown absorber layer. In the case of nonequilibrium growth, both metalorganic chemical vapor deposition (MOCVD) and molecular beam epitaxy (MBE) were getting started. The major issue for both MOCVD and MBE has been a problem in how to handle the Mercury. Brian Mullen addressed the problem with MOCVD by instituting the IMP process to decouple the mercury deposition from the other growth constituents.[2] In the case of MBE, both Jean Paul Faurie and Jan Schetzina addressed the mercury sticking coefficient issue and were the major contributors to the design of MBE facilities for mercury based alloy systems.

IR Sensors based upon the principles of photodiodes and photocapacitors were starting. We were just beginning to get some insight to the defect structure of mercury cadmium telluride and the role that the weak mercury bond plays in this understanding. I have never forgotten a point that Bill Spicer made to me at an early DARPA review regarding the direct communication between bulk and the surface. It said to me that there should be minimal stress in regions of high mercury content, be it chemical, mechanical, electrical or other environmentally induced stress, and that the design of such sensors and the processes selected in realizing them must be acutely sensitive to this issue.

In the last ten years since the founding of the workshop series we have seen steady, continuous progress. Liquid phase epitaxy has become the baseline growth method. The cadmium telluride substrates used in epitaxial growth have steadily improved in quality and increased in size. The addition of zinc to the CdTe substrate, initially carried out at NVEOL to reduce/eliminate the lattice mismatch with the mercury cadmium telluride, showed an immediate benefit in improvement in the surface morphology of the grown layer and its sensitivity to misorientation of the substrate surface. An added benefit to this addition of zinc has been the increased strength of the normally fragile substrate wafers, a point that has been substantiated theoretically by Arden Sher and his co-workers.[3] This paved the way for further substrate quality and size advances. Alternate options for the substrates have been the use of Sapphire, GaAs, and Silicon. The general rational for these options has been the availability of large, quality wafers at lower costs. Work is continuing on these options. The issues and/or limitations are different for each option and all the answers are not yet in. In the area of Passivation, SiO_2 with or without a ZnS layer, anodic sulfide and the growth of CdTe (or CdZnTe) on the MCT surface have been used. Each has its own advocates. A very significant advance has been the growth of heterostructures, particularly heterojunctions, and the doping control that this entails. Work on superlattices by MBE has been started. There has been steady progress in the area of defect reduction/purity improvements, careful annealing processes, understanding of clustering of defects/impurities and a growing understanding of the relationship of defects to $1/f$ noise, uniformity and long term stability. There is more to be done.

In the next ten years, we can expect improved tools both in processing and characterization. These tools will provide the control necessary to reproducibly carry out experiments that lead to a better understanding of this material system. As a result of the processing tool improvements, nonequilibrium growth will be the baseline. This includes MOCVD, MBE and their offspring, MOMBE. We will see significant advances in heterostructures including superlattices. We will also see an emergence of patterned growth in situ.

In the area of characterization improvements, we can expect to have the ability to conduct characterization during processing. This will be made possible because nonequilibrium growth methods generally have the growth surface exposed to the characterization probes. With the advances in nonintrusive characterization techniques, the growth process should be minimally perturbed by such probes. We can expect steady improvements in resolution of important parameters, such as trap levels or defect/impurity clusters in the material being grown, and in sensitivity of measurements such as impurity concentrations of various species, both intentional and unintentional.

The ability to characterize the growth surface in this way will provide the key to control of the growth in real time. This ability and the reproducibility of experiments that it affords, will enhance the theoretical correlation to experiment and vice versa that advances in understanding will require.

There is a challenge to the unique position of the HgCdTe material system for IR sensing. This challenge comes from the options afforded by non II–VI superlattice structures (quantum wells, strained layers). There is considerable activity both theoretically and experimentally. How it will fare as a challenge is not known. A number of questions must be resolved regarding, for example, quantum efficiency, dark current, and noise (both temporal and spatial). There is important science to be done here, and we do not fully know where this work will lead. Measurements are crucial before we can make predictions about what these options will offer.

Finally, in the words of a former Director of the Night Vision Laboratory and a good friend of mine, Donald J. Looft, I pass on the thought: "Know what it is that you are trying to do. The rest is easy."

[1] W. D. Lawson, S. Nielson, E. H. Putley, and A. S. Young, J. Phys. Chem. Sol. **9**, 325 (1959).
[2] S. J. C. Irvine, J. Tunnicliffe, and J. B. Mullin, Mater. Lett. B **2**, 304 (1984).
[3] A. Sher, A. B. Chen, W. E. Spicer, and C. K. Shih, J. Vac. Sci. Technol. A **3**, 105 (1985).

Diffusion mechanisms in mercury cadmium telluride

D. A. Stevenson and M-F. S. Tang
Department of Materials Science and Engineering, Stanford University, Stanford, California 94305

(Received 2 October 1990; accepted 18 December 1990)

The key diffusion quantities for component diffusion in mercury cadmium telluride are reviewed: interdiffusion; intrinsic diffusion; and tracer self-diffusion. The interrelation between these quantities is described as well as their relation to practical diffusion problems. A review of the experimental information for these diffusion quantities is presented in the framework of diffusion theory and diffusion mechanisms. In particular, the interrelation between the fundamental diffusion quantities is used to assign diffusion values. The interpretation of marker motion and Kirkendall effect experiments is presented in the context of theory and experimental diffusion information in this system.

I. INTRODUCTION

Semiconductor devices rely on the ability to introduce controlled concentrations of electrically active species at precise locations in a semiconductor chip. For elemental semiconductors, this is achieved by introducing acceptor or donor dopant atoms by growth, diffusion, or ion implantation. Diffusion is of interest not only as a method for introducing dopants, but also for dopant redistribution during subsequent processing steps which require annealing, such as activation of implants, oxidation, and during the growth of heterojunctions. Binary compound semiconductors usually rely upon dopant atoms to achieve desired properties, but there is the additional complication of defects arising from nonstoichiometry. The nonstoichiometry is usually accomodated with vacancies and interstitials of the component atoms, which are referred to as native defects (self-interstitials and self-vacancies) which are usually electrically active. For example, excess metal can be accommodated by metal interstitials or nonmetal vacancies, both of which are donor species. In the common III–V compounds (GaAs and InP), the energies of formation of these native defects are quite large resulting in relatively low concentrations and dopant atoms are the predominant electrically active species. In the II–VI compounds, however, the native defect concentration is appreciable and can compete with the dopant atoms for controlling the electrical properties, leading to the phenomenon of self-compensation which limits many of them to one carrier type. Ternary semiconductors, such as mercury cadmium telluride (MCT; $Hg_{1-x}Cd_xTe$), are of interest since the band gap can be varied by controlling X, the mole fraction of CdTe in the ternary compound. Consequently, the overall composition is a major concern. Because of the difficulties in growing bulk crystals of uniform composition, ternary semiconductors are usually prepared by heteroepitaxial growth methods [such as liquid phase epitaxy (LPE), metalorganic chemical vapor deposition (MOCVD), or molecular-beam epitaxy (MBE)] on a binary semiconductor substrate. The resulting layers are therefore subject to composition change arising from interdiffusion between the layer and the substrate. This concern is particularly important for very thin films and for superlattices.

Native defects, such as mercury vacancies and interstitials, play a dominant role in the electrical properties of MCT. The component ambient pressures during high temperature annealing in a well defined ambient exert a dominant influence on the concentrations of these point defects. Brebrick has shown that for pseudobinary systems such as MCT, the chemical potentials of HgTe and CdTe are independent of the stoichiometry and are only functions of the temperature and the X value.[1] The partial pressure of Cd or Hg for any composition X in MCT is related to the vapor pressure of Te by the chemical potential of the corresponding binary.[2] Therefore, fixing P_{Hg} for a given X value fixes all the other component pressures at the annealing temperature.[1] There are several experimental studies and theoretical calculations that have established the influence of composition (X) and stoichiometry (P_{Hg}) on the concentration of electrically active defects in MCT.[3-12] These results have led to proposed defect equilibria (Fig. 1) and a log P_{Hg} versus $1/T$ "road map" for the electrical properties (Fig. 2). Based on this information, one can establish general trends for defect concentrations with the variation in stoichiometry (P_{Hg}), composition (X), and temperature (T), as is shown schematically in Fig 3.

As a consequence of the dependence of the band gap on

DEFECT CHEMISTRY IMPLIED FROM ELECTRICAL PROPERTY MEASUREMENTS

NATIVE DONOR: Hg_i (mercury interstitial)

NATIVE ACCEPTOR: V_{Hg} (mercury vacancy)

DEFECT EQUILIBRIA:

$Hg_{(v)} = Hg_i^{\bullet\bullet} + 2e'$; $K = [Hg_i^{\bullet\bullet}]n^2/P_{Hg}$

$$[Hg_i^{\bullet\bullet}] \sim P_{Hg}$$

$Hg_{Hg} = V_{Hg}'' + 2h^{\bullet} + Hg_{(v)}$; $K = [V_{Hg}'']p^2P_{Hg}$

$$[V_{Hg}''] \sim P_{Hg}^{-1}$$

FIG. 1. Defect equilibria established from electrical property measurements.

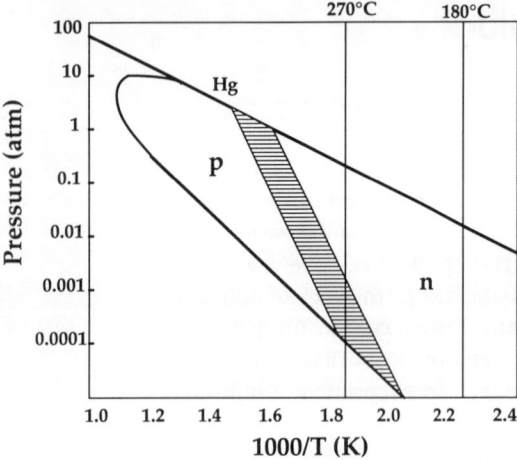

FIG. 2. The mercury pressure-temperature diagram showing the influence of annealing on the electrical properties (after C. Jones et al., Ref. 12).

component composition and the influence of native defects on electrical properties, the practical interest for diffusion in MCT has focused on component diffusion, specifically, the interdiffusion of the components and the change in stoichiometry by annealing in Hg vapor. These processes relate to the diffusion quantities: the interdiffusion coefficient (D); and the intrinsic diffusion coefficient (D_i). These diffusion quantities, along with the tracer self-diffusion coefficient (D_i^*) comprise the three fundamental diffusion quantities for component diffusion. For binary systems, these are interrelated by the Darken Equation and a thermodynamic factor,[13,14] whereas for a ternary system such as MCT, they are interelated by a more complex relation.[15,16] To clarify these relationships, we review the fundamental diffusion quantities and the interrelation between these quantities.

TRENDS ESTABLISHED FROM ELECTRICAL
PROPERTY MEASUREMENTS AND THEORETICAL
CALCULATIONS

STOICHIOMETRY; TEMPERATURE, X VALUE

- HIGHER P_{Hg} FAVORS Hg_i
  LOWER P_{Hg} FAVORS V_{Hg}

- HIGHER T FAVORS V_{Hg} RELATIVE TO Hg_i

- HIGHER X VALUE LOWERS [V_{Hg}] AND FAVORS Hg_i RELATIVE TO V_{Hg}

V_{Hg} <---------------------> Hg_i
                P_{Hg}
           ------------>
                 X
           ------------>
                 T
           <------------

FIG. 3. Trends established from electrical property measurements and theoretical calculations.

II. BASIC DIFFUSION QUANTITIES

The most easily conceptualized diffusion quantity is the tracer self-diffusion coefficient D_i^*. This is measured by first equilibrating the diffusion host with precisely the same ambient conditions as for the diffusion anneal with a nontracer source of components, then replacing this with a tracer source for a given diffusion anneal at a fixed time and temperature. Since the preanneal establishes uniform concentrations of atoms and native defects, there is no thermodynamic driving force other than the ideal entropy of mixing of the isotopic species of the components in question and the completely random walk of these atoms produces a tracer profile. From the resulting tracer profile and the relevant solution to Fick's law for diffusion, a D_i^* value is calculated. Since there is no concentration change throughout the sample during the diffusion anneal, D_i^* is a constant independent of distance and time. In the case of MCT, this is achieved by an initial anneal under well defined mercury overpressure until equilibrium is established and then replacing this mercury ambient with a radio tracer mercury ambient. If the diffusion host were first annealed under different conditions (for example, tellurium-rich conditions), prior to introducing a mercury-rich mercury tracer source, the diffusion would be driven by a thermodynamic driving force in addition to random walk. The resulting diffusion coefficient under these conditions is the intrinsic mercury diffusion coefficient D_{Hg} (or D_i in general). Since the composition is changing with position and time during the anneal, D_i may not be a constant. For a binary system:

$$D_i = D_i^*[\text{TF}], \qquad (1)$$

where TF is the thermodynamic factor defined as

$$[\text{TF}] = \partial \ln a_i / \partial \ln X_i, \qquad (2)$$

where a_i is the thermodynamic activity and X_i the mole fraction of one of the components i.

When there is a sharp junction between two compositions, such as a MCT/CdTe, junction, then intermixing occurs by interdiffusion and the resulting diffusion coefficient is the interdiffusion coefficient D. For a binary system, the Darken Equation relates D, D_i, and D_i^* as follows[13,14]:

$$D = (X_A D_B + X_B D_A) = (X_A D_B^* + X_B D_A^*)[\text{TF}]. \qquad (3)$$

For ternary systems, such as MCT, there are additional constraints on the system, such as the Gibbs–Duhem equation for a ternary system, electrical neutrality, and conservation of lattice sites. This leads to a more complex relation between these three quantities,[15] which, for MCT, can be written:

$$D = \{D_{Hg}^* D_{Cd}^* + D_{Te}^*[XD_{Hg}^* + (1-X)D_{Cd}^*]\}[\text{TF}]/$$
$$[(1-X)D_{Hg}^* + XD_{Cd}^* + D_{Te}^*], \qquad (4)$$

where X is the mole fraction of CdTe in MCT.

In a pseudobinary system such as MCT, the interdiffusion coefficient is related to the intrinsic diffusion coefficients by the same expression as for a binary diffusion couple[15]:

$$D = XD_{Hg} + (1-X)D_{Cd}. \qquad (5)$$

There are two important limiting forms for Eq. (4). When

$D_{Te}^* \ll D_{Cd}^*$, and D_{Hg}^*, the limiting form is a Nernst–Planck type of equation:

$$D = \{D_{Cd}^* D_{Hg}^* / [XD_{Cd}^* + (1-X)D_{Hg}^*]\}[TF]. \quad (6)$$

The Nernst–Planck equation was originally developed for the transport of ionic species, such as in an aqueous liquid junction between two different electrolytes. When one cation diffuses more rapidly than the other, the ionic charge leads to an electrical potential gradient, with the potential gradient retarding the diffusion of the faster cation and enhancing the slower diffusing cation. The interdiffusion is then rate limited by the slower cation and for the case of MCT[16]:

$$D \sim D_{Cd} \sim D_{Hg}. \quad (7)$$

If the intrinsic charge density is large compared to the charge density of the mobile defect, as is the case for MCT at higher temperatures, or if the mobile defects are neutral, then an opposing electric field is not developed. However, there is the constraint of metal to nonmetal stoichiometric ratio that produces a retarding force for the faster diffusing species. As the metal to nonmetal ratio becomes larger on one side of the couple and smaller on the other as a result of the unequal transport of the metal species, a gradient in the chemical potential develops that opposes the initial driving force for Hg. If this did not occur, then the metal and nonmetal composition on the two sides of the couple would exceed the solubility limit and precipitation of Hg-rich and Te-rich phases would occur on the two sides.

If $D_{Te}^* \gg D_{Hg}^* > D_{Cd}^*$, then Te can diffuse along with Hg, and no electric field or disparity in the metal/nonmetal ratio would develop. For this case, one may view the diffusion as a true pseudobinary diffusion couple between HgTe and CdTe species. The limiting form of the general diffusion is the same as for a binary system, namely a Darken type of equation[15]:

$$D = [(1-X)D_{Cd}^* + XD_{Hg}^*][TF]. \quad (8)$$

With these quantities defined, we proceed with a review of some of the experimental work on interdiffusion and self-diffusion in MCT, discussing the extent of agreement between different investigations, the proposed mechanisms, and any unusual features of the experimental results.

III. INTERDIFFUSION

There are several studies of interdiffusion in MCT, as outlined in Table I. Above 450 °C, there is remarkably good agreement between the results of these studies, in spite of the use of different methods and different types of samples. Most of these studies have used CdTe/HgTe diffusion couples, with the diffusion profiles analyzed with the Boltzmann Matano method, or the analysis of the growth kinetics for isothermal vapor phase epitaxial growth (ISOVPE) in the diffusion controlled regime. There are rather well behaved trends at temperatures above 450 °C, which will be discussed initially. Typical behavior is shown in Figs. 4 and 5. The important observed trends are: D is a strong function of X, changing by ~ two orders of magnitude over the entire X range, with higher values for lower X values; D is independent of the ambient Hg pressure; and D is not significantly different for polycrystalline or single crystalline samples or influenced by the orientation for single crystals.

Analytical expressions for $D(X,T)$ are given by several authors, for example[17]:

$$D(\text{cm}^2/\text{s}) = 300\exp(-7.53X)\exp(-1.92\text{eV}/kT)$$
$$(\text{for } X < 0.6 \text{ and } T > 450\,°C), \quad (9)$$

or by an Arrhenius expression with the activation energy a function of X:[20,21]

$$D = D_0 \exp(-\Delta E_a/kT); \quad (\text{with } \Delta E_a = a + bX)$$

For example, for $X < 0.5$ and $T > 450\,°C$:

$$D(X,T) = 1.0\exp[-(1.53 + 0.51X)/kT]. \quad (10)$$

These results are consistent with the dependence of interstitial and vacancy concentrations upon X, as determined by experiment[3–6] and by theory.[11] There is a tendency for higher defect densities in HgTe than in CdTe, due to the lower cohesive energy of HgTe. Thus, one expects D to increase as X decreases, as is observed. The dominant native defects used to model the electrical properties are mercury interstitials (Hg_i) and vacancies (V_{Hg}) as donors and acceptors, respectively. In the temperature range of these earlier diffusion studies, the electrical property road map (Fig. 2) describing the influence of component pressure anneals on the electrical properties, shows a transition from n to p type within the stability blade of the $\ln P_{Hg}$ versus $1/T$ diagram (Fig. 2). This implies that Hg_i species predominate on the Hg-saturated side and V_{Hg} on the Te-saturated side. The independence of D on pressure can then be explained by either of two models: a dual vacancy-interstitial mechanism, with comparable mobilities for both species; or a V_{Hg}–Hg_i complex that is approximately independent of component pressure when there are comparable concentrations of the two native defects. The latter mechanism is proposed for the tracer diffusion of Cd in CdTe, which is independent of the Cd ambient pressure.[18,19]

The trend from p-type to n-type behavior of lower temperatures indicates that interstitals become the dominant native defect at lower temperatures. This indicates two trends: a lower dependence of D on X, since the interstitial defect formation energy is less dependent on X than for the vacancy; and a P_{Hg} dependence developing at lower temperatures. There is interest in interdiffusion at lower temperatures (~ 150–300 °C) in order to predict the sharpness of junctions for MBE and MOCVD heteroepitaxy. Interdiffusion studies have been made in this temperature region for that reason.

Using a junction couple and angle lapping methods to increase the spatial resolution, diffusion measurements were extended down to 300 °C.[20,21] The results, shown in Fig 5, show the following noteworthy trends as the temperature is lowered below 450 °C: (i) the activation energy decreases; (ii) D is less dependent on X; and (iii) D depends on P_{Hg}, increasing with an increase in the Hg overpressure (especially at the higher X values). These results and trends are consistent with the defect models proposed from electrical characterization. At lower temperature, mercury interstitials increase in concentration leading to n-type behavior. They also become the dominant mobile defect, particularly at higher X values where the vacancy formation energy is rela-

TABLE I. Summary of interdiffusion studies.

Sample and preannealing treatment	Temperature	Ambient	Remarks	References
Parallelepipeds of $10\times4\times2$ mm HgTe and CdTe crystals	400–600 °C			Rodot & Henoc (38)
(111) single crystal of Bridgman grown CdTe anneal with HgTe source	400–600 °C	Hg overpressure	A high temperature (~600 °C), D decreases with increasing P_{Hg}, propose vacancy mechanism	Bailly, Cohen, Solal and Marfaing (39,40,41)
Bridgman grown HgTe, CdTe. Cleaved surfaces. Preannealed HgTe, CdTe (metal, nonmetal saturated and P_{min} conditions). Diffusion couple was enclosed in HgTe polycrystalline matrix	500–600 °C	HgTe powder, pretreated under Hg,Te saturated and P_{min} condition	D is independent of P_{Hg}. D varies with dopants (Ag,In). use electron microprobe analysis (EMA) and modified Boltzman–Matano analysis	Leute, Schmidtke, Stradenmann and Winking (42)
(111) CdTe crystals (Bridgman) use $Hg_{1-x}Cd_xTe$ equilibrium source, grow epitaxial layer on CdTe	450–700 °C	Te-saturated MCT Source	EMA to obtain profile. Analyze the kinetics of isothermal growth process to obtain: $D = 300\exp(-7.53X)\exp(-1.92\text{ eV}/kT)$ limited D values: $X=0.1$–0.6, for $T=450$–550 °C $X=0.3$–0.6, $T>550$ °C	Fleming and Stevenson (17)
$Hg_{1-x}Cd_xTe$ multilayer grown by LPE, LPE, MOCVD, ultrahigh vacuum (UHV)	400–600 °C		EMA, Auger electron spectroscopy (AES), x-ray photoelectron spectroscopy (XPS), analyze the profile $D = 3.15\times10^{10}\times10^{-3.53X}\times\exp(-2.24\times10^4/kT)$	Zanio and Massopust (22)
(111) CdTe, HgTe(111) and polycrystalline HgTe to form diffusion couple. Preanneal CdTe in metal and Te saturated condition	300–600 °C	preannealed HgTe powder under Hg and Te saturated condition.	EMA $D = D_0 - \exp(-\Delta E_a/kT)$ $\Delta E_a = a + bX$ $D_0 = 1.0\times\exp[(-1.53+0.51X/kT)]$ for $x<0.5$, $T>450$ °C	Tang and Stevenson (20,21)
MBE grown HgTe–CdTe superlattices	185 °C	He	Assuming D is independent of X. Analyzing the decay intensities of satellite peaks (of x-ray) $\Delta E = 0.9, 0.72, 0.5$ eV	Arch, Staudemann and Faurie (23)
MBE grown MCT ($x=0.27$)/MCT ($x=0.83$) Superlattices	200–265 °C	Ar-filled furnace	Chemical lattice imaging technique. D depends on junction depth (z) $D = D_z\exp(-0.3/kT)\times\exp[-0.5(1-X)/kT]$	Kim, Qurmazd and Feldman (24)

tively higher than for lower X values. This leads to an obvious dependence upon P_{Hg}.

There are other studies of the interdiffusion at lower temperatures. Zanio and Massopust established experimental profiles using electron microprobe, Auger (cross section and depth profiling) and x-ray photoelectron spectroscopy with their results in general agreement with other data.[22] In addition, they described methods to model junction profiles obtained during low temperature heterojunction growth. Arch et al. explored the intensity of x-ray side band peaks to determine the interdiffusion in superlattices at 185 °C.[23] Their results agree with the extrapolation of the data and the diffusion model of Tang and Stevenson.[21] Kim et al.[24] employed chemical lattice imaging mode of high resolution transmission electron microscopy in combination with vector pattern recognition to determine profiles at low temperatures at superlattice junctions. The interdiffusion coefficient was found to depend on the depth beneath the surface; at 200 °C, D

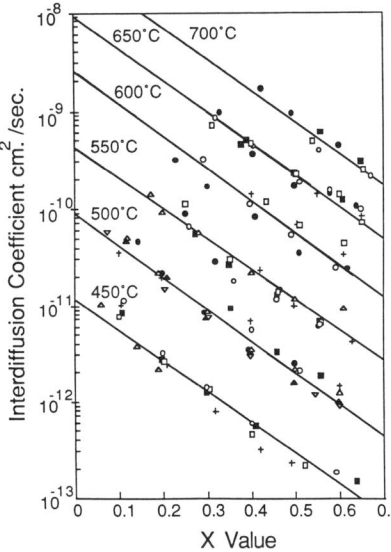

FIG. 4. A plot of the interdiffusion coefficient as a function of X and T (after Fleming and Stevenson, Ref. 17).

increases by two orders of magnitude as the junction depth varies from 7000 to 100 Å.

In summary, at higher temperatures (> 400 °C), there are several studies of D over the entire X range of composition, with generally good agreement between the different studies. An exponential dependence of D on X reflects the lower defect concentrations with increasing X, due to the stronger Cd–Te bonds. D is independent of P_{Hg} over the entire range of composition from Te rich to Hg rich. A mixed vacancy-interstitial or a vacancy-interstitial complex is proposed as the diffusing species. At lower temperatures and higher X values, the activation energy decreases, the dependence on X decreases, and D dependence on P_{Hg} arises, with D increasing with increasing P_{Hg}, and is consistent with a tendency toward metal interstitials as the dominant mobile species.

IV. SELF-DIFFUSION STUDIES

Extensive information on the influence of mercury annealing on the concentration of electrically active native defects has established mercury interstitials and vacancies as the dominant defects at different temperatures and stoichiometry (P_{Hg}). Do these defect models also explain the self-diffusion behavior? For MCT, self-diffusion studies performed as a function of the Hg component pressure provide valuable insight into the dominant mobile defects. Furthermore, self-diffusion coefficients are related to the interdiffusion coefficient using the basic diffusion equation for a ternary diffusion system [Eq. (4)].

For MCT, there are convenient radio tracers for all three components. There are several studies of self-diffusion, mostly for Hg, which are summarized in Table II. An interesting characteristic of most of these studies is the observation of complex diffusion profiles for the metal components, as illustrated in Fig. 6. Typically, there are two branches that are associated with slower and faster diffusion mechanisms ($D^*_{Hg,f}$ and $D^*_{Hg,s}$). (Note: Although D^*_i is a symbol ordinarily used for the true self-diffusion coefficient, we use the symbol initially to signify the tracer diffusion coefficient of species, prior to explicit assignment.) Complex profiles, with fast and slow diffusion branches, may arise from diffusion short circuits (such as subgrain boundaries or dislocations) or incomplete preannealing so that isoconcentration conditions do not exist initially during the diffusion anneal. However, for the MCT system, the multiple branches are observed both with and without preannealing and they are observed for a wide range of samples with different orientation and substructure. Furthermore, the faster component for $D^*_{Hg,f}$ shows a dependence on P_{Hg}, characteristic of lattice diffusion, and $D^*_{Hg,f}$ and $D^*_{Hg,s}$ are both related to the interdiffusion coefficient using fundamental equations for bulk diffusion. We believe, therefore, that at least two components represent lattice diffusion.

Our discussion concerns mainly the results of Chen et al.,[25,26] and Tang and Stevenson[27,28] for $X = 0.2$ MCT, since these two studies include all three self-diffusion coefficients and are in good agreement with one another. A typical tracer profile shows two branches (Fig. 6) and is represented with a superposition of two error functions with associated diffusion coefficients: $D^*_{Hg,f}$ and $D^*_{Hg,s}$. Figure 7 shows diffusion isotherms—log D^*_{Hg} versus log(P_{Hg})— for $D^*_{Hg,f}$ and $D^*_{Hg,s}$. The $D^*_{Hg,f}$ isotherm shows a change in slope from -1

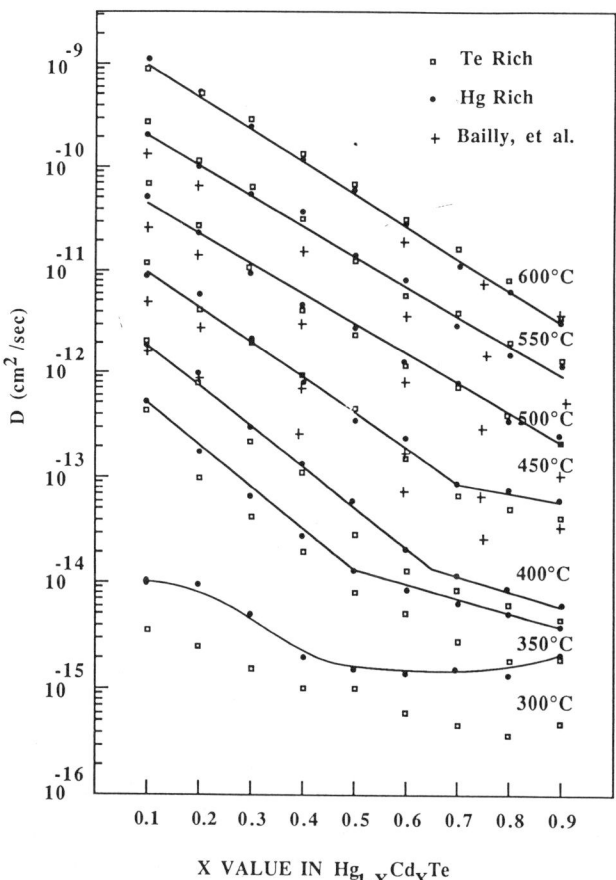

FIG. 5. A plot of the interdiffusion coefficient as a function of X and T (after Tang and Stevenson, Ref. 21).

TABLE II. Summary of tracer diffusion studies in $X = 0.2$ MCT.

Tracer	Sample pretreatment	Temperature	Ambient	Remarks	References
Hg^{203}	Preannealed with Hg-saturated condition	225 °C	Hg saturated	3 branches found preannealed sample has similar result to that of nonpreannealed one	Brown and Willoughby (33,34)
	(111) polished single crystal preannealed	400, 450, 490 °C	Hg Over pressure	2 components found (fast and slow) $D_{f,v}^* = 4.7 \times 10^5 \exp(-2.40 \text{ eV}/kT) P_{Hg}^{-1}$ at low P_{Hg}—vacancy $P_{f,i}^* = 1.1 \times 10^{-7} \exp(-0.54 \text{ eV}/kT) P_{Hg}$ at high P_{Hg}—interstitial $D_s = 9.9 \times 10^{-1} \exp(-1.50 \text{ eV}/kT)$—independent of P_{Hg} assign to exchange of Hg isotope with Hg in the diffusion host	Chen (25,26)
	Single crystal preannealed	I. 400, 450 °C II. 200, 250 °C	Hg over pressure	2 components found f, s both show pressure dependence, but the slope of diffusion isotherms changed at two temperature regions. At 450 °C, D_f and D_s are faster than the values of Chen's result by 2 orders of magnitude	Gorshkov and Zaitov (43,44)
	(111) polished single crystal preannealed	350, 400 450, 500 °C	Hg over pressure	2 components found (f,s) $D_{f,v}^* = 4.87 \times 10^4 \exp(-2.10 \text{ eV}/kT) P_{Hg}^{-1}$ at low P_{Hg}—vacancy $D_{f,i}^* = 5.5 \times 10^{-7} \exp(-0.61 \text{ eV}/kT) P_{Hg}$ at high P_{Hg}—interstitial $D_{Hg,s}^* = 6.07 \exp(-1.70 \text{ eV}/kT)$ independent of P_{Hg} vacancy, interstitial in series mechanism. Assign s component to intrinsic diffusion coefficient	Tang and Stevenson (27,28)
	Bridgman grown single crystals ($X=0.16$) LPE ($X=0.23$)	300–400 °C	Hg over pressure	2 components found for both samples	Archer and Palphrey (45)
Cd^{109}	(111) Single crystal, preannealed	400, 450, 490 °C	Hg over pressure	$D_f^* = 1.0 \exp(-1.50 \text{ eV}/kT)$ $D_s^* = 0.7 \exp(-1.55 \text{ eV}/kT)$ both independent of P_{Hg} assign, f, s to the exchange of Cd tracer with Cd and Hg	Chen (25,26)
	Preannealed	230–500 °C	Hg over pressure	Single component is found At $T = 280$ °C, D_{Cd}^* independent of P_{Hg}, except for Hg-rich, $D_{Cd}^* \propto P_{Cd}^{1/2}$. At $T > 280$ °C, $D_{Cd}^* \propto P_{Cd}^{1/2}$.	Shaw (31)
	(111) polished single crystal preannealed	350–500 °C	Hg over pressure	2 components found, both are independent of P_{Hg} $D_f = 10.9 \exp(-1.70 \text{ eV}/kT)$ $D_s = 1.48 \exp(-1.65 \text{ eV}/kT)$ propose-vacancy, interstitial in series mechanism f—true self diffusion s—intrinsic diffusion coefficient	Tang and Stevenson (27,28)
Te^{123m}	(111) single crystal preannealed with Hg overpressure	400, 450, 490 °C	Hg over pressure	$D_{Te}^* = 47.8 \exp(-2.12 \text{ eV}/kT) P_{Hg}^{-1}$—interstitial mechanism	Chen (25,26)
	Preannealed with Hg overpressure	450 °C	Hg over pressure	Diffusion isotherm changes the across the P_{Hg} field indicating a change of mechanism from vacancy to interstitial as P_{Hg} increases	Gorshkov and Zaitov (43)

TABLE II. (continued).

(111) single crystal preannealed with Hg overpressure	400–500 °C	Hg over-pressure	After preannealing, only one component is found. $D^*_{Tc} = 88.6 \exp(-2.15 \text{ eV}/kT) P_{Hg}^1$ interstitial mechanism	Tang and Stevenson (27,28)

to $+1$ as the stoichiometry proceeds from Te rich to Hg rich, whereas the $D^*_{Hg,s}$ isotherm shows a weaker dependence on P_{Hg}, but with a maximum rather than a minimum towards the center of the phase field. The diffusion isotherm for $D^*_{Hg,f}$ is an exact parallel to the n and p isotherms established from the defect equilibria in Fig. 1, and both are consistent with a defect model with V_{Hg} and Hg_i as the dominant mobile defects at low and high P_{Hg}, respectively.[4–6] (Since $X = 0.2$ MCT is intrinsic at the diffusion temperatures, one can not determine the charge state; however, according to the literature, the respective ionization energies are low and it is reasonable to assume that both defects are doubly ionized.[8–10]) It is proposed that both defects have comparable mobilities and can diffuse in parallel, independent of one another. We refer to this as the V–I in parallel model. There is a minimum value for $D^*_{Hg,f}$ when the stoichiometry of the crystal is close to the transition from n to p type and there is a minimum in the sum of Hg_i (native donor) and V_{Hg} (native acceptor) species. With this model, one associates the diffusion in the two branches with diffusion of Hg_i and V_{Hg}, respectively, and can express the diffusion coefficient for vacancies and interstitials in Arrhenius form:

$$D^*_{Hg,f,V} = 4.9 \times 10^4 \exp(-2.1 \text{ eV}/kT) P_{Hg}^{-1};$$
at $P_{Hg} = 0.1$ atm, (11)

$$D^*_{Hg,f,i} = 5.5 \times 10^{-7} \exp(-0.61 \text{ eV}/kT) P_{Hg};$$
at $P_{Hg} = 1$ atm. (12)

The $D^*_{Hg,s}$ isotherm shows a weaker P_{Hg} dependence, but with a maximum in the center of the phase field, rather than a minimum. An explanation of this behavior is a vacancy–

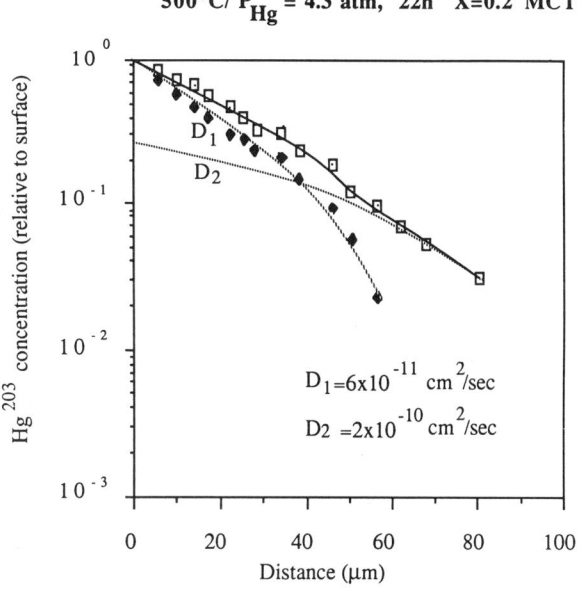

FIG. 6. Hg tracer diffusion profile for $X = 0.2$ MCT (500 °C, Hg = 4.3 atm, 22 h) (after Tang, Ref. 16).

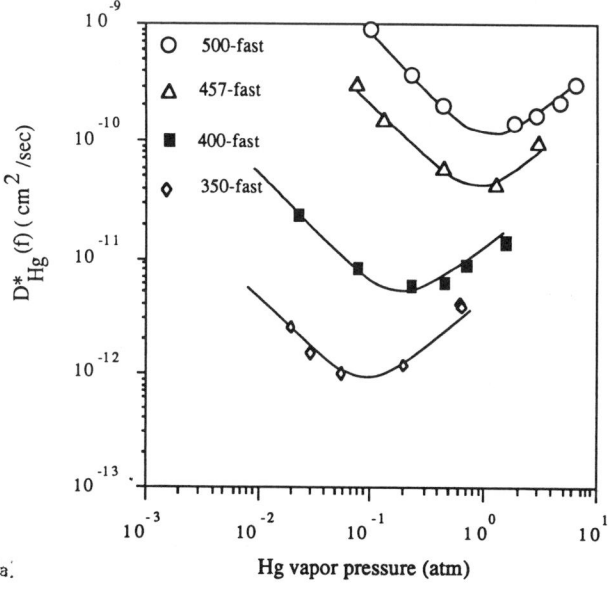

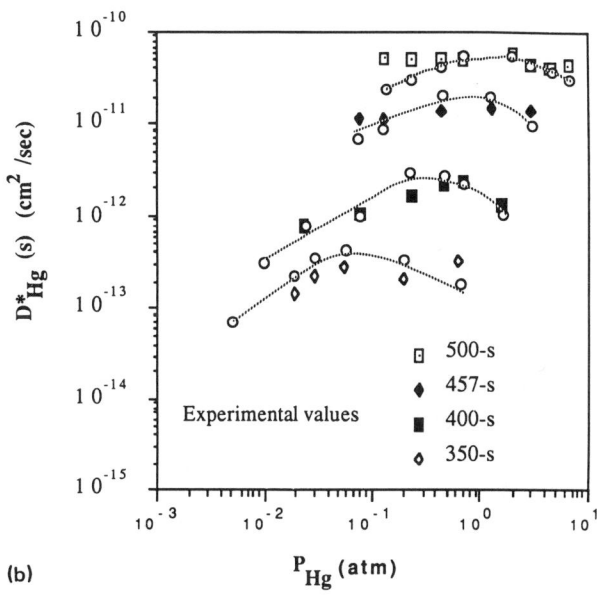

FIG. 7 (a) $D^*_{Hg,f}$ vs P_{Hg}, for $X = 0.2$ (b) $D^*_{Hg,f}$ vs P_{Hg}, for $X = 0.2$; experimental values (○) and calculated values, assuming a V–I in series mechanism. (···) (after Tang and Stevenson, Ref. 27).

interstitial in series mechanism (*V–I* in series). For this mechanism, the jump of a substitutional Hg atom is promoted by an interstitialcy step; instead of a direct jump from a substitutional site to an adjacent vacancy, a Hg_i first displaces the diffusing atom into an interstitial site (an intestitialcy mechanism) en route to its final jump into the vacancy. The *V–I* in parallel and *V–I* in series paths are schematically illustrated in Fig 8. For the series mechanism, the diffusion process is limited by interstitials at low P_{Hg} and by V_{Hg} at high P_{Hg} and is a maximum for comparable concentrations for the two species somewhere in the center of the phase field. Analysis of the *V–I* in series process[27,29,30] leads to the following relation between $D^*_{Hg,s}$, $D^*_{Hg,V}$, and $D^*_{Hg,i}$:

$$1/D^*_{Hg,s} = 1/2(1/D^*_{Hg,V} + 1/D^*_{Hg,i}). \quad (13)$$

Values of $D^*_{Hg,s}$ are calculated using $D^*_{Hg,V}$, and $D^*_{Hg,i}$ from Eqs. (11) and (12) and are plotted on Fig 7(b), showing good agreement with experiment. Although the mathematical fit to the model of *V–I* in parallel for the fast component and *V–I* in series for the slow component is very good, we do not explain why these two paths should be sufficiently independent to establish two distinct profiles.

Tracer-diffusion profiles of Cd also show two branches with corresponding fast and slow self-diffusion coefficients, $D^*_{Cd,f}$ and $D^*_{Cd,s}$, with both coefficients almost independent of P_{Hg} and with values lower than respective values for mercury (Fig. 9).[27,28,31] Our proposed interpretation of these two values is based on a comparison of all three component self-diffusion coefficients with the interdiffusion coefficients using the basic diffusion equation, as is discussed below.

The Te self-diffusion profiles of preannealed samples of MCT show only a single branch and a single D^*_{Te} which decreases with P_{Hg} in accord with $D^*_{Te} \sim P_{Hg}^{-1.26-28}$. The pressure dependence is consistent with an interstitial mechanism, similar to the behavior in CdTe described in Refs 18 and 19, but the charge state can not be established because the diffusion is performed at temperatures where MCT is intrinsic. The most significant aspect of the D^*_{Te} behavior is that it is substantially lower than either of the cation tracer diffusion coefficients. This leads to a simplification in the basic diffusion equation that we employ to identify the fast and slow components of the metal tracer diffusion. A summary of the trends and values for the tracer diffusion coefficients are given in Fig. 10.

Thus far, we propose that the fast and the slow component of the mercury tracer diffusion corresponds respectively to a vacancy–interstitial in parallel and in series mechanism without specific assignment of these two values. Further interpretation of these values and those for cadmium tracer diffusion is provided by relating these values to the interdiffusion coefficient. As was mentioned, the relative values of the tracer diffusion coefficients ($D^*_{Te} \ll D^*_{Cd}$ and D^*_{Hg}) leads to a simplification of the general diffusion equation [Eq. (4)], specifically to a Nernst–Planck type of equation [Eq.

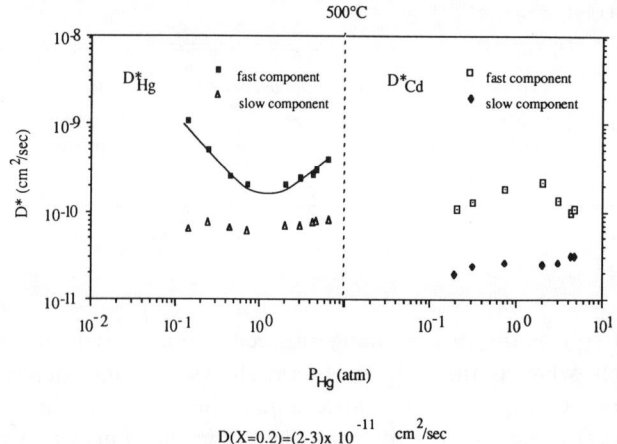

FIG. 9. Comparison between D and D^*_i at 500 °C. $D(X=0.2) = (2-3) \times 10^{-11}$ cm^2/s (after Tang and Stevenson, Ref. 28).

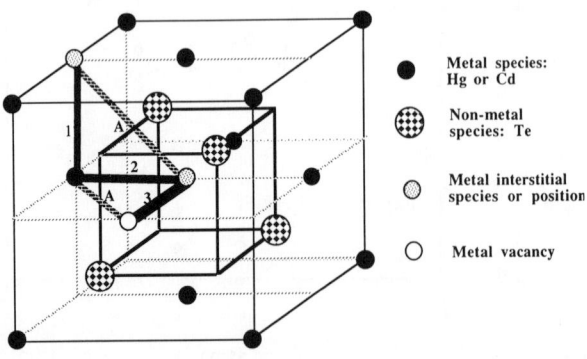

FIG. 8. Display of the metal diffusion paths: *V–I* in parallel indicated by path A; and *V–I* in series by the sequence 1–2–3 (after Tang and Stevenson, Ref. 27).

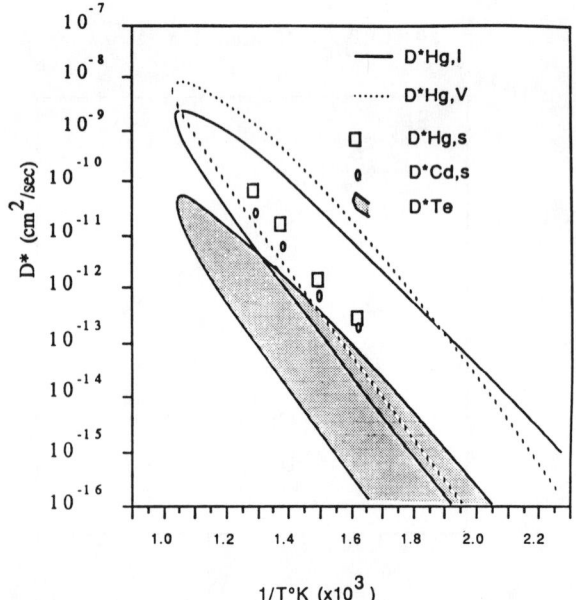

FIG. 10. Tracer diffusion coefficients of Hg (fast and slow components), Cd (slow component) and Te vs temperature for the whole stoichiometric region for $X = 0.2$ MCT (after Tang and Stevenson, Ref. 28).

(6)]. For $X = 0.2$ and for $D^*_{Cd} \sim D^*_{Hg}$, Eq. (6) is further simplified to:

$$D \sim D^*_{Cd} [TF] \tag{14}$$

which is an intrinsic diffusion coefficient for Cd. Comparison of D^*_{Cd} with D for several temperatures shows that $D = D^*_{Cd,s}$, thus justifying an assignment of $D^*_{Cd,s}$ to D_{Cd}, the intrinsic diffusion coefficient of Cd, and $D^*_{Cd,f}$ is assigned to the self-diffusion coefficient of Cd. This latter assignment is further justified by calculating the thermodynamic factor [TF] from the data of Tung et al.[2] This calculation leads to values of [TF] < 1 (corresponding to a positive enthalpy of mixing), with values such that: $D_{Cd} = D^*_{Cd,s} = D^*_{Cd,f} [TF]$. Thus, the fast and slow components of the Cd tracer diffusion are assigned to the self-diffusion and the intrinsic diffusion coefficient of Cd, respectively. Having identified the intrinsic and the self-diffusion coefficients for Cd, they may be introduced into Eqs. (5) and (6), respectively. The intrinsic and self-diffusion coefficients of Hg can be identified by noting that $D^*_{Hg,s}$ leads to the correct value of D in Eq. (5), and is therefore an intrinsic diffusion coefficient, D_{Hg}. Likewise, $D^*_{Hg,f}$ leads to the correct value of D in Eq (6), and is therefore the Hg self-diffusion coefficient, D^*_{Hg}. Thus, for both Cd and Hg, the fast tracer diffusion coefficients are the self-diffusion coefficients and the slower are the intrinsic tracer diffusion coefficients.

One may ask the question, why should one have an intrinsic diffusion coefficient in MCT if it has been preannealed? The tracer diffusion of one of the metal components in MCT presents two possibilities. A tracer Hg atom may exchange with a nontracer Hg atom, in which case there is no change in chemistry arising from that exchange and it is therefore true self-diffusion. If, however, a tracer Hg exchanges with a Cd atom, there is a local change in chemistry and the associated diffusion coefficient corresponds to intrinsic diffusion. One notes that this intrinsic diffusion results in a local change in the X value, but not in stoichiometry.

Another form of intrinsic diffusion in MCT corresponds to a change in stoichiometry, such as the low temperature mercury annealing used to convert from p to n type. The kinetics of low temperature Hg annealing made by Schaake et al.[32] was modeled with a diffusion coefficient close to the value for the slow component of the tracer diffusion coefficient reported by Brown and Willoughby,[33,34] and the latter is consistent with our $D^*_{Hg,f}$ extrapolated to low temperature. The model used to analyze Hg annealing by Schaake et al.[32] involves vacancies and interstitials and is consistent with the proposed V–I in parallel self-diffusion model.

Why should the self-diffusion of mercury proceed by a V–I in parallel process whereas the intrinsic diffusion coefficient of Hg proceeds by a V–I in series process? The latter diffusion process involves the displacement of Cd, which forms a stronger bond with Te than does Hg. In the V–I in parallel mechanism the diffusing atom surmounts a higher energy barrier arising from a closer approach to Te than for the V–I in series mechanism. Thus, the V–I in series process is preferred for jumps involving Cd in order to avoid the higher energy barrier resulting from a closer Cd–Te distance.

V. MARKER MOTION AND THE KIRKENDALL EFFECT IN MCT

A difference in the mobility of two components in a binary diffusion couple can produce motion of inert markers that are embedded in the diffusion zone, as was first established by Smigelskas and Kirkendall for binary metal system.[35] For ideal conditions, the markers move with individual lattice planes. If there is no volume change upon mixing, the marker velocity is the difference in the mobility of the two components. Darken pointed out that only if markers of different materials and size move at the same rate, does the marker motion correspond to the velocity of specific planes.[13] For the ideal Kirkendall effect, there is local defect equilibrium; for a vacancy diffusion mechanism, sources and sinks of vacancies are sufficient to maintain local vacancy equilibrium. When this is not the case, a supersaturation of vacancies results in the formation of voids, referred to as "Kirkendall voids," and there is not a "pure" Kirkendall effect in which the markers are fixed to lattice planes with the marker velocity equal to the difference in diffusion of the two species.

In quasi-binary systems, the Kirkendall effect is more complex, since there are three atomic species to be considered, in addition to point defects such as vacancies. For MCT the self-diffusion coefficient of Hg (D^*_{Hg}) is somewhat larger than that for Cd (D^*_{Cd}), which leads to the consensus opinion that Hg diffuses more rapidly than Cd. This fact leads one to conclude that there should be a Kirkendall effect in this system. Marker motion has indeed been observed in the HgTe–CdTe diffusion system and interpreted as a Kirkendall effect.[36]

Diffusion studies in the MCT system have established that the self-diffusion coefficient of Te (D^*_{Te}) is significantly smaller than for either Hg or Cd. This observation, along with an analysis of the interrelation between self-diffusion, intrinsic diffusion and interdiffusion for such a system, leads to the conclusion that Hg and Cd are constrained to diffuse at the same rate [Eq. (7)], and there should not be a true Kirkendall effect. Marker motion is indeed observed but the velocity depends on the marker size and type and does not represent the motion of lattice planes.[37]

From an analysis of the data of Tung et al.,[2] it can be shown that the activity of Te decreases as the X value in MCT increases. Hence, if there is a positive gradient in X, there will be a negative gradient for Te. Hence, there is a tendency for Te to diffuse from lower X to higher X MCT. However, at the diffusion temperatures at which Marker motion experiments are performed, the Te sublattice is essentially immobile. However, in the presence of pores or markers (in the form of Al_2O_3 or W particles), the void space or the particle interface provides a short circuit for Te transport (through the vapor for pores or along the interface for particles), thereby causing the void or particle to move. This mechanism is illustrated in Fig. 11. As a consequence, there is no true Kirkendall effect. The marker motion is explained by the lack in local point defect equilibrium in the diffusion zone, and the vapor transport and/or surface diffusion of Te along the markers.

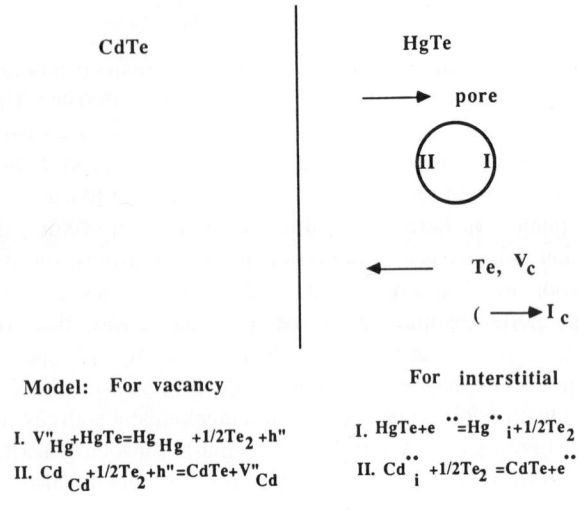

FIG. 11. Model for pore and/or marker movement in a MCT interdiffusion couple.

VI. SUMMARY

The influence of composition (X value in MCT), stoichiometry (P_{Hg}) and temperature on the component diffusion in MCT is reviewed in the context of the known defect chemistry models of MCT. Ionized mercury interstitials and vacancies are the dominant electrically active defects and also play a dominant role in the diffusion behavior. In general, vacancies become relatively more important compared to interstitials with increasing temperature, lower P_{Hg}, and lower X value and vice versa.

The interdiffusion coefficient at $T > 400\,°C$ decreases exponentially with X and is independent of P_{Hg} pressure. The model of a dual vacancy-interstitial (V–I) or a vacancy interstitial complex diffusing species is proposed to explain the participation of vacancies and interstitials, with D independent of the P_{Hg} pressure. For lower temperatures, the X dependence and the activation energy decrease and there is a dependence of D on the P_{Hg} overpressure with increasing P_{Hg} pressure. An interstitial is proposed as the dominant mobile defect in this temperature regime.

The tracer diffusion of Hg and Cd both show two branches with a corresponding fast and slow diffusion coefficient. The fast component for mercury is modeled with a V–I parallel process, and shows a dependence on the mercury pressure. The other diffusion coefficients are explained by a V–I in series mechanism which is more effective in surmounting the larger activation barrier for jumps involving the Cd atoms. The tellurium tracer diffusion shows only a single branch with values much smaller than the tracer diffusion of either Hg or Cd. Inserting the relative values into a basic diffusion equation develops a Nernst–Planck type of equation as the guiding equation interrelating self-diffusion coefficients to interdiffusion coefficients in the MCT system. Self-consistency with these basic diffusion equations is obtained by assigning the fast component for both Cd and Hg to the true self-diffusion coefficients and the slow components to the intrinsic diffusion coefficients of Hg and Cd.

[1] R. F. Brebrick, J. Phys. Chem. Solids **40**, 177 (1979).
[2] T. Tung, L. Golonka, and R. F. Brebrick, J. Electrochem. Soc. **128**, 451 (1981).
[3] J. Strauss and R. F. Brebrick, J. Phys Chem. Solids **31**, 2283 (1970).
[4] H. R. Vydyanath, J. Electrochem. Soc. **128**, 260 (1981).
[5] H. R. Vydyanath, J. C. Donovan, and D. A. Nelson, J. Electrochem. Soc. **128**, 2685 (1981).
[6] J. Nishizawa and K. Suto, J. Phys. Chem. Solids **37**, 33 (1976).
[7] H. F. Schaake, J. Electron. Mater. **14**, 513 (1985).
[8] A. Kobayashi, O. F. Sankey, and J. D. Dow, Phys. Rev. B **25**, 6367 (1982).
[9] C. A. Swarts, M. S. Daw, and T. C. McGill, J. Vac. Sci. Technol. A **1**, 198 (1982).
[10] C. L. Wang, S. Wu, and D. S. Pan, J. Vac. Sci. Technol. A **1**, 1631 (1983).
[11] C. G. Morgan-Pond and R. Raghavan, Phys. Rev. B **31**, 6616 (1985).
[12] C. L. Jones, M. J. T. Quench, P. Cooper, and J. J. Gosney, J. Appl. Phys. **53**, 9080 (1982).
[13] L. S. Darken, Trans. AIME **175**, 184 (1948).
[14] J. R. Manning, *Diffusion Kinetics for Atoms in Crystals* (Van Nostrand, Princeton, NJ, 1968).
[15] M-F. S. Tang and D. A. Stevenson, J. Phys. Chem. Solids **51**, 563 (1990).
[16] M-F. S. Tang, Ph.D. thesis, Stanford University 1987.
[17] J. G. Fleming and D. A. Stevenson, Phys. Status Solidi A **105**, 77 (1987).
[18] P. M. Borsenberger and D. A. Stevenson, J. Phys. Chem. Solids **29**, 1277 (1968).
[19] S. S. Chern and F. A. Kroger, J. Solid State Chem. **14**, 299 (1975).
[20] M-F. S. Tang and D. A. Stevenson, Appl. Phys. Lett. **50**, 1272 (1987).
[21] M-F. S. Tang and D. A. Stevenson, J. Vac. Sci. Technol. A **5**, 3124 (1987).
[22] K. Zanio and T. Massopust, J. Electron. Mater. **15**, 103 (1986).
[23] D. K. Arch, J. L. Staudenmann, and J. P. Faurie, Appl. Phys. Lett. **48**, 1588 (1986).
[24] Y. Kim, A. Ourmazd, and R. D. Feldman, J. Vac. Sci. Technol. A **8**, 1117 (1990).
[25] J. S. Chen, F. A. Kroger, and W. L. Ahlgren, Extended Abstract, U. S. Workshop on the Physics and Chemistry of Mercury Cadmium Telluride, 1984 (unpublished), p. 109.
[26] J. S. Chen, Ph.D. thesis, University of Southern California, 1985.
[27] M-F. S. Tang and D. A. Stevenson, J. Vac. Sci. Technol. A **7**, 544 (1989).
[28] M-F. S. Tang and D. A. Stevenson, J. Vac. Sci. Technol. A **6**, 2650 (1988).
[29] R. Kikuchi and H. Sato, J. Chem. Phys. **51**, 161 (1969).
[30] R. Kikuchi, J. Chem. Phys. **53**, 2702 (1970).
[31] D. Shaw, Philos. Mag. A **53**, 727 (1986).
[32] H. F. Schaake, J. H. Tregilgas, J. D. Beck, M. A. Kinch, and B. E. Gnade, J. Vac. Sci. Technol. A **3**, 143 (1985).
[33] M. Brown and A. F. Willoughby, J. Cryst. Growth **59**, 27 (1982).
[34] M. Brown and A. F. Willoughby, J. Vac. Sci. Technol. A **1**, 1641 (1983).
[35] A. D. Smigelskas and E. Q. Kirkendall, Trans. AIME **171**, 130 (1949).
[36] V. Leute and W. Stratmann, Z. Physik. Chem. **90**, 172 (1974).
[37] M-F. S. Tang and D. A. Stevenson (to be published).
[38] H. Rodot and J. Henoc, Compt. Rend. Acad. Sci. **256**, 1954 (1963).
[39] F. Bailly, G. Cohen-Solal, and Y. Marfaing, Compt. Rend. Acad. Sci. **277**, 103 (1963).
[40] F. Bailly, Compt. Rend. Acad. Sci. **262**, 635 (1966).
[41] L. Svob, Y. Marfaing, R. Triboulet, F. Bailly, and G. Cohen-Solal, J. Appl. Phys. **46**, 4251 (1975).
[42] V. Leute, H. M. Schmidtke, W. Stratmann, and W. Winking, Phys. Status Solidi A **67**, 183 (1981).
[43] A. V. Gorshkov, F. A. Zaitov, S. B. Shangin, G. M. Shalyapina, I. N. Petrov, and I. S. Asaturova, Sov. Phys. Solid State **26**, 1787 (1984).
[44] A. Zaitov, A. V. Gorshkov, and G. M. Shalyapina, Sov. Phys. Solid State **21**, 112 (1979).
[45] N. Archer and H. Palfrey, presented at the Electronic Materials Conference, Santa Barbara, CA, June 27–29, 1990 (unpublished).

Selective annealing for the planar processing of HgCdTe devices

K. K. Parat, H. Ehsani, I. B. Bhat, and S. K. Ghandhi
Electrical, Computer and Systems Engineering Department, Rensselaer Polytechnic Institute, Troy, New York 12180

(Received 2 October 1990; accepted 19 October 1990)

$Hg_{1-x}Cd_x Te$ layers grown by organometallic vapor phase epitaxy are p type with carrier concentrations around $4 \times 10^{16}/cm^3$ due to group II vacancies. Following a Hg saturated anneal at 220 °C for 25 h, these layers become n type with carrier concentrations around $5 \times 10^{14}/cm^3$. However, the presence of a 0.5–0.8 μm thick CdTe cap inhibits the annealing of the underlying $Hg_{1-x}Cd_x Te$ layer, since it acts as a barrier for Hg diffusion. By opening windows in this cap, the underlying $Hg_{1-x}Cd_x Te$ layer can be annealed and converted to n type in a selective manner. P–N junction photodiodes were fabricated using this planar technique. Some of these diodes employed the CdTe cap itself as the surface passivant; in others, the CdTe cap was stripped and anodic sulfide was used as the junction passivant. In both the cases, diodes had $R_0 A$ values comparable to the best values reported in literature. N-channel enhancement mode metal–insulator semiconductor field effect transistors were also fabricated using anodic sulfide as the surface passivant. Here, the n-type source and drain regions were formed by selectively annealing the as grown p-type $Hg_{1-x}Cd_x Te$ layer.

I. INTRODUCTION

Fabrication of semiconductor devices requires the formation of p-type and n-type regions in the material. This can be achieved using extrinsic dopants or intrinsic defects. In case of $Hg_{1-x}Cd_x Te$, group II vacancies are shallow acceptors, thus p-type $Hg_{1-x}Cd_x Te$ can be achieved using these vacancies.[1,2] It is a common feature of $Hg_{1-x}Cd_x Te$ layers grown by various techniques that once the group II vacancy concentration in the layer is reduced below a certain level by suitable annealing techniques, the layer becomes n type due to the residual impurities in the layer which are often donors.[1-6] In addition, lattice defects, such as ion implantation induced damage act as donors in $Hg_{1-x}Cd_x Te$.[7] Thus it is possible to achieve both n-type and p-type $Hg_{1-x}Cd_x Te$ using native defects alone. Fabrication of p–n junctions in $Hg_{1-x}Cd_x Te$ is often based on this technique of forming p and n regions using native defects.[8-11]

In this paper we report on the fabrication of $Hg_{1-x}Cd_x Te$ devices using a selective annealing technique. $Hg_{1-x}Cd_x Te$ layers grown by organometallic vapor phase epitaxy (OMVPE) are p type due to group II vacancies. The layers can be converted to n type by annealing under Hg overpressure.[6] However, the presence of a 0.5–0.8 μm thick CdTe cap prevents the underlying $Hg_{1-x}Cd_x Te$ from getting annealed. This is due to the lower diffusion coefficient of Hg in CdTe, as compared to that in $Hg_{1-x}Cd_x Te$.[1,12] By opening windows in the CdTe cap, the underlying $Hg_{1-x}Cd_x Te$ layer can be annealed in a selective manner, thus allowing the formation of n-type regions in a p-type layer in a planar fashion. P–N junction photodiodes were fabricated by this technique, where the p region was vacancy doped, and the n regions were formed by annealing. Photodiodes passivated using anodic sulfide as well as the CdTe cap layer were fabricated, and the performance of these devices are comparable to the best values reported for $Hg_{1-x}Cd_x Te$ photodiodes fabricated by other techniques.[8-11] n-channel metal–insulator semiconductor field effect transistors (MISFETs) were also fabricated using anodic sulfide for gate passivation. Here, the n-type source and drain regions were formed by selective annealing. The details of annealing techniques and the performance of the devices are described.

II. EXPERIMENTAL

$Hg_{1-x}Cd_x Te$ layers used in this study were grown at 370 °C in a vertical reactor by the direct alloy growth process involving the simultaneous pyrolysis of elemental mercury, dimethylcadmium, and diisopropyltelluride.[13] (100) CdTe, misoriented 4° to 8° towards (110) were used as the substrates. Typically 10–12 μm thick undoped $Hg_{1-x}Cd_x Te$ layers were grown, followed by a 0.5–0.8 μm thick undoped CdTe cap. This cap prevents any unintentional or partial annealing[14] of the $Hg_{1-x}Cd_x Te$ layer during cool down after the layer growth. It is subsequently used as the diffusion barrier for Hg during selective annealing of the underlying $Hg_{1-x}Cd_x Te$.

The thickness and alloy composition of the $Hg_{1-x}Cd_x Te$ layer were determined using Fourier transform Infrared spectrometer. Here, the energy corresponding to an absorption coefficient of 500 cm^{-1} was used for calculating the alloy composition.[15] Annealings were carried out in a sealed quartz ampoule, along with 99.99999% pure Hg. Some layers were annealed with the CdTe cap intact, and in other cases the cap was removed prior to annealing. For selective annealing, windows were opened in the CdTe cap using photolithography and chemical etching. Annealings were done at 220 °C for 25 h, with the samples kept about 2 °C warmer than the Hg reservoir to avoid Hg from condensing on the samples. After annealing, the ampoule was pulled out of the furnace and quenched in air. Some of the annealed samples were characterized by Hall effect measurements using the van der Pauw technique.[16]

To make p–n junction diodes and MISFETs, the CdTe cap and about 0.5 μm of the underlying $Hg_{1-x}Cd_xTe$ was removed from the samples after annealing, and the surface was passivated using 200 Å of anodic sulfide.[17] Next, 2500 Å of thermally evaporated ZnS was deposited. Windows were opened in this insulator and evaporated indium was used for making contacts to the n region. In the case of MISFETs, indium was also used as the gate metal. The insulator was removed from regions close to the edges of the sample, and evaporated gold was used for forming ohmic contacts to the underlying p-$Hg_{1-x}Cd_xTe$.

In some cases, the CdTe cap itself was used as the junction passivant for the p–n junction diodes. Here, ohmic contacts were formed on the n and p regions of the $Hg_{1-x}Cd_xTe$ following annealing, and the intermediate step of removing the CdTe cap and passivating the surface was omitted. Finished devices were bonded in flatpacks with front and back illumination facility.

III. RESULTS AND DISCUSSIONS

The as grown $Hg_{1-x}Cd_xTe$ layers which are p type with carrier concentrations around 4×10^{16}/cc due to group II vacancies, become n type with carrier concentrations around 5×10^{14}/cc following a Hg-rich anneal. Full annealing and conversion to n type could be achieved by a single heat treatment at 220 °C. It was not necessary to subject the layers to a higher annealing temperatures, as in previous reports by us[6] as well as by others.[18]

Some of our earlier layers showed only partial annealing at 220 °C.[6] There it was seen that annealings at temperatures in the range of 200–230 °C, were inadequate for achieving conversion to n type. Even though the vacancy concentrations in the layers reduced to the low 10^{14}–10^{15}/cc range following the annealing, they still remained p type, Such annealing behaviours have been observed by other workers as well.[18,19] These layers could be converted to n type following a two step anneal which involved an initial high temperature anneal at 290 °C followed by another lower temperature anneal at 220 °C.[6] However, our recent growth process results in layers which become n type following a single low temperature anneal at 220 °C. The results described in this paper are based on these layers.

Figure 1 shows the temperature dependence of the Hall coefficient of two layers annealed at 220 °C for 25 h under Hg saturated conditions. The alloy composition of these layers was 0.23, and the thickness was 11.0 μm. The layers were grown with a 0.5 μm thick CdTe cap. For sample A, the cap was left intact, whereas the cap was removed prior to annealing sample B. It can be seen from the Hall data that the capped sample remains p type, whereas the uncapped sample has become n type following the anneal. At 77 K, the measured hole concentration in the p-type layer was 4.9×10^{16}/cm^3, whereas the electron concentration in the n-type layer was 6.5×10^{14}/cm^3. Table I lists the Hall data on a number of as grown layers and annealed layers together with the annealing conditions. The measured carrier concentration at 77 K is quoted in all the cases. It can be seen from these data that the CdTe cap acts as an effective barrier against the diffusion of Hg, so that the original vacancy concentration in the underlying $Hg_{1-x}Cd_xTe$ layer remains unchanged.

TABLE I. 77 K carrier concentration in as-grown and annealed $Hg_{1-x}Cd_xTe$ layers.

No.	Annealing conditions	x	Carrier concentration (/cm^3)	
1	As grown	0.267	5×10^{16}	(p type)
2	As grown	0.228	3.1×10^{16}	(p type)
3	Uncapped anneal 220 °C/25 h	0.229	7.8×10^{14}	(n type)
4	Uncapped anneal 220 °C/25 h	0.293	6.8×10^{14}	(n type)
5	Capped anneal 220 °C/25 h	0.230	4.9×10^{16}	(p type)
6	Capped anneal 220 °C/25 h	0.293	5×10^{16}	(p type)

IV. p–n JUNCTION PHOTODIODES

In order to make p–n junction photodiodes, an array of windows, 600 μm in diameter and spaced 1000 μm apart, were opened in the CdTe cap, using 2:2:1 by volume ratio of HBr, H_3PO_4, and 1N–$K_2Cr_2O_7$. This etchant was used instead of Br-methanol, because it does not attack the positive photoresist. During annealing, the entire $Hg_{1-x}Cd_xTe$ lay-

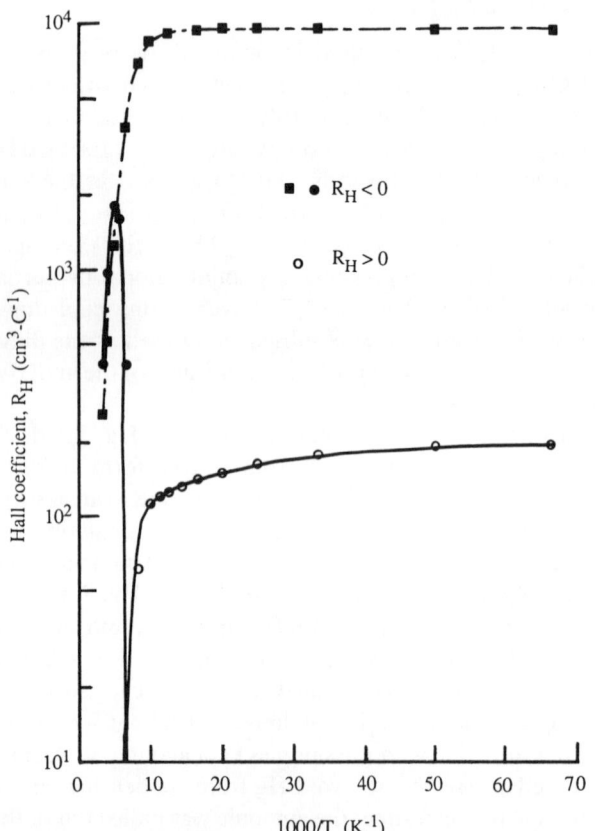

FIG. 1. R_H vs 1000/T for two $Hg_{0.77}Cd_{0.23}Te$ layers annealed at 220 °C for 25 h. Sample A (●, ○): capped layer. Sample B (■): uncapped layer.

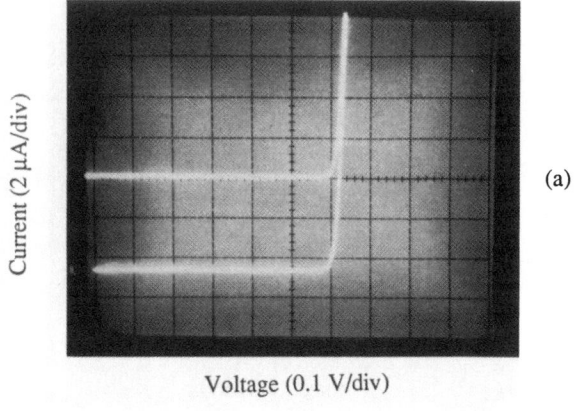

(a)

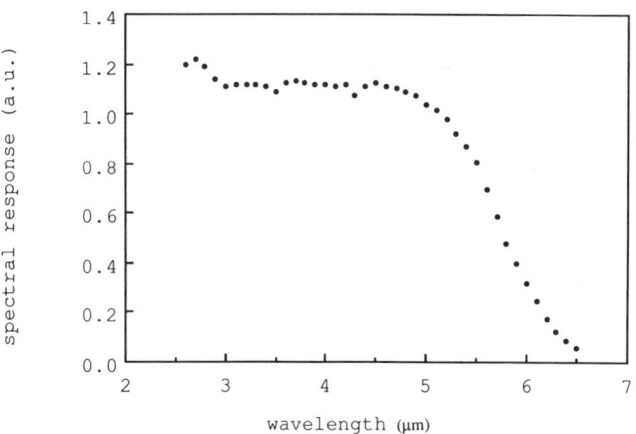

FIG. 3 Spectral response of the photodiode of Fig. 2 at 77 K.

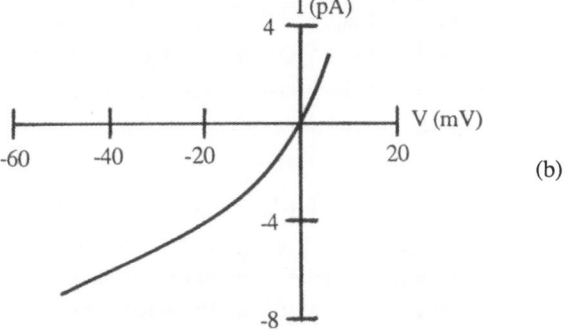

(b)

FIG. 2 I–V characteristics of a $Hg_{0.715}Cd_{0.25}$Te photodiode at 77 K. (a): Full characteristics under 77 K (thermal equilibrium) background, and under 295 K (room temperature) background. (b): detailed characteristics around zero bias under thermal equilibrium background at 77 K. x axis: 20 mV/div. y axis: 4 pA/div.

er under the open window is converted to n type, whereas rest of the region remains p type. Thus the p–n junction formed by this technique is of a cylindrical nature with vertical junctions, and has a junction area of πdt, where d is the diameter of the n-type region, and t is the thickness of the $Hg_{1-x}Cd_x$Te layer. The contacts to the n-type regions were formed using evaporated indium dots, 400 μm in diameter.

Figure 2 shows the I–V characteristic of a $Hg_{0.715}Cd_{0.285}$Te photodiode formed by the above technique. The CdTe cap was 0.8 μm thick and the annealing was carried out at 220 °C for 24 h. Here the thickness of the $Hg_{0.715}Cd_{0.285}$Te layer in the finished device was 9.4 μm, which results in a junction area of 1.8×10^{-4} cm^2. The junctions were passivated using anodic sulfide. Figure 2(a) shows the I–V characteristic of the photodiode at 77 K, under 77 K (thermal equilibrium) background and under 295 K (room temperature) background. The sensitivity of the device to the background infrared (IR) radiation is seen in the induced photocurrent of 4.6 μA and photovoltage of 121 mV. Figure 2(b) shows the detailed I–V characteristic of the same diode around zero bias, under thermal equilibrium background. The zero bias resistance of the device is 3×10^9 Ω, resulting in an R_0A product of $5.4 \times 10^5 \Omega$ cm^2.

Figure 3 shows the relative spectral response of this photodiode at 77 K under constant photon flux, for the case of backside illumination. Due to the vertical junction, the photodiode operates in a lateral collection mode, and the carriers generated within one diffusion length of the depletion layer will be collected. In such a case, the photocurrent is not a direct function of the area of the diode, and thus does not give the quantum efficiency of the device. Consequently, the spectral response is plotted in arbitrary units. It is nearly flat over the wavelength range of 2.5 to 5 μm, and falls to 50% of this value at a wavelength of 5.7 μm, which is considered to be the cutoff wavelength of this device.

V. MISFETs

In order to fabricate MISFETs, annealing of the layers were done after opening rectangular windows in the CdTe cap. Annealing converts the region below the open windows to n type to provide the source and drain regions, while the rest remains p type. After removing the CdTe cap and passivating the surface, indium was deposited between the source and drain to form the gate metal. Indium was also used for forming contacts to the source and drain.

Figure 4 shows the source drain I–V characteristics of a MISFET at 77 K for different values of gate voltage. The alloy composition of the $Hg_{1-x}Cd_x$Te was 0.30 for this device. The source drain spacing was 300 μm and the gate width was 500 μm. Figure 4(a) shows the complete I–V characteristics of the device including the linear as well as the saturation regime. Figure 4(b) shows the detailed I–V characteristics of the device in the linear regime.

From Fig. 4 it can be seen that the device exhibits classical MISFET characteristics with clearly defined ohmic and saturation regimes. The threshold voltage of the device is about 1.4 V. Due to the high p-type doping in the substrate, as well as the low density of fixed charges in the anodic sulfide, the p-type region below the insulator is not inverted at zero gate voltage. As a result the n-channel MISFETs fabricated using anodic sulfide insulator are enhancement mode devices. From the linear region of the source-drain I–V, the mobility of electrons in the inversion layer was estimated to be 6,700 cm^2/V s using the following relation[20]

$$I_D = C_i \mu_n (Z/L)(V_G - V_T)V_D,$$

where I_D is the drain current, C_i is the insulator capacitance

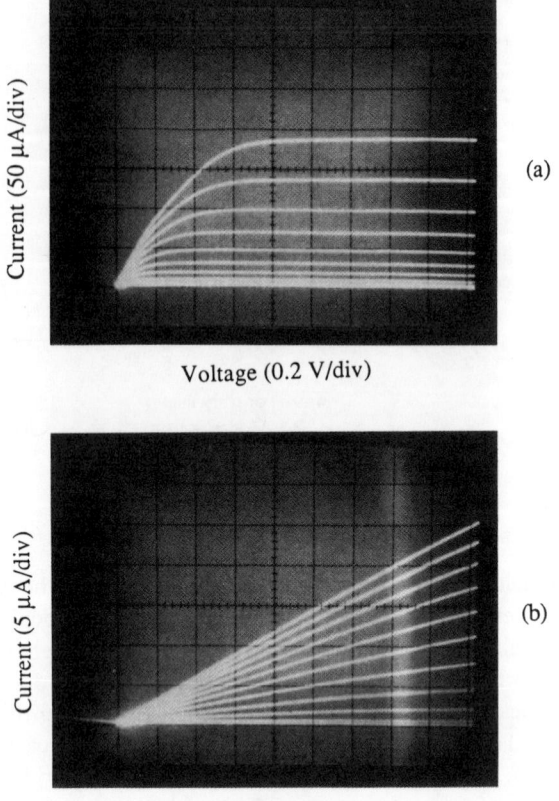

FIG. 4 *I-V* characteristic of an *n*-channel $Hg_{0.7}Cd_{0.3}Te$ MISFET. Gate voltage V_{GS}): 1.2, 1.4, 1.6,..., 3.2 V. (a): Full characteristics. (b): Characteristics in the linear region.

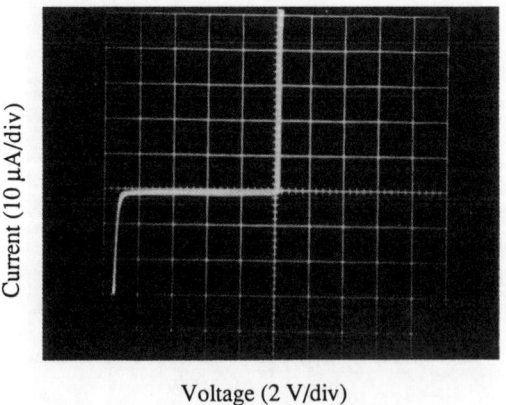

FIG. 5 *I-V* characteristics of a $Hg_{0.68}Cd_{0.32}Te$ photodiode at 77 K under thermal equilibrium background.

per unit area, V_G is the gate voltage, V_T is the threshold voltage, μ_n is the mobility of electrons in the inversion layer, Z is the gate width, L is the gate length, and V_D is the drain voltage. For this device C_i was 3×10^{-8} F/cm^2, Z was 500 μm, and L was 300 μm. The typical bulk electron mobility in *n*-type $Hg_{0.7}Cd_{0.3}Te$ at 77 K is around 30 000 cm^2/V s. The lower mobility of electrons in the inversion layer can be attributed to surface scattering, and is observed by other workers as well.[21]

VI. PHOTODIODES USING CdTe AS THE JUNCTION PASSIVANT

Some photodiodes were also fabricated using the CdTe cap itself as the junction passivant.[22] Figure 5 shows the *I-V* characteristics of one such diode fabricated in a $Hg_{0.68}Cd_{0.32}Te$ layer. The *I-V* was taken at 77 K under thermal equilibrium background.

The cutoff wavelength of this diode is 4.5 μm, and the R_0A product is 8.8×10^7 Ω cm.^{2} The reverse breakdown voltage of the diode is about 9 V. The sharp and high breakdown voltage is a result of the excellent junction passivation provided by the CdTe cap.

CONCLUSIONS

The effect of a CdTe cap on the annealing behaviour of $Hg_{1-x}Cd_xTe$ was described. It was shown that a 0.5–0.8 μm thick CdTe cap can be used as an effective barrier against mercury diffusion during annealing of epitaxial $Hg_{1-x}Cd_xTe$ at 220°C. This allows the selective formation of *n*-type regions in a *p*-type layer, by annealing. These regions are relatively undamaged, compared to those formed by the conventional approach of using ion implantation for this purpose.[7] As a result *p–n* junctions formed by this technique provide means for characterizing the as grown material without the introduction of further defects into the material. This is reflected in the excellent electrical performance of the devices made using this technique. The technique is also relatively straightforward and provides for planar processing of devices.

ACKNOWLEDGMENTS

The authors would like to thank J. Barthel for technical assistance on this program and P. Magilligan for manuscript preparation. CdTe substrate material was kindly supplied by C.J. Johnson of II–VI, Inc., Saxonburg, PA. Partial funding for this program, from the Raytheon Corporation, is hereby acknowledged. This work was sponsored by the Defense Advanced Research Projects Agency (Contract No. N-00014-85-K-0151), administered through the Office of Naval Research, Arlington, VA. This support is greatly appreciated.

[1] H. R. Vydyanath, J. Electrochem. Soc. **128**, 2609 (1981).
[2] J. L. Schmit and E. L. Stelzer, J. Electron. Mater. **7** 65 (1978).
[3] C. L. Jones, M. J. T. Quelch, P. Capper, and J. J. Gosney, J. Appl. Phys. **53**, 9080 (1982).
[4] H. R. Vydyanath and C. H. Hiner, J. Appl. Phys. **65**, 3080 (1989).
[5] H. F. Schaake, J. H. Tregilgas, J. D. Beck, M. A. Kinch, and B. E. Gnade, J. Vac. Sci. Technol. A **3**, 143 (1985).
[6] K. K. Parat, N. R. Taskar, I. B. Bhat, and S. K. Ghandhi, J. Cryst. Growth, **102**, 413 (1990).
[7] L. O. Bubulac, J. Cryst. Growth **86**, 723 (1988).
[8] M. Lanir and K. J. Riley, IEEE Trans. Electron Dev. ED **29**, 274 (1982).
[9] E. R. Gertner, S. H. Shin, D. D. Edwall, L. O. Bubulac, D. S. Lo, and W. E. Tennant, Appl. Phys. Lett. **46**, 851 (1985).
[10] J. B. Mullin, S. J. C. Irvine, J. Geiss, J. S. Gough, A. Royle, and M. C. L. Ward, SPIE Proc. **1106**, 17 (1989).

[11] A. Rogalski, Infrared Phys. **28**, 139 (1988).
[12] K. Takita, K. Murakami, H. Otake, K. Masuda, S. Seki, and H. Kudo, Appl. Phys. Lett. **44**, 996 (1984).
[13] S. K. Ghandhi, I. B. Bhat, and H. Fardi, Appl. Phys. Lett. **52**, 392 (1988).
[14] S. J. C. Irvine, J. S. Gough, J. Riess, M. J. Gibbs, A. Royle, C. A. Taylor, G. T. Brown, A. M. Keir, and J. B. Mullin, J. Vac. Sci. Technol. A **7** 285 (1989).
[15] G. L. Hansen, J. L. Schmit, and T. N. Casselman, J. Appl. Phys. **53**, 7099 (1982).
[16] L. J. van der Pauw, Phillips Res. Rep. **13**, 1 (1958).
[17] Y. Nemirovsky, L. Burstein, and I. Kidron, J. Appl. Phys. **58**, 366 (1985).
[18] P. Capper, B. C. Easton, P. A. C. Whiffin, and C. D. Maxey, J. Cryst. Growth **79**, 508 (1986).
[19] B. Pelliciari, G. L. Destefanis, and L. DiCioccio, J. Vac. Sci. Techol. A **4**, 27 (1986).
[20] S. M. Sze, *Physics of Semiconductor Devices*, 2nd ed. (Wiley, New York, 1980).
[21] G. M. Williams and E. R. Gertner, Electron. Lett. **16**, 839 (1980).
[22] S. K. Ghandhi, K. K. Parat, H. Ehsani, and I. B. Bhat, Appl. Phys. Lett. **58**, 828 (1991).

The metalorganic chemical vapor deposition growth of HgCdTe on GaAs at 300°C using diisopropyltelluride

R. Korenstein, P. Hallock, and B. MacLeod
Raytheon Company, Research division, Lexington, Massachusetts 02173

(Received 1 October 1990; accepted 17 December 1990)

HgCdTe was grown by the metalorganic chemical vapor deposition alloy process on (111)B CdTe/GaAs and (111)B CdTe substrates at 300 °C using diisopropyltelluride, dimethylcadmium, and elemental mercury. Excellent surface morphology and compositional uniformity were obtained at this temperature. Hall effect measurements indicate that n-type (111) HgCdTe with carrier concentration below 10^{15} cm^{-3} and good mobilities at 77 K can be obtained on GaAs substrates.

I. INTRODUCTION

Hg$_{1-x}$Cd$_x$Te (MCT) grown on GaAs substrates is an important material for advanced infrared detector devices. Compared to CdTe substrates, GaAs offers the advantages of high crystalline perfection, low price, ruggedness, and the availability in large area. Typically, MCT is deposited by metalorganic chemical vapor deposition (MOCVD) on (100) oriented GaAs at temperatures in the range of 360–420 °C.[1,2,3] Relatively low doping of the MCT layer by either gallium or arsenic from the GaAs substrate is reported for the (100) orientation. For example, MCT grown at 370 °C by the MOCVD alloy process is found to be n type with a carrier concentration of 3.5×10^{15} cm^{-3} at 77 K.[1] However, significant gallium incorporation ($>10^{17}$ gallium atoms cm^{-3}) is found when MCT is grown in the (111)B orientation at a temperature of 360 °C on GaAs substrates.[4] A corresponding carrier concentration in excess of 1×10^{17} cm^{-3} at 77 K as measured by Hall effect is observed for this material. The dependence of gallium incorporation on the crystallographic orientation can be explained in terms of the chemical reactivity of each surface.[4]

MCT grown in the (111)B orientation by MOCVD has a much smoother surface morphology than MCT grown in the (100) orientation which is characterized by microterraces and hillocks.[2] Thus the (111)B orientation would be preferred if the gallium concentration could be greatly reduced. A lower growth temperature is expected to reduce the chemical reactions between the metalorganic compounds and the GaAs substrate believed to be responsible for gallium incorporation in MCT grown at high temperatures.[4]

In the MOCVD growth of MCT the lower limit to the growth temperature is governed by the pyrolyzing efficiency of the tellurium source. Typical growth temperatures for MCT grown by MOCVD are between 360–400 °C using diisopropyltelluride (DIPTe) as the tellurium alkyl.[1,3,4] Although other tellurium precursors such as ditertiarybutyltelluride[5] and diallyltelluride[6] have been used to grow HgTe and CdTe below 300 °C, the purity of these tellurium alkyls is still questionable. On the other hand, DIPTe is currently available in very pure form as evidenced by the very low n-type carrier concentration (low 10^{14} cm^{-3}) of MCT grown with this compound.[7] For this reason we chose to use DIPTe in this work even though very low MCT growth rates were expected. This paper will describe the growth of MCT at lower temperatures, namely 300 °C on GaAs substrates. It will be shown that gallium incorporation in MCT can be reduced by approximately two orders of magnitude by growing at this reduced temperature and that 300 °C is a viable temperature for the MOCVD growth of MCT on both GaAs and CdTe substrates.

II. EXPERIMENT

MCT was grown by the MOCVD alloy process in a horizontal reactor at 300 °C. The substrates were rotated about a plane parallel to the gas flow during the growth process in order to improve the compositional uniformity of the grown MCT. For comparison, MCT was also grown at 360 °C under conditions given in Ref. 4. The metalorganic reagents were DIPTe and dimethylcadmium. Elemental Hg, used as the mercury source, was contained in a heated reservoir external to the growth chamber. Pd-diffused H$_2$ was used as the carrier gas at a total flow rate of 2.7 ℓ/min. MCT was grown at 300 °C both on (111)B CdTe and (100) GaAs substrates. CdTe substrates purchased from II–VI, Inc. were chemo-mechanically polished before growth. GaAs substrates were purchased polished in the standard misorientation of 2° towards (110). (111) CdTe buffer layers were deposited onto GaAs substrates at 300 °C by hot wall epitaxy.[8] The polarity of these buffer layers was determined to be (111)B using the Nakagawa etch.[9]

The Cd mole fraction in the layers, x, was determined from infrared (IR) transmission measurements performed on a Perkin Elmer 983G IR spectrophotometer. Film thicknesses were obtained from cleaved cross sections. Dislocation densities in the layers were measured by using a K$_2$Cr$_2$O$_7$/HNO$_3$/HCl/H$_2$O etch.[10] The transport properties of the material were deduced from Van der Pauw Hall effect measurements either as a function of temperature or at room temperature and 77 K. Gallium and arsenic incorporation in the layers were determined by secondary ion mass spectrometry (SIMS) performed at Charles Evans and Associates.

III. RESULTS AND DISCUSSION

MCT was grown at 300 °C both on (111)B CdTe/GaAs and (111)B CdTe substrates. The DIPTe mole fraction was set at 0.003 in order to obtain a reasonable growth rate on the order of 1.5 μm/h at a total flow rate of 2.7 ℓ/min. The resulting film thicknesses ranged between 5 and 7 μm. The surface morphology of MCT grown on (111)B CdTe/GaAs is very smooth and highly specular as exhibited in Fig. 1. As expected for this orientation, the surface is devoid of the microterraces and hillocks which are characteristic of the (100) orientation.

A considerable improvement in the compositional uniformity was observed for layers grown at 300 °C compared to those grown at 360 °C. This is illustrated in Fig. 2 which depicts the compositional uniformity of two 15 mm × 20 mm wafers grown at 300 °C and 360 °C. The compositional uniformity as defined by the maximum variation in x divided by the mean value is typically between 2–4% for layers grown at 300 °C compared to 7–10% for layers grown at 360 °C. The improvement in uniformity at the lower growth temperature could be related to a decrease in the gas phase parasitic reactions.

The defect microstructure of MCT grown on (111)B CdTe/GaAs was revealed by chemical etching and is illustrated in Fig. 3. Typical etch pit densities (EPD) are in the $1–3 \times 10^6$ cm^{-2} range. These values which are similar to what has been observed for MCT grown on GaAs at 380–400 °C by the MOCVD interdiffused multilayer process[3] indicate that the crystalline quality of this material is comparable to that grown at higher temperatures. Twins which are typically seen for the (111) orientation were not present in this material or in the MCT grown at 360 °C on (111)B CdTe/GaAs. On the other hand, twins are observed on MCT grown on (111)B CdTe substrates at both temperatures. The absence of twins in GaAs-based material is probably related to the misorientation of the GaAs. It has been shown that substrate misorientation can result in an absence of twins for MCT grown on misoriented CdTe.[2]

Table I lists the transport properties of MCT grown in this study. In order to eliminate Hg vacancies, all samples were annealed in a Hg ambient for 15 h at 220 °C before Hall effect measurements. The low carrier concentration and high electron mobility for sample 1256 which was grown at 360 °C on CdTe are indicative of good n-type, low-doped material. Sample 891B grown at 360 °C on (111)B CdTe/GaAs is also n-type but the carrier concentration is 2.4×10^{17} cm^{-3} at 77 K. This sample consists of a 7-μm-thick MCT layer grown on a 5-μm-thick CdTe buffer layer. From the SIMS profile shown in Fig. 4, it can be seen that the corresponding gallium concentration in the MCT layer is also approximately 2×10^{17} cm^{-3}. Carrier concentrations in the low 10^{17} cm^{-3} with correspondingly high levels of gallium doping are char-

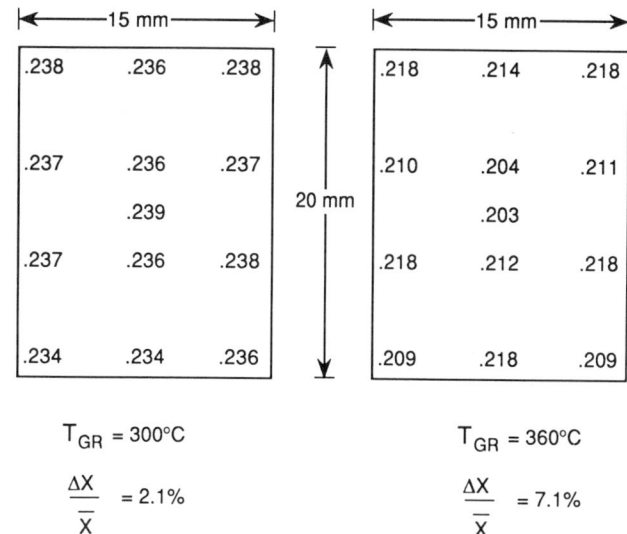

FIG. 2. Effect of growth temperature on compositional uniformity. The compositional uniformity is defined as the maximum variation in x divided by the mean.

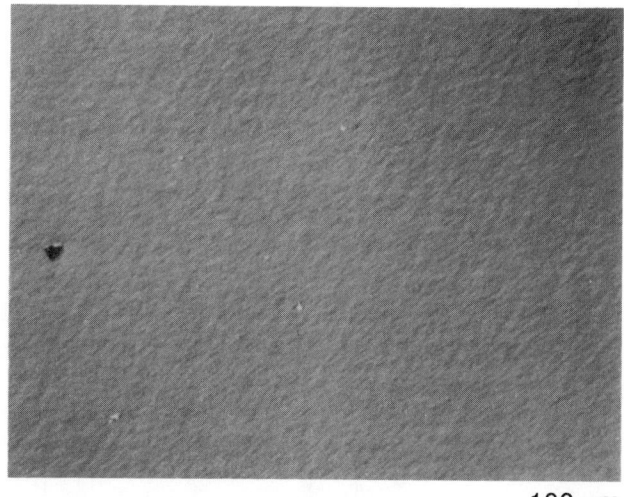

FIG. 1. Surface morphology of MCT grown on (111)B CdTe/GaAs at 300 °C.

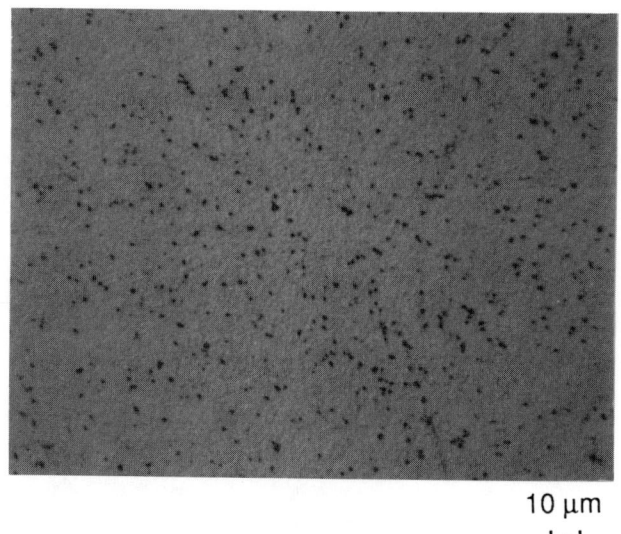

FIG. 3. Etch pit density of MCT grown on (111)B CdTe/GaAs at 300 °C.

TABLE I. Transport properties of MCT.

Sample	Substrate	T growth (°C)	x	n_H 77 K ($\times 10^{14}$ cm^{-3})	μ_H 77 K ($\times 10^3$ cm^2/V-sec)
1256	(111)B CdTe	360	0.218	6.6	94
891B	(111)B CdTe/GaAs	360	0.260	2400	33
1233	(111)B CdTe/GaAs	300	0.220	9.0	90
1262	(111)B CdTe/GaAs	300	0.209	9.1	217
1264	(111)B CdTe/GaAs	300	0.223	5.5	97
1271	(111)B CdTe/GaAs	300	0.225	3.7	111
1172	(111)B CdTe	300	0.224	2.9	103
1174	(111)B CdTe	300	0.210	7.5	92

acteristic of all layers grown on (111)B CdTe/GaAs at 360 °C. The SIMS results confirm that gallium, which is expected to be an n-type dopant in the II–VI lattice, is responsible for the high doping levels measured by Hall effect.

On the other hand, as can be seen from Table I, MCT grown on (111)B CdTe/GaAs at 300 °C has considerably lower carrier concentrations. These layers are all n type with carrier concentrations below 1×10^{15} cm^{-3} at 77 K. A SIMS profile of sample 1271 shown in Fig. 5 also confirms that the MCT grown at 300 °C has considerably less gallium (1×10^{15} cm^{-3}). It should be noted that SIMS analysis did not find any arsenic above the arsenic detection limit of 1×10^{15} arsenic atoms/cm^3. Hence, the low carrier concentrations of the layers reported in Table I are a direct result of the decrease in the gallium concentration in these layers due to the lower growth temperature. The temperature dependence of the Hall coefficient, carrier concentration and mobility for sample 1233 is shown in Fig. 6. The classical n-type behavior (net donor level, N_D-N_A, of 9.0×10^{14} cm^{-3} and electron mobility in excess of 10^5 cm^2/V s at low temperatures) is another indication of the good n-type material that can be grown at low temperature. For purposes of comparison, MCT was also grown on (111)B CdTe substrates at 300 °C (sample 1174). As can be seen from Table I, MCT grown on GaAs at 300 °C has comparable low doping levels to that of MCT grown on CdTe both at 300 °C and 360 °C. Reducing the growth temperature decreases gallium incor-

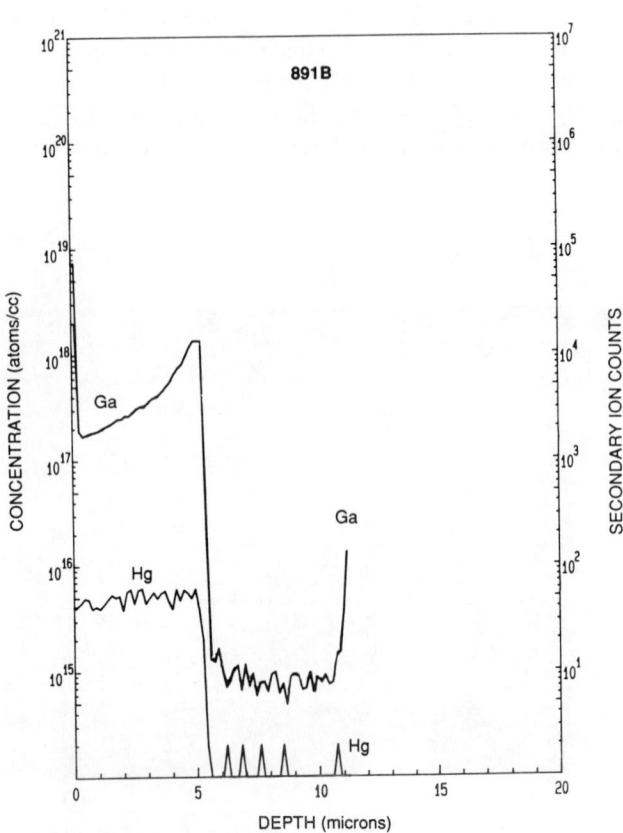

FIG. 4. SIMS profile through a 7-μm-thick MCT layer grown on (111)B CdTe/GaAs at 360 °C. The CdTe buffer layer is 5 μm thick.

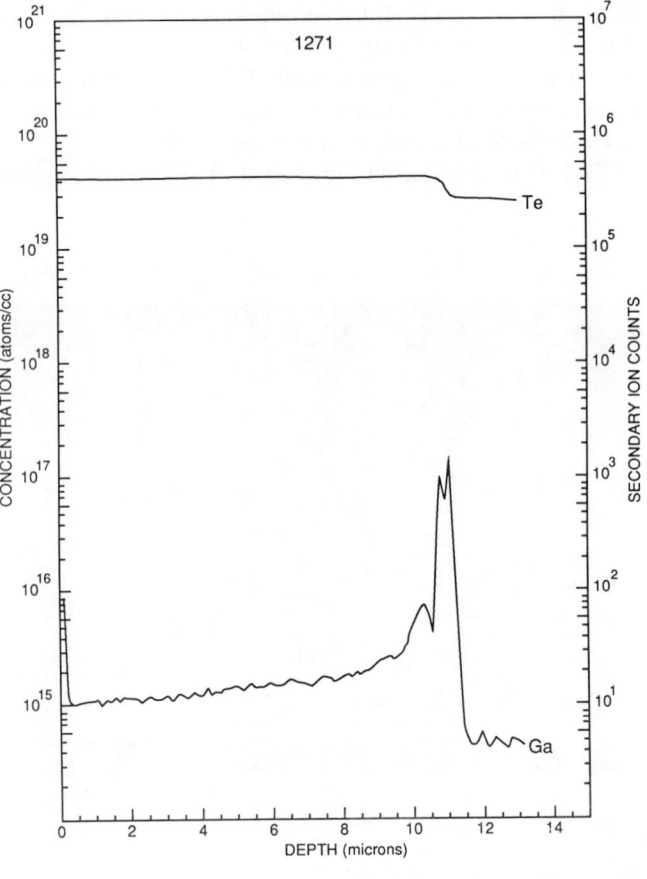

FIG. 5 SIMS profile through a 10-μm-thick MCT layer grown on (111)B CdTe/GaAs at 300 °C.

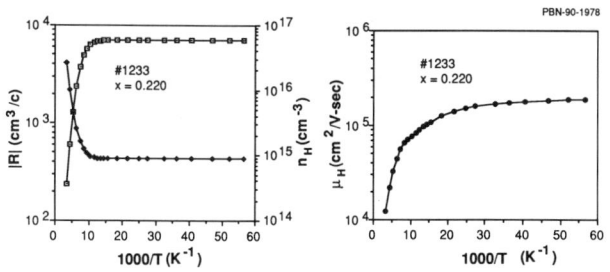

FIG. 6 Temperature dependence of Hall coefficient, carrier concentration, and mobility for $x = 0.220$ MCT layer grown on (111)B CdTe/GaAs at 300 °C.

poration by approximately two orders of magnitude and is an effective way of producing low doped n-type (111) MCT grown on GaAs substrates.

IV. CONCLUSIONS

We have shown the feasibility of growth at 300 °C on (111)B CdTe/GaAs substrates. Excellent surface morphology and compositional uniformity can be obtained at this temperature. Etch pit density measurements indicate that the MCT has comparable microstructure to that grown at higher temperature. Twins, which are commonly observed in MCT grown in (111) substrates, were not observed when MCT was grown on (111)B CdTe/GaAs. A significant reduction in gallium incorporation is observed when MCT is grown on GaAs substrates at this reduced temperature. n-type MCT layers with carrier concentrations below 10^{15} cm^{-3} and good mobilities can be grown on both (111)B CdTe/GaAs and CdTe substrates at 300 °C.

ACKNOWLEDGMENT

The authors thank P. Madison for the growth of (111)B CdTe layers on GaAs substrates by hot wall epitaxy.

[1] N. R. Taskar, I. B. Bhat, K. K. Parat, D. Terry, H. Ehsani, and S. K. Ghandi, J. Vac. Sci. Technol. A 7, 281 (1989).
[2] P. Capper, C. D. Maxey, P. A. C. Whiffin, and B. C. Easton, J. Cryst. Growth 96, 519 (1989).
[3] D. D. Edwall, J. Bajaj, and E. R. Gertner, J. Vac. Sci. Technol. A 8, 1045 (1990).
[4] R. Korenstein, P. Hallock, B. MacLeod, W. Hoke, and S. Oguz, J. Vac. Sci. Technol. A 8, 1039 (1990).
[5] W. E. Hoke and P. J. Lemonias, Appl. Phys. Lett. 48, 1669 (1986).
[6] R. Korenstein, W. E. Hoke, P. J. Lemonias, K. T. Higa, and D. C. Harris, J. Appl. Phys. 62, 4929, (1987).
[7] S. Oguz, R. J. Olson, D. L. Lee, L. T. Specht, and V. G. Kreismanis, SPIE Proc. Electro-Optical Materials for Switches, Coatings, Sensor Optics and Detectors, Vol. 1307 (1990).
[8] R. Korenstein and B. MacLeod, J. Cryst. Growth 86, 382 (1988).
[9] K. Nakagawa, K. Maeda, and S. Takeuchi, Appl. Phys. Lett. 34, 574 (1979).
[10] J. S. Chen and S. L. Johnston, Presented at the 1989 U.S. Workshop on the Physics and Chemistry of Mercury Cadmium Telluride and Related II-VI Compounds, San Diego, CA, Oct. 1989.

The effect of growth orientation on the morphology, composition, and growth rate of mercury cadmium telluride layers grown by metalorganic vapor phase epitaxy

G. Cinader, A. Raizman, and A. Sher
Solid State Physics Department, Soreq Nuclear Research Center, Yavne 70600, Israel

(Received 2 October 1990; accepted 15 January 1991)

We have measured the growth rate dependence of CdTe and HgTe layers in the growth surface direction. The growth mechanism of HgTe and CdTe were concluded to be heterogeneous, surface kinetic limited. The morphology of CdTe and mercury cadmium telluride interdiffusion multilayer process layers is found to be strongly effected by the large anisotropy of the CdTe growth rate. The best morphologies were found to be on surfaces having directions of maximum CdTe growth rates, with respect to surfaces of other crystallographic directions. A model for hillocks formation is given, based on the CdTe growth rate anisotropy. The different nature of twins stacking faults in (111)A and (111)B layers is explained by the different morphologies of these layers.

I. INTRODUCTION

The structural and morphological defects, common in mercury cadmium telluride (MCT) layers, are major limiting factors for device fabrication. The nature and appearance of these defects are determined by the specific growth method, growth conditions, substrate type, and its crystallographic orientation. Among these defects, microtwins and hillocks are typical of layers grown by molecular-beam epitaxy (MBE) and metalorganic vapor phase epitaxy (MOVPE) gas phase methods. The presence of microtwins was found to reduce the MCT carrier mobility.[1] Hillocks create morphological disturbances, serve as sources for crystalline defects, and are associated with large local variations of composition.[2] In this work we studied the effect of the growth orientation on the morphology, composition, and growth rate of $Hg_{1-x}Cd_xTe$ (MCT) grown by MOVPE, using the interdiffusion multilayer process (IMP).[3]

The most common crystallographic orientations used, so far, for MCT layer growth, have been (111)B and (001), on either CdTe or GaAs substrates. A very smooth morphology is usually obtained for the (111)B growth orientation.[4] However, lamellar twins structure is revealed using transmission electron microscopy (TEM).[5] On the other hand, for growth in the (111)A direction, under similar growth conditions, a rough morphology, consisting of pyramidal structures, accompanied by a double positioning twinned configuration, has always been obtained.[4] CdTe or MCT layers of the (001) orientation were found to be free of microtwins but usually exhibited a high density of hillocks.[5] All the hillocks present in a particular layer are of the same shape, size, and orientation. In general, the hillocks start to grow at the substrate layer interface and have (111) or vicinal underlay facets.[6] Recently, several groups have investigated the growth on surfaces of other crystallographic directions. It was found that there are directions, such as (112)B, which yield layers of very flat morphologies, free of hillocks and microtwins.[7-10] The results of the present investigation relate the growth rate and structural features of the epilayers with the growth surface crystallographic orientation. We have found that for CdTe and MCT layers grown by MOVPE, the preferred orientations for obtaining layers free of microtwins and hillocks are those which have high growth rates and deviate from (111)B.

II. EXPERIMENTAL

We have used a Cambridge Instruments MR-100 MOVPE system with a novel horizontal reaction cell. The source materials were diisopropyltelluride (DiPTe), dimethylcadmium (DMCd), and metallic mercury. The MCT layers were grown by the IMP method. The growth conditions were in the following range: substrate temperature 340–380 °C, hydrogen flow 0.6–4 SLPM, flow rates of 5.2×10^{-5}, 4.7×10^{-5}, and 8.6×10^{-4} mol/min for DiPTe, DMCd, and Hg, respectively. The total layer thickness d and the interdiffused layer composition x were deduced from the transmission curves measured by a Fourier transform infrared (FTIR) spectometer. The growth rates R_c and R_m of the individual binary CdTe and HgTe layers, respectively, are derived from the layer composition, thickness, and total growth times of CdTe and HgTe. The substrates were usually of CdTe or of lattice matched CdZnTe. Each epitaxial growth run was usually performed simultaneously on several substrates of different crystallographic orientations. Double crystal diffractometry was used for assessment the presence of twins in the layers.[4] Optical microscopy was used for the examination of hillocks and morphology.

III. RESULTS

Under the above growth conditions, the morphology, growth rate, and composition of the epilayers were found to be strongly dependent on two major factors: substrate orientation and substrate polarity. The influence of these factors on the resulting epilayers is depicted in Figs. 1–3. The effect of the substrate orientation on the epilayers morphology is revealed in Fig. 1. Here, the epilayer was grown on a multigrain substrate with arbitrary crystallographic directions. The differences in the morphology of three nearby domains

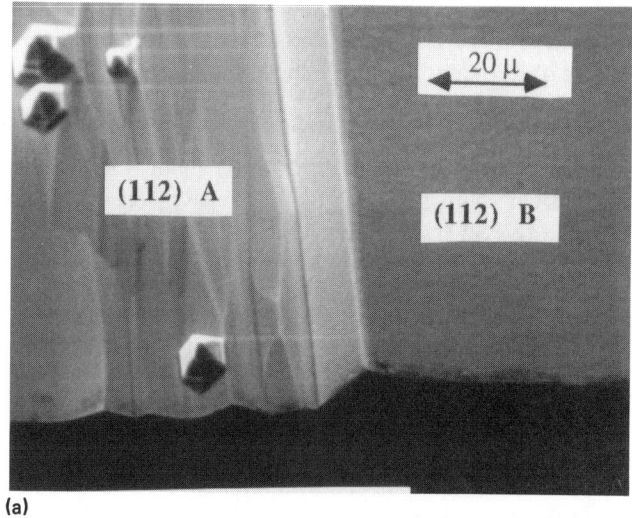

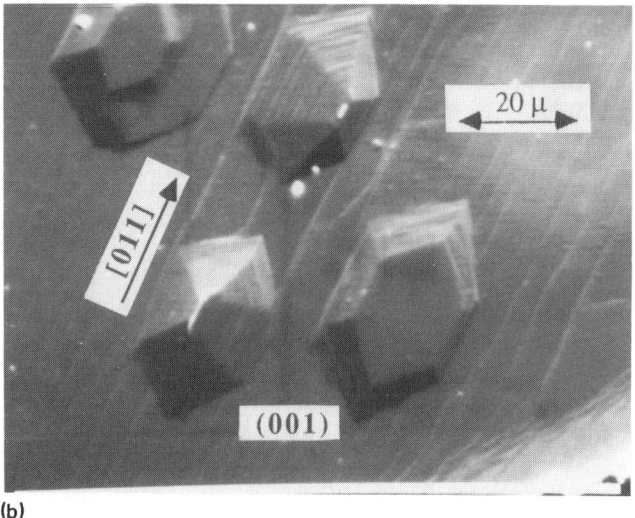

FIG. 1. (a) Typical morphology of an MCT layer grown on a substrate having domains of different crystallographic orientations. (b) A magnified hillock micrograph in the (001) domain.

FIG. 3. (a) Morphology and (b) cross section of a (112) MCT layer with two domains of A and B faces.

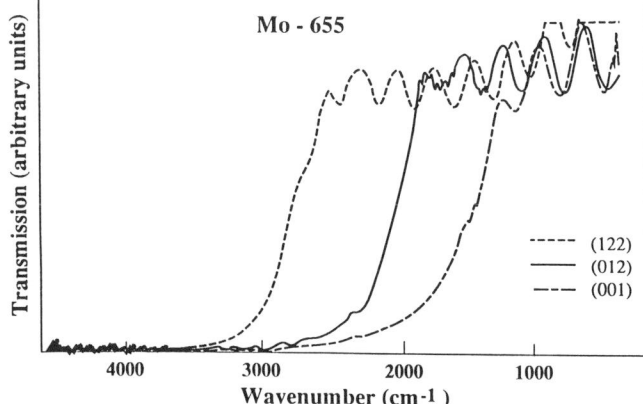

FIG. 2. Transmission curves of an MCT layer, in three domains of different crystallographic orientations.

are clearly seen. The variations of epilayers thickness and compositions, due to the different substrate orientations are revealed through the infrared (IR) transmission measurements.

The three IR spectra, presented in Fig. 2, were taken on the same layer, consisting of three domains, having different crystallographic orientations. The composition and the thickness of the layer in each domain are derived from the wavelength cutoff values and the interference fringes, respectively. Most significant variations of these parameters were obtained.

The polarity effect is well illustrated in Fig. 3 where a plane view and a cleaved cross section of a layer, grown on a twinned (112) substrate, is shown. In this case, a (111) twinning plane in the substrate, perpendicular to the (112) surface, yielded (112)A and (112)B adjacent domains. The A and B face notations designate that the growth plane is in the vicinity of (111)Cd and (111)Te planes, respectively. The B polarity is determined by the shiny surface in a Nakagawa etch. The layer, the morphology of which is shown in Fig. 3(a), grew coherently to the substrate. The (112)A domain exhibits a very rough and faceted morphology compared with the smooth morphology obtained in the (112)B orientation domain. Similar polarity dependence has been obtained for the (111) growth direction.[4] The overall growth rate in the (112)A direction is about 3 to 4 times smaller than that in the (112)B direction as is reflected from the respective layers thickness [Fig. 3(b)].

In order to have a better understanding of the IMP results we have looked at the morphology of layers of the binary compounds, CdTe and HgTe. CdTe and MCT–IMP layers free of hillocks were obtained only when grown on substrates which could be categorized as "B face." Epilayers grown on "A face" exhibited high density of hillocks, while those grown on (001) revealed moderate density of hillocks (Fig. 1). On the other hand, HgTe layers were in general smooth, except for A face growth, and free of hillocks on B face and on (001) surfaces.

The growth rates R_c and R_m for layers of different crystallographic directions, deviating from the (111)B direction toward either (001) or (011) directions, under identical growth conditions, are given in Table I. To study further the factors governing the growth process, we have grown MCT on charge transfer (CT) (112)B substrates under various mass flow rates of DiPTe and DMCd. The resulting growth rates are presented in Table II.

Finally we have studied the influence of the growth surface crystallographic orientation on the appearance of microtwins in the layers. Using x-ray diffraction, we have found that in layers tilted about 3° from the (111)B direction toward (001) the fractional volume of twins is $\approx 0.7\%$. This is compared with a value of 50% for an exact (111)B orientation. A similar reduction in the twins fractional volume was observed for layers tilted toward the (011) direction. No twins were observed in (112) CdTe and MCT (IMP) layers. The results are summarized in Table III. Our twin detection sensitivity is about 0.05%.

TABLE I. Growth rates dependence on growth surface direction.

Sample no.	Growth direction[a]	R_c (μm/h)	R_m (μm/h)
652.1	(012)	2.77	2.34
652.2	(112)	4.49	2.76
654.1	(112)	4.1	2.65
654.2	(111)	3.62	2.81
655	(001)	2.12	2.56
655	(012)	2.79	2.56
655	(122)	3.95	2.66
657.1	(001)	1.77	2.14
657.2	(123)	4.24	2.60
664.1	(223)	5.14	2.83
664.2	(112)	4.34	2.88
686.1	(255)	4.34	2.42
682.2	(112)	4.26	2.62

[a] The polarity of all growth faces is B.

TABLE II. Growth rates dependence on precursors flow rates.[a]

DiPTe (10^{-5} mol/min)	DMCd (10^{-5} mol/min)	R_c (μm/h)	R_m (μm/h)
5.2	4.7	4.3	2.4
5.2	3.1	3.7	2.4
3.5	4.7	4.4	1.8

[a] All growth is in (112)B direction. Mercury flow rate is 8.6×10^{-4} mol/min.

IV. DISCUSSION

A. Growth mechanism

Examining the results presented in Table I, it is clearly seen that, under the above mentioned growth conditions R_m, the growth rate of HgTe, is almost independent of the growth surface direction for all B faces and (001) planes. However, for A face planes, a drastic decrease in R_m is observed. The behavior of R_c, the CdTe growth rate, is quite different. It was found that R_c is strongly dependent on the crystallographic growth direction. The results summarized in Table I and the morphology of the corresponding layers indicate the existance of a correlation between the relative values of R_c and the abundancy and size of the hillocks: layers grown in directions of higher R_c yield smoother morphologies. In particular, crystallographic directions having nearly maximum values of R_c yielded layers entirely free of hillocks. A good example is the (112)B orientation. The results presented in Table II elucidate another experimental finding: R_c was found to be, to first order, practically independent of the partial pressure P of DiPTe. This is in spite of the fact that $P(\text{DiPTe}) \approx P(\text{DMCd})$ and DiPTe is a more stable molecule than is DMCd and hence was expected to be the rate limiting species. The sensitivity of R_c to the crystallographic orientation and its independence of $P(\text{DiPTe})$ lead to the conclusion that the growth rate of CdTe is controlled by surface catalytic processes. Sublimation effects are excluded due to the very low sublimation rates of CdTe under the relevant growth conditions. Our conclusion is in agreement with the model suggested by Bhat et al.,[11] who proposed surface catalytic pyrolysis of DiPTe by chemically adsorbed Cd. Moreover, our findings exclude the applicability of a previous model,[12] explaining the enhancement of DiPTe pyrolysis by an homogeneous adduct formation reaction in the gas phase.

TABLE III. Twins concentration dependence on growth surface direction.

Sample no.	Deviation of growth surface from (111)B	Twins concentration (%)
654	< 1°	50
693.2	3°	0.7
694.2	6°	0.45
685.2	9.4°	0.1
685.1	19.4°	< 0.05

The growth process of HgTe is different. We see in Table I that it is quite isotropic, namely nearly independent of the growth surface direction for (111)B and vicinal planes, as well as for (001). On the other hand R_m is very strongly effected by the polarity of the growth surface, as can be seen in Fig. 3(b). Hence we have to conclude that the HgTe growth process is heterogeneous too, probably due to autocatalytic pyrolysis of the Te precursors.[13] The effect of the unisotropic sublimation was measured and found to be negligible, under our growth conditions. The temperature dependence of R_m was measured in the range 340–380 °C, and found to be of the Ahrenious type with an activation energy of $E_a \approx 35$ kcal/mole. Such a temperature dependence excludes a mass transfer growth mechanism for HgTe, under our experimental conditions.

Finally, we have noticed that the orientational dependence of R_c is similar to what was earlier found for GaAs.[14,15] The A and B face polarities in CdTe growth play a similar role to the B and A face polarities in GaAs growth, respectively. This might be related to the higher volatility of Cd and As in the two systems. The similarity of these systems further extends the understanding of the CdTe growth mechanism, due to the extensive investigation already done on GaAs.

B. Features morphology

Once it is understood that the nature of the surface is a dominant factor in controlling the CdTe rate of growth, then the occurrence of some morphological features in the IMP structures may be explained. It has already been pointed out that the hillocks start to grow at the substrate-layer interface. It is reasonable to assume that the substrate surface has facets of different crystallographic planes, which could have been created by thermal and/or chemical etching prior to growth. The highly defected grown layer, near the interface, might be very susceptible to formation of thermal etch pits too. Among the exposed planes are some of high R_c, such as the (111)B and (112)B planes, and would give rise to a locally faster growth in these directions, leading to the formation of faceted hillocks, along such prefered crystallographic directions. Hence hillock formation results from the unisotropy of R_c. This explains the completely different morphology of (111)A, a low growth rate direction, and of (111)B face, high growth rate direction, of CdTe, HgTe and MCT layers. This model of hillock formation predicts that in MCT layers every hillock will be associated with a local increase in the CdTe concentration (x), as it results from the faster growth rate of CdTe on the hillock facets directions, while the growth rate of HgTe remains nearly constant. Such an effect was actually observed by Capper et al.[2] This feature is detrimental for device fabrication. Mouruzi–Khorasani et al.[16] have observed by TEM lamella twins inside a hillock cross section, in a (001) MCT layer. It is known that twins can not be formed in (001) growth.[10] Hence it has to be concluded that at least one facet of this hillock was oriented in the (111)B direction during the growth process. According to Million et al.[6] such facets are very common.

In view of this behavior, a practical conclusion can be drawn: the preferred growth direction for obtaining epilayers free of hillocks should be those of highest R_c. While on (100) and on B face surfaces the morphology features are dominated by the anisotropy of R_c, on A face substrates the anisotropy of R_m is also an important factor. The absence of hillocks in (001) HgTe layers results from the nearly maximum value of R_m there.

C. Twins configuration

In a previous study[4] we have shown that under our growth conditions, microtwins are common defects in CdTe layers grown in (111) orientation, while no twin defects were found in HgTe layers.[17] The twin defects present in (111)B CdTe layers exhibit a lamellar structure with equal population of both twin orientations (namely coherent and 60° rotated with respect to the substrate). In (111)A layers, the twins were of the double positioning (DP) type and have a much smaller fractional volume, compared with that of (111)B layers.[4] We claim that the different twin structure found in (111)A and (111)B layers result from their different morphologies. While the morphology of (111)B layers is flat, the morphology of (111)A layers is full of faceted hillocks, and the actual local growth surfaces significantly deviate from the (111) direction. In the present work we have found that the total volume of the twinned structures is drastically reduced by increasing the tilt angle of the growth surface from the (111)B orientation, as can be seen in Table III. Similar findings were reported by Schaake and Koestner[10] for MBE grown HgTe layers.

We propose the following process for the formation of DP twin configuration in (111)A layers: In the early stage of growth, when the morphology is still smooth and the growth surface is in the (111)A direction, some lamella of twins are formed (covering a few percents of the growth area), exactly as in (111)B layers. As soon as the surface begins to roughen, and the actual local direction of the growth surface deviate more than a few degrees from the (111) direction, no further twins can be formed, and the initial configuration is freezed out. Any further growth continues coherently. This is exactly a DP configuration.

V. CONCLUSIONS

In this investigation we have looked for the origin of various structural phenomena, concerned with MCT layers grown by MOVPE, using the IMP method. We have observed that under similar growth conditions, the composition, total thickness, morphology, and crystallinity of the epilayers were all strongly dependent on the epilayer growth orientation. We could explain all these findings in view of the dependence of the growth rate on the crystallographic surface orientation. We have found that the growth rate of HgTe depend strongly on the growth surface polarity but otherwise is quite isotropic. The growth reaction is heterogeneous with activation energy of about 35 kcal/mole. The growth rate of the CdTe constituent of the IMP is surface kinetic limited, leading to significant growth rate dependence on the growth surface crystallographic orientation. The growth reaction of CdTe is heterogeneous with catalytic pyrolysis of the Te precursors by adsorbed Cd. The formation of hillocks was related to the large anisotropy of R_c, and hillocks exist only on surface orientations of low CdTe growth rates. The difference in the twin structures of (111)A and (111)B layers is explained by the difference in the morphology of these layers.

ACKNOWLEDGMENTS

We are indebted to B. Bezerano for his outstanding technical help and fruitful ideas and to H. Shacham for the x-ray measurements.

[1] R. D. Feldman, S. Nakahara, R. F. Austin, T. Bonne, R. L. Opila, and A. S. Wynn, Appl. Phys. Lett. **51**, 1239 (1987).
[2] P. Capper, C. D. Maxey, P. A. C. Whiffin, and B. C. Easton, J. Cryst. Growth **96**, 519 (1989).
[3] S. J. Irvine, J. Tunnicliffe, and J. B. Mullin, Mater. Lett. **2**, 305 (1984).
[4] M. Oron, A. Raizman, Hadas Shtrikman, and G. Cinader, Appl. Phys. Lett. **52**, 1059 (1988).
[5] P. D. Brown, J. E. Hails, G. L. Russel, and J. Woods, Appl. Phys. Lett. **50**, 1114 (1987).
[6] A. Million, L. Di Cioccio, J. P. Gailliard, and J. Piaguat, J. Vac. Sci. Technol. A **6**, 2813 (1988).
[7] R. J. Koestner, H. Y. Liu, and H. F. Schaake, J. Vac. Sci. Technol. A **7**, 517 (1989).
[8] W. L. Ahlgren, S. M. Johnson, E. J. Smith, R. P. Ruth, B. C. Johnston, M. H. Kalisher, C. K. Cockrum, T. W. James, D. L. Arney, C. K. Ziegler, and W. Lick, J. Vac. Sci. Technol. A **7**, 331 (1989).
[9] J. M. Arias, S. H. Shin, J. G. Pasko, R. E. DeWames, and E. R. Gertner, J. Appl. Phys. **65**, 1747 (1989).
[10] H. F. Schaake and R. J. Koestner, J. Cryst. Growth **86**, 452 (1988).
[11] I. B. Bhat, N. R. Taskar, and S. K. Ghandhi, J. Electrochem. Soc. **1987**, 195.
[12] J. B. Mullin, S. J. C. Irvine, and D. J. Ashen, J. Cryst. Growth **55**, 92 (1981).
[13] M. R. Czerniak and B. C. Easton, J. Cryst. Growth **68**, 128 (1984).
[14] L. Hollan and C. Schiler, J. Cryst. Growth **13/14** (1972).
[15] Don W. Shaw, J. Cryst. Growth **31**, 130, 325 (1975).
[16] A. Nouruzi-Khorasami, I. P. Jones, P. S. Dobson, Y. Etem, D. J. Williams, M. G. Astles, C. Ard, and G. Coates, J. Cryst. Growth **102**, 819 (1990).
[17] G. Cinader, A. Raizman, and M. Oron, J. Cryst. Growth **101**, 167 (1990).

Growth and carrier concentration control of Hg$_{1-x}$Cd$_x$Te heterostructures using isothermal vapor phase epitaxy and vapor phase epitaxy techniques

S. B. Lee, D. Kim, and D. A. Stevenson
Department of Materials Science and Engineering, Stanford University, Stanford, California 94305

(Received 6 November 1990; accepted 26 December 1990)

We report on the growth and characterization of mercury cadmium telluride (MCT) heterostructures grown by isothermal vapor phase epitaxy (ISOVPE) and vapor phase epitaxy (VPE). This report consists of the use of both ISOVPE and VPE methods for growing wider band gap MCT on narrow band gap MCT, the compositional and structural characterization of as-grown layers, and the use of two zone Hg ampoules for carrier concentration control. The crystal perfection of some heterostructures is as good as that of the ISOVPE-grown layers. The carrier concentrations and mobilities of the ISOVPE-grown layers are substantially improved by a two zone Hg annealing technique.

I. INTRODUCTION

Narrow band gap mercury cadmium telluride (MCT) is widely used as a long wavelength ($>10\mu$m) infrared detecting material. Application of this material to the infrared detecting devices such as photodiodes and metal–insulator semiconductor (MIS) devices suffers from an increase in thermally generated minority carrier processes and tunneling processes with narrower band gaps. Heterostructures (wide band gap material on the surface of a narrow band gap material) are proposed as a way of overcoming some of the problems inherent in single layer narrow band gap MCT devices.[1] Advantages of heterostructure MCT include: the improvement in MIS performance by increasing well capacity and reducing dark current; and the integration of associated signal processing circuitry on-chip in the wide band gap region of MCT. The requirements for these applications are:

(i) In order to increase the well capacity and signal-to-noise ratio, a 1 to 5 μm thick surface layer of wide band gap MCT is required. Precise control of composition is not required for this purpose and graded junction profiles are acceptable.

(ii) For on-chip fabrication of signal processing circuitry, thicker layers ($>10\mu$m) are required and there is no preference about the junction shape as long as the composition profile at the surface region is flat. We describe a study employing isothermal vapor phase epitaxy (ISOVPE) and vapor phase epitaxy (VPE) for the growth of wider band gap MCT on narrow band gap MCT, along with the electrical and compositional control and characterization of the layers.

II. EXPERIMENTS

We have studied six conversion techniques using ISOVPE and VPE. Solid state recrystalized (SSR) single crystal narrow band gap MCT ($x=0.2$) measuring $4\times10\times0.5$ mm^3 is used as a starting material to fabricate the heterostructures. These SSR MCT substrates are not oriented to a specific direction and are mechanically polished with 1 μm alumina powder, then are chemomechanically polished with 2% bromine in ethylene glycol. Each process is discussed in the following sections.

A. ISOVPE conversion

Conversions using a one zone furnace (source and substrate are located in the same temperature regime) is the simpler method. We have studied four techniques in this category and we will call them ISOVPE conversions followed by the source types. The geometry of each conversion ampoule and the vapor pressure of each species are illustrated in Fig. 1.

The ISOVPE conversion process with MCT ($x=0.3$) source is the reverse of the ISOVPE growth process.[2] The Te-rich MCT ($x=0.3$) source is fabricated by the method described in Ref. 3. As shown in Fig. 1(a), vapor pressures of Hg and Te are higher at the specimen side because the mole fraction of the CdTe is lower at the specimen than at the source; hence Hg and Te are transported from the specimen to the source, thus increasing the x value of the surface of the sample. The higher x source material in this conversion technique maintains the ambient vapor pressure inside the conversion ampoule and controls the surface concentration of the specimen.

If Cd vapor pressure is higher in the ambient than the equilibrium Cd pressure of the specimen, the surface of the lower x value MCT can be converted to the higher x value MCT. Figure 1(b) shows an ISOVPE conversion with a Cd source. Small pieces of Cd and substrate MCT are separately placed in an evacuated quartz ampoule and are kept at 500 °C in a furnace. In this ampoule, this conversion condition is satisfied. Mercury vaporizes out from the surface of substrate MCT due to its higher equilibrium vapor pressure at these temperatures and the higher Cd vapor pressure in the ambient deposits Cd atoms into the Te-rich surface of the substrate MCT forming a CdTe surface layer. The surface of the substrate MCT is converted into CdTe and can be polished up to the region of graded concentration. Alternatively, MCT with the desired composition can be grown on the surface by the ISOVPE growth technique.

The other two techniques in the ISOVPE conversion cate-

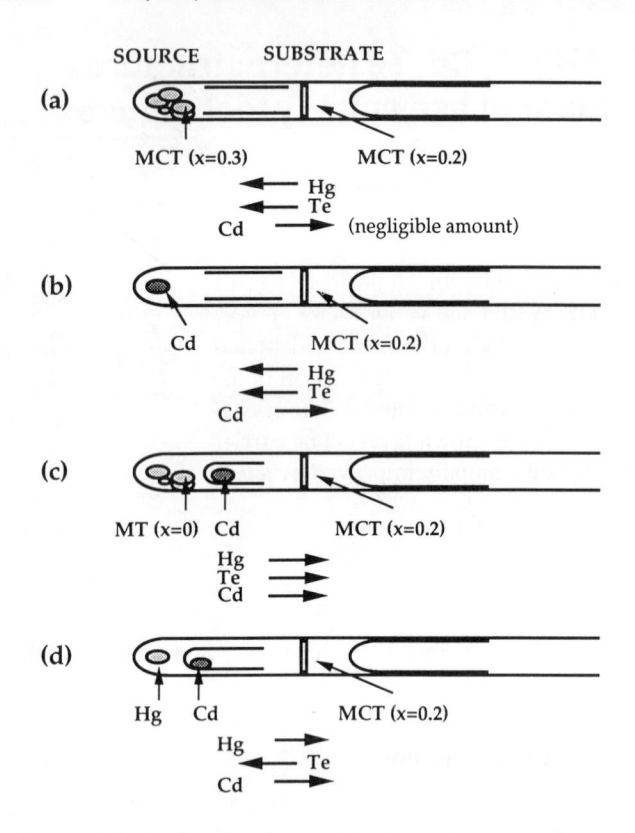

FIG. 1. The geometry of ISOVPE conversion ampoules. Typical processing temperature and time for each conversion technique is: (a) ISOVPE conversion with MCT ($x = 0.3$) source, 450 °C, 24 h, (b) ISOVPE conversion with Cd source, 500 °C, 2–50 h, (c) ISOVPE conversion with HgTe and Cd sources, 450 °C, 5 h, (d) ISOVPE conversion with Hg and Cd sources, 375–400 °C, 5–15 h.

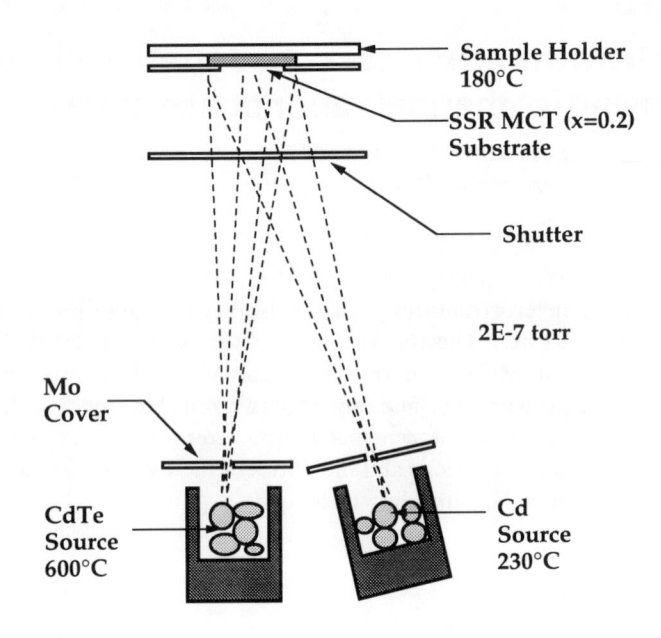

FIG. 2. Deposition of CdTe layers in vacuum evaporation chamber.

gory are conducted using either HgTe and Cd as sources or Hg and Cd as sources [Figs. 1(c) and 1(d)]. These methods are basically the same as the ISOVPE conversion technique with a Cd source [Fig. 1(b)] except that the fast loss of Hg from the surface of the substrate MCT is prevented by controlling the ambient Hg pressure with Te-rich HgTe or Hg sources in addition to the Cd source. The conversion process with HgTe and Cd sources is conducted at 450 °C for 5 h, but the conversion process with Hg and Cd sources is conducted at the lower temperatures (375 and 400 °C) to prevent high Hg pressure in the annealing ampoule.

B. VPE conversion

Another possible method to fabricate the heterostructures is to use two zone or multizone VPE techniques. We have studied two techniques in this category: the growth of CdTe on the surface of substrate MCT using a conventional vacuum evaporation chamber; and the growth of CdTe layers in the evacuated quartz ampoule using a two zone furnace.

The schematic diagram of the vacuum evaporation chamber used for the growth of the CdTe layer is shown in Fig. 2. The Te-rich CdTe source and the Cd source are placed in the source holder made of alumina, which is wrapped with the heating wire. The outgoing vapors are aimed to go directly to the MCT substrate and the substrate is held in a heating stage. The vacuum level inside the chamber is 2×10^{-7} Torr and the MCT substrate, CdTe source, and Cd source are heated to 180, 600, and 230 °C, respectively. We fix the temperature of the MCT substrate at 180 °C, because 180 °C is a reasonable temperature for MCT ($x = 0.2$) without severe loss of Hg in the evaporation chamber. The temperatures of the CdTe and Cd sources are fixed too, and the deposition rate of the CdTe is controlled by covering the source with molybdenum foil with a small hole. Growth rates of 1 and 2.5 μm/h are used for the CdTe layers. The role of Cd source is to compensate for the lower sticking coefficient of Cd than that of Te.[4]

As shown in Fig. 3, VPE conversion with a CdTe source is

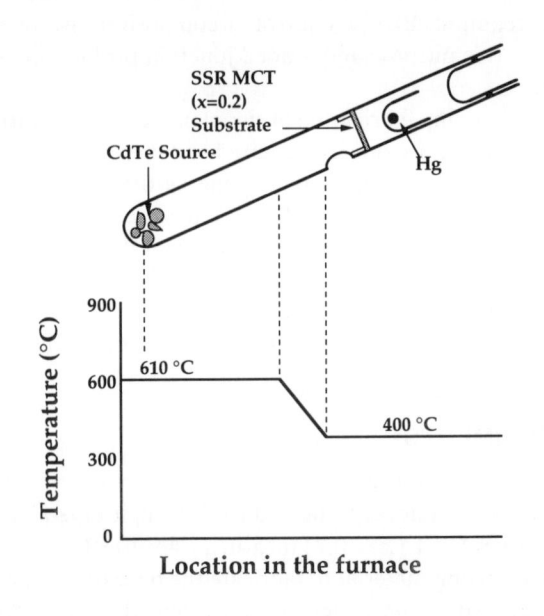

FIG. 3. Geometry of VPE conversion ampoule and temperature profile of the two zone furnace.

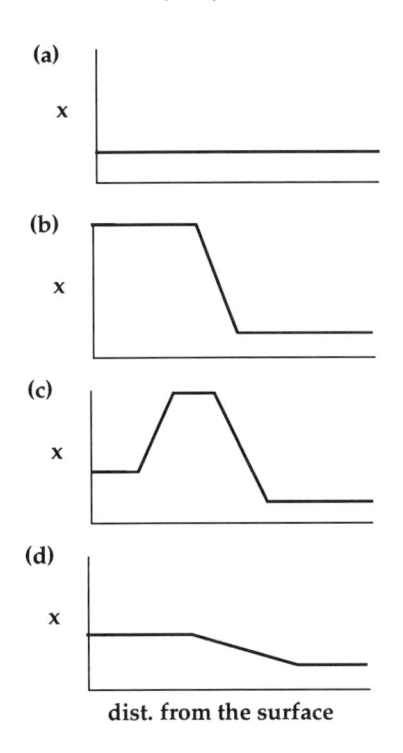

FIG. 4. Two stage conversion mechanism Stage 1: (a) SSR MCT ($x = 0.2$) starting material, (b) Formation of CdTe Layer (either by VPE or ISOVPE) on SSR MCT; Stage 2: (c) Wider band gap MCT ($x = 0.3$) grown on CdTe interlayer by ISOVPE, (d) CdTe Interlayer Disappeared by Hg–Cd Interdiffusion for longer ISOVPE growth.

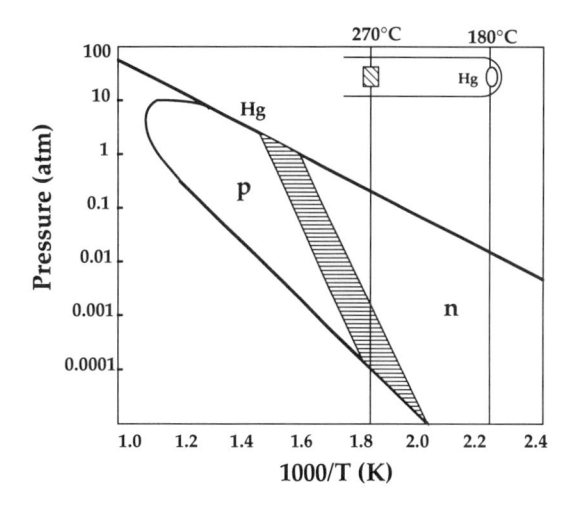

FIG. 5. Carrier concentration control using two zone Hg- annealing (graph is referred from the work of Jones et al. Ref. 5).

conducted using a two zone furnace. The Te-rich CdTe source is located at the higher temperature (610 °C) and the MCT substrate is located at the lower temperature (400 °C). In order to avoid the condition where the deposition of CdTe occurs before the vapors reach the substrate, the substrate is placed at the border of the lower temperature region where the temperature changes from 400 °C to a higher value. To prevent severe Hg loss at 400 °C, a small amount of Hg is added to the ampoule. The amount of the Hg is calculated to maintain the Hg pressure of the ampoule as 2–3 atm at 400–600 °C.

C. Two stage conversion

In order to get relatively thick MCT layers with the desired compositions ($0.3 < x < 0.7$), a two stage conversion is proposed: in its first stage, CdTe is grown on top of narrow band gap MCT and in the second stage, the desired composition is grown on a CdTe layer by the ISOVPE technique. The mechanism of this two stage conversion process is shown in Fig. 4. This technique provides a flat surface composition profile thicker than 10 μm, which is preferred for on-chip fabrication of signal processing circuitry.

D. Two zone mercury annealing

As-grown MCT is usually p type due to retrograde Te solubility but the material can be converted to n type by low temperature anneals in Hg ambient. The material is converted to n type either as a result of Hg interstitials or as the result of donor impurities which become dominant as the metal vacancies are annihilated. In either case, lowering the Hg pressure in the annealing ampoule without changing the specimen temperature is a reasonable approach to lower the carrier concentration of post-growth annealed MCT layers, while maintaining the MCT specimen temperature at 270 °C, Hg pressure in the annealing ampoule is lowered to one tenth of its value at 270 °C by lowering the temperature of the Hg reservoir to 180 °C using a two zone furnace. The annealing condition is shown schematically in Fig. 5.[5] Two zone Hg annealing has been conducted on the ISOVPE-grown MCT layers (30–50 μm thick) for 16–24 h to investigate the controllability of carrier concentration using this technique.

E. Characterization of heterostructures

The composition profiles of the heterostructures are measured by electron probe microanalysis (EPMA) with the wavelength dispersive spectroscopic mode. The standards used for the analysis are mercury sulphide (HgS) and elemental tellurium and cadmium. An accuracy of ± 0.005 of the x value can be obtained with EPMA. The surface morphology of the heterostructures is examined by optical microscopy. The thickness of the CdTe layer grown in the vacuum evaporation chamber is measured using an alpha step apparatus. The Laue method and Read camera are used for investigating the crystal quality of the converted layers. Electrical characterization of the post-growth treated ISOVPE-grown layers is performed by Hall measurements at 77 K using the van der Pauw technique.

III. RESULTS AND DISCUSSION

A. Chemical composition profiles of the heterostructures

The total thickness of the converted region and the junction profiles of the heterostructures are listed as Table I with the conversion conditions. Figure 6 shows the actual composition profiles of the heterostructures measured using EPMA. Five heterostructures converted by the ISOVPE technique show graded junction profiles compared to the abrupt profile of the evaporated CdTe layer. Roughly 1 to 2 μm thick wide band gap MCT grown by molecular-beam epitaxy (MBE) technique is reported to increase the well capacity of the MIS structure by improving the breakdown electric field.[1] However, graded junction profiles are also

TABLE I. The thickness of the converted layers and the junction profiles of the heterostructures for various conversion conditions.

Conversion technique	Source material	Source temp. (°C)	Substrate temp. (°C)	Conversion rate (μm/h)	Thickness of converted layer (μm)	Junction shape
ISOVPE	MCT (0.3)	450	450	0.2	5	Graded
ISOVPE	Cd	500	500	3–10	20–150	Graded
ISOVPE	Cd HgTe	450 450	450	2	10	Graded
ISOVPE	Cd Hg	400 400	400	0.5	8	Graded
ISOVPE	Cd Hg	375 375	375	0.9	4	Graded
VPE	CdTe Cd	600 230	180	2.5	10	Abrupt
VPE	CdTe Cd	600 230	180	1	8.5	Abrupt
VPE	CdTe Hg	610 400	400	N/A	N/A	N/A

acceptable for this purpose and the precise control of composition is not necessary. ISOVPE converted MCT is a good candidate for fabricating enhanced MIS detectors.

The optimum time and temperature for the ISOVPE conversion with MCT ($x = 0.3$) source are 24 h at 450 °C. For the higher conversion temperature or longer conversion time, surface morphology is worse and for the lower temperature, conversion is too slow without improving the surface morphology.

A composition profile of the heterostructure converted with the Cd source at 500 °C for 2 h is shown in Fig. 6 (profile B). The converted layer is CdTe at the surface and the thickness of the CdTe depends on the conversion time and temperature as well as the free volume of the ampoule. For the ampoule with the larger free volume, more mercury can leave the specimen before the vapor pressure of mercury reaches the equilibrium vapor pressure of the specimen; hence, there is more chance for cadmium to be substituted into mercury vacancy site. Thick CdTe layers (> 10 μm) can be formed by this method.

The composition profiles of the heterostructures converted by the Hg compensated ISOVPE technique with a Cd source are shown in Fig. 6 (profiles C, D, and E). The surface compositions of the C, D, and E heterostructures are 0.9, 0.8, and 0.4, respectively, in terms of x value, and these profiles are appropriate for improved MIS device fabrication after some polishing.

The VPE conversion process using a vacuum chamber produces an abrupt junction profile as profile F in Fig. 6, but the Hg compensated VPE conversion process (Fig. 3) deposits Te-rich (85–90 at. % Te) layers on the MCT substrate. Currently we don't have an explanation for this unexpected result.

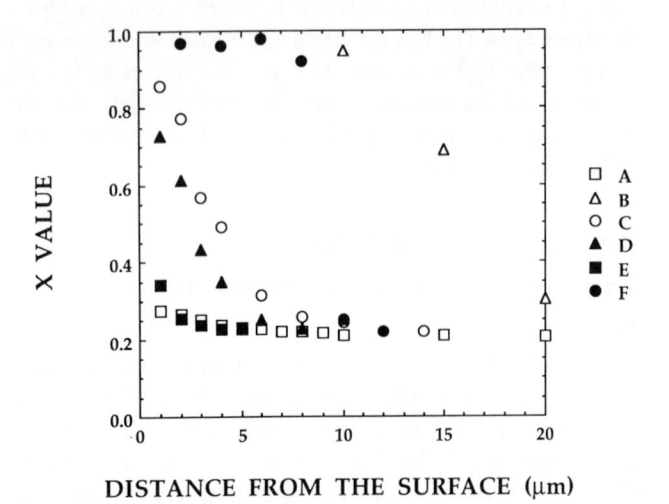

FIG. 6. Composition profiles of the heterostructures converted with following conditions. (measured by EPMA). A: ISOVPE with MCT ($x = 0.3$) source, 450 °C, 24 h; B: ISOVPE with Cd source, 500 °C, 2 h; C: ISOVPE with Cd and HgTe source, 450 °C, 5 h; D: ISOVPE with Cd and Hg source, 400 °C, 15 h; E: ISOVPE with Cd and Hg source, 375 °C, 5 h; F: VPE in vacuum evaporation chamber, 1 μm/h.

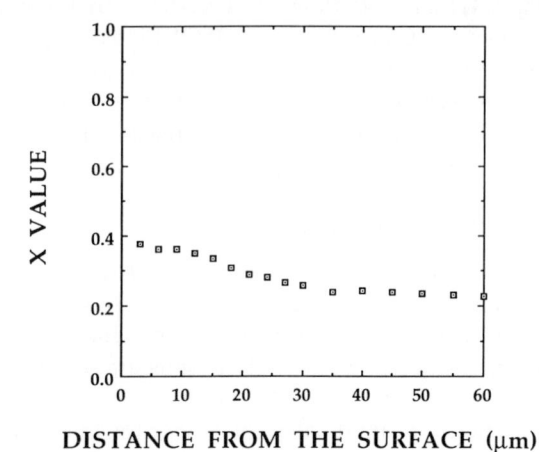

FIG. 7. Composition profile of two stage converted heterostructure starting material: SSR MCT ($x = 0.2$). First stage: ISOVPE conversion with Cd source, 450 °C, 2 h; second stage: ISOVPE growth with MCT ($x = 0.3$) source, 550 °C, 5 h.

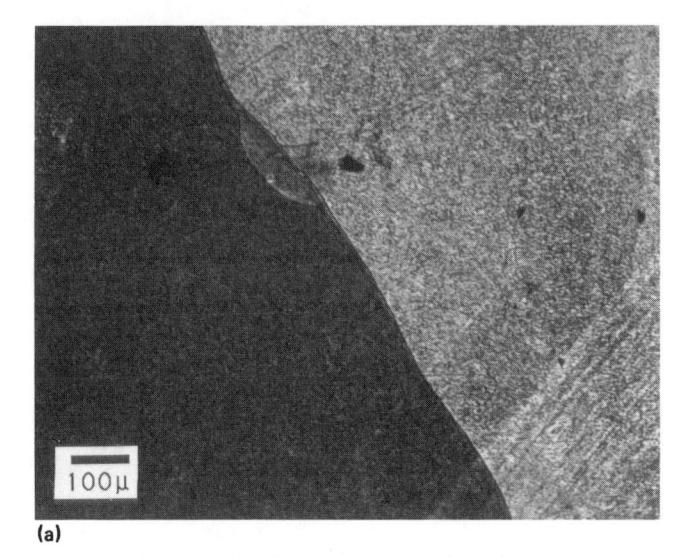

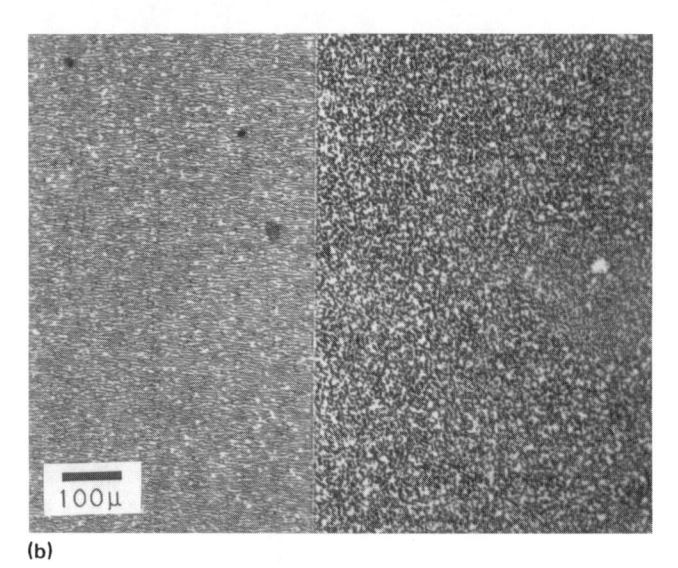

FIG. 8. The dependence of surface morphology to the orientation of the starting materials. (a) Sample with a grain boundary, (b) Sample with a twin boundary. (a) & (b) are converted by ISOVPE method with MCT ($x = 0.3$) source at 450 °C, 24 h.

FIG. 9. The surface morphology of the ISOVPE converted MCT with MCT ($x = 0.3$) source is porous with some cracks. (450 °C, 50 h).

The trial two stage conversion process was conducted using an ISOVPE converted layer with a Cd source and the composition profile of this heterostructure is shown in Fig. 7. This profile is similar to Fig. 4(d).

B. Surface morphology of heterostructures

The surface morphology depends both on orientation of the MCT substrate and the conversion time for the layers converted by the ISOVPE method with a MCT ($x = 0.3$) source. Two converted specimens with a grain boundary and a twin boundary show different surface morphology at each single crystal region (Fig. 8). Even with the optimum condition (450 °C, 24 h), the surface needs further polishing for device application as shown in Fig. 8. Longer conversion (50 h) at 450 °C results in a rougher surface morphology which is not shiny and looks frosted to the bare eye and is porous with some cracks under the microscope, while the junction profile remains the same (Fig. 9).

The ISOVPE conversion process with a Cd source produces a smooth looking surface compared with the surfaces in Figs. 8 and 9, although not as shiny as the polished surface or the ISOVPE-grown layers. As shown in Fig. 10, further polishing is required to get a shiny surface.

The Hg compensated ISOVPE conversion processes with Cd sources produce better quality surfaces than the noncompensated ISOVPE converted layers with a Cd source. The specimen E in Fig. 6 (converted with the Hg–Cd source) shows a mirror-like surface as shown in Fig. 11. No subsequent polishing is required for device application on this specimen.

The surfaces of some CdTe layers evaporated in the vacuum chamber are as shiny as a polished surface and no polishing is required for the further growth of wider band gap MCT on it. Two surfaces evaporated in the vacuum chamber are shown in Fig. 12.

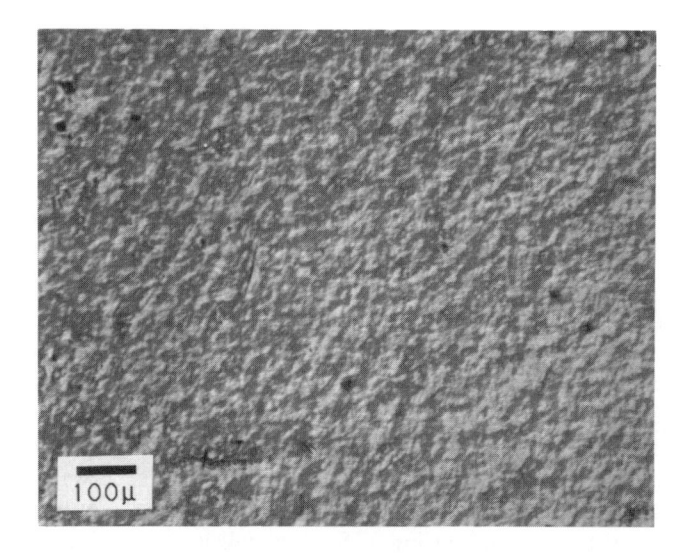

FIG. 10. The surface morphology of the ISOVPE converted MCT with Cd source. Smooth looking surface to bare eyes, although not as shiny as the polished surface, is rough under the microscope. (500 °C, 5 h).

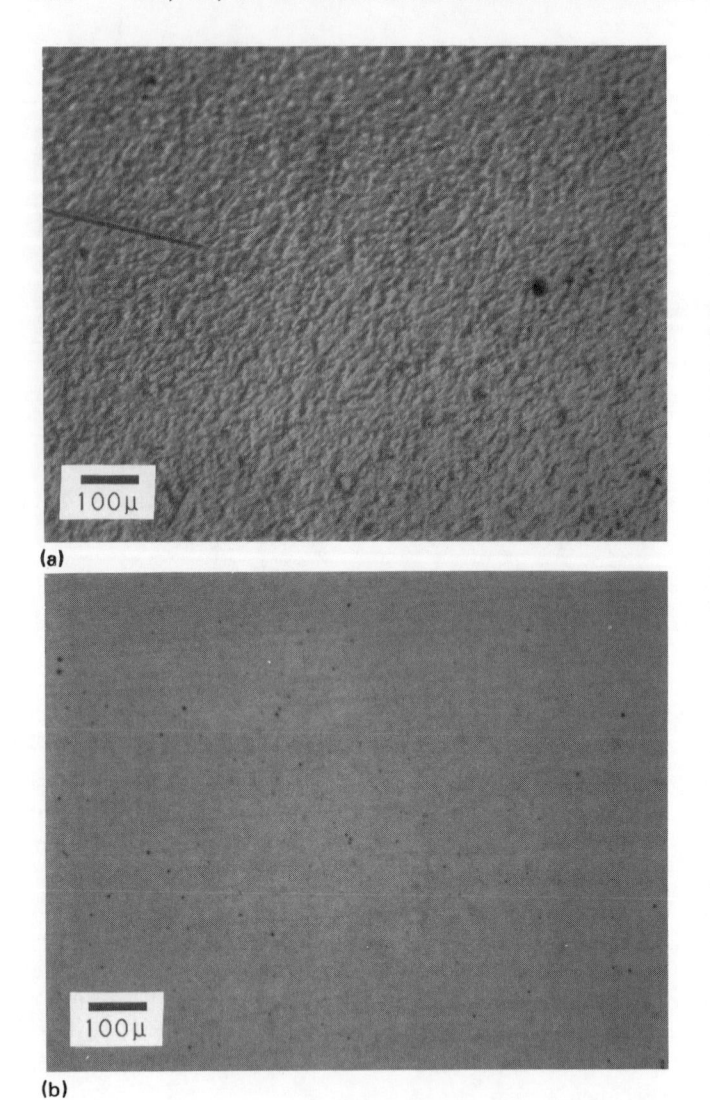

FIG. 11. The surface morphology of the Hg-compensated ISOVPE converted MCT. (a) Sample with a Cd–HgTe source, 450 °C, 5 h, (b) Sample with a Cd–Hg source, 375 °C, 5 h, [(b) shows mirror-like surface after conversion].

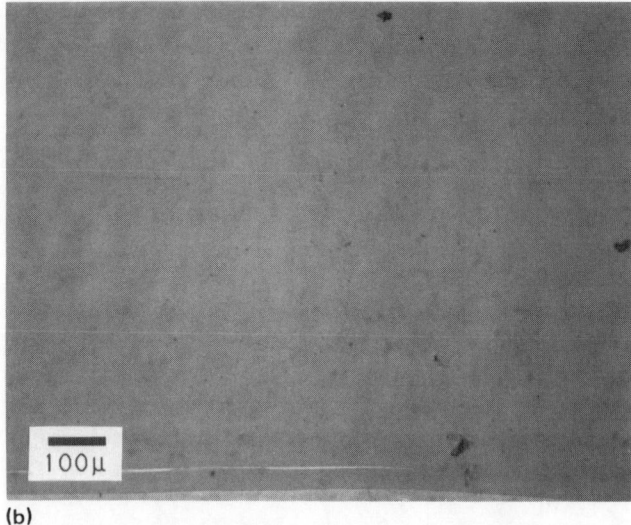

FIG. 12. The surface morphology of the CdTe layers evaporated in the vacuum chamber. (a) Sample G in Table II, growth rate: 2.5 μm/h, (b) Sample H in Table II, growth rate: 1 μm/h (sample holder: 180 °C, CdTe source: 600 °C, Cd source: 230 °C).

C. Crystal quality of heterostructures

Crystal quality of the converted layers is evaluated by optical microscopy and x-ray techniques, and listed in Table II. Layers B, F, and G are not single crystals, but the other layers are. Many Kirkendall voids have been found on layer B and some voids are found under optical microscopy at X500 magnification for specimens C and D. The hazy Laue spots of specimen C and D may be explained by these voids and a locally distorted lattice arising from these voids.

Both layers A and E show acceptable crystalline quality for device application when examined by optical microscopy and the Laue method. There are no voids and no obvious defects in the converted region of these specimens. If other factors are also concerned for device applications, E has several obvious advantages. Surface polishing is not required for E but is required for A. Surface polishing reduces the thickness of the converted layers and also reduces the x value at the surface. The conversion time is shorter for E and a Hg–Cd source is simpler to prepare than a MCT ($x = 0.3$)

TABLE II. Crystal quality of the converted heterostructures.

Conversion condition[a]	Single crystal	Converted layer thickness (μm)	Used x-ray technique	Kirkendall voids
A	Yes	5	Laue	No
B	No	20	Laue	Many
C	Yes[b]	10	Laue	Few
D	Yes[b]	8	Laue	Few
E	Yes	4	Laue	No
F	No	8.5	Laue	No
G	No	10	Laue	No
H	Yes	2	Read	No

[a] A: ISOVPE with MCT ($x = 0.3$) source, 450 °C, 24 h; B: ISOVPE with Cd source, 500 °C, 2 h; C: ISOVPE with Cd and HgTe source, 450 °C, 5 h; D: ISOVPE with Cd and Hg source, 400 °C, 15 h; E: ISOVPE with Cd and Hg source, 375 °C, 5 h; F: VPE in vacuum evaporation chamber, 1 μm/h; G: VPE in vacuum evaporation chamber, 2.5 μm/h; H: VPE in vacuum evaporation chamber, 1 μm/h.

[b] Laue spot is not as bright as other single crystal layers, but doesn't show any elongated shape.

source. The composition profile of E is more abrupt than that of A, as shown in Fig. 6. An abrupt junction profile is not necessary for the MIS application, but is preferred.

D. Improved electrical properties of ISOVPE-grown MCT

Two zone mercury annealing (the Hg reservoir at a lower temperature than the annealed sample) has been conducted for the ISOVPE-grown MCT ($x = 0.2$) layers and is effective to lower the carrier concentration. Carrier concentrations of 5×10^{14}/cm^3 to 1×10^{15}/cm^3 and mobilities of 70,000 to 90,000 cm^2/V s are measured reproducibly for the ISOVPE-grown MCT layers. These are remarkably improved electrical properties compared with the properties of the one zone (270 °C) Hg annealed samples (Carrier concentrations: 1×10^{15}/cm^3 to 1×10^{16}/cm^3, mobilities: 15 000–30 000 cm^2/V s).

When a p-type MCT is converted to an n-type MCT by Hg annealing, if Hg vacancies are still the dominant native defects over Hg interstitials and the type conversion is achieved by n-type impurities which become comparable to Hg vacancies in number as Hg vacancies are reduced by Hg annealing, the only way to reduce the carrier concentration and to increase the mobility of n-type MCT is reducing the concentration of n-type impurities. If this statement is true and two zone Hg annealing (lower Hg pressure than one zone annealing) reduces the n-type carrier concentration by allowing more vacancies which can compensate n-type impurities, improved carrier mobility is hard to explain because there is no decrease in charged impurity scattering.

If Hg interstitials are the dominant defects which convert p-type MCT to n type, improved carrier mobility with decreased carrier concentration is easily explained. Currently we do not have information about n-type impurity concentrations, but improved carrier mobility with decreased carrier concentration indicates that Hg interstitials could be the dominant factor for type conversion of ISOVPE-grown layers.

The surface conversion methods mentioned above produce p-type heterostructures and two zone mercury annealing can also be effective to lower carrier concentration of heterostructures MCT.

IV. SUMMARY

The main achievements of the present work are: the growth of heterostructures with wider band gap MCT at the surface region for better performance of MIS devices and on-chip signal processing, utilizing relatively easy and inexpensive methods like the ISOVPE and VPE techniques; and the development of a two zone mercury annealing method to control the carrier concentration.

(1) Six methods are studied as the heterostructure fabricating techniques on the SSR MCT ($x = 0.2$) substrates. All four ISOVPE conversion processes produce heterostructures with graded composition profiles, which are suitable for enhanced performance MIS devices. The VPE conversion process produces an abrupt junction profile. The heterostructure formed by the two stage conversion technique shows a flat composition profile which is adequate for on-chip fabrication of the signal processing circuitry.

(2) The surface morphology and the crystal quality of the converted heterostructures are evaluated and some layers have mirror-like surface morphology and good crystal quality. The Hg compensated atmosphere and lower conversion temperature greatly improve the surface morphology and crystal quality.

(3) The electrical properties of the ISOVPE-grown layers are improved by using the two zone Hg annealing technique. Both the carrier concentration and the mobility are improved and this indicates that the dominant defect for the type conversion could be Hg interstitials. The same method can be used to control the carrier concentration of heterostructures.

ACKNOWLEDGMENTS

The authors would like to thank Dr. J. H. Tregilgas and Dr. H. F. Schaake of Texas Instruments, Incorporated for helpful discussions and encouragement.

[1] M. W. Goodwin, M. A. Kinch, and R. J. Koestner, J. Vac. Sci. Technol. A **8**, 1226 (1990).
[2] S. B. Lee, L. K. Magel, M. F. S. Tang, D. A. Stevenson, J. H. Tregilgas, M. A. Goodwin, and R. L. Strong, J. Vac. Sci. Technol. A **8**, 1098 (1990).
[3] J. G. Fleming and D. A. Stevenson, J. Cryst. Growth **82**, 621 (1987).
[4] J. M. Arias, S. H. Shin, D. E. Cooper, M. Jandian, J. G. Pasko, E. R. Gertner, R. E. DeWames, and J. Singh, J. Vac. Sci. Technol. A **8**, 1025 (1990).
[5] C. L. Jones, M. J. T. Quelch, P. Capper, and J. J. Gosney, J. Appl. Phys. **53**, 9080 (1982).

Dislocation density reduction by thermal annealing of HgCdTe epilayers grown by molecular beam epitaxy on GaAs substrates

J. M. Arias, M. Zandian, S. H. Shin, W. V. McLevige, J. G. Pasko, and R. E. DeWames
Rockwell International Science Center, 1040 Camino dos Rios, Post Office Box 1085, Thousand Oaks, California 91358

(Received 3 October 1990; accepted 26 November 1990)

Post growth thermal annealing has been used to reduce the threading dislocation density of $Hg_{1-x}Cd_xTe$ ($0.20 \leqslant x \leqslant 0.28$) epilayers grown on (211)B GaAs substrates by molecular beam epitaxy. Etch pit density studies indicate an order of magnitude reduction on the surface threading dislocations after annealing at 490 °C for 30 min. The dislocation density at the HgCdTe surface on this highly mismatched system is only a factor of 2–6 times higher than the best values (1×10^5 cm^{-2}) we have obtained using CdZnTe bulk lattice-matched substrates. The reduction of dislocations may be due to enhanced dislocation movement and their annihilation and coalescence at Hg vacancies point defect pinning centers introduced during the annealing process.

I. INTRODUCTION

The use of GaAs and Si substrates for HgCdTe epitaxial growth is highly desirable for the manufacturing of large focal plane arrays (greater than 128×128) using the hybrid approach. These HgCdTe multilayer structures make it conceivable to integrate signal processing and infrared detection in a monolithic structure. One of the major problems in obtaining high quality HgCdTe films on GaAs is the high defect generation that takes place because of the large lattice mismatch (14.5%) between the CdTe buffer layer and the GaAs substrate.[1] A threading dislocation density in the mid 10^6 cm^{-2} range is obtained for molecular beam epitaxy (MBE) HgCdTe epilayers grown on (211)B GaAs.[2] Dislocations affect the HgCdTe infrared detector dark currents for devices operating at low temperatures[2,3] ($\leqslant 77$ K) because they are thought to produce mid-gap states in the bandgap; therefore it is highly desirable to reduce their density.

We have used post-growth thermal annealing to consistently reduce the threading dislocation density by an order of magnitude for large area HgCdTe epilayers grown on (211)GaAs and GaAs/Si substrates. Etch pit densities (EPD) as low as 2×10^5 cm^{-2} have been obtained. This is a significant result since the EPD values of these highly mismatched systems are close to the MBE HgCdTe epilayers grown on lattice-matched CdZnTe substrates.[4,5] In this paper, we present the thermal annealing studies followed to achieve the reduction of the dislocation density. We also discuss a possible annealing mechanism for the annihilation of dislocations.

II. EXPERIMENT

The selection of the (211)B orientation was motivated by our work on (211)B CdZnTe substrates,[4] where very good surface morphology and diode performance were obtained on epilayers grown on this orientation. The surface was smooth and free of hillocks, oval defects, and antiphase domains[6,7] (laminar twins) which are common growth defects present on other orientations [i.e., (100), (111)B and (110)] and degrade diode performance.

The epilayers were grown by MBE on (211)B bulk GaAs substrates and epitaxial (211)B GaAs grown on bulk Silicon substrates. The (211)B GaAs/Silicon substrates were purchased from the Kopin Corporation; metal-organic chemical vapor deposition (MOCVD) was the method used to grow the undoped GaAs thin films (2 μm thick) on silicon. Substrate preparation was carried out by etching the GaAs epilayer in a $H_2SO_4/H_2O_2/H_2O$ (5/1/1) chemical solution at room temperature to remove about 0.5 μm of the layer. In-situ thermal cleaning was carried out in the growth chamber by heating the substrate to 595 °C for 5 min to desorb the protective oxide layer on the GaAs surface.[2] To avoid both Ga outdiffusion from the GaAs and an extremely high dislocation density in the HgCdTe, a 4–8-μm-thick CdTe buffer layer was grown at 300 °C under Te-stabilized growth conditions.[8,9]

The system and procedures employed here have been previously described.[4] The growth of the $Hg_{1-x}Cd_xTe$ epilayer was carried out without substrate rotation with effusive sources of Hg, CdTe, and Te$_2$, and an initial Hg/Te$_2$ relative flux ratio of about 85, as measured with a nude ion gauge in the growth position and corrected for atomic and molecular weight and ionization efficiency. The minimum Hg flux necessary to maintain monocrystalline HgCdTe growth under these conditions was determined by reflection high-energy electron diffraction (RHEED). Growth rates were about 5 μm/h and the thickness of the epitaxial heterostructures was in the 15–25 μm range. The growth of HgCdTe was carried out at 190–200 °C.

The relatively high temperature ($T > 410$ °C) annealing of the HgCdTe/CdTe/GaAs and HgCdTe/CdTe/Si epilayers was carried out in a high pressure system to control the Hg vapor pressure.[10] Annealing at $T < 410$ °C was carried out in a closed ampoule under Hg environment. After the high temperature annealing process was completed, the epilayers were *n* type annealed at 250 °C under a Hg environment to remove nonstoichiometric Hg vacancies induced by the annealing process.[11]

(211)B CdTe epilayer threading dislocations were investigated with the EPD technique using the lactic

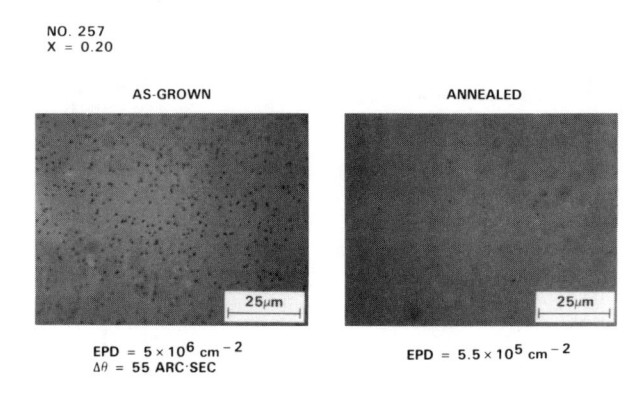

FIG. 1. Surface EPD photos of as-grown and 490 °C annealed (211)B Hg$_{0.80}$Cd$_{0.20}$Te/CdTe/GaAs structure (No. 257) grown by MBE.

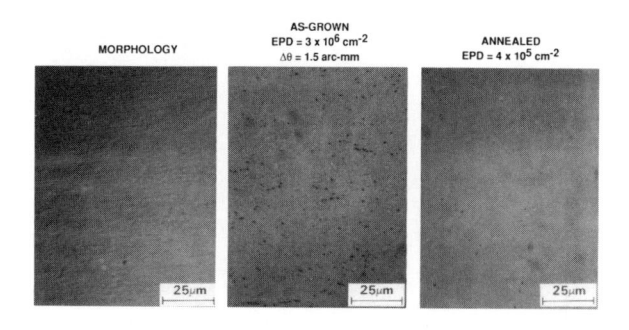

FIG. 2. Surface morphology and surface EPD photos of as-grown and 490 °C annealed (211)B Hg$_{0.75}$Cd$_{0.25}$Te/CdTe/GaAs/Si structure (No. 269) grown by MBE.

acid/HNO$_3$/HF (20 mℓ/4 mℓ/1 mℓ) chemical solution; this etch also reveals pits on the (111)B orientation. (211)B HgCdTe epilayer EPD threading and misfit dislocations values were obtained using the HNO$_3$/K$_2$Cr$_2$O$_7$/HCl/H$_2$O chemical solution.[12]

III. RESULTS

In Table I, we have summarized the (211)B Hg$_{1-x}$Cd$_x$Te/CdTe/GaAs as-grown and high temperature annealed material characteristics. Sample 269 was grown on (211)B GaAs/Si. HgCdTe alloy cadmium compositions were in the 0.20–0.28 range. The epitaxial surface morphology of the as-grown and annealed samples were smooth and free of common macroscopic defects such as: hillocks, oval defects, and antiphase domains.

Typical full width at half-maximum double-crystal x-ray rocking curve measurements for the Hg$_{1-x}$Cd$_x$Te/CdTe layers grown on the (211)B GaAs substrates are in the 1.0–1.5 arc min range. The full width at half maximum ($\Delta\theta$) for the purchased GaAs on Silicon was 2.4 arc min. After a CdTe buffer layer and a HgCdTe layer (17 μm thick) were grown, the $\Delta\theta$ was reduced to 1.5 arc min which clearly indicates an improvement in the crystallinity. This value is comparable to those obtained by growing HgCdTe on GaAs bulk substrates.

Typical EPD threading dislocations values in the 10^6 cm^{-2} range were measured in the as-grown MBE HgCdTe epilayers grown on CdTe(buffer layer)/GaAs as shown in Table I. In this table, we also illustrate the reduction of EPD values on a set of HgCdTe epilayers grown on GaAs and GaAs/Si (No. 269) obtained after annealing at 490 °C for 30 min and 390 °C for 120 min. A reduction at the surface of about one order of magnitude is obtained after the 490 °C annealing.

In Fig. 1, we show a typical photo of a (211)B HgCdTe/CdTe/GaAs (No. 257) surface after chemical etching has been carried out to reveal threading dislocations in an as-grown and annealed HgCdTe layer. An EPD value of 5 $\times$ 10^6 cm^{-2} was measured on the as-grown material, but after the high temperature annealing process was carried out, the surface EPD value dropped by one order of magnitude to 5.5 $\times$ 10^5 cm^{-2}. In Fig. 2, we also illustrate the reduction of EPD values by thermal annealing on a HgCdTe epilayer grown on (211)B CdTe/GaAs/Si (No. 269). A reduction of about one order of magnitude (from 3 $\times$ 10^6 to 4 $\times$ 10^5 cm^{-2}) is also obtained. Fig. 3 shows the correlation between the surface EPD in the as-grown and thermal annealed HgCdTe/CdTe/GaAs samples. The electrical properties (Hall mobilities and minority carrier lifetimes), carrier concentrations and surface morphologies of these annealed epilayers are those of normal HgCdTe epilayers grown by other epitaxial techniques.[13]

TABLE I. MBE Hg$_{1-x}$Cd$_x$Te/CdTe/GaAs as-grown and annealed material properties.

Layer No.	Cd composition (x)	Thickness (μm)	As-grown EPD (cm^{-2})	490 °C annealed EPD (cm^{-2})	390 °C annealed EPD (cm^{-2})
257	0.200	16.2	5.4 $\times$ 10^6	3.0 $\times$ 10^5	
260	0.202	25.0	4.4 $\times$ 10^6	3.0 $\times$ 10^5	
235	0.210	15.0	6.0 $\times$ 10^6	7.4 $\times$ 10^5	1.6 $\times$ 10^6
259	0.222	15.2	2.4 $\times$ 10^6	2.4 $\times$ 10^5	
258	0.227	16.0	2.4 $\times$ 10^6	2.4 $\times$ 10^5	
231	0.248	19.5	1.2 $\times$ 10^7	1.6 $\times$ 10^6	
238	0.248	16.3	6.9 $\times$ 10^6	7.7 $\times$ 10^5	2.1 $\times$ 10^6
269 (Si)	0.256	17.0	3.1 $\times$ 10^6	3.0 $\times$ 10^5	
239	0.278	21.0	3.1 $\times$ 10^6	3.0 $\times$ 10^5	

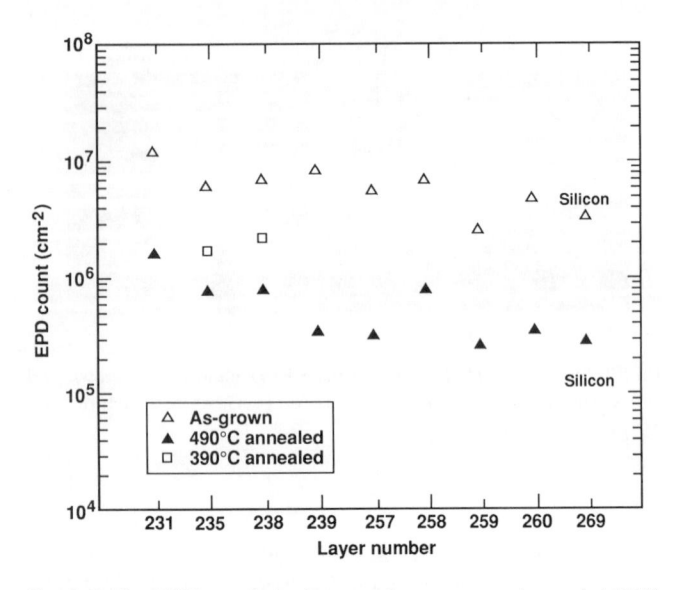

FIG. 3. Surface EPD correlation between the as-grown and annealed (211) HgCdTe epilayers grown by MBE on GaAs substrates.

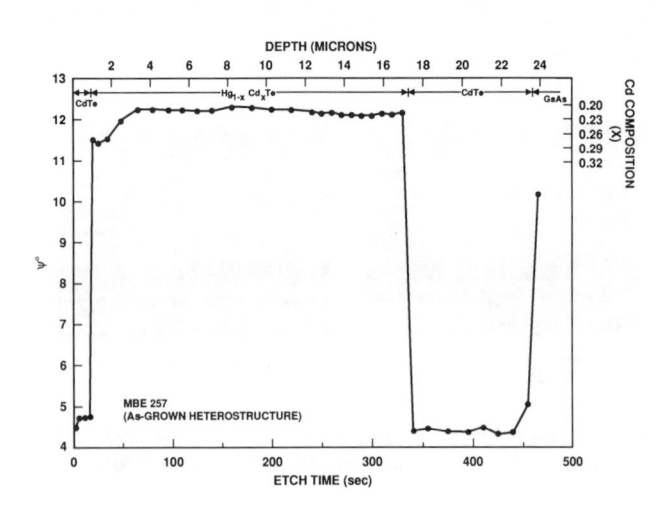

FIG. 5. Compositional depth profile (measured with an ellipsometer) of an as-grown HgCdTe/CdTe heterostructure (No. 257) grown by MBE on a (211)B GaAs substrate.

In Fig. 4, we show a scanning electron microscope (SEM) photo of a cleaved and etched cross-sectional (110) plane of an as-grown (211)B Hg$_{0.77}$Cd$_{0.23}$Te/CdTe/GaAs epilayer. Many misfit dislocations are confined to the Hg$_{0.77}$Cd$_{0.23}$Te/CdTe interface as shown in the figure. This result shows that the lattice mismatch ($\sim 2.3 \times 10^{-3}$) is effectively accommodated by the introduction of misfit dislocations.[14] The misfit dislocation density of this HgCdTe epilayer is estimated to be in the high 10^8 cm^{-2} range near the CdTe interface. As seen in the figure the misfit dislocations are well confined to a very narrow area, which is a result of the very abrupt compositional transition between the HgCdTe epilayer and the CdTe buffer layer. This very abrupt transition is illustrated in Fig. 5 where the compositional depth profile of an as-grown heterostructure (No. 257) was measured with an ellipsometer. In this heterostructure the cadmium composition at the end of the run was continuously graded to $x \sim 0.30$. Step etching was carried out using a Br$_2$–methanol chemical solution.

As-grown HgCdTe epilayers grown by MBE on lattice-matched (211)B CdZnTe substrates have EPD values of at least one order of magnitude lower[4,5] than those grown on GaAs. We believe the higher value of the HgCdTe grown on GaAs is due to threading dislocations propagated into the HgCdTe epilayer from the CdTe buffer (4–8 μm thick) layer, and also probably due to a fraction of threading dislocations which were not effectively bent out at the abrupt interface.

A thick CdTe buffer acts as an intermediate layer to avoid Ga outdiffusion into the HgCdTe, and also reduces the very high misfit dislocation density (10^9–10^{11} cm^{-2}) formed during initial CdTe/GaAs nucleation and growth.[1] A growth of 4 μm or more of (211)B CdTe buffer layer reduces the EPD value to the 10^6 cm^{-2} range. This reduction is primarily produced by the annihilation of dislocations by interaction. In Fig. 6 we show typical SEM pictures of the surface of (211)B CdTe buffer layers grown on GaAs (No. 257, 5 μm thick) and GaAs/Si (No. 269, 8 μm thick) after chemical etching was carried out to reveal threading dislocations in the as-grown material. As illustrated in the figure, well-de-

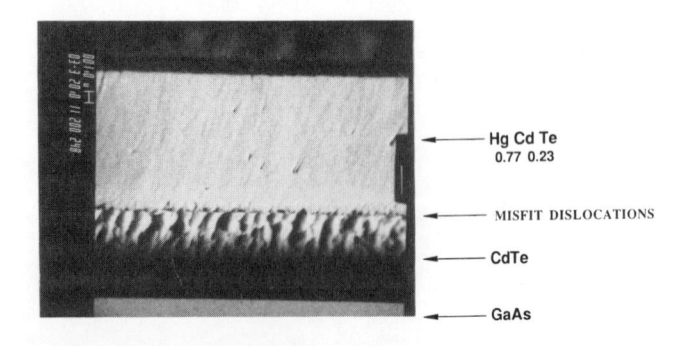

FIG. 4. SEM EPD cross section photo of a (110) cleaved plane of an as-grown MBE (211)B Hg$_{0.77}$Cd$_{0.23}$Te/CdTe/GaAs structure.

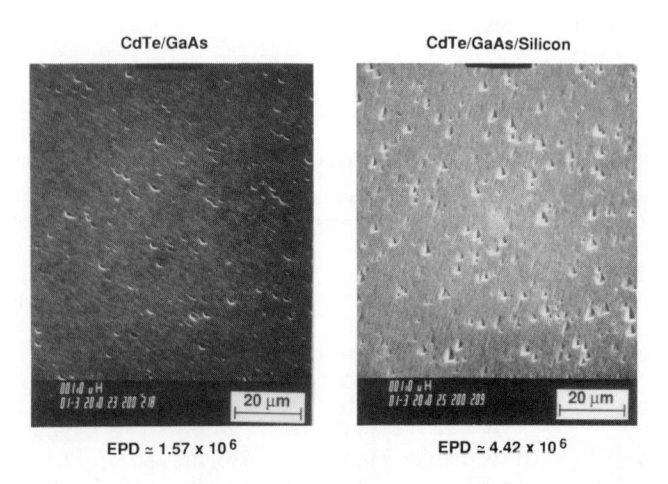

FIG. 6. Surface EPD SEM photos of (211)B CdTe buffer layers grown at 300 °C by MBE on GaAs (No. 257) and silicon (No. 269) substrates.

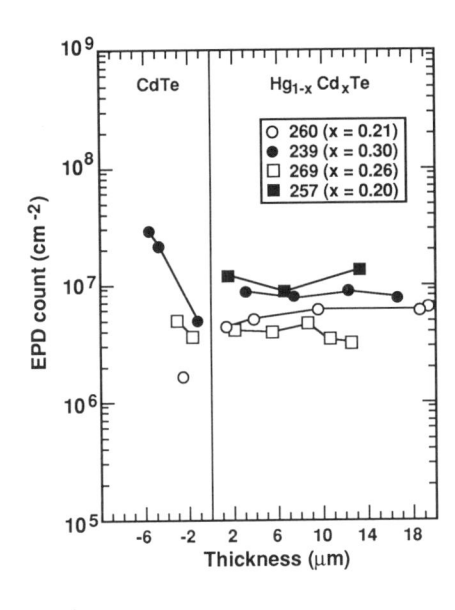

FIG. 7. Surface EPD count as a function of epilayer thickness for as-grown (211)B HgCdTe/CdTe layers grown by MBE on GaAs and GaAs/Si (269) substrates. The solid line connecting the data point is being used as an aid to the eye.

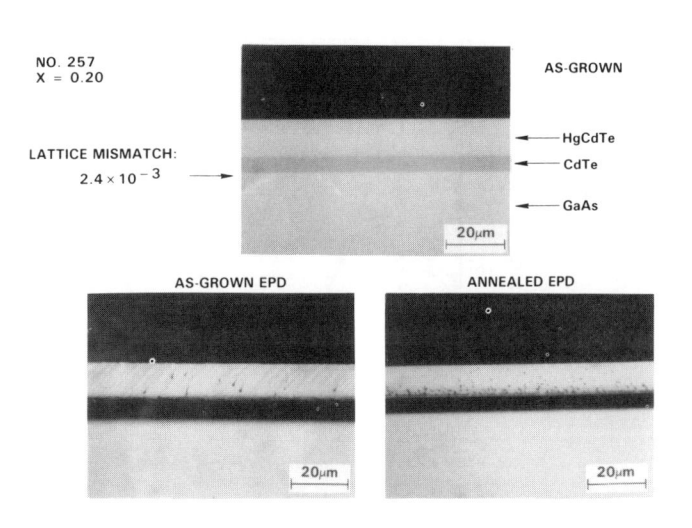

FIG. 8. Photos of the cross-sectional (110) plane of an as-grown layer and the etched EPD dislocation of the as-grown and 490 °C annealed (211)B $Hg_{0.80}Cd_{0.20}Te/CdTe/GaAs$ epilayer (No. 257).

fined pits are obtained using the lactic acid/HNO_3/HF chemical solution. To carry out the EPD analysis of these buffer layers the HgCdTe epitaxial layers were first removed using Br_2–methanol.

In Fig. 7, we have plotted the EPD counts as a function of epilayer thickness for HgCdTe/CdTe layers grown on GaAs substrates. From this figure we observe that the as-grown HgCdTe EPD values are fairly constant in depth. We also observed that the EPD values of the HgCdTe epilayers are higher than those of their corresponding CdTe buffer layers at the interface. If both EPD chemical solutions are revealing all of the threading dislocations in the epilayers, then this indicates that a fraction of the expected misfit dislocations (based on the lattice mismatched) are not being effectively bent out at the abrupt interface, and therefore they thread to the surface. Transmission electron microscopy studies must be carried out to confirm this observation.

In Fig. 8, we show photos of the cross-sectional (110) plane of an as-grown layer and the EPD dislocation etching of the as-grown and annealed (211)B $Hg_{0.80}Cd_{0.20}Te/CdTe/GaAs$ epilayer (No. 257). Notice that the annealed layer has a high density of misfit dislocations concentrated near the HgCdTe/CdTe interface, and a very low density at the HgCdTe surface which is consistent with the measured surface EPD values shown in Fig. 1 and Table I. The same behavior is observed for the (211)B HgCdTe/CdTe/GaAs/Si.

In Fig. 9, we have plotted the surface EPD count for several annealed samples as a function of epilayer thickness. As seen from these EPD surface profiles, a reduction of about one order of magnitude with respect to the as-grown epilayers was obtained at the surface. This is a very significant result because formation of conventional p-on-n electrical junctions in HgCdTe detector occurs near the surface where the dislocation is reduced. The high concentration of dislo-

cations observed near the annealed HgCdTe/CdTe interface is consistent with Fig. 8. The increase in the width of this high density misfit area, as compared to the as-grown layer, indicates that the HgCdTe/CdTe interface has been compositionally graded due to interdiffusion occurring during the annealing process. We have analyzed this interdiffusion effect by ellipsometry. In Fig. 10, the compositional depth profile of a heterostructure (No. 257) after high temperature post growth annealing is illustrated. A graded transition region at the interface of about 3 μm was found.

Thermal annealing is one of the most effective processes for annihilating threading dislocations because it increases dislocation motion with subsequent interaction, annihilation, and formation of looping networks which effectively pin threading dislocations.[15] The reduction of dislocations in MBE HgCdTe/GaAs could probably occur by pinning dislocations at Hg-vacancies point defects formed during growth and high temperature annealing. At thermodynamic equilibrium during the annealing process the net Hg vacancy concentration[16] of a $Hg_{0.80}Cd_{0.20}Te$ layer at 490 °C equilibrated under Hg-saturated conditions ($P_{Hg} = 7.4$ atm) is 7×10^{16} cm^{-3}. MBE growth of $Hg_{0.80}Cd_{0.20}Te$ epilayers at 200 °C is carried out under Te-saturated conditions, and in this case, the Hg vacancy concentration[17] is about 4×10^{15} cm^{-3}; these Hg vacancy concentrations have been reported in the as-grown MBE HgCdTe layers.[2,18]

The Hg vacancies may assist in enhancing dislocation climb, resulting in the reduction of the dislocation density. This annealing dislocation reduction mechanism by enhanced dislocation climb caused by an increase in the vacancy cation concentration has been recently proposed for the GaAs/Si system.[19,20] In this model the excess number of dislocations on one slip plane can climb to another plane where the total number of dislocations can be reduced by further annihilation. During the HgCdTe annealing process the increase in Hg vacancies may enhance the probability of dislocation climb and annihilation by interacting with mov-

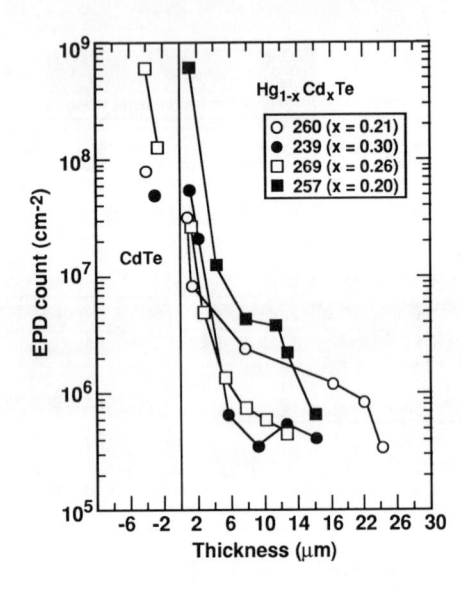

FIG. 9. Surface EPD count as a function of epilayer thickness for 490 °C annealed (211)B HgCdTe/CdTe samples grown by MBE on GaAs and GaAs/Si (269) substrates. The solid line connecting the data point is being used as an aid to the eye.

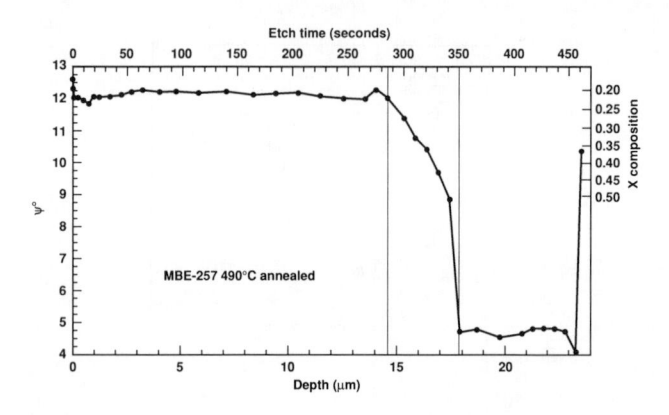

FIG. 10. Compositional depth profile (measured with an ellipsometer) of a 490 °C annealed HgCdTe/CdTe heterostructure (NO. 257) grown by MBE on a (211)B GaAs substrate.

ing dislocations. Supplemental studies specially TEM studies are needed to test this interpretation.

IV. SUMMARY

Post growth thermal annealing has been used to reduce the threading dislocation density of HgCdTe epilayers grown on (211)B CdTe/GaAs by MBE. EPD studies indicate an order of magnitude reduction on the surface threading dislocations. This is a very significant result because dislocations affect HgCdTe infrared detector dark currents for devices operating at low temperatures[2,3] ($T \leqslant 77$ K). The dislocation density at the surface on these highly mismatched systems is only a factor of 2–6 times higher than the best values obtained by using CdZnTe lattice-matched substrates. A model to explain the dislocation reduction in the MBE HgCdTe material has been suggested. In this model the annealing process creates an excess concentration of Hg vacancies in the epilayer, which may act as pinning centers for dislocations, resulting then either directly in annihilation of the dislocations, or increasing dislocation motion and more extensive mutual annihilation of dislocations.

Further improvements in HgCdTe EPD values on GaAs can be obtained by multiple thermal annealing cycles or by reducing the threading dislocation density of the buffer layer (CdTe or CdZnTe) from 10^6 to the 10^5 cm^{-2} range.

ACKNOWLEDGMENTS

We are grateful to E. R. Gertner and J. S. Chen for helpful technical discussions. This work was supported by Rockwell International IR&D funds and by DARPA/CNVEO under contract No. DAAL03-87-C-0014.

[1] J. T. Cheung and T. Magee, J. Vac. Sci. Technol. A **1**, 1604 (1983).
[2] J. M. Arias, R. E. DeWames, S. H. Shin, J. G. Pasko, J. S. Chen, and E. R. Gertner, Appl. Phys. Lett. **54**, 1025 (1989).
[3] J. H. Tregilgas, T. L. Polgreen, and M. C. Chen, J. Cryst. Growth **86**, 460 (1988).
[4] J. M. Arias, S. H. Shin, J. G. Pasko, R. E. DeWames, and E. R. Gertner, J. Appl. Phys. **65**, 1747 (1989).
[5] K. A. Harris, T. H. Myers, R. W. Yanka, L. M. Mohnkern, R. W. Green, and N. Otsuka, J. Vac. Sci. Technol. A **8**, 1013 (1990).
[6] R. D. Feldman, S. Nakahara, R. F. Austin, T. Boone, R. L. Opila, and A. S. Wynn, Appl. Phys. Lett. **51**, 1239 (1987).
[7] R. J. Koestner and H. F. Schaake, J. Vac. Sci. Technol. A **6**, 2834 (1988).
[8] J. D. Benson, B. K. Wagner, A. Torabi, and C. J. Summers, Appl. Phys. Lett. **49**, 1034 (1986).
[9] J. M. Arias and J. Singh, Appl. Phys. Lett. **55**, 1561 (1989).
[10] C. C. Wang, S. H. Shin, M. Chu, M. Lanir, and A. H. B. Vanderwyck, J. Electrochem. Soc. **127**, 175 (1980).
[11] G. L. Destefanis, J. Cryst. Growth **86**, 700 (1988).
[12] J. S. Chen, U.S. Patent No. 4,897,152.
[13] J. M. Arias, M. Zandian, J. G. Pasko, S. H. Shin, L. O. Bubulac, R. E. DeWames, and W. E. Tennant, J. Appl. Phys. **69**, 2143 (1991).
[14] M. Yoshikawa, K. Maruyama, T. Saito, T. Maekawa, and H. Takigawa, J. Vac. Sci. Technol. A **5**, 3052 (1987).
[15] J. W. Lee, H. Shichijo, H. L. Tsai, and R. J. Marti, Appl. Phys. Lett. **50**, 31 (1987).
[16] T. Tung, M. H. Kalisher, A. P. Stevens, and P. E. Herning, Mat. Res. Soc. Symp. Proc. **90**, 321 (1987).
[17] R. J. Koestner and H. F. Schaake, Mat. Res. Soc. Symp. Proc. **90**, 321 (1987).
[18] R. J. Koestner and H. F. Schaake, J. Vac. Sci. Technol. A **6**, 2834 (1988).
[19] D. G. Deppe, N. Holonyak, Jr., K. C. Hsieh, D. W. Nam, W. E. Plano, R. J. Matyi, and H. Shichijo, Appl. Phys. Lett. **52**, 1812 (1988).
[20] J. L. Lee, H. Kobayashi, S. Tanigawa, and M. Kawabe, Jpn. J. Appl. Phys. **29**, L860 (1990).

Molecular-beam epitaxy of CdTe on large area Si(100)

R. Sporken, M. D. Lange, and J. P. Faurie
Microphysics Laboratory, Department of Physics, University of Illinois at Chicago, P.O. Box 4348, Chicago, Illinois 60680

J. Petruzzello
Philips Laboratories, North American Philips Corporation, Briarcliff Manor, New York 10510

(Received 2 October 1990; accepted 19 October 1990)

We have grown CdTe directly on 2- and 5-in. diam Si(100) by molecular-beam epitaxy and characterized the layers by *in situ* reflection high-energy electron diffraction, double crystal x-ray diffraction, scanning electron microscopy, transmission electron microscopy, and low-temperature photoluminescence. The films are up to 10-μm thick and mirror-like over their entire surface. Even on 5-in. diam wafers, the structural and thickness uniformity is excellent. Two domains, oriented 90° apart, are observed in the CdTe films on oriented Si(100) substrates, whereas single-domain films are grown on Si(100) tilted 6° or 8° toward [011]. The layers on misoriented substrates have better morphology than those on oriented Si(100), and the substrate tilt also eliminates twinning in the CdTe layers. First attempts to grow HgCdTe on Si(100) with a CdTe buffer layer have produced up to 10-μm thick layers with cutoff wavelengths between 5 and 10-μm and with an average full width at half-maximum of the double-crystal x-ray diffraction peaks of 200 arc s.

I. INTRODUCTION

Recently, there has been considerable interest in the use of silicon as a substrate for the epitaxial growth of CdTe[1-4] and HgCdTe.[3] There are several reasons why silicon is a particularly attractive substrate for CdTe and HgCdTe. So far, CdTe and CdZnTe single crystals are the most widely used substrates for epitaxial growth of HgCdTe, but they are very expensive and difficult to produce with large area. CdTe and CdZnTe substrates are also very fragile and difficult to handle and to process. On the other hand, silicon substrates are available in large area and are inexpensive compared to CdTe, and silicon is much stronger than CdTe. GaAs has also received much attention[5-9] as a substrate for CdTe and HgCdTe, but where price, size and strength are concerned, its properties are only intermediate between CdTe or CdZnTe and silicon. Most of all, if good HgCdTe and CdTe can be grown directly on silicon, it could eventually become possible to produce large area monolithic infrared (IR) focal-plane arrays, combining HgCdTe IR detectors with Si integrated circuits for signal processing. Another approach to the same problem is to use GaAs/Si substrates,[10,11] but this is clearly more expensive and complicated than the direct growth of HgCdTe and CdTe on silicon.

There are several difficulties associated with the growth of CdTe on silicon. The large lattice mismatch (19%) between CdTe and silicon generates a high density of dislocations at the interface, which must be prevented from travelling into the active part of the layer. Also, when the temperature of the sample is changed, the thermal mismatch between CdTe and silicon creates additional strain and dislocations. The first difficulty is reduced by the fact that, if Si(100) substrates are used, CdTe grows in the (111)B orientation, and this reduces the lattice mismatch to 3.4% along the CdTe[$\bar{2}$11] axis. However, the overall mismatch remains high and it will be necessary to use dislocation reduction techniques such as selected area epitaxy[12] or strained layer superlattices[13] in order to achieve material properties that will allow the fabrication of high-quality devices. Furthermore, two epitaxial relationships are possible between CdTe(111)B and Si(100), corresponding to the CdTe[$\bar{2}$11] axis parallel to either the Si[011] or the Si[0$\bar{1}$1] axis. This leads to the formation of two domains in the CdTe layers, but this problem can be avoided if the silicon substrates are cut off-axis.[3]

In this paper, we will describe the present status of our work on the molecular-beam epitaxy (MBE) growth of CdTe on silicon. Special emphasis will be placed on the growth of CdTe on large-area substrates as well as on the influence of the substrate tilt on the structural characteristics of the layers.

II. EXPERIMENTAL

The MBE growth was done in OPUS 45, which is a prototype multiwafer and large-area MBE system manufactured by ISA RIBER. The system consists of three ultrahigh vacuum chambers for loading and unloading, substrate outgassing, and MBE growth, respectively. Up to three 2-in. wafers or one 3 or 4-in. wafer deposited on a molybdenum platen or one 5-in. wafer (platen-free) can be transferred and processed in a horizontal face-down position during all steps. This configuration was chosen to avoid particle contamination as much as possible. The system is designed for In-free mounting of the substrates, but mounting with indium can also be used if necessary. In the growth chamber, the mercury cell is in a central position, as close as possible to the wafers in order to limit the Hg consumption. The other eight cell ports face the wafer at 45° and are regularly positioned around the chamber. All the cells use a dual-zone filament, except for the Hg cell. In the work described here, CdTe was evaporated from a single CdTe cell.

All the substrates used for this work were Si(100), with a 0°, 6°, or 8° misorientation toward [011]. 2-in. as well as 5-in. diam. substrates were used, with different conduction types and doping levels, which will be specified later. Immediately before loading into the MBE system, the wafers were cleaned using either the standard RCA cleaning method[14] or a method described by Ishizaka and Shiraki.[15] Although the latter method gave results similar to the RCA technique when it was successful, we found that the success rate was much higher for the RCA method. Both methods start with a degreasing step, followed by a wet chemical etch. At the end, the substrate is passivated with a thin oxide layer. The substrates are outgassed in situ at 500 °C, and the oxide is removed at 850 °C.

We have grown up to 10-μm thick CdTe layers at a growth rate between 1 and 2 Å/s. The sample temperature is measured by a thermocouple placed about 5 mm behind the sample. This thermocouple was calibrated against a temporarily installed thermocouple which was touching a 2-in. silicon wafer. The temperature at which the oxide desorbs from the silicon wafer, and the temperature where tellurium starts to stick on CdTe grown on silicon were also used for calibration. The CdTe layers were grown by a two-step method, similar to what is used for the growth of GaAs on silicon.[16] First, a layer about 50–100 nm thick is grown at 220–250 °C, followed by a 10 min anneal at 360 °C. The remainder of the layer is then grown at 300–320 °C.

III. RESULTS

A. Growth on large area

Recently, we reported the MBE growth of CdTe on 5-in. diam Si(100) substrates.[4] In this paper, we describe the MBE growth of CdTe on large-area Si substrates and we address the differences that exist between growth on 2-in. and 5-in. substrates. In these experiments, the 5-in. substrates were p type with 0.01–0.02 Ω cm resistivity and the 2-in. substrates were p type with 15–30 Ω cm.

The CdTe layers were shiny and mirror-like over their entire area. Streaky reflection high-energy electron diffraction (RHEED) patterns were observed from the beginning of the growth, indicating that smooth, two-dimensional growth occurs. This is in contrast to the direct growth of CdTe on Si(100) without using a two-step process. In that case, some roughness was observed even for layer thickness up to 1 μm.[3] The reconstruction is $(2\sqrt{3} \times 2\sqrt{3}) R\, 30°$, which is typical for CdTe(111)B grown by MBE when the substrate is at 275–310 °C and a single CdTe effusion cell is used.[17] Because of the existence of two domains in the CdTe(111)B layers, the RHEED pattern for the e beam along Si[011] is the superposition of the CdTe[01$\bar{1}$] and [$\bar{2}$11] diffraction patterns.

The thickness of these layers was measured from the interference fringes in the IR transmission spectra between 4000 and 5000 cm^{-1}. Figure 1 shows the thickness variation for a CdTe layer on a 5-in. Si(100) substrate. The average thickness of this layer is 6 μm. The standard deviation is 2.3% across the 5-in. sample, whereas the maximum thickness variation is $(d_{max} - d_{min})/d_{av} = 6.5\%$. For the same layer,

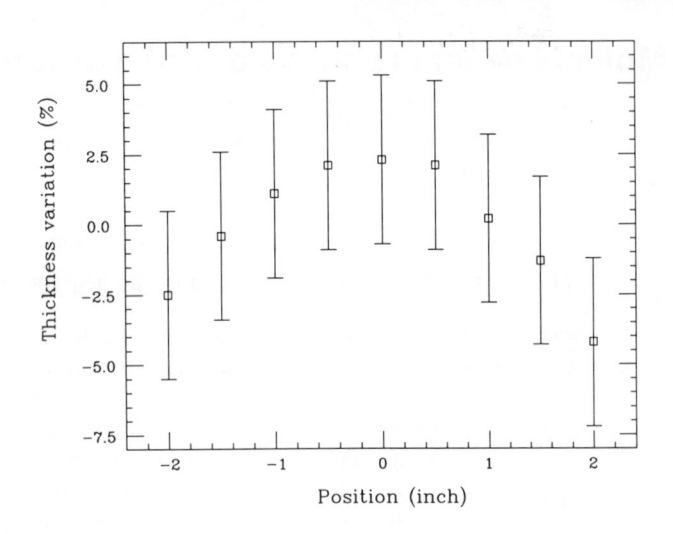

FIG. 1. Thickness variation of a CdTe layer grown by MBE on a 5-in. diam Si(100) wafer. The average layer thickness is 6 μm.

the maximum thickness variation along a 2-in. diam in the center of the sample is 1.4%. These results were obtained when the CdTe cell was oriented so that the cell axis points midway between the center and the edge of the sample. For best thickness uniformity on a 2-in. wafer centered on a 5-in. platen, the CdTe cell axis should aim at the center of the substrate. This configuration gave a standard deviation of 0.3% across a 2-in. diam in the center of a 5-in. sample, with a maximum thickness variation of 1%. However, the maximum thickness variation over the full 5-in. wafer is 12% in this case.

The crystalline quality and structural uniformity of the CdTe layers was checked by x-ray diffraction. Double-crystal rocking curves (DCRCs) of the CdTe(333) reflection have an average full width at half-maximum (FWHM) of 514 arc s on a 5-in. sample, with a standard deviation of only 3.3%. The best value obtained so far for CdTe(111)B grown on 5-in. Si(100) is 460 arc s. On 2-in. substrates, we found x-ray DCRCs with 570 arc s FWHM. We think that there are two reasons why the layers grown on 5-in. Si(100) are slightly better than those on 2-in. substrates. First, the 5-in. substrates do not need a sample holder, whereas the 2-in. substrates are deposited on a 5-in. molybdenum platen. Therefore, it is likely that the vacuum conditions in the vicinity of the sample are better in the case of a 5-in. substrate during the oxide removal at 850 °C. Second, because of the higher doping level of the 5-in. substrates, less heating power is needed to reach a substrate temperature of 850 °C, and this also improves the vacuum conditions close to the sample.

B. Influence of the substrate tilt

In order to avoid the formation of two domains in the CdTe films, all single-atomic steps on the Si(100) surface must be eliminated so that only double-atomic steps remain. This situation can be achieved by tilting the substrate surface away from an exact (100) orientation or by preheating the substrate at high temperature (above 900 or 1000 °C) for a long time.[18] In our case, heating the substrate higher than 850–900 °C is not practical, especially for a long time. We therefore used Si(100) substrates tilted 6° or 8° toward [011]. These substrates were 2-in. n type, with 10–30 Ω cm resistivity. After cleaning at 850 °C, all the steps should be two atomic layers high, and all the Si dangling bonds should then be parallel to only one of the Si⟨011⟩ directions, and this imposes a single orientation for the CdTe(111)B layer.

We have indeed observed by RHEED that single-domain CdTe(111)B films are obtained on Si(100) tilted 6° or 8° toward [011]. These layers have a very different surface morphology, as is seen from the scanning electron microscope (SEM) images, shown in Fig. 2. While the layers grown on misoriented substrates are featureless at this magnification, layers on oriented Si(100) substrates show a very characteristic domain structure, with a domain size of 1–2 μm. A comparison of the cross section of these layers by transmission electron microscopy (TEM) reveals that these domains are columns extending from the CdTe/Si interface to the surface of the layer [Fig. 3(a)]. Such columns are not observed in the layers grown on misoriented substrates [Fig. 3(b)]. Electron diffraction experiments performed in the transmission electron microscope show that these columns are in fact the domains which are observed by RHEED during the growth of these films. For the dark columns in Fig. 3(a), the CdTe [1$\bar{1}$0] axis is parallel to the Si[011], whereas the CdTe [11$\bar{2}$] axis is parallel to the Si[011] for the light ones. A closeup view of the boundary between these domains shows that twinning occurs inside the domains (Fig. 4), while such twinning is not observed in the films on misoriented substrates. On oriented substrates, twinning is also seen by RHEED during the initial stages of the growth. This remarkable agreement between RHEED and TEM confirms the power of RHEED as an *in situ* analysis tool during MBE growth.

Low-temperature photoluminescence (PL) spectra were measured on epitaxial CdTe layers in order to investigate the influence of the substrate tilt on the quality of the layers. The spectra were excited with the 514.5 nm line of a 50 mW Ar ion laser. Figure 5(a) shows the PL spectrum measured on a 10-μm thick layer of CdTe(111)B on Si(100). The narrow peak at 1.593 eV is related to bound exciton (BE) recombi-

FIG. 2. Scanning electron microscope image of the surface of a 9-μm thick CdTe layer on Si(100) (a) and a 4-μm thick CdTe layer on Si(100) tilted 8° toward [011] (b).

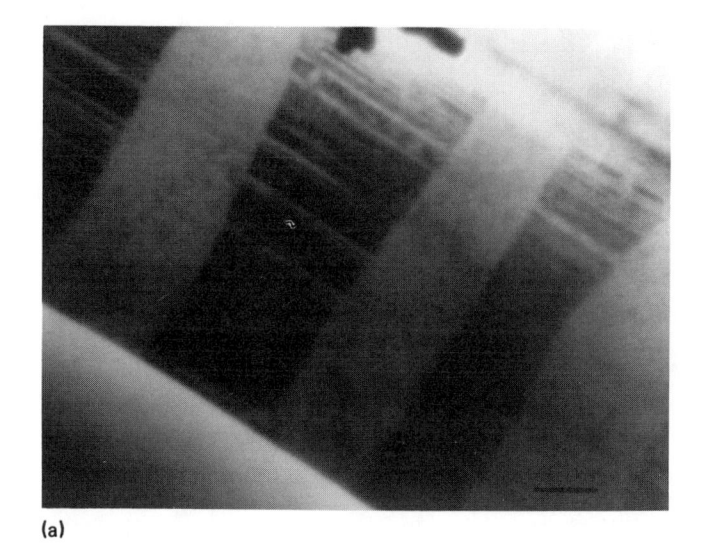

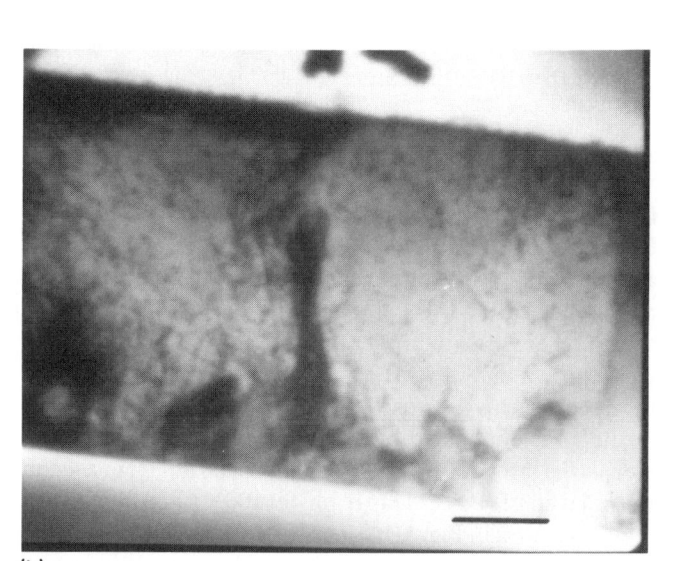

FIG. 3. Transmission electron micrograph of the cross section of a 4.7-μm thick CdTe layer on Si(100) (a) and a 4-μm thick CdTe layer on Si(100) tilted 8° toward [011] (b). Markers are 1 μm.

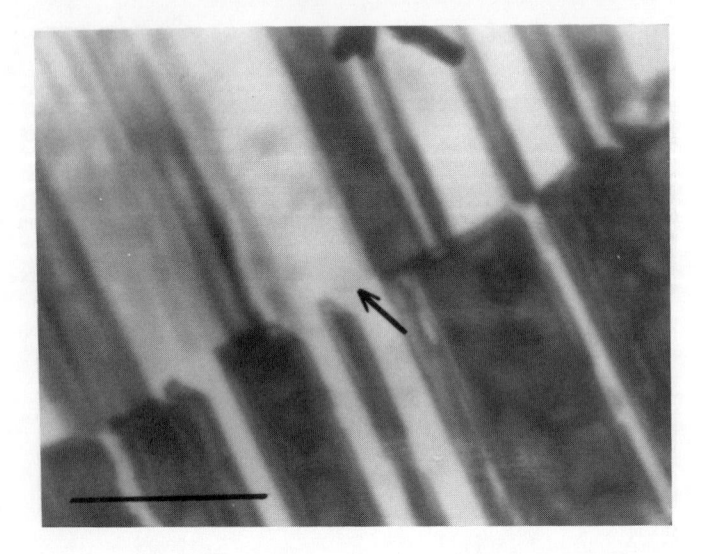

FIG. 4. Transmission electron micrograph of the domain boundary in CdTe(111)B on Si(100). The arrow indicates the domain boundary, and the dark and light bands are twins, within the domains. Marker is 0.1 μm.

nations, whereas the other features are due to defects in the CdTe. The BE peak has a FWHM of 4–4.5 meV, which is already significantly lower than the values of 20, 12, and 5–6 meV reported for CdTe grown on Si by MOCVD,[2] CdTe grown on Si with a (Ca,Ba) F$_2$ buffer layer by MBE[19] and CdTe grown on Si by MBE,[3] respectively. However, when the laser power is decreased, the BE peak disappears, and the spectrum is dominated by the defect-related features, as is seen in the inset of Fig. 5(a). The PL spectrum from a 4-μm thick layer of CdTe(111)B on Si(100) tilted 8° toward [011] is shown in Fig. 5(b). The BE peak has shifted to 1.58 eV, which could be due to strain in this layer. It is indeed possible that for CdTe on nominal Si(100), most of the strain is released at the boundaries of the domains, which are absent in CdTe layers on misoriented Si(100). In contrast to the case of CdTe on nominal Si(100), the BE peak is now the dominant feature in the PL spectrum, even at very low laser power. Furthermore, the FWHM of the BE peak is 3 meV, which is the best value we have measured so far for CdTe on silicon. We conclude that CdTe grown on Si(100) tilted 8° toward [011] is of better quality than CdTe on nominal Si(100).

IV. CONCLUSIONS

In this paper, we have shown that CdTe(111)B can be grown by MBE with excellent structural and thickness uniformity on Si(100) substrates with up to 5-in. diam. Better substrate cleaning and the use of a two-step growth technique have led to improved layer quality, compared to earlier results. Very recently, we have grown HgCdTe on CdTe/Si(100) with 3- and 5-in. diam. These very first attempts have produced up to 10-μm thick layers of HgCdTe(111)B with x-ray DCRCs with an average FWHM of 200 arc s for the best layer obtained so far. This is a significant improvement over the CdTe DCRCs, indicating that CdTe is indeed an efficient buffer layer for the growth of HgCdTe on silicon. We anticipate further improvements of the CdTe, especially through the use of dislocation reduction techniques, as well as significant improvements of the quality of the HgCdTe after the growth conditions have been optimized. This will bring the fabrication of HgCdTe IR detectors on silicon within reach. We are currently investigating the effect of selected area epitaxy on the density of dislocations. A method to prepare patterned Si(100) substrates for MBE growth of CdTe has been developed, and CdTe was grown successfully on such substrates. A complete characterization of these layers is expected to be available soon.

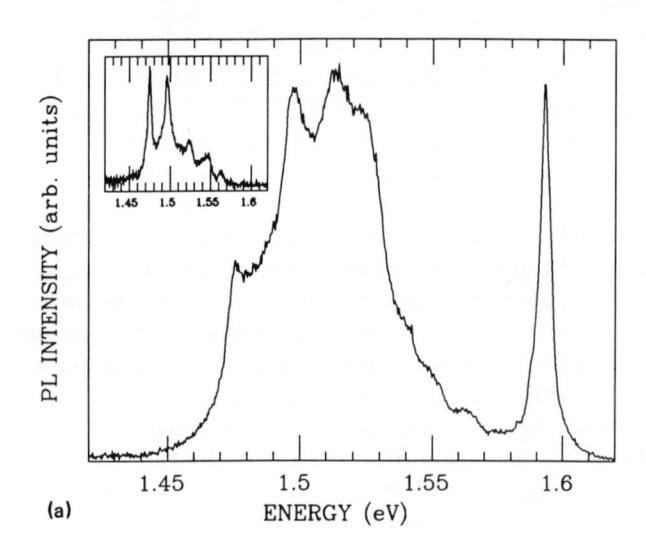

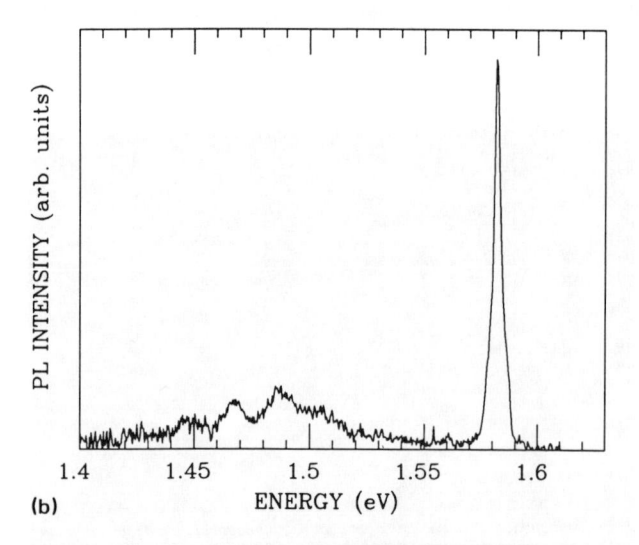

FIG. 5. Low-temperature photoluminescence spectrum of CdTe(111)B on nominal Si(100) (a) and Si(100) tilted 8° toward [011] (b). The approximate excitation powers are 50 W/cm^2 (a), and 0.025 W/cm^2 (b). The inset in Fig. 5(a) shows the same spectrum measured with 0.5 W/cm^2 excitation power.

On nominal Si(100) substrates, the CdTe layers consist of two types of domains, corresponding to the CdTe [11$\bar{2}$] axis parallel to the Si[011] or Si[01$\bar{1}$] axis. These domains are 1 to 2 μm wide columns, extending from the CdTe/Si interface to the layer surface. This domain structure, as well as the twinning which occurs within the domains, is avoided when misoriented Si(100) substrates are used. The improved quality of the CdTe layers grown under such conditions is confirmed by low-temperature PL spectroscopy.

ACKNOWLEDGMENTS

This work was supported by DARPA and monitored by AFOSR Contract No. 49620-87-C-0021 and SDIO Sensors Office, coordinated through WRDC/MLPO, WPAFB, in conjunction with NRL Contract No. N00014-89-J-2030. One of us (R.S.) is a senior research assistant of the Belgian National Fund for Scientific Research. We are grateful to Dr. K. Mahavadi for the photoluminescence measurements, to Dr. S. Sivananthan and Dr. M. Boukerche for their help and suggestions, and Z. Ali for technical assistance.

[1] Y. Lo, R. N. Bicknell, T. H. Myers, J. F. Schetzina, and H. H. Stadelmaier, J. Appl. Phys. **54**, 4238 (1983).
[2] R. L. Chou, M. S. Lin, and K. S. Chou, Appl. Phys. Lett. **48**, 523 (1986).
[3] R. Sporken, S. Sivananthan, K. K. Mahavadi, G. Monfroy, M. Boukerche, and J. P. Faurie, Appl. Phys. Lett. **55**, 1879 (1989).
[4] R. Sporken, M. D. Lange, C. Masset, and J. P. Faurie, Appl. Phys. Lett. **57**, 1449 (1990).
[5] J. P. Faurie, S. Sivananthan, M. Boukerche, and J. Reno, Apply. Phys. Lett. **45**, 1307 (1984).
[6] J. M. Arias, R. E. DeWames, S. H. Shin, J. G. Pasko, J. S. Chen, and E. R. Gertner, Appl. Phys. Lett. **54**, 1025 (1989).
[7] J. L. Reno, P. L. Gourley, G. Monfroy, and J. P. Faurie, Appl. Phys. Lett. **53**, 1747 (1988).
[8] R. D. Feldman and R. F. Austin, Appl. Phys. Lett. **49**, 954 (1986).
[9] J. Cibert, Y. Gobil, K. Saminadayer, S. Tatarenko, A. Chami, G. Feuillet, Le Si Dang, and E. Ligeon, Appl. Phys. Lett. **54**, 828 (1989).
[10] S. M. Johnson, M. H. Kalisher, W. L. Ahlgren, J. B. James, and C. A. Cockrum, Appl. Phys. Lett. **56**, 946 (1990).
[11] K. Zanio, R. Bean, R. Mattson, P. Vu, S. Taylor, D. McIntyre, C. Ito, and M. Chu, Appl. Phys. Lett. **56**, 1207 (1990).
[12] E. A. Fitzgerald, G. P. Watson, R. E. Proano, D. G. Ast, P. D. Kirchner, G. D. Pettit, and J. M. Woodall, J. Appl. Phys. **65**, 2220 (1989).
[13] J. Petruzzello, D. Olego, X. Chu, and J. P. Faurie, J. Appl. Phys. **66**, 2980 (1989).
[14] W. Kern and D. A. Puotinen, RCA Rev. **31**, 187 (1970).
[15] A. Ishizaka and Y. Shiraki, J. Electrochem. Soc. **1333**, 666 (1986).
[16] W. I. Wang, Appl. Phys. Lett. **44**, 1149 (1984).
[17] S. Sivananthan, Ph.D. thesis, Chicago, 1988.
[18] H. Noge, H. Kano, M. Hashimoto, and I. Igarashi, J. Appl. Phys. **64**, 2246 (1988).
[19] H. Zogg and S. Blunier, Appl. Phys. Lett. **49**, 1531 (1986).

Characterization of CdTe, HgTe, and Hg$_{1-x}$Cd$_x$Te grown by chemical beam epitaxy

B. K. Wagner, D. Rajavel, R. G. Benz II, and C. J. Summers
Physical Sciences Laboratory, Georgia Tech Research Institute, Atlanta, Georgia 30332

(Received 2 October 1990; accepted 19 December 1990)

Detailed characterization of chemical beam epitaxially (CBE) grown CdTe and Hg$_{1-x}$Cd$_x$Te layers are reported. These characterizations include photoluminescence, infrared transmission, energy dispersive x-ray analysis, and variable temperature (10–300 K) Hall effect and resistivity measurements. The results indicate that high quality HgCdTe layers can be grown by CBE.

I. INTRODUCTION

Over the last decade the molecular-beam epitaxial (MBE) growth of Hg$_{1-x}$Cd$_x$Te has seen rapid progress.[1,2] Very homogeneous and high quality alloy epitaxial layers as well as complex superlattice (SL) structures for enhanced device performance can be reproducibly obtained. In fact, the low growth temperatures made possible by this technique make it the only growth process capable of producing HgCdTe SL and multiquantum well (MQW) device structures. However, despite its success, MBE still has some limitations. The degree of flux control, stability, and reproducibility available with the thermal effusion cell is inadequate for the growth of long-wavelength alloy material with a high degree of compositional uniformity. Furthermore, the evaporation of group V elements for *p*-type doping produces tetramers and solid Te produces Te dimers, respectively, which can not be incorporated effectively on the growth surface. On the other hand, the use of gas sources whose flow rates can be accurately and reproducibly controlled with precision flow controllers make this option an attractive alternative to conventional thermal sources. This has lead to the evolution of the chemical beam epitaxy (CBE) technique in which hydrocarbon vapor compounds of the constituent elements are delivered through flow controllers which regulate the flux of the species incident upon the growth surface. In this manner, monomer species can be provided, as required, to enhance growth kinetics.[3] CBE also allows premixing of the constituent elements, thereby minimizing the number of gas injectors and achieving greater compositional uniformity.[4] Several III–V and Zn(Se, Te) CBE systems are in operation today, but very little work has been reported on the CBE growth of (Hg,Cd)Te alloys. The first such system dedicated to the growth of (Hg,Cd,Zn)Te alloys was developed at Georgia Tech in 1989.[5–7]

We report in this paper the properties of CdTe, HgTe, and HgCdTe layers grown on GaAs, CdTe, and ZnCdTe substrates by this technique.

II. EXPERIMENTAL PROCEDURES

The CBE system used for these studies has been described previously[6,7] and so will not be discussed at length. The system consists of a Varian GEN II MBE which has been modified to handle gas sources. These consist of MKS Instruments 1150B pressure flow controllers for diisopropyltelluride (DipTe), diethylcadmium (DeCd), and diethylzinc (DeZn) source gases which have been obtained from Air Products Inc., as well as a Hg-pressure controlled vapor source (Hg-PCVS) developed at Georgia Tech.[5] The choice of DeCd, DeZn, and DipTe was based on previously reported work on these or similar precursors. In the growth of GaAs by CBE, layers grown using triethylgallium showed considerably less carbon incorporation that those grown under similar conditions using trimethlygallium.[8] This was attributed to the stronger chemical bond between gallium and the carbon complex in trimethlygallium as opposed to triethylgallium. The bond strength of diethylcadmium (26.5 kcal/mol) is considerably less than that of dimethylcadmium (33.5 kcal/mol) and so should reduce the amount of carbon incorporated in the layers and also be easier to pyrolize. Additionally, because the ethyl complex is larger than the methyl complex, it is expected that it will produce less carbon incorporation in the film. Unfortunately, higher order Cd-alkyl complexes which are expected to crack at lower temperatures and produce even lower probabilities of carbon incorporation of the growing film are either not available commercially, or in high enough purity. Similar arguments were used for the selection of DipTe which is the largest hydrocarbon-Te complex readily available in high purity. The CBE system also has precision flow controllers for arsine and ethyliodide for *p*- and *n*-type doping, respectively. The group II and VI gases are precracked in separate high temperature injectors to eliminate gas phase reactions. The system also has conventional MBE CdTe, Cd, Zn, and Te solid sources. Finally, the pumping system consists of a Varian cryopump and a Balzers MBE series turbomolecular pump for handling the high gas loads. The gases exhausted or purged from the pumps are routed through an EMCORE scrubber in order to remove the toxic gases.

Growth of (001) oriented CdTe, HgTe, and Hg$_{1-x}$Cd$_x$Te layers was carried out on (001) GaAs, CdTe, and ZnCdTe substrates. The GaAs substrates were prepared with a standard degreasing and 3:1:1 (H$_2$SO$_4$:H$_2$O:H$_2$O$_2$) free etch procedure while the II–VI substrates were degreased and then chemomechanically polished with a 1% bromine:methanol solution. The substrates were then mounted onto Mo blocks with a colloidal graphite suspension and placed into the CBE load lock where they were outgassed at low temperatures. After the oxide (GaAs) or amorphous Te (CdTe and ZnCdTe) layer was desorbed, as determined by reflection high-energy diffraction

(RHEED), CdTe growth was initiated on the II–VI substrates at 300 °C using the solid CdTe source. The GaAs substrates were cooled from the oxide desorption temperature to approximately 375 °C under a combination of Zn and CdTe fluxes to ensure (001) growth. Once the (001) orientation was identified by RHEED, the substrate temperature was again lowered to 300 °C and CdTe growth continued. After approximately 5 min of further growth, a temperature calibration was carried out using a Te condensation method described previously.[9,10] Growth of the buffer layer at the corrected 300 °C temperature was then continued by depositing a further 1 and 0.2 μm of CdTe on the GaAs and CdTe/ZnCdTe substrates, respectively. CBE growth was then initiated at growth rates of 0.5 to 1 μm/h and substrate temperatures of 200 to 250 °C for CdTe, and 120 to 185 °C for the Hg-based layers. CdTe layers were n type as grown. Figures 1 and 2 show typical RHEED patterns observed during the growth of HgTe and HgCdTe, respectively. Using the pressure controlled sources to optimize the ratios of the constituent fluxes, the conditions for high quality single crystal HgCdTe growth were determined as assessed by RHEED. For both CdTe and HgTe growth the DipTe injector was operated at conditions estimated to completely crack the DipTe and to give a flux of 80% Te monomers. The DeCd injector was also operated so as to completely dissociate the compound. After growth, the Hg-based layers were capped with a thin (< 200 Å) MBE CdTe layer. The epitaxial layers were then characterized by optical and electron microscopy, photoluminescence (PL), room temperature Fourier transform infrared (FTIR) transmission, Hall effect, and resistivity measurements.

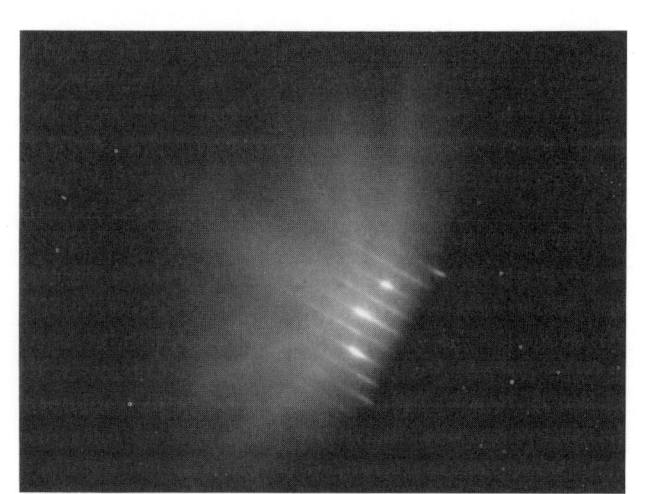

FIG. 1. RHEED patterns of CBE HgTe: (a) [$\bar{1}10$] azimuth; (b) [110] azimuth.

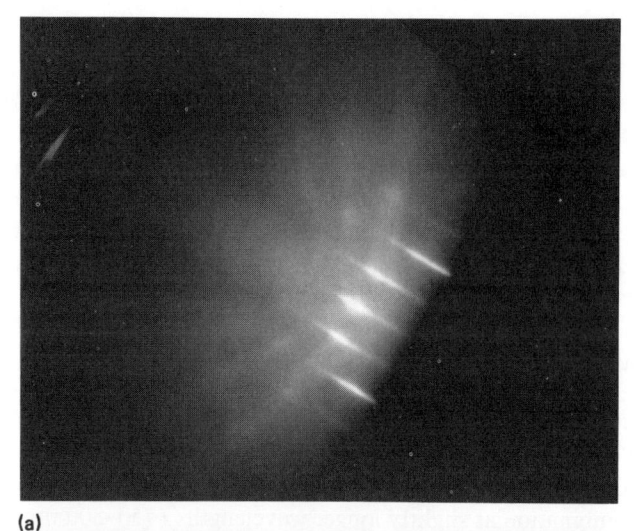

FIG. 2. RHEED patterns of CBE HgCdTe: (a) [$\bar{1}10$] azimuth; (b) [110] azimuth.

III. RESULTS AND DISCUSSION

Immediately following growth, the structural properties of the epitaxial HgCdTe layers were examined by optical microscopy and scanning electron microscopy (SEM). This study showed the presence of small hillocks on the surfaces, with the best sample having a hillock density of 10^5 cm^{-2} and the worst of 10^6 cm^{-2}. Although no optimization was performed to lower the hillock density, these values are competitive with MBE grown layers. SEM and energy dispersive x-ray analysis (EDAX) were also performed to estimate the composition of the layers using reference samples of CdTe,

HgTe, and bulk $Hg_{0.79}Cd_{0.21}Te$ obtained from Cominco. This data indicated that samples with x values between 0.16 and 0.4 were grown. The growth rates of CdTe, HgTe, and HgCdTe were determined by measurement of the thicknesses of the epilayers grown on GaAs substrates. This was accomplished by selectively etching the epilayers, as described by Leech et al.[11] and then using profilometry to measure the etch step. Long wavelength FTIR measurements were also used to determine the thicknesses.

The PL data obtained for CdTe grown by CBE on GaAs substrates at various DeCd/DipTe ratios are shown in Fig. 3. The magnitude of the near-band edge emission is large and is comparable to that measured for MBE grown material. The spectra show features that can be assigned to donor and acceptor bound excitons as well as donor–acceptor pair recombination at slightly longer wavelengths (750–800 nm). In the 1.45 eV region (830–880 nm), the broad band which has been attributed to Te-defect complexes is absent or very weak for these samples. Sample A90-15 which exhibits the stongest acceptor bound exciton luminescence also has the lowest carrier mobility due to compensation effects.

Figure 4 shows the room temperature FTIR transmission spectra from 2–18 μm for four samples grown on GaAs substrates. These thin (2–5 μm) layers exhibit relatively sharp cutoffs with 50% transmission band gaps corresponding to x values of 0.19, 0.21, 0.24, and 0.37. The position of the cutoff moves to longer wavelengths in approximate agreement with the x value determination from the EDAX measurements. The slope of the cutoff feature also becomes less abrupt with decreasing x value both because of the decreasing absorption coefficient and the increased distribution of thermally excited conduction electrons in smaller band gap alloys. In fact the Moss–Burstein effect in n-type material shifts the apparent absorption edge to shorter wavelengths and thus makes it difficult to obtain precise x-value determinations by this technique without extensive analysis. Ideally, x-value determinations require temperature dependent transmission data.

X-ray characterization of the HgCdTe layers has not yet been performed, however, previous studies on CdTe grown on GaAs substrates exhibited double crystal rocking curve full widths at half-maximum of 60–200 arc s.

In Fig. 5 the dependence of the electron mobility on temperature is shown for two HgTe layers grown by CBE at 165 °C on a (001) CdZnTe and a (001) GaAs substrate, respectively. For growth on GaAs, a 1-μm thick CdTe buffer layer was first grown. As shown, very comparable mobility values were found for both samples, with the HgTe/CdZnTe layer showing slightly higher mobilities at the higher temperatures (150–300 K). The mobilities increase rapidly from 2.5×10^4 cm^2 V^{-1} s^{-1} at 300 K to about 1.1×10^5 cm^2 V^{-1} s^{-1} at 20 K due to the reduction in longitudinal optical (LO)-phonon scattering. As shown, the mobility values and the temperature dependence are in very good agreement with the theory[12] and show that high quality HgTe can be grown by CBE.

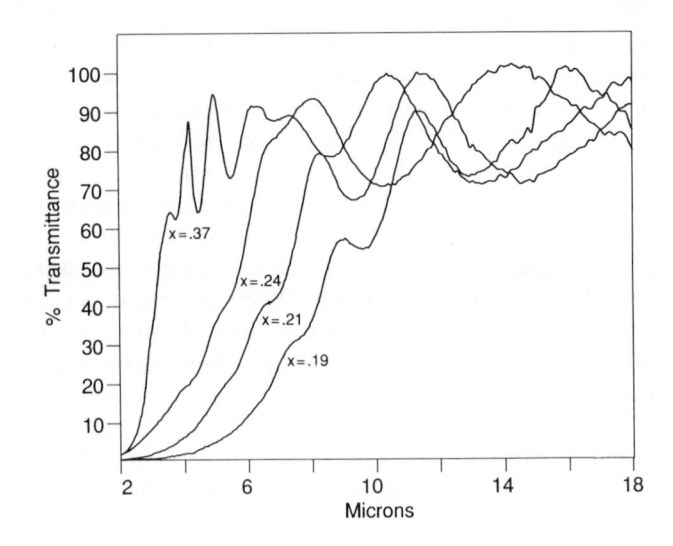

FIG. 4. Room temperature FTIR spectra of four HgCdTe/GaAs layers with various x values.

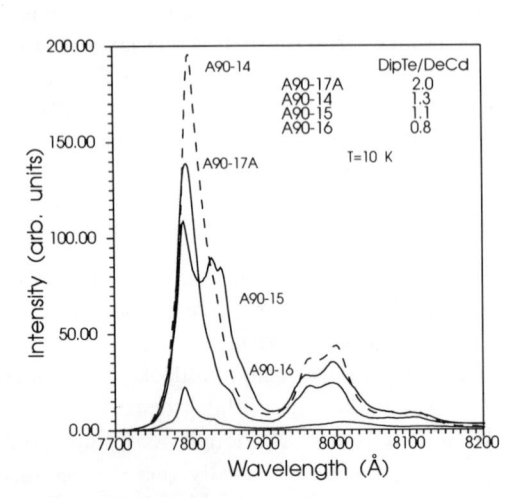

FIG. 3. Photoluminescence spectra of CBE CdTe grown at 250 °C with various DeCd/DipTe ratios.

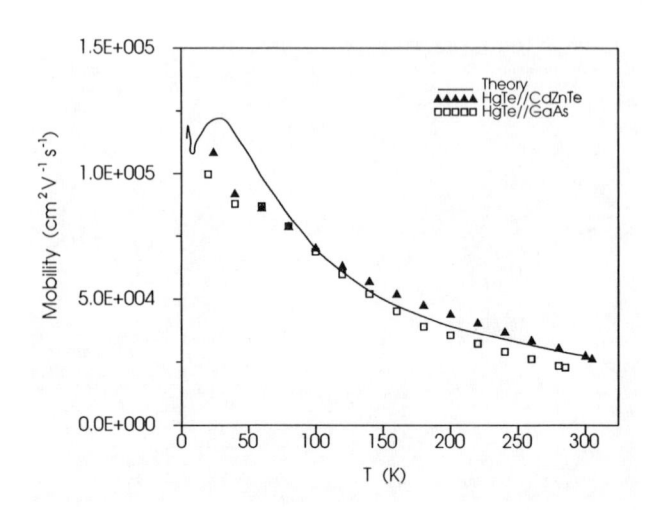

FIG. 5. Electron mobility vs temperature for two HgTe layers; solid line represents theory from Ref. 12.

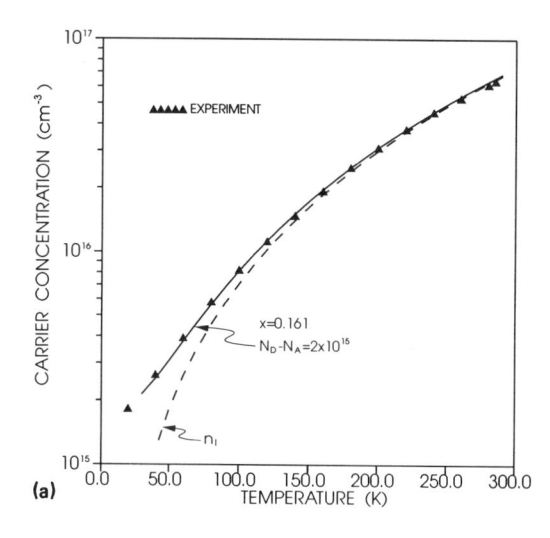

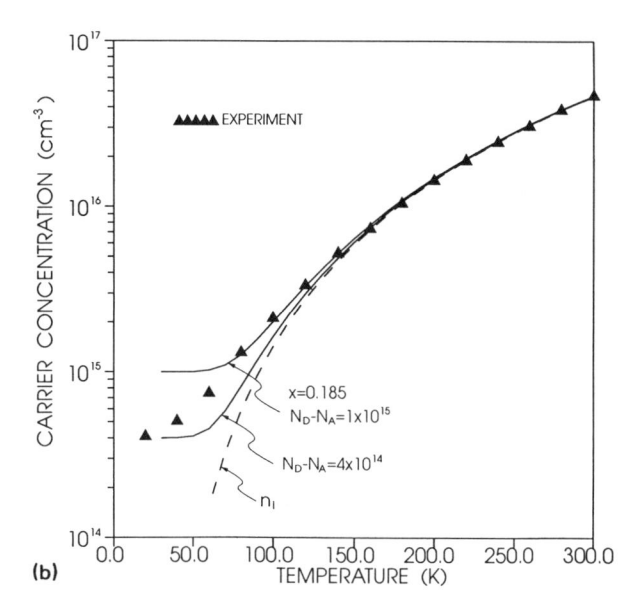

FIG. 6. Electron concentration vs temperature for (a) $x = 0.161$ HgCdTe layer; (b) $x = 0.185$ HgCdTe layer. Dashed and solid lines depict empirical fitting of intrinsic and extrinsic carrier concentrations, respectively.

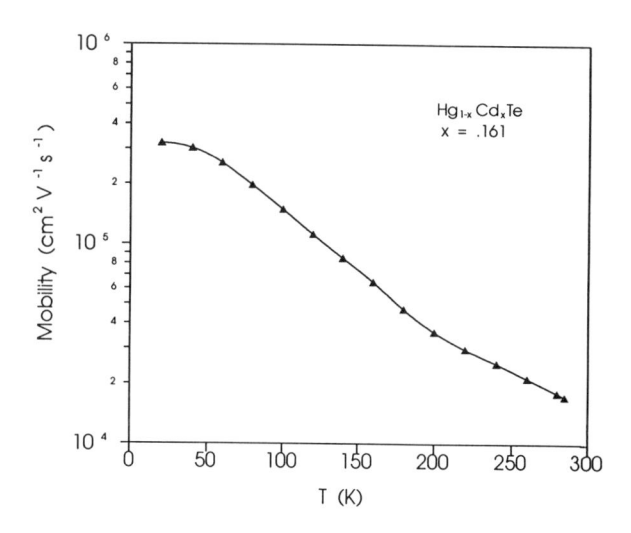

FIG. 7. Electron mobility vs temperature for a $Hg_{0.84}Cd_{0.16}Te$ layer.

The dependence of the electron concentration on temperature for a HgCdTe sample is shown in Fig. 6(a). The rapid decrease in electron concentration between 300 and 100 K is a direct measure of the number of electrons thermally excited across the band gap. Thus by fitting this variable temperature Hall data with an empirical expression relating the intrinsic carrier concentration to temperature and alloy composition an accurate x value for the sample can be found.[13] For this procedure the net background donor concentration $(N_D - N_A)$ assumed here to be completely ionized was also incorporated in the fitting routine, and was accounted for in the total electron concentration of the material via the charge neutrality criterion. The excellent fit obtained to the temperature dependence of the electron concentration over the whole temperature range is obtained for an x value of 0.161 ± 0.002 and a value of $N_D - N_A = 2 \times 10^{15}$ cm^{-3}, and reflects the high uniformity and quality of this layer. A very similar fit is also shown in Fig. 6(b) for a second sample using the fitting parameters $x = 0.185 \pm 0.002$ and $N_D - N_A = 4 \times 10^{14}$ cm^{-3}. The higher carrier concentration measured between 40–100 K is attributed to the presence of a second or extended donor state of concentration 7×10^{14} cm^{-3} approximately 2.5 meV below the conduction band. In fact using a model which incorporates two donor states and the full Kane band structure for these alloys computed with Fermi–Dirac statistics, a very good fit can be obtained to this curve. Similar but slightly less exact fits have also been obtained for samples with x values of 0.21 ± 0.005 ($N_D - N_A = 3.1 \times 10^{14}$ cm^{-3}) and 0.33 ± 0.005 ($N_D - N_A = 10^{13}$ cm^{-3}). For these last two samples the mobility data indicate significant compensation which could cause variations from the expected temperature dependence of the electron concentration. An example of the mobility data is shown in Fig. 7 for a $Hg_{0.84}Cd_{0.16}Te$ layer. As shown, the mobility increases by an order of magnitude as the temperature is reduced from 300–100 K, attributed to decreasing LO-phonon scattering. At lower temperatures the mobility plateaus at a value of 3.2×10^5 cm^2 V^{-1} s^{-1}. Table I lists the electrical properties at 20 K for several CBE layers. These electrical data indicate that high quality HgCdTe layers can be grown by CBE and that carbon incorporation does not appear to be a significant problem for low temperature CBE growth.

IV. SUMMARY

Chemical beam epitaxy using precracked DeCd and DipTe and a Hg-PCVS has been used to grow CdTe, HgTe, and low x valued $Hg_{1-x}Cd_xTe$ alloys ($x = 0.16 - 0.22$) as needed for the important infrared detector range of 10–15 μm. Structural, optical and electrical measurements show that high quality epitaxial HgTe and HgCdTe layer growth can be achieved by this technique at a low growth temperature of 165 °C. This data suggests that the carbon contamination is low and does not significantly affect the electrical

TABLE I. Electrical properties of some CBE $Hg_{1-x}Cd_xTe$ layers at 20 K.

Sample	X	Substrate	T_{GROWTH} (°C)	$N_D - N_A$ (cm^{-3})	μ_H (cm^2 V^{-1} s^{-1})
30 A	0.0	(001) GaAs	120	2×10^{16}	7.2×10^4
1	0.0	(001) GaAs	165	3×10^{16}	1×10^5
11	0.0	(001) ZnCdTe	165	1×10^{16}	1.1×10^5
18	0.14	(001) GaAs	165	2.3×10^{15}	8.7×10^5
4	0.16	(001) GaAs	165	2×10^{15}	3.2×10^5
16	0.18	(001) GaAs	165	3.6×10^{14}	7.5×10^4
7	0.21	(001) GaAs	165	3.1×10^{14}	2.2×10^4

properties and also that the stainless steel Hg-source enclosure has little effect on the material properties.

ACKNOWLEDGMENTS

This work was supported by the Wright Research Development center under Contract No. F33615-89-C-1066, and by the Internal Research Program of the Georgia Tech Research Institute.

[1] S. Sivananthan, M. D. Lange, G. Monfroy, and J. P. Faurie, J. Vac. Sci. Technol. B 6, 788 (1988).
[2] T. H. Myers, R. W. Yanka, K. A. Harris, A. R. Reisinger, J. Han, S. Hwang, Z. Yang, N. C. Giles, J. W. Cook, J. F. Schetzina, R. W. Green, and S. McDevitt, J. Vac Sci. Technol. A 7, 300 (1989).
[3] C. J. Summers, R. G. Benz II, B. K. Wagner, J. D. Benson, and D. Rajavel, SPIE Proc. 1106, 2 (1989).
[4] W. T. Tsang, J. Cryst. Growth 95, 121 (1989).
[5] B. K. Wagner, R. G. Benz II, and C. J. Summers, J. Vac. Sci. Technol. A 7, 295 (1989).
[6] R. G. Benz II, B. K. Wagner, and C. J. Summers, J. Vac. Sci. Technol. A 8, 1020 (1990).
[7] R. G. Benz II, B. K. Wagner, D. Rajavel, and C. J. Summers, J. Cryst. Growth (to be published).
[8] H. Heinecke, K. Werner, M. Weyers, H. Luth, and P. Balk, J. Cryst. Growth 81, 270 (1987).
[9] D. Rajavel, F. Mueller, J. D. Benson, B. K. Wagner, R. G. Benz II, and C. J. Summers, J. Vac. Sci. Technol. A 8, 1002 (1990).
[10] D. Rajavel, F. Mueller, J. D. Benson, B. K. Wagner, R. G. Benz II, and C. J. Summers, J. Vac. Sci. Technol. B 8, 191 (1990).
[11] P. W. Leech, P. W. Gwynn, and M. H. Kibel, Appl. Surf. Sci. 37, 291 (1989).
[12] J. R. Meyer, C. A. Hoffman, F. J. Bartoli, J. M. Perez, J. E. Furneaux, R. J. Wagner, R. J. Koestner, and M. W. Goodwin, J. Vac. Sci. Technol. A 6, 2775 (1988).
[13] W. M. Higgins, G. N. Pultz, R. G. Roy, R. A. Lancaster, and J. L. Schmit, J. Vac. Sci. Technol. A 7, 271 (1989).

Low-temperature growth of midwavelength infrared liquid phase epitaxy HgCdTe on sapphire

S. Johnston, E. R. Blazejewski, J. Bajaj, J. S. Chen, L. Bubulac, and G. Williams

Rockwell International Science Center, Thousand Oaks, California 91360

(Received 29 October 1990; accepted 21 December 1990)

Consistently low dislocation density midwavelength infrared ($x = 0.31$) liquid phase epitaxy HgCdTe epitaxial layers with excellent morphology were grown on 2 in. sapphire substrates (PACE-1) using a new low-temperature (420 °C) Te melt process. Surface etch pit densities (EPDs) between 5×10^5 and 9×10^5 cm^{-2} were revealed using a previously reported chemical etch (J. S. Chen, US Patent No. 4897152) on a 20-layer sample set. Cross-sectional EPD profiles reveal a more rapid decrease of defects from the CdTe buffer layer interface as compared to conventionally grown (500 °C) material. X-ray rocking curve widths from 43 to 66 arcsec were routinely observed. 77 K electron mobilities as high as 51 000 cm^2/V-s were measured. Secondary-ion mass spectrometry profiles show a minimum of impurity gettering at the HgCdTe/CdTe buffer layer interface.

I. INTRODUCTION

$Hg_{1-x}Cd_xTe$ has become the preferred material for advanced infrared (IR) focal plane arrays operating in regions of the IR spectrum ranging from 1 to 12 μm. In addition to high quantum efficiency HgCdTe has the desirable feature of possessing an approximately linear variation of band gap with composition across the pseudobinary range from mercury telluride to cadmium telluride. This property implies complete miscibility of the two compounds in the solid state and a relatively simple phase diagram. The variable bandgap capability of HgCdTe permits an application specific choice of material cutoff wavelength through proper selection of material composition.

The difficulties associated with bulk crystal growth, such as high Hg vapor pressure and noncongruent solidification, and the desire for large-area HgCdTe of good crystalline quality have stimulated rapid progress in epitaxial approaches. These approaches include vapor phase epitaxy (VPE), metalorganic chemical vapor phase epitaxy (MOCVD), liquid phase epitaxy (LPE), and molecular beam epitaxy (MBE). The relative cost and simplicity with which a LPE growth apparatus can be assembled has resulted in rapid maturation of this technique and its introduction into the production environment.

A major impediment to the producibility and cost of epitaxial HgCdTe has been the lack of large high quality substrates. The metallurgical compatibility of HgCdTe with CdTe or CdZnTe implies these are natural choices for substrate material. The advantages of these II–VI substrates, however, are offset by the difficulties in obtaining low-cost, large, and rugged single-crystal wafers.

These bulk II–VI substrate difficulties have prompted extensive research in the area of alternate substrates for the epitaxial growth of HgCdTe. Previous work has included growth of CdTe by MBE, MOCVD, and laser evaporation onto substrates of Al_2O_3 (sapphire),[2,3] GaAs, InP, and Si. The demonstration and realization of the full potential of utilizing sapphire alternate substrates for the growth of device quality LPE HgCdTe has been discussed by a number of authors.[4–9] Rockwell's current midwavelength infrared (MWIR) detector array production process makes extensive use of 2 in. sapphire substrates in conjunction with the LPE growth of HgCdTe. The term associated with this unique growth process is PACE-1 which stands for Producible Alternative for CdTe Epitaxy.

In previous works[4–9] a high growth temperature of ~500 °C was used to grow epitaxial films of $x = 0.31$ HgCdTe on hybrid CdTe/sapphire substrates. One paper reported limited success in the low-temperature growth of $x = 0.2$ HgCdTe on CdTe substrates.[10] The present paper discusses the electrical and structural properties of epitaxial films of $x = 0.31$ HgCdTe grown by LPE at 420 °C on sapphire substrates. The advantages anticipated by using a lower growth temperature included a decrease in interdiffusion of the matrix elements at the interface, reduced dislocation density in the HgCdTe removed from the interface, Te precipitate reduction, impurity reduction, and subsequent interface gettering and improved morphology due to some elevation of the surface tension of the Te-rich solution.

II. EXPERIMENTAL

A. LPE growth

The key steps in the growth of the MWIR LPE HgCdTe on hybrid CdTe/sapphire substrates (PACE-1) are described in this section and a cross section is shown in Fig. 1(a). The sapphire substrates are suitably prepared and screened. A thin (5–7 μm) single-crystal layer of CdTe is grown by MOCVD on a (0001) oriented single crystal 2 in. sapphire wafer using dimethyl tellurium and dimethyl cadmium in a H_2 carrier gas. The active LPE HgCdTe material is grown on these hybrid CdTe/sapphire substrates using growth conditions similar to those used for bulk CdTe substrates. LPE growth is carried out in a horizontal graphite boat using Te-rich liquids at growth temperatures of both 500 and 420 °C. The technique utilizes the horizontal tipping apparatus illustrated in Fig. 2(a).

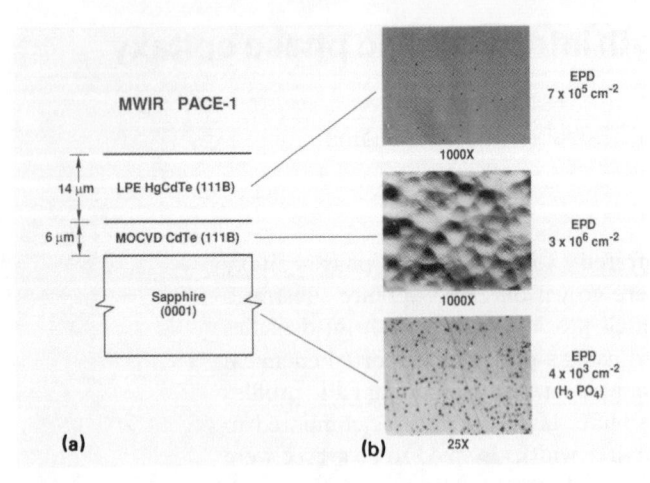

FIG. 1. (a) Cross section of the HgCdTe/CdTe/sapphire (PACE-1) material system. (b) Etch pit density distributions of LPE HgCdTe, MOCVD CdTe, and bulk sapphire components.

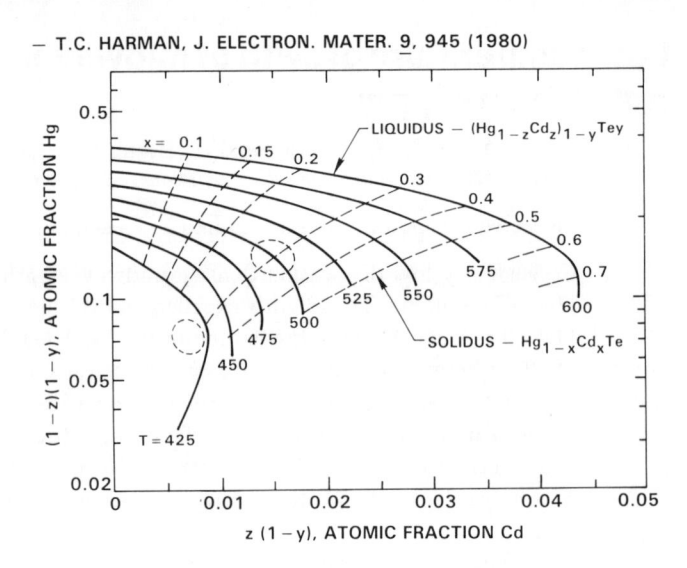

FIG. 3. Temperature-composition phase diagram for the Hg–Cd–Te system (reproduced from Harman, see Ref. 11).

The time-temperature sequence of growth is shown in Fig. 2(b). After proper weighing of the charge the boat is heated to an equilibration temperature of around 550 °C. The solution remains at this temperature long enough to ensure formation of a homogeneous growth solution. The temperature is then reduced to the growth temperature which in this work was either 500 or 420 °C. Growth is initiated by tipping the furnace in the direction that rolls the liquid melt onto the CdTe/sapphire substrate. Cooling then continues at rates between 0.01 and 0.2 °C. Depending on the growth parameters and layer thickness desired, the substrate and solution are kept in contact for periods of approximately 70 min. The furnace is then tipped back and the solution is rolled off. The furnace is then allowed to cool slowly to room temperature.

B. HgCdTe phase diagram considerations

The LPE growth of MWIR HgCdTe ($x = 0.31$) occurred using a Te-rich melt that has the advantages of reduced Hg vapor pressure and increased Cd solubility. The composition-temperature phase diagrams both experimental[11] and theoretical[12,13] have been published for the Te-rich region. An excellent summary of the preferred numerical relations describing the Cd-Hg-Te phase diagram was published by Brice.[14] Figure 3 shows the experimental Te-rich phase diagram by Harman.[11] The two regions of interest for this work are indicated by the dotted regions located on the isotherms at 500 and 420 °C. The 500 °C growth region is well established while that around 420 °C is not well characterized.

Since a semiopen tube apparatus was employed in this work some ambiguity in the final Hg content of the liquidus just prior to layer growth initiation prevented the inclusion of an exact data point on the phase diagram although general agreement is observed. Better agreement was obtained when current data was compared to the theoretical phase diagram recently published by J. Sanz-Maudes et al.[13] This new diagram moves the solidus lines to a lower cadmium to mercury ratio than Harman.

Growth temperature was an important concern. Theoretically, the growth of an epitaxial layer may occur at a tem-

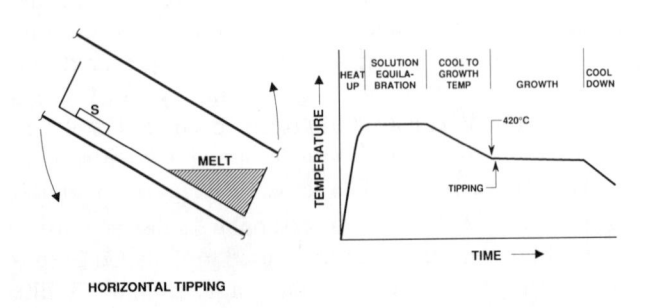

FIG. 2. Illustration of the LPE tipping apparatus and a schematic of the temperature-time growth cycle for Te-rich LPE HgCdTe Growth.

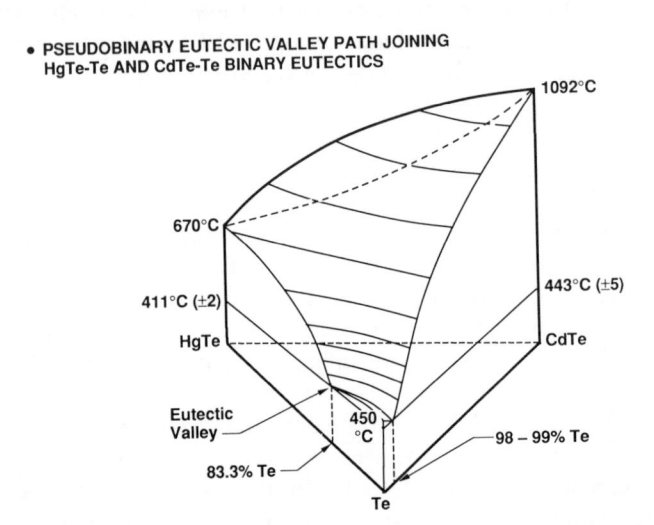

FIG. 4. Three-dimensional illustration of the Hg-Cd-Te phase diagram emphasizing the Te corner. HgTe–Te and CdTe–Te binary eutectic temperatures are indicated.

perature below the eutectic point of the growth solution as long as the solution has maintained its liquid state metastably. The Te-rich solution used in this work yielded epitaxial growths of HgCdTe even at a temperature around 400 °C which is at least 11 °C below the eutectic point of HgTe. These layers, however, had large melt spots.

Figure 4 shows a three-dimensional construction of the Hg-Cd-Te phase diagram.[15] The eutectic temperatures were taken from a numerical review of the phase diagram authored by Brice.[14] The left side of the three-dimensional Gibbs triangle shows the HgTe-Te eutectic occurring at ~411 °C ($\pm$ 2 °C). The right side of the figure shows the CdTe-Te eutectic occurring at ~443 °C ($\pm$ 5 °C). The path of the pseudobinary eutectic valley line which joins the binary HgTe-Te and the CdTe-Te eutectic points is also indicated. A growth temperature of 420 °C should lie on the phase diagram above the ternary eutectic depending on liquidus composition. All the low-temperature layers reported in this paper were grown at 420 °C. The presence of second phase Te was not observed by either optical or scanning electron microscope (SEM) techniques.

C. Morphology and composition

All layers grown at both 500 and 420 °C routinely yielded good morphology. Figure 5 shows the dependence of layer morphology on substrate orientation for the 420 °C growths of $x = 0.31$ HgCdTe on the CdTe/sapphire substrate. A natural progression of increased surface features is observed with increased substrate misorientation. The 0.2° misorientation example exhibits excellent featureless morphology. The contrast of this photograph was enhanced using a Nomarski attachment.

The composition of the layers was determined at room temperature by IR transmission and verified later by actual device cutoff wavelength measurements. Figure 6 is a example of an IR transmission spectrum of sample J1-48. The cuton is sharp and intersects the x axis at a value of 3.76 μm. Interference fringes indicating a reasonably planar interface are observed. A thickness of 15.2 μm was determined by calculations using the fringes. These values typically correlate with thicknesses determined by direct optical measurement of a cross section of the epilayers.

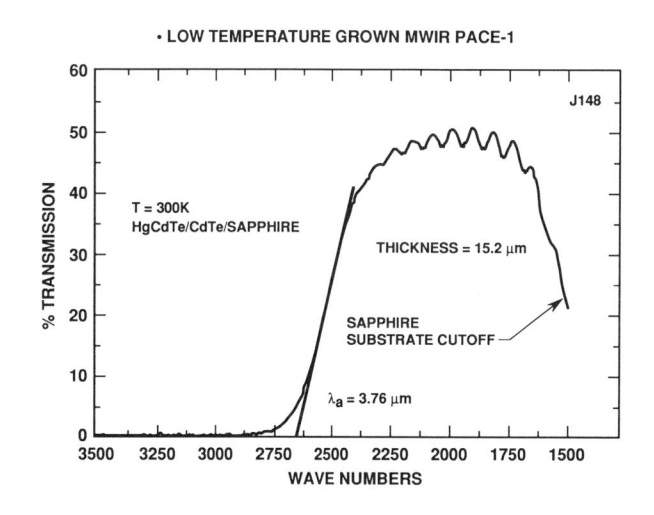

FIG. 6. IR transmission spectrum of a 420 °C grown PACE-1 MWIR HgCdTe layer.

Compositional depth profiles of both the 500 and the 420 °C grown epilayers were obtained using energy dispersive x-ray analysis (EDAX). Figure 7 indicates the results. Both layers are approximately the same composition. The high-temperature grown layer exhibits compositional grading as indicated by the more rounded Hg and Cd curves around 4 μm. The low-temperature grown layer exhibits a sharper transition from the CdTe buffer layer to the HgCdTe layer. The reduced compositional grading observed in the 420 °C layer is consistent with a reduction of matrix interdiffusion associated with a reduction in temperature.

D. Crystallinity

The crystallinity of the low-temperature layers was examined by the double-crystal x-ray rocking curve technique. Excellent full width at half maximum (FWHM) values of between 40 and 59 arcsec were obtained as indicated in Fig. 8. These values which are comparable to LPE HgCdTe grown on CdTe substrates indicate the quality of HgCdTe that can be obtained using the lattice-mismatched CdTe/

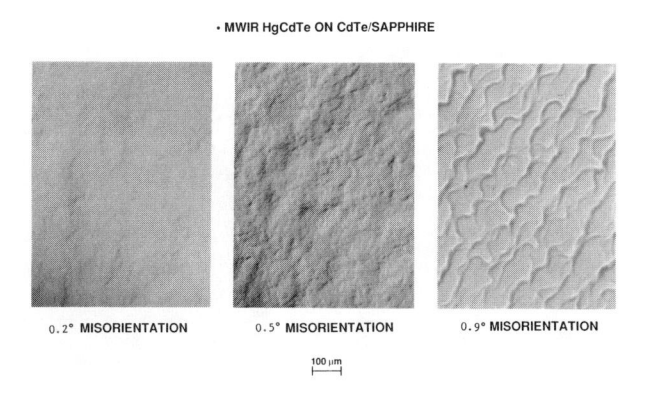

FIG. 5. Photomicrographs of the (111)B surface of MWIR HgCdTe grown at 420 °C on CdTe/sapphire (PACE-1) substrates as a function of substrate misorientation.

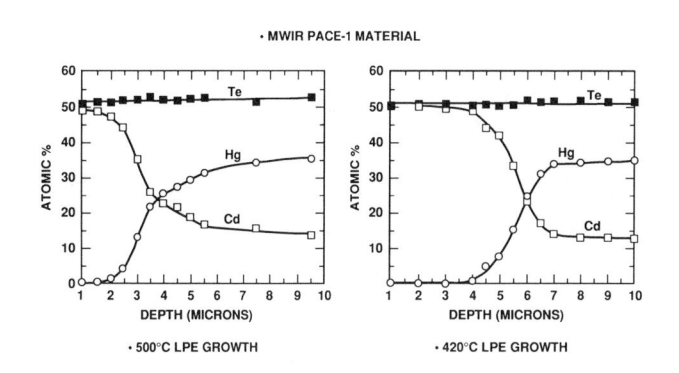

FIG. 7. EDAX compositional profile of MWIR PACE-1 HgCdTe material grown from Te-rich solutions at 500 and 420 °C temperatures.

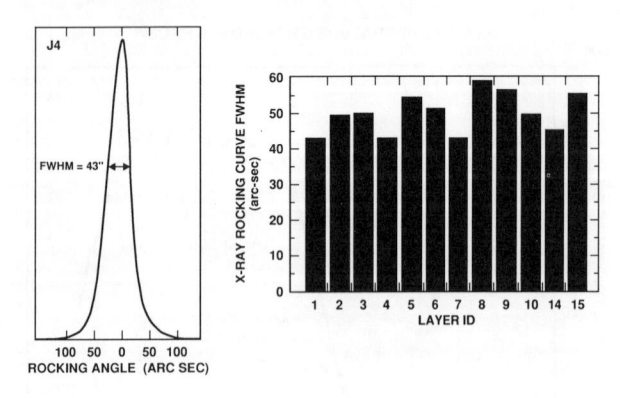

FIG. 8. Double-crystal x-ray rocking curves for 420 °C grown PACE-1 material.

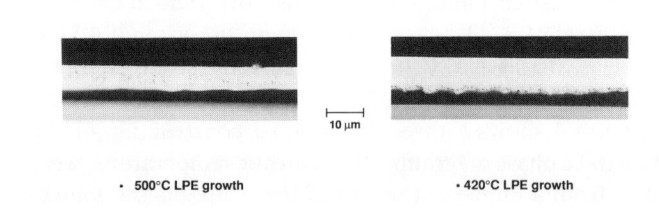

FIG. 10. Photomicrographs of the cross sections of 500 and 420 °C MWIR PACE-1 layers after exposure to a dislocation etch.

sapphire system. Rocking curves were taken on a series of 15 PACE-1 low-temperature layers. The results were consistent as shown in Fig. 8(b).

Further studies of the crystallinity were performed using a new chemical etch which can reveal dislocations on various crystalline planes.[1] Since PACE-1 is a lattice-mismatched material system, higher dislocation densities are expected. A desire for dislocation reduction was one of the motivations for this study of low-temperature growth. Recent work by other authors has indicated a relationship between compositional grading of HgCdTe and misfit dislocation generation.[16] As observed in Fig. 7 the low-temperature growth technique results in a more abrupt HgCdTe/CdTe profile. This abruptness is expected to yield reduced dislocations.

A general description of the (111B)HgCdTe/sapphire material system is shown in cross section in Fig. 1 with accompanying photomicrographs indicating the etch pit density distributions observed on each of the constituent layers. The bottom picture indicates the results of etching the sapphire substrate with hot phosphoric acid to reveal basal plane (0001) dislocations.[17] EPDs of 4×10^3 cm^{-2} were observed. The (111B) surfaces of both the CdTe buffer layer and the $Hg_{0.69}Cd_{0.31}Te$ layer were treated by the new $K_2Cr_2O_7/HNO_3/HCl/H_2O$ etch and produced triangular etch pits.[1] EPD values of 3×10^6 cm^{-2} were revealed in the MOCVD CdTe buffer layer after a light etch of the faceted (111B) surface which was ~6 µm from the CdTe/sapphire interface. EPD values of 7×10^5 cm^{-2} were revealed at the surface of the 420 °C grown LPE layer of HgCdTe which was ~12 µm from the HgCdTe/CdTe interface. Etch pit densities increased with depth for both layers with the CdTe EPD reaching ~10^7 cm^{-2} approximately 3 µm from the CdTe/sapphire interface and the HgCdTe EPD reaching 7×10^6 cm^{-2} approximately 1 µm from the HgCdTe/CdTe interface. Layer depths were determined by optical cross section.

SEM micrographs of the etch pit density (EPD) etched (111B) surfaces of 500 and 420 °C grown MWIR HgCdTe layers are shown in Fig. 9 for the same magnification. The etch pit count is significantly reduced for the 420 °C grown layer. The pits are triangular as expected although a number of them have tails. It is postulated that these tails are replications of the features left from etching the same dislocation in material slightly above the current surface.

The EPD depth dependence and the misfit dislocation density at the CdTe/HgCdTe interface were studied further by EPD etching cross sections of a high temperature (500 °C) and a low temperature (420 °C) grown layer. Figure 10 shows optical micrographs of these two layers. The CdTe/HgCdTe interface of the high temperature layer exhibits reduced dislocations while the low temperature layer shows a high dislocation density directly at the interface.

These two layers were examined in more detail by SEM as shown in Fig. 11. Figure 11 clearly shows the relationship of

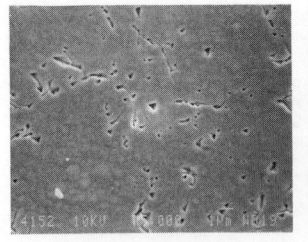

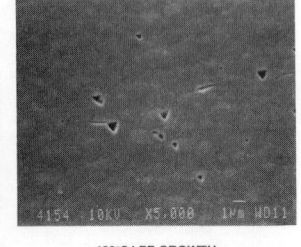

FIG. 9. SEM micrographs of 500 and 420 °C grown PACE-1 MWIR HgCdTe epilayers after exposure to a dislocation etch.

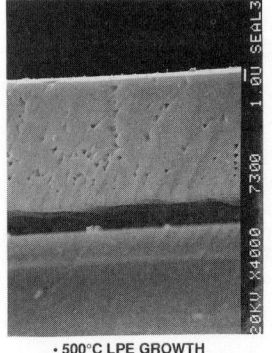

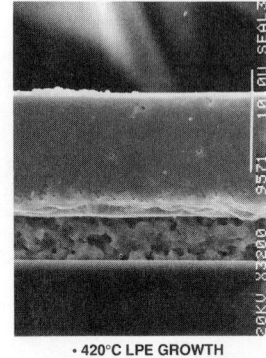

FIG. 11. SEM micrographs of the cross section of 500 and 420 °C MWIR PACE-1 layers after exposure to a dislocation etch.

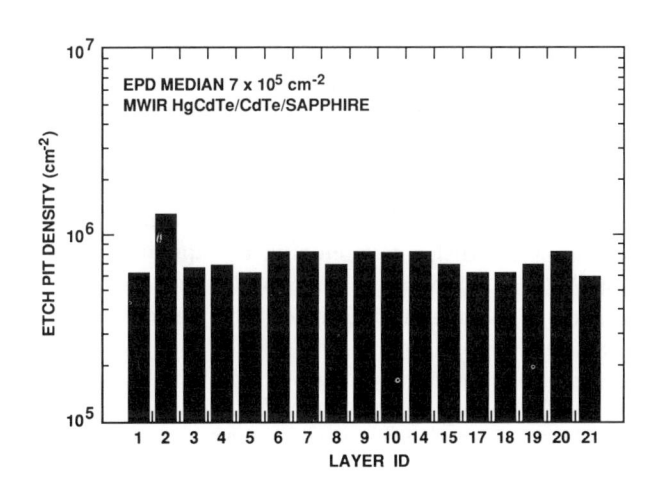

FIG. 12. Etch pit density results for numerous MWIR HgCdTe PACE-1 420 °C grown epilayers as revealed by chemical etch.

EPD distribution as a function of depth for the two growth temperatures. The high-temperature layer shows dislocations distributed throughout the crystal while the low-temperature layer shows a high dislocation count at the interface but a significantly reduced count throughout the bulk of the layer. The dislocation etch appears to deferentially attack CdTe. This property results in the removal of material into the plane of the picture as seen near the shaded strip in the middle of the photographs. Some features resulting from cleaving are also observed on the layer cross sections.

The samples used for compositional profiling shown in Fig. 7 were used in the cross-sectional EPD examination of Fig. 11. Comparison indicates the data is in qualitative agreement with that of Yoshikawa[16] and can be explained as follows. The reduction in growth temperature results in less interdiffusion of HgCdTe and CdTe at the buffer layer interface and therefore less compositional grading. This interdiffusion occurs during the growth and subsequent cooldown. Less compositional interdiffusion requires that the lattice mismatch that occurs between HgCdTe and CdTe must be accommodated over a shorter distance. This accommodation results in increased misfit dislocations confined near the interface. Any additional dislocations appearing in the film are attributed to threading dislocations. The consequence is that for low-temperature grown layers the EPD is high at the interface and much reduced in the bulk of the layer. Higher temperature material with increased compositional grading forces misfit dislocations throughout the material.

The use of the low-temperature growth technique has resulted in an order of magnitude reduction of EPD for the HgCdTe/CdTe/sapphire material system. Data from 17 layers examined by using the new EPD etch is shown in Fig. 12. Layers were found to yield consistently lower EPD with a median value of 7×10^5 cm^{-2}.

E. Impurity content and Hall data

The impurity content of both high-temperature and low-temperature grown PACE-1 layers was compared by secondary-ion mass spectrometry (SIMS). Figure 13 shows impurity profiles for both 500 °C and 420 °C growths. It is difficult to say anything concerning the absolute values of the impurities observed since some quantitative variations were observed from layer to layer. However, the presence of impurity gettering at the HgCdTe/CdTe buffer layer interface is observed. Li, Na, Al, and Ga are peaked at the interface for the 500 °C layer. Little impurity build up is observed for the 420 °C case.

The 77 K Hall characteristics of the 420 °C grown layers were excellent. As-grown p-type carrier concentrations of $\sim 2 \times 10^{16}$ cm^{-3} were routinely observed with excellent hole mobilities ranging from 300 to 400 cm^2/V-s at 77 K. N-type characteristics were also obtained by closed tube Hg anneal. Electron mobilities as high as 51 000 cm^2/V-s were measured.

III. SUMMARY

The feasibility of growing the lattice-mismatched system Hg$_{0.69}$Cd$_{0.31}$Te/CdTe/sapphire (PACE-1) at low tempera-

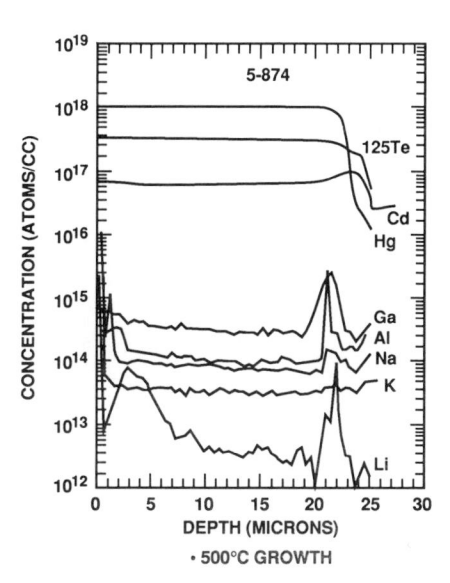

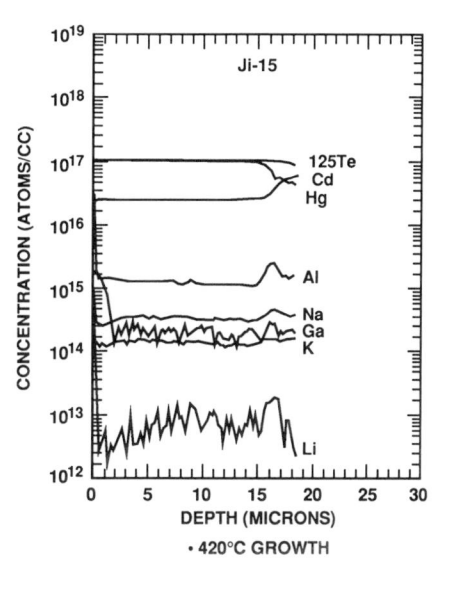

FIG. 13. Impurity distributions of both 500 and 420 °C MWIR HgCdTe PACE-1 epilayers as revealed by SIMS.

tures (420 °C) was demonstrated. Liquidus and solidus data was found to be in general agreement with existing phase diagrams. The low-temperature grown material was characterized in detail and found to be of excellent quality. X-ray rocking curves were found to be narrow and consistent with values obtained for MWIR LPE HgCdTe on CdTe. The etch pit density of the material was examined both on the (111)B surface and on the (110) cross section. Median EPD counts of 7×10^5 cm^{-2} were routinely observe using a new dislocation etch. These values are equivalent to counts routinely observed from LPE (111)B HgCdTe material grown on CdTe substrates.

The compositional profiles of both high-temperature and low-temperature grown material were obtained by EDAX and compared to EPD cross section SEM micrographs. The less interdiffused 420 °C grown layer corresponded to a high interface EPD count and a reduced HgCdTe epilayer EPD count. The more interdiffused 500 °C grown layer corresponded to a reduced interface EPD count and an increased HgCdTe layer EPD count.

The electrical properties of the 420 °C layers yielded consistent as-grown p-type carrier concentrations of 2×10^{16} cm^{-3} and hole mobilities as high as 400 cm^2/V-s at 77 K. Electron mobilities as high as 51 000 cm^2/V-s were measured after a Hg saturated low-temperature anneal. Excellent MWIR photodiode performance was observed with devices exhibiting diffusion limited performance at temperatures typically $\geqslant 120$ K.

The most significant result of this work is that the lattice-mismatch system, Hg$_{0.69}$Cd$_{0.31}$Te/CdTe/sapphire (PACE-1), can be grown under conditions that produce high quality material with characteristics (x-ray rocking curves, Hall data, and EPD) that mimic those obtained using the more lattice-matched HgCdTe/CdTe system.

ACKNOWLEDGMENTS

The authors thank G. Bostrup for growth of some of the high-temperature MWIR layers, and M. Diem for his discussions of the dislocation etching of sapphire. We also thank W. Tennant for his guidance and original suggestions and both S. Irvine and E. Gertner for their useful discussions.

[1] J. S. Chen, United States Patent No. 4897152.
[2] T. H. Myers, Y. Lo, R. N. Bickness, and J. F. Schetzina, Appl. Phys. Lett. 42, 247 (1983).
[3] H. S. Cole, H. H. Woodbury, and J. F. Schetzina, J. Appl. Phys. 55, 3166 (1984).
[4] E. R. Gertner, W. E. Tennant, J. D. Blackwell, and J. P. Rode, J. Cryst. Growth 72, 461 (1985).
[5] D. D. Edwall, E. R. Gertner, and W. E. Tennant, Proceed. of the IRIS Specialty Group on Infrared Devices, Aug. 1985.
[6] W. E. Tennant, Techn. Digest, IDEM (IEEE, New York, 1983) p. 704.
[7] J. P. Rode, Proc. SPIE 443, 120 (1983).
[8] R. A. Riedel, E. R. Gertner, D. D. Edwall, and W. E. Tennant, Appl. Phys. Lett. 46, 64 (1985).
[9] E. R. Gertner, Annu. Rev. Mater. Sci. 15, 303 (1985).
[10] C. D. Chaing and T. B. Wu, J. Cryst. Growth 94, 499 (1989).
[11] T. C. Harman, J. Electron. Mater. 9, No. 6 (1980).
[12] Tse Tung, Ching-Hua Su, Pok-Kai Liao, and R. F. Brebrick, J. Vac. Sci. Technol. 21, May/June (1982).
[13] J. Sanz-Maudes, J. Sangrador, and T. Rodriguez, J. Cryst. Growth 102, 1065 (1990).
[14] J. C. Brice, Prog. Cryst. Growth Charact. 13, 39 (1986).
[15] D. J. Williams, J. Cryst. Growth 58, 657 (1982).
[16] M. Yoshikawa, K. Maruyama, T. Saito, T. Maekawa, and H. Takigawa, J. Vac. Sci. Technol. A 5, No. 5. Sept./Oct. (1987).
[17] D. L. Stephens and W. J. Alford, J. Am. Ceram. Soc. 47, No. 2, 81 (1963).

A review of impurity behavior in bulk and epitaxial Hg$_{1-x}$Cd$_x$Te

P. Capper

Philips Infrared Defence Components Ltd., Southampton, Hants, SO9 7QG, United Kingdom

(Received 2 October 1990; accepted 29 October 1990)

Extrinsic doping using elements which produce stability of electrical properties will become increasingly important in future infrared device structures based on Hg$_{1-x}$Cd$_x$Te (MCT). This paper reviews the incorporation and activation of dopants in the most widely used bulk and epitaxial growth techniques. Stoichiometry at the growth temperature is demonstrated to be the critical factor which affects dopant activation. A number of factors, including stoichiometry, can affect the as-grown electrical properties of MCT and the importance of determining the type of conduction in the as-grown state, if successful extrinsic doping is to be accomplished, is stressed. The minimum criterion for confirmation of dopant activity is established as agreement between electrical and chemical data on the same low temperature Hg-annealed sample. At low concentrations of dopants, an additional requirement is to confirm the absence of other potential impurity dopants at equivalent levels. Most elements are active dopants in accordance with their relative position in the periodic table but several important exceptions exist, notably group V elements in Te-rich material. Slow-diffusing dopants are preferred and techniques are described which produce stable doped/undoped heterostructures, using As as the acceptor element in metalorganic vapour phase epitaxy growth. Data on dopant segregation behavior, in growth from liquids, acceptor ionization energies and minority carrier lifetimes are presented and their importance is discussed. Ionization energies can be used to differentiate doped from undoped material, providing the degree of compensation is known. Doping using extrinsic acceptors has been shown to improve minority carrier lifetimes in material grown by certain techniques but is unsuccessful in other types of material. Some recent data on annealing undoped Te-rich liquid phase epitaxial material will be presented which suggests that higher minority carrier lifetimes can be achieved purely by defect control.

I. INTRODUCTION

Establishing the behavior of extrinsic dopants in Hg$_{1-x}$Cd$_x$Te, (MCT), is important for two very different reasons. Small concentrations of unintentionally added elements can, if electrically active, affect the properties of the material and influence the performance of devices. More importantly, the need to produce photodiodes with improved performance may ultimately require intentionally doped material with known and controllable concentrations of both donor and acceptor elements on both sides of the junction. It is necessary, therefore, to determine the effects of various elements contained within the particular type of material, i.e., bulk, epitaxial, etc., being considered. Once this information has been obtained, methods for incorporating useful dopants at the correct levels and eliminating undesirable impurities can be developed.

There is considerable evidence that both donor and acceptor behavior is affected by the growth technique employed and by stoichiometry. In addition to the improved control of devices, increases in both minority carrier lifetime and mobility are expected in extrinsically doped p-type MCT due to a reduction in the number of native defects through the law of mass action. It is generally recognized that both chemical and electrical data on a specific element, preferably within the same low temperature Hg-annealed sample, should correlate before electrical activity can be assigned to an added dopant. At low dopant concentrations, a third criterion, namely the absence of active accidental dopants at equivalent concentrations within the sample, can be considered to be important.

In order fully to characterize the behavior of a particular dopant it may be necessary to establish the extent of its segregation behavior (especially relevant to bulk crystals), its ionization energy and its effect, if any, on carrier lifetime. It is also self evident that the nature of undoped material, both in the as-grown and low temperature Hg-annealed states, grown by a technique must be determined before the effect of deliberate doping can be understood. This aspect becomes increasingly important in epitaxial processes where growth temperatures are lower and the material produced changes from native defect controlled p-type to n-type or p-type governed by native defects or background impurities. Only when all of this relevant data for the chosen elements in the particular growth technique are established can controllable doping be achieved and, in the case of epitaxial processes, conditions for the growth of extrinsically doped heterostructures be determined for use in advanced device structures (e.g., those aimed at raising the operating temperature).

This review presents results for the various bulk processes [solid state recrystallized (SSR), Bridgman and Travelling heater method (THM)] before reviewing the state of extrinsic doping in the three most common epitaxial techniques [i.e., liquid phase epitaxy (LPE), metal-organic vapour phase epitaxy (MOVPE) and molecular beam epitaxy (MBE)]. This is a rapidly developing field in the epitaxial processes, bulk material having been extensively studied already, and this review should be treated as a snapshot of current thinking. A discussion is developed throughout the paper comparing the behavior of dopants in these different growth techniques with particular emphasis on the effect of the stoichiometry of conditions at growth.

II. BULK-GROWN MATERIAL

A. Undoped material—electrical behavior

This section is included as a base-line to which the doped material can be compared. In SSR growth the material is equilibrated at a high temperature and the conduction is p-type, on cooling to ambient, due to Hg vacancies. Several groups[1-10] have studied the behavior of as-grown and Hg-annealed SSR material and the picture is one of high p-type concentrations in the as-grown state with low background n-type levels ($1-10\times 10^{14}$ cm^{-3}) produced by low temperature (<300 °C) Hg annealing which fills Hg vacancies. A more complicated mechanism has emerged from the work in Refs. 6-9 in which Hg diffusing into the material gives rise to an n-type surface layer and which also dissolves Te precipitates. This removal of precipitates drives the major impurities (Cu, Ag) ahead of the Hg leaving the core p type. On further annealing, these impurities redistribute throughout the slice and the whole sample reverts to p type. This is a clear example of the interactive nature of impurities and defects in MCT and emphasizes the need to understand undoped material prior to investigating extrinsic doping.

Work on Bridgman material[2,11] has shown that as the crystal is cooled slowly to ambient temperature, n-type conduction results. The background donor level ($<1\times 10^{15}$ cm^{-3}) is believed to be due to unintentionally added impurities rather than Hg interstitials or other native defects. Annealing at high temperatures, with Hg, or two-temperature annealing can convert such material to p type through the introduction of Hg vacancies. Figure 1 shows the extent of the p- and n-type regions as determined for bulk SSR and Bridgman material. Evidence exists[12,13] that SSR material can reconvert to p type after extended storage at ambient temperature but such a mechanism does not occur in Bridgman samples.[14,15] A recent improvement has been made to the basic Bridgman process by the addition of the accelerated crucible rotation technique, (ACRT).[16-18] This resulted in improved x uniformity but also gave p-type conduction as-grown for $x<0.3$.[19] As x decreased from 0.3 to 0.2 the p-type carrier concentration increased from 2×10^{16} cm^{-3} to 5×10^{17} cm^{-3} but such material was readily converted to n type after a conventional low temperature Hg anneal. ACRT material is thought to equilibrate along the Te-saturated solidus (Fig. 1), unlike normal Bridgman growth which is metal saturated by nature.

Reducing the growth temperature (to 600–550 °C) using the THM process still produces p-type, $x = 0.2$ material as-grown ($1-2\times 10^{17}$ cm^{-3}), from the Te-rich solvent. This material can be converted to low ($1-4\times 10^{14}$ cm^{-3}) n-type by suitable Hg annealing.[20-22]

It can be concluded, therefore, that SSR, THM, and ACRT Bridgman ($x<0.3$) material are examples of Te-rich growth while normal Bridgman produces Hg-rich material

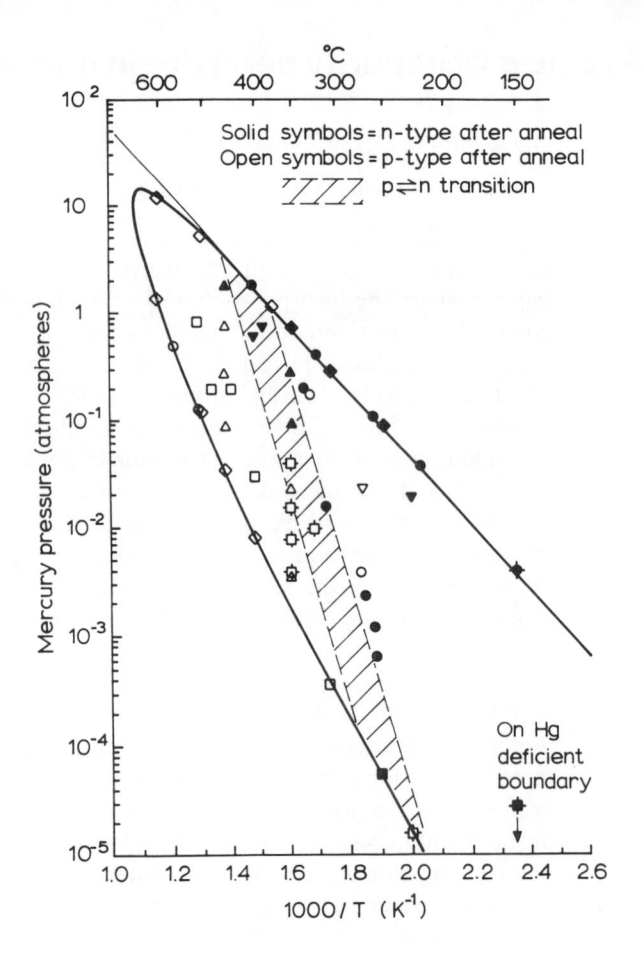

FIG. 1. Hg pressure vs temperature for $x = 0.2$–0.3 material (from Ref. 11).

in the as-grown state which is n-type. These differences must be considered in any studies of doped material.

B. Electrical activity of dopants

Table I shows the expected behavior of dopants compared to their relative positions in the periodic table and the relevant substitutional site.[23] Group I and III elements should be acceptors and donors, respectively, on the metal sites while groups V and VII elements should be acceptors and donors on Te sites. Group IV elements may be amphoteric in nature while group VIII elements may behave as donors on Te sites. Difficulties arise if dopants collect in precipitates (probably Te) rather than distributing through the lattice or if they are complexed with native defects to give neutral or charged species.

The first systematic study of extrinsic doping was carried out by Johnson and Schmit[24] in SSR material ($0.2<x<0.4$). Both diffusion after growth (for Li, Cu, Ag, Al, Ga, Si, Sn, P, As, and Br) and direct addition to melts (for Cu, Ag, and Si) were used and a summary of their results is presented in Table I, along with other data from work cited below. Acceptors were found to be Li, Cu, Ag, P, and As (the latter two only after a high temperature, 650 °C, diffusion) while donors included Al, Ga, In, Si, and Br (the latter only after a 650 °C diffusion). Tin was found to be inactive and was assumed to be isoelectronic with Hg and Cd (valency of 2). Elements substituting on metal sites were

TABLE I. Impurity behavior in bulk material.

Element	Group	Expected behavior[a]	Actual behavior[b] SSR (Te)	I.I. (Te)	Brid. (Hg)	ACRT (Te)	ACRT (Hg)
Li	IA	$A(m)$	A		A		A
Cu	IA	$A(m)$	A	A	A	A	A
Ag	IB	$A(m)$	A	A	A	A	A
Au	IB	$A(m)$	A		I	A	A
Zn	IIB	$I(m),D(i)$		D		I	I
Hg	IIB	$I(m),D(i)$		D			
B	IIIB	$D(m)$	D	D			
Al	IIIB	$D(m)$	D	D	D		
Ga	IIIB	$D(m)$	D				
In	IIIB	$D(m)$	D	D	D	D	D
Si	IVB	$D(m),A(t)$	D		D		
Ge	IVB	$D(m),A(t)$	I				
Sn	IVB	$D(m),A(t)$	I				
Pb	IVB	$D(m),A(t)$	I				
P	VB	$A(t)$	$I/A+$	A	A		
As	VB	$A(t)$	$I/A+$	A	A	I	A
Sb	VB	$A(t)$	I		A	I	A
Bi	VB	$A(t)$				I	I
Cr	VIA	$I(t)$	I				
O	VIB	$D(i),I(t)$	D				
F	VIIB	$D(t)$		D			
Cl	VIIB	$D(t)$	D		D	I	D
Br	VIIB	$D(t)$	D				
I	VIIB	$D(t)$	D				
Fe	VIII	$I(m),D(t)$			I		
Ni	VIII	$I(m),D(t)$	I				

[a] (m) = on metal sites, (t) = on tellurium sites, (i) = on interstitial sites.
[b] A = acceptor, D = donor, I = inactive, $I/A+$ = conflicting results, I.I. = ion implantation.

found to be rapid diffusers while those on Te sites were slow diffusers.

The activities of Cu and Ag were confirmed by Lin.[25] Vydyanath[3] found Cu to be an acceptor but precipitation occurred at high concentrations leading to a saturation of activity. This same author concluded that In was only partially active as a donor, the remainder being present as In_2Te_3. Iodine was found to be a donor in $x = 0.25$ SSR material.[26] Vydyanath et al.[27] noted the amphoteric nature of P as a dopant. The annealing behavior of undoped SSR material was explained by Tregilgas et al.[8] by a mechanism involving the redistribution of an acceptor impurity, identified as Cu. Dingrong et al.[28] confirmed In as a partially active donor but concluded that the inactive In was present as In–Hg vacancy pairs. Finkman and Nemirovsky[29] show Au to be an acceptor in SSR material although they suggest that its energy level is co-incident with the lower energy level of the doubly ionized Hg vacancy, suggesting the Au may be complexed with the metal vacancy. Studies of Cl in SSR crystals[30,31] show it to be a donor with $\approx 16\%$ activity at 1×10^{17} cm^{-3} decreasing to $\approx 3\%$ at 4×10^{14} cm^{-3}.[31]

Ion implantation has been widely used to introduce dopant species into SSR material with dopant concentrations confirmed by secondary ion mass spectrometry (SIMS). Early work (Ref. 32 and references therein) suggested n-type conduction was produced for B, Al, In, Hg, and Zn and even when using Au or P. However, Ryssel et al.,[33] found that p-type doping could be obtained, albeit with low efficiency. Bahir and Kalish[34] showed how laser annealing could produce p-type and n-type conduction when P (acceptor) and B and In (donors) were used. Baars et al.[35] found that a high temperature anneal (≈ 400 °C) was necessary to activate As after implantation. Table I summarizes the ion implantation results.

Direct addition to Bridgman melts was used by Capper.[23] Samples for electrical assessment (Hall measurements) and chemical analysis [atomic absorbtion spectrometry (AAS)] were taken from the same region in each crystal (i.e. $x \approx 0.2$–0.3) to avoid problems of dopant and matrix element segregation. Table I presents the results for the donors Si, Al, In, and Cl and the acceptors Li, Cu, Ag, P, As, and Sb, all of which were $\sim 100\%$ electrically active. Both Au and Fe showed low electrical activity. A more detailed study of Au[36] showed it to reside on nonsubstitutional sites and it was concluded that while not suitable as a controllable acceptor it could be useful as a contact metal at temperatures below 150 °C. An extension to the acceptor doping of Bridgman material was reported by Capper et al.[37] Controlled doping over the range 1×10^{16}–1×10^{18} cm^{-3} for Ag and Sb was demonstrated.

In the case of ACRT Bridgman material Capper et al.[19] demonstrated that the stoichiometry at growth was different from normal Bridgman and that this had a marked effect on dopant behavior. For material with $x < 0.3$ (i.e. Te-rich growth conditions) annealing in Hg was necessary to remove Hg vacancies and reveal the true nature of the added dopant. In this way Cu was seen to be a fully active acceptor while Ag and Au were $\approx 50\%$ and 10% active, respectively, following annealing. Prior to this anneal both Ag and Au levels, determined by AAS, agreed with the Hall data, suggesting full activity as the Hg vacancy concentration before annealing was below the dopant concentration. This contrasts with Bridgman growth where Au was essentially inactive. Indium was found to be a donor while Cl was seen to have an activity of only a few percent for $x < 0.3$ but $\approx 100\%$ activity for $x > 0.3$. This relatively low efficiency in material with $x < 0.3$ agrees with the behavior in SSR material.[31] Similar behavior to Cl was seen for As and Sb, i.e. active in that section of the crystal with $x > 0.3$ and inactive for $x < 0.3$. Chemical analysis showed high levels of incorporation for all of the Group V elements but n-type concentrations (after annealing) were $\leq 5 \times 10^{15}$ cm^{-3}. This demonstrates that the Group V elements were inactive electrically and not behaving amphoterically (i.e. as donors on metal sites) as seen in Te-rich LPE (see Sec. III B). In this case Bi was electrically inactive. Clearly for elements which are expected to substitute onto Te sites (groups V and VII) electrical activity is only likely for material grown under Hg-rich conditions, i.e. where Te lattice sites are available for the substitutional atoms. This may also account for the amphoteric nature of P in SSR material (Te rich) as observed by Vydyanath.[27] Those elements which substitute on metal sites (groups I and III) appear to be active dopants irrespective of growth stoichiometry, although Ag, Au, and Li[19] are seen to diffuse on annealing even in metal-rich material. These results are also given in summary in Table I which has been taken from p. 70 of Ref. 38 and updated.

C. Dopant segregation

Macroscopic segregation of dopants is expected to be small in SSR crystals where rapid cooling from the growth temperature is used. By contrast, the slow-grown Bridgman, ACRT, and THM crystals exhibit equilibrium (or near-equilibrium) segregation on the macroscopic scale in both radial and axial directions.

Johnson and Schmit[24] and Lin[26] found segregation coefficients, k, of 1.0 for Cu, Ag, and Si in SSR material, as expected. Dingrong et al.[28] established $k = 1.0$ for In. Tregilgas and Gnade[39] found that Ag diffused in SSR material, which was heavily doped with Au, at room temperature in ambient light. The Dember effect was thought to be responsible for this movement of Ag, on interstitial sites, towards the sample surface. Several other interstitial impurities behaved similarly. Talasek et al.[40] reported on the electromigration of Cu in SSR crystals at room temperature. Conversion to p type was observed.

Triboulet et al.[20] describe results for a Au-doped THM crystal with $x \approx 0.7$. A segregation coefficient of 0.003 was obtained. This demonstrates the purification which zoning produces.

Marked segregation of dopants occurs in Bridgman crystals. Capper[23] studied both singly and multi-doped crystals and used Pfann's mixing equation to determine k values. Concentrations of dopants were measured by AAS and spark source mass spectrometry (SSMS). Table II shows the results for the elements studied. It was found in Ref. 23 that k's decreased as the cube of the atomic radius. In the majority of cases ionic radii were used but for the more electronegative elements (Cl, Br, S) tetrahedral radii fitted the trend line more closely, supporting the view that binding in CMT is mixed covalent and ionic.

The marked difference in matrix element segregation between ACRT and normal Bridgman suggested that dopant segregation would also differ. This was found to be the case[41] and Table II summarizes the findings. Some elements (Au, As, Cl, and Sb) showed an increase in k towards 1.0 as growth proceeded. This region corresponded to, though extended beyond, the constant x region seen in ACRT crystals,[16] indicating that material can be grown uniform in both x and dopant concentration by ACRT. This behavior could provide a reproducible doping process for these elements. Table II also suggests that, in general, k's are lower in ACRT crystals. This may be due to increased stirring causing Te-rich conditions at the interface which reduces the boundary layer thickness[42,43] and the number of sites available for those elements which substitute on the Te sites (i.e., As, Sb, Bi, Cl, and Br). Alternatively the mechanism based on the "facet" effect of Ref. 44 might also play a role.

TABLE II. Segregation coefficients in bulk material.

Element	Group	Bridgman[a] (Hg)	ACRT (Te)	THM (Te)
Li	IA	0.4[a]	0.7	
Na	IA	1.0	0.04	
K	IA	2.1	<0.01	
Cu	IB	0.2[a]	0.5[a]	
Ag	IB	0.1[a]	0.3[a]	
Au	IB	0.02[a]	0.03[a]	0.003[a]
Mg	IIA	2.2		
Ca	IIA	2.0		
Al	IIIB	2.8[a]	1.0[a]	
Ga	IIIB	1.8	0.5	
In	IIIB	2.0	1.6	
C	IVB	0.5	0.6	
Si	IVB	2.4	2.3	
Sn	IVB	<0.01		
Pb	IVB	<0.01		
P	VB	1.0[a]		
As	VB	0.1[a]	0.01[a]	
Sb	VB	<0.01[a]	<0.01[a]	
Bi	VB	<0.01	<0.01[a]	
Cr	VIA	0.15	<0.01	
Mo	VIA		1.0	
O	VIB	<0.01	0.04	
S	VIB	1.0		
Mn	VIIA	1.6		
Cl	VIIB	0.4[a]	0.02[a]	
Br	VIIB	0.5	<0.01	
Fe	VIII	0.5[a]	0.5	
Co	VIII	0.15		

[a] From singly doped crystals.

D. Carrier lifetimes and ionization energies

1. n-type material

A review of carrier lifetime in *n*-type material has been published recently [p. 130 of Ref. (38)]. Essentially $x \approx 0.2$ material grown by SSR is dominated by Auger 1 recombination,[45] where lifetimes varied from 0.01 to 70 μs as *n*-type levels changed from 1×10^{16} to 1×10^{14} cm^{-3}, as is $x \approx 0.2$ THM material.[46] For Bridgman material, including ACRT, Shockley-Read (S-R) recombination was proposed to explain the results for both $x \approx 0.2$ and 0.3 samples,[14,15] where lifetimes varied from 0.1 to 10 μs as *n*-type levels decreased from 5×10^{15} to 4×10^{14} cm^{-3} for $x = 0.23$ (the corresponding values for $x = 0.35$ samples were 0.7–100 μs for 1×10^{16}–4×10^{14} cm^{-3}). In Ref. 15 doping with the donors Si, Cl, and In was shown to have little effect on lifetime but Al doping did reduce lifetime as did addition of a high concentration of inactive Fe.[23] When inactive Au was added, however, there was no reduction in lifetime. Jones (p. 150 of Ref. 38) has reviewed the ionization energies of donors in MCT. Monovalent donors are always fully ionized and Hall measurements on annealed SSR, or as-grown Bridgman, confirm that donor ionization energy, E_d, is close to zero. Crystals doped with In, Al, or Cl also showed $E_d \approx 0$.

2. p-type material

Lifetimes in *p*-type material were reviewed in Ref. 38 p. 139. For $x \approx 0.3$ most studies find undoped SSR and Bridgman material to be dominated by S-R recombination although one group suggests radiative recombination due to *n*-type channels in a *p*-type matrix in their Bridgman material.[47] Lanir et al.[48] have proposed radiative recombination in their SSR material and also show that Au doping reduces lifetime, believed to be via the introduction of S-R centers. Doping with As was found to increase lifetime in $x = 0.3$ SSR crystals[49] from 500 to 750 ns at 1×10^{16} cm^{-3}. The S-R center is generally thought to be the Hg vacancy but Jones et al.[50] argue that a defect or impurity (antisite Te and O, C, Si, or Cl impurity) on the metal site could also explain the results.

For $x \approx 0.2$ material S-R recombination appears most often as the explanation for the variation of lifetime with temperature in both SSR[51,52] and Bridgman,[53] but Schacham and Finkman[54] suggest the Auger 7 mechanism for their Au-doped SSR samples. Lacklison and Capper[55] studied both undoped and doped Bridgman samples with $x \approx 0.3$–0.21. They concluded that Auger 7 was less important than S-R recombination. There was no marked dependence of lifetime on x or dopant species (Ag, As, or Sb), but the lifetime decreased as the acceptor concentration, N_a, increased. This lack of improvement in lifetime for doped material contrasts with the As-doped data given in.[49] The Bridgman material studied in[55] was as-grown ACRT and two-temperature annealed Bridgman, both believed to be *p* type due to Hg vacancies, and showed lower lifetimes than SSR material of similar x and N_a. There is clearly a difference in the defect levels between the two growth techniques. More recently, Nemirovsky and Rosenfeld[56] have observed a difference between Au-doped SSR and undoped "slush"-grown (Cominco) material of a factor of ≈ 2, with the doped samples having the higher lifetimes. Fastow and Nemirovsky[57] point out that steady state lifetimes rather than transient values give a better measure of material quality. Both As and Au doping were shown to increase lifetime in $x = 0.22$ SSR and THM samples. The use of transient lifetimes as a measure rather than the more usual steady state values may explain why no differences were seen between doped and undoped material in Ref. 55.

A great deal of work has been carried out to understand acceptor ionization energies[58–67] and reviews have been published recently (Ref. 58 and p. 151 of Ref. 38). Figures 2 and 3 show the data for undoped and doped Bridgman (plus ACRT) material found by Kenworthy et al.[58] It can be seen that acceptor ionization energy, E_a, increases with x and decreases with $N_a^{1/3}$. Lower values are reported in the literature for compensated material, suggesting that values for x, N_a, and degree of compensation should be given when quoting a figure for E_a. For material doped with Cu, Ag, Au, As, and Sb values for E_a are lower than equivalent undoped samples, as shown in Figs. 2 and 3.[58] Doped samples were given a low temperature Hg anneal prior to measurement to remove the Hg vacancies. For narrow ranges of N_a (2.5–3.5$\times 10^{17}$ and 0.8–1.5$\times 10^{17}$ cm^{-3} for undoped and doped samples, respectively) and x (0.2–0.24) Kenworthy et al.[58] quote:

$$E_a (\text{undoped}) = 91x + 2.66 - 1.42 \times 10^{-5} N_a^{1/3},$$

$$E_a (\text{doped}) = 42x + 1.36 - 1.40 \times 10^{-5} N_a^{1/3}.$$

This difference in E_a values provides a valuable diagnostic technique for differentiating between doped and undoped *p*-type material. The technique is not sufficiently refined, however, to differentiate between the various acceptor dopants. Detailed chemical analysis was used to determine dopant concentrations to compare to N_a values, in Ref. 58, unlike most of the earlier reports where assumptions were made concerning the incorporation of dopants which may not have been valid (see Sec. II B above).

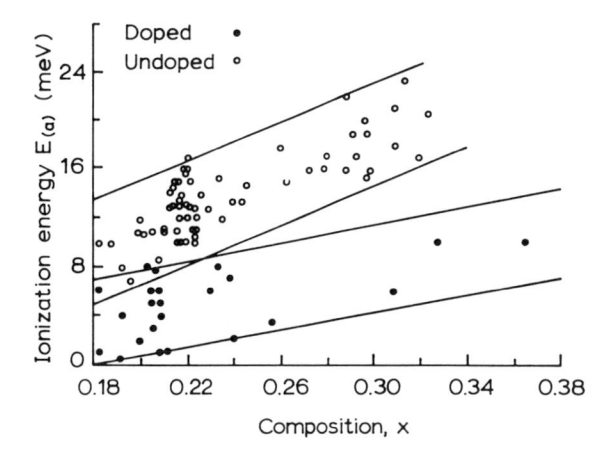

FIG. 2. Acceptor ionization energy vs x for undoped and doped Bridgman and ACRT material (from Ref. 58).

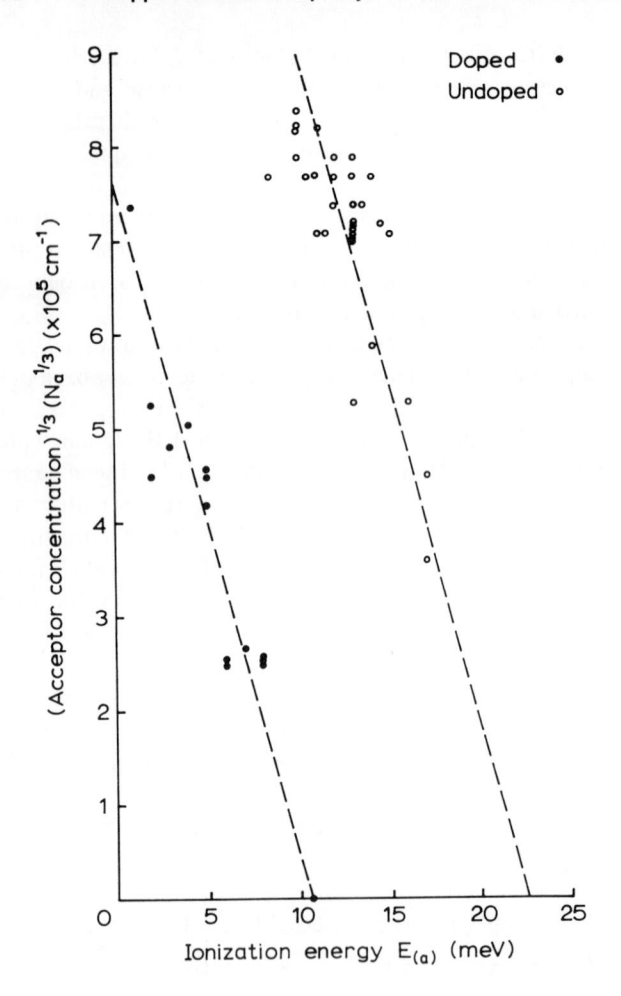

FIG. 3. Acceptor ionization energy vs $N_a^{1/3}$ for undoped and doped Bridgman and ACRT material (from Ref. 58).

III. LPE MATERIAL

A. Undoped material—electrical behavior

Both Te-rich and Hg-rich growth techniques have been used to grow LPE layers. The majority of studies have employed Te-rich conditions at temperatures above 450 °C and hence p-type material is produced (see Fig. 1). Annealing in Hg at temperatures below 300 °C fills in these vacancies[68–78] and produces n-type material due to residual donor impurities ($1-50\times10^{14}$ cm^{-3}). Several authors (Refs. 74, 76, 78, and 136) have suggested that LPE material converts to n type more slowly than bulk material, as determined by Jones et al..[11] A model to explain this was proposed in[78] based on the lack of grain structure in LPE layers. Brown and Willoughby[79] found that Hg diffusion into bulk material proceeded via a combination of slow and fast components.

Vydyanath and Hiner[80] have annealed their Te-rich LPE layers ($x\approx0.2$) at temperatures between 150–400 °C in both Hg and Te-rich ambients and find that the equilibrium defect levels are comparable with equivalent bulk samples. n-type carrier concentrations of $\approx 6\times10^{14}$ cm^{-3} were seen in annealed samples and they concluded that this was due to impurities and not native donor defects, e.g. Te antisites. Astles et al.[81,82] have shown how the use of a Bi wash-melt prior to growth can remove impurities from the substrate/layer interface and reduce the background n-type level from 2×10^{15} cm^{-3} to 1×10^{14} cm^{-3}. They found from SIMS analyses that Si and Cl were the only likely donor dopants present in the annealed samples. Brice et al.[83] used laser source mass spectrometry (LSMS)[84] survey analysis on their Te-rich LPE samples and found significant levels of Al, Si, O, and Cl. After annealing in Hg such layers converted to $5-50\times10^{14}$ cm^{-3} n type, from $2\times10^{18}-2\times10^{16}$ cm^{-3} p type for x between 0.2 and 0.4.

Growth from Hg-rich melts[85,86] occurs at temperatures between 450–550 °C and can produce n-type material directly. Figure 1 shows, however, that even at 475 °C on the Hg-rich solidus p-type material should be obtained. Sangha et al.[86] comment that their as-grown n-type results could be due to a degree of annealing on slow cool-down to room temperature in a Hg ambient. Tung and co-workers[87,88] report that their $x\approx0.2$ Hg-rich LPE is indeed p type as-grown at $\approx 1-2\times10^{17}$ cm^{-3}. This level, however, is well below the equivalent figure for Te-rich LPE with $x\approx0.2$ of $>1\times10^{18}$ cm^{-3}.

B. Electrical activity of dopants

The most comprehensive study of dopant behavior has been carried out by Kalisher[89] in Hg-rich LPE. A total of 13 elements were added to the growth melts and Hall measurements together with SIMS analyses were used to determine dopant activity. Table III lists the activities found. In most cases substitution on active sites can be seen, the exception being Zn, as it is isoelectronic with Cd and Hg. Ge is a donor dopant, indicating that it substitutes on metal sites as did Si in Hg-rich Bridgman material.[23] The only discrepancy with the behavior found in Hg-rich Bridgman is that of Au, which is an acceptor in the LPE layers. Tung and co-workers[87,88] have given a full account of the growth and doping of Hg-rich LPE layers. They show that As and In are acceptors and donors over at least two orders of magnitude of dopant concentration and note that if the desired dopant concentration is less than the native defect level then an anneal is required to eliminate Hg vacancies.

Bubulac et al.[90] found Li to be an acceptor in Te-rich LPE ($x = 0.3$–0.2) and showed that it decorated defects. Movement of Li on annealing in Hg was seen in[90] and has been confirmed by Astles,[82] who also found Na to be mobile. It has been proposed[82] that Li moves via the Dember effect as found for Ag in SSR material at room temperature.[39] Yoshikawa et al.[75] reduced their background n-type level by lowering the oxygen concentration in their Te-rich LPE layers. In this way they were able to establish controlled n-type doping using In as the donor impurity. Edwall et al.[76] showed that Cu acts as an acceptor in Te-rich LPE ($x = 0.32$) although the material was thought to be highly compensated.

Destefanis[91] used ion implantation to introduce In into his Te-rich LPE layers. He confirmed that the n-type conduction was due to added donor impurities by implanting with Xe and measuring p-type conduction. Most of the In was electrically active even at high implant doses. Bubulac et al.[92] have shown that both B and In (10% activity) are donors when implanted into $x = 0.3$ Te-rich LPE layers.

TABLE III. Impurity behavior in epitaxial material.

			Actual behavior				
Element	Group	Expected behavior[a]	LPE (Hg)	LPE (Te)	MOVPE (Hg)	MOVPE (Te)	MBE (Te)
Li	IA	$A(m)$		A			A
Cu	IB	$A(m)$	A	A			
Ag	IB	$A(m)$	A			A	A
Au	IB	$A(m)$	A				
Zn	IIB	$I(m),D(i)$	I	I			
B	IIIB	$D(m)$		D			
Al	IIIB	$D(m)$	D	D		D	D
Ga	IIIB	$D(m)$	D	D		D	
In	IIIB	$D(m)$	D	D	D	D	D
Tl	IIIB	$D(m)$	D				
Si	IVB	$D(m),A(t)$		$D?$		D	D
Ge	IVB	$D(m),A(t)$	D				
Sn	IVB	$D(m),A(t)$	$D?$				
P	VB	$A(t)$		$I(A)$	A	I	
As	VB	$A(t)$	A	$I(A)$	A	I	D/A
Sb	VB	$A(t)$	A	$I(A)$	A		D
Bi	VB	$A(t)$		I			
O	VIB	$D(i),I(t)$		D			
Mn	VIIA	$I(t)$	$D?$				
Cl	VIIB	$D(t)$		$D?$			
I	VIIB	$D(t)$		D		D	
Fe	VIII	$I(m),D(t)$	I				

[a] (m) = on metal sites, (t) = on Te sites, (i) = on interstitial sites.
[b] A = acceptor, D = donor, I = inactive, $I(A)$ = requires high temperature anneal to activate dopant, D/A = conflicting results from different centers.

A detailed study of the behavior of Group V elements in Te-rich LPE ($x \approx 0.24$–0.31) was reported by Vydyanath et al.[93] In the as-grown state P, As, Sb, and Bi were inactive. When annealed at 200 °C layers converted to n type as did undoped layers of similar x. However, when a preanneal at 500 °C (in Hg) was used before the 200 °C anneal, p-type conduction was obtained for P, As, and Sb but not Bi. This behavior was explained by the dopants occupying metal sites in the as-grown state and being moved onto active Te sites by the 500 °C anneal. This may also explain the need to introduce P and As into SSR bulk material via a high temperature (650 °C) diffusion[24] and the 400 °C anneal needed to activate the As implanted into SSR material.[35]

Chen and Dodge[94] suggest that Sb is an active acceptor in $x \approx 0.22$ Te-rich LPE following a 270 °C Hg anneal. No comparisons were given with similarly treated undoped samples, however, nor were any confirmatory chemical analysis data presented. Recent work by Sarusi et al.[95] with Te-rich LPE layers ($x \approx 0.2$) doped with Cu and Ga shows low doping efficiencies for both elements. Annealing for 24 h at 270 °C reduced the electrically active fraction of Cu. Layers doped with Ga were all n type. An increase in activity was observed after a 3 h/390 °C Hg anneal and this was compared to the behavior of group V acceptors in Te-rich LPE as reported by Vydyanath et al.[93] Chen[96] has studied Cu and As-doped LPE layers and found that both produced p-type conduction. No growth or annealing details were given, however, making it impossible to assign doping efficiencies or to categorize the material as Te or Hg rich. In a conference report Cirlin et al.[97] show that Cu is an active acceptor but that Fe is inactive in $x = 0.3$ epitaxial layers. It was not stated which process was used for the growth but as the samples were provided by the Santa Barbara Research Center group they were probably Hg-rich LPE.

Table III summarizes the activities found in both Hg- and Te-rich LPE processes.

C. Segregation of dopants

Segregation of added elements occurs to some degree in both Hg- and Te-rich LPE processes. There have been only a limited number of systematic studies in this area despite its importance in achieving reproducible and controlled doping.

Kalisher[89] determined segregation coefficients (k) for 13 elements in his Hg-rich LPE material. The k's decreased with increasing atomic number, within a group, as seen by Capper[23] in Bridgman material. Kalisher also found that k decreased as the dopant concentration increased, for Ga, In, and As. Lapides et al.[98] used the segregation data of Kalisher to grow double layer heterostructures, in which In and As controlled the n- and p-type conduction, respectively. The dopant distributions and electrical activities were stable during device manufacture demonstrating the advantage of having reliable dopant segregation data in a particular growth system.

Several studies have been carried out in Te-rich LPE material, either looking at single elements added deliberately or at a range of residual impurity dopants. The former group include work with Li,[90] In,[75] Sb,[94] and P, As, Sb, and Bi.[93] Annealing at 400 or 250 °C caused Li to move to the surface.[90] Vydyanath et al.[93] compared their data with those in Hg-rich LPE and bulk material and showed that k's in Te-rich material were lower (i.e., further away from unity). Studies of residual impurity segregation behavior include those of Astles[82] and unpublished work of Brice et al.[99] The k values found for Al, Si, K, and Cr were in reasonable agreement. For Li and Na Astles et al.[82] point out that for the higher growth temperature of 500 °C used by Brice et al. both elements could diffuse to the surface during the cooldown to room temperature, giving an apparent low k value. They also suggest that Li could diffuse to the surface, against the concentration gradient, via the Dember effect. Care must be exercised when carrying out comparisons based on residual levels of impurities as elemental sensitivities in mass spectrometric analyses vary, especially for low dopant concentrations.

D. Carrier lifetimes and ionization energies

There have been few systematic investigations of minority carrier lifetime in LPE material. For undoped n-type (i.e. after a Hg anneal) material with $x = 0.2$, Bajaj et al.,[100] Doyle et al.,[101] and Amingual et al.[102] quote values which are comparable with or slightly below those found in equivalent bulk samples. The latter group points out that measurements made on thin layers may not be representative of the true bulk lifetime but rather reflect the surface recombination velocity. Bajaj et al.[100] report lifetime versus temperature measurements which indicate that both S-R recombination and surface related effects can dominate the measurements at low temperatures.

For $x \approx 0.22$ undoped Te-rich LPE material which was p type, either as-grown or after a suitable Hg anneal, Chen et al.[103] report the values shown in Fig. 4. They suggest that the reason these values are below those of equivalent bulk samples (line 1, Fig. 4) is recombination either at the surface or the substrate/layer interface. Measurements of lifetime versus temperature and the dependence of lifetime on N_a (i.e., $N_a^{-1.85}$) indicate that the Auger recombination mechanism is dominant at low temperatures in their material. Nemirovsky et al.[104] give lifetime values for 2 p-type samples which agree with the trend shown in Fig. 4, from Ref. 103. Unpublished work in this author's laboratory on undoped Te-rich LPE layers are also shown in Fig. 4. Values for the three layers studied show a large variation with some close to the Rockwell data[103] while others are within the SBRC[87] range. The differences seen within each layer are due to different ex-situ anneal treatments indicating that control of defects can lead to high lifetimes in Te-rich LPE. Measurements against temperature indicate S-R recombination is dominant, at 77 K, in most of the samples.

While there are no reports of lifetimes in doped p-type Te-rich LPE material Tung et al.[87] present data for As-doped Hg-rich LPE layers with $x = 0.2$ and 0.3. The trend lines for samples with these two x values are also shown in Fig. 4. By comparison with undoped Te-rich LPE (from Ref. 103) and bulk samples (see Sec. II D 2 above) these lifetimes are ex-

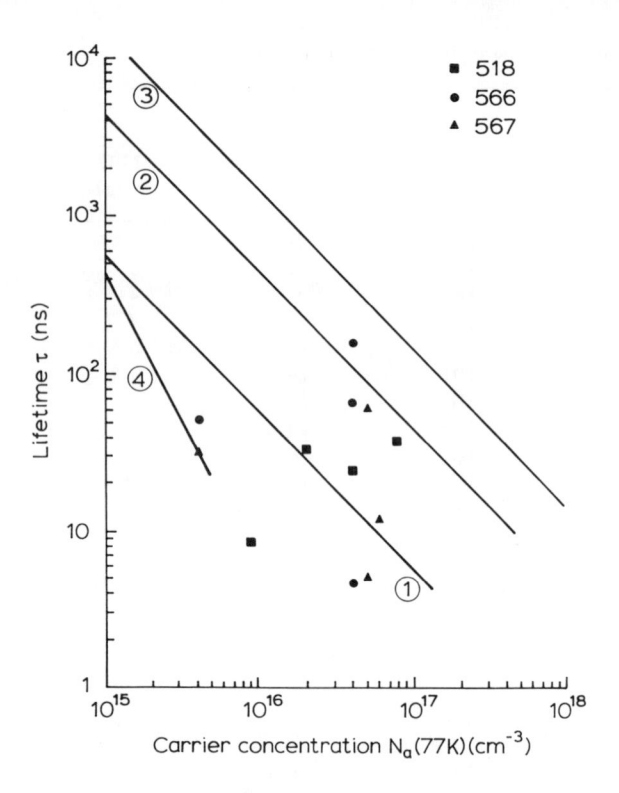

FIG. 4. Minority carrier lifetime vs N_a for 1—bulk samples (from Ref. 55), 2 and 3—$x = 0.2$ and 0.3 As-doped Hg-rich LPE (from Ref. 87), and 4—undoped Te-rich LPE (from Ref. 103). Data points for samples from three undoped Te-rich LPE layers after various post-growth anneal treatments (from Ref. 136)

tremely long. Tung et al.[87] do not quote values for equivalent undoped Hg-rich LPE layers but a private communication from the author indicates that lifetimes are lower by a factor of ≈ 2–5 in undoped material of similar x and p-type level. This increase in lifetime is thought[87] to arise from a reduction in S-R centers in the doped material as the added As decreases the native Hg vacancy level via the law of mass action. In Bridgman material no such increase in lifetime was found in doped samples.[55]

Reports of values for E_a in LPE layers are limited. Values in undoped material agree with those seen in bulk samples (see Figs. 2 and 3). In Te-rich LPE layers Edwall et al.[76] found a low value for Cu, in agreement with similar bulk samples. Chen and Dodge[94] quote low values for Sb in their Te-rich LPE while Chen[96] gives low figures for As and Cu, although whether the material is Hg or Te-rich is not stated. Kalisher[89] reports values for a range of As concentrations in his Hg-rich LPE layers, while Lou and Frye[105] give a low value for As in similar material. Elements from both groups I and V produce acceptor levels which are clearly, therefore, determined by the host lattice.

IV. MOVPE MATERIAL

A. Undoped material—electrical behavior

Early work on MOVPE-grown material showed samples with $x = 0.2$–0.25 grown at ≈ 400 °C to be n-type. This was attributed to contamination by In[106] and by Al and Ga.[107] More recent studies by several groups[78,108,109] using the interdiffused multilayer process (IMP) technique have produced p-type material, as predicted by Fig. 1. n-type as-grown material can be caused by other mechanisms besides contamination. In their IMP-grown layers Whitely et al.[110] report material with $x = 0.2$–0.3 to be p type (at mid-10^{16} cm^{-3}) after removal of an n-type surface layer. A similar phenomenon was seen by Bhat and Ghandhi[111] who demonstrated that when the Hg was left at 230 °C during cool-down from 400–430 °C to room temperature their $x = 0.19$ layers were n type (3×10^{15} cm^{-3}) but if layers were removed at ≈ 350 °C p-type (1×10^{16} cm^{-3}) material was produced. The latter group uses alloy growth but Irvine et al.[112] have shown that cool-down conditions can lead to n-type surface layers in IMP material grown at 350 °C. They described how the $p \leftrightarrow n$ transition band of Fig. 1 was crossed during their cool-down schedule.

Bhat et al.[113] grew $x = 0.2$ layers onto GaAs substrates at 415 °C by the alloy growth technique. They found n-type conduction at 1–30×10^{16} cm^{-3}, depending on the thickness of the CdTe buffer layer, which they attributed to strain at the substrate/layer interface and not to diffusion of Ga from the substrate. Gertner et al.,[114] however, report that layers with $x = 0.25$–0.35 grown at 400 °C onto CdTe buffered GaAs were p type (1–3×10^{16} cm^{-3}) as-grown. Taskar et al.[115] suggest that their $x = 0.3$ layers grown at 370 °C by the alloy technique are either weakly p type or weakly n type (3×10^{15} cm^{-3}) depending on several growth parameters, e.g. substrate type, Hg pressure, reactor design, chemical purity, etc. On lattice-matched CdTeSe substrates Bhat et al.[116] obtained p-type material if a 0.5 μm CdTe cap was grown on to their alloy-grown MCT but n-type material resulted if a 1 h/300 °C Hg anneal was used after growth.

A systematic study of IMP material grown at 410–350 °C has been carried out by Capper et al.[117] All of the layers grown at 400 °C were p type irrespective of substrate orientation, in contrast to the behavior seen in[108] where layers on (111) substrates were often anomalous electrically. Samples were etched prior to Hall assessment in[117] to remove any n-type surface layers. The p-type levels were below those found in Te-rich LPE layers grown at 500 °C in the same laboratory,[83] as expected. For growth at 350 °C Thompson et al.[118] found lower p-type levels as-grown, as expected. In the case of the (111) layer grown at 350 °C[117] n-type conduction was obtained. Chemical analysis by LSMS showed insufficient donor impurities to account for this and thinning revealed n-type conduction throughout the layer. It was concluded in Ref. 117 that twins were responsible for the n-type behavior. This may also explain the n-type (9×10^{15} cm^{-3}) results obtained by Hoke and Lemonias[119] in their growth of $x = 0.2$ layers at 350 °C, using di-iso propyl telluride (DIPT), on (111) CdTe. Young et al.[120] found similar behavior in (111) material grown at 350 °C on CdTe buffered GaAs. This group found that for (100) growth with a 10-min post-growth anneal, n-type material was produced while a 30-min post-growth heat treatment gave p-type conduction. Additionally, all the n-type layers moved towards p type on storage at room temperature. This was assigned to out-diffusion of donor defects rather than in-diffusion of acceptors or any impurity movement. This may be a similar phenomenon to that seen in SSR material after long periods of storage.[12,13]

Edwall et al.[121] have grown p-type (1–15×10^{17} cm^{-3}), $x = 0.22$–0.32 layers on GaAs and GaAs/Si substrates at 380–400 °C by IMP, although some n-type layers were produced. Low pressure growth using methylallyltelluride (MATe) at 320 °C produced n-type (2×10^{15} cm^{-3}) $x = 0.23$ material.[122] Reasons proposed for the n-type conduction included the low growth temperature, defects from lattice-mismatch (although a CdTe substrate was used) and impure MATe. In view of other reported results the final possibility seems most likely.

Photon-enhanced MOVPE has been employed by Ahlgren and co-workers[123,124] to produce layers at 280 °C which were n-type as-grown (7×10^{16} cm^{-3}), as expected from the data in Fig. 1, changing to 2×10^{16} cm^{-3} after low temperature Hg annealing. In Ref. 124 this group shows that superlattices of HgTe–CdTe grown at 180–240 °C are n-type as-grown and after annealing, which they ascribe to the shunting effect of the HgTe layers. Precracking of the alkyls was used in Ref. 125 to grow $x = 0.3$ layers at 225 °C. The material was n type at $\approx 1 \times 10^{17}$ cm^{-3} when grown on (100) CdTe substrates. In a later paper[126] this technique was shown to produce $x = 0.23$ layers which were n type at $\approx 5 \times 10^{15}$ cm^{-3}.

In summary, p-type behavior is expected for growth at temperatures above ≈ 300 °C, depending on Hg pressure, but n-type conduction can be caused by a number of effects. These include contamination,[106,107] surface layers formed during cool-down[110–112] or oxidation,[109] strain on GaAs substrates,[113] twins in (111) growth[117,119,120] and perhaps other parameters.[115,121,122] Low temperature growth will also obviously lead to n-type material.[123–126] It is vital therefore to establish the as-grown electrical properties of MOVPE before proceeding with doping studies. This is even more critical than in bulk and LPE processes in view of the variety of causes of "unexpected" n-type behavior. The superlattices of Ahlgren et al.[124] form a special category and will require a different approach to doping from the cases of alloy and IMP growth.

In Ref. 117 samples of both $x < 0.27$ and $x > 0.27$ were annealed in the same runs as similar LPE material. Conversion to n type occurred less often for MOVPE than for LPE layers and for $x > 0.27$ layers of either origin, agreeing with the view expressed in Ref. 78 with regard to the possible decrease in Hg diffusion rate in epitaxial material. Contamination during annealing was ruled out in Ref. 117, as samples annealed together did not always behave in the same manner. n-type concentrations between 5 and 10×10^{15} cm^{-3} were found in Ref. 117 for MOVPE layers. Annealing of (100) layers in Ref. 108 produced n-type conduction but (111) material produced anomalous results. Whiteley et al.[110] found n-type levels of 5–10×10^{14} cm^{-3} ($x = 0.27$) on annealing at 260/250 °C for 16 h. Edwall et al.[127] have compared LPE material to both IMP and alloy-grown MOVPE layers of similar x. They found that three IMP layers were p type as-grown but that most alloy layers were n-type and in annealed material n-type levels were highest in the alloy MOVPE samples followed by IMP material with LPE layers giving the lowest values (5×10^{14} cm^{-3}). Capper et al.[117]

have, however, pointed out the problems which can arise in such comparisons if it is assumed that all epitaxial material converts to n type at the same rate as bulk material. When the slower conversion for LPE layers was taken into account, the n-type levels in LPE layers were found to be comparable to those in the thinner MOVPE layers.[117]

B. Electrical activity of dopants

Table III includes the data for dopants in both conventional Te-rich and metal-rich MOVPE. The donor nature of the group III elements has been described above[106,107] in Te-rich material. The data for In and I are from Easton et al.[128] while the acceptor nature of Ag in Te-rich material is from unpublished work by the same group. Whiteley et al.[110] also found In to be a donor although whether Te or metal-rich conditions applied is not stated.

The difficulty of achieving acceptor doping with slow-diffusing group V elements in Te-rich LPE was outlined in Sec. III B above. The same basic problem occurs in MOVPE material. The first report of successful doping using phosphine to introduce P was given in Ref. 78. Both P and As were shown to behave similarly by Capper et al.[129] in $x = 0.26$–0.3 layers. Under normal Te-rich conditions both elements were electrically inactive but when alkyl ratios were changed to produce metal-rich conditions both were seen to be $\approx 100\%$ active at levels of 6×10^{16} cm^{-3} and 1–2×10^{17} cm^{-3}, respectively. Undoped LPE and MOVPE control samples were annealed in the same run and converted to 2×10^{14} cm^{-3} and 2×10^{15} cm^{-3} n type, respectively, indicating that the anneal had filled Hg vacancies and had not introduced any extraneous impurities. Analysis by LSMS[84] confirmed the absence of other potential acceptors at levels approaching that of the added As, suggesting the electrical activity was due solely to the As. For an $x = 0.2$ As-doped layer the p-type level was $\approx 3 \times 10^{17}$ cm^{-3} as-grown, reducing to $\approx 1 \times 10^{17}$ cm^{-3} after a Hg anneal (see Fig. 5). Analysis by SIMS shows As present at 1.5×10^{17} cm^{-3} in both samples. The decrease in p-type level was due to removal of the Hg vacancies. No high temperature "activation" anneals were required to produce active acceptor behavior, unlike the work of Vydyanath et al.[93] in Te-rich LPE material.

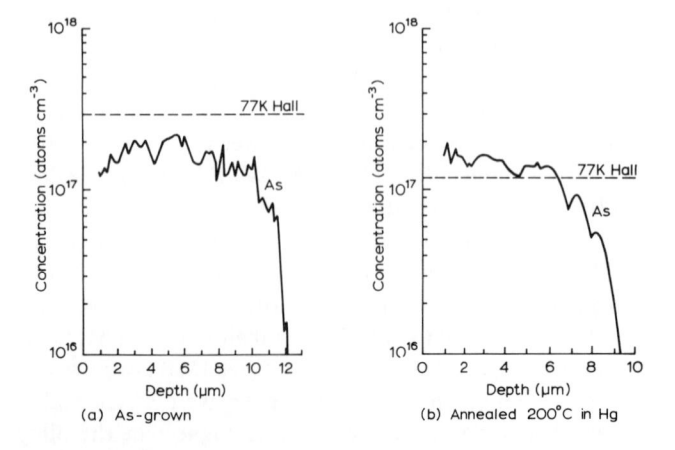

FIG. 5. SIMS profiles and carrier concentrations a) before and b) after Hg annealing in an $x = 0.2$ As-doped MOVPE layer (from Ref. 130).

This, together with the lack of diffusion of As during annealing, opens up the possibility of heterostructure growth with better defined junctions; see below.

Subsequently, Whiteley et al.[110] and Taskar et al.[115] confirmed activity for As and Sb in IMP growth and As in alloy growth, respectively. In neither case were any chemical data given for the doped samples and so it is not possible to assign efficiencies to the activity of the dopants.

A full description of doping IMP layers with P and As has been given by Capper et al.[130] Control over the range 5×10^{15}–3×10^{17} cm^{-3} (saturation) was established for growth at 410 °C on CdTe substrates. When growth was at 350 °C the As level increased, believed to be due to a reduction in depletion of the dopant-alkyl adduct upstream of the substrate at the lower temperature. p-type conduction was obtained for dopant concentrations of 6×10^{16} cm^{-3} for P and 2×10^{16}–8×10^{17} cm^{-3} for As. The importance of maintaining metal-rich conditions during the CdTe IMP cycle was stressed in Ref. 130. An alkyl ratio of < 0.9 was found to be necessary for $x = 0.2$ layers. It was also stressed in Ref. 130 that to assign a p-type level unequivocally to an added dopant requires that a survey analysis reveals no other potentially active acceptors at levels near to the dopant concentration. This was found to be the case for 1×10^{17} cm^{-3} As but as the level decreased to 5×10^{15} cm^{-3} other unintentionally added elements were present at similar concentrations.

As the arsenic is incorporated only during the CdTe IMP cycle and because it is a slow diffuser, a potential problem arises if the As does not diffuse into the HgTe IMP layers sufficiently to dominate the electrical properties. This aspect has been addressed by Capper et al.[131] and Maxey et al.[132,133] Profiling As distributions, by SIMS, in test structures added to theoretical modelling resulted in a figure for the diffusion coefficient (D_{As}) of $\approx 1 \times 10^{-14}$ cm^2 s^{-1} in $x = 0.5$ material. For typical IMP thicknesses of $\geqslant 500$ and 2000 Å (CdTe and HgTe) problems were found at mid-10^{16} cm^{-3} As levels. Increasing the post-growth anneal times removed the anomalous behavior and p-type levels corresponding to the As levels were obtained; i.e., As diffusion had proceeded to a level sufficient to outweigh the residual background donor level. Thus it was possible to achieve active acceptor doping controllably down to 1×10^{16} cm^{-3}. In addition, doped/undoped heterostructures were grown which, after Hg annealing, produced a n/p junction with the As controlling the p-type, high-x side of the junction,[133] see Fig. 6. These were the first reports of such doped structures prepared by MOVPE, although several reports of undoped heterostructure growth have been made.[134,135]

C. Ionization energies and carrier lifetimes

Capper et al.[78] found two levels in their P-doped layers. The higher value quoted was assigned to the Hg vacancies and agrees with the range found for undoped bulk material of similar x- and p-type concentration (see Figs. 2 and 3). The lower value in Ref. 78 is also commensurate with those seen for other group V acceptors in bulk material with similar properties. The values for As from[129,132] fall within the

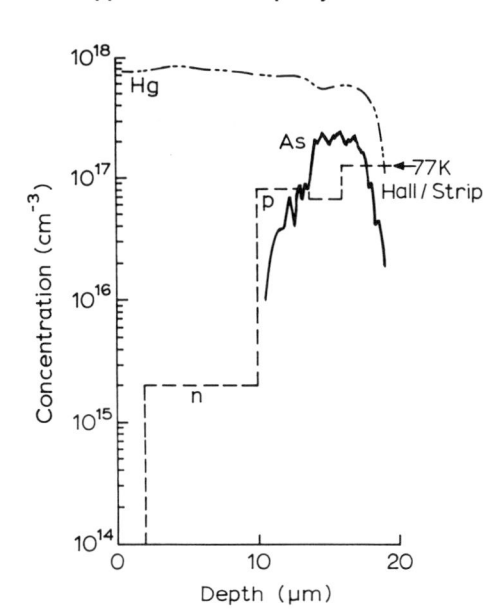

FIG. 6. SIMS and Hall/strip profiles of a doped/undoped MOVPE heterostructure, after Hg annealing (from Ref. 133).

range of those measured in As-doped bulk material and As-doped Hg-rich LPE layers. Taskar et al.[115] show an inverse linear dependence of E_a on N_a in the range studied, and their values agree very well with those of Kalisher[89] and Lou and Frye[105] in LPE layers grown from Hg-rich solutions.

There have been few reports of lifetime measurements on undoped MOVPE material. This may be due to the difficulty of measuring meaningful lifetimes in such thin (15 μm) layers as described by workers in the LPE field.[102,103] Unpublished work[136] on material grown in this author's laboratory suggest a wide range of values for lifetime in as-grown p-type layers with the best figures in undoped material being similar to bulk values. Edwall et al.[137] quote a value for an $x = 0.27$ IMP layer grown on a (221) GaAs substrate of 150 ns, which is also equivalent to bulk values at 5×10^{15} cm^{-3} (see Fig. 4). In addition, they give a value of 1550 ns for an annealed $x = 0.24$ layer which did not convert to n-type. This is close to the values reported for As-doped Hg-rich LPE[87] and, as no conversion occurred, might indicate that this sample was doped inadvertently. Maxey et al.[133] have recently published data for As-doped MOVPE layers ($0.21 < x < 0.35$) which were comparable to equivalent bulk material.[55] For the higher x samples the lifetimes were still below those reported by Tung[88] in his Hg-rich LPE layers. Whether this represents a fundamental difference between these two types of material or is due to problems in measuring very thin MOVPE layers is not known at present.

V. MBE MATERIAL

A. Undoped material—electrical behavior

Electrical properties of undoped MBE material have received increasing attention in recent years because of the need to dope the material successfully with both acceptor and donor impurities. In their recent review of doping MBE layers Boukerche et al.[138] summarise the situation for as-grown undoped material. A complex picture of behavior is apparent and factors which influence electrical properties include substrate orientation, substrate temperature, Hg flux, and the x value of the layer. They comment that a few °C change in substrate or Hg cell temperature can switch the type of the material being grown, hence the need to stabilize properties with extrinsic dopants. In (111) growth increasing Hg flux or decreasing substrate temperature leads to the formation of twins which degrade electrical properties. These are thought to be anti-phase domains and can result in n-type material. This behavior mirrors that seen in MOVPE layers as discussed by Capper et al.[117] For (111) MBE layers which are p type, excess Te, either interstitial or at antiphase boundaries, controls the p-type level. Growth on (100) produces n-type layers when $x < 0.35$ and this may be due to antisite Te donors. For substrate temperatures above 200 °C p-type material can be obtained on (100) but not for $x < 0.24$. For these reasons and because more Hg is needed for (100) growth the preferred orientation is the (111)B.

Arias et al.[139] have grown 10–20-μm-thick layers on (211)B CdZnTe substrates at 195 °C and again produced either n-type or p-type material depending on substrate temperature and Hg flux. Annealing, at 200 °C for 24 h, produced conversion to 7×10^{14} cm^{-3} n-type, indicating that Hg vacancies were responsible for the p-type level. Improvements have resulted in intrinsic levels being reduced to $<1 \times 10^{16}$ cm^{-3}.[140] Temofonte et al.[141] report their undoped (100) $x < 0.3$ layers are n type at $\geqslant 1 \times 10^{15}$ cm^{-3}. For $x \approx 0.26$ material p-type levels of 6×10^{15} cm^{-3} were seen on occasions.

Undoped HgTe–CdTe superlattices grown at ≈ 160 °C on (100) CdTe/GaAs substrates are also n type (5–10×10^{15} cm^{-3}) as-grown.[142] This is as expected if the HgTe layers are acting as shunting layers, as was found by Ahlgren et al.[123,124]

B. Electrical activity of dopants

The first successful extrinsic doping of MBE material was reported by Boukerche and co-workers[143,144] who grew layers at ≈ 195 °C on (111) CdTe//(100) GaAs substrates and used In as the n-type dopant. Doping efficiencies of 10–60% were found at doping levels of 2×10^{17}–1×10^{18} cm^{-3} for x between 0.2 and 0.55. Wroge et al.[142] report the first successful p-type doping of superlattices, using Sb, but the doping process was not reproducible. In alloy growth Wijewarnasuriya et al.[145] describe how Li can be used to dope (100) layers with $x \approx 0.17$–0.25 p type. Both incorporation and electrical activity were found to be close to 100%. Growth of a doped/undoped sample, followed by SIMS analysis, showed Li to be a fast diffuser, even at the growth temperature of ≈ 190 °C, limiting its usefulness as an acceptor dopant.

Peterman et al.[146] report on the alternative of doping layers with Ag. Growth at ≈ 165 °C onto both (100) and (111) substrates resulted in a doping range of 1×10^{16}–1×10^{18} cm^{-3} in $x \approx 0.18$–0.24 layers. SIMS data in an undoped/doped/undoped structure revealed minimal diffusion, from a level of 4×10^{17} cm^{-3} Ag, into the undoped regions. In a later paper a diffusion coefficient for Ag of $\approx 7\times 10^{-14}$ cm^2 s^{-1} at 165 °C was given.[147] Heterojunctions, where In controlled the n-type levels in both the $x = 0.21$ and 0.3 regions, were grown at 190 °C by Boukerche et al.[148] These authors also report that silicon can be used as an n-type dopant in a p-n homojunction structure. The p-type level of 2×10^{16} cm^{-3} was fixed by stoichiometric adjustment while the Si was incorporated at $\approx 5\times 10^{16}$ cm^{-3} in this $x\approx 0.27$ layer.

In their review of doping in MBE material Boukerche et al.[138] give more detail of their own doping studies. It is apparent that In doping results in a memory effect within the growth kit. In the case of Li doping, anneals in Hg produced n-type material, presumably caused by removal of Li to the surface. This has been observed previously in bulk and LPE material (see earlier sections). Storage at room temperature also resulted in changing electrical properties, mirroring the behavior seen by Young et al.[120] in MOVPE material. For the alternative acceptor elements, these authors report that both As and Sb are donors on the metal sites. Isothermal annealing at 200–250 °C produced n-type material despite the presence of the Group V element. Neither laser nor UV illumination during growth produced any dopant activation.

An extension to their work on Ag doping was reported recently by Peterman et al.[149] Superlattices of HgTe–CdTe grown at ≈ 160 °C were doped with Ag in the range 1×10^{16}–1×10^{18} cm^{-3}. n-type doping using In was also reported in these superlattices. Diffusion of As at 380–450 °C from a Hg solution containing As was used by Arias et al.[139] to produce p^+/n homojunction diodes in their MBE growth on (211)B CdZnTe substrates. Presumably the high temperature diffusion has moved the As onto active Te sites in this Te-rich material as found necessary by Vydyanath et al.[93] in Te-rich LPE.

Boukerche et al.[140] report that improvement of the control of stoichiometry has now resulted in low 10^{16} cm^{-3} In levels being obtained reproducibly. This implied that the memory effect noted in Ref. 134 can be reduced and they also note that doping efficiencies for In can be close to 100% for concentrations of 3×10^{16}–1×10^{17} cm^{-3}. A further extension of their doping studies in superlattices has been published by Wroge et al.[150] Doping with In in the range 1×10^{16}–1×10^{18} cm^{-3} was achieved with no memory effect. A structure was grown in which Ag and In doping were used to control the p and n sides, respectively, of a junction.

Temofonte et al.[141] used InP as the source of In, rather than elemental In, and saw 50% doping efficiencies in the range 1–10$\times 10^{15}$ cm^{-3}. These are lower values than previously reported. Doping with Ag was achieved in the range 5×10^{15}–5×10^{16} cm^{-3}. Modulation doping via thin Ag-doped layers incorporated into thicker layers produced hole concentrations of $\approx 7\times 10^{14}$ cm^{-3}. The problems in measuring such low acceptor levels were outlined in Ref. 141, where the importance of variable magnetic field Hall measurements (up to 5 Tesla) was stressed. Sou et al.[151] have shown how In is triply ionized and predominantly incorporated interstitially in their (100) MBE layers, supporting the Te antisite model for this orientation.[152] Normal incorporation of In on the metal sites was found for (111) growth. Lower compensation levels ($\sim 1\times 10^{16}$ cm^{-3}) were achieved in these layers.

Han and co-workers[153,154] have followed an alternative approach to doping in MBE layers. Thin CdTe layers heavily doped with As are grown in the MCT layer using photo-assisted MBE (PAMBE). These 50-Å layers, with 1000-Å spacings, provide holes into the narrow band-gap material. No acceptor freeze-out is thus seen as no acceptor ions are present in the narrow gap regions. Layers with "x" ≈ 0.18–0.26 were produced with p-type levels of 5×10^{16}–1×10^{18} cm^{-3}. Doping with In in the range 1×10^{17}–3×10^{18} cm^{-3} was reported in Ref. 150. New properties were predicted for this "quantum alloy" of MCT. Arias et al.[155] have grown doped superlattices at 155 °C with 12.6 μm cut-off response and stress the importance of establishing cation-stabilized conditions, mirroring the work on MOVPE.[130] Low activity of As was seen in as-grown samples increasing to 50% after a 250 °C Hg anneal, while undoped samples were n-type as-grown and after Hg annealing.

In their recent review of MBE, Summers et al.[156] suggest that group V elements may be incorporated interstitially rather than substitutionally, due to their size and low surface mobility. They then develop a convincing argument for a photo-assisted chemical beam epitaxy approach using groups V and III hydrides or metal-organic compounds, for acceptor and donor doping, respectively. Wu et al.[157] have produced (111) layers at 170–180 °C which were p-type as-grown due to As doping, although as Summers et al.[156] point out such layers with $x \geqslant 0.3$ are normally p type due to stoichiometry. Annealing the doped layers in Hg did not convert layers to n-type, suggesting active As, but no chemical analysis data were presented in either as-grown or annealed samples, nor were any results on annealing similar undoped material. In (100) layers no acceptor activity was seen for As, samples were n type as-grown and after annealing—as seen in undoped (100) material.

Table III lists the activities found in MBE material.

C. Carrier lifetimes and ionization energies

Few studies of acceptor ionization energies (E_a) have been made with MBE material. Boukerche et al.[143] show a range of values from 0–8 meV in undoped, compensated material with $x\approx 0.21$–0.38. These values are similar to those for compensated bulk material of corresponding x and N_a. Wijewarnasuriya et al.[145] find a value of 8.3 meV for Li in $x\approx 0.26$ compensated material with a p-type level of $\approx 2\times 10^{16}$ cm^{-3}. This value is a little high for such doped material, when compared to those for bulk material.

Lifetime data on MBE layers are equally scarce. In an undoped layer with $x\approx 0.4$ and p-type level of 2.5×10^{14} cm^{-3} a value of 1 μs was quoted by Faurie et al.[158] A later report by the same group[159] gives a value of 100 ns for an $x\approx 0.34$ layer with $N_a \approx 3.6\times 10^{15}$ cm^{-3}. Both of these values are slightly below those expected from bulk material of similar x (see line 1 in Fig. 4). A lifetime value in a Ag-doped MBE layer was reported by Wroge et al.[147] For a layer of

$x \approx 0.25$ with $N_a \approx 2 \times 10^{16}$ cm^{-3} a value of <25 ns was found at 77 K. This is again low compared to bulk material and clearly less than the As-doped Hg-rich LPE of Tung.[88] Fits to lifetime-versus-temperature curves could be obtained either using radiative and Auger recombination or radiative and S-R recombination. No conclusions with regard to mechanisms could thus be made in Ref. 147. Auger recombination was cited in[138] for MBE layers with $x > 0.3$ while S-R recombination was reported as dominant in Ref. 152—clearly the situation is still uncertain. A value of 4 μs was given in Ref. 157 for a layer of $x = 0.33$ and $N_a = 5 \times 10^{15}$ cm^{-3} doped with As and annealed. This is equivalent to that seen by the same group in Hg-rich LPE doped with As,[87] although no comments were made regarding the recombination mechanism in Ref. 157.

VI. CONCLUSIONS

In order to establish a reproducible doping process in a particular growth system it is necessary first to assess the undoped material produced by that technique. For material equilibrated at high temperatures (≥ 300 °C), p-type conduction results due to Hg vacancies which can be removed by annealing in Hg. Bridgman growth produces n-type material as-grown as can Hg-rich LPE, if the cool-down conditions are favorable. n-type conduction in MOVPE growth can occur due to contamination, annealing during cooldown, surface layers due to oxidation, twins and strain. In MBE growth factors governing as-grown n-type conduction include Hg flux, substrate temperature, substrate orientation, and x value.

The minimum requirement to confirm doped behavior is established as a correlation between electrical data, on an annealed sample, and chemical data from AAS or SIMS. At low concentrations it should also be established whether any other potentially active dopant impurities are present at significant levels. Most elements are electrically active in accordance with their relative position in the periodic table. This is particularly true of material grown under Hg-saturated conditions (e.g. Bridgman and Hg-rich LPE). In Te-rich LPE and MOVPE material acceptor and donor activity, from groups I and III, respectively, is largely unchanged but group V acceptor elements require a high temperature "activation" anneal. Even this type of anneal does not produce acceptor activation in MBE layers; the group V elements occupy metal sites as donors, and resort has been made to the faster diffusing group I elements, although one recent report has indicated active acceptor doping using a group V element. Changing conditions from Te rich to metal rich in MOVPE growth gives acceptor activity using group V elements with no need for "activation" anneals and opens up the potential for stable doped heterostructures. There is some evidence of movement of group I elements at temperatures as low as room temperature in bulk, LPE, and MBE material. The implications of using a slow diffusing acceptor impurity in MOVPE layers produced by IMP have been outlined. The new "quantum alloy" of modulation-doped PAMBE material may offer unique possibilities for novel devices.

In addition to the basic electrical activity of dopants this review has also highlighted the importance of other features. Knowledge of dopant segregation, particularly in growth from liquids, is also critical in achieving controlled doping. Acceptor ionization energies can be used to differentiate doped from undoped material but not, as yet, between dopants. There is some evidence that acceptor doping with As and Au can improve lifetimes in Hg-rich LPE and SSR material, respectively, but this increase is not universally observed in other types of material, e.g. Bridgman and MBE. Defect control in undoped Te-rich LPE material can also lead to improved lifetimes.

ACKNOWLEDGMENTS

The author would like to thank Dr. C. L. Jones and Dr. I. M. Baker, J. J. G. Gosney, I. Kenworthy, and C. Ard (Philips, Southampton), and Dr. J. C. Brice, J. A. Roberts, B. C. Easton, P. A. C. Whiffin, and C. D. Maxey (PRL, Redhill) for their help in this work during the past several years and for critically reading the manuscript. This work was carried out with the support of Procurement Executive, Ministry of Defence, sponsored by RSRE Malvern.

[1] J. L. Schmit and E. L. Stelzer, J. Electron. Mater. **7**, 65 (1978).
[2] B. E. Bartlett, P. Capper, J. E. Harris, and M. J. T. Quelch, J. Cryst. Growth **49**, 600 (1980).
[3] H. R. Vydyanath, J. Electrochem. Soc. **128**, 2609 and 2619 (1981).
[4] H. R. Vydyanath, J. C. Donovan, and D. A. Nelson, J. Electrochem. Soc. **128**, 2625 (1981).
[5] A. J. Syllaios and M. J. Williams, J. Vac. Sci. Technol. **21**, 201 (1982).
[6] H. F. Schaake and J. H. Tregilgas, J. Electron. Mater. **12**, 931 (1983).
[7] H. F. Schaake, J. H. Tregilgas, A. J. Lewis, and P. M. Everett, J. Vac. Sci. Technol. A **1**, 1625 (1983).
[8] J. H. Tregilgas, J. D. Beck, and B. E. Gnade, J. Vac. Sci. Technol. A **3**, 150 (1985).
[9] H. F. Schaake, J. H. Tregilgas, J. D. Beck, M. A. Kinch, and B. E. Gnade, J. Vac. Sci. Technol. A **4**, 143 (1985).
[10] J. Yang, Z. Yu, and D. Tang, J. Cryst. Growth **72**, 275 (1985).
[11] C. L. Jones, M. J. T. Quelch, P. Capper, and J. J. Gosney, J. Appl. Phys. **53**, 9080 (1982).
[12] G. Nimtz, B. Schlicht, and R. Dornhaus, Appl. Phys. Letts. **34**, 490 (1979).
[13] W. F. H. Micklethwaite and R. F. Redden, Appl. Phys. Lett. **36**, 379 (1980).
[14] R. G. Pratt, J. Hewett, P. Capper, C. L. Jones, and M. J. T. Quelch, J. Appl. Phys. **54**, 5152 (1983).
[15] R. G. Pratt, J. Hewett, P. Capper, C. L. Jones, and N. Judd, J. Appl. Phys. **60**, 2377 (1986).
[16] P. Capper, J. J. G. Gosney, and C. L. Jones, J. Cryst. Growth **70**, 356 (1984).
[17] P. Capper, J. J. G. Gosney, C. L. Jones, and E. J. Pearce, J. Electron. Mater. **15**, 361 (1986).
[18] P. Capper, J. J. G. Gosney, C. L. Jones and I. Kenworthy, J. Electron. Mater. **15**, 371 (1986).
[19] P. Capper, J. A. Roberts, I. Kenworthy, C. L. Jones, J. J. G. Gosney, C. K. Ard, and W. G. Coates, J. Appl. Phys. **64**, 6227 (1988).
[20] R. Triboulet, T. Nguyen Duy, and A. Durand, J. Vac. Sci. Technol. A **3**, 95 (1985).
[21] T. Nguyen Duy, A. Durand, and J. L. Lyot, Mater. Res. Symp. Proc. **90**, 81 (1987).
[22] L. Colombo, R. R. Chang, C. J. Chang, and B. A. Baird, J. Vac. Sci. Technol. A **6**, 2795 (1988).
[23] P. Capper, J. Cryst. Growth **57**, 280 (1982).
[24] E. S. Johnson and J. L. Schmit, J. Electron. Mater. **6**, 25 (1977).
[25] J. W. Lin, J. Solid State Chem. **15**, 96 (1975).

[26] H. R. Vydyanath and F. A. Kroger, J. Electron. Mater. **11**, 111 (1982).
[27] H. R. Vydyanath, R. C. Abott, and D. A. Nelson, J. Appl. Phys. **54**, 1323 (1983).
[28] Q. Dingrong, T. Wenguo, S. Jie, C. Junhao, and Z. Guozhen, Solid State Comms. **56**, 813 (1985).
[29] E. Finkman and Y. Nemirovsky, J. Appl. Phys. **59**, 1205 (1986).
[30] M. A. Marais, H. J. Strydom, J. H. Basson, D. E. C. Rogers, and H. Booyens, J. Cryst. Growth **88**, 391 (1988).
[31] A. P. Botha, H. J. Strydom, and M. A. Marais, Nucl. Instrum. Methods Phys. Res. B **35**, 420 (1988).
[32] S. Margalit, Y. Nemirovsky, and I. Rotstein, J. Appl. Phys. **50**, 6386 (1979).
[33] H. Ryssel, G. Lang, J. P. Biersack, K. Muller, and W. Kruger, IEEE Trans. Electron Devices **ED-27**, 58 (1980).
[34] G. Bahir and R. Kalish, Appl. Phys. Letts. **39**, 730 (1981).
[35] J. Baars, H. Seelewind, C. Fritzsche, U. Kaiser, and J. Ziegler, J. Cryst. Growth **86**, 762 (1988).
[36] C. L. Jones, P. Capper, M. J. T. Quelch, and M. Brown, J. Cryst. Growth **64**, 417 (1983).
[37] P. Capper, J. J. G. Gosney, C. L. Jones, I. Kenworthy, and J. A. Roberts, J. Cryst. Growth **71**, 57 (1985).
[38] *Properties of Mercury Cadmium Telluride*, edited by J. C. Brice and P. Capper (IEE, Hitchin, United Kingdom 1987).
[39] J. H. Tregilgas and B. Gnade, J. Vac. Sci. Technol. A **3**, 156 (1985).
[40] R. T. Talasek, M. J. Ohlson, and A. J. Syllaios, J. Electrochem. Soc. **133**, 230 (1986).
[41] P. Capper, I. G. Gale, F. Grainger, J. A. Roberts, C. L. Jones, J. J. G. Gosney, I. Kenworthy, C. K. Ard, and W. G. Coates, J. Cryst. Growth **92**, 1 (1988).
[42] J. A. Burton, R. C. Prim, and W. P. Slichter, J. Chem. Phys. **21**, 1987 (1953).
[43] J. A. Burton, E. D. Kolb, W. P. Slichter, and J. D. Struthers, J. Chem. Phys. **21**, 1991 (1953).
[44] J. B. Mullin and K. F. Hulme, J. Phys. Chem. Solids **17**, 1 (1960).
[45] M. A. Kinch, M. J. Brau, and A. Simmons, J. Appl. Phys. **44**, 1649 (1973).
[46] J. Calas and J. Allegre, Phys. Status Solidi B **112**, 179 (1982).
[47] N. L. Bazhenov, V. I. Ivanov-Omskii, and V. K. Ogorodnikov, Sov. Phys. Semicond. **18**, 911 (1984).
[48] M. Lanir, A. H. B. Vanderwyck, and C. C. Wang, J. Appl. Phys. **49**, 6182 (1987).
[49] C. E. Jones, K. James, J. Merz, R. Braunstein, M. Burd, M. Eetemadi, S. Hutton, and J. Drumheller, J. Vac. Sci. Technol. A **3**, 131 (1985).
[50] C. E. Jones, V. Nair, J. Lindquist, and D. L. Polla, J. Vac. Sci. Technol. **21**, 187 (1982).
[51] D. L. Polla, S. P. Tobin, M. B. Reine, and A. K. Sood, J. Appl. Phys. **52**, 5182 (1981).
[52] D. L. Polla, R. L. Aggarwal, D. A. Nelson, J. F. Shanley, and M. B. Reine, Appl. Phys. Letts. **43**, 941 (1983).
[53] A. V. Voitsekhovskii and Yu. V. Lilenko, Sov. Phys. Semicond. **15**, 845 (1981).
[54] S. E. Schacham and E. Finkman, J. Appl. Phys. **57**, 2001 (1985).
[55] D. E. Lacklison and P. Capper, Semicond. Sci. Technol. **2**, 33 (1987).
[56] Y. Nemirovsky and D. Rosenfeld, J. Appl. Phys. **63**, 2435 (1988).
[57] R. Fastow and Y. Nemirovsky, J. Vac. Sci. Technol. A **8**, 1245 (1990).
[58] I. Kenworthy, P. Capper, C. L. Jones, J. J. G. Gosney, and W. G. Coates, Semicond. Sci. Technol. **5**, 854 (1990).
[59] M. C. Gold and D. A. Nelson, J. Vac. Sci. Technol. A **4**, 2040 (1986).
[60] F. J. Bartoli, C. A. Hoffman, and J. R. Meyer, J. Vac. Sci. Technol. A **4**, 2047 (1986).
[61] M. A. Kinch, J. Vac. Sci. Technol. **21**, 215 (1982).
[62] W. Scott, E. L. Stelzer, and R. J. Hager, J. Appl. Phys. **47**, 1408 (1976).
[63] Yu. G. Arapov, B. B. Ponikarov, I. M. Tsidil'kovskii, and I. M. Nesmelova, Sov. Phys. Semicond. **13**, 409 (1979).
[64] O. Caporaletti and W. F. H. Micklethwaite, Phys. Lett. **89A**, 151 (1982).
[65] A. I. Elizarov, L. P. Zverev, V. V. Kruzhaev, G. M. Min'kov, and O. E. Rut, Sov. Phys. Solid State **25**, 156 (1983).
[66] E. Janik and R. Triboulet, J. Phys. D. **16**, 2333 (1983).
[67] M. C. Chen and J. H. Tregilgas, J. Appl. Phys. **61**, 787 (1987).
[68] J. E. Bowers, J. L. Schmit, C. J. Speerschneider, and R. B. Maciolek, IEEE Trans. Electron Devices **ED-27**, 24 (1980).
[69] J. L. Schmit and J. E. Bowers, Appl. Phys. Lett. **35**, 457 (1979).
[70] M. Chu, J. Appl. Phys. **51**, 5876 (1980).
[71] S. H. Shin, M. Chu, A. H. B. Vanderwyck, M. Lanir, and C. C. Wang, J. Appl. Phys. **51** 3772 (1980).
[72] Y. Nemirovsky, S. Margalit, E. Finkman, Y. Schacham-Diamand, and I. Kidron, J. Electron. Mater. **11**, 133 (1982).
[73] J. Bajaj, S. H. Shin, G. Bostrup, and D. T. Cheung, J. Vac. Sci. Technol. **21**, 244 (1982).
[74] K. Nagahama, R. Ohkata, K. Nishitani, and T. Murotani, J. Electron. Mater. **13**, 67 (1984).
[75] M. Yoshikawa, S. Ueda, K. Maruyama, and H. Takigawa, J. Vac. Sci. Technol. A **3**, 153 (1985).
[76] D. D. Edwall, E. R. Gertner, and W. E. Tennant, J. Electron. Mater. **14**, 245 (1985).
[77] S. H. Shin, M. Khoshnevisan, C. Morgan-Pond, and R. Raghavan, J. Appl. Phys. **58**, 1470 (1985).
[78] P. Capper, B. C. Easton, P. A. C. Whiffin, and C. D. Maxey, J. Cryst. Growth **79**, 508 (1986).
[79] M. Brown and A. F. W. Willoughby, J. Vac. Sci. Technol. A **1**, 1641 (1983).
[80] H. R. Vydyanath and C. H. Hiner, J. Appl. Phys. **65**, 3080 (1989).
[81] M. Astles, G. Blackmore, V. Steward, D. C. Rodway, and P. Kirton, J. Cryst. Growth **80**, 1 (1987).
[82] M. Astles, H. Hill, G. Blackmore, S. Courtney, and N. Shaw, J. Cryst. Growth **91**, 1 (1988).
[83] J. C. Brice, P. Capper, B. C. Easton, J. L. Page, and P. A. C. Whiffin, Semicond. Sci. Technol. **2**, 710 (1987).
[84] F. Grainger and J. A. Roberts, Semicond. Sci. Technol. **3**, 802 (1988).
[85] C. A. Castro and R. Korenstein, Proc. SPIE **317**, 262 (1981).
[86] S. P. S. Sangha, L. M. Rinn, and R. E. Nicholls, J. Cryst. Growth **88**, 107 (1988).
[87] T. Tung, M. H. Kalisher, A. P. Stevens, and P. E. Herning, Mater. Res. Soc. Symp. Proc. **90**, 321 (1987).
[88] T. Tung, J. Cryst. Growth **86**, 161 (1988).
[89] M. H. Kalisher, J. Cryst. Growth **70**, 365 (1984).
[90] L. O. Bubulac, W. E. Tennant, R. A. Riedel, J. Bajaj, and D. D. Edwall, J. Vac. Sci. Technol. A **1**, 1646 (1983).
[91] G. L. Destefanis, J. Vac. Sci. Technol. A **3**, 171 (1985).
[92] L. O. Bubulac, D. S. Lo, W. E. Tennant, D. D. Edwall, and J. C. Robinson, J. Vac. Sci. Technol. A **4**, 2169 (1986).
[93] H. R. Vydyanath, J. A. Ellsworth, and C. M. Devaney, J. Electron. Mater. **16**, 13 (1987).
[94] M. C. Chen and J. A. Dodge, Solid State Comms. **59**, 449 (1986).
[95] G. Sarusi, A. Zemel, and D. Eger, J. Appl. Phys. **65**, 672 (1989).
[96] M. C. Chen, J. Appl. Phys. **65**, 1571 (1989).
[97] E. -H. Cirlin, S. Shin, R. DeWames, E. Gertner, and D. Edwall, *Extended Abstracts* (MCT Workshop, New Orleans, 1987).
[98] L. E. Lapides, R. L. Whitney, and C. A. Crosson, Mater. Res. Soc. Symp. Proc. **48**, 365 (1985).
[99] J. C. Brice, J. L. Page, P. A. C. Whiffin, and J. A. Roberts (unpublished results).
[100] J. Bajaj, S. H. Shin, J. G. Pasko, and M. Koshnevisan, J. Vac. Sci. Technol. A **1**, 1749 (1983).
[101] O. L. Doyle, J. A. Mroczkowski, and J. F. Shanley, J. Vac. Sci. Technol. A **3**, 259 (1985).
[102] D. Amingual, G. L. Destefanis, S. Guillot, J. L. Ouvrier-Buffet, S. Paltrier, and D. Zenatti, SPIE **659**, 85 (1986).
[103] J. S. Chen, J. Bajaj, W. E. Tennant, D. S. Lo, M. Brown, and G. Bostrup, Mater. Res. Soc. Symp. **90**, 287 (1987).
[104] Y. Nemirovsky, S. Margalit, E. Finkman, Y. Shacham-Diamand, and I. Kidron, J. Electron. Mater. **11**, 133 (1982).
[105] L. F. Lou and W. N. Frye, J. Appl. Phys. **56**, 2253 (1984).
[106] S. J. C. Irvine and J. B. Mullin, J. Crystal Growth **55**, 107 (1981).
[107] W. E. Hoke and R. Traczewski, J. Appl. Phys. **54**, 5087 (1983).
[108] V. Vincent, C. Wilson, and J. M. Lansdowne, Proc. SPIE **659**, 55 (1986).
[109] J. B. Mullin, J. Giess, S. J. C. Irvine, J. S. Gough, and A. Royle, Mater. Res. Soc. Symp. Proc. **90**, 367 (1987).
[110] J. S. Whiteley, P. Koppel, V. L. Conger, and K. E. Owens, J. Vac. Sci. Technol. A **6**, 2804 (1988).

[111] I. B. Bhat and S. K. Ghandhi, J. Cryst. Growth **75**, 241 (1986).
[112] S. J. C. Irvine, J. B. Mullin, J. Giess, J. S. Gough, A. Royle, and G. Crimes, J. Cryst. Growth **93**, 732 (1988).
[113] I. B. Bhat, N. R. Taskar, and S. K. Ghandhi, J. Vac. Sci. Technol. A **4**, 2230 (1986).
[114] E. R. Gertner, S. H. Shin, D. D. Edwall, L. O. Bubulac, D. S. Lo, and W. E. Tennant, Appl. Phys. Lett. **46**, 851 (1985).
[115] N. R. Taskar, I. B. Bhat, K. K. Parat, D. Terry, H. Ehsani, and S. K. Ghandhi, J. Vac. Sci. Technol. A **7**, 281 (1989).
[116] I. B. Bhat, H. Fardi, S. K. Ghandhi, and C. J. Johnson, J. Vac. Sci. Technol. A **6**, 2800 (1988).
[117] P. Capper, C. D. Maxey, P. A. C. Whiffin, and B. C. Easton, J. Cryst. Growth **96**, 519 (1989).
[118] J. Thompson, P. Mackett, L. M. Smith, D. J. Cole-Hamilton, and D. V. Shenai-Khatkhate, J. Cryst. Growth **86**, 233 (1988).
[119] W. E. Hoke and P. J. Lemonias, Appl. Phys. Lett. **46**, 398 (1985).
[120] M. L. Young, J. Giess, and S. J. C. Irvine, paper given at EW-MOVPE-III, Montpellier, France, (1989).
[121] D. D. Edwall, J. Bajaj, and E. R. Gertner, J. Vac. Sci. Technol. A **8**, 1045 (1990).
[122] I. B. Bhat, H. Ehsani, and S. K. Ghandhi, J. Vac. Sci. Technol. A **8**, 1054 (1990).
[123] W. L. Ahlgren, R. H. Himoto, S. Sen, and R. P. Ruth, paper given at IC-MOVPE-III, Los Angeles, (1986).
[124] W. L. Ahlgren, E. J. Smith, J. B. James, T. W. James, R. P. Ruth, E. A. Patten, R. D. Knox, and J. -L. Staudenmann, J. Cryst. Growth **86**, 198 (1988).
[125] P. -Y. Lu, C. -H. Wang, L. M. Williams, S. N. G. Chu, and C. M. Stiles, Appl. Phys. Lett. **49**, 1372 (1986).
[126] P. -Y. Lu, L. M. Williams, S. N. G. Chu, and M. H. Ross, Appl. Phys. Lett. **54**, 2021 (1989).
[127] D. D. Edwall, E. R. Gertner, and L. O. Bubulac, J. Cryst. Growth **86**, 240 (1988).
[128] B. C. Easton, C. D. Maxey, P. A. C. Whiffin, J. A. Roberts, I. G. Gale, F. Grainger, and P. Capper (unpublished).
[129] P. Capper, P. A. C. Whiffin, B. C. Easton, C. D. Maxey, and I. Kenworthy, Mat. Lett. **6**, 365 (1988).
[130] P. Capper, C. D. Maxey, P. A. C. Whiffin, and B. C. Easton, J. Cryst. Growth **97**, 833 (1989).
[131] P. Capper, C. D. Maxey, P. A. C. Whiffin, B. C. Easton, I. Gale, J. B. Clegg, and A. Harker, paper given at EW-MOVPE-III, Montpellier, France, (1989).
[132] C. D. Maxey, P. Capper, P. A. C. Whiffin, B. C. Easton, I. Gale, J. B. Clegg, and A. Harker, Mater. Lett. **8**, 190 (1989).
[133] C. D. Maxey, P. Capper, P. A. C. Whiffin, B. C. Easton, I. Gale, J. B. Clegg, A. Harker, and C. L. Jones, J. Cryst. Growth **101**, 300 (1990).
[134] P. A. C. Whiffin, B. C. Easton, P. Capper, and C. D. Maxey, J. Cryst. Growth **79**, 935 (1986).
[135] S. J. C. Irvine, J. Giess, J. S. Gough, G. W. Blackmore, A. Royle, J. B. Mullin, N. G. Chew, and A. G. Cullis, J. Cryst. Growth **77**, 437 (1986).
[136] C. L. Jones, C. Ard, A. McAllister, A. Clark, and S. Barton (unpublished).
[137] D. D. Edwall, J. S. Chen, J. Bajaj, and E. R. Gertner, Semicond. Sci. Technol. **5**, S221 (1990).
[138] M. Boukerche, P. S. Wijewarnasuriya, S. Sivananthan, I. K. Sou, Y. J. Kim, K. K. Mahavadi, and J. P. Faurie, J. Vac. Sci. Technol. A **6**, 2830 (1988).
[139] J. M. Arias, S. H. Shin, J. G. Pasko, R. E. DeWames, and E. R. Gertner, J. Appl. Phys. **65**, 1747 (1989).
[140] M. Boukerche, S. Sivananthan, P. S. Wijewarnasuriya, I. K. Sou, and J. P. Faurie, J. Vac. Sci. Technol. A **7**, 311 (1989).
[141] T. A. Temofonte, A. J. Noreika, M. J. Bevan, P. R. Emtage, C. F. Seiler, and P. Mitra, J. Vac. Sci. Technol. A **7**, 440 (1989).
[142] M. L. Wroge, D. J. Leopold, J. M. Ballingall, D. J. Peterman, B. J. Morris, J. G. Broerman, F. A. Ponce, and G. B. Anderson, J. Vac. Sci. Technol. B **4**, 1306 (1986).
[143] M. Boukerche, J. Reno, I. K. Sou, C. Hsu, and J. P. Faurie, Appl. Phys. Lett. **48**, 1733 (1986).
[144] M. Boukerche, P. S. Wijewarnasuriya, J. Reno, I. K. Sou, and J. P. Faurie, J. Vac. Sci. Technol. A **4**, 2072 (1986).
[145] P. S. Wijewarnasuriya, I. K. Sou, Y. J. Kim, K. K. Mahavadi, S. Sivananthan, M. Boukerche, and J. P. Faurie, Appl. Phys. Lett. **51**, 2025 (1987).
[146] D. J. Peterman, M. L. Wroge, B. J. Morris, D. J. Leopold, and J. G. Broerman, J. Appl. Phys. **63**, 1951 (1988).
[147] M. L. Wroge, D. J. Peterman, B. J. Morris, D. J. Leopold, J. G. Broerman, and B. J. Feldman, J. Vac. Sci. Technol. A **6**, 2826 (1988).
[148] M. Boukerche, S. Yoo, I. K. Sou, M. DeSouza, and J. P. Faurie, J. Vac. Sci. Technol. A **6**, 2623 (1988).
[149] D. J. Peterman, M. L. Wroge, B. J. Morris, D. J. Leopold, and J. G. Broerman, J. Appl. Phys. **65**, 1550 (1989).
[150] M. L. Wroge, D. J. Peterman, B. J. Feldman, B. J. Morris, D. J. Leopold, and J. G. Broerman, J. Vac. Sci. Technol. A **7**, 435 (1989).
[151] I. K. Sou, P. S. Wijewarnasuriya, M. Boukerche, and J. P. Faurie, Appl. Phys. Lett. **55**, 954 (1989).
[152] M. Boukerche and J. P. Faurie, Inst. Phys. Conf. Ser. **95**, 351 (1989).
[153] J. W. Han, S. Hwang, Y. Lansari, R. L. Harper, Z. Yang, N. C. Giles, J. W. Cook, Jr., and J. F. Schetzina, Appl. Phys. Lett. **54**, 63 (1989).
[154] J. W. Han, S. Hwang, Y. Lansari, R. L. Harper, Z. Yang, N. C. Giles, J. W. Cook, Jr., J. F. Schetzina, and S. Sen, J. Vac. Sci. Technol. A **7**, 305 (1989).
[155] J. M. Arias, S. H. Shin, D. E. Cooper, M. Zandian, J. G. Pasko, E. R. Gertner, R. E. DeWames, and J. Singh, J. Vac. Sci. Technol. A **8**, 1025 (1990).
[156] C. J. Summers, R. G. Benz II, B. K. Wagner, J. D. Benson, and D. Rajavel, Procs. SPIE **1106**, 1 (1989).
[157] O. K. Wu, G. S. Kamath, W. A. Radford, P. R. Bratt, and E. A. Patten, J. Vac. Sci. Technol. A **8**, 1034 (1990).
[158] J. P. Faurie, J. Reno, S. Sivananthan, I. K. Sou, X. Chu, M. Boukerche, and P. S. Wijewarnasuriya, J. Vac. Sci. Technol. A **4**, 2067 (1986).
[159] J. P. Faurie, S. Sivananthan, M. Lange, R. E. DeWames, A. M. B. Vanderwyck, G. M. Williams, D. Yamini, and E. Yao, Appl. Phys. Lett. **52**, 2151 (1988).

Impurities and metal organic chemical-vapor deposition growth of mercury cadmium telluride

B. C. Easton, C. D. Maxey, P. A. C. Whiffin, J. A. Roberts, I. G. Gale, and F. Grainger
Philips Research Laboratories, Redhill, Surrey, RH1 5HA United Kingdom

P. Capper
Philips Infra-red Defence Components, Southampton, SO9 7BH United Kingdom

(Received 3 October 1990; accepted 18 December 1990)

The doping behavior of indium and iodine have been investigated for mercury cadmium telluride (MCT, $Hg_{1-x}Cd_xTe$) layers deposited by the interdiffused multilayer process procedure at 400 °C using diethyl tellurium and dimethyl cadmium. Trimethyl indium and solid iodine were used as dopant sources. Both elements exhibited donor behavior under the conditions employed. Secondary ion mass spectrometry profile analysis was used to demonstrate that indium required a relatively long period during growth to attain an equilibrium concentration in the layer; in addition a significant reactor system memory was observed allied to a relatively fast diffusion rate. Iodine showed encouraging dopant properties at low concentration levels, the chemical concentration of iodine was in good agreement with the free donor level from Hall measurements. The advantage and applicability of each of the three chemical analysis procedures used in this work are discussed together with comments on the residual impurity content and electrical properties of undoped layers.

I. INTRODUCTION

The ability to prepare doped p–n epitaxial mercury cadmium telluride (MCT) structures is an essential requirement if some of the new developments in infrared photovoltaic detectors are to be realized in practice. The location of the doped regions must be well defined within a structure and the concentrations of the active impurities must be adequately controlled. The criteria for a "well behaved" dopant impurity are:

(i) It should have a slow diffusion rate at the growth temperature so that sufficiently abrupt and stable concentration profiles can be produced and maintained.

(ii) The dopant should be active in the MCT layer in the as-grown state and require no high temperature anneal treatments.

(iii) The way in which the dopant is introduced should be compatible with the deposition process, i.e. it should not have a significant influence on layer composition or growth rate.

In previously published work[1–3] we have studied arsenic as an acceptor impurity in MCT layers deposited by the interdiffused multilayer process (IMP)[4] technique. We have shown that As fits the above criteria providing due consideration is given to the chemical stoichiometry during deposition, specifically during the CdTe phase of the IMP cycle. In work reported here we discuss the extension of our doping studies to donor doping, using indium and iodine.

In all our investigations chemical analysis has made an important contribution to understanding incorporation mechanisms and dopant behavior, it has also been essential for establishing unequivocally whether or not the observed electrical behavior was due to the particular dopant being studied. In addition to the application of chemical analysis to the donor doping work we also report some further aspects of the residual impurity situation in nominally undoped material.

II. EXPERIMENT

MCT layers were deposited at 400 °C on CdTe substrates, oriented 2° or 4° off (100), using alkyls diethyl tellurium and dimethyl cadmium from Epichem Ltd. Details of the reactor system and its operation have been given elsewhere,[5] a schematic of the reactor tube assembly is shown in Fig. 1. A particular feature, considered relevant to doping experiments, is the use of a central "injector" tube down which the Cd alkyl and the dopant are introduced. The hydrogen carrier gas flow rate is constant during the HgTe and the CdTe phases of the IMP growth. For indium doping studies trimethyl indium was employed as the source material, dispensed from a temperature controlled stainless steel "bubbler" with a mass flow controlled supply of hydrogen as carrier gas. In the case of iodine the free solid element was used in a similar manner but contained in a glass vessel. All the layers discussed here were grown under "cadmium rich" conditions, found to be necessary for achieving electrical activity for As, with "x" values in the region 0.20–0.23.

Any investigation of doping behavior in MCT must take into account the role of metal vacancies. Undoped layers, deposited under conditions described above and taken straight from our particular growth system, are invariably p type with acceptor concentrations in the range $1–2\times10^{17}$ cm^{-3}, this we associate with the metal vacancy concentration. These layers, after annealing isothermally in mercury vapour at 200 °C, are converted to "low" n type, consistent with the removal of the vacancy defects. Such an anneal treatment is a standard procedure applied during the study of all impurity doping.

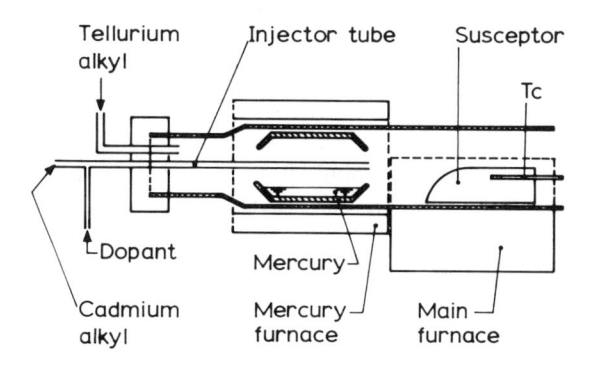

FIG. 1. Reactor tube assembly, showing the "injector" tube used to introduce the cadmium alkyl and the dopant.

Three analytical techniques are employed for determining impurity concentrations and distribution profiles. Secondary ion mass spectrometry (SIMS) analysis is carried out using an Atomika α-DIDA system. This ultrahigh vacuum (UHV) system has a high sensitivity for most elements (typically 5×10^{14} atoms cm^{-3}) combined with a depth resolution down to 10 nm. SIMS is particularly important for studying the diffusion characteristics and distribution of dopants through a layer structure. Laser scan mass spectrometry (LSMS) is a technique which has been developed at our laboratory and has been described previously.[6] The basis of the procedure depends on the ablation and ionization of the analysis sample in a pulsed laser beam (1064 nm) in combination with a rastered movement of the sample holder, thereby allowing thin sections to be removed and analyzed. The system is developed from an AEI Ltd. MS702 equipment with a magnetic sector spectrometer. LSMS is particularly suited to the survey analysis of MCT, it has a sensitivity comparable to the SIMS, however unlike the latter the sensitivity factors are close to unity for most elements. As a consequence of this LSMS has higher sensitivities for certain elements, similar benefits arise from the superior mass resolution of the LSMS system. The depth resolution for profile analysis is ~ 2 μm. Graphite furnace atomic absorption spectrometry (GFAAS) is a well established procedure[7] which is used to provide standards for the spectrometric techniques, it is also invaluable for carrying out whole layer impurity analyses of certain impurities at levels below or in the region of the SIMS and LSMS detection limits.

In this work 77 K Hall measurements (0.32 Tesla field strength) have been used to establish electrical behavior. These measurements have been combined with standard etch/strip procedures for investigating depth uniformity characteristics. Selected samples were also examined over a temperature range (liquid He to room temperature) in order to validate conclusions drawn from the 77 K measurements. Layer composition and thickness are measured using a Bio-Rad FTS-40 Fourier transform infrared spectrometer, fitted with a microscope attachment.

III. RESULTS AND DISCUSSION

A. Indium doping

The first experiments were aimed at investigating the way In was incorporated during growth of MCT, HgTe, and CdTe. Figure 2 is a SIMS profile through a HgTe/CdTe/MCT multilayer structure. Trimethyl indium was introduced ($\sim 10^{-5}$ mole fraction over the substrate) part way through the growth of the MCT layer, adjacent to the substrate. The gas-phase concentration was doubled at the commencement of growth of the second HgTe layer. The SIMS profile indicates the concentration of In through the structure and the Hg profile (not quantitative) is included in order to permit the different parts of the structure to be identified. It is apparent that there is a significant delay (~ 6 min) before In is indicated in the deposit, this is not thought to be related to delays in the gas system but to surface adsorption effects. Such a view is supported by the high concentration at the surface when the dopant was turned off with the termination of growth. Indium is incorporated at greater concentration in the HgTe layer, unlike As which is predominantly incorporated in the CdTe. The increase in the In concentration in HgTe is considerably more than doubled on doubling the concentration of InMe$_3$, this may also be related to the time taken for the In concentration to reach an equilibrium level during deposition of a layer.

Figure 3 is a SIMS profile of an In MCT doped layer ([InMe$_3$] 10^{-5} mole fraction). Again the rise time for the In signal is slow (total growth time 60 min), supporting the previous evidence that the dopant requires a considerable time to reach an equilibrium level. GFAAS analysis of the average In content of the layer gave 3×10^{19} atoms cm^{-3}, this is in good agreement with the SIMS profile. Hall measurements showed that only about 20% of the In was active. Figure 4 is the SIMS profile of a nominally undoped layer grown after the last experiment. A system "memory" effect from the previous doping experiments is indicated, in addition it is clear that there is a significant back diffusion of In

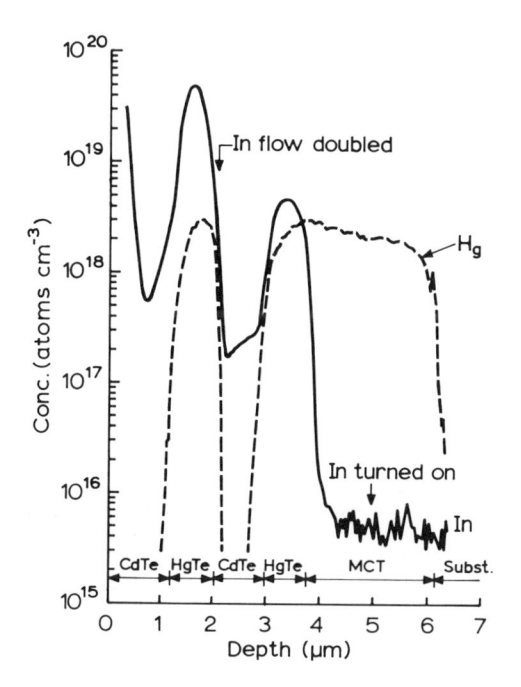

FIG. 2. SIMS profile of a MCT/HgTe/CdTe multilayer structure showing the difference in the incorporation of In in the different regions.

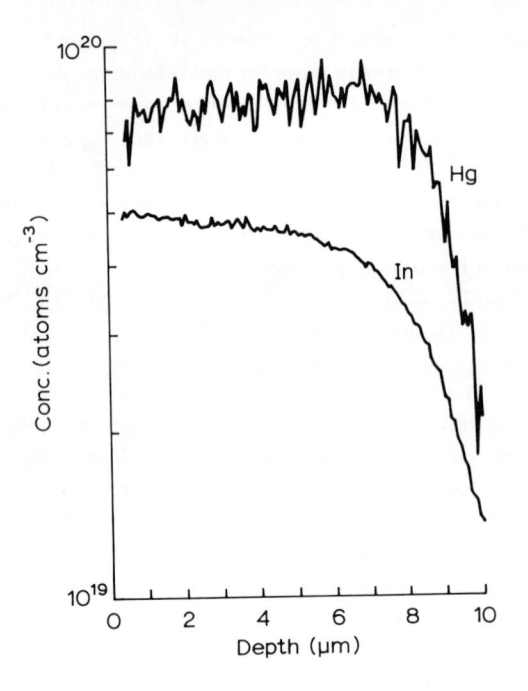

FIG. 3. SIMS profile of MCT layer doped at high concentration with In. Showing the significant time during growth to attain an equilibrium level.

B. Iodine doping

From our observations it was considered that In would be unlikely to fulfil the requirements for a satisfactory donor dopant for growth with DET at 400 °C. As an alternative it was decided that iodine should be examined. The solid has a relatively low vapor pressure that might be suitable as a source material. In addition it was expected that the diffusion rate for iodine would be low due to substitution on the Te site (as was found for As).

The first experiment was encouraging and showed that iodine could be incorporated in the mid 10^{15} atoms cm^{-3} region (determined by LSMS using I^{2+} species) for calculated concentration of ~ 5 vpma over the substrate, Hall measurements of the Hg annealed layers showed that the layer was n type with a free donor concentration which agreed with the chemical analysis. A series of experiments was carried out to produce a range of dopant concentrations using a combination of different temperatures and carrier gas flow rates. Figure 5 shows the results obtained as a plot of calculated reactor gas concentration versus free donor concentration (after annealing in Hg). Mobility values were in the region 8×10^4–10^5 cm^2 V^{-1} s^{-1} at 77 K. Before anneal treatment the layers generally showed mixed conduction behavior, one layer was n type as grown with a carrier concentration close to the annealed sample. Such results may be associated with the behavior of iodine in bulk material, reported by Vydyanath,[8] however it is difficult to explain how such low iodine concentrations used in the present study can influence the as-grown vacancy concentration to such an extent. Iodine is a particularly difficult element to analyze by SIMS because of interference effects of the Te isotopes, as a

into the CdTe substrate. GFAAS analysis of this layer showed 1.4×10^{16} In atoms cm^{-3} to be present. Hall measurements of the annealed/etched layer (after removal of $\sim 4 \mu$m from the surface) indicated $N_{D-A} = 4 \times 10^{16}$ cm^{-3}. Such a carrier concentration is consistent with the SIMS profile and shows that the doping activity for In is close to 100% at this concentration.

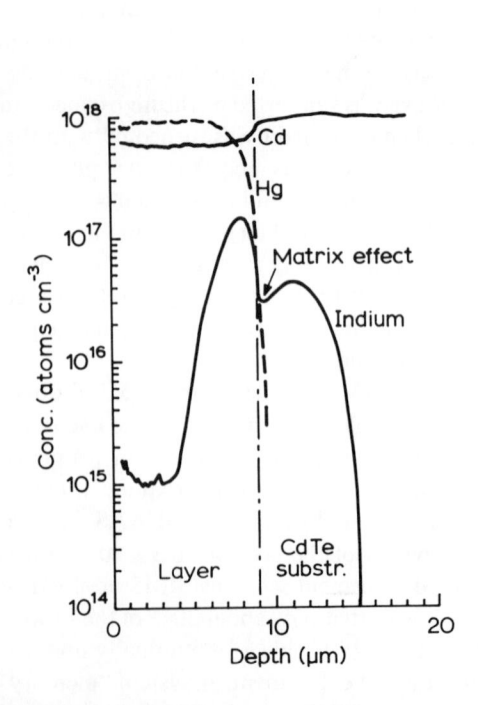

FIG. 4. SIMS indium profile of an unintentionally doped MCT layer, indicating the residual dopant memory effect and diffusion into the substrate.

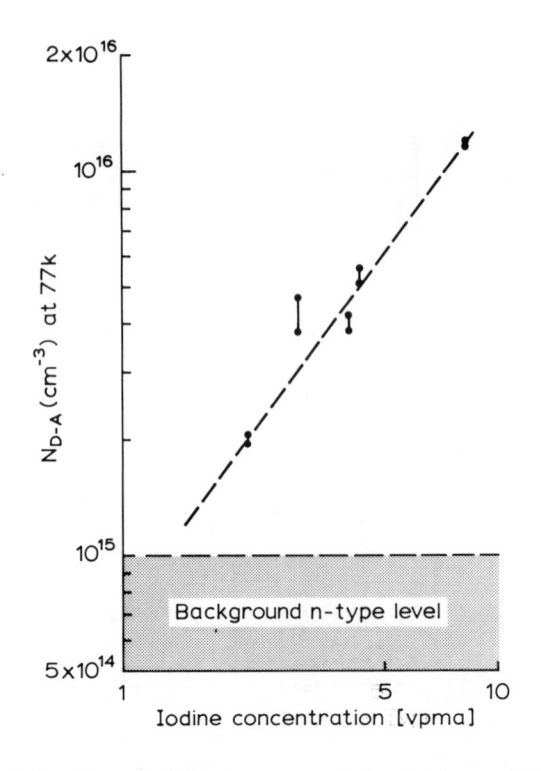

FIG. 5. Plot of measured free donor concentration in Hg annealed MCT layers vs iodine gas phase concentration.

result it has not been possible to carry out profile analyses to investigate the mode of incorporation (as reported for In). From Hall measurements however it does appear that the iodine content may increase with increasing x of the layer (this is shown in Fig. 5 where the point off the line represents a layer with $x = 0.23$ compared with 0.22 for the other layers). More work is required to confirm this aspect. The fact that iodine is expected to react directly with the dimethyl cadmium does not appear to be a disadvantage.

C. Residual impurities

In earlier work the free donor level in undoped MCT layers ($x = 0.20$–0.23) was generally in the region 2–3×10^{15} cm^{-3} in more recent material this has been reduced to 5×10^{14}–10^{15} cm^{-3}. It is not clear whether this improvement is due to an improvement in the procedures and the reactor system or to cleaner starting materials, e.g., the alkyls. Table I shows an LSMS analysis of a recent undoped layer 15 μm thick, it shows the concentrations of elements present above the technique detection limit. Analyses are shown for four regions in the structure, i.e., the top surface—where contamination effects primarily arise from handling after growth, the "bulk" or center region of the layer, the interface and the CdTe substrate. This particular layer had a free carrier concentration of 4.8×10^{14} cm^{-3} and a mobility of 1.1×10^{5} cm^2 V^{-1} s^{-1} at 77 K after annealing in Hg. Away from the surface region C and O are the major impurities, reducing significantly on reaching the substrate. Experience indicates that the concentrations of these elements may show quite large variations without exerting a significant influence on the Hall measurements. It is presumed that aluminium is rendered inactive by the presence of oxygen. No particular elements(s) can be assigned as being responsible for the measured free donor concentration. There is evidence of high impurity levels for some elements in the region of the interface but the substrate has impurities at levels below the detection limits in most cases.

Sodium is a common impurity in as-grown MCT layers, it is also a potential acceptor impurity. Figure 6 is a SIMS profile for an as-grown layer which was also part doped with As. The Na concentration is in the mid 10^{15} atoms cm^{-3} region in the undoped part of the layer. Figure 7 is a second profile of the same layer after it was annealed in Hg, the Na concentration in the layer is reduced to a low level by the anneal treatment with a reduced interface peak remaining. It appears that the Hg displaces the Na to the layer surface during the anneal, further that the As doping minimizes the uptake/incorporation of sodium. It has been observed that substrate cleaning treatments using nominally high purity organic solvents can prove a source of contamination in this case.

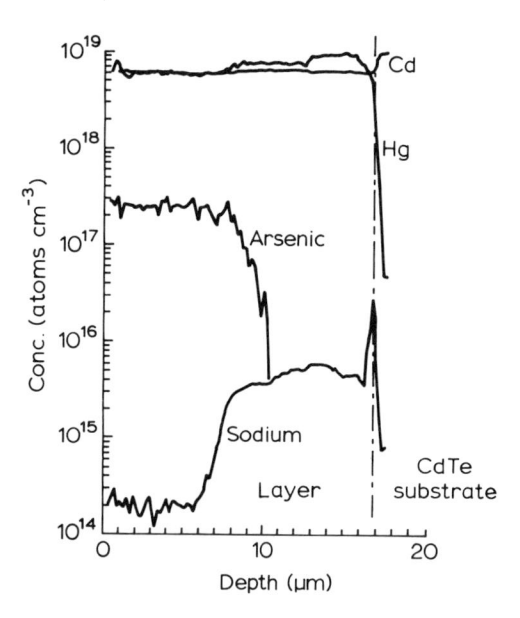

FIG. 6. SIMS profile of a layer part doped with As and contaminated with Na.

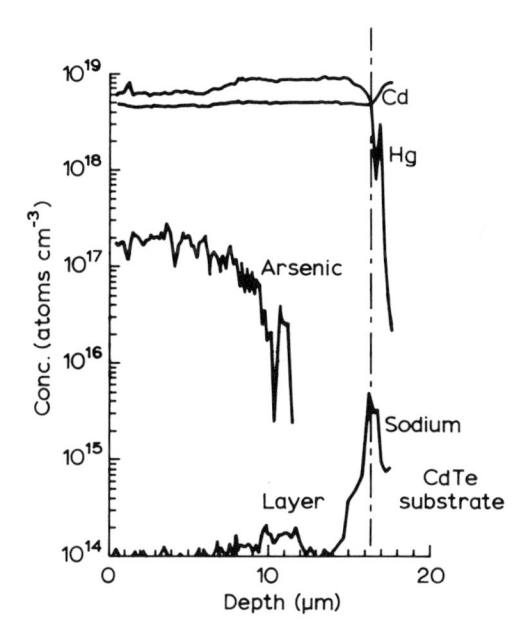

FIG. 7. Profile of layer in Fig. 6 after an anneal in Hg vapor at 200 °C.

TABLE I. LSMS analysis of undoped MCT layer ($\times 10^{15}$ atoms cm^{-3}).

Element	Surface	Bulk	Interface	Substrate
C	3000	3000	3000	60
O	3000	30	600	0.6
F	9	0.9	1.5	<0.2
Na	9	0.9	1.5	6[a]
Al	6	3	0.6	0.3
Si	60	<0.9	1.5	<1.5
P	0.9	<0.2	0.3	<0.2
S	150	0.6	15	0.6
Cl	90	0.3	30	<0.3
K	6	0.9	0.6	6[a]
As	0.3	<0.2	0.6[a]	<0.2
Se	60	60	<0.3	<0.6
Br	15	<0.3	<0.3	<0.6

[a] Heterogeneous distribution.

IV. CONCLUSIONS

Indium has been shown to behave as a donor in this work. When introduced as the trimethyl alkyl it is incorporated more strongly during deposition of HgTe than CdTe. This is probably related to the reduced competition for metal sites exerted by Hg as compared to Cd. Indium shows a considerable doping inertia effect, both with respect to the time taken to achieve an equilibrium level and the system memory effect. This, combined with the fast diffusion behavior (assuming that the diffusion rate in MCT is at least as high as that in CdTe), renders In unsuitable as a dopant where sharp junctions are required. It may however be suitable for establishing a higher background doping level should this be required. One might expect the memory effect to be greater in a reactor which did not incorporate the injector tube used in this work.

Iodine shows promise as a donor impurity at low levels. The element provides a convenient dopant source with a satisfactory vapor pressure at room temperature. Because of the difficulty in analyzing iodine it has not been possible to use SIMS to determine whether or not the iodine is preferentially incorporated in HgTe or CdTe and confirm that the diffusion is low, as expected for an element substituted on a Te site. More work is required to prove satisfactory behavior at high doping levels.

Deposition of MCT layers at 400 °C using DET can provide material of consistently high electron mobility and low carrier concentration (following anneal treatment in Hg), carbon and oxygen are the major impurities however no single element can be identified as being responsible for the consistent n-type behavior. Sodium has been shown to be an extremely mobile impurity in MCT during anneal treatments. Although the Na remains within a layer during the relatively high temperature deposition stage it moves to the surface at a lower temperature (200 °C) when a high mercury vapor pressure is maintained, i.e. the conditions used for removing the metal vacancies. As a result the potential acceptor doping behavior is not observed following an anneal treatment.

ACKNOWLEDGMENTS

The authors thank A. M. Cole for the FTIR measurements. This work was carried out with the support of the Procurement Executive, Ministry of Defence.

[1] P. Capper, P. A. C. Whiffin, B. C. Easton, C. D. Maxey, and I. Kenworthy, Mat. Lett. **6**, 356 (1988).
[2] C. D., Maxey, P. Capper, P. A. C. Whiffin, B. C. Easton, I. Gale, and J. B. Clegg, Mat. Lett. **8**, 385 (1989).
[3] C. D. Maxey, P. Capper, P. A. C. Whiffin, B. C. Easton, I. G. Gale, J. B. Clegg, A. Harker, and C. L. Jones, J. Cryst. Growth. **101**, 300 (1990).
[4] J. B. Mullin, S. J. C. Irvine, and J. Tunnicliffe, J. Cryst. Growth **68**, 214 (1984).
[5] P. A. C. Whiffin, B. C. Easton, P. C. Capper, and C. D. Maxey, J. Cryst. Growth **79**, 935 (1986).
[6] F. Grainger and J. A. Roberts, Semicond. Sci. Technol. **3**, 802 (1988).
[7] F. Grainger and I. G. Gale, J. Mat. Sci. **14**, 1370 (1979).
[8] H. R. Vydyanath and F. A. J. Kroger, Elect. Mat. **11**, 111 (1982).

The growth and properties of In-doped metalorganic vapor phase epitaxy interdiffused multilayer process (HgCd)Te

J. S. Gough, M. R. Houlton, S. J. C. Irvine, N. Shaw, M. L. Young, and M. G. Astles
RSRE, St. Andrews Road, Malvern, Worcestershire, WR14 3PS United Kingdom

(Received 3 October 1990; accepted 9 November 1990)

Indium doping of $Hg_{1-x}Cd_xTe$ layers grown by metalorganic vapor phase epitaxy using the interdiffused multilayer process has been investigated using trimethyl indium as the dopant source. The epitaxial growth was performed onto 2° off ⟨100⟩ CdTeSe substrates at ~350 °C using dimethyl cadmium (DMCd) and di-isopropyl telluride as the alkyl sources. Doping in the range 10^{17}–2×10^{18} cm^{-3} was achieved. By comparing Hall effect measurements of carrier concentration with secondary ion mass spectrometry analyses of the In concentrations in the layers, it was found that the In was only 30% active in the as-grown layers but ~100% active after Hg-rich isothermal annealing at 250 °C for 48 h. At In concentrations greater than 2×10^{18} cm^{-3}, the carrier concentration levels off, probably due to the solubility limit of In being reached. The annealed doped layers show slightly higher carrier mobilities than as-grown layers for the same carrier concentration. An apparent shift of the absorption edge to shorter wavelength with increasing donor concentration is thought to be due to the Moss–Burstein effect rather than a change in alloy composition x. A significant "memory" effect has been found with trimethyl In which persists for several runs and is probably due to strong adsorption onto the stainless steel surfaces in the growth system. This can be overcome by vacuum baking of the pipework. The growth of a heterostructure using In doping has shown that there is no serious diffusion problem with In under the growth conditions used.

I. INTRODUCTION

The growth of epitaxial $Hg_{1-x}Cd_xTe$ (MCT) by metal-organic vapor phase epitaxy (MOVPE) using the interdiffused multilayer process (IMP) is now well established and has demonstrated a capability for producing large areas of uniform MCT on a variety of substrates with good structural and electrical properties.[1-3] Most as-grown material to date has had electrical properties dominated by grown-in mercury vacancies which act as acceptors and under normal growth conditions are present in concentrations of 10^{16}–10^{17} cm^{-3}. To convert undoped material to n-type, layers are usually isothermally annealed under Hg-rich conditions at temperatures ~200–250 °C to reduce the Hg vacancy concentration below the residual electrically active donor impurity concentration, which is typically ~10^{15} cm^{-3}. Unfortunately the control of electrical properties by means of native defects is difficult particularly in the regions of n/p less than 10^{16} cm^{-3}. This, coupled with the belief that for p-type material, extrinsically doped MCT should have better minority-carrier lifetimes than vacancy-doped material[4] has led to an interest in both donor and acceptor doping of MCT by the MOVPE process. Good progress has been made by workers in the UK and USA on arsenic doping for p-type material.[5,6] At RSRE, we have recently been looking at donor doping with Al and In, and this paper presents our results on In doping.

II. EXPERIMENT

The epitaxial MCT layers were grown at ~350 °C by the IMP process of MOVPE[2] using the precursors dimethyl cadmium and di-isopropyl telluride with trimethyl In (TMIn) as the dopant source, all of which were supplied by Epichem Ltd. (UK). The respective bubbler temperatures of the alkyl sources were 25, 25, and 10 °C, and the partial pressure of the TMIn was varied in the range 7×10^{-6}–2.7×10^{-5} atm. The MCT layers were deposited onto CdTeSe ⟨100⟩ 2° off substrates supplied by II–VI Inc. and all samples were buffered and capped with a thin layer of CdTe. The x values of the layers were mainly in the range of 0.25–0.28. The In doping was carried out during the CdTe part of the IMP cycle since the In is incorporated at a lower level in the CdTe layers than in the HgTe and hence provides better control particularly at the lower doping levels. Pieces of all the layers were isothermally annealed in a Hg-rich atmosphere in evacuated silica ampoules for 48 h at 250 °C in order to remove the mercury vacancies in the as-grown material and activate the In dopant. The layers were assessed by (i) infrared (IR) absorption to obtain the value of x from the wavelength at which the absorption coefficient $\alpha = 500$ cm^{-1} using the relationship of Finkman and Schacham;[7] (ii) Hall effect measurements at 77 K as a function of magnetic field B over the range 0.01–1 T, both in the as-grown state and after isothermal annealing; (iii) secondary ion mass spectrometry (SIMS) using a CAMECA IMS 3f machine with 12.5 keV O_2^+ primary beam focussed to a 20–30 μm spot giving an erosion rate of ~4 μm/h^{-1}. Calibration was carried out using doped specimens of bulk MCT analyzed by spark-source mass spectrometry or atomic absorption spectroscopy;[8] (iv) Rutherford backscatter (RBS) has been used to assess the alloy composition x of one of the doped layers to compare with the IR absorption value. The RBS data was obtained using ^{4}He ions in the energy range 2.0–2.5 MeV.

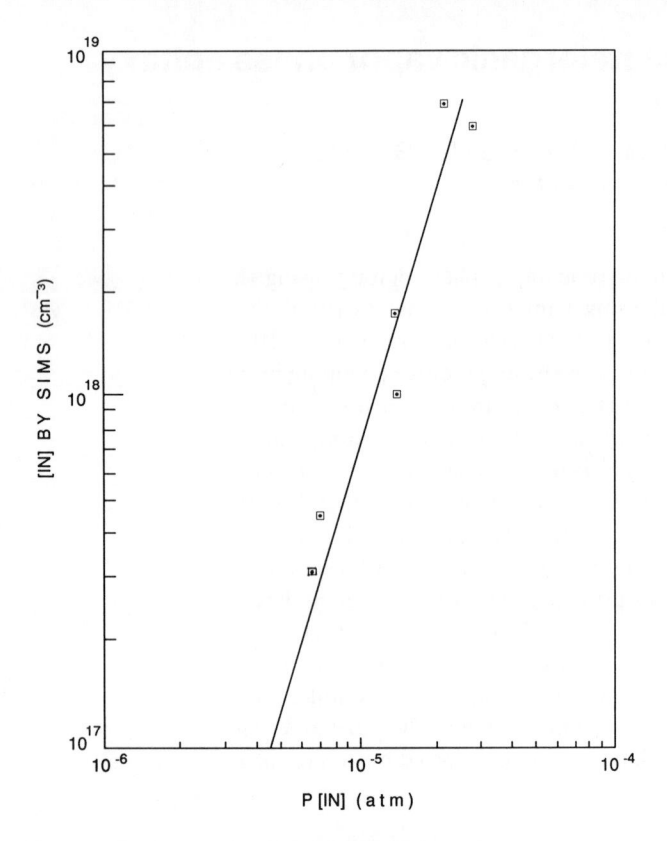

FIG. 1. Atomic concentration of indium in the doped ($Cd_xHg_{1-x}Te$) layers measured by SIMS (atoms cm^{-3}) against partial pressure of TMIn in the vapor phase (atmospheres) for growth at 350 °C and $x \sim 0.28$.

III. RESULTS

The variation of the atomic concentration of In in the doped layers as measured by SIMS with the partial pressure of TMIn is shown in Fig. 1 covering a range from $[In] = 2 \times 10^{17} - \sim 8 \times 10^{18}$ cm^{-3}. There is some scatter on the results and more data points are required. The scatter is probably due to the long time constant to reach a steady state In concentration in the layers. A typical SIMS profile is shown in Fig. 2 where it can be seen that there is a delay before the (In) reaches a constant doping level in the layer. This may be due to the time taken for TMIn adsorption on the walls of the pipework and reactor to reach saturation. This interpretation is supported by the observation of a significant "memory" effect with TMIn, whereby subsequent, nominally undoped layers show In incorporation by SIMS. Careful baking of the pipework under vacuum is needed to remove this contamination. This has been confirmed by SIMS analysis which shows that after baking, levels of In down to $\sim 10^{15}$ cm^{-3} can be obtained which represents the background level in the SIMS machine.

The undoped layers grown in the kit prior to In doping were p type with 77 K hole concentration of $\sim 4 \times 10^{16}$ cm^{-3}, consistent with the expected Hg vacancy concentration for the growth conditions. On In doping the layers converted to n type when the chemical concentration of In in the layers was greater than 1×10^{17} cm^{-3}. At these In concentrations the layers were uniformly n type and had flat 77 K carrier profiles, no significant magnetic field or depth dependence of the Hall coefficient being found. The variation of the measured n-type carrier concentration with the chemical concentration of In in the layers is shown in Fig. 3 for both as-grown and annealed layers. It can be seen that in the as-grown conditions the electrical activity of the In is only $\sim 30\%$ whereas after annealing the activity increases to

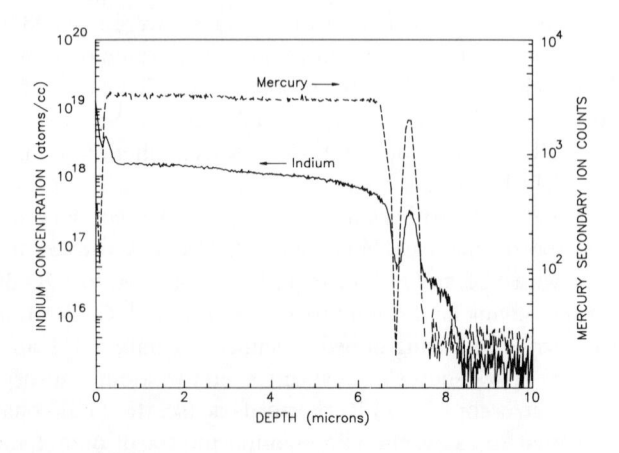

FIG. 2. SIMS depth profile of typical indium-doped $Cd_xHg_{1-x}Te$ layer grown on a CdTeSe substrate at a partial pressure of TMIn of 1.4×10^{-5} atm.

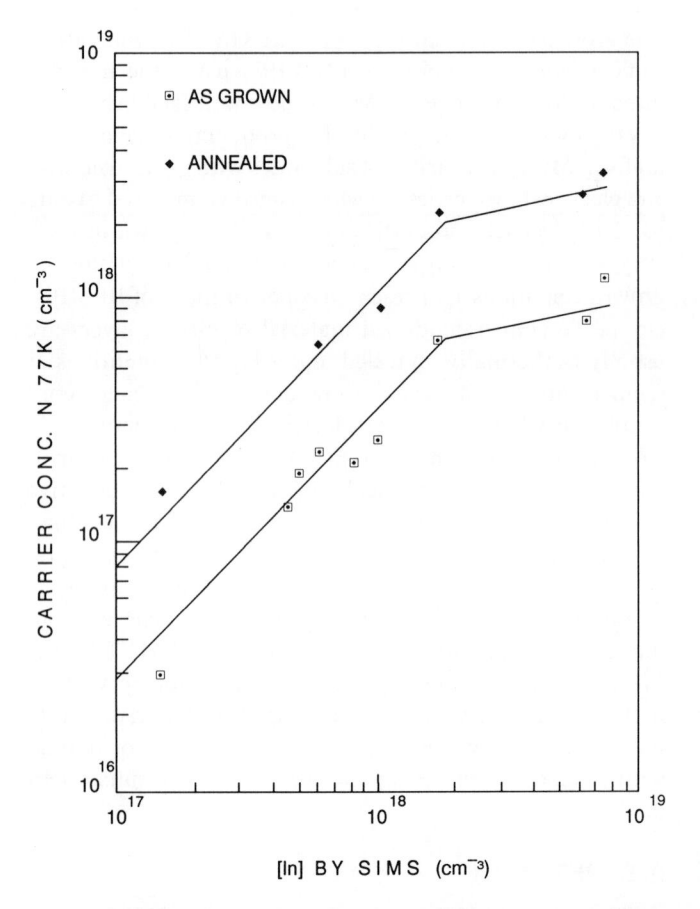

FIG. 3. Electron concentration measured at 77 K (cm^{-3}) for as-grown and annealed In-doped $Cd_xHg_{1-x}Te$ layers against the concentration of In in the layers measured by SIMS (cm^{-3}).

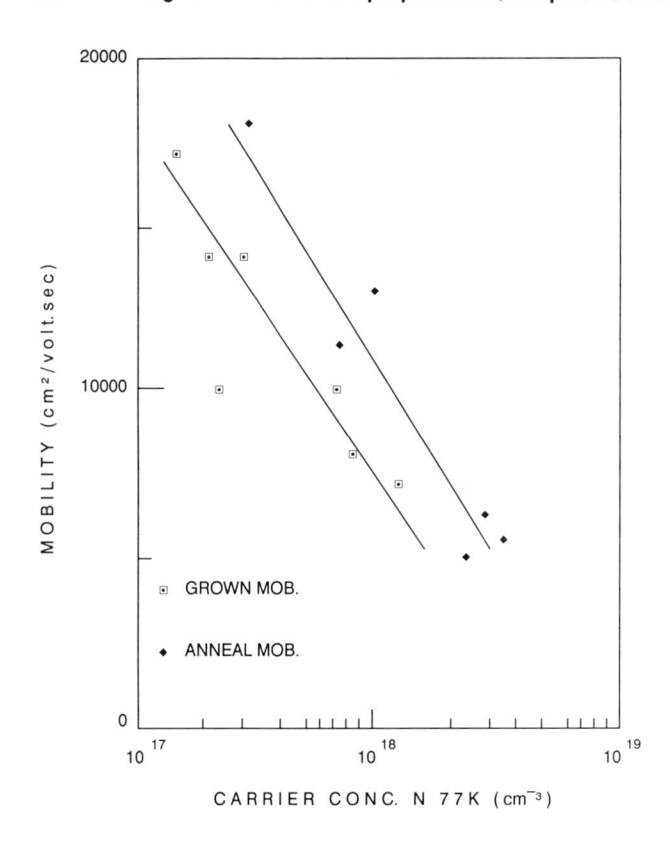

FIG. 4. Hall mobility at 77 K (cm² V⁻¹ s⁻¹) as a function of electron concentration (cm⁻³) for as-grown and annealed In-doped $Cd_xHg_{1-x}Te$ layers ($x = 0.28$).

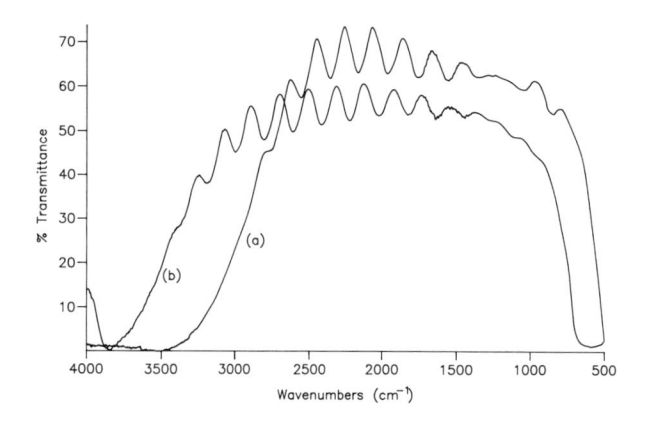

FIG. 5. IR transmission spectrum for In-doped $Cd_xHg_{1-x}Te$ layer with 7×10^{18} cm⁻³ In (SIMS) (a) as-grown and (b) after isothermal annealing.

annealed specimen shows a reduction in transmission at low wavenumber (long wavelength) due to the shift in the plasma edge. The variation of the shift in the apparent x value (compared to what would be expected from an undoped layer grown under identical conditions) as a function of carrier concentration is shown in Fig. 6. Measurements of x by RBS have confirmed that at the higher doping levels this effect is not due to a change in alloy composition, but almost certainly to the Moss–Burstein effect. In fact, no significant effect of

100% at concentrations up to $\sim 2 \times 10^{18}$ cm⁻³. At the low end of the doping range the low activity of the as-grown layers might be expected, due to compensation by the residual background of 4×10^{16} cm⁻³ Hg vacancy acceptors. However, it also occurs at higher doping levels, which is similar to the effect seen in our work on In doping of liquid-phase epitaxy (LPE) material. Although the low as-grown activity could be due to In incorporation on electrically inactive sites, it seems more likely to be due to autocompensation, with the In doping generating additional compensating acceptor defects, most probably Hg vacancies. Figure 4 shows a plot of 77 K Hall mobility as a function of carrier concentration for as-grown and annealed material. There appears to be a small but significant increase in the mobility for annealed material of the same carrier concentration. This would be consistent with the annealing reducing the concentration of compensating acceptor defects rather than redistribution of In onto active sites. We note that in Fig. 3 there is a leveling-off of the carrier concentration above 2×10^{18} cm⁻³. This has been seen in In doping work on LPE material and is probably due to the solubility limit for In being reached with the formation of In-containing precipitates. TEM work is in hand to test this hypothesis.

Significant changes were observed in the IR transmission spectrum upon annealing, consistent with the increase in the n-type carrier concentration. Figure 5 shows spectra before and after annealing a layer containing 7×10^{18} cm⁻³ In. Not only is there a shift in the position of the band edge but the

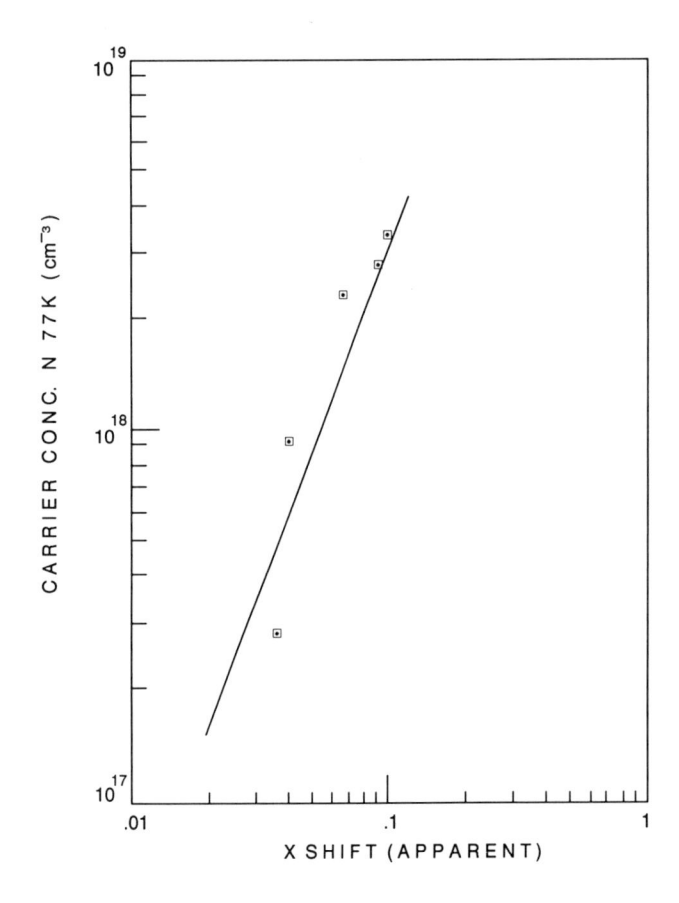

FIG. 6. Apparent shift in x from IR transmission measurements against electron concentration (cm⁻³) for In-doped $Cd_xHg_{1-x}Te$ layers.

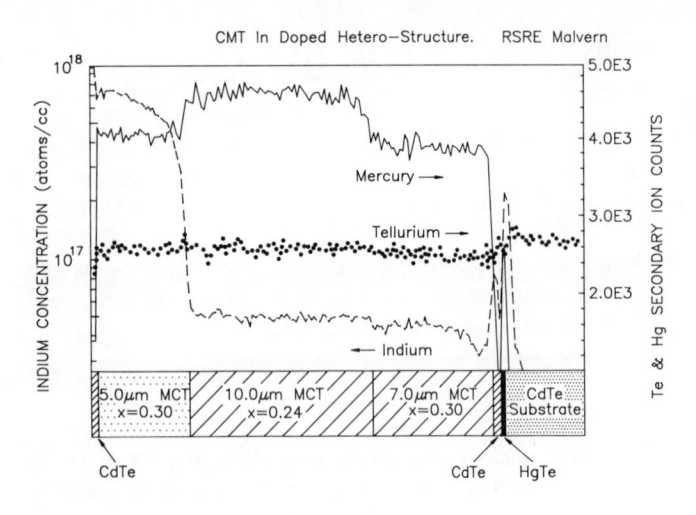

FIG. 7. SIMS depth profiles for In, Hg, and Te for structure consisting of 0.1 μm CdTe (buffer)/7 μm $x=0.30$ (undoped)/10 μm $x=0.24$ (undoped)/5 μm $x=0.30$ (In doped)/0.1 μm CdTe (cap) grown on CdTe substrate using ^{200}Hg, ^{125}Te, and ^{115}In species.

TMIn partial pressure on the alloy composition was found in this work.

To demonstrate the potential for growing doped heterostructures using In as the n-type dopant and to assess any problems with diffusion of In a heterostructure was grown consisting of 0.1 μm HgTe (nucleation)/0.5 μm CdTe (buffer)/7 μm $x=0.30$ (undoped)/10 μm $x=0.24$ (undoped)/5 μm $x=0.3$ (In doped)/0.1 μm CdTe (Cap). Figure 7 shows a SIMS depth profile of the structure for the Hg, Te, and In species. It should be noted that this structure was grown with an In memory level of $\sim 5\times 10^{16}$ cm^{-3} in the growth apparatus. The sharp transition in the In signal between the undoped (nominally) and the deliberately doped layer incidates that there is no major problem with In diffusion at the growth temperature used. Also, it can be seen that during the growth of the initial HgTe nucleation layer and the CdTe buffer layers, the incorporation of the In from the background TMIn is greater in the HgTe than in the CdTe layer by a factor of ~ 4, in qualitative agreement with the observations of Easton et al.[9]

IV. CONCLUSIONS

Doping of IMP MOVPE grown layers of Cd$_x$Hg$_{1-x}$Te has been successfully achieved using TMIn as the dopant source. The doping is well controlled in the range 10^{17}–$\sim 2\times 10^{18}$ cm^{-3}. Further work is needed to extend the doping range to lower carrier concentration. At high dopant concentrations the electrical activity saturates probably due to a solubility limit being reached. The In dopant is 30% active as-grown and is $\sim 100\%$ active after isothermal annealing. It has been found that there is a significant memory effect probably due to the adsorption of the alkyl on the internal surfaces of the growth system which currently makes it impossible to grow p-n structures with the n-type layer grown first. The adsorbed material can be removed by vacuum baking of the mixing manifold between runs. A more direct technique for introducing the dopant such as the injection tube used by Whiffen et al.[10] has been shown to give a much reduced "memory" effect. We have shown that the shift of the band edge seen in the doped layers is not due to a change of alloy composition but probably to the Moss-Burstein effect. The potential of In for use in doped MCT heterojunctions has been demonstrated.

ACKNOWLEDGMENTS

The authors are grateful to Dr. J. Hails and J. Giess (RSRE) for technical discussions on the MOVPE growth, to A. Horsfall (RSRE) for the Hall effect measurements, to B. Clarke and C. Burford (RSRE) for the annealing work and D. Jones for the preparation of the substrates. We also acknowledge the work of D. Diskett and A. Avery at the Royal Military College of Science, Shrivenham, in providing the RBS result.

[1] J. Thompson, P. Hackett, L. M. Smith, D. J. Cole-Hamilton, and D. V. Shenai-Khatkhate, J. Cryst. Growth 86, 233 (1988).
[2] S. J. C. Irvine, J. B. Mullin, J. Giess, J. S. Gough, and A. Royle, J. Cryst. Growth 93, 732 (1988).
[3] D. D. Edwall, J. J. Bajaj, and E. R. Gertner, J. Vac. Sci. Technol. A 8, 1045 (1990).
[4] T. Tung, J. Cryst Growth 86, 161 (1988).
[5] P. Capper, P. A. C. Whiffen, B. C. Easton, C. D. Maxey, and I. Kenworthy, Mater. Lett. 6, 356 (1988).
[6] S. K. Ghandhi, N. R. Taskar, K. K. Parat, D. Terry, and I. B. Bhat, Appl. Phys. Lett. 53, 1641 (1988).
[7] E. Finkman and S. E. Schacham, J. Appl. Phys. 56, 2896 (1984).
[8] R. Holland and G. Blackmore, Surf. Interface. Anal. 4, 174 (1982).
[9] B. C. Easton, C. D. Maxey, P. A. C. Whiffen, J. A. Roberts, I. Gale, and F. Grainger, J. Vac. Sci. Technol. B 9, 1682 (1991).
[10] P. A. C. Whiffen, B. C. Easton, P. C. Capper, and C. D. Maxey, J. Cryst. Growth 79, 935 (1986).

Arsenic doping in metalorganic chemical vapor deposition $Hg_{1-x}Cd_xTe$ using tertiarybutylarsine and diethylarsine

D. D. Edwall, J.-S. Chen, and L. O. Bubulac
Rockwell International Science Center, Thousand Oaks, California 91360

(Received 29 October 1990; accepted 12 December 1990)

P-type arsenic doped epitaxial layers of HgCdTe have been grown by metalorganic chemical vapor deposition using two alkyl sources, tertiarybutylarsine and diethylarsine. Data are presented on Hall characteristics and arsenic concentration profiles. High activation efficiencies and hole mobilities have been obtained over the range mid-10^{15} to low-10^{17} cm^{-3}.

I. INTRODUCTION

The rationale for developing metalorganic chemical vapor deposition (MOCVD) grown $Hg_{1-x}Cd_xTe$ (MCT) is based on MOCVDs dual capability in high throughput and sophisticated material/device engineering. Early results of MOCVD MCT have confirmed the throughput capabilities by demonstrating large area uniformity[1] and material quality compatible with device fabrication.[2] Considerations for higher performance, process simplification, and producibility are leading to device structures based on *in situ* grown junctions or diffused junctions from *in situ* deposited diffusion sources which require the incorporation and activation of extrinsic dopants during the MOCVD growth cycle. Reported here are our results for *p*-type layers extrinsically doped with arsenic.

II. EXPERIMENTAL

The interdiffused growth method[3] was used to grow (100) MCT layers in MOCVD reactors operating at atmospheric pressure. Layers were grown on (100) GaAs substrates misoriented towards the nearest (110). The following metalorganic alkyls were used: dimethylcadmium (DMCd) and dimethyltellurium (DMTe) for growth of CdTe; diisopropyltellurium and elemental Hg for growth of HgTe; and tertiarybutylarsine (TBAs) and diethylarsine (DEAs) for the arsenic dopant sources. Successful *p*-type doping of MOCVD MCT using arsine was reported by Maxey *et al.*[4,5] and Prof. Ghandhi's group,[6,7] but the use of metalorganic sources offers a less toxic alternative to the use of arsine. Doping using the metalorganic sources trimethylarsenic and trimethylantimony has been reported by Whitely *et al.*[8]

MCT growth temperatures were < 400 °C with HgTe–CdTe period thicknesses of typically 1000–1100 Å. Further growth details can be found in Refs. 1,9,10. The arsenic alkyls were injected into the reactors during the CdTe growth cycles under Cd-rich (relative to Te) vapor concentration conditions in order to promote arsenic substitution on the Te sublattice, consistent with the work of others in both MOCVD[11] and molecular-beam epitaxy MBE growth.[12]

Layer characterization techniques included room-temperature infrared transmission to determine composition x and layer thickness, Hall measurements at 2 kG magnetic field to determine electrical carrier concentration and mobility, and secondary ion mass spectroscopy (SIMS) performed at Charles Evans Associates to determine arsenic concentrations. All electrical measurements were carried out after post-growth *n*-type annealing of the layers under Hg-saturated conditions at 250 °C for 15–20 h to remove Hg vacancies. After such annealing, undoped layers typically showed mixed conduction at 77 K or were lightly *p* type with carrier concentrations less than 5×10^{15} cm^{-3}, due to residual background impurities.

III. RESULTS

A. Doping using TBAs

Table I shows results for MCT layers doped using TBAs. Data are indicated for composition x, Hall characterization at 77 K, and arsenic concentrations determined by SIMS. For all three layers the hole mobilities are good considering both the x values and carrier concentrations. The arsenic concentrations determined by SIMS agree well with the 77 K carrier concentrations. A sample of layer 246 yielded the same Hall parameters after annealing at 400 °C for 2 h followed by the standard 250 °C anneal, demonstrating no increase in activation efficiency of the arsenic acceptors.

Figure 1 shows SIMS arsenic profiles versus depth from the surface for two of the Table I layers. For both layers the arsenic concentrations are reasonably flat with depth. For undoped layers the arsenic concentrations are at the SIMS detection limit of approximately 1×10^{15} cm^{-3} or below, well below the levels in Fig. 1. For layer 260 the region from approximately 5 to 7 μm represents the (uncalibrated) arsenic concentration in the CdTe buffer layer; due to the difference in ion yield between CdTe and MCT, we estimate that the actual arsenic concentration in the CdTe layer is ~5–6 times lower than indicated in the SIMS profiles.

TABLE I. Characteristics of layers doped using TBAs.

Layer	x	p (cm^{-3})	μ (cm^2/V s)	As, SIMS (cm^{-3})
		77K Hall		
423	0.23	8.5×10^{15}	606	2.5×10^{16}
246	0.31	1.9×10^{17}	220[a]	2×10^{17}
260	0.27	2.3×10^{17}	277	2×10^{17}

[a] Same Hall parameters obtained after 400 °C (2 h)/250 °C anneal.

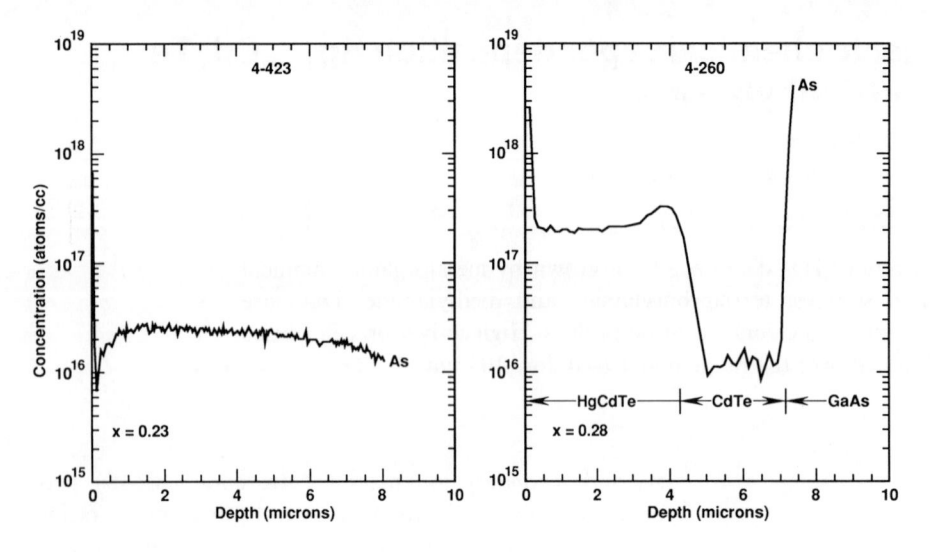

FIG. 1. SIMS arsenic profiles for two layers doped using TBAs.

Therefore, the arsenic diffusion into the CdTe buffer layer is small, near or below the detection limit.

Figure 2 shows the SIMS arsenic profile for a doped heterostructure. The region greater than approximately 6 μm from the surface represents an undoped MCT layer with composition $x = 0.2$. On top of this layer was grown an arsenic doped layer with $x = 0.3$. The arsenic concentration in the doped layer is reasonably flat. Most of the arsenic profile change between the two layers occurs over a distance of approximately 1 μm; the arsenic concentration drops by almost two orders of magnitude from the doped to undoped regions. Because the SIMS analysis was carried out with cesium ion bombardment for maximum arsenic yield, we do not have markers for Cd or Hg to locate the position of the heterointerface, nor the point in time when the arsenic source was turned on. Consequently, only an estimate of the diffusion coefficient can be given. An analysis[13] of the profile yields a diffusion coefficient of $\sim (1-2) \times 10^{-13}$ cm^2 s^{-1}. This value lies in the range of $(1-20) \times 10^{-14}$ cm^2 s^{-1} reported by Maxey et al.[14] for arsenic diffusion in MOCVD layers grown at 410 °C, and slightly better than the value of 2×10^{-12} cm^2 s^{-1} obtained at 385 °C in liquid phase epitaxy (LPE) material for an arsenic doped MCT layer grown from Hg solution on a layer grown from Te solution.[15]

B. Doping using DEAs

Table II shows 77 K Hall data for several layers doped using DEAs. The acceptor concentrations range from mid-10^{15} to mid-10^{16} cm^{-3}. The mobilities are similar to those measured in our Hg vacancy-doped layers.

Figure 3 shows variable temperature Hall data for one of the Table II layers. The carrier concentration shows the expected carrier freeze-out dependence on decreasing temperature. Measuring the slope over the temperature range $\sim$25–70 K yields an acceptor ionization energy of 4.4 meV. Both this value and the general shapes of both carrier concentration and mobility are very similar to data published by Capper et al.[16] for their layers doped using arsine. The decrease in low temperature mobility is probably caused by a relatively large ionized center scattering component due to the relatively high carrier concentration (6×10^{16} cm^{-3} at 77 K). At higher temperature the carrier concentration begins to saturate at approximately 6×10^{16} cm^{-3} before the onset of intrinsic-dominated conduction. Because the SIMS arsenic concentration for this layer was 1×10^{17} cm^{-3}, the arsenic acceptor activation efficiency is approximately 60%. The DMCd/DMTe vapor phase ratio was 1.05 for this layer;

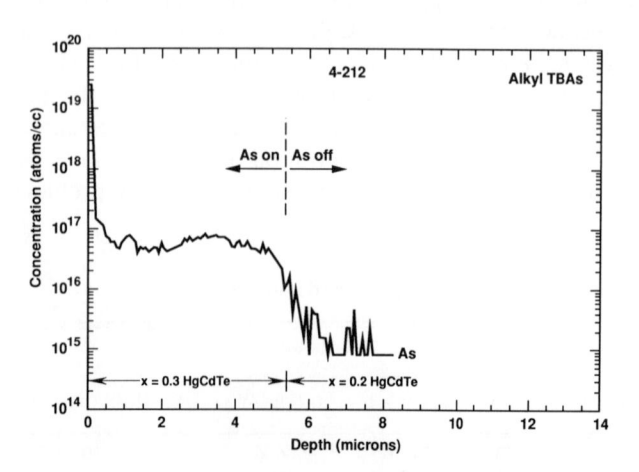

FIG. 2. SIMS arsenic profile for TBAs-doped heterostructure No. 4-212. Arsenic doped $x = 0.3$ layer grown on undoped $x = 0.2$ layer in one growth run.

TABLE II. Characteristics of layers doped using DEAs.

		77 K Hall	
Layer	x	μ (cm^2/V s)	p (cm^{-3})
1285	0.30	353	6.1×10^{16}
1287	0.32	501	1.4×10^{16}
1302	0.26	351	2.7×10^{16}
1316	0.28	320	4.0×10^{15}
1322	0.25	346	1.6×10^{16}

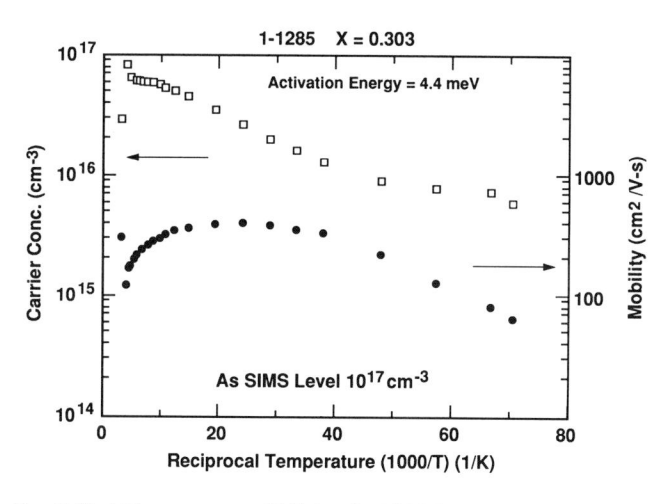

FIG. 3. Variable temperature Hall data for MCT layer doped using DEAs. Activation efficiency is approximately 60%.

we believe that a higher ratio might result in a higher activation efficiency due to more incorporation of arsenic atoms on Te sites.

Figure 4 shows SIMS arsenic profiles for two of the Table II layers. Again the arsenic concentrations are reasonably flat with depth.

C. Activation efficiency

Figure 5 shows the relationship between 77 K hole concentration and SIMS arsenic concentration for all the layers in Tables I and II. The data demonstrate a high activation efficiency for both doping sources from approximately mid-10^{15} to 2×10^{17} cm^{-3} doping levels. Because of hole freezeout at lower temperature, the actual activation efficiencies are higher than those implied by the Fig. 5 data. Increasing the activation efficiency by optimizing the DMCd/DMTe ratio has not yet been studied.

IV. SUMMARY

MOCVD MCT layers have been successfully doped with arsenic from mid-10^{15} to low-10^{17} cm^{-3} levels using two metalorganic sources, TBAs and DEAs. While acceptable results have been obtained using both sources, for low doping level applications the use of DEAs may be preferable due to its much lower vapor pressure. Good 77 K hole mobilities and high activation efficiencies have been obtained. The arsenic appears to be electrically stable even after 400 °C annealing for 2 h. An adequately sharp arsenic diffusion profile has been measured in a heterostructure grown by this method. Minority carrier lifetime results on some of these layers can be found in Ref. 17.

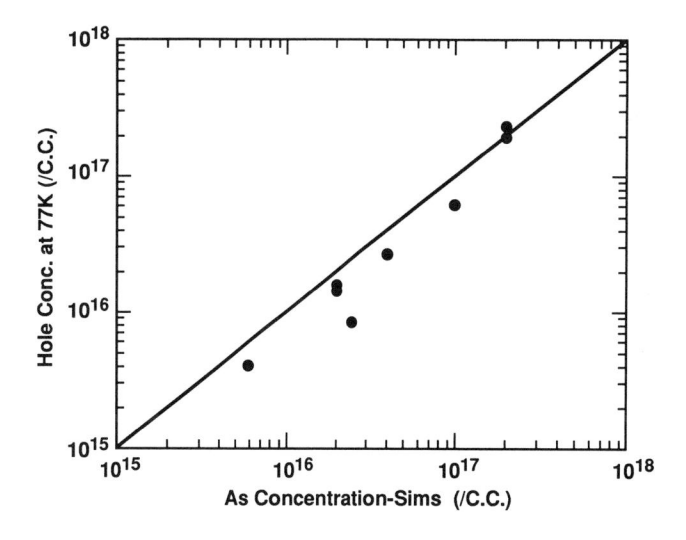

FIG. 5. Relationship between 77 K hole concentration and SIMS arsenic concentration, demonstrating high arsenic activation efficiency.

ACKNOWLEDGMENTS

The authors gratefully acknowledge the financial support of SDIO/TNS through AFSTC/SWS, managed by WRDC/MLPO Contract No. F33615-89-C-5557, Lynn Brown technical monitor.

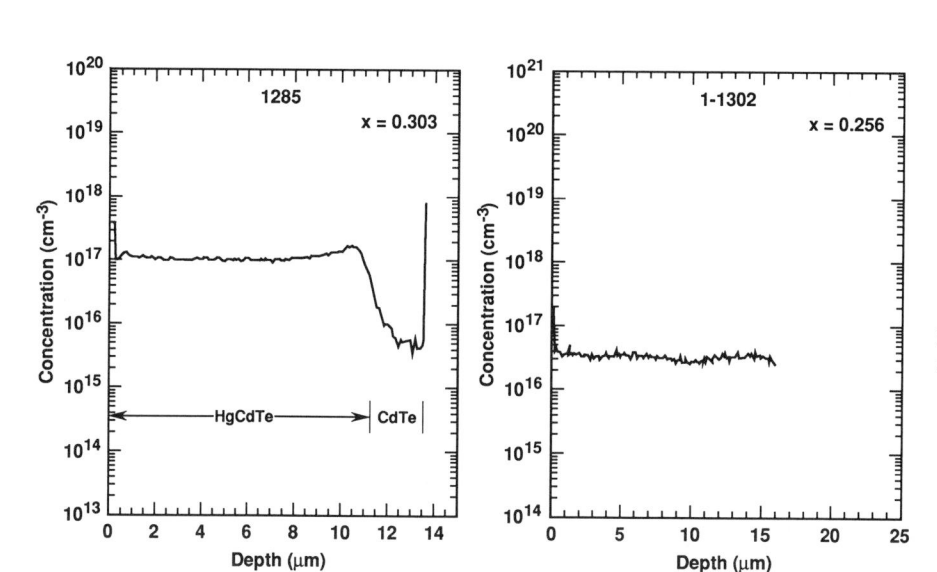

FIG. 4. SIMS arsenic profiles for two layers doped using DEAs.

[1] D. D. Edwall, J. Bajaj, and E. R. Gertner, J. Vac. Sci. Technol. A 8, 1045 (1990).
[2] L. O. Bubulac, D. D. Edwall, D. McConnell, R. E. DeWames, E. R. Blazejewski, and E. R. Gertner, Semicond. Sci. Technol. 5, S45 (1990).
[3] J. Tunicliffe, S. J. C. Irvine, O. D. Dosser, and J. B. Mullin, J. Cryst. Growth 66, 245 (1984).
[4] C. D. Maxey, P. Capper, P. A. C. Whiffin, B. C. Easton, I. Gale, and J. B. Clegg, Mater. Letters 8, 385 (1989).
[5] C. D. Maxey, P. Capper, P. A. C. Whiffin, B. C. Easton, I. Gale, and J. B. Clegg, A. Harker, and C. L. Jones, J. Cryst. Growth 101, 300 (1990).
[6] N. R. Taskar et al., J. Vac. Sci. Technol. B 9, 1705 (1991).
[7] S. K. Ghandhi, N. R. Taskar, K. K. Parat, D. Terry, and J. B. Bhat, Appl. Phys. Lett. 53, 1641 (1988).
[8] J. S. Whitely, P. Koppel, V. L. Conger, and K. E. Owens, J. Vac. Sci. Technol. A 6, 2804 (1988).
[9] D. D. Edwall, E. R. Gertner, and L. O. Bubulac, J. Cryst. Growth 86, 240 (1988).
[10] D. D. Edwall, J.-S. Chen, J. Bajaj, and E. R. Gertner, Semicond. Sci. Technol. 5, S221 (1990).
[11] P. Capper, P. A. C. Whiffin, B. C. Easton, C. D. Maxey, and I. Kenworthy, Mater. Letters 6, 356 (1988).
[12] J. M. Arias, S. H. Shin, D. E. Cooper, M. Zandian, J. G. Pasko, E. R. Gertner, R. E. DeWames, and J. Singh, J. Vac. Sci. Technol. A 8, 1025 (1990).
[13] L. O. Bubulac et al., J. Vac. Sci. Technol. B 9, xxxx (1991).
[14] C. D. Maxey, P. Capper, P. A. C. Whiffin, B. C. Easton, I. Gale, J. B. Clegg, and A. Harker, Mater. Letters 8, 190 (1989).
[15] J.-S. Chen (private communication).
[16] P. Capper, C. D. Maxey, P. A. C. Whiffin, and B. C. Easton, J. Cryst. Growth 97, 833 (1989).
[17] R. Zucca et al., J. Vac. Sci. Technol. B 9, xxxx (1991).

Dynamics of arsenic diffusion in metalorganic chemical vapor deposited HgCdTe on GaAs/Si substrates

L. O. Bubulac and D. D. Edwall
Rockwell International Science Center, 1049 Camino Dos Rios, Thousand Oaks, California 91360

C. R. Viswanathan
University of California at Los Angeles, Electrical Engineering Department, Los Angeles, California 90024

(Received 19 November 1990; accepted 28 January 1991)

This paper describes the behavior of arsenic when diffused from an ion implanted source in $Hg_{1-x}Cd_xTe$ ($x \simeq 0.19$–0.23) grown by metalorganic chemical vapor deposition (MOCVD) on GaAs/Si substrates. The results are compared with the diffusion of arsenic from a grown source in the same material. Diffusion mechanisms of arsenic were studied from chemical analysis (secondary ion mass spectroscopy) and theoretical modeling, combined with techniques for radiation damage study [transmission electron microscopy (TEM)] for revealing the defect structure of material [chemical defect etching-etch pit density (EPD), TEM], and for electrical junction determination (electron beam induced current). The main experimental observation was that arsenic redistributes during thermal anneal by a multicomponent mechanism in which the tail-components differ depending on the EPD of the material. We propose a model that explains the four components observed: (1) a surface retarded diffusion component; (2) an atomic vacancy-based component in which arsenic starts on Te sublattice, well described by the theoretical Gaussian solution to Fick's second law of diffusion with a constant diffusion coefficient ($D_{(400°C)} \simeq 3 \times 10^{-14}$ cm^2/s and $D_{(450°C)} \simeq 2 \times 10^{-13}$ cm^2/s); (3) an atomic vacancy-based component in which arsenic starts on metal sublattice and undergoes a diffusion process enhanced by a chemical nonequilibrium dependent on EPD of the material; and (4) a short circuit diffusion component. The same model applies to diffusion from a grown source. To control the electrical junction of a photovoltaic device, the concentration of arsenic in the tail components should be reduced to values lower than the n-type background. We show that the primary requirement to obtain this control is reducing the dislocation density of the material. Proper choice of implant/anneal conditions helps in reducing the tail of the arsenic profiles for a given EPD in the material. No diffusion tails were observed in MOCVD material with EPD in the low 10^6/cm^2 range.

I. INTRODUCTION

The approach of p-on-n junction in a heterostructure configuration, with extrinsic doping, for photovoltaic devices has taken an increasingly critical role in the long-wavelength IR (LWIR) technology based on HgCdTe material. To date the majority of these devices have been formed in material grown by liquid phase epitaxy (LPE). More recently MOCVD growth of HgCdTe has matured, allowing fabrication of p-on-n photodiodes using ion implantation techniques.

Acceptor extrinsic doping of HgCdTe has been the subject of study of various groups attempting to replace the native acceptor doping defects by extrinsic acceptor dopants. Column V dopants, in particular As and Sb, are the preferred choices because, due to their large atom size, they have low diffusivity. These elements when substituted into Te-sublattice site will behave, p type,[1] on metal-sublattice site, n type,[1] and when an As-acceptor is associated to an As-donor, neutral.[2] Also they may undergo a lattice-site transfer in a certain thermal anneal.[1]

High performance p-on-n devices have been obtained by ion implantation in a heterostructure MOCVD material on GaAs substrates.[3] However, the electrical junctions of these devices are controlled by the tailing component of the arsenic diffusion profile rather than by a mechanism representing the true volume diffusion. Electrical activity of these components can vary (n, p type, or neutral) depending on the nature of the diffusion mechanism which generated them. Good control and quality of electrical p-on-n junctions require understanding of the dopant diffusion mechanism, i.e., nature, stability, and the electrical activity of the dopant. Diffusion study of As from any source requires the consideration of point defects and of the extended defects in the starting material.

We have selected to study As behavior in $Hg_{1-x}Cd_xTe$ ($x = 0.19$–0.23) grown by MOVPE on GaAs/Si substrates when diffused from a shallow source formed by ion implantation and to compare the results with As diffusion in a grown p-on-n heterostructure. Based on the knowledge gained in this study we propose a model of As diffusion behavior.

II. EXPERIMENT

$Hg_{1-x}Cd_xTe$ material used in this work is epitaxially grown by metalorganic chemical vapor deposition (MOCVD) on GaAs/Si substrates by an interdiffused growth method. The advantages of GaAs/Si substrates for HgCdTe focal plane technology are the size and ruggedness of Si and the ease of nucleation on GaAs. GaAs/Si substrates are currently the most direct path toward very large hybrid focal plane arrays (overcoming thermal mismatch of substrate and Si multiplexer) and the monolithic integration of IR detectors and readout electronics. The interdiffused growth method was used to grow HgCdTe (MCT) layers in an atmospheric-pressure horizontal reactor chamber. The growth technique is described in Refs. 4 and 5. (100) GaAs/Si substrates were purchased from the Kopin Corporation. The layers used throughout this work had a Cd molar fraction $x = 0.19$–0.23. The EPD in these layers ranged from ~ 0.3 to $6 \times 10^7/cm^2$. Occasionally we measured lower EPD. The ion implantation and anneal parameters used in the study were implantation doses of 2 and $10 \times 10^{14}/cm^2$; ion energies of 50 and 100 keV; implant temperatures of 23 and 100 °C. Postimplant anneals were performed in a sealed ampoule at 400 and 450 °C for 2 h in a Hg atmosphere. To compare the results with arsenic diffusion from a grown source, a double layer structure was grown in which the top As doped layer had a larger bandgap ($x \simeq 0.3$).[4]

Various layers with different EPD were investigated, and the implant/anneal experiments were repeated between 1 and 7 times on each sample to assess the reproducibility of As redistribution. For each studied case, 3–5 profiles were taken on the sample. Two layers with a difference of an order of magnitude in EPD ($2.6 \times 10^6/cm^2$ and $2.7 \times 10^7/cm^2$) were studied in detail for As redistribution. The profiles presented in this paper are typical.

The experimental technique used throughout this study was secondary ion mass spectroscopy (SIMS) to monitor the depth distribution of the As-atom concentration. This work was carried out at Charles Evans and Assoc. Corp. in a Cameca IMS-3F machine. The C_s^+ primary ion current was 300 mA, scanned over an area of 500 μm or 250 μm square. The craters obtained in this process had sharp edges and a flat bottom. The analyzed area was restricted to diameters of 150 or 85 μm. The detection limit and/or instrumental background was between 5×10^{14} and $1 \times 10^{15}/cm^3$.

Transmission electron microscopy (TEM) was used to analyze the defect structure of the starting material and of the radiation induced damage in plan view prepared samples from the as grown and from the implanted/annealed layers, respectively. Wedge-shaped samples for TEM experiments were prepared by chemical-jet thinning from the silicon substrate followed by an iodine-ion milling.

Electron beam induced current (EBIC) analysis was used to locate the electrical junction and to correlate it with the As-depth profile. A chemical etch[6] was used to reveal the etch pit density (EPD).

III. ARSENIC DIFFUSION FROM AN ION IMPLANTED SOURCE

A. Observations

The typical redistribution of As from an implanted source ($2 \times 10^{14}/cm^2$, 100 keV) during the postimplant thermal anneal (450 °C/2 h + 250 °C/20 h) of MOCVD-HgCdTe/GaAs/Si devices is shown in Fig. 1. The SIMS instrumental background was $7 \times 10^{14}/cm^3$. The main experimental observations are:

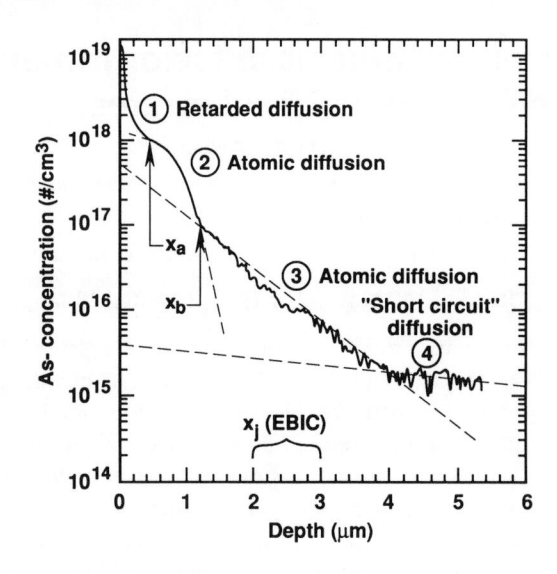

FIG. 1. Typical As redistribution profile from an ion implanted source in a typical MOCVD-HgCdTe/Si material (EPD = $2.7 \times 10^7/cm^2$)—full line; proposed model for As diffusion from an ion implanted source: (1) retarded diffusion; (2) atomic diffusion—As starts on Te sublattice; (3) atomic diffusion—As starts on metal sublattice; and (4) short circuit diffusion. Implant/anneal: $2 \times 10^{14}/cm^2$, 100 keV; 450 °C/2 h + 250/20 h.

(1) The arsenic redistribution from the ion implanted source typically has a multicomponent behavior in which the tail component varies from sample to sample.

(2) The multicomponent distribution differs depending on dislocation density in the material. This is illustrated in Figs. 2(a)–2(c) for material with various EPD. In materials with relatively lower EPD ($\sim 2.6 \times 10^6/cm^2$) no tail component was observed down to the SIMS detection limit [Fig. 2(a)].

(3) The dominant electrical activity of the As tail component is p type as deduced from junction location determined by EBIC analysis.

B. Proposed model

The multicomponent behavior of the typical depth profile of the As concentration is indicative of a complex mechanism by which As diffuses from the implanted source (Fig. 1). We propose a model that explains the nature of these components, based on four diffusion mechanisms. In the shallow region close to the surface, retarded diffusion occurs (component 1). Deeper in the material atomic diffusion is dominant. As diffusion proceeds on two independent networks: As starts either on the anion (Te) sublattice (component 2); or it starts on the cation (metal) sublattice where enhanced atomic diffusion occurs, forming an extended tail (component 3). This component is associated with the radiation induced damage of ion implantation as well as with the defect density in the starting material. And, finally, a fast diffusing tail (component 4) is due to a short circuit diffusion mechanism. In the following sections we discuss each diffusion component separately.

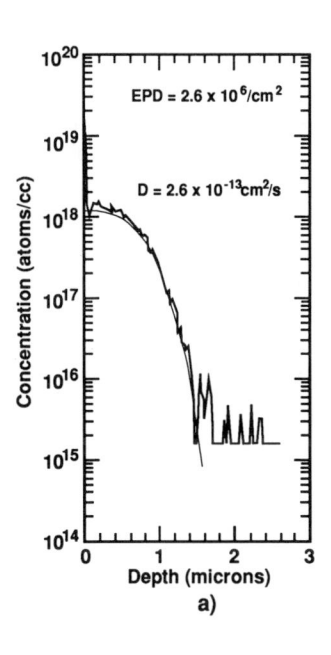

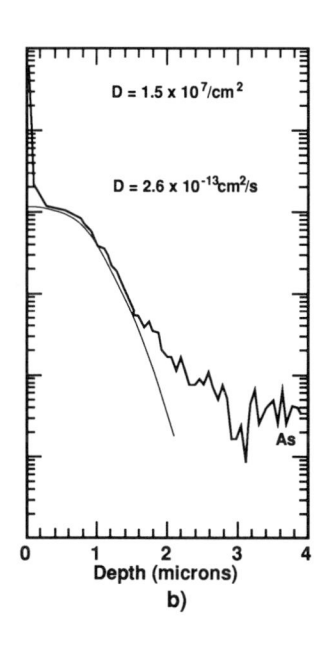

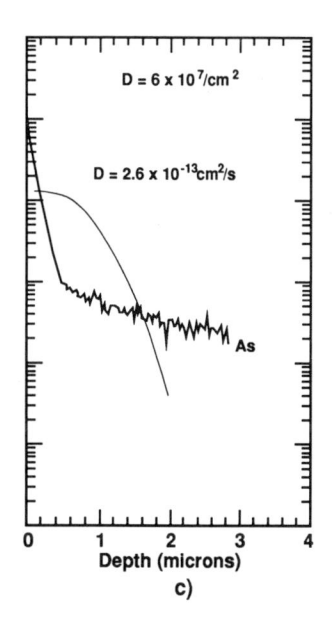

FIG. 2. Typical As redistribution profiles from an ion implanted source in materials with various EPD: (a) EPD $= 2.6\times10^6/\text{cm}^2$; (b) EPD $= 1.5\times10^7/\text{cm}^2$; and (c) EPD $= 6\times10^7/\text{cm}^2$. A Gaussian curve with $D = 2.6\times10^{-13}$ cm^2/s is plotted as a visual aid to show the multicomponent nature of As redistribution.

C. Results and discussion

1. The arsenic diffusion source

The region of lattice disorder induced by the irradiation process of ion implantation is the source for all the migrating species during the postimplant anneal. The ion implantation in HgCdTe material forms a disordered region with high defect concentration which, however, retains the crystalline structure.[7] The amount of the radiation damage (surface damage) and the specific nature of the damage depend on ion species, energy, dose, and implant temperature. The chemical composition and the defect structure of the starting material also play a role. The dominant type of disorder consists of extended defects. In addition, a large number of point defects result, e.g., interstitials consisting of "knock-out" atoms, vacancies, and antisites (e.g., Te$_{\text{Hg}}$).[8,9] The implanted atoms are introduced into the lattice at various lattice sites. Complex types of defects also result. After undergoing free migration some of the point defects coalesce and form a large number of dislocation loops of vacancy[10] and interstitial[11] types; some of the interstitials become attached to or associated with vacancy complexes.[10] The work of various groups has shown that in HgCdTe the depth of disordered region is considerably greater than the projected range of the implanted ions.[7,10,12,13] A known effect of thermal annealing is complete or partial dissociation of dislocation loops, resulting in the release of point defects and perhaps complexes, and their migration into the material.[7,10,14] Thus, study of the dynamics of implanted arsenic and of the nature of various mechanisms requires consideration of irradiation-induced point defects, especially vacancies and interstitials, and of extended defects, as well as of the defect structure of the starting material.

2. Component (1)

Component (1) of the As profile peaks at the surface and decreases rapidly in depth. The extent of this component is defined by its intersection with the deeper Gaussian-like component (2) (x_a in Fig. 1). We propose that the As distribution in the near-surface region occurs due to a retarded diffusion mechanism operating in the radiation damage region of the sample. One possible explanation for the retardation effect is the complexing of diffusing of As atoms with the disorder in this region during the annealing process (e.g., trapping in the extended defects). The increase in the total amount of radiation damage has been shown to be associated with the deeper extent of the damage in the as-implanted material.[12] In the postimplant anneal process a migration of damage occurs at distances depending on their amount. A measure of the remaining disorder in the near-surface region after annealing is the depth of this region as determined from the SIMS As profile. This is supported by TEM analysis of the depth distribution of the radiation induced damage. The plan-view images taken at various depths showed that the radiation induced dislocation loops extended to about 0.3 μm. The image showing the disappearance of the damage at this depth is presented in Fig. 3(a). This depth agrees, with a good approximation, with the extent of component (1) in the As SIMS profile. Furthermore, a sample with relatively low EPD ($2.6\times10^6/\text{cm}^2$), in which the radiation damage annealed completely, as shown in Fig. 3(b) by the plan-view image taken at ~ 0.1 μm, showed no component (1) in the corresponding As profile [Fig. 2(a)]. The dependence of component (1) on the EPD of the starting material and on implant/anneal conditions is shown in Table I. In each portion of this table the depth of surface damage is indicated for different EPDs, ion implantation conditions or postimplant anneal conditions. Of all these variables, the one that shows a strong correlation with the extent of surface damage is EPD. As the EPD decreases from $6\times10^7/\text{cm}^2$ to $2.6\times10^6/\text{cm}^2$ the extent of component (1) decreases from 0.45 to 0.05 μm. Because of the interaction of irradiation process of ion implantation with the defect structure of the material[15] the surface radiation damage is enhanced by dislocations in the starting material.

The As of this component, being complexed with the damage, should be electrically inactive. This interpretation is

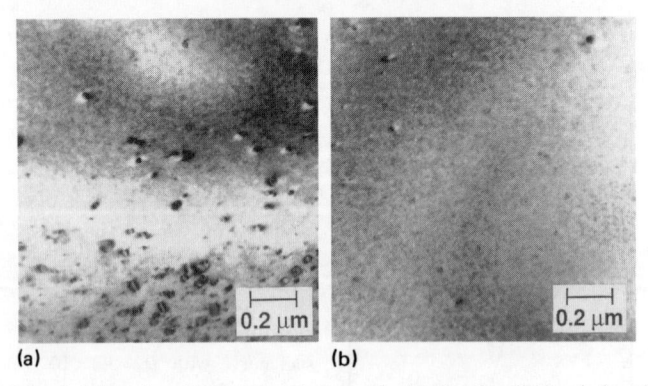

FIG. 3. Plan view TEM microphotograph showing irradiation induced damage in MOCVD-HgCdTe/Si ion implanted and annealed materials (2×10^{14}/cm^2, 100 keV; 450 °C/2 h + 250 °C/20 h) with various EPD: (a) EPD = 2.7×10^7/cm^2, depth = 0.3 µm and (b) EPD = 2.6×10^6/cm^2, depth = 0.1 µm.

suggested by the differential Hall measurement performed on As-implanted/annealed samples from LPE grown material.[16]

3. Component (2)

Deeper in the material a second As-diffusion component with a Gaussian-like distribution can be identified (component 2—Fig. 1). The mechanism we propose for this component is atomic vacancy-based diffusion in which As starts on the Te sublattice in the surface damage region. Because of a large number of Hg interstitials generated in this region as a result of knock-out Hg atoms in the irradiation process of ion implantation,[14] the conditions under which As is introduced in the lattice approximates a "Hg-rich" condition, resulting in a larger fraction of As incorporation on the Te sublattice versus the Hg sublattice.

We propose for this component an atomic diffusion at defect equilibrium (i.e., uniform native defect concentration), and a constant diffusion coefficient (independent on position, time, defect concentration, and on impurity concentration). The As diffusion in this component can be described by the Fick's law with a constant diffusion coefficient D. The solution to Fick's equation with a limited source as the boundary conditions is a Gaussian distribution.[17] The following experiments and theoretical calculations confirm this assumption.

Typical samples with EPD $\sim 10^7$/cm^2 were studied. Examples of SIMS As-redistribution profiles are presented in Figs. 4(a) and 4(b) annealed at 400 and 450 °C respectively. The Gaussian solution with diffusion coefficients of $D = 2.8\times10^{-14}$ cm^2/s for 400 °C and $D = 1.9\times10^{-13}$ cm^2/s for 450 °C provided good fit to the experimental data up to the depth of ~ 1.0 µm for 400 °C [Fig. 4(a)] and ~ 1.5 µm for 450 °C [Fig. 4(b)] when the tail component becomes dominant.

Since the EPD of the material is one of the most sensitive parameters in affecting the radiation damage, as discussed

TABLE I. Dependence of As concentration in the tail component (3) on the surface damage when EPD of the material and implant/anneal conditions are varied.

Dislocation density (EPD) As, 2×10^{14}/cm^2, 100 keV; 450 °C/2 h + 250 °C/20 h					Ion implantation conditions EPD = 2.7×10^7/cm^3			Post-implant anneal conditions EPD = 2.7×10^7/cm^3		
EPD (cm^{-2})	2.6×10^6	2×10^7	2.7×10^7	6×10^7	Dose (cm^{-2})	2×10^{14}	1×10^{15}	Tann (°C)	400	450
Sf. damage depth (µm)	0.05	0.27	0.15	0.45	Sf. damage depth (µm)	0.15	0.7	Sf. damage depth (µm)	0.15	0.27
As-conc (cm^{-3})	1×10^{15}	5×10^{16}	1×10^{17}	1×10^{18}	As-conc (cm^{-3})	1×10^{17}	2×10^{18}	As-conc (cm^{-3})	3×10^{16}	3×10^{17}
					Energy keV	50	100	Δ Tann $(T_s - T_{Hg})$ (°C)	0	30
					Sf. damage depth (µm)	0.15	0.27	Sf. damage depth (µm)	0.05	0.27
					As-conc (cm^{-3})	1×10^{17}	3×10^{17}	As-conc (cm^{-3})	4×10^{16}	3×10^{17}
					T_{impl} (°C)	23	100			
					Sf. damage depth (µm)	0.27	0.1			
					As-conc depth (cm^{-3})	3×10^{17}	1×10^{16}			

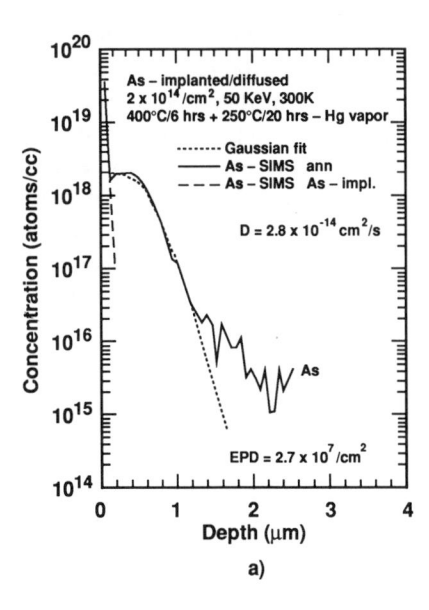

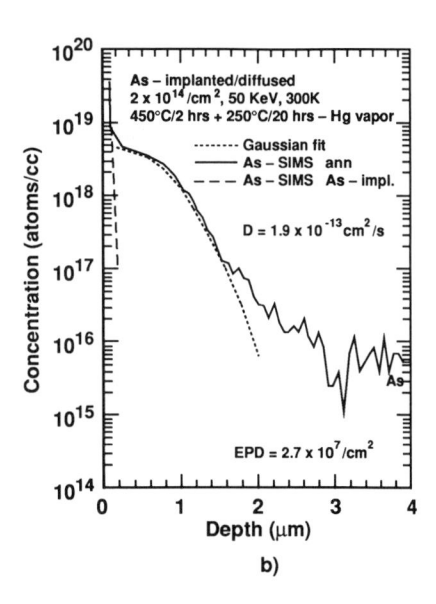

FIG. 4. Examples of SIMS As diffusion profiles together with theoretical Gaussian model for a typical EPD ($2.7 \times 10^7/\text{cm}^2$) obtained at two different temperatures: (a) 400 °C/6 h + 250 °C/20 h and (b) 450 °C/2 h + 250 °C/20 h. Ion implantation: $2 \times 10^{14}/\text{cm}^2$, 100 keV.

for component (1), we investigated the diffusion coefficient of component (2), in a material with low EPD ($2.6 \times 10^6/\text{cm}^2$) and compared it with the typical case of $10^7/\text{cm}^2$ EPDs. Examples of the As profiles after 400 and 450 °C anneals for 2 h for the low EPD sample are shown in Figs. 5(a) and 5(b). The arsenic concentration profile exhibits mainly the Gaussian component down to SIMS detection limit. The diffusion coefficients calculated from the theoretical Gaussian model which fits the experimental data down to the SIMS detection limit, were $D = 1.8 \times 10^{-14}$ cm^2/s for 400 °C and $D = 2.0 \times 10^{-13}$ cm^2/s for 450 °C. These values are very close to those obtained with higher EPD material, and indicate that the diffusion coefficient is independent of the EPD at both temperatures investigated.

A further confirmation of the classical diffusion of this component was obtained from a detailed experiment on MOCVD HgCdTe in which the time of a postimplant anneal at 450 °C was varied. By fitting the SIMS profiles with the theoretical Gaussian model for each anneal time, we obtained the diffusion length and plotted it as a function of the square root of time (Fig. 6). The linear dependence observed confirms the Gaussian redistribution of arsenic from the ion implanted source.

This classical component is always present. The depth in which it is observed depends on the magnitude of the other components as shown in Fig. 2.

The calculated value of diffusion coefficient is in agreement with the range of values presented in the literature at 400 °C; Falconer et al.[18] showed values of $(8 \pm 3) \, 10^{-15}$ cm^2/s for LPE grown material when arsenic was diffused from an external source in Hg vapor at 400 °C, and Maxey et al.[4] showed $D = (1-20) \times 10^{-14}$ cm^2/s in MOCVD-HgCdTe material when arsenic was diffused from a grown source at 410 °C.

The excellent reproducibility of the diffusion coefficient on different samples is evidence that this diffusion mechanism is a true representation of the bulk diffusion. This component is p-type electrically active because it starts and diffuses in Te sublattice.[1] This is the most desirable mechanism for a controlled junction formation process.

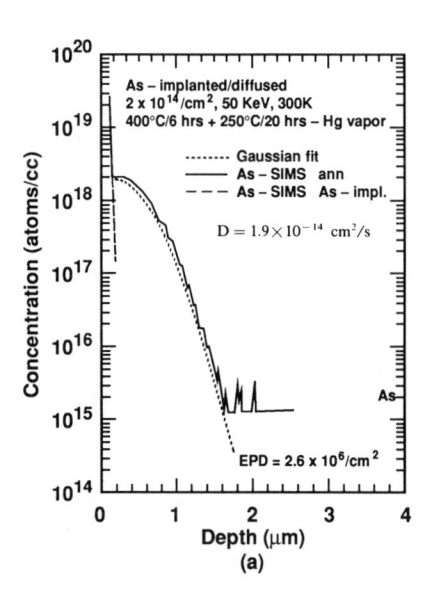

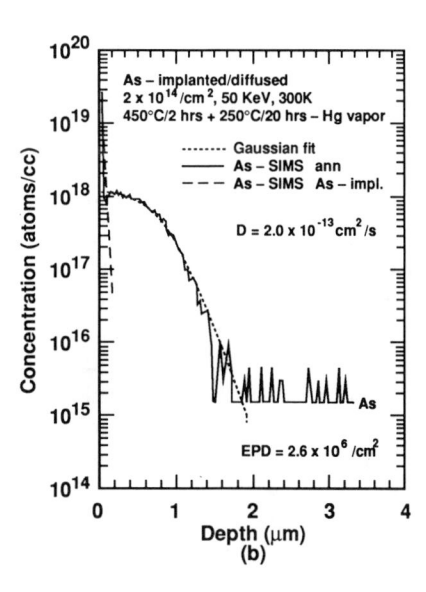

FIG. 5. Examples of SIMS As diffusion profiles together with theoretical Gaussian model for a material with low EPD ($2.6 \times 10^6/\text{cm}^2$) obtained at two different temperatures in Hg vapor: (a) 400 °C/6 h + 250 °C/20 h and (b) 450 °C/2 h + 250 °C/20 h. Ion implantation: $2 \times 10^{14}/\text{cm}^2$, 50 keV.

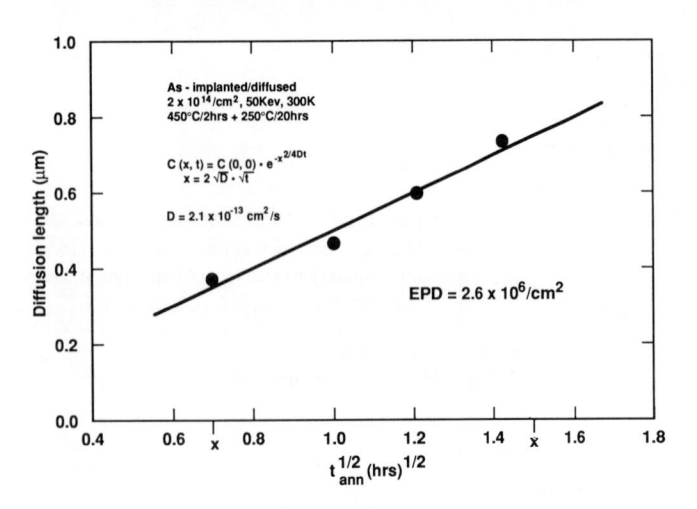

FIG. 6. Diffusion length of As vs square root of annealing time. Implant/anneal: $2\times10^{14}/cm^2$, 50 keV; 450 °C + 250 °C.

4. Component (3)

In ion implantation process a fraction of As atoms can also be introduced on metal sublattice sites. As on the metal sublattice diffuses by an enhanced vacancy-based mechanism, which is possible as a consequence of a departure from the defect equilibrium within the atomic diffusion zone. Among all the radiation induced vacancies, Hg vacancies have the highest diffusion rate[19] and these vacancies we will consider in the model. Assuming that the starting material was in a defect equilibrium state at the conditions of the thermal anneal, and, moreover, the defect diffusion was faster than the As-atomic diffusion, a constant defect population has still not been established throughout the atomic diffusion zone during the postimplant thermal anneal. (The rate at which defect equilibrium is established[7] is $D_{200\,°C} = 2\times10^{-11}$ cm^2/s, orders of magnitude higher than As diffusion coefficient at the anneal temperatures.) This is because during the restructuring process in the anneal, the dislocation loops in the surface damage region dissociate to a sufficient rate to maintain a continuous metal vacancy flux to the diffusion zone and to cause a gradient of vacancy concentration. Since vacancies enhance the substitutional diffusion, the atomic diffusion in component (3) (Fig. 1) is affected by the gradient of vacancy distribution and thus the resulted diffusion coefficient will be dependent on the position in the respective diffusion zone.

In this proposed mechanism the diffusion process will be affected by the same parameters that affect the radiation induced damage (surface damage). Thus, we have conducted extensive experiments to study the dependence of component (3) on the surface damage. The approach we have taken was to deliberately alter the surface damage by varying the parameters known to affect this region; thus we varied EPD, and ion implant and postimplant anneal conditions. We then observed the changes in the surface damage monitoring the extent of component (1) and the corresponding response in the component (3) at its intersection with the Gaussian component (2) (x_b in Fig. 1). The results from these experiments are compiled in Table I.

a. Component (3) dependence on EPD. Samples for these experiments were chosen to have a large spread in EPD (from 2.6×10^6 to $6\times10^7/cm^2$). The data in the left part of Table I clearly show that an increase in EPD coupled with an increase in the depth of surface damage causes an increase in arsenic concentration in component (3). Thus, an increase of about one order of magnitude in EPD (from 2.6×10^6 to $2.7\times10^7/cm^2$) causes an increase in surface damage depth by a factor of 9 (from 0.05 to 0.45 μm) and an increase of about 2 orders of magnitude in the arsenic concentration (from 10^{15} to $10^{17}/cm^3$). At the highest EPD [Fig. 2(c)] the component (1) dominates.

b. Component (3) dependence on implant/anneal parameters. Material with typical EPD ($2.7\times10^7/cm^2$) was studied for a correlation between surface damage and arsenic concentration in component (3) by varying independently each implantation and postimplant anneal parameter. By decreasing the implantation dose the amount and extent of surface damage increases.[12,20] When increasing the dose in these experiments from $2\times10^{14}/cm^2$ to $1\times10^{15}/cm^2$ (central part of Table I) the extent of surface damage increases from 0.15 to 0.7 μm and the arsenic concentration in component (3) increases also from $1\times10^{17}/cm^3$ to $2\times10^{18}/cm^3$. Increasing implant energy causes an increase of the total damage, with a corresponding shift of the maximum damage deeper into the semiconductor.[12] Varying the energy from 50 to 100 keV in these experiments caused an increase in surface damage depth from 0.15 to 0.27 and in the As concentration of component (3) from $1\times10^{17}/cm^3$ to $3\times10^{17}/cm^3$, as shown in the central part of Table I. Another parameter used to alter the radiation induced damage was the implant temperature. The surface damage decreased at higher implant temperature due to some annealing effects during the implantation. In our experiments we changed the implant temperature from room temperature to 100 °C. As a result the surface damage extent decreased from 0.27 to 0.1 μm and the arsenic concentration in the tail decreased from $3\times10^{17}/cm^3$ to $1\times10^{16}/cm^3$ as seen in the center part of Table I.

Modification of surface damage can also be accomplished by varying the postimplant annealing conditions. In one experiment we varied the postimplant annealing temperature from 400 to 450 °C for 2 h. In another experiment we changed the difference between sample and Hg temperatures in the anneal ($\Delta T = T_s - T_{Hg}$) from 0 to 30 °C. The results showed that when the damage extent increases, the As concentration in component (3) increases in both cases by about one order of magnitude from $3-4\times10^{16}/cm^3$ to $3\times10^{17}/cm^3$ (right-hand side of Table I).

The results of these experiments support the proposed model in that the dependence of component (3) on the defect structure of the material and on the radiation induced damage is shown. To minimize component (3) and still have freedom to choose the appropriate annealing regime for forming the junction on the controllable Te-sublattice component, the principal consideration is to reduce the dislocation density in the material [Fig. 2(a)]. Other parameters should be optimized toward lower dose (as low as possible), lower energy, higher implant temperature, and lower tem-

The above described component (3) was observed to control the electrical junction of photovoltaic devices by determining the position of the junction in the material. This component is electrically complex, consisting of n- and p-type active As, and perhaps neutral As as well.[2] A site transfer with the associated change in electrical activity is possible, depending on subsequent thermal treatment and phase equilibria. To control the junction performance, component (3) should be minimized. The principal factor is EPD. For a EPD = low $10^6/cm^2$ in this material, component (3) was minimal.

5. Component (4)

For the deepest component of As diffusion, [component (4) in Fig. 1] we propose a short circuit mechanism due to the presence of isolated and nested dislocations in the starting material.[21-23] A typical example of dislocations revealed by chemical etching is shown in Fig. 7(a) by an SEM microphotograph. The area of the pits in this example varies from a $\sim 0.1^2$ to $\sim 1.0^2$ μm^2, suggesting that they reveal both isolated and nested dislocations. The distribution of both is random. A detailed image of dislocations taken by TEM on a similar sample is shown in Fig. 7(b). A rich substructure observable in this TEM microphotograph, in an area similar to the large pit in Fig. 7(a), suggests that the large pits indeed result from nested or clustered dislocations. This type of density and distribution of dislocations can result in paths for short circuit diffusion.

We have conducted experiments to determine when the influence of short circuit paths on the diffusion profile becomes apparent. Thus, parameters such as EPD, implant and postimplant anneal conditions have been changed successively. The results (Table II) showed that for the same range of EPD as in the component (3) study, the As concentration in component (4) varies by about 1 order of magnitude (from $< 10^{15}$ to $1 \times 10^{16}/cm^2$). For a EPD

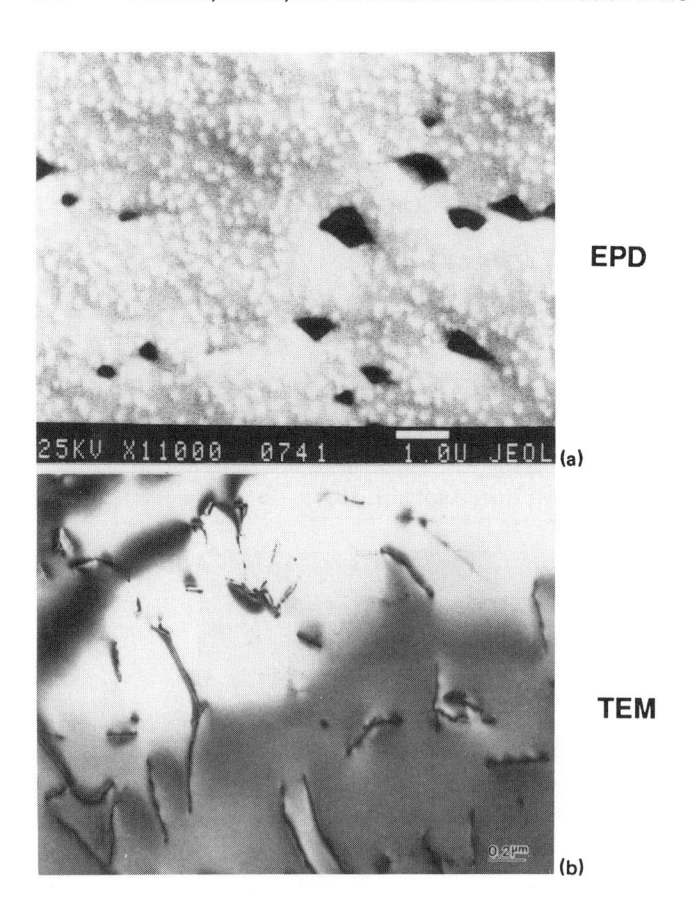

FIG. 7. Examples of dislocation revealing in a material with typical EPD, $2.7 \times 10^{17}/cm^2$, by (a) etch pitting technique (EP)—detailed SEM microphotograph and (b) plan-view TEM image. Isolated dislocations and nestings are observed that account for multicomponent As diffusion.

perature (and possibly time) in annealing. The data above show a relationship between component (3) and the EPD. However, the details of the dislocation distribution may still affect the result.

TABLE II. Dependence of As concentration in the tail component (4) on the surface damage when EPD of the material and implant/anneal conditions are varied.

Dislocation density (EPD) As, $2 \times 10^{14}/cm^2$, 100 keV: 450 °C/2 h + 250°C/20 h				Ion implantation conditions EPD = $2.7 \times 10^7/cm^3$			Postimplant anneal conditions EPD = $2.7 \times 10^7/cm^3$		
EPD (cm^{-2})	2.6×10^6	2.6×10^7	2.6×10^7	Dose (cm^{-2})	2×10^{14}	1×10^{15}	Tann (°C)	400	450
As-conc (cm^{-3})	$< 10^{15}$	4×10^{15}	1×10^{16}	As-conc (cm^{-3})	4×10^{15}	1×10^{16}	As-conc (cm^{-3})	3×10^{15}	1×10^{16}
				Energy (keV)	50	100	Δ Tann $(T_s - T_{Hg})$ (°C)	0	30
				As-conc (cm^{-3})	3×10^{15}	1×10^{16}	As-conc (cm^{-3})	2×10^{15}	4×10^{15}
				T_{impl} (°C)	23	100			
				As-conc (cm^{-3})	1×10^{16}	2×10^{15}			

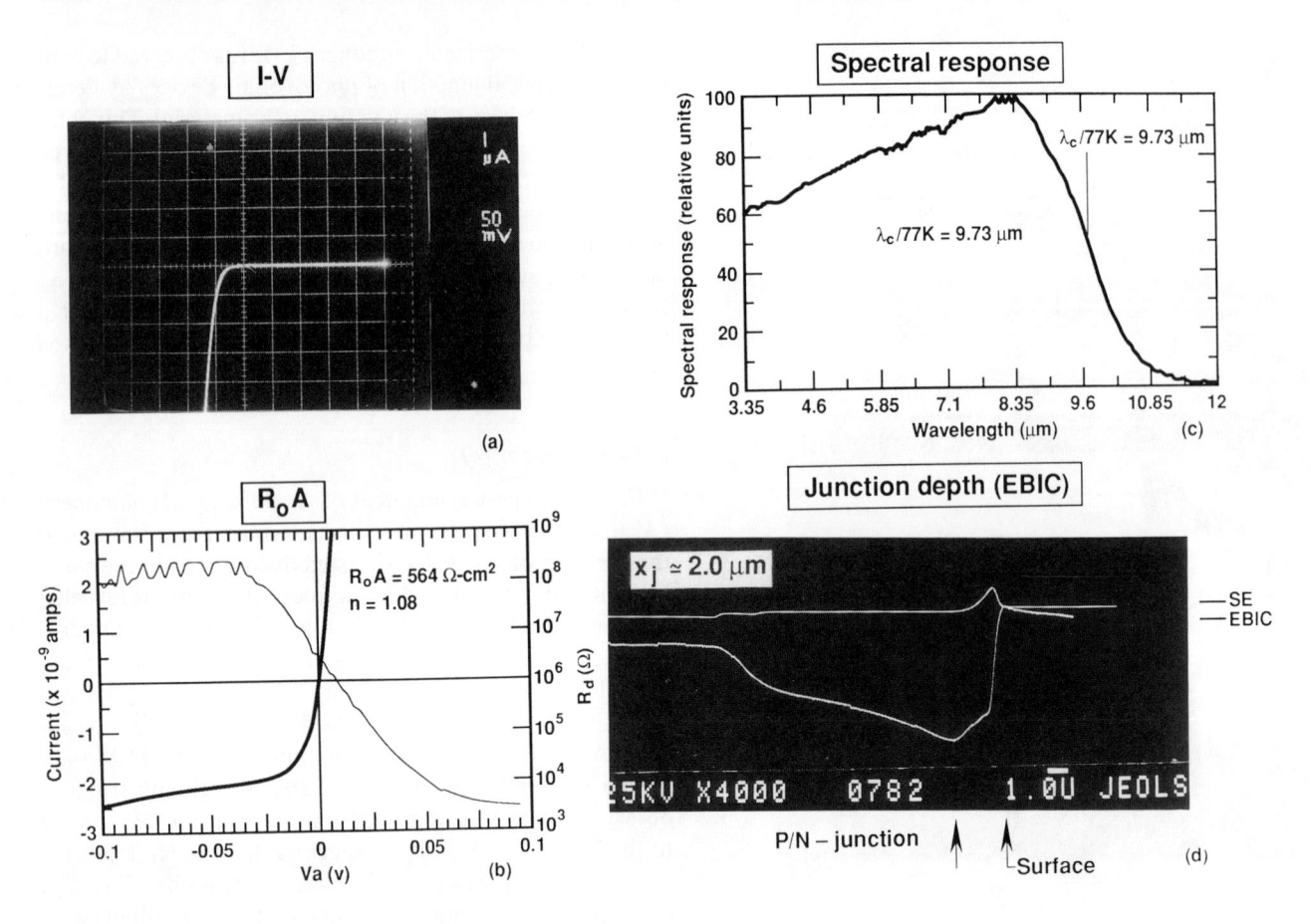

FIG. 8. Performance at 77 K of a *p*-on-*n* device obtained by As implantation/diffusion process in MOCVD HgCdTe/Si: (a) curve tracer *I-V*; (b) detailed *I-V* and differential resistance; (c) junction depth (EBIC); and (d) spectral response.

$= 2.6 \times 10^6/\text{cm}^2$, the As tail was not observed ($< 10^{15}/\text{cm}^3$). The implant/anneal conditions seem to affect the number of As atoms that diffuse by this mechanism. The specifics of the thermal anneal determine the electrical activity of As in this component. We have observed *p*-type activation [by a site transfer mechanism[1] as in component (3)] causing electrical shorts or leakage paths in the junction region. The most effective way to minimize component (4) is to reduce the EPD of the material. Proper choice of implant/anneal, as discussed in component (3), is also helpful.

6. Device results

High performance was achieved with *p*-on-*n* LWIR photovoltaic devices on MOCVD-HgCdTe material grown on Si substrates by an As-implant/diffusion process. An example of a good device is shown in Fig. 8. The curve tracer [Fig. 8(a)] and the detailed *I-V* at 77 K, together with the differential resistance [Fig. 8(b)] indicated a high performance device (zero-bias resistance multiplied to the area, $R_0A = 546 \, \Omega \, \text{cm}^2$) operating in a diffusion limited regime, with an ideality factor determined from forward characteristics, $n = 1.08$. The cutoff wavelength for this device was $\lambda_c = 9.73 \, \mu\text{m}$ at 77 K [Fig. 8(c)]. The junctions in these devices were located at depths between 2.0 and 3.0 μm as shown by the EBIC data in Fig. 8(d).

IV. ARSENIC DIFFUSION FROM A GROWN SOURCE

We have evidence that As diffusion from grown and implanted sources behaves similarly. An example of the As redistribution profile from a grown source after a heat treatment at 400 °C (similar to a postimplant anneal) is shown in Fig. 9. The source for this diffusion is an As doped layer grown *in situ* on a smaller bandgap *n*-type layer. As was introduced in the CdTe cycles, under Cd-rich conditions, to enhance As incorporation on the Te-sublattice sites.

As in the ion implantation case, the As distribution exhibits a multicomponent behavior. The shallowest component, component (1), is caused by a deposition of residual alkyls in the reactor at the end of the growth and a gettering mechanism in the surface defect layer. This defect layer can be seen in the SIMS measurements when the As profile is taken in conditions that expand the surface region. Component (2) occurs as a result of an atomic diffusion in which As starts on the Te sublattice. The diffusion coefficient in this case, where the source of As is a thick film, is concentration dependent. By using Boltzmann–Matano analysis,[9,24] we determined the value as $D_{400\,°C} \simeq 5 \times 10^{-12} \, \text{cm}^2/\text{s}$. The behavior of component (3) varies from sample to sample. Similar to ion implantation, we propose an atomic diffusion process in which As started on metal sublattice. This component originates in the surface defect region and/or in the As doped layer due to

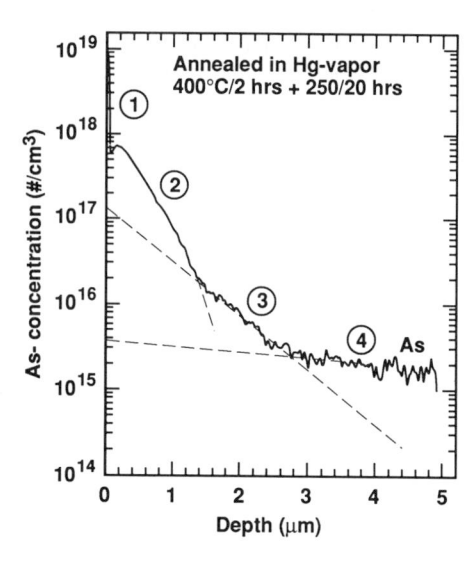

FIG. 9. Proposed model for As diffusion from a grown source: (1) gettering; (2) atomic diffusion—As start on Te sublattice; (3) atomic diffusion—As starts on metal sublattice; and (4) short circuit diffusion. Postgrowth anneal at 400 °C/2 h + 250 °C/20 h in Hg vapor.

the fraction of As atoms not incorporated on Te sublattice which occupy the metal-sublattice sites. For the fast diffusing component (4) we propose a short circuit diffusion mechanism similar to that described in the implantation case. In the grown junctions the growth temperature and time can be lower than those of the postimplant anneal. Consequently, the tailing components can be made smaller for the same EPD.

V. SUMMARY

We have observed a multicomponent diffusion behavior of As from an implanted source. As diffusion tails differ depending on the dislocation density in the material for the same implant-anneal conditions. We propose a model that explains the nature of various components: all the components originate in the radiation damage region; in component (1), As is retarded in the diffusion process within the surface damage region; in component (2), As starts on the Te sublattice and undergoes a classical vacancy-based diffusion process; in component (3), As starts on the metal sublattice and moves by an enhanced vacancy-based diffusion mechanism due to defect nonequilibrium; a gradient in the Hg-vacancy distribution in the diffusion zone is maintained by a continuous flux of Hg-vacancies freed in the annealing process; in component (4), As diffusion is further enhanced due to isolated and nested dislocations that provide paths for short circuit diffusion.

The Te-sublattice component (component 2) is reproducible from sample to sample and therefore is representative of the true volume diffusion mechanism. Diffusion coefficients of $D_{400\,°C} = 3 \times 10^{-14}$ cm^2/s and $D_{450\,°C} = 2 \times 10^{-13}$ cm^2/s were calculated from the data. The electrical activity of As in this component is p type. The metal-sublattice component [component (3)] is characterized by a position-dependent diffusion coefficient and is dependent on the defect structure of the material (dislocation density). Arsenic in this component should behave as n type; however, at the annealing temperatures used for As implants it can activate to p type by a site transfer mechanism. As in this component can also be electrically inactive if an As-acceptor complexes with an As donor. The short circuit component (4) is mainly EPD dependent. Implant/anneal conditions also affect its behavior. The electrical activity of As in component (4) can be similar to that in component (3).

The experimental data support the proposed model: the extent of the retarded diffusion (component 1) is within the surface damage region (TEM results); the Te-sublattice component (component 2) is very reproducible from sample to sample and has a Gaussian behavior; the metal-sublattice component [component (3)] is dependent on the amount of surface damage and on the EPD of the starting material, supporting the proposal that it originates in the radiation damage region; high EPD and presence of dislocation nestings (EP, TEM results) seem to be responsible for the short circuit diffusion, and implant/anneal conditions seem to affect the number of As atoms that diffuse by this mechanism. Vital to a controllable As-diffusion mechanism, and therefore to controllable junction formation, are the dislocation density and its distribution in the material. Implant/anneal conditions are also important. As diffusion from a grown source behaves similarly to As diffusion from an ion implanted source in equivalent thermal treatment conditions.

ACKNOWLEDGMENTS

Two of the authors (L. O. B. and D. D. E.) thank Dr. W. E. Tennant and E. R. Gertner for their support, advice and criticism; and Dr. Gayle Lux of Charles Evans and Associates who worked with us in acquiring SIMS data and whose expertise was valuable for the SIMS analysis.

[1] H. R. Vydyanath, A. Ellsworth, and C. M. Devaney, J. Electron. Mater. **16** (1), 13 (1987).
[2] H. R. Vydyanath (private communication).
[3] L. O. Bubulac, D. D. Edwall, D. McConnell, R. E. DeWames, E. R. Blazejewski, and E. R. Gertner, Semicond. Sci. Technol. **5**, 545 (1990).
[4] C. D. Maxey, P. Capper, P. A. C. Whiffin, B. C. Easton, I. Gale, and J. B. Clegg, J. Harker, and C. L. Jones, J. Cryst. Growth, **101**, (1-4), 300 (1990).
[5] D. D. Edwall, J. Bajaj, and E. R. Gertner, J. Vac. Sci. Technol. A **8**, 1045 (1990).
[6] J. S. Chen, United States Patent No. 4897152.
[7] G. L. Destefaris, J. Cryst. Growth **86** 700, (1988).
[8] L. O. Bubulac, W. E. Tennant, S. H. Shin, C. C. Wang, M. Lanir, E. R. Gertner, and E. D. Marshall, Jpn. J. Appl. Phys. **19**, 495 (1980).
[9] S. M. Sze, *VLSI Technology* (McGraw-Hill, New York, 1983).
[10] L. O. Bubulac, W. E. Tennant, R. A. Riedel, and T. J. Magee, J. Vac. Sci. Technol. **21**, 251 (1982).
[11] H. F. Schaake, J. Vacuum Sci. Technol. A **4**, 2174 (1986).
[12] Yu V. Lilenco, K. V. Shastov, A. S. Petrov, A. V. Voitsekhovskii, V. S. Kulikauskas, N. V. Kuznetsov, and A. P. Mamontov, Phys. Status Solidi A **113**, 285 (1989).
[13] G. Bahir and R. Kalish, J. Appl. Phys. **54**, 313 (1983).
[14] L. O. Bubulac and W. E. Tennant, Appl. Phys. Lett. **51**, 355 (1987).
[15] L. O. Bubulac, Appl. Phys. Lett. **46**, 976 (1985).
[16] L. O. Bubulac, D. S. Lo, W. E. Tennant, D. D. Edwall, J. C. Chen, and J. Ratusnik, Appl. Phys. Lett. **50**, 1586 (1987).

[17] A. S. Grove, *Physics and Technology of Semiconductor Devices* (Wiley, New York, 1979).
[18] J. E. Falconer and H. D. Palfrey, J. Cryst. Growth **100**, 275 (1990).
[19] J. S. Chen, Ph.D. dissertation, University of Southern California, 1985.
[20] H. Ryssel, G. Lang, J. P. Biersack, and K. Muller, IEEE Trans. Electron. Devices **ED-27**, 58 (1980).
[21] A. D. LeClaire and A. Rabinovich, J. Phys. C. **14**, 3863 (1981).
[22] A. D. LeClaire and A. Rabinovich, J. Phys. C. **16**, 2087 (1983).
[23] D. Shaw, *Atomic Diffusion in Semiconductor* (Plenum, New York, 1973).
[24] D. Shaw, J. Cryst. Growth **86**, 778 (1988).

Extrinsic p-doped HgCdTe grown by direct alloy growth organometallic epitaxy

N. R. Taskar,[a] I. B. Bhat, K. K. Parat, and S. K. Ghandhi
Electrical, Computer and Systems Engineering Department, Rensselaer Polytechnic Institute, Troy, New York 12180

G. J. Scilla
IBM Thomas J. Watson Research Center, Yorktown Heights, New York 10598

(Received 3 October 1990; accepted 17 December 1990)

This paper describes the doping behavior of arsenic in HgCdTe, grown by organometallic epitaxy using the direct alloy growth process. It is shown that arsenic readily incorporates into HgCdTe during this growth process, to a doping concentration of 1×10^{17} cm^{-3}. Secondary-ion mass spectroscopy (SIMS) data clearly establishes the presence of arsenic in these layers. Moreover, excellent stability under annealing conditions (16 h at 270 °C followed by 10 h at 220 °C) indicates the suitability of arsenic as an extrinsic dopant source for HgCdTe. The net acceptor concentration is shown to be linearly proportional to the Hg overpressure. This fact, combined with mobility values which are comparable to those of bulk grown material, indicates that these layers are relatively uncompensated. High mobility values are preserved well into the saturated region of the acceptor concentration versus arsine flow rate characteristic. This is explained by the fact that the doping concentration is limited by surface coverage of the As species, and not by arsenic incorporation into compensating or inactive sites. Evidence for this process is presented by the SIMS data, which shows that the total arsenic concentration is linearly related to the net acceptor concentration over the entire doping range.

I. INTRODUCTION

Mercury cadmium telluride layers, grown by organometallic vapor phase epitaxy (OMVPE), have shown considerable promise for use in far infrared detector applications in recent years.[1–4] Layers with uniformity in composition and thickness, that are required for present day device structures, can be grown by this method.[5,6] The fabrication of stable p-n junctions in these layers requires slow diffusing p- and n-type dopants that will not move during subsequent device processing. Of these, n doping can be carried out with indium as the impurity, and we have shown that this can be achieved readily in OMVPE-grown material, using trimethylindium as the doping source.[7]

HgCdTe layers grown by OMVPE are generally p type due to the presence of Hg vacancies, but these are fast diffusers. Hence, it is necessary to use external impurities to obtain stable p-type layers. Here, Column V dopants are the preferred choice because they have low thermal diffusivity.[8,9] These elements, when substituted into Te sites, will behave p type, but will be donor-like if incorporated on Column II sites. Hence, it is of concern that they may substitute Column II sites and compensate the Te-substituted acceptors. In fact, doping with arsenic (as well as other Column V species) has been unsuccessful in the liquid phase epitaxy of HgCdTe grown from a Te-rich melt, because of its low distribution coefficient and also because some arsenic incorporates in Hg sites, resulting in compensated layers.[10] In molecular beam epitaxy, doping using solid sources of As and Sb has not resulted in p-type layers.[11] However, we have shown that it is possible to dope HgCdTe using arsine,[12] even though growth by OMVPE occurs under Hg-deficient conditions with a high concentration of Hg vacancies (mid 10^{15}–mid 10^{16} cm^3). In this paper, we present new results on the doping behavior of arsenic in HgCdTe grown by OMVPE, using the direct alloy growth (DAG) process, and propose a model for its incorporation.

II. EXPERIMENTAL

HgCdTe layers were grown at 370 °C in an atmospheric pressure, horizontal reactor by the DAG process, involving the simultaneous introduction of elemental mercury, dimethylcadmium (DMCd), and diisopropyltelluride (DIPTe) into the reaction chamber. Substrates were 2° misoriented (100)Cd$_{0.96}$Zn$_{0.04}$Te, which is closely lattice matched to the epilayer. The thickness of Hg$_{1-x}$Cd$_x$Te layers reported here was ~7 μm, and the alloy composition x was in the range 0.27–0.31. A cap of CdTe layer, approximately 1 μm thick, was grown on the HgCdTe as a passivant layer. A 500 ppm mixture of arsine in hydrogen was used for doping purposes.

HgCdTe layers were also grown on GaAs substrates for the purpose of comparison. Here, a 2-μm-thick layer of CdTe was first grown on the GaAs, and served as a buffer to relieve the effects of the 14.6% lattice mismatch between the GaAs and the HgCdTe.

Hall mobility and resistivity measurements were made in cloverleaf-patterned samples with a magnetic field strength of 0.5 to 6 kG, and over the temperature range from 10 to 300 K. Measurement over this temperature range is necessary for p-type layers, in order to determine the presence of any surface inversion.[13] In those cases where a surface inversion layer was present, especially in low-doped annealed layers without a cap, a two-layer model was used to curve fit the experimental points to obtain the true hole concentration.[14]

Secondary-ion mass spectrometry (SIMS) was employed to determining depth distribution profiles. A CAMECA 4f instrument was used in this study. Primary $^{32}O_2^+$ ions, accelerated to 8.5 keV, were used for sputtering the samples, and positive secondary ions were monitored. Depth calibration was carried out by determining sputtering rates from crater depth measurements, made on a Tencor Instruments Alpha Step 200.

III. RESULTS AND DISCUSSION

Figure 1 shows the acceptor concentration measured as a function of arsine flow, for arsenic-doped layers grown lattice matched on CdZnTe substrates. The measured doping concentration is seen to increase from 8×10^{15} to 9×10^{16} cm^{-3} as the arsine flow was increased from 10 to 250 sccm. Increasing the arsine flow beyond 250 sccm did not increase the doping concentration. Also shown is the doping characteristic when GaAs substrates were used, with a 2-μm-thick CdTe buffer layer. Consistently, a factor of 2 to 4 higher doping level was achieved when lattice-matched CdZnTe was used as the substrate. We believe that HgCdTe grown on GaAs has more defects than material grown on lattice-matched substrates, and that As segregation in these defect sites is the main reason for this difference.

Arsenic in HgCdTe was found to be very stable, after low-temperature isothermal annealing in a Hg-rich ambient. Measurements of the Hall coefficient were made as a function of temperature for pairs of samples, grown side by side, during the same run. One was an as-grown layer, while the other was annealed under Hg overpressure at 270 °C for 16 h, followed by 220 °C for 10 h. This temperature should be more than sufficient to convert the entire epitaxial layer to n type[15] if the p doping observed is due to mercury vacancies alone. In each case, as seen in Fig. 2, both samples had almost identical behavior, which suggests that arsenic is stable under this heat treatment process.

Annealing experiments, carried out on GaAs substrates

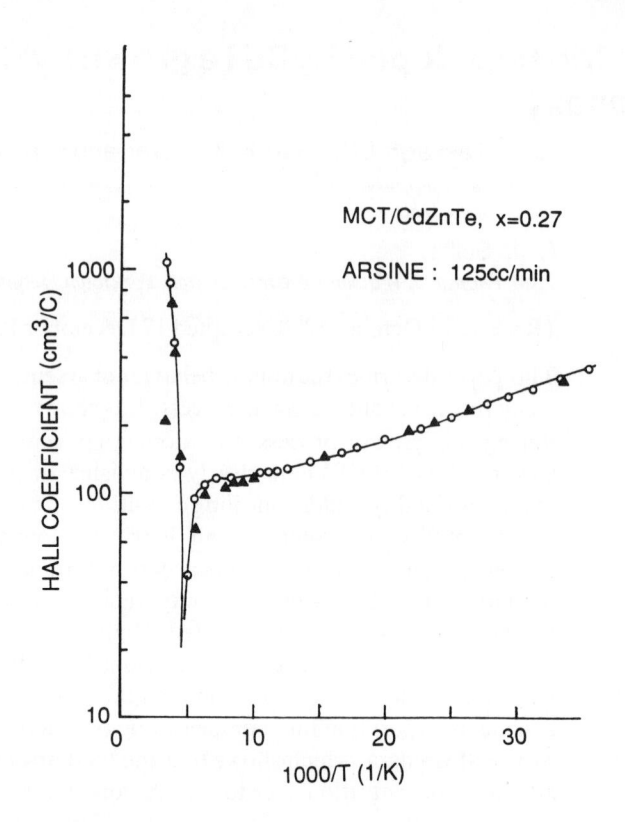

FIG. 2. Hall coefficient for HgCdTe layers (O) as grown, and ($\triangle$) after an isothermal anneal at 270 °C for 16 h, followed by 220 °C for 10 h.

with a CdTe buffer for the conditions given above, gave inconclusive results. In general, annealing resulted in a significant increase in the hole concentration, and varied from sample to sample. We believe that this result is a direct consequence of the fact that significant inactive arsenic is incorporated into defect sites (as seen from Fig. 1), and that some of this arsenic is released during the annealing step. Further studies of this problem are being undertaken at the present time.

SIMS measurements were made in order to establish the presence of As in the layer. Figure 3 shows a typical SIMS profile of a HgCdTe layer grown with 50 sccm of arsine flow, and shows clearly the presence of incorporated arsenic. The hole concentration in this layer, as measured by Hall effect, was 4.5×10^{16} cm^{-3}.

Figure 4 shows the low-temperature (40 K) mobility for layers as a function of the arsine flow. The mobility values shown here are comparable to values obtained in bulk p-type layers with similar doping concentration and composition, and indicate that the layers are relatively uncompensated.

Site incorporation of arsenic in HgCdTe was studied by growing doped layers at different Hg partial pressures (0.008 and 0.03 atm), with all other conditions being held constant. Measurements of the net acceptor concentration were made after these layers were fully annealed under Hg-rich conditions. This was performed in a sealed tube, and consisted of a 270 °C anneal of 16 h, followed by 10 h at 220 °C.

If we assume[16] that $V_{Te} \propto P_{Hg}$ and that $V_{Hg} \propto 1/P_{Hg}$, and that arsenic incorporates into these sites, its relative site oc-

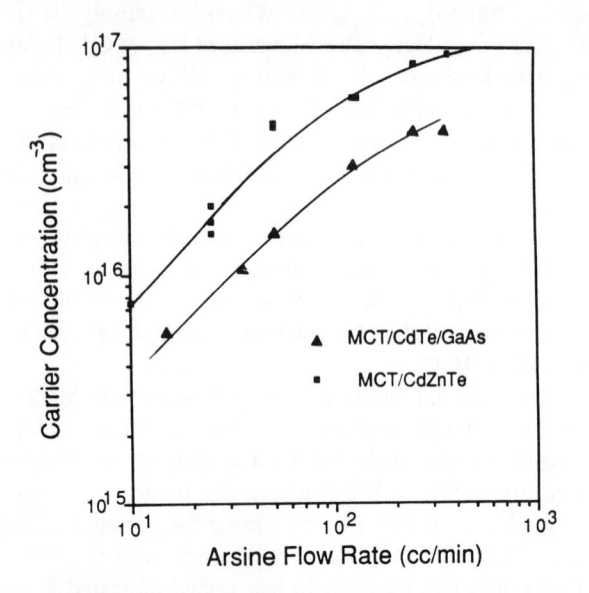

FIG. 1. Net acceptor concentration as a function of arsine flow ( ) CdZnTe substrates ($\triangle$) GaAs substrates ($x = 0.27$–0.31).

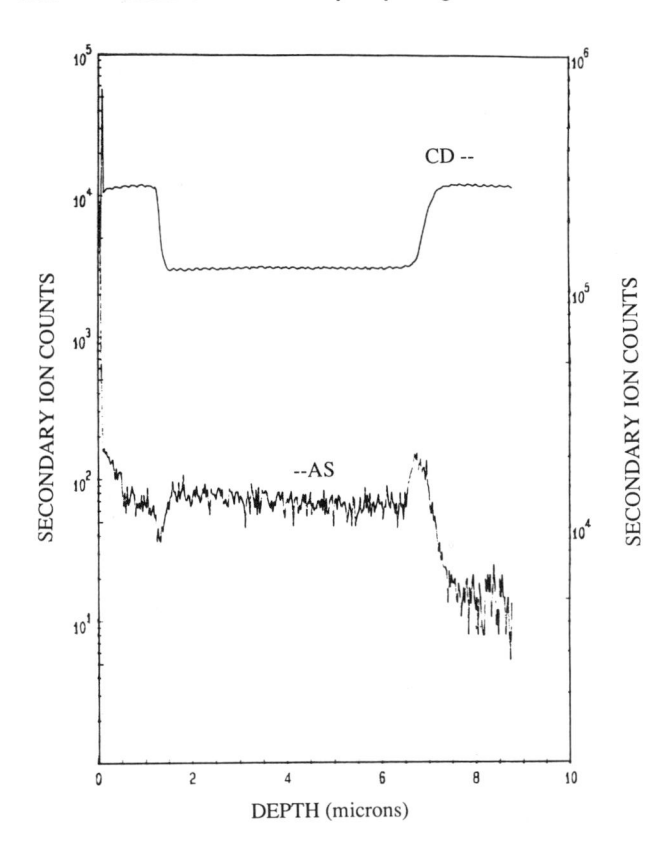

FIG. 3. SIMS profile for arsenic in a HgCdTe layer.

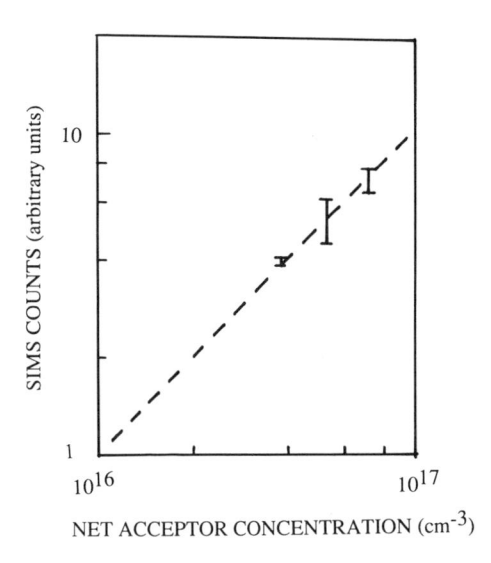

FIG. 5. SIMS count vs net acceptor concentration for a series of heavily doped HgCdTe layers.

cupation can be determined from the net acceptor concentration. For samples grown with a net arsine flow of 125 sccm, a factor of 3.7 increase in P_{Hg} resulted in an increase in $(N_A - N_D)$ by a factor of 3.9. For this situation, and on the premise that the arsenic is either p or n type, depending on its site occupation, we calculate a compensation ratio of less than 0.2 for these samples.

The high-mobility values measured on HgCdTe layers grown in the saturated region of the doping characteristic of Fig. 1 are very unusual, since doping into this region is commonly accompanied by a sharp falloff in mobility, due to dopant incorporation into inactive or compensating sites. This result can be explained with reference to Fig. 5, which shows a plot of the SIMS arsenic count, (which is directly proportional to the *incorporated* arsenic) versus the net acceptor concentration. Here, the linear correspondence between the SIMS count and the net acceptor concentration indicates that saturation of the doping characteristic is related to a saturation of dopant surface coverage as the arsine flow is increased, and not to increased incorporation of arsenic into inactive or compensating sites.

The growth of As-doped HgCdTe with low compensation comes about by the selective incorporation of this dopant into Column VI sites. We believe that this selective incorporation, even in the presence of a high concentration of Column II vacancies, can be explained by the following model. DMCd and AsH_3 are relatively strong Lewis acids and Lewis bases, respectively. Thus, it is probable that an adduct reaction occurs in the gas phase, with the transport of a Cd-

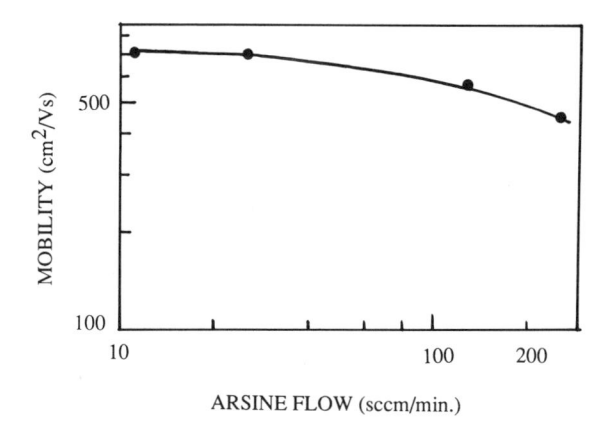

FIG. 4. Hole mobility at 40 K as a function of arsine flow ($x = 0.27$–0.31).

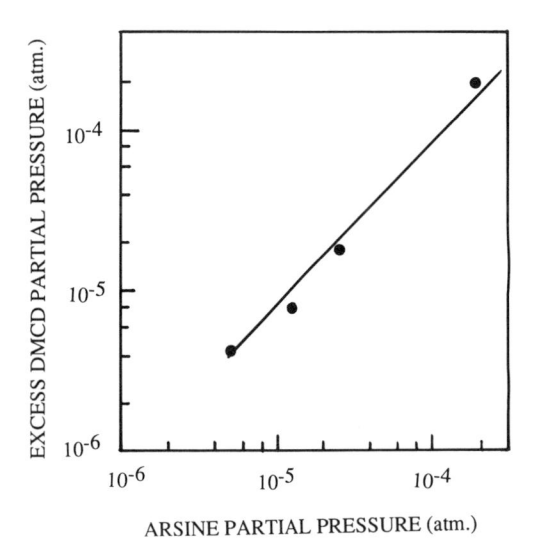

FIG. 6. DMCd partial pressure in excess of that required for undoped layers ($x = 0.27$) plotted as a function of arsine partial pressure.

As complex to the substrate. Incorporation of As into the growing HgCdTe would then come about by its location in sites contiguous to those occupied by Cd, i.e., into Te sites. We estimate an incorporation efficiency of about 0.5–1% for this process, so that much of this adduct is desorbed from the HgCdTe surface. Following this model, we can expect the introduction of AsH_3 to deplete the DMCd which is available for growth, and the value of x should fall with As doping. This has indeed been observed in our experiments.

Figure 6 shows the results of a series of doping experiments where the partial pressure of DMCd was adjusted to keep the x value constant with doping. All other parameters were held constant in these experiments. Here, it is seen that the excess DMCd partial pressure is linearly proportional to the arsine partial pressure, and is of comparable magnitude. This supports our proposed model for As doping of HgCdTe.

IV. CONCLUSIONS

We have shown that arsenic is readily incorporated into HgCdTe which is grown by the direct alloy growth (DAG) process, and doping to $1\times10^{17}/cm^3$ can be achieved in this manner. This arsenic is shown to be stable under subsequent heat treatments. High values of hole mobility indicate a very low compensation in these doped layers. These values remain high, even for layers grown in the saturated region of the doping characteristic. This unusual result has been explained by SIMS measurements, which indicate that saturation in the carrier concentration is due to saturation of the surface coverage of the arsenic species, which in turn limits the incorporation of this acceptor into HgCdTe.

ACKNOWLEDGEMENT

The authors would like to thank J. Barthel for technical assistance on this program and P. Magilligan for manuscript preparation. CdZnTe substrate material was kindly supplied by C. J. Johnson of II–VI, Inc., Saxonburg, PA. This work was sponsored by the Defense Advanced Research Projects Agency (Contract No. N-00014-85-K-0151), administered through the Office of Naval Research, Arlington, VA, and by a grant from the Raytheon Corporation. This support is greatly appreciated.

[a] Presently at Philips Research Labs, Briarcliff Manor, New York.

[1] S. J. C. Irvine and J. B. Mullin, J. Cryst. Growth **55**, 107 (1981).
[2] S. K. Ghandhi and I. B. Bhat, Appl. Phys. Lett. **44**, 7779 (1984).
[3] W. E. Hoke, P. J. Lemonias, and R. Taczewski, Appl. Phys. Lett. **45**, 1092 (1984).
[4] J. L. Schmit, J. Vac. Sci. Technol. A **3**, 89 (1985).
[5] S. K. Ghandhi, I. B. Bhat, and H. Fardi, Appl. Phys. Lett. **52**, 392 (1988).
[6] J. Thompson, P. Mackett, L. M. Smith, D. J. Cole-Hamilton, and D. V. Shenai-Khatkhate, J. Cryst. Growth **86**, 233 (1988).
[7] S. K. Ghandhi, N. R. Taskar, K. K. Parat, and I. B. Bhat, Appl. Phys. Lett. **57**, 252 (1990).
[8] E. S. Johnson and J. L. Schmit, J. Electron. Mater. **6**, 25 (1977).
[9] M. Brown and A. F. W. Willoughby, J. Cryst. Growth **59**, 27 (1982).
[10] H. R. Vydyanath, J. A. Ellsworth, and C. M. Devaney, J. Electron. Mater. **16**, 13 (1987).
[11] P. S. Wijewarnasuriya, I. K. Sou, Y. J. Kim, K. K. Mahavadi, and S. Sivananthan, Appl. Phys. Lett. **51**, 2026 (1987).
[12] N. R. Taskar, I. B. Bhat, K. K. Parat, D. Terry, H. Ehsani, and S. K. Ghandhi, J. Vac. Sci. Technol. A **7**, 281 (1989).
[13] A. Zemel, A. Sher, and D. Eger, J. Appl. Phys. **62**, 1861 (1987).
[14] L. F. Lou and W. H. Frye, J. Appl. Phys. **56**, 2253 (1984).
[15] C. L. Jones, M. J. T. Quelch, P. Capper, and J. J. Gosney, J. Appl. Phys. **53**, 9080 (1982).
[16] H. R. Vydyanath, J. Electrochem. Soc. **128**, 2610 (1981).

Determination of acceptor densities in p-type Hg$_{1-x}$Cd$_x$Te by thermoelectric measurements

J. Baars, D. Brink, and J. Ziegler,[a]
Fraunhofer-Institut für Angewandte Festkörperphysik, Tullastr. 72, D-7800 Freiburg, Germany

(Received 2 October 1990; accepted 9 November 1990)

The differential thermoelectric voltage of p-type bulk samples and epitaxial layers of Hg$_{1-x}$Cd$_x$Te (MCT) (0.2 < x < 0.25) in the temperature range from 20 to 300 K is measured using two different experimental techniques, the hot point method, and the lateral gradient method. The samples were also examined by Hall effect and conductivity measurements. In addition the Seebeck coefficient of p-MCT for acceptor densities $10^{14} < N_A < 10^{17}$ cm^{-3} is calculated employing empirical relations for the energy gap, the intrinsic carrier density, the carrier mobilities, and the LO phonon frequencies. By fitting the calculated temperature dependence of the thermoelectric voltage to the experimental one, the lateral gradient method proved to be an adequate tool for determining the effective acceptor density in p-type MCT including surface inversion. The hot point method is found to be insensitive to surface inversion. It may be used for determining the temperature of zero thermoelectric power which directly yields a good estimate of the acceptor density.

I. INTRODUCTION

There is an urgent need for reliable techniques to characterize p-type Hg$_{1-x}$Cd$_x$Te (MCT) of CdTe mole fractions 0.2 < x < 0.25. The determination of acceptor densities and compensation in p-MCT by Hall effect requires considerable effort in measurement and evaluation.[1-5] Moreover, the interpretation of the measured data is often complicated by surface inversion layers which obscure the bulk properties. In particular, this applies to epitaxial layers.[6-9] In view of these difficulties alternative methods to characterize p-type MCT have been discussed recently. The conductivity at 77 K was found to be an adequate parameter for routine characterization of p-type bulk samples.[10] The magnetic field dependence of the magnetoconductivity was shown to provide detailed information on the densities and mobilities of all the carriers involved in the conduction.[11] Infrared reflectance analyses in the plasmon–phonon range proved to be a reliable method for characterization of p-type bulk crystals and epitaxial layers.[12]

In this study we measured the differential thermoelectric voltage (DTV) of p-type bulk samples and of epitaxial layers of MCT (0.2 < x < 0.25) in the temperature range from 20 to 300 K using two different experimental techniques, namely thermoelectric probing—well known as hot point method—and the lateral gradient method. The samples were also examined by Hall effect and conductivity measurements. In addition we calculated the Seebeck coefficient of p-MCT for acceptor densities $10^{14} < N_A < 10^{17}$ cm^{-3} employing empirical relations for the energy gap, the intrinsic carrier density, the carrier mobilities, and the LO phonon frequencies. By fitting the calculated temperature dependence of the thermoelectric voltage to the experimental one, the lateral gradient method proved to be an adequate tool for determining the effective acceptor density in p-type MCT including surface inversion. The hot point method was found to be insensitive to surface inversion. It may be used for determining the temperature of zero thermoelectric power, which directly yields an estimate of the acceptor density.

II. EXPERIMENT

For this study p-type MCT bulk crystals and epitaxial layers of CdTe mole fractions 0.2 < x < 0.25 were used. The bulk crystals were grown by the solid-state recrystallization method.[13] The epitaxial layers were deposited from Te-rich solutions on (111)B surfaces of (Cd,Zn)Te crystals using the dipping technique.[14] The effective acceptor density $N_A - N_D$ of the as-grown material was reduced to a level between 10^{14} and 10^{17} cm^{-3} by annealing 0.5-mm-thick slices cut from bulk crystals, and epitaxial layers in sealed quartz ampules under saturated Hg vapor pressure at temperatures of about 300 °C for some weeks. The samples were mechanically polished and subsequently etched in a methanol solution containing about 0.5% of bromine. The annealed samples were of uniform composition and acceptor density, as verified from room temperature infrared transmission[25] and reflectance measurements.[12] In addition, the uniformity of annealed bulk samples was verified by a sequence of temperature dependent Hall measurements performed on samples which were systematically thinned between measurements. Further annealing at 225 °C results in uniform n-type material with donor densities between 10^{14} cm^{-3} and 10^{15} cm^{-3} corresponding to the residual donor density of the p-type samples used in this study.

Prior to the thermoelectric and Hall-effect measurements the samples were subjected to a special electrochemical etch.[15] XPS examinations proved this final treatment to produce clean surfaces of stoichiometric composition.

For the conductivity and Hall-effect measurements we employed the van der Pauw method using Au leads bonded to Au/In contact pads evaporated amid the edges of the square-shaped samples of about 8 × 8 mm^2. The electrical properties of the contacts at temperatures between 15 and 320 K were found to be ohmic. The Hall effect was measured using magnetic fields of 0.3 kG and 1 kG.

Figure 1 shows the Hall coefficents versus reciprocal temperature for p-type MCT specimens used in this study. The plots are typical of mixed electron and hole conduction at

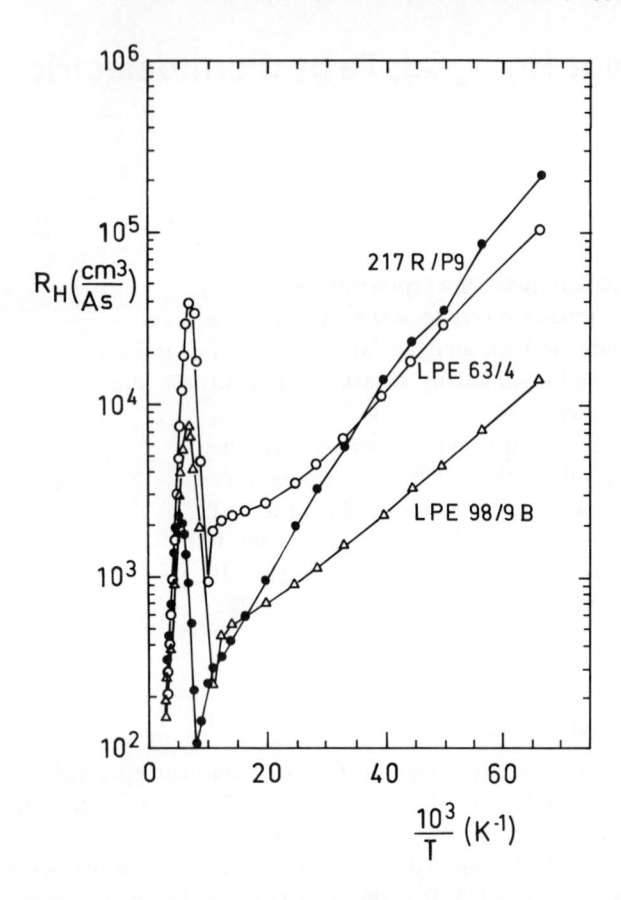

FIG. 1. Hall coefficient R_H vs reciprocal temperature of a MCT bulk crystal (217R/P9, $x = 0.215$, $d = 0.3$ mm) and of two MCT epitaxial layers on (Cd,Zn)Te substrates (LPE 63/4, $x = 0.226$, $d = 8.8$ μm; LPE 98/9B, $x = 0.216$, $d = 30$ μm). A magnetic field of 0.3 kG was used for these measurements. Prior to the Hall measurements the samples were subjected to a special electrochemical etch to avoid surface inversion.

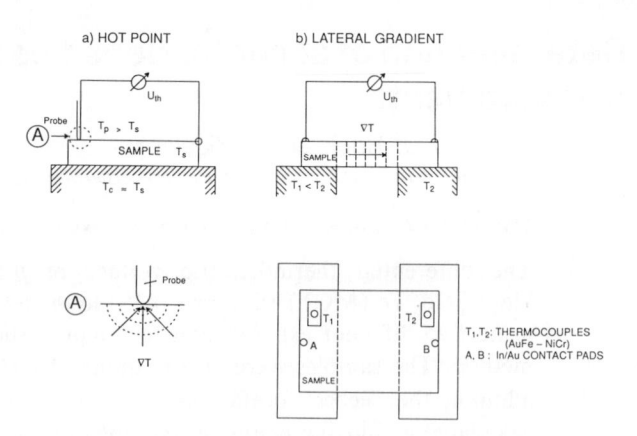

FIG. 2. Schematic arrangement for measuring the thermoelectric power of MCT by (a) the hot point technique and (b) the lateral gradient technique. In contrast to the lateral gradient technique, the temperature gradient produced by the hot probe is almost perpendicular to the sample surface (A).

higher temperatures, and they clearly show continuous freeze-out of holes at low temperatures. The temperature dependence of the Hall coefficient does not exhibit an anomalous second sign reversal at low temperatures. A second sign reversal is found for samples with oxidized surfaces only.

The Hall contacts also served the DTV measurements when the lateral temperature gradient technique was employed. Figure 2 illustrates the two arrangements used for measuring the thermoelectric power of MCT, (a) the hot point technique and (b) the lateral gradient technique.

For hot point thermoelectric probing the specimen was placed on the cold finger of a liquid nitrogen cryostate using vacuum grease as adhesive. NiCr–NiAl thermocouples (Philips Thermocoax CI3S) sheathed with 0.5-mm-diam covers of an Ni–Cr–Fe alloy (INCONEL) were used for controlling the temperatures of the specimen and of the hot probe. The thermocouple sheaths served as probes for the DTV measurements. The sheath proved to be a good thermoelectric probe for MCT, producing thermoelectric signals about two orders in magnitude smaller than MCT. The DTV was recorded while the MCT specimen was touched by the hot probe for one second, applying a fixed pressure and a 5 K temperature difference between the hot probe and the specimen. Although these hot point measurements turned out to be fairly reproducible, showing variations of 10% only, their absolute accuracy is much less, due to the uncertainty in determining the actual temperature difference generated by the hot probe. It should be noted that thermoelectric probing causes some damage. Microscopic examinations revealed small scratches on the surface due to probing. The depth of the scratches was found to be less than 0.5 μm (Sloan Dektak II). The range of crystalline damage, however, may be much larger, and the electrical properties might be considerably altered in the damaged region, though the total conductivity proved to be unaffected by probing. By reducing the load, scratches can be avoided at the expense of reproducibility.

For the lateral gradient technique a Sterling type cooling engine was used allowing DTV measurements between 20 K and 300 K. To achieve uniform lateral temperature gradient over the entire width of the specimen, together with good electrical insulation, two sapphire plates were attached to the specimen covering one quarter each and leaving a gap in the middle as shown in Fig. 2(b). The specimen and the sapphire plates were joined with a low-temperature glue (GE 7031) to provide a good thermal contact. The sample was placed in the cooling engine, one sapphire plate providing the thermal contact to the cold end, and the other to an electric heating foil. A copper cap at sample temperature served as radiation screen. AuFe–NiCr thermocouples, electrically insulated from the specimen by thin ceramic plates, were used to control the temperatures on both sides of the gap. To provide good thermal contact the thermocouples were glued to the specimen. A temperature difference of 2 K across the gap was chosen for DTV measurements between 20 and 50 K, and a difference of 5 K at higher temperatures. This setup was also found to be capable of recording the DTV continuously versus temperature. Starting the cooling engine at 300 K and keeping the heating power constant, the temperature difference typically decreased from 10 K at 300 K to 0.5 K at 20 K. Owing to the small temperature differences at low temperatures, the DTV determination below 50 K becomes increasingly uncertain.

In contrast to the lateral gradient technique, the temperature gradient produced by the hot probe is almost perpendicular to the surface of the specimen. Thus, the DTV caused by the hot probe in thin surface layers will be negligibly small, whereas the DTV generated by lateral temperature gradients in surface layers might very well compare in magnitude with the DTV of the bulk.

Since the low-temperature glue is soluble in methanol, the ceramic plates with the thermocouples and the sapphire plates can be removed without damage. Thus the lateral gradient technique is nondestructive.

III. THERMOELECTRIC MODEL FOR p-TYPE MCT

The differential thermoelectric voltage or Seebeck coefficient of a cubic semiconductor is equivalent to the voltage produced by two generators in parallel, the admittances of which are proportional to the conductivity of electrons and holes, respectively[16]:

$$\alpha = (\alpha_h p \mu_h + \alpha_e n \mu_e)(p \mu_h + n \mu_e)^{-1}. \quad (1)$$

Here, n and p are the densities, μ_e and μ_h are the mobilities of electrons and holes, respectively. The relaxation time approximation of the Boltzmann equation yields the thermoelectric coefficients α_e and α_h of electrons and holes, respectively, owing to charge carrier diffusion in a temperature gradient.

$$\alpha_e = -\frac{k}{q}\left(\frac{\langle \tau_e \epsilon \rangle}{\langle \tau_e \rangle} - \eta\right); \quad \alpha_h = \frac{k}{q}\left(\frac{\langle \tau_h \epsilon \rangle}{\langle \tau_h \rangle} - \eta'\right),$$

where

$$\epsilon = \frac{E}{kT}; \quad \epsilon_g = \frac{E_g}{kT}; \quad \eta = \frac{E_F}{kT}; \quad \eta' = -\eta - \epsilon_g.$$

The kinetic term $\langle \tau \epsilon \rangle / \langle \tau \rangle$ depends on the charge carrier scattering mechanisms involved. τ is the relaxation time, E the energy measured from the edge of the conduction band, E_g the band gap energy, E_F the Fermi energy, T the temperature, k the Boltzmann constant, and q the elementary charge.

In general, however, the generation of thermoelectricity in semiconductors involves charge carrier diffusion and drag of the carriers by phonons as well. Up to now only limited information on the thermoelectric properties of MCT has been available. Most of the published data refer to HgTe. The Seebeck coefficient of intrinsic HgTe at temperatures between 5 and 300 K was studied by Sologub et al.[17] and Dubowski et al.[18] They showed that the thermoelectric power of HgTe at temperatures above 50 K is controlled by diffusion of charge carriers which are predominantly scattered by longitudinal optical phonons. Phonon drag of holes was found to be effective at temperatures below 50 K only.

Recently, Höschl et al.[19] measured the thermoelectric power of p-type MCT samples of CdTe mole fractions close to $x = 0.21$ in the temperature range from 77 to 300 K using the hot point technique. The experimental DTV data was found to agree with calculations based on pure optical mode scattering, assuming identical kinetic terms for electrons and holes. The same kinetic term is given by Devlin[20] for optical mode scattering.

The numerical calculations of the thermoelectric coefficient of MCT presented here are based on the diffusion mechanism of charge carrier transport employing Devlin's kinetic term for pure optical mode scattering:

$$\frac{\langle \tau_h \epsilon \rangle}{\langle \tau_h \rangle} = \frac{\langle \tau_e \epsilon \rangle}{\langle \tau_e \rangle} = 2.5 + 0.5 e^{-(\theta/T)^2}$$
$$- 0.86 \left(\frac{\theta}{T}\right)^{25/9} e^{-10\theta/(9T)}. \quad (3)$$

For the Debye temperature θ the average of the two LO phonon frequencies of MCT is used:

$$\theta = (\omega_{LO_1} + \omega_{LO_2}) 1.24 \times 10^{-4} (2k)^{-1}. \quad (4)$$

The thermoelectric model used is confined to nondegenerate electron and hole densities. This assumption holds good for p-type MCT in the temperature range of interest, provided $x > 0.18$. In addition, the heavy hole band is assumed to be parabolic, $N_v = 2(2\pi m_{hh}^* kT/h^2)^{2/3}$ being its effective density-of-states. Light holes are not considered. Hence, in this approximation the Fermi energy is given by

$$E_F = kT \ln(N_v/p) - E_g. \quad (5)$$

Furthermore, the model is restricted to a single discrete acceptor level of density N_A at an energy E_A above the valence band edge, and to donors N_D which are assumed to be fully ionized even at low temperatures,

$$N_D^+ = N_D; \quad N_A^- = N_A(1 + 4e^{(E_A + E_F')/(kT)})^{-1}. \quad (6)$$

This is in agreement with variable temperature Hall effect measurements performed on the samples used in this study. The donor densities were found to be in the order of 10^{14} cm^{-3}.

The Hall coefficient for low magnetic fields ($\mu B \ll 1$) used for evaluation is given by

$$-q R_H = (n \mu_e^2 - p \mu_h^2)(n \mu_e + p \mu_h)^{-2}. \quad (7)$$

Equation (7) proved to be a good approximation for p-type MCT and magnetic fields as low as 0.3 kG.

Finally the following empirical relations are used for the intrinsic carrier density[21] n_i, the energy gap[22] E_g, the LO phonon frequencies,[23] and the electron-to-hole mobility ratio:

$$n_i = [5.585 - 3.82x + (1.753 \times 10^{-3} T) - (1.364 \times 10^{-3} xT)] 10^{14} \, E_g^{3/4} T^{3/2} e^{-E_g/(2kT)}, \quad (8)$$

$$E_g = -0.302 + 1.93x + (5.35 \times 10^{-4})(T^3 - 1822) \times (1 - 2x)(T^2 + 255.2)^{-1} - 0.81x^2 + 0.832x^3, \quad (9)$$

$$\omega_{LO_1} = [13 + (2 \times 10^{-2} T)]x - [(2 \times 10^{-2} T) + 155]; \quad (10)$$

$$\omega_{LO_2} = -[10 - 1 \times 10^{-2} T]x - (1 \times 10^{-2} T) + 140,$$

$$\frac{\mu_e}{\mu_h} = \frac{610}{(E_g T)^{1/2}}. \quad (11)$$

Equation (11) is based on experimental data from various sources.[2-6] This relation turned out to be a good approximation for p-type MCT at temperatures between 40 and 300 K, though both the electron and the hole mobility depend on the effective acceptor density, and there is a considerable variance among the experimental data. The effective mass of the heavy hole band[24] is assumed to be $m_{hh}^* = 0.5\, m_0$ where m_0 is the electron rest mass.

Now, the electron and hole densities can be determined from the charge neutrality equation and their common relation to the intrinsic carrier density n_i:

$$p + N_D^+ = n + N_A^-; \quad np = n_i^2. \tag{12}$$

Once p and n are known, the Seebeck coefficient is obtained from Eq. (1) in combination with Eqs. (2)–(6).

When the lateral gradient method is employed, possible surface inversion layers will contribute to the thermoelectric voltage as well. In this case the total thermoelectric coefficient is given by

$$\alpha = \alpha_v \frac{\sigma_v}{\sigma} + \alpha_s \frac{\sigma_s}{\sigma} \frac{d_s}{d_v}, \tag{13}$$

where $\sigma_v = q(\mu_e n + \mu_h p)$; $\sigma_s = q\mu_{es} n_s$; $\sigma = \sigma_v + \sigma_s d_s/d_v$. α_v, α_s, σ_v, σ_s, d_v, and d_s are the thermoelectric coefficient, the conductivity, and the thickness of the bulk (index v) and of the surface inversion layer (index s), respectively. n_s (cm^{-3}) is the inversion layer electron concentration. It is equivalent to the ratio of fixed surface charges caused by oxidation, and the thickness of the inversion layer d_s, which is assumed to be twice the Debye length[6]

$$d_s = 2 L_D = 2\{kT/[q^2(n+p)]\}^{1/2}. \tag{14}$$

The thermoelectric coefficient for the completely degenerate electron gas of the inversion layer is given by[16]

$$\alpha_s = -\frac{\pi^2}{3}\left(r + \frac{3}{2}\right) \frac{k^2}{q} \frac{T}{E_{F_s}}. \tag{15}$$

In the $k \times p$ approximation the Fermi energy for degenerate conduction electrons and spherical Fermi surfaces is given by[24]

$$E_F = -E_g/2 + [E_g^2/4 + (2/3)P^2(3\pi n_s)^{2/3}]^{1/2}, \tag{16}$$

where $P^2 = (18 + 3x)\hbar^2/(2m_0)$.

The thermoelectric contribution of the inversion layer electrons will dominate at low temperatures only, when the hole density in the bulk has decreased due to freeze out. At low temperatures ionized defects will be the most efficient scatterers for the electrons in the inversion layer. Therefore, r in Eq. (15) was chosen to be 3/2. The mobility of the inversion layer electrons is assumed to be independent of temperature, $\mu_{es} = 1\times 10^4$ cm^2/(V s), in agreement with low temperature Hall measurements performed on oxidized p-type samples.[7,8]

IV. RESULTS AND DISCUSSION

The temperature dependence of the differential thermoelectric voltage (DTV) calculated according to Eqs. (1)–(6) and (12) for p-type MCT of various CdTe mole fractions is presented in Fig. 3(a). Also shown is experimental data obtained from DTV measurements on a MCT epitaxial layer using the lateral gradient method. The DTV or Seebeck coefficient α is determined by the ratio of the measured thermoelectric voltage U_{th} to the temperature difference $T_2 - T_1$ applied to the sample [Fig. 2(b)]:

$$\alpha = U_{th}/(T_2 - T_1). \tag{17}$$

The calculation is based on the empirical relations for MCT given by Eqs. (8)–(11) including the effective acceptor density and the activation energy of the specimen determined by Hall effect measurements (Fig. 1). The small donor density of approximately 5×10^{14} cm^{-3} was found to be ineffective. The best fit is obtained for $x = 0.225$ which is close to the mole fraction $x = 0.226$ found by transmission measurements at room temperature.[25] Figure 3(b) presents a set of curves calculated for a fixed CdTe mole fraction $x = 0.225$ and for various values of the acceptor density N_A. Again the experimental data is shown for convenience.

Figure 3 clearly illustrates the thermoelectric features of p-type MCT, and shows the degree of accuracy achievable in the evaluation of the effective acceptor density by the lateral gradient method.

At high temperatures the thermoelectric power is determined by the intrinsic charge carrier density. At intermediate temperatures the curves for different CdTe mole fractions spread and change sign due to the influence of the doping in relation to the intrinsic carrier density which is higher for lower x. When the intrinsic carrier density has further decreased with temperature, and the thermoelectric power is completely controlled by doping, the curves join at a plateau which depends on the acceptor density only. The

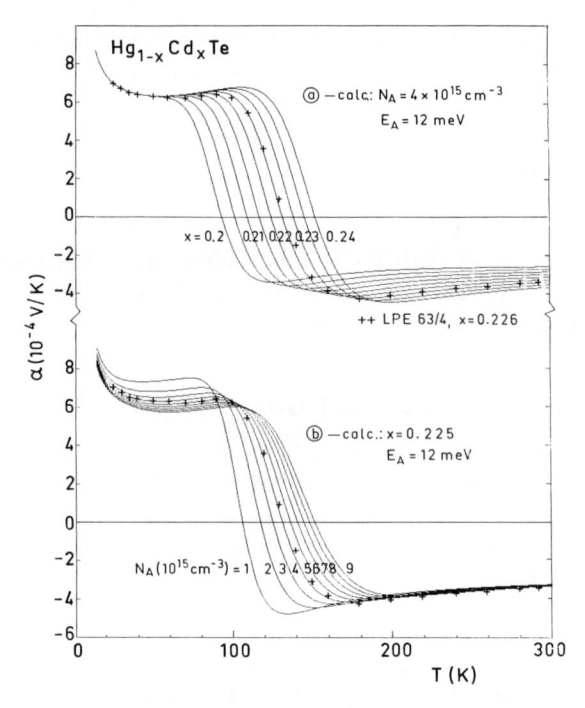

FIG. 3. Differential thermoelectric voltage vs temperature calculated a) for MCT $0.2 < x < 0.24$ with fixed values of N_A and E_A ($N_A = 4 \times 10^{15}$ cm^{-3}, $E_A = 12$ meV), and b) for $x = 0.225$, $1 \times 10^{15} < N_A < 9 \times 10^{15}$ cm^{-3}, and $E_A = 12$ meV. The experimental data (+) is determined from measurements of the thermoelectric power of a MCT epitaxial layer (LPE 63/4, $x = 0.226$, $d = 8.8$ μm) using the lateral gradient technique. The calculations are based on empirical relations of the properties of MCT. The values of the effective acceptor density $N_A = 4 \times 10^{15}$ cm^{-3}, and of the acceptor energy $E_A = 12$ meV used in the calculations are obtained from sample LPE 63/4 by Hall effect measurements.

subsequent increase of the thermoelectric power at low temperatures is caused by the freeze-out of the holes. The experimental data presented in Fig. 3 was obtained after the specimen was given a special electrochemical etch treatment to avoid surface inversion.[15]

The effect of surface inversion was studied on p-type MCT samples covered with anodically grown oxide layers. Native oxides are known to produce surface inversion in p-type MCT. Figures 4 and 5 show plots of the Hall coefficient against reciprocal temperature for a bulk sample and an epitaxial layer with and without anodic oxides. The values of the acceptor densities and energies presented in Figs. 4 and 5 were obtained by fitting the Hall coefficients of the nonoxidized samples with Eq. (7). The best fits were obtained for donor densities less than 10^{15} cm^{-3}. The plots for the oxidized samples are characteristic of inverted surfaces. The Hall coefficient of the oxidized bulk sample exhibits a transition from positive to negative values at 60 K, and that of the oxidized epitaxial layer, due to its much smaller thickness, is negative in the entire temperature range. The Hall coefficient of thin epitaxial layers as shown in Fig. 5 can be considerably affected by inversion layers, and attempts to fit its temperature dependence by using a two-layer model including surface inversion often lead to misinterpretation.

The results of the DTV measurements performed on the same samples using the lateral gradient technique are presented in Figs. 6 and 7. A set of curves calculated for various degrees of surface inversion according to Eqs. (13)–(16) is shown for comparison. The sheet electron density normalized to the thickness of the sample is used as a parameter.

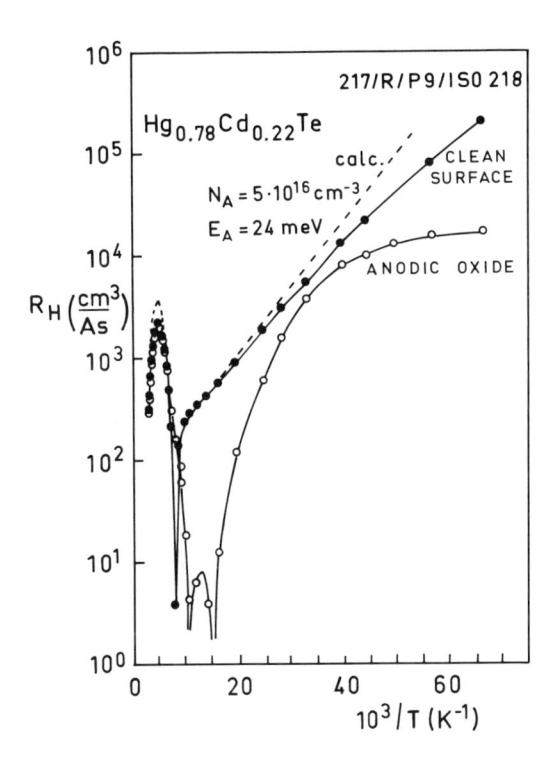

FIG. 4. Hall coefficient R_H vs reciprocal temperature determined before and after anodic oxidation of a MCT bulk crystal (217R/P9, $d = 301$ μm). Fitting the Hall coefficient of the nonoxidized sample (clean surface) with $-qR_H = (n\mu_e^2 - p\mu_h^2)(n\mu_e + p\mu_h)^{-2}$ yields $N_A = 5\times 10^{16}$ cm^{-3}, $N_D < 1\times 10^{15}$ cm^{-3}, and $E_A = 24$ meV.

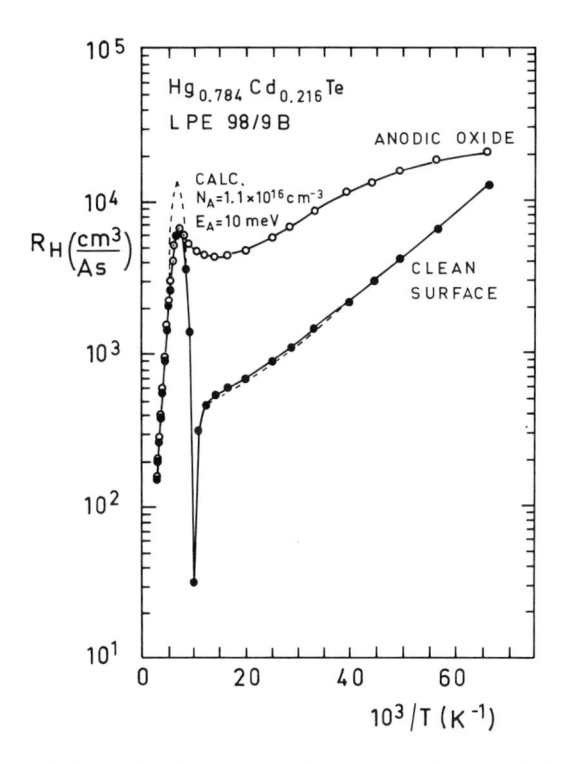

FIG. 5. Hall coefficient R_H vs reciprocal temperature determined before and after anodic oxidation of a MCT epitaxial layer (LPE 98/9B, $d = 30$ μm). Fitting the Hall coefficient of the nonoxidized sample (clean surface) with $-qR_H = (n\mu_e^2 - p\mu_h^2)(n\mu_e + p\mu_h)^{-2}$ yields $N_A = 1.1\times 10^{16}$ cm^{-3}, $N_D < 1\times 10^{15}$ cm^{-3}, and $E_A = 10$ meV.

The acceptor densities and energies which were obtained from the Hall effect analyses of the nonoxidized samples are used for the calculation. The donor density is too small to be effective, and therefore is not considered.

The data obtained from DTV measurements performed on the samples before oxidation agrees very well with that calculated for clean surfaces ($N_s = 0$). The temperature dependence of the Seebeck coefficient measured on the oxidized samples is in fair agreement with curves calculated for $N_s = 2\times 10^{11}$ cm^{-2} which is typical of p-MCT MIS devices with anodic oxides.

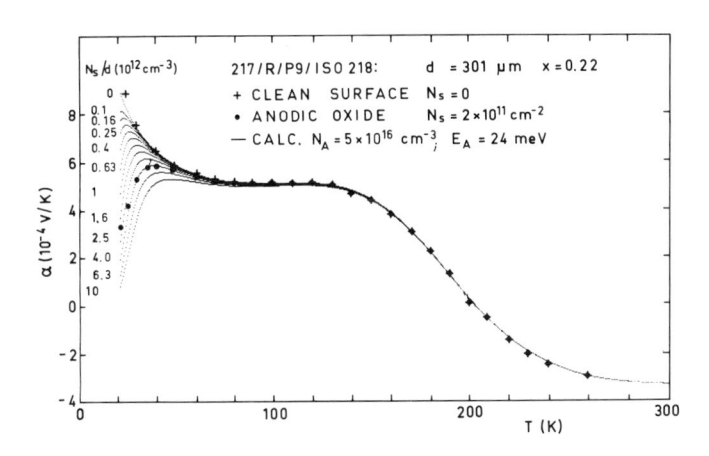

FIG. 6. Differential thermoelectric voltage vs temperature of a MCT bulk crystal (217R/P9) determined before (+) and after (●) anodic oxidation using the lateral gradient technique. The set of curves represents the thermoelectric voltage calculated for various degrees of surface inversion due to anodic oxidation. The sheet electron density N_s normalized to the thickness d of the MCT specimen is used as parameter.

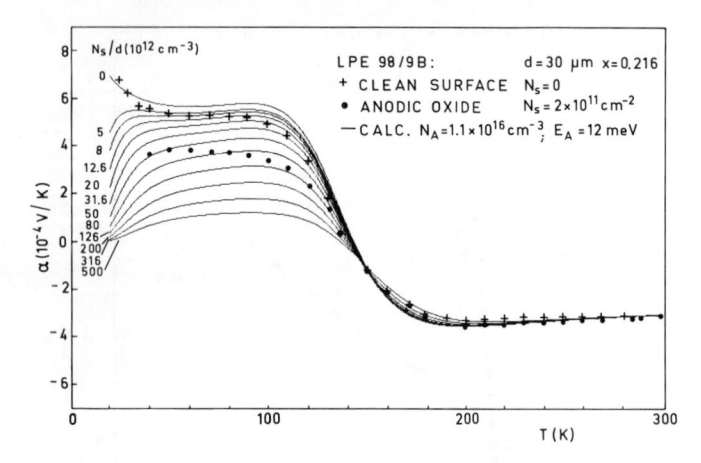

FIG. 7. Differential thermoelectric voltage vs temperature of a MCT epitaxial layer (LPE 98/9B) determined before (+) and after (●) anodic oxidation using the lateral gradient technique. The set of curves represents the thermoelectric voltage calculated for various degrees of surface inversion due to anodic oxidation. The sheet electron density N_s normalized to the thickness d of the MCT specimen is used as parameter.

Hot point thermoelectric probing was found to be quite reproducible. Its absolute accuracy, however, turned out to be quite low owing to the uncertainty regarding the measurements of the actual temperature difference generated by the hot probe. The sign reversal at the temperature of zero thermoelectric power is determined with high accuracy, though. This is illustrated in Fig. 8. The effective acceptor density is plotted against the sign reversal temperature of the thermoelectric voltage calculated according to Eqs. (1)–(12) for MCT with $0.2 < x < 0.24$. The experimental data shown was obtained from thermoelectric measurements performed on various p-type MCT samples using both the hot point and the lateral gradient techniques. The CdTe mole fractions,

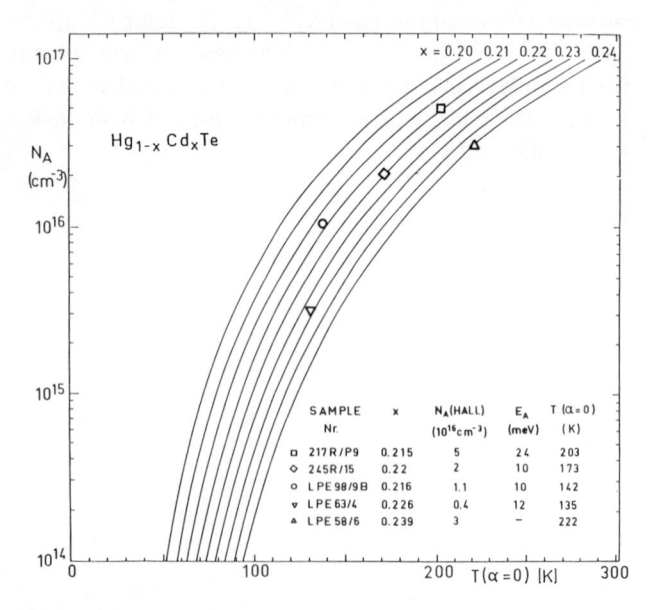

FIG. 8. Effective acceptor density vs temperature of sign reversal of the thermoelectric voltage calculated for MCT $0.2 < x < 0.24$, and obtained from thermoelectric measurements using both the hot point and the lateral gradient techniques. The inserted table lists the CdTe mole fraction x, the acceptor densities N_A and energies E_A determined by Hall effect analyses, and the sign reversal temperatures of zero thermoelectric power for the samples used in this study.

the acceptor densities and energies, and the sign reversal temperatures of zero thermoelectric power for the samples used are listed in the table inserted in Fig. 8. The effective acceptor densities obtained from the plots compare very well with the values determined by Hall effect analyses.

Since hot point thermoelectric probing can easily be applied at temperatures between 80 and 300 K, where the acceptors in MCT are almost completely ionized, it is a useful technique for examining p-type MCT with respect to acceptor densities ranging from 10^{14} to 10^{17} cm^{-3}. Thermoelectric probing was found to be particularly suited for the characterization of p-type MCT layers on n-type MCT substrates for which examinations by Hall effect measurements fail.[26] p-type layers, about 2 μm thick, were formed in n-type MCT wafers by As implantation and annealing. The temperature dependence of the thermoelectric voltage produced by hot point probing these p-type layers was found to be characteristic of uniform p-type MCT samples. There was no apparent effect due to the n-type substrate, though it is expected to shunt the p-type layer at temperatures beyond about 200 K, when the p-n junction is flooded. The sign reversal temperature of zero thermoelectric power was used to determine the effective acceptor density which compared well with the value obtained by infrared reflectance analysis.[12]

Also the lateral gradient method has a high potential for characterizing layer structures like p-on-n homo- and heterojunctions. The total thermoelectric voltage for such a structure is the sum of the thermoelectric voltages of the individual layers weighed by their conductance G_i relative to the total conductance G of the structure

$$U_{\text{th}} = U_1 \, G_1/G + U_2 \, G_2/G \qquad (18)$$

where $G_i/G = (\sigma_i/\sigma) d_i/d$. Thus the contributions of the individual layers are proportional to their conductivities σ_i and to their thicknesses d_i. Since the contribution of an individual layer to the total Hall coefficient is a quadratic function of the layer conductivity[27] the lateral gradient technique has a great advantage over standard Hall effect measurements versus temperature or versus magnetic field. A thin p-type layer of low conductivity which is difficult to assess by Hall effect measurements if it is grown on a highly conducting n-type substrate may be very well characterized by the lateral gradient technique.

V. CONCLUSIONS

A simple thermoelectric model based on the diffusion of electrons and holes in a temperature gradient was found to be sufficient to fit the temperature dependence of the thermoelectric coefficient of p-type MCT samples including surface inversion.

The lateral gradient technique proved to be an adequate tool for determining the effective acceptor density in p-type MCT.

Hot point thermoelectric probing turned out to be a reliable technique for determining the sign reversal temperature of zero thermoelectric power which directly yields a good estimate of the acceptor density.

[a] AEG, Theresienstr. 2, D-7100 Heilbronn, Germany.
[1] W. Scott, E. L. Stelzer, and R. J. Hager, J. Appl. Phys. **47**, 1408 (1976).
[2] D. D. Edwall, E. R. Gertner, and W. E. Tennant, J. Electron. Mat. **14**, 245 (1985).
[3] D. Eger, A. Zemel, D. Mordowicz, and A. Sher, Appl. Phys. Lett. **46**, 989 (1985).
[4] M. C. Gold and D. A. Nelson, J. Vac. Sci. Technol. A **4**, 2040 (1986).
[5] S. E. Schacham and E. Finkman, J. Appl. Phys. **60**, 2860 (1986).
[6] L. F. Lou and W. H. Frye, J. Appl. Phys. **56**, 2253 (1984).
[7] M. C. Chen, J. Appl. Phys. **65**, 1571 (1989).
[8] P. R. Emtage, T. A. Temofonte, A. J. Noreika, and C. F. Seiler, Appl. Phys. Lett. **54**, 2015 (1989).
[9] P. S. Wijewarnasuriya, M. Boukerche, and J. P. Faurie, J. Appl. Phys. **67**, 859 (1990).
[10] P. N. J. Dennis and C. T. Elliott, Infrared Phys. **22**, 167 (1982).
[11] D. L. Leslie-Pelecky, D. G. Seiler, M. R. Loloee, and C. L. Littler, Appl. Phys. Lett. **51**, 1916 (1987).
[12] J. Baars, V. Hurm, T. Jakobus, H. Seelewind, and J. Ziegler, SPIE **659**, 44 (1986).
[13] H. Maier, J. Hesse, *Crystals Growth, Properties and Applications* (Springer, Berlin, 1980), Vol. 4, pp. 195–199
[14] C. Geibel, H. Maier, and J. Ziegler, SPIE **659**, 110 (1986).
[15] M. Seelmann-Eggebert, D. Brink, Patent applied.
[16] R. A. Smith, *Semiconductors* (Cambridge University Press, Cambridge, 1964), p. 173.
[17] V. V. Sologub, V. I. Ivanov-Omskii, V. M. Muzhdaba, and S. S. Shalyt, Sov. Phys. Solid State **13**, 1452 (1971).
[18] J. J. Dubowski, T. Dietl, W. Szymanska, and R. R. Galazka, J. Phys. Chem. Solids **42**, 351 (1981).
[19] P. Höschl, P. Moravec, E. Belas, J. Franc, and R. Grill, Phys. Status Solidi B **147**, 621 (1988).
[20] S. S. Devlin, *Physics and Chemistry of II–VI Compounds*, edited by M. Aven and J. S. Prener (North-Holland, Amsterdam, 1967), p. 562.
[21] G. L. Hansen and J. L. Schmit, J. Appl. Phys. **54**, 1639 (1983).
[22] D. G. Seiler, J. R. Lowney, C. L. Littler, and M. R. Loloee, J. Vac. Sci. Technol. A **8**, 1237 (1990).
[23] J. Baars and F. Sorger, Solid State Commun. **10**, 875 (1972).
[24] R. Dornhaus and G. Nimtz, *Springer Tracts in Modern Physics*, (Springer, Berlin, 1983), Vol. 98, p. 217.
[25] E. Finkman and S. E. Schacham, J. Appl. Phys. **65**, 2896 (1984).
[26] J. Baars, H. Seelewind, Ch. Fritzsche, U. Kaiser, and J. Ziegler, J. Cryst. Growth **86**, 762 (1988).
[27] R. L. Petritz, Phys. Rev. **110**, 1254 (1958).

Mechanisms of incorporation of donor and acceptor dopants in (Hg,Cd)Te alloys

H. R. Vydyanath
Aerojet Electronic Systems Division, 1100 W. Hollyvale Street, Azusa, California 91702

(Received 25 January 1991; accepted 11 February 1991)

The usefulness of the quasichemical approach in predicting the defect structure of undoped and doped (Hg,Cd)Te crystals is emphasized. To begin with, the defect structure of undoped (Hg,Cd)Te crystals is established along with mass action constants for the intrinsic excitation process and the formation of the dominant native acceptor defects. Subsequently, the defect structures of donor doped and acceptor doped (Hg,Cd)Te crystals are established. With the values of the mass action constants established for the undoped crystals along with those for the doped crystals, the concentrations of defects present at the high temperature are calculated along with the carrier concentrations expected in the cooled crystals. The results of the carrier concentration calculations are compared with the experimental values to verify the validity of the incorporation mechanisms established for the doped crystals.

I. INTRODUCTION

The importance of (Hg,Cd)Te as an infrared detector material needs no elaboration.[1] With the expanding need for large density two dimensional arrays, epitaxial growth techniques such as metalorganic chemical vapor deposition (MOCVD), and molecular-beam epitaxy (MBE) are playing an increasingly significant role along with liquid phase epitaxy (LPE). Current and future focal plane performance requirements for various sensor systems have created a need to develop novel device structures in the form of heterojunctions and superlattices. These classes of devices require extrinsic doping with a minimum concentration of native defect related Shockley–Read centers which degrade the carrier lifetime in the material below the fundamental limit.

The issue of understanding the behavior of extrinsically doped (Hg,Cd)Te alloys has been pursued for a number of years by various researchers.[2-15] The mechanisms of incorporation of dopants in (Hg,Cd)Te alloys are far more complicated than in Silicon or GaAs. The quasichemical approach pioneered by Kröger and Vink[16] serves as an elegant tool in unraveling the defect structures of undoped and doped elemental and binary compound crystals. Numerous examples of the results of defect analysis in many elemental and binary compound systems are cited in Kröger's book.[17]

The paper will begin with a description of the quasichemical approach followed by an analysis of the electrical characteristics of undoped (Hg,Cd)Te alloys. The nature of the dominant native defects along with the mass action constants associated with the formation of these defects, will be discussed. Next, the electrical data in donor doped and acceptor doped (Hg,Cd)Te crystals will be reviewed followed by a discussion of the mechanisms of incorporation of these dopants.

Although the discussions in the paper focus on the experimental data mostly in bulk grown (Hg,Cd)Te material and to a much smaller extent in LPE material, the conclusions regarding the incorporation mechanisms should be valid for all (Hg,Cd)Te material irrespective of the method of growth such as bulk, LPE, vapor phase epitaxy (VPE), MOCVD, or MBE.

II. DEFECT EQUILIBRIA AND QUASICHEMICAL APPROACH

The quasichemical approach[16] has proven to be very successful in predicting the defect structures of many binary compound crystals. In this approach electronic and atomic defects are treated as chemical species and appropriate formation reactions along with respective mass action constants are formulated. The electroneutrality condition and the dopant balance equation (for doped crystals) are then written down. All the electronic and atomic defect species in the electroneutrality condition and the dopant balance equation are then expressed in terms of the relevant mass action constants and the concentration of one defect species. This results in equations containing the various mass action constants, concentration of one defect species and the activity or partial pressure of one of the components (p_M or p_{x_2}) of the binary compound MX (M—metal, X—nonmetal). In the case of the doped crystals, the equations will also contain the partial pressure (p_F), activity (a_F) or concentration of the foreign dopant [F]. Numerical solution of the electroneutrality and the dopant balance equations will then yield the concentration of one defect species for given values of the mass action constants with p_M or p_{x_2} and p_F, a_F, or [F] as variables. Once the concentration of one defect species is known, the concentrations of all the other defect species can be immediately calculated. Approximation of the electroneutrality and the dopant balance equations by only the dominant members results in simple expressions for the defects in terms of the component partial pressure p_M or p_{x_2} and the concentration of the dopant (for the doped crystals). Concentration $\propto p_M^r [F_{\text{tot}}]^s$ where r and s are small integers or fractions.

In deducing the high temperature defect state of the crystals from measurements on the crystals cooled to room temperature, the assumptions made are that electrons and holes recombine during cooling, that all the atomic defects at the high temperature are frozen in. Additionally, in the case of measurement of the electrical characteristics of the cooled crystals, the temperature of measurement should be low enough to ensure that the crystals are extrinsic with the Fermi level pinned at the defect level. The carrier concentration then gives the concentration of the atomic defects corresponding to the high temperature equilibrium.

In the deduction of the high temperature defect from measurements on the cooled crystals, the assumption of a near perfect quenching from the high temperature is often impractical leading to the precipitation of the atomic defects during the quench and hence errors in the evaluation of the high temperature defect state.

The subsequent sections in this paper will illustrate the applications of the quasichemical approach to specifically discuss the native defect structures of undoped (Hg,Cd)Te alloys and incorporation mechanisms of donor and acceptor dopants in these alloys.

III. ILLUSTRATION OF THE QUASICHEMICAL APPROACH FOR UNDOPED $Hg_{0.8}Cd_{0.2}Te$ AND $Hg_{0.6}Cd_{0.4}Te$

Assuming the presence of only electrons (e'), holes ($h^\cdot$), singly ionized and doubly ionized vacancies of Hg (V'_{Hg} and V''_{Hg}) we can formulate the following defect formation reactions along with the respective mass action constants.

(The defect notations used in this paper are those of Kröger and Vink[16,17] according to which subscripts denote the lattice site occupied by the defect and superscripts denote the charge, a dash (') standing for an effective negative charge, a dot (·) for an effective positive charge and a cross (x) for an effective neutral charge. Square brackets [ ] indicate concentrations of defects in terms of site fractions—# cm^{-3} of the defect ÷ # cm^{-3} of the total number of lattice sites.)

Defect formation reactions	Mass action constants
1. $0 \to e' + h^\cdot$	$K_i = [e'][h^\cdot]$
2. $Hg^x_{Hg} \to V'_{Hg} + h^\cdot + Hg(g)$	$K'_{VHg} = \dfrac{[V'_{Hg}][h^\cdot]p_{Hg}}{[Hg^x_{Hg}]}$
3. $Hg^x_{Hg} \to V''_{Hg} + 2h^\cdot + Hg(g)$	$K''_{VHg} = \dfrac{[V''_{Hg}][h^\cdot]^2 p_{Hg}}{[Hg^x_{Hg}]}$
Noting $[Hg^x_{Hg}] \approx 1.0$	$K'_{VHg} = [V'_{Hg}][h^\cdot]p_{Hg}$
And	$K''_{VHg} = [V''_{Hg}][h^\cdot]^2 p_{Hg}$

Electroneutrality condition

$$[e'] + [V'_{Hg}] + 2[V''_{Hg}] = [h^\cdot]$$

Or in terms of relevant mass action constants

$$\frac{K_i}{[h^\cdot]} + \frac{K'_{VHg}}{[h^\cdot]}p_{Hg}^{-1} + \frac{2K''_{VHg}}{[h^\cdot]^2}p_{Hg}^{-1} = [h^\cdot]$$

Three different approximations to the electroneutrality condition can be considered.

A. Crystal is intrinsic with

$$[e'] = [h^\cdot] = \sqrt{k_i} \gg [V'_{Hg}] \text{ or }$$

$$[V''_{Hg}] \text{ at the high temperature (HT)}$$

Then $[V'_{Hg}] = \dfrac{K'_{VHg}}{\sqrt{k_i}} p_{Hg}^{-1}$ and $[V''_{Hg}] = \dfrac{K''_{VHg}}{k_i} p_{Hg}^{-1}$

In cooled crystals at 77 K:

$$[e']_{77\,K} \ll [h^\cdot]_{77\,K} = [V'_{Hg}] + 2[V''_{Hg}]$$

$$= \left(\frac{K'_{VHg}}{\sqrt{k_i}} + 2\frac{K''_{VHg}}{k_i}\right) p_{Hg}^{-1}$$

B. Crystal is extrinsic with

$$[e'] \ll [h^\cdot] = [V'_{Hg}] \gg [V''_{Hg}] \text{ at high temperature}$$

Then $K'_{VHg} = [V'_{Hg}]^2 p_{Hg}$ or $[V'_{Hg}] = K'_{VHg}{}^{1/2} p_{Hg}^{-1/2}$

For crystals at 77 K

$$[h^\cdot]_{77\,K} = [V'_{Hg}]_{HT} = (K'_{VHg})^{1/2} p_{Hg}^{-1/2}$$

C. Crystal is extrinsic with

$$(e') \ll [h^\cdot] = 2[V''_{Hg}] \gg [V'_{Hg}] \text{ at high temperature}$$

Then $K''_{VHg} = [V''_{Hg}](2[V''_{Hg}])^2 p_{Hg}$

Or $[V''_{Hg}]_{HT} = \left(\dfrac{K''_{VHg}}{4}\right)^{1/3} p_{Hg}^{-1/3}$

For crystals at 77 K:

$$[h^\cdot]_{77\,K} = 2[V''_{Hg}]_{HT} = \left\{2\left(\frac{K''_{VHg}}{4}\right)^{1/3} p_{Hg}^{-1/3}\right\} HT$$

Actual experimental results as shown in Fig. 1 indicate that the 77 K hole concentration in $Hg_{0.8}Cd_{0.2}Te$ crystals varies inversely with p_{Hg} indicating that the crystals are intrinsic at the high temperature[3,18] corresponding to the electroneutrality approximation A. However, the low temperature (192 K) hole concentration in $Hg_{0.6}Cd_{0.4}Te$ crystals varies as $p_{Hg}^{-1/3}$ (Fig. 2 and Ref. 19) indicating these crystals to be extrinsic with the dominance of doubly ionized Hg vacancies corresponding to the electroneutrality approximation C. It should be noted here that the results would be identical if tellurium interstitials Te'_i, Te''_i were considered instead of V'_{Hg} and V''_{Hg}.

The intrinsic behavior of undoped $Hg_{0.6}Cd_{0.4}Te$ can be accounted for, by noting that the band gap of $Hg_{0.8}Cd_{0.2}Te$ (0.1 eV) is only a quarter of the band gap of $Hg_{0.6}Cd_{0.4}Te$ (0.4 eV) at 77 K. Although electrical measurements cannot differentiate between Hg vacancies and Te interstitials, recent diffusion data[20] indicate the native acceptor defects to be Hg vacancies. Because of the deeper energy level of de-

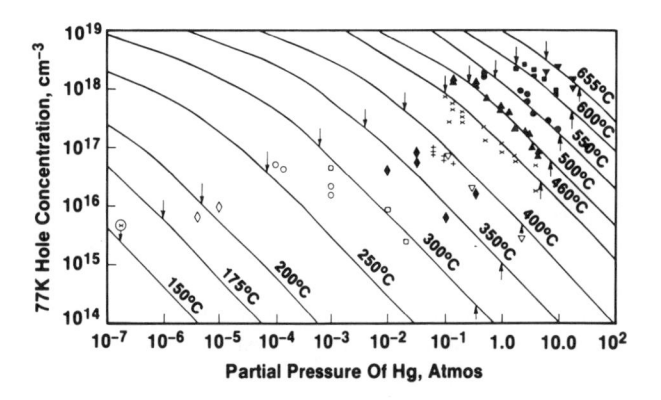

FIG. 1. 77 K hole concentration data for $Hg_{0.8}Cd_{0.2}Te$ crystals (bulk—400 to 655 °C, Epi—150 to 400 °C) as a function of the partial pressure of Hg showing p_{Hg}^{-1} dependence of the native acceptor defect concentration, solid lines indicate calculations using K_i and K''_{VHg} values discussed in Sec. III C of the text (Ref. 18).

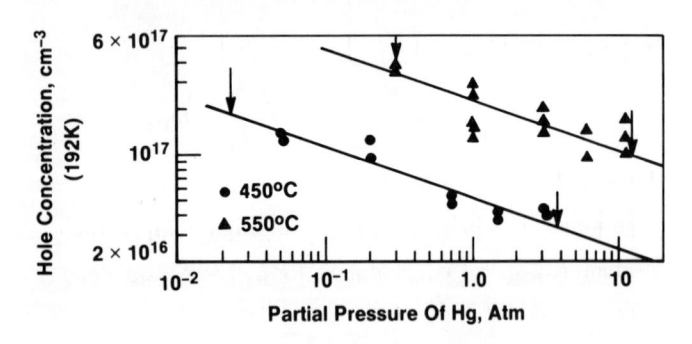

FIG. 2. 192 K hole concentration data for bulk $Hg_{0.6}Cd_{0.4}Te$ crystals as a function of the partial pressure of Hg showing $p_{Hg}^{-1/3}$ dependence of the native acceptor defect concentration (Ref. 19).

fects in the larger band gap $Hg_{0.6}Cd_{0.4}Te$ crystals a higher measurement temperature of 192 K is used to assure complete ionization of all the native defects. Experimentally, the native acceptor defects were found to be completely ionized at 77 K in $Hg_{0.8}Cd_{0.2}Te^{3,18}$ and at 192 K in $Hg_{0.6}Cd_{0.4}Te.^{19}$

A. Degree of ionization of the dominant native acceptor defects in undoped (Hg,Cd)Te

Although the native acceptor defects in $Hg_{0.6}Cd_{0.4}Te$ crystals could be confirmed to be doubly ionized by the $p_{Hg}^{-1/3}$ dependence of the hole concentration in the cooled crystals, the p_{Hg}^{-1} dependence of the 77 K hole concentration in $Hg_{0.8}Cd_{0.2}Te$ crystals does not establish whether these defects are singly ionized or doubly ionized.

To resolve the question of the degree of ionization of the native acceptor defects in $Hg_{0.8}Cd_{0.2}Te$ crystals the present author resorted to the hole mobility measurements in undoped crystals and in crystals doped with a dopant such as copper which was thought to be p type and singly ionized. For comparable hole concentrations, the experimentally measured hole mobility in the copper doped samples was found to be higher than in the undoped samples;[3,18] further, the experimentally measured mobility values were in better agreement with calculations for singly ionized centers in copper doped crystals and for doubly ionized centers in undoped crystals (Appendix A of Ref. 3). Based on these inferences the native acceptor defects in $Hg_{0.8}Cd_{0.2}Te$ also were established to be doubly ionized.

B. Lack of dominance of native donor defects in (Hg,Cd)Te alloys and origin of p to n conversion

Although undoped $Hg_{0.8}Cd_{0.2}Te$ and $Hg_{0.6}Cd_{0.4}Te$ crystals were found to be p type after equilibration at high temperatures in excess of 400 °C under all partial pressures of $Hg^{3,18,19}$ within the phase boundary limits of partial pressure of Hg, some of the crystals turn n type after anneals at temperatures below 400 °C. In those crystals where p to n conversion occurs, the conversion temperature is different for different samples and the electron concentration is independent of temperature with further decrease in the equilibration temperature. Additionally, the electron concentration is also independent of the partial pressure of Hg and the concentration is different in different samples. Finally, many samples do not turn n type at all even after equilibration at very low temperatures (< 200 °C) under Hg saturated conditions. Although independence of electron concentration on temperature does not by itself rule out the presence of native donor defects [Ref. 3, Appendix B], the independence of electron concentration on the partial pressure of Hg, the differing conversion temperature for different samples and the different electron concentrations for the different samples can only be explained if the p to n conversion occurs due to the presence of residual foreign donors and such a conversion occurs only in samples where the residual donor concentration is in excess of the residual acceptor concentration and the native acceptor defect concentration is well below the residual donor concentration. Samples with higher residual donor concentrations turn n type at higher temperatures and show higher electron concentrations. This explains why different samples turn n type at different temperatures and have different electron concentrations. Based on the results on undoped $Hg_{0.8}Cd_{0.2}Te^{3,18}$ and $Hg_{0.6}Cd_{0.4}Te^{19}$ it appears that the native donor concentration is at least below 10^{15} cm^{-3} at temperatures of 400 °C.

Absence of compensation in the undoped crystals has also been inferred[3,19] based on the lack of dependence of hole mobility on the partial pressure of Hg or annealing temperature but only on the hole concentration. If there was compensation due to Frenkel or Schottky defects, their concentration would be exponentially temperature dependent on the annealing temperature and dependence of hole mobility on the annealing temperature should have been inferred, in contrast to the actual experimental results.

C. Mass action constants K_i and K''_{VHg} for undoped $Hg_{0.8}Cd_{0.2}Te$

The two mass action constants K_i and K''_{VHg} which can predict the defect concentration in $Hg_{0.8}Cd_{0.2}Te$ from a temperature ranging from 150 to 655 °C are given by

$$K_i = [e'][h^\cdot] = 5.8 \times 10^{-4} \exp(-0.57 \text{ eV}/kT) \text{Site Fr}^2$$
$$= 9.2 \times 10^{40} \exp(-0.57 \text{ eV}/kT) \text{cm}^{-6}$$

and

$$K''_{VHg} = [V''_{Hg}][h^\cdot]^2 p_{Hg}$$
$$= 790 \exp(-2.24 \text{ eV}/kT) \text{ (Site Fr)}^3 \text{ atm}$$
$$= 1.6 \times 10^{69} \exp(-2.24 \text{ eV}/kT) \text{cm}^{-9} \text{ atm}.$$

The calculated hole concentrations (or Hg vacancies) in $Hg_{0.8}Cd_{0.2}Te$ are shown as solid lines in Fig. 1.

IV. INCORPORATION OF DONOR DOPANTS IN (HG,CD)TE ALLOYS

Donor behavior is expected of elements from group III B on metal lattice sites and in interstitial sites where as elements from group VII B substituting Te lattice sites would be expected to behave as donors and as acceptors in interstitial sites.

There is evidence of the donor behavior of group III B elements B, Ga, Al, and In and of group VII B elements Br, Cl, and I.[2,5,8,9] However, to the author's knowledge, a detailed investigation of the mode of incorporation has been pursued only for indium and iodine[4,6] and incorporation behavior of these elements will be presented in more detail below.

A. Indium doped Hg$_{0.8}$Cd$_{0.2}$Te

Because of the high solubility and moderately high diffusivity of indium in (Hg,Cd)Te it is relatively easy to prepare samples with high concentrations of indium via diffusion in relatively short periods of time. In the indium doped crystals of Hg$_{0.8}$Cd$_{0.2}$Te, the electron concentration at 77 K is found to increase with increase in the concentration of indium and with increase in the partial pressure of Hg at the temperature of anneal [Figs. 3, 4 and Ref. 4]. Additionally, the concentration of electrons [e'] is found to be much less than the total indium concentration, and [e']$_{77\,K} \propto$ [In$_{Tot}$]$^{1/2}$, and $p_{Hg}^{1/2}$ [Figs. 3, 4]. The incorporation mechanism which explains these data most satisfactorily is the one where indium acts as a single donor occupying metal lattice sites and at concentrations of indium in excess of 10^{18} to 10^{19} cm^{-3} at the annealing temperatures of 500 to 600 °C, a major fraction of the indium appears to be incorporated as (In$_{Hg}$V$_{Hg}$In$_{Hg}$)^x which is identical to In$_2$Te$_3$(s) dissolved in (Hg,Cd)Te. Despite the presence of large concentrations of indium, the crystals appear to behave intrinsic at 500 to 600 °C because of the presence of most of the indium as neutral triplets of (In$_{Hg}$V$_{Hg}$In$_{Hg}$)^x. The dependence of 77 K electron concentration on [In$_{Tot}$] and p_{Hg} can be inferred via the incorporation reaction:

$$2e' + 2\text{In}_{Hg}^{\bullet} + \text{Hg}_{Hg}^x \rightarrow (\text{In}_{Hg}V_{Hg}\text{In}_{Hg})^x + \text{Hg}(g).$$

If the crystals are intrinsic at the temperature of anneal, [e'] = [$h^{\bullet}$] = $K_i^{1/2} \gg$ [In$_{Hg}^{\bullet}$]; with [(In$_{Hg}$V$_{Hg}$In$_{Hg}$)^x] $\sim$ [In$_{Tot}$]/2 $\gg$ [In$_{Hg}^{\bullet}$]

then $K_{(\text{In}_{Hg}V_{Hg}\text{In}_{Hg})^x} = \dfrac{[\text{In}_{Tot}]/2\, p_{Hg}}{K_i [\text{In}_{Hg}^{\bullet}]^2}$

or [In$_{Hg}^{\bullet}$] $\propto$ [In$_{Tot}$]$^{1/2} p_{Hg}^{1/2}$.

In the crystals cooled to 77 K, the Fermi level is pinned at the donor level and [e']$_{77\,K}$ = [In$_{Hg}^{\bullet}$] minus 2[V_{Hg}'']; if [In$_{Hg}^{\bullet}$] $\gg$ [V_{Hg}''] [e']$_{77\,K} \propto$ [In$_{Tot}$]$^{1/2} p_{Hg}^{1/2}$.

Figure 5 shows the defect isotherm for indium doped Hg$_{0.8}$Cd$_{0.2}$Te at an annealing temperature of 500 °C and partial pressure of Hg, p_{Hg} = 3 atm.

B. Iodine doped Hg$_{0.8}$Cd$_{0.2}$Te

The behavior of iodine in Hg$_{0.8}$Cd$_{0.2}$Te crystals[6] is somewhat similar to that of indium, with the 77 K electron concentration increasing with increase in p_{Hg} in Iodine doped crystals annealed at temperatures of 450 to 600 °C at various partial pressures of Hg (Fig. 6 and Ref. 6). Just as in the case of the indium doped crystals, iodine doped crystals are intrinsic at the temperatures of anneal. The electrical data can

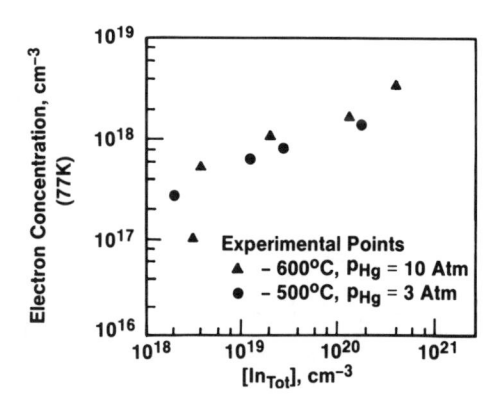

FIG. 3. 77 K electron concentration as a function of the total indium concentration for Hg$_{0.8}$Cd$_{0.2}$Te crystals which were quenched to room temperature subsequent to anneal at 500 and 600 °C at the indicated partial pressures of Hg; the electron concentration roughly increases as [In$_{Tot}$]$^{1/2}$ (Ref. 4).

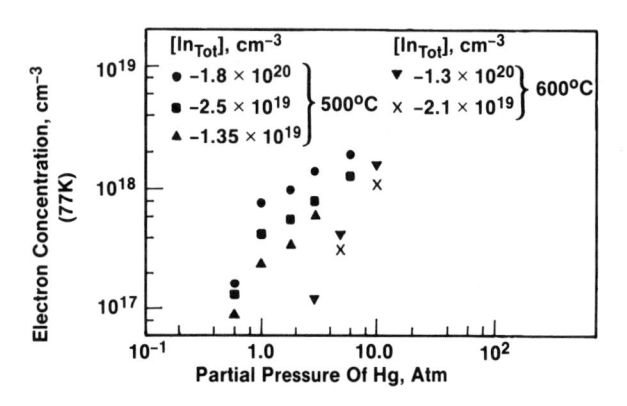

FIG. 4. 77 K electron concentration as a function of the partial pressure of Hg for Hg$_{0.8}$Cd$_{0.2}$Te crystals which were doped with various total concentrations of indium, annealed at 500 and 600 °C at various partial pressures of Hg and quenched to room temperature; the electron concentration roughly increases as $p_{Hg}^{1/2}$ in a region away from the p–n transition. (Ref. 4).

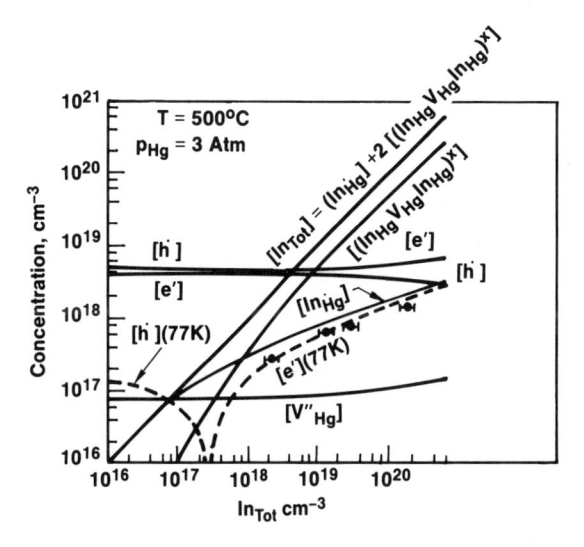

FIG. 5. Calculated defect concentrations ([e'], [$h^{\bullet}$], [V_{Hg}''], [In$_{Hg}^{\bullet}$], and [(In$_{Hg}$V$_{Hg}$In$_{Hg}$)^x]) as a function of the total indium concentration in the Hg$_{0.8}$Cd$_{0.2}$Te crystals at 500 °C and p_{Hg} = 3 atm; the calculated 77 K carrier concentrations in the crystals quenched to room temperature are shown as dashed lines along with experimentally measured electron concentrations. Calculations (detailed in Ref. 4) assume the values for the mass action constants K_i and $K_{V_{Hg}''}''$ given in Sec. III C of the text.

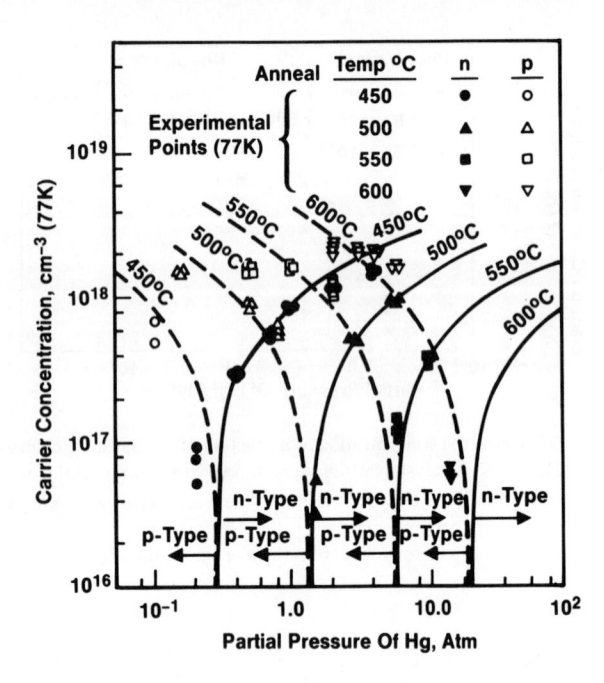

FIG. 6. Carrier concentration at 77 K as a function of the partial pressure of Hg for iodine doped $Hg_{0.8}Cd_{0.2}Te$ crystals annealed at various temperatures and quenched to room temperature; solid lines show the calculated electron concentrations while the dashed lines show the calculated hole concentrations (Ref. 6).

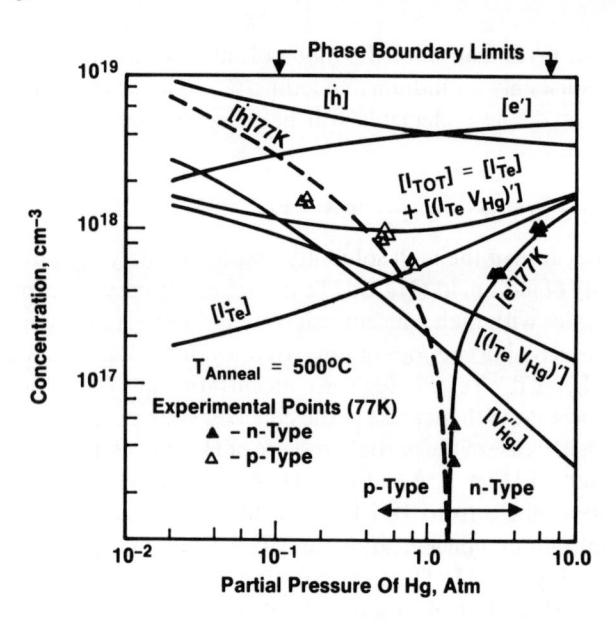

FIG. 7. Calculated defect concentrations ($[e']$, $[h^\cdot]$, $[I_{Te}^\bullet]$, $[(I_{Te}V_{Hg})']$ and $[V_{Hg}'']$) as a function of the partial pressure of Hg for iodine doped $Hg_{0.8}Cd_{0.2}Te$ samples which were annealed at 500 °C at various partial pressures of Hg and quenched to room temperature; calculated electron and hole concentrations at 77 K indicated by solid and dashed lines, respectively, are also shown along with the experimental points. Calculations (detailed in Ref. 6) assume the values for the mass action constants K_i and $K_{V_{Hg}}''$ given in Sec. III C of the text.

be explained on the basis of a defect model where the (Hg,Cd)Te crystals are saturated with $(Hg,Cd)I_2$ with the iodine being present as donors on Te lattice sites ($I_{Te}^\bullet$) and a fraction being present as negatively charged $(I_{Te}V_{Hg})'$ pairs. The incorporation reaction of interest is

$$Hg(g) + (Hg_{0.8}Cd_{0.2})I_2(s)$$
$$\rightarrow 2I_{Te}^\bullet + 2e' + 1.8Hg_{Hg}^x + 0.2Cd_{Hg}^x$$

$$K_{I_{Te}} = \frac{[I_{Te}^\bullet]^2[e']^2[Hg_{Hg}^x]^{1.8}[Cd_{Hg}^x]^{0.2}}{p_{Hg} a_{(Hg_{0.8}Cd_{0.2})I_2}},$$

where $a_{(Hg_{0.8}Cd_{0.2})I_2}$ is the activity of $(Hg_{0.8}Cd_{0.2})I_2$. If the crystals are intrinsic $[e'] = [h^\cdot] = K_i^{1/2} \gg [I_{Te}^\bullet]$. If $(Hg_{0.8}Cd_{0.2})I_2$ exists as a pure second phase, $a_{(Hg_{0.8}Cd_{0.2})I_2} = 1.0$, then $[I_{Te}^\bullet] \propto p_{Hg}^{1/2}$. Similarly, it can be shown that $[(I_{Te}V_{Hg})'] \propto p_{Hg}^{-1/2}$. Figure 6 shows the calculated carrier concentration at 77 K along with experimental points. Figure 7 shows the defect isotherm at an annealing temperature of 500 °C.

V. INCORPORATION OF ACCEPTOR DOPANTS IN (Hg,Cd)Te ALLOYS

Elements of group I are expected to behave as acceptors substituting metal lattice sites and as donors interstitially. Elements of group V are expected to behave as acceptors interstitially and substituting Te lattice sites whereas they are expected to behave as donors substituting metal lattice sites.

Although Ag, Cu, and Au have been found to behave as acceptors in $(Hg,Cd)Te$[2] the behavior of copper appears to be most straightforward.[3] It has a high solubility in excess of 10^{19} cm^{-3} and an extremely high diffusivity. Copper has been found to behave as a single acceptor occupying metal lattice sites and there is no evidence of its incorporation in interstitial sites.[3]

A. Phosphorus doped $Hg_{0.8}Cd_{0.2}Te$

Elements from group V B are found to be very useful as p type dopants in (Hg,Cd)Te.[11,14,15] The extremely low diffusivity of these elements makes them very attractive for fabricating stable p–n junctions. P, As, and Sb have been used extensively as dopants to make photovoltaic devices. Of these elements, the mode of incorporation of phosphorus has been studied in greater detail[7] than of the others.

Figure 8 shows the 77 K hole concentration in P doped

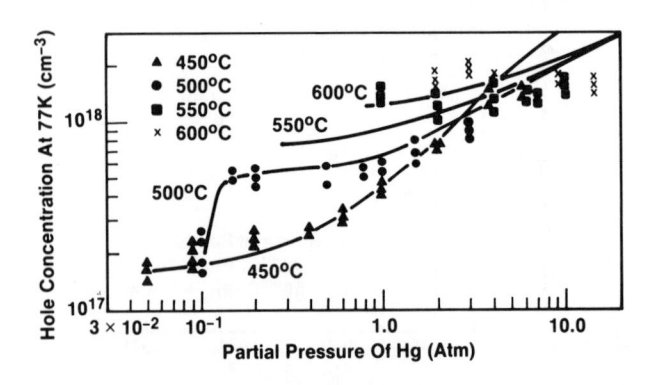

FIG. 8. Hole concentration at 77 K as a function of the partial pressure of Hg for the phosphorus doped (10^{19} cm^{-3}) $Hg_{0.8}Cd_{0.2}Te$ samples which were quenched to room temperature after annealing at different temperatures and partial pressures of Hg. The solid curves show the calculated hole concentrations to be expected in the samples on the basis of the defect model detailed in Ref. 7.

TABLE I. Defect formation reactions, mass-action relations, electroneutrality condition, and phosphorous balance equation.

Reaction	Mass-action relation
(1) $0 \rightarrow e' + h^{\cdot}$;	$K_i = [e'][h^{\cdot}]$
(2) $Hg^x_{Hg} \rightarrow V''_{Hg} + 2h^{\cdot} + Hg(g)$;	$K''_{v_{Hg}} = [V''_{Hg}][h^{\cdot}]^2 p_{Hg}$
(3) $2h^{\cdot} + (P_i P_{Hg})^x + Hg^x_{Hg} \rightarrow 2P^{\bullet}_{Hg} + Hg(g)$;	$K_{P_{Hg}} = [P^{\bullet}_{Hg}]^2 p_{Hg}/[h^{\cdot}]^2[(P_i P_{Hg})^x]$
(4) $(P_i P_{Hg})^x + 3Hg(g) \rightarrow 3Hg^x_{Hg} + 2P'_{Te} + 2h^{\cdot}$;	$K_{P_{Te}} = [P'_{Te}]^2[h^{\cdot}]^2/[(P_i P_{Hg})^x]p^3_{Hg}$
(5) $P'_i + P^{\bullet}_{Hg} \rightarrow (P_i P_{Hg})^x$;	$K_P(P_i P_{Hg})^x = [(P_i P_{Hg})^x]/[P'_i][P^{\bullet}_{Hg}]$
(6) $P^{\bullet}_{Hg} + V''_{Hg} \rightarrow (P_{Hg} V_{Hg})'$;	$K_P(P_{Hg} V_{Hg})' = [(P_{Hg} V_{Hg})']/[P^{\bullet}_{Hg}][V''_{Hg}]$
(7) $(P_i P_{Hg})^x + 2h^{\cdot} + 3Hg^x_{Hg} \rightarrow 2(P_{Hg} V_{Hg})^{\bullet} + 3Hg(g)$;	$K(P_{Hg} V_{Hg})^{\bullet} = [(P_{Hg} V_{Hg})^{\bullet}]^2 p^3_{Hg}/[(P_i P_{Hg})^x][h^{\cdot}]^2$

Electroneutrality condition
$[e'] + 2[V''_{Hg}] + [P'_i] + [P'_{Te}] + [(P_{Hg} V_{Hg})'] = [h^{\cdot}] + [P^{\bullet}_{Hg}] + [(P_{Hg} V_{Hg})^{\bullet}]$

Phosphorus balance equation
$[P'_i] + [P^{\bullet}_{Hg}] + [P'_{Te}] + [(P_{Hg} V_{Hg})'] + [(P_{Hg} V_{Hg})^{\bullet}] + 2[(P_i P_{Hg})^x] = [P_{Tot}]$

$p(77K) = [P'_i] + [P'_{Te}] + [(P_{Hg} V_{Hg})'] + 2[V''_{Hg}] - [P^{\bullet}_{Hg}] - [(P_{Hg} V_{Hg})^{\bullet}]$.

bulk $Hg_{0.8}Cd_{0.2}Te$ crystals subsequent to anneals at various temperatures and partial pressures of Hg. All the samples are found to be p type with the hole concentration increasing with increase in p_{Hg} in contrast to the behavior of undoped $Hg_{0.8}Cd_{0.2}Te$ crystals (Fig. 1 and Refs. 3, 18) where the hole concentration decreases with increase in p_{Hg}. Further, it is to be noted that the hole concentration in the P doped crystals is higher than that of the undoped crystals at high p_{Hg} and lower than that of the undoped crystals at low p_{Hg} indicating that P acts as an acceptor at high p_{Hg} and as a donor at low p_{Hg}. The concentration of holes is also found to be much less than that of P in the crystals. All these inferences can be explained on the basis of a defect model according to which P behaves amphoterically in $Hg_{0.8}Cd_{0.2}Te$ acting as a single acceptor interstitially (P'_i) and on Te lattice sites (P'_{Te}) at high Hg pressures and as a single donor on metal lattice sites ($P^{\bullet}_{Hg}$) at low Hg pressures. The defect model also predicts a majority of the P to be present as neutral pairs $(P_i P_{Hg})^x$ at intermediate Hg pressures along with a large fraction being present as pairs $(P_{Hg} V_{Hg})^{\bullet}$ and $(P_{Hg} V_{Hg})'$ at low Hg pressures. Table I lists the various defect formation reactions along with respective mass action constants which are relevant to the incorporation of P in $Hg_{0.8}Cd_{0.2}Te$; the table also includes the electroneutrality and the P dopant balance equation needed to calculate the defect concentrations as $f(p_{Hg})$ at the annealing temperature, the table also lists the expression for the 77 K hole concentration in the crystals in terms of the defect concentrations at the high temperature. Table II lists the values of the mass action constants for the various defect formation reactions listed in Table I. With the values of these mass action constants, the electroneutrality condition and the P dopant balance equation (shown in Ta-

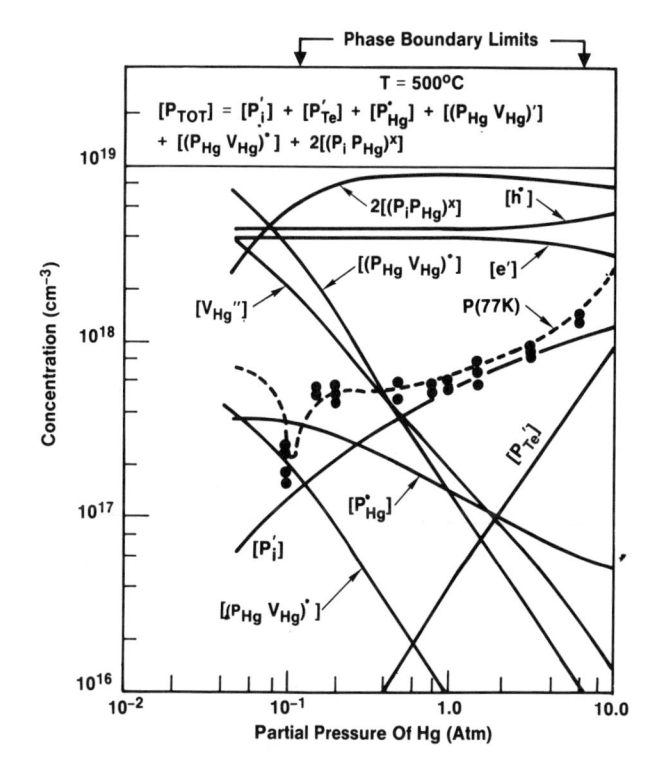

TABLE II. Values of the parameters for the equilibrium constants, $K = K_0 \exp(-H/KT)$ defined in Table I.

Equilibrium constant	K_0 (site fr., atm)	H (eV)	Source
1 K_i	5.77×10^{-4}	0.57	Ref. 3, 18
2 $K''_{v_{Hg}}$	7.9×10^2	2.24	
3 $K_{P_{Hg}}$	1.36×10^{-2}	-0.360	
4 $K_{P_{Te}}$	1.81×10^{-26}	-1.73	
5 $K_P(P_i P_{Hg})^x$	1.76×10^4	-0.252	Ref. 7
6 $K_{P(P_{Hg} V_{Hg})'}$	17	-0.36	
7 $K_{(P_{Hg} V_{Hg})^{\bullet}}$	2.1×10^{28}	4.29	

There are 1.26×10^{22} molecules/cm^3 in $Hg_{0.8}Cd_{0.2}Te$.

FIG. 9. Calculated defect concentrations ($[e']$, $[h^{\cdot}]$, $[V''_{Hg}]$, $[P'_i]$, $[P'_{Te}]$, $[P^{\bullet}_{Hg}]$, $[(P_{Hg} V_{Hg})^{\bullet}]$, $[P_{Hg} V_{Hg})']$, and $[(P_i P_{Hg})^x]$) as a function of the partial pressure of Hg at 500 °C; the expected hole concentration in the crystals cooled to 77 K is indicated as a dashed curve along with the experimental results. Calculations detailed in Ref. 7 are based on the values of the mass action constants listed in Table II.

TABLE III. Evidence of amphoteric behavior of $Hg_{1-x}Cd_xTe$ films grown from Te-rich solutions doped with group V elements and annealed at 200 °C under Hg saturated conditions with and without a 500 °C Hg saturated preanneal.

Film number	Composition	Dopant and concentration in the Te rich growth solution, At. %	200 °C Hg saturated anneal $\|N_A - N_D\|$ cm^{-3} (Variable Temp R_H)	77 K Hall mobility cm^2/V s	200 °C Hg saturated anneal with a 500 °C Hg saturated preanneal $\|N_A - N_D\|$ cm^{-3} (Variable temp R_H)	77 K Hall mobility cm^2/V s
109	0.31	P, 0.8	4×10^{16} p type	186		
133	0.30	P, 0.09	1.7×10^{15} n type	3.3×10^4	1×10^{16} p type	111
134	0.30	P, 0.2	2×10^{15} p type	328	4×10^{16} p type	210
100	0.29	As, 1	1.5×10^{14} n type	1.8×10^3	4.4×10^{15} p type	176
121	0.31	As, 10	6×10^{14} n type	1.4×10^4	1.2×10^{17} p type	292
103	0.30	Sb, 0.7	2×10^{13} n type	1.2×10^4	4.5×10^{14} n type	3.7×10^3
117	0.30	Sb, 1.0	1.6×10^{15} p type	172	1×10^{16}	304
118	0.33	Sb, 10.0	4×10^{14} n type	3.6×10^3	1.3×10^{17} p type	196
107	0.29	Bi, 0.95	7×10^{14} n type	4.4×10^4	6×10^{14} n type	2.2×10^4
132	0.27	Bi, 8.7	1×10^{16} n type	2.2×10^4	1.6×10^{15} n type	7.8×10^4
146	0.24	Bi, 0.006	2×10^{15} n type	7.3×10^4	1×10^{15} n type	4.5×10^4
147	0.28	Bi, 0.06	1.6×10^{15} n type	4.2×10^4	9×10^{14} n type	3.4×10^4

ble I) can be solved in terms of the concentration of one of the defect species with p_{Hg}, T, and $[P_{Tot}]$ as variables. Once the concentration of one defect species is calculated, the concentrations of all the other defect species can be immediately determined. With the knowledge of the concentration of all the defect species at the annealing temperature as a function of p_{Hg} and T, the hole concentration at 77 K is also easily calculated.

We summarize the results of defect model calculations for P doped $Hg_{0.8}Cd_{0.2}Te$ in Fig. 9 where the concentration of all the defects is shown as a function of the partial pressure of Hg at $T = 500$ °C and $[P_{Tot}] = 10^{19}$ cm^{-3}. The calculated hole concentration in the crystals cooled to 77 K is shown as a dotted line along with experimental points. The details of the defect model calculations can be found in Ref. 7.

B. Group V element doping issues in (Hg,Cd)Te grown liquid phase epitaxially from Te rich solutions

Because of the increasingly significant role of the epitaxial growth technology needed in fabricating novel multicompositional heterostructures, it is appropriate to discuss the group V element doping issues in epitaxially grown (Hg,Cd)Te. Although there is a vast amount of group V element doping work, currently in progress, in MBE and MOCVD grown (Hg,Cd)Te, we will restrict our discussions to LPE grown material; details of doping work in MBE and MOCVD material can be found in the excellent review article by Capper.[21]

Vydyanath et. al.[12] have confirmed the amphoteric behavior of the group V dopants P, As, and Sb in epitaxial $Hg_{0.7}Cd_{0.3}Te$ films grown from Te rich solutions. In a series of anneals at 200 °C under Hg-saturated conditions with and without a 500 °C Hg-saturated preanneal, they observed n to p conversion in many of the films [see Table III]. The site transfer reactions associated with transfer of the group V elements from metal lattice sites to interstitial or Te lattice sites are given by

$$D^{\bullet}_{Hg} + 2Hg(g) \rightarrow D'_{Te} + 2h^{\bullet} + 2Hg^x_{Hg}$$

$D \equiv$ P, As, Sb, or Bi

$$K_{D^{\bullet}_{Hg}D'_{Te}} = \frac{[D'_{Te}][h^{\bullet}]^2}{[D^{\bullet}_{Hg}]P^2_{Hg}}$$

$$D^{\bullet}_{Hg} + Hg(g) \rightarrow D'_i + 2h^{\bullet} + Hg^x_{Hg}$$

$D \equiv$ P, As, Sb, or Bi

$$K_{D^{\bullet}_{Hg}D'_i} = \frac{[D'_i][h^{\bullet}]^2}{[D^{\bullet}_{Hg}]p_{Hg}}.$$

The 500 °C preanneal under Hg rich conditions provides the thermal energy required for the site transfer. Vydyanath[15] has calculated the values of the constants

$K_{D^{\bullet}_{Hg}D'_{Te}}$ and $K_{D^{\bullet}_{Hg}D'_i}$

for P, As, and Sb from the electrical data in epitaxial $Hg_{0.7}Cd_{0.3}Te$ films (Table IV). Finally, the distribution coefficients of these elements for growths from Te rich solutions have been evaluated[12] and are shown in Fig. 10 along

TABLE IV. Values of mass action constants $K_{D^{\bullet}_{Hg}D'_{Te}}$ and $K_{D^{\bullet}_{Hg}D'_i}$ for group V dopants ($D \equiv$ P, As, and Sb) in epitaxial $Hg_{0.7}Cd_{0.3}Te$ films grown from Te-rich solutions (Ref. 12).

Dopant	Temperature (°C)	$K_{D^{\bullet}_{Hg}D'_{Te}}$ (cm^{-6} atm^{-2})	$K_{D^{\bullet}_{Hg}D'_i}$ (cm^{-6} atm^{-1})
Phosphorus	500	7×10^{36} (5.7×10^{36})[a]	1.2×10^{37} (7×10^{37})[a]
Arsenic	500	2.4×10^{38}	4×10^{38}
Antimony	500	3.7×10^{38}	6.3×10^{38}

[a] Bulk $Hg_{0.8}Cd_{0.2}Te$ (Ref. 7).

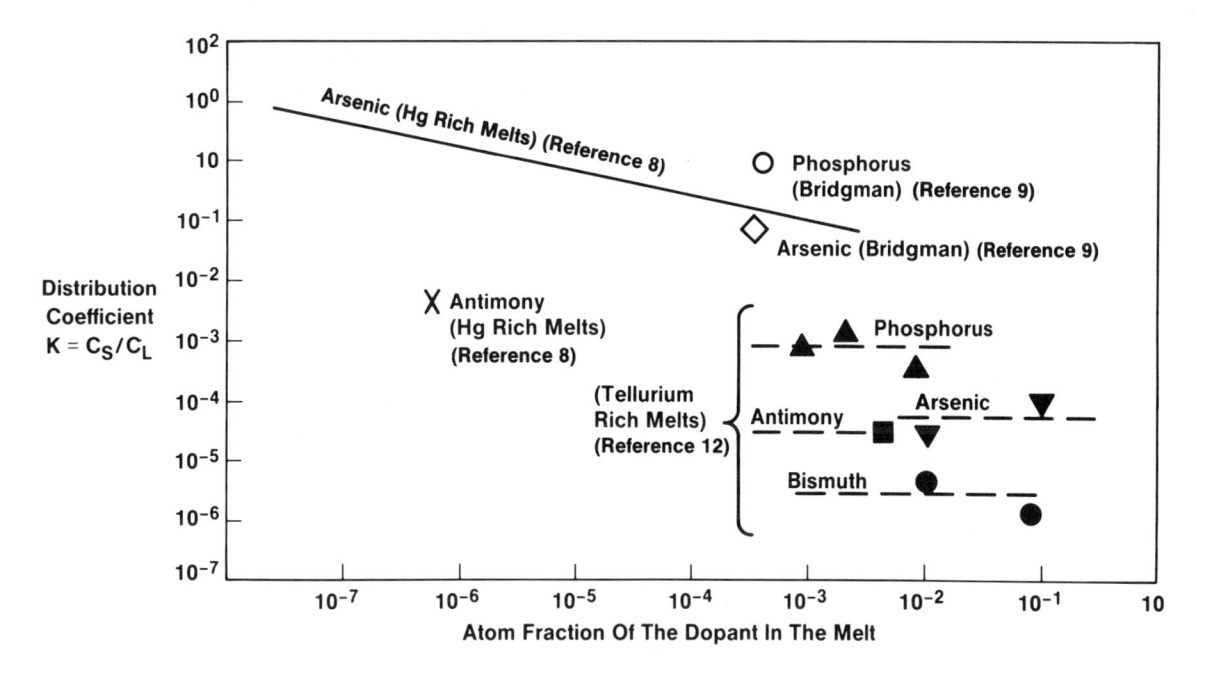

FIG. 10. Distribution coefficient of the group V elements in $Hg_{1-x}Cd_xTe$ as a function of the dopant concentration in the melt for Hg rich, Te rich and pseudobinary growths. (Ref. 12).

with the data for growths from Hg rich[8] and pseudobinary solutions.[9] From the results shown in the figure, it can be noted that the distribution coefficient of group V elements for growths from Te rich solutions is orders of magnitude lower than that for growths from Hg rich solutions and has been attributed to the activity coefficient of the group V elements being orders of magnitude lower in Te rich solutions than in Hg rich solutions.[12]

VI. SUMMARY

The defect structure of undoped $Hg_{0.8}Cd_{0.2}Te$ and $Hg_{0.6}Cd_{0.4}Te$ is established experimentally using the quasichemical approach developed by Kröger and Vink.[16,17] The dominant native defects in (Hg,Cd)Te crystals have been shown to be acceptor in nature and to be doubly ionized. The native donor defect concentration is established to be negligible and the p to n conversion has been shown to result from the presence of residual donors. Equilibrium constants have been arrived at, for the intrinsic excitation process and the formation of the native acceptor defects. The mechanisms of incorporation of $Hg_{0.8}Cd_{0.2}Te$ crystals doped with indium, iodine, phosphorus and other group V B elements are discussed.

With the equilibrium constants established for the formation of defects in the undoped and doped crystals, the defect concentrations have been calculated along with carrier concentrations expected in the cooled crystals as a function of the thermodynamic conditions of preparation of the crystals at the high temperature.

[1] *Semiconductors and Semimetals*, edited by R. K. Willardson and A. C. Beer (Academic, New York, 1981), Vol. 18.
[2] E. S. Johnson and J. L. Schmit, J. Electron. Mater. **6**, 25 (1977).
[3] H. R. Vydyanath, J. Electrochem. Soc. **128**, 2609 (1981).
[4] H. R. Vydyanath, J. Electrochem. Soc. **128**, 2619 (1981).
[5] P. Capper, J. Cryst. Growth **57**, 280 (1982).
[6] H. R. Vydyanath and F. A. Kröger, J. Electron. Mater. **11**, 111 (1982).
[7] H. R. Vydyanath, R. C. Abbott, and D. A. Nelson, J. Appl. Phys. **54**, 1323 (1983).
[8] M. H. Kalisher, J. Cryst. Growth **70**, 365 (1984).
[9] P. Capper, J. G. Gosney, C. L. Jones, and I. Kenworthy, J. Cryst. Growth **71**, 57 (1985).
[10] C. E. Jones, K. James, J. Merz, R. Braunstein, M. Burd, M. Eetemadi, and J. Drumheller, J. Vac. Sci. Technol. A **3**, 131 (1985).
[11] L. O. Bubulac, D. S. Lo, W. E. Tennant, D. D. Edwall, J. C. Chen, and J. Ratusnik, Appl. Phys. Lett. **50**, 22 (1987).
[12] H. R. Vydyanath, J. A. Ellsworth, and C. M. Devaney, J. Electron. Mater. **16**, 13 (1987).
[13] G. L. Destafanis, J. Cryst. Growth **86**, 700 (1988).
[14] L. O. Bubulac, J. Cryst. Growth **86**, 723 (1988).
[15] H. R. Vydyanath, Semicond. Sci. Technol. **5**, S213 (1990).
[16] F. A. Kröger and H. J. Vink, *Solid State Physics III*, edited by F. Seitz and D. Turnbull (Academic, New York, 1956), p. 307.
[17] F. A. Kröger, *The Chemistry of Imperfect Crystals*, 2nd ed. (North Holland, Amsterdam, 1974), Vol. 2.
[18] H. R. Vydyanath and C. H. Hiner, J. Appl. Phys. **65**, 3080 (1989).
[19] H. R. Vydyanath, J. C. Donovan, and D. A. Nelson, J. Electrochem. Soc. **128**, 2625 (1981).
[20] M. S. Tang and D. A. Stevenson, J. Vac. Sci. Technol. A **7**, 544 (1989).
[21] P. Capper, J. Vac. Sci. Technol. B **9**, 1667 (1991).

Growth and characterization of P-on-n HgCdTe liquid-phase epitaxy heterojunction material for 11–18 μm applications

G. N. Pultz, Peter W. Norton, E. Eric Krueger, and M. B. Reine
Loral Infrared and Imaging Systems, Lexington, Massachusetts 02173

(Received 18 December 1990; accepted 11 February 1991)

The capability of growing long-wavelength infrared HgCdTe liquid-phase epitaxy P-on-n heterojunction films with state-of-the-art photodiode performance, with excellent thickness uniformity ($\pm 10\%$), and with excellent cutoff wavelength uniformity (e.g., 10.5 ± 0.1 μm) across 2.5 cm $\times$ 4.0 cm wafers has been demonstrated. In addition, we have extended the region of HgCdTe photodiode operation to wavelengths of 18–19 μm at 80 K. Both measured carrier lifetime and photodiode data show that the n-type HgCdTe base layers are of excellent quality, with 77 K carrier lifetimes at the calculated Auger-1 limit for film carrier concentrations above 4×10^{14} cm^{-3}. The R_0A products for large-area diodes (10^{-3} cm^2) with cutoff wavelengths of 11–19 μm are consistent with n-side diffusion current calculated using the film Auger-1 lifetime. Smaller diodes of area 1×10^{-5} cm^2 have typical R_0A values of 12 Ω cm^2 at 80 K for a 12.2 μm cutoff wavelength. Large area diodes with an 80 K cutoff wavelength of 18–19 μm have R_0A products of 0.14 Ω cm^2 at 80 K and 0.4 Ω cm^2 at 70 K. Quantum efficiency values of 60% are observed with no antireflection coating (corresponding to 75% internal quantum efficiency) and the spectral response data are classical.

I. INTRODUCTION

Meeting the specifications for long-wavelength infrared (LWIR) HgCdTe photodiodes in the next generation of satellite instruments currently being designed will require improvements in HgCdTe material in order to extend detection out to the 18 μm range. While detectors in previous instruments have been based on either bulk grown HgCdTe[1] or single layer Liquid phase epitaxy (LPE) HgCdTe films that are ion implanted to form n-on-p homojunctions, it has become apparent that future LWIR needs will only be met by taking advantage of epitaxially grown P-on-n heterostructures (here the uppercase P denotes the region of wider energy band gap). To date, high quality LWIR HgCdTe heterojunction photodiodes grown by LPE in both the P-on-n and N-on-p configurations have been reported[2–6] with detector cutoff wavelengths as long as 11.5 μm. Alternatives such as metalorganic chemical vapor deposition offer the hope of more sophisticated multiple layer structures in the future.

Growth of single layer HgCdTe by both Te-rich and Hg-rich LPE has been demonstrated by many investigators.[7–10] However, the growth of multiple layer P-n junction heterostructures adds significant complexity. When compared to the III–V materials grown by LPE, it would seem that Te-rich rather than Hg-rich growth would be the preferred approach, since its typical format of using a horizontal slider boat is most similar to the Ga-based growth of GaAs/GaAlAs multiple layer structures which have served the optoelectronics industry for so long. Additionally, the Te-rich dipper has the advantages of high Cd solubility at temperatures of 450–500 °C allowing good control of x value, a low Hg overpressure (about 1 atm), and the use of indium as a donor dopant. Unfortunately, the high growth temperatures for Te-rich growth (420–490 °C) result in significant interdiffusion of HgCdTe layers, and data for p-type films grown by Te-rich LPE show that acceptor dopants are difficult to activate.[11–12] Hg-rich LPE systems, on the other hand, are basically high-pressure dipper systems which are not readily applicable for multiple layer, sequentially grown structures. However, acceptor dopants (Sb and As) are readily activated in Hg-rich LPE,[2–5] and low temperature growth can maintain layer integrity if growth times are minimized. We have taken advantage of the known strong points of both the Te-rich LPE and Hg-rich LPE growth formats, and combined them into a sequential two-step process capable of producing high-yield, high-performance HgCdTe heterostructure detectors and have extended the 80 K wavelength range out to 19 μm.

This paper will report on the state-of-the-art performance achieved in variable-area photodiode arrays fabricated from films grown with this process.

II. EXPERIMENT

A. Film growth and characterization

The first step of the growth process was to grow an indium-doped n-type base layer of Hg$_{1-x}$Cd$_x$Te, 15 to 22 μm thick, using a Te-rich horizontal slider. CdTe substrates of $\langle 111 \rangle$B orientation were used. Base layer doping was achieved by forming an intermediate dilution ingot and was easily controlled for resultant carrier concentrations in the low-10^{14} cm^{-3} to mid-10^{15} cm^{-3} range. The carrier concentrations were measured from patterned samples on an annealed sacrificial piece of the base layer using the Van der Pauw-Hall technique. Additionally, capacitance–voltage (C–V) measurements were made on large diodes for comparison to the Hall measurements. The base layer thickness was determined from interference fringes in the room-temperature transmittance data measured on a Bio-Rad FTL40 Fourier Transform Spectrometer. The x value of the base was also determined from the transmittance spectra. Minority carrier lifetime values are measured in the annealed n-type base layer piece from either Loral's Optically Modulated Absorption technique[13] or photoconductive (PC) frequency response measurements at 85 K.

After growth of the base layer, the base layer is loaded into the vertical slider, Hg-rich system, and a 1–2-μm-thick p-type cap layer is grown, As-doped (5×10^{16} cm^{-3} to 1×10^{18} cm^{-3}), at a growth temperature around 400 °C. The patterned cap layer sample for the Van der Pauw–Hall measurement comes from a sacrificial CdTe piece fitted in the graphite boat set during cap growth. Thickness of the cap layer film is measured from an optical photograph of the cross section of the sacrificial CdTe piece. The cap layer x value was measured by surface reflectance spectroscopy[14] in the visible wavelength region.

The resulting double-layer films are then annealed under Hg-saturation conditions at a temperature of 240–260 °C to reduce the native defect acceptor concentration to values well below the indium donor concentration in the base layer, thereby converting the base layer from p-type to n-type.

B. Photodiode characterization

To characterize the grown junction, arrays of 43 ZnS-passivated circular backside-illuminated mesa photodiodes of various areas (0.13×10^{-4} to 9.6×10^{-4} cm^2) were fabricated and bump mounted to circuit boards. Figure 1 depicts a cross section of a double layer film after photodiode fabrication. The relative spectral response was measured for representative photodiodes on each array at a temperature of 80 K, from which the cutoff wavelength (defined as that wavelength at which the response has dropped to 50% of its peak value) was determined. The zero bias impedance R_0 and quantum efficiency (QE) were measured for all photodiodes at 80 K and 70 K. The QE was measured by flood illumination from a chopped blackbody source; either a 500 K blackbody and an f/5 cold aperture or a 1000 K blackbody with a 4-μm-spike cold filter was used for the measurements. Current versus voltage (I–V) measurements were made on selected diodes of each size at biases from -300 to $+180$ mV. Care was taken to reduce the background photon flux on the photodiode, by means of either the f/5 cold aperture or the 4-μm-spike cold filter, so that the measured photodiode resistances were not affected by the background-induced shunt resistance effect reported by Rosbeck et al.[15]

III. RESULTS AND DISCUSSION

A. Film characterization

The Hall coefficient R_H for the n-type base layer samples was measured at 300 K and at 77 K for magnetic field strengths of 0.1–8.0 kG. Typically there was no significant field dependence of the Hall coefficient. Figure 2 shows the results at 4 kG of these measurements for films with $x = 0.21$ at the three doping levels that were grown for this study. With the exception of a few of the growths, the Hall mobility ranged between 1×10^5 and 3×10^5 cm^2/V s, independent of doping level. The room-temperature resistivity and intrinsic carrier concentration for the LPE base layers agree closely with the published values for high-quality bulk-grown material.[16] The carrier concentration values obtained from the large-area diode C–V measurements at reverse biases of greater than 100 mV agree with the Hall measurements. A typical plot of carrier concentration vs depth determined from C–V data is shown in Fig. 3. The disagreement below 100 mV have a number of causes. The indium could be diffusing out of the base layer into the cap during cap growth, and/or the arsenic in the cap could be compensating the indium. Additionally, the band gap is changing in this region.

A typical room temperature transmission spectrum is shown in Fig. 4, from which the film thickness was found to be 19.5 μm. The prominent interference fringing is indicative of a flat CdTe/HgCdTe interface.

Photodiode cutoff wavelength data at 80 K have been related to the room-temperature transmission cutoff wavelength (defined as the wavelength having a transmission value corresponding to an absorption coefficient of 500 cm^{-1}), as shown in Fig. 5. This relation allows prediction of the final device cutoff with a simple room-temperature measurement on the base layer. It can be seen from the curve that the final device 80 K cutoff wavelength can be predicted from the room-temperature transmittance spectra with an accuracy of better than $\pm 0.3 \mu$m in the 10–12 μm range. Because the films are grown by LPE with a gradient and with thicknesses that are on the order of the minority carrier wavelength, it is not expected that the relation between room-temperature facing target sputtering measurements and 80 K device cutoff wavelengths be exactly that of Hansen, Schmit, and Casselman[17] which is plotted as the solid line in Fig. 5. This is indeed apparent particularly in the longer wavelength region ($> 13 \mu$m).

The lifetime data for the n-type LWIR base layers are plotted in Fig. 6 versus the Hall carrier concentration measured at 77 K. Also shown in Fig. 6 for comparison are pho-

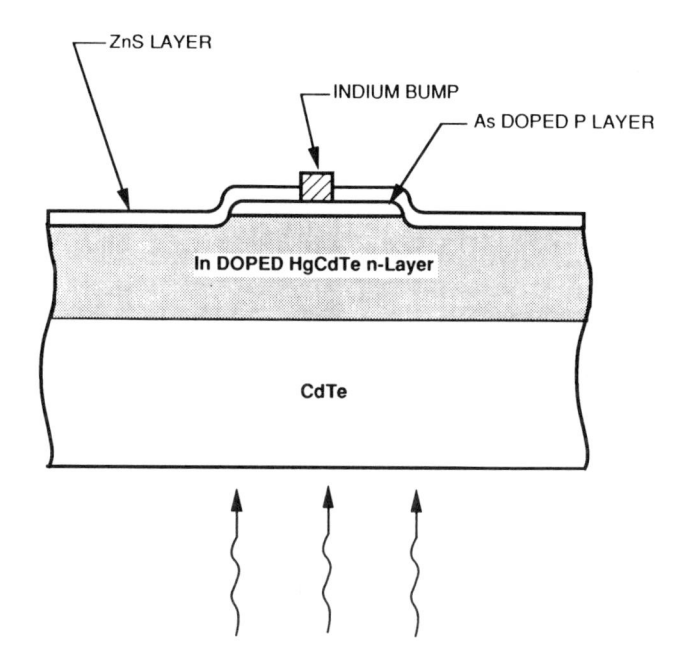

FIG. 1. Cross section of P-on-n heterojunction film after mesa photodiode fabrication.

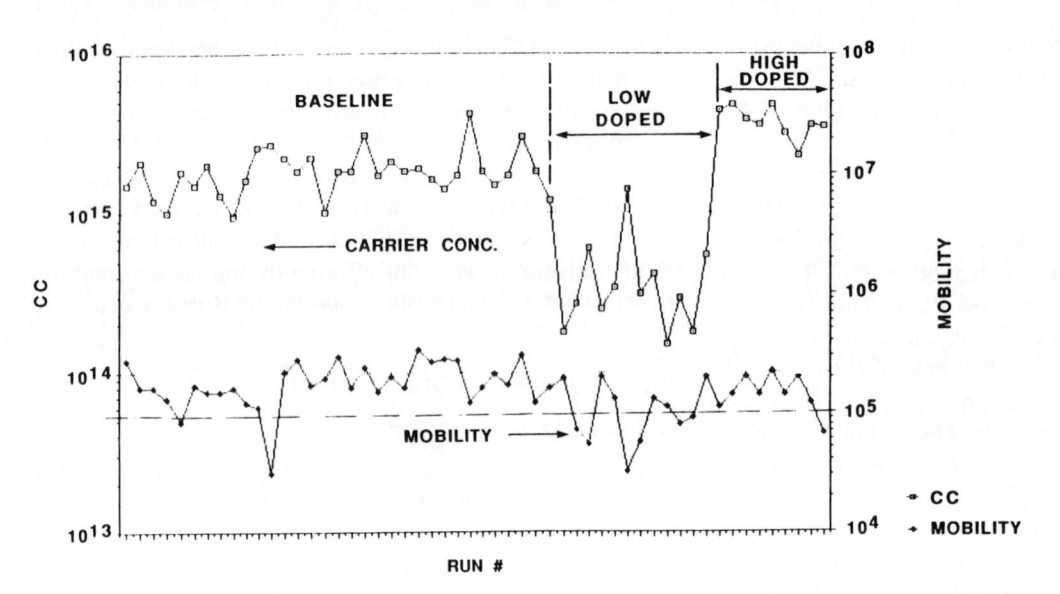

FIG. 2. Base layer (n-type, x = 0.21) carrier concentrations and Hall mobilities at 77 K for the three doping levels used in this study. The n-type dopant was indium. The magnetic field was 4 kG.

toconductive decay lifetime data at 77 K measured on high-quality Loral bulk-grown[16] n-type $Hg_{0.794}Cd_{0.206}Te$ samples.[18] Both the LPE and the bulk sample data show a rapid decrease in lifetime with increasing carrier concentration, typical of the Auger-1 recombination mechanism that is known to determine carrier lifetime in well behaved n-type $Hg_{1-x}Cd_xTe$ at these temperatures.[19,20] At carrier concentrations above 4×10^{14} cm^{-3} both LPE and bulk data agree very well with the calculated Auger-1 lifetime for extrinsic n-type $Hg_{1-x}Cd_xTe$ shown in Fig. 6 by the solid line. This line is consistent with Blakemore's equations[21] for the Auger-1 lifetime with the overlap integral term $|F_1F_2|^2$ set equal to 0.041. For carrier concentrations below 4×10^{14} cm^{-3} the LPE data fall below the Auger-1 limit, which we attribute to Shockley–Read recombination or to background-generated increase in the carrier concentration.

Measurements of the Hall coefficient of the As-doped p-type cap layer samples grown on CdTe yielded measured carrier concentrations ranging from 5×10^{16} cm^{-3} to 10^{18} cm^{-3}. Over the magnetic field range of 0.1 to 8 kG mixed conduction usually was not observed. The thicknesses of the cap layers were typically 1–2 μm. The $Hg_{1-x}Cd_xTe$ alloy composition increase Δx from base to cap layer was typically 0.03. The composition change between the cap layer and base layer was minimized to ensure low p-side contact and series resistance and to reduce the possibility of valence band barrier formation on the n side. Simple models comparing Auger-1 limited diffusion currents from the cap and base

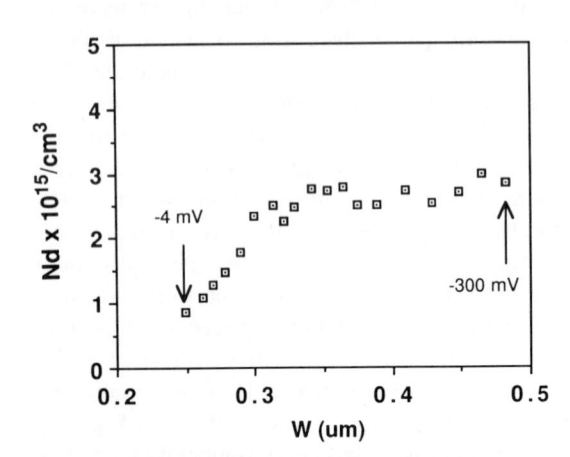

FIG. 3. Carrier concentration vs depletion width values obtained from C–V measurements are in close agreement with Hall data of 2.1×10^{15} cm^{-3} at reverse biases greater than 100 mV. Cutoff wavelength for film was 12.2 μm at 80 K; the diode area was 9.6×10^{-4} cm^2.

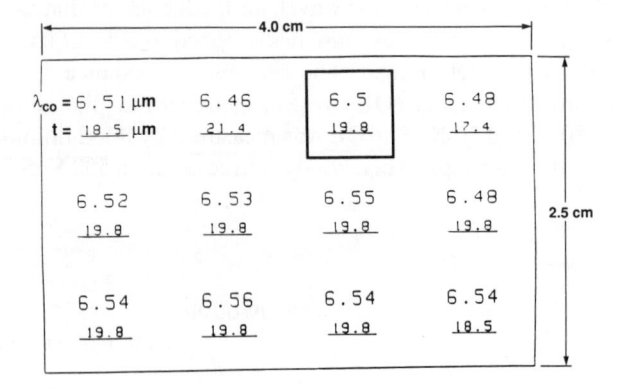

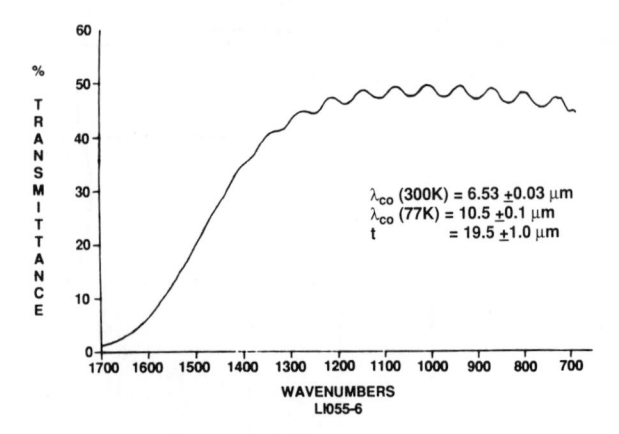

FIG. 4. Transmission map over an LPE LWIR HgCdTe 2.5×4.0 cm^2 base layer. Data summary: L_{co}(300 K) = 6.53 ± 0.03 μm → L_{co}(80 K calc.) = 10.5 μm; Thickness = 19.5 ± 1.0 μm. Transmission = 50%.

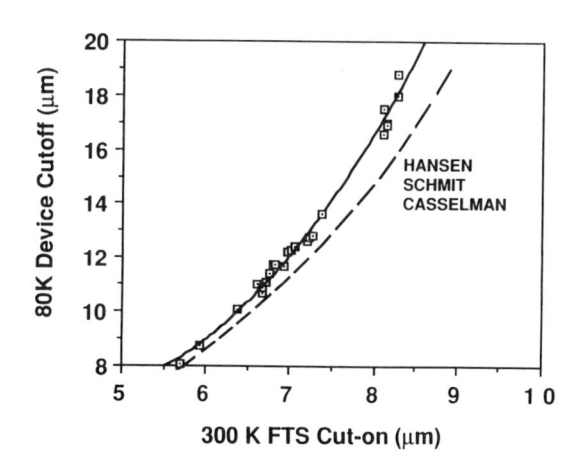

FIG. 5. Room temperature FTS cut-on wavelengths (before cap growth) vs device 80 K cut-on wavelength.

layers of diodes have shown that the small change in x-value is large enough to ensure that the base layer diffusion current dominates the performance of the diode.[22]

B. Photodiode characterization

The goal of the diode data analysis was to obtain a value of R_0A which was characteristic of the film and not of diode geometry or fabrication steps which affect the surface. To this end, simple geometric models were used to extract "infinite area" or bulk quantum efficiency and R_0A values from the variable area optical response and resistance data. These values are then easily compared to film electrical properties because one-dimensional calculations can be used with no assumptions about surface and geometric effects on the diode performance.

1. Impedance measurements

The electrical characteristics of the LPE films were deduced from the measured R_0A values for the different area diodes as reported by Briggs et al.[23] and by Chung et al.[24] Values of $1/R_0A$, which are proportional to the dark currents, were plotted versus the perimeter to junction area ratio P/A_j. The R_0A value, denoted by $(R_0A)_{bulk}$, obtained from the intercept at $P/A_j = 0$, approximates the performance of an infinite-area diode with no geometric effects. Additional information about the performance limiting mechanisms of smaller diodes can be obtained from the dependence of $1/R_0A$ values on the P/A_j ratio. Although the following two examples have similar $(R_0A)_{bulk}$ values, the R_0A values for smaller diodes have very different dependences on the P/A_j ratio and hence surface affects and/or lateral collection.

The $1/R_0A$ versus P/A_j plot shown in Fig. 7 is for a film which has a base layer carrier concentration of 2.5×10^{15} cm^{-3} and a 12.2 μm cutoff wavelength at 80 K. The data from each size diode are closely distributed and can be fit to a straight line. The dependence of $1/R_0A$ on the P/A_j ratio is given by

$$\frac{1}{(R_0A)_{meas}} = \frac{1}{(R_0A)_{bulk}} + \frac{qn_i W s_0}{V_{bi}} \frac{P}{A_j}.$$

The value of R_0A_{meas} is the product of the measured impedance and the junction area. The surface recombination velocity s_0 was deduced from the slope of the $1/R_0A$ versus P/A_j plot by calculating the intrinsic carrier concentration n_i, the depletion width W, and the built in potential V_{bi}. The value for $(R_0A)_{bulk}$, which is independent of diode geometry and any lateral effects, is found from the intercept of the best fit line. This value can be plotted as a function of temperature and cutoff wavelength to determine the quality of the base layer and the junction in the absence of geometrical and surface effects. For the particular array in Fig. 7, $s_0 = 10^4$ cm/s and $(R_0A)_{bulk} = 23$ Ω cm^2.

These plots are easily analyzed when the hole diffusion lengths are relatively short ($<10\,\mu$m) with respect to diode radii. This corresponds to a base layer carrier concentration of $\approx 1\times10^{15}$ cm^{-3} for the lifetime dependence on carrier concentration shown in Fig. 6. However, when the hole diffusion length L_h was long relative to diode radii, the simple linear trend was no longer observed because a significant

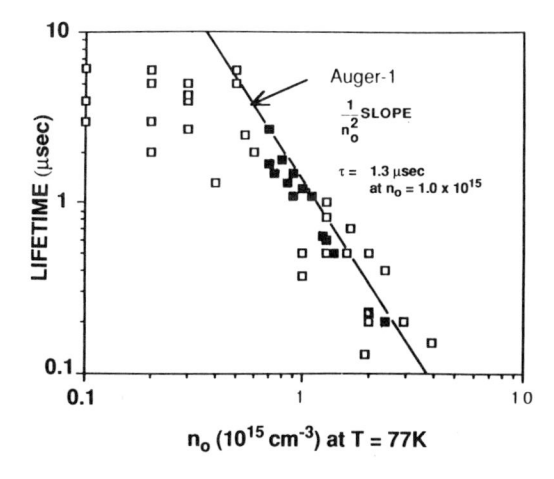

FIG. 6. n-type base layer minority lifetime measured by optically modulated absorption or photoconductive frequency response, plotted vs 77 K Hall carrier concentration. Solid points are from high quality bulk crystal material ($x = 0.206$) measured at 77 K, open symbols are from LPE films ($x = 0.21$) grown in Te-rich slider measured at 85 K. The solid line is the calculated Auger-1 lifetime at 77 K using $|F_1 F_2|^2 = 0.041$ and $x = 0.206$.

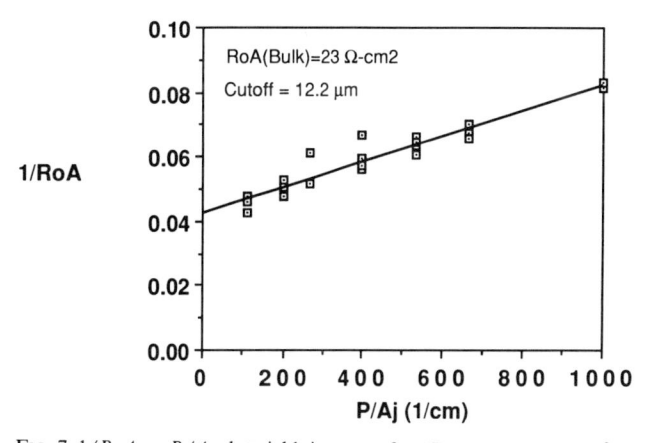

FIG. 7. $1/R_0A_j$ vs P/A_j plot yields intercept for $(R_0A)_{bulk} = 23$ Ω cm^2 and $s_0 = 10^4$ cm/s for film with 80 K cutoff wavelength of 12.2 μm and $N_d = 2.5\times10^{15}$ cm^{-3}. Impedance measurements made at 80 K.

component of the dark current was due to lateral diffusion. The data were analyzed by adding a lateral diffusion term to the previous equation

$$\frac{1}{(R_0A)_{\text{meas}}} = \frac{1}{(R_0A)_{\text{bulk diff}}}\left[1 + L_{\text{opt}}\frac{P}{A_j} + \frac{L_{\text{opt}}^2}{4}\left(\frac{P}{A_j}\right)^2\right]$$
$$+ \frac{1}{(R_0A)_{\text{bulk }G-R}} + \frac{qn_iWs_0}{V_{\text{bi}}}\frac{P}{A_j},$$

where L_{opt} is a lateral optical collection length which is proportional to the minority carrier diffusion length. The value of $(R_0A)_{\text{bulk}}$ is again found from the intercept which is equal to the sum of $1/(R_0A)_{\text{bulk diff}}$ and $1/(R_0A)_{\text{bulk }G-R}$. An example of this type of behavior is shown in Fig. 8. The base layer carrier concentration was 2×10^{14} cm^{-3}, and consequently the expected hole diffusion length was greater than 30 μm. The magnitude of the effect of lateral diffusion current on the diode performance depends on the size of the detector. We expect small diodes (with a radius comparable to the optical path length L_{opt}) to have a higher proportion of lateral diffusion current than larger diodes. Alternatively, once an optical path length L_{opt} is known from the signal data, analysis can be done as in the first example except plotting $1/(R_0A_0)$ versus P/A_j. Where R_0A_0 is the product of the junction impedance and the optical area.

It is also interesting to note that the change in R_0A from 70 to 80 K was dependent on the carrier concentration. For low doped films (2×10^{14} cm^3) a proportionally higher G–R current was observed due to a larger depletion region width (about a factor of three for an abrupt junction approximation). Correspondingly, the R_0A increase from 70 to 80 K in the larger diodes for low doped material was a factor of 3, whereas the R_0A increase for higher doped material was a factor of 7 on average.

The $(R_0A)_{\text{bulk}}$ values from a subset of the films grown with base layer thicknesses of $20\pm2\ \mu$m and n-type doping (indium) to a carrier concentration of $(2.0\pm0.5)\times10^{15}$ cm^{-3} are shown in Fig. 9 as a function of cutoff wavelength. The data closely follow an Auger-1, one-dimensional, n-side

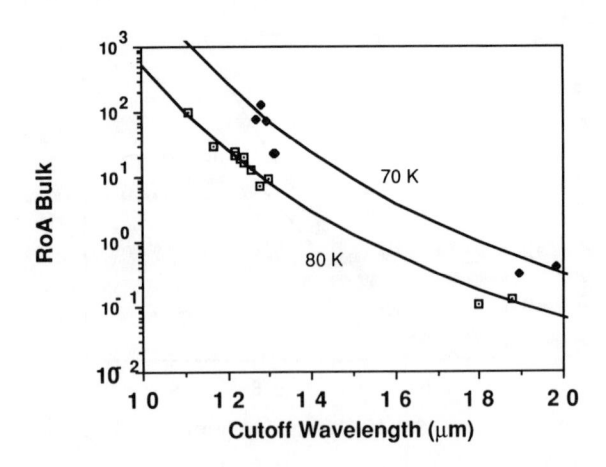

FIG. 9. $(R_0A)_{\text{bulk}}$ vs cutoff wavelength. Open symbols are 80 K data and closed symbols are 70 K data. Solid lines are calculated from an Auger-1, one-dimensional, n-side, diffusion model for 80 and 70 K. The calculations were made using a base layer thickness of 20 μm and carrier concentration of 2×10^{15} cm^{-3}.

diffusion limit calculated assuming the above carrier concentration and thickness.[25] The model is represented by solid lines which are plotted for 70 and 80 K temperatures. Note that the $(R_0A)_{\text{bulk}}$ values are at the Auger-1 diffusion limit, indicating that the junctions are of very high quality. This is a significant finding as the performance limiting mechanism, once surface and geometric effects are removed, is the base layer diffusion current (which is Auger-1 limited) and not tunneling or G–R currents in the junction.

Many applications require the photodiodes to be operated at a small but significant reverse bias (e.g., -20 to -40 mV) to improve multiplexer input coupling efficiency and to decrease detector thermal noise. The dynamic impedance data for devices with 80 K cutoff wavelengths of 12.7 and 18.8 μm are shown in Figs. 10 and 11. The junction quality is evident in both cases, as substantial increases in the dynamic impedances were observed at reverse bias. At 40 mV reverse bias the dynamic resistance of the 12.7 μm diode was roughly 20 times R_0. Similarly, at 40 mV reverse bias, the dynamic

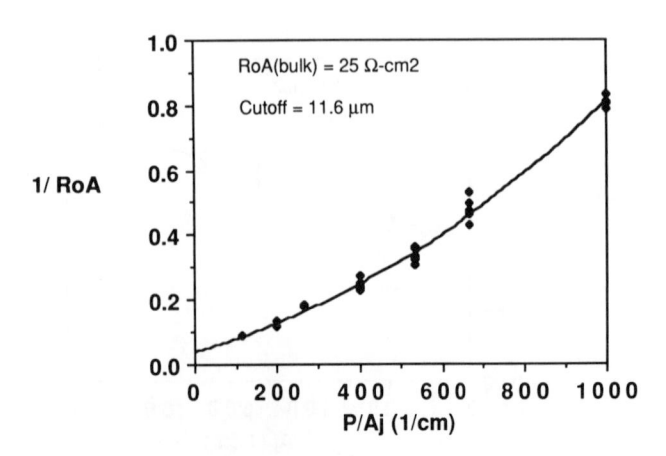

FIG. 8. $1/R_0A_j$ vs P/A_j plot yields an intercept of $(R_0A)_{\text{bulk}} = 25\ \Omega$ cm^2 for film with 80 K cutoff wavelength of 11.6 μm and $N_d = 2\times10^{14}$ cm^{-3}. Impedance measurements made at 80 K.

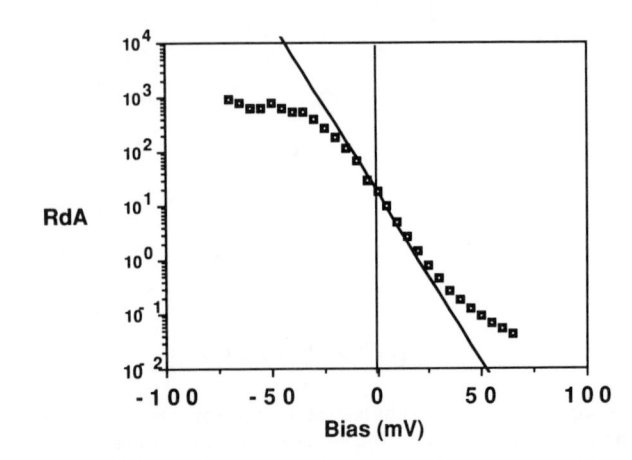

FIG. 10. The 80 K dynamic impedance-area product, R_dA, vs reverse bias for a film with an 80 K cutoff wavelength 12.7 μm, 80 K $R_0A = 23\ \Omega$ cm^2 and $A_j = 3.14\times10^{-4}$ cm^2.

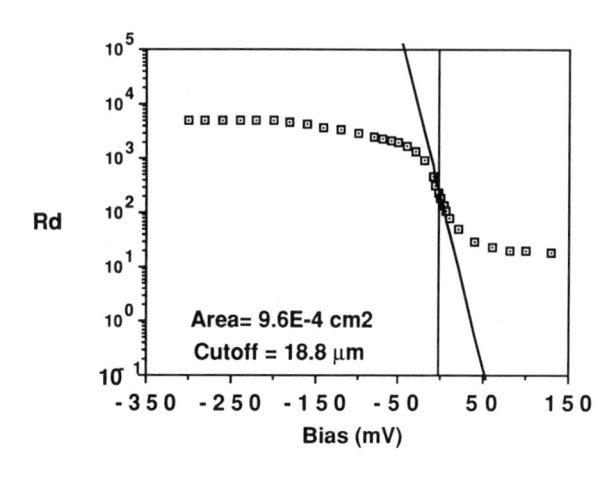

FIG. 11. The dynamic impedance, R_0, vs reverse bias for a film with an 80 K cutoff wavelength 18.8 μm, 80 K $R_0 A = 25$ Ω cm^2 and $A_j = 9.62 \times 10^{-4}$ cm^2.

resistance of the 18–19 μm diodes were roughly a factor of 10 times R_0 for larger diodes (as shown in Fig. 11) and a factor of roughly 3 in the small diodes. The solid straight lines are plots of $R_0 \exp(eV/kT)$, which is dynamic impedance due to diffusion current alone, normalized to the measured R_0 at $V = 0$, and show what the dynamic impedance would be if diffusion current were the only current present.

2. Optical characterization

The variable-area array was designed to have well-separated circular mesa diodes. Therefore the diodes have optical collection areas that are larger than the junction areas. To obtain the meaningful quantum efficiency values, the "measured" quantum efficiency QE$_{meas}$ for each diode was calculated using the junction area. Using a very simple model where the optical collection area is calculated by adding an optical length L_{opt} to the junction radius r, the "measured" quantum efficiency QE$_{meas}$ is given by

$$QE_{meas} = QE_{inf}\left(1 + \frac{L_{opt}}{r}\right)^2,$$

where L_{opt} is the lateral optical collection length and QE$_{inf}$, obtained from the intercept of a plot of the square root of QE$_{meas}$ versus $1/r$, is a good approximation to the quantum efficiency that an infinite area diode, would have.

Figure 12 contains variable area array data from two films with different base layer thicknesses and carrier concentrations. Data from a film with a 15 μm thick base layer and a low carrier concentration of 2.0×10^{14} cm^{-3} (open symbols) and data from a film with a 20 μm thick base layer and a carrier concentration of 2.0×10^{15} cm^{-3} (solid symbols) are plotted together with best fit lines. The derived infinite area quantum efficiency values are displayed as well as the optical lengths. The fact that the infinite area quantum efficiency values were comparable with such different optical lengths can be explained by both a thinner base layer and an effective electric field resulting from the small but significant composition gradient in the base layer which enhances the optical collection efficiency.

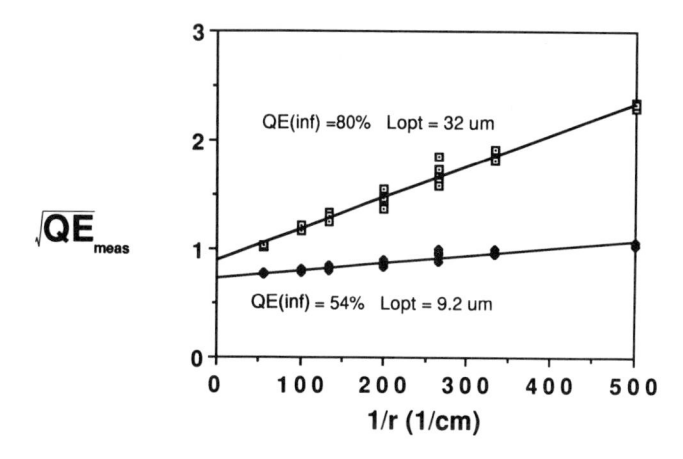

FIG. 12. Square root of measured quantum efficiency vs $1/r$ for two arrays from films with carrier concentrations of (open symbols) $N_d = 2 \times 10^{14}$ cm^{-3} and (closed symbols) $N_d = 2 \times 10^{15}$ cm^{-3} which are 15 and 20 μm thick, respectively. QE and optical path length values for the $N_d = 2 \times 10^{14}$ cm^{-3} are 80% and 32 μm. QE and optical path length values for the $N_d = 2 \times 10^{15}$ cm^{-3} are 54% and 9 μm.

Figure 13 shows values of L_{opt} for a range of base layer carrier concentrations. The theoretical minority carrier diffusion length L_h shown by the straight line was calculated for a hole mobility μ_h of 450 cm^2/V s and a calculated Auger-1 minority carrier lifetime

$$L_h = \sqrt{\frac{KT}{q}\mu_h \tau}.$$

The measured L_{opt} values are in close agreement with the L_h calculated values. The correlation of electrical measurements on final devices to base layer doping level and minority carrier lifetime is evidence that the excellent base layer quality was maintained during the cap layer growth process.

Typical relative spectral response and relative quantum efficiency curves for a test device with a cutoff wavelength of 18.8 μm at 80 K are shown in Fig. 14. The spectral response curve is classical with a sharp cutoff. The quantum efficiency obtained for the 18.8 μm device was 20%. This low quantum

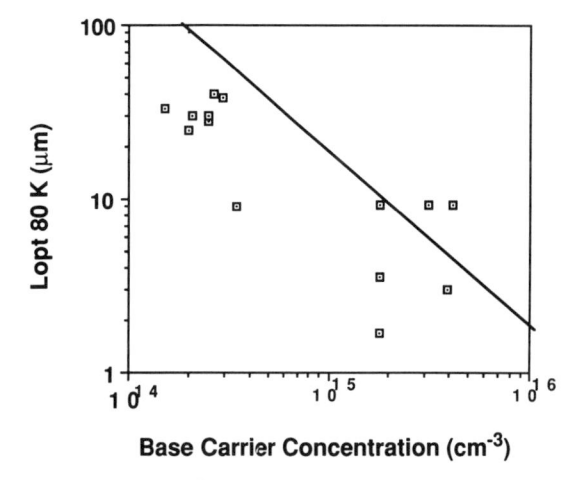

FIG. 13. Lateral optical collection length L_{opt} vs 77 K Hall carrier concentration. Solid line is a calculated hole diffusion length using Auger-1 lifetime and hole mobility of 450 V cm^2/s.

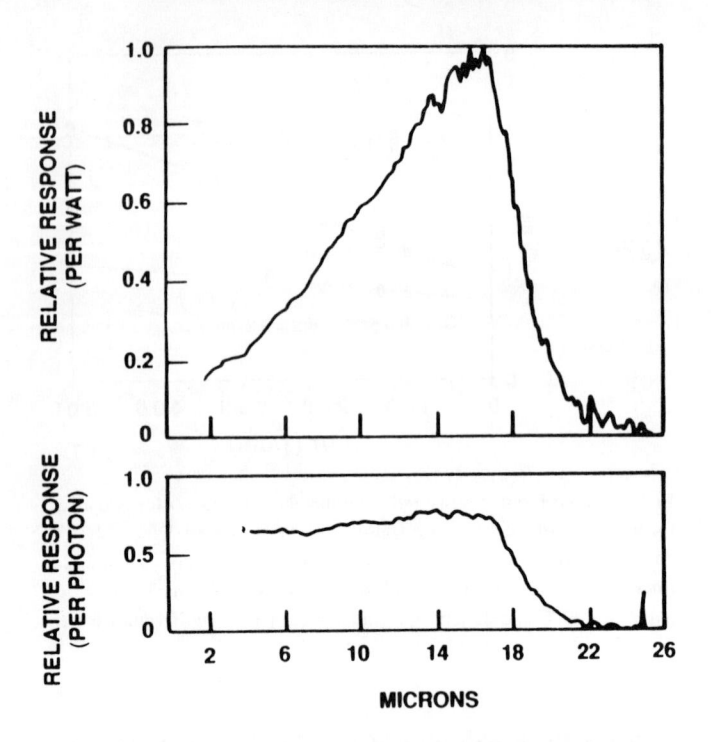

FIG. 14. Relative spectral response data (per watt in upper curve and per photon in lower curve) at 80 K on a two-layer LPE film with 18.8 μm cutoff wavelength.

efficiency is consistent with a 22 μm thick base layer having a short diffusion length calculated for an n-type base layer with a carrier concentration of 2×10^{15} cm^{-3}. Lower doping and thinner base should increase the quantum efficiency to values of 65%–80% that are typically seen on two-layer LPE films with cutoff wavelengths in the 11–13 μm range.

IV. SUMMARY & CONCLUSIONS

We have demonstrated the capability of growing improved double-layer P-on-n LPE HgCdTe heterojunction films with excellent thickness uniformity ($\pm 10\%$) and excellent cutoff wavelength uniformity (e.g., 10.5 ± 0.1 μm) across wafers of up to 2.5×4.0 cm^2 in area. Data show that the n-type base layers are of excellent quality with carrier lifetimes which agree with bulk material values and vary with carrier concentration with an Auger-1 dependence. Furthermore, large diode R_0A data follow a diffusion limited dependence with an Auger-1 lifetime. Quantum efficiency values of up to 80% for cutoff wavelengths of 11–13 μm and 20% at cutoff wavelengths in the 18–19 μm range are observed (with no antireflection coating) and the spectral response curves are classical. These data show the potential for extending the region of operation for photovoltaic $Hg_{1-x}Cd_xTe$ devices to wavelengths of 18–19 μm at 80 K.

ACKNOWLEDGMENTS

This work was supported by Loral Internal Funds and NASA/JPL Contract No. 958606. The authors acknowledge the efforts and contributions of M. Cody, L. Firth, and T. Dunning (LPE growth and characterization); J. Mroczkowski and B. White (lifetime measurements), P. O'Dette (device fabrication); D. King (device testing); and M. H. Weiler, N. Hartle, R. Briggs, K. Maschhoff, M. Krueger, J. Marciniec, and P. Zimmermann (technical discussions).

[1] P. W. Norton, P. H. Zimmermann, R. J. Briggs, and N. M. Hartle, Proc. SPIE **686**, 138 (1986).
[2] T. Tung, M. H. Kalisher, A. P. Stevens, and P. E. Herning, Mater. Res. Soc. Symp. Proc. **90**, 321 (1987).
[3] S. M. Johnson, M. H. Kalisher, W. L. Ahlgren, J. B. James, and C. A. Cockrum, Appl. Phys. Lett. **56**, 946 (1990).
[4] S. M. Johnson, W. L. Ahlgren, J. B. James, W. J. Hamilton Jr., M. H. Kalisher, and M. Ray, Mater. Res. Soc. Symp. Proc. (to be published).
[5] T. Tung, J. Cryst. Growth **86**, 161 (1988).
[6] C. C. Wang, Fermionics Corporation, Simi Valley, California (private communication).
[7] J. A. Mroczkowski and H. R. Vydyanath, J. Electrochem. Soc. **128**, 655 (1981).
[8] E. R. Gertner, Mater. Res. Soc. Symp. Proc. **90**, 357 (1987).
[9] T. C. Harman, J. Electron. Mater. **10**, 1069 (1981).
[10] Y. Nemirovsky, S. Margalit, E. Finkman, Y. Shacham-Diamand, and I. Kidron, J. Electron. Mater. **11**, 133 (1982).
[11] H. R. Vydyanath, J. Vac. Sci. Technol. **9**, 1711 (1991).
[12] P. Capper, J. Vac. Sci. Technol. **9**, 1662 (1991).
[13] O. L. Doyle, J. A. Mroczkowski, and J. F. Shanley, J. Vac. Sci. Technol. A **3**, 259 (1985).
[14] M. Grimbergen and A. Szilagyi, Mater. Res. Soc. Symp. Proc. **69**, 257 (1986).
[15] J. P. Rosbeck, R. E. Starr, S. P. Price, and K. J. Riley, J. Appl. Phys. **53**, 6430 (1982).
[16] W. M. Higgins, G. N. Pultz, R. G. Roy, R. A. Lancaster, and J. L. Schmit, J. Vac. Sci. Technol. A **7**, 271 (1989).
[17] G. L. Hansen, J. L. Schmit, and T. N. Casselman, J. Appl. Phys. **53**, 7099 (1982).
[18] D. A. Nelson, Loral Infrared & Imaging Systems (unpublished work, 1981).
[19] M. A. Kinch, M. J. Brau, and A. Simmons, J. Appl. Phys. **44**, 1649 (1973).
[20] P. Capper, in *Properties of Mercury Cadmium Telluride*, edited by J. Brice and P. Capper (INSPEC, IEE, London, 1987), Chap. 5.7.
[21] J. S. Blakemore, *Semiconductor Statistics* (Dover, New York, 1987), Equations 620.1 and 620.2.
[22] M. H. Weiler, Loral Infrared & Imaging Systems (unpublished work, 1989).
[23] R. J. Briggs, J. W. Marciniec, P. H. Zimmermann and A. K. Sood, IEEE International Electron Devices Meeting Technical Digest, 1980 (unpublished), p. 496.
[24] H. K. Chung, M. A. Rosenberg, and P. H. Zimmermann, J. Vac. Sci. Technol. A **3**, 189 (1985).
[25] M. B. Reine, A. K. Sood and T. J. Tredwell, in *Semiconductors and Semimetals*, edited by R. K. Willardson and A. C. Beer (Academic, New York, 1981), Vol. 18, Chap. 6.

Improved breakdown voltage in molecular beam epitaxy HgCdTe heterostructures

R. J. Koestner, M. W. Goodwin, and H. F. Schaake
Texas Instruments Inc., Central Research Laboratories, Dallas, Texas 75265

(Received 28 November 1990; accepted 7 February 1991)

We present recent progress in the development of metal–insulator–semiconductor (MIS) heterostructure detectors grown by molecular beam epitaxy (MBE); these heterostructure films should significantly improve the available well capacity for MIS long wavelength infrared (LWIR) detectors. Recent MBE grown HgCdTe(112)Te heterostructures show a 4× increase in available charge capacity and a 2× decrease in net donor density relative to our MBE HgCdTe(001) layers. To obtain full advantage of the MIS heterostructure however, another 2.5× increase in the MWIR breakdown field (0.8–2.0 V/μm) and a 2× decrease in the MWIR donor density (1×10^{15}–5×10^{14} cm^{-3}) is necessary. Since the MWIR breakdown field does not appear to be limited by compositional micro-inhomogeneities or by compound twin defects in the MBE HgCdTe layers, we suspect either dopant micro-inhomogeneities or native point defects are responsible for the significant tunnel currents measured in our MWIR MBE HgCdTe.

I. INTRODUCTION

HgCdTe heterostructures consisting of a thin *n*-type widegap (250 meV or 5 μm cutoff) layer deposited on an *n*-type narrow gap (100–125 meV or 10–13 μm cutoff) layer offer the promise of very high performance metal–insulator–semiconductor (MIS) photocapacitors for long wavelength infrared (LWIR) detection.[1] To date, HgCdTe(001) heterostructures grown by molecular beam epitaxy (MBE) have not achieved the predicted MIS performance due to premature breakdown in the widegap layer. In this paper, we examine the improved breakdown voltages measured in widegap HgCdTe(112)Te relative to HgCdTe(001) layers.

MIS breakdown voltages measured for MBE grown HgCdTe(001) layers with a cutoff wavelength varying from 4–11 μm is found to be ~0.2 V. In contrast, our recent MBE HgCdTe(112)Te layers show an increase in breakdown voltage from 0.2 V to 0.8 V over the same cutoff wavelength range. The 4× improvement in our middle wavelength IR (MWIR) MBE breakdown voltage has led to the first demonstration of a 10.2 μm cutoff MBE heterostructure with greater well capacity (30–35 nC/cm^2) than that for high quality bulk HgCdTe with the same cutoff (25–30 nC/cm^2). At another laboratory,[2] LWIR photodiodes fabricated on MBE grown HgCdTe(112)Te films also yielded R_0A products comparable to that obtained from the more mature liquid phase epitaxy (LPE) growth technology.

Nonetheless, a well capacity of 100 nC/cm^2 should be obtained in an MIS heterostructure if the MWIR layer breakdown field is increased from our present 0.8 V/μm to the optimal value of 2.0 V/μm found in high quality MWIR bulk HgCdTe. The premature breakdown in MBE HgCdTe epilayers may be due (1) to a compositional nonuniformity on a 1000 Å scale that we have observed in earlier MBE HgCdTe samples[3] by cross-sectional transmission electron microscopy (XTEM), (2) to the presence of highly defective pyramidal hillocks[4] or other compound twins under the MIS gate, (3) or finally to dopant micro-inhomogeneities or native point defects within the HgCdTe film.

With a 4–5 h post-growth anneal at 340–360 °C followed by a 24 h stoichiometric anneal at 200 °C under Hg saturated conditions, we should have interdiffused the compositional nonuniformity observed in earlier films;[5] however, the four MWIR HgCdTe(001) and the three MWIR HgCdTe(112)Te annealed in this manner did not show any increase in available breakdown field.

In addition, we have reduced the compound twin density in our MBE HgCdTe layers from the *low* 10^4 cm^{-2} to the *low* 10^3 cm^{-2} range by passivating the internal surfaces of our Hg source. These compound twins always lead to surface pyramids on HgCdTe(001) layers, but are usually only observed after defect etching HgCdTe(112)Te films. With the 10× reduction in surface pyramids, we have isolated MIS gates on a MBE HgCdTe(001) layer without any underlying hillocks and only find a small increase in the MWIR breakdown voltage.

At this point, we believe either a dopant micro-inhomogeneity or high native point defect concentration is responsible for the premature breakdown of our MWIR MBE HgCdTe. High local electric fields associated with n^+ regions in MBE HgCdTe will lead to higher tunnel currents. Also, point defects such as Te antisites may be kinetically trapped in MBE HgCdTe due to the low growth temperature; and these point defects may then lead to mid-bandgap states that enhance trap-assisted tunneling at the moderate electric fields produced during MIS detector operation.

II. EXPERIMENTAL

The HgCdTe epilayers were deposited in a RIBER 2300P MBE system on CdZnTe(112)Te substrates held at 185–200 °C; the substrates were prepared with a standard Br in MeOH chemimechanical polish. A Hg beam equivalent pressure of 0.5 and 6×10^{-4} Torr was employed for the (112)Te and (001) layers, respectively. The Te and CdTe effusion cell temperatures were adjusted to achieve a Hg$_{1-x}$Cd$_x$Te composition of $x = 0.22$–0.30 with a growth rate in the 1.5–5.0 μm/h range. A CdTe buffer layer of

~2000 Å was first deposited at 300–350 °C before the growth of a ~5 μm thick HgCdTe layer.

III. RESULTS AND DISCUSSION

The MIS HgCdTe heterostructure device concept is schematically presented in Fig. 1. The heterostructure is grown with a thin (1–3 μm) wide-band gap (MWIR) layer on top of the normal narrow-band gap (LWIR) detecting layer. Both layers are n-type with net donor densities at or below 1×10^{15} cm^{-3}. The LWIR layer is grown sufficiently thick (~5 μm) to achieve a high quantum efficiency, while the MWIR thickness is chosen to produce simultaneous breakdown in both layers at the maximum bias applied in deep depletion. The critical advantage to this device structure is that the MWIR layer is inserted in the high electric field region of the depletion volume which substantially reduces tunnel and generation-recombination dark current in the detector. This allows larger biases to be applied to the MIS gate and hence more charge to be stored in the inversion layer.

The operation of the MIS heterostructure as illustrated in Fig. 1 is very similar to the simple integrating MIS detector. By applying a voltage pulse large enough to drive it into deep depletion and also large enough to remove any potential barriers at the heterointerface, the heterostructure device will allow the optically generated minority carriers to flow into the well. As the charge fills the inversion layer, the surface potential will decrease until the electric field at the MWIR/LWIR HgCdTe interface is zero. At this point, any further decrease of the surface potential at the insulator/HgCdTe interface will create a potential barrier at the heterointerface and prevent further integration of the optically generated LWIR layer minority charge. The size of the voltage pulse needed for deep depletion is determined by the breakdown electric field in the MWIR layer. Measured breakdown fields for high quality SSR MWIR and LWIR HgCdTe are 2.0 and 0.6 V/μm, respectively. For the maximum MIS well capacity, the thickness of the top layer should be that value which will produce breakdown electric fields in both the top and bottom layers at the same gate voltage. Using the above mentioned breakdown fields and a MWIR layer donor density of 1×10^{15} cm^{-3}, we calculate an optimal MWIR layer thickness of 1.4 μm. Due to its fine control in layer thickness, we chose MBE as the primary growth technique for this structure.

Figure 2 illustrates the improvement in available well capacity possible with a MIS heterostructure detector. The calculation assumes a net donor density of 5×10^{14} cm^{-3} in both the MWIR and LWIR layers. High quality bulk HgCdTe with a 10 μm cutoff has a breakdown field of 0.6 V/μm; this leads to an available charge capacity of 28 nC/cm^2 as shown in Fig. 2 by the horizontal line. In the MIS heterostructure however, the well capacity for the 10 μm cutoff detecting layer can be improved to as much as 110 nC/cm^2 if the MWIR layer shows a breakdown field of 2.0 V/μm which is found in high quality bulk HgCdTe.

For the first time, we demonstrate an MBE HgCdTe heterostructure that has a well capacity (30–35 nC/cm^2) greater than that for high quality bulk HgCdTe with a 10 μm cutoff (28 nC/cm^2). This became possible with the higher breakdown fields and lower donor densities observed in our HgCdTe(112)Te films relative to HgCdTe(001). This HgCdTe(112)Te heterostructure (run 676) consists of a 4.5 μm cutoff wide-band gap layer (0.7 μm thick) deposited on a 10.2 μm cutoff narrow-band gap layer (5.0 μm thick). The net donor density at 80 K is $6-8 \times 10^{14}$ cm^{-3} for the detectors tested. A breakdown field of 0.8 V/μm is found for this heterostructure by biasing the metal gate until a total minority carrier current of 2 mA/cm^2 is obtained. (This minority carrier current is representative of the current created by a tactical flux level.) The well capacity of 30–35 nC/cm^2 is then calculated from this breakdown field and the measured MIS donor density.

To increase the well capacity in the MIS heterostructure further, we need to improve the breakdown voltage of our MBE MWIR HgCdTe films. Table I summarizes the best

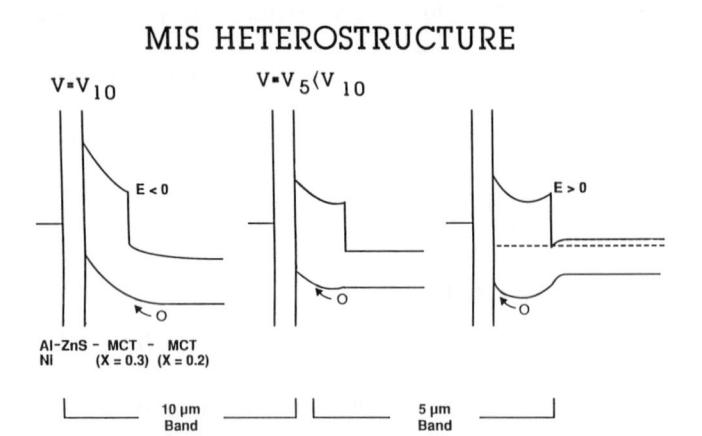

FIG. 1. A schematic of an MIS heterostructure is illustrated. A MWIR HgCdTe layer is inserted into the high electric field region of the depletion volume to substantially reduce tunnel and generation-recombination dark current in the LWIR HgCdTe detecting layer.

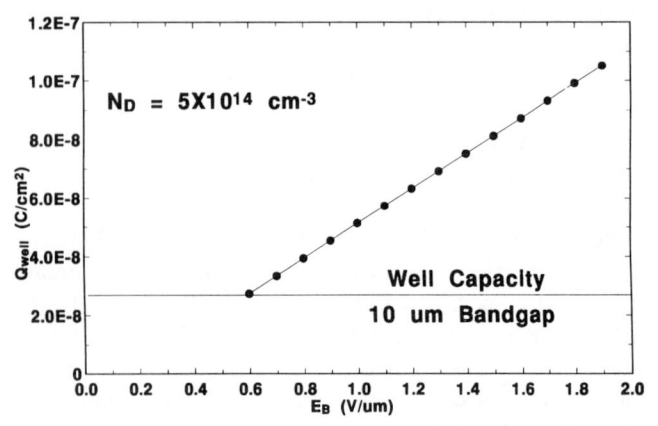

FIG. 2. The well capacity in a LWIR MIS heterostructure is calculated as a function of the breakdown field in the MWIR HgCdTe layer. The donor density in both MWIR and LWIR layers is set at 5×10^{14} cm^{-3} and the breakdown field in the LWIR layer is assumed to be 0.6 V/μm.

TABLE I. Comparison of MWIR MBE HgCdTe(001) and (112)Te.

	(112)		(001)Te	
	Best	Typical	Best	Typical
$N_d - N_a$ (cm^{-3})	2×10^{14}	8×10^{14}	1×10^{15}	2×10^{15}
$V_{\text{breakdown}}$	0.8	0.5	0.2	0.15
Q_{well} (nC/cm^2)	35	20	8	6

and typical values we find in MBE MWIR HgCdTe(112)Te and (001) films. For MWIR HgCdTe(112)Te, we measure donor densities as low as 2×10^{14} cm^{-3} and breakdown voltages as high as 0.8 V, while typical values are 8×10^{14} cm^{-3} and 0.5 V, respectively. Donor densities in the high 10^{14} cm^{-3} range for MBE HgCdTe(112)Te films have also been reported by another laboratory.[6] For MWIR HgCdTe(001) on the other hand, we find donor densities as low as 1×10^{15} cm^{-3} and breakdown voltages as high as 0.2 V, while typical values are 2×10^{15} cm^{-3} and 0.15 V, respectively.

The improved performance of MBE HgCdTe(112)Te films is shown graphically in Fig. 3. The breakdown voltage for MBE HgCdTe(112)Te layers is seen in panel (a) to increase from 0.3 to 0.8 V as the 80 K cutoff decreases from 9 to 5 μm. Our MBE HgCdTe(001) films given in panel (b), on the other hand, show a constant breakdown voltage of 0.2 V over the same wavelength range. Nonetheless, the MBE HgCdTe(112)Te layers illustrated in panel (a) still fall short of high quality bulk HgCdTe which shows an increase from 0.3 to 1.5 V as the 80 K cutoff decreases from 10 to 4 μm.

In order to improve the breakdown characteristics in our MBE MWIR HgCdTe films further, we have tried to identify the defect responsible for the higher measured tunnel currents. Figure 4 illustrates one possible cause of premature breakdown in MBE HgCdTe. The bright-field cross-sectional transmission electron micrograph (XTEM) shows a compositional inhomogeneity at a scale of 500 Å with an x-value variation at or above 0.1.[3] The HgCdTe(112)Cd layer was grown near 200 °C and has an average x value of 0.31. A HgTe rich band travels diagonally through the film and is clearly associated with the ridging observed on the sample surface. The ridging develops early in growth, and the resulting (001) and (111)Cd microfacets on the growth front have different HgTe growth rates. The dark diagonal band in Fig. 4 accompanies the (001) microfacet since this orientation has a higher HgTe growth rate than the (111)Cd. We have also observed this compositional inhomogeneity in our MBE HgCdTe(001)0,5° layers with a scale of 1000 Å, but this micro-inhomogeneity, if present in our MBE HgCdTe(112)Te films, falls below the detection limit for XTEM.

The x-value variation of ~0.1 in Fig. 4 can be removed by a 4–5 h anneal at 340–360 °C under Hg saturated conditions since the expected interdiffusion length is ~0.5 μm.[5] Because the compositional inhomogeneity is less pronounced

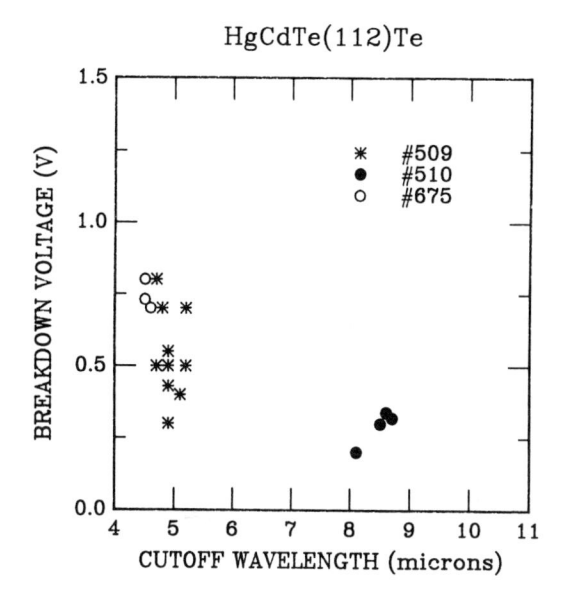

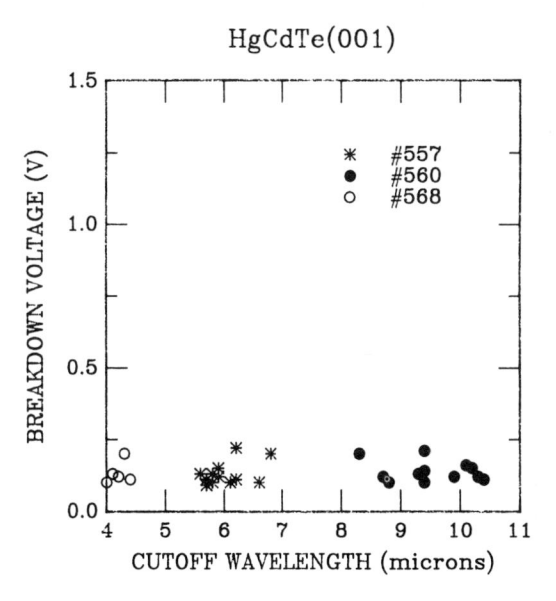

FIG. 3. The measured breakdown voltage for (a) MBE HgCdTe(112)Te and (b) (001) films is plotted as a function of the 80 K cutoff wavelength. The breakdown voltage in high quality bulk HgCdTe increases from 0.3 to 1.5 V as the cutoff is decreased from 10 to 5 μm.

with our MBE HgCdTe(001) and not even detectable for our MBE HgCdTe(112)Te, this interdiffusion anneal should be more than sufficient for all three oriented HgCdTe films.

However, Fig. 5 compares the breakdown voltage and donor density for two HgCdTe(112)Te samples; one sample (run 509) was stoichiometrically annealed at 200 °C, and the other (run 664) was annealed at 360 and then 200 °C. There is in fact a significant decrease rather than an increase in the measured breakdown voltage of the HgCdTe(112)Te layer after the high temperature annealing process. This result is supported by the four HgCdTe(001) and the other two HgCdTe(112)Te MWIR samples that were high temperature annealed as well. The breakdown voltage was less

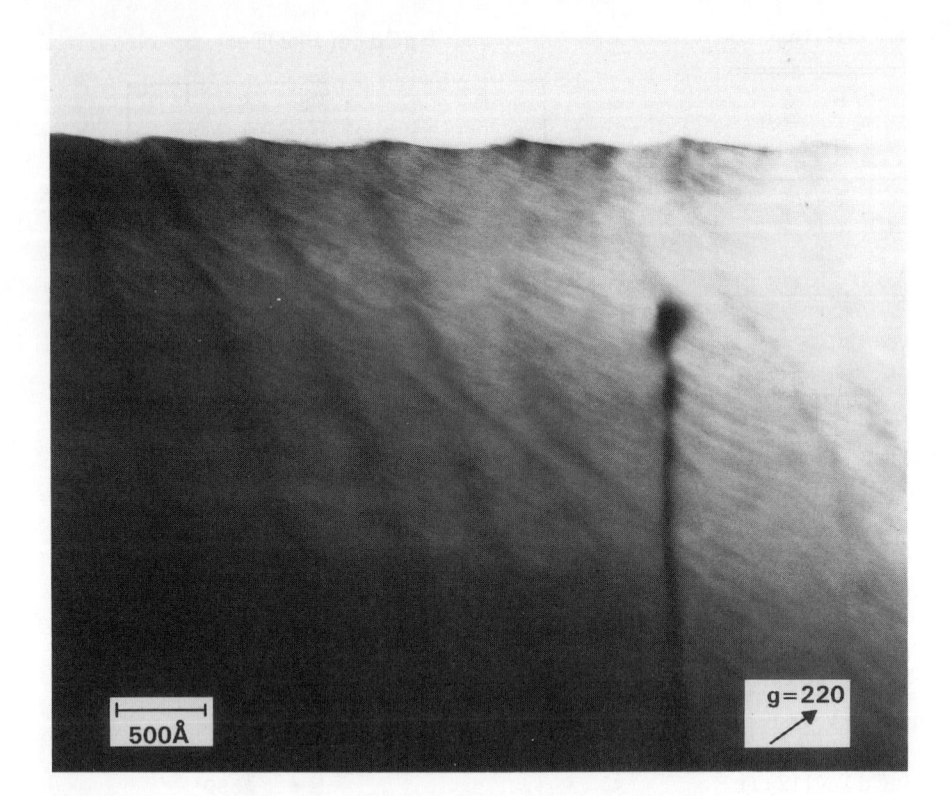

FIG. 4. A bright field XTEM micrograph of a 1.6 μm HgCdTe(112)Cd layer is illustrated. The fine-scale ridging observed at the surface of the film is associated with a compositional inhomogeneity in the layer [$x = 0.30$; P_{Hg}(meas) $= 9.0 \times 10^{-4}$ Torr].

than 0.1 and 0.2 V for the HgCdTe(001) and (112)Te layers, respectively, which is significantly less than the average values listed in Table I that were obtained after only a low temperature (200 °C) postgrowth anneal under Hg saturated conditions. On the other hand, the donor density remained similar; and the MIS storage time actually improved after the high temperature anneal. Typical MWIR full-well storage times of 1–10 ms improved to 40–80 ms with the high temperature postgrowth anneal.

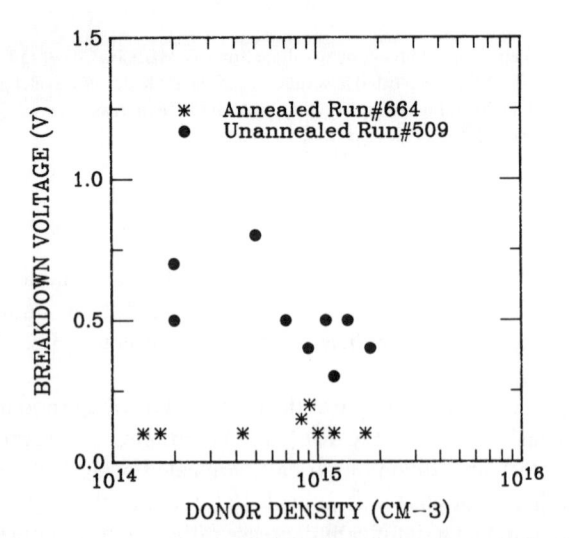

FIG. 5. The breakdown voltage for two MWIR HgCdTe(112)Te layers is plotted as a function of the MIS donor density. Run 664 was annealed under Hg saturated conditions at 360 °C for 4–5 h followed by a 200 °C for 24 hs; run 509 was only stoichiometrically annealed under Hg saturated conditions at 200 °C for 24 hs.

Since the high temperature (340–360 °C) postgrowth annealing unfortunately decreased the measured MIS breakdown voltage in our MBE MWIR HgCdTe(112)Te and (001) layers, we next examined the possible influence of compound twins within the depletion volume on the breakdown voltage. These compound twins always lead to pyramidal hillocks on MBE HgCdTe(001) films, while the twin volume is found by XTEM and chemical etching to contain a very high dislocation density. By reducing the twin density into the 10^3–10^4 cm^{-2} range and also by reducing the MIS gate area from the standard 10×15 to 5×5 mil^2, we have fabricated some MIS gates on a HgCdTe(001) layer that have no compound twins underneath them.

Figure 6 shows the measured MIS breakdown voltage for this MWIR HgCdTe(001) film where 10×15 and 5×5 mil^2 MIS gates were fabricated in a side-by-side fashion. The six large area gates resulted in measured breakdown voltages of 0.1–0.2 V, while the small area gates yielded breakdown voltages as high as 0.3–0.4 V. At the measured hillock density of 3×10^3 cm^{-2}, one hillock will on average be present under every other small area (5×5 mil^2) gate. Of the seven small area gates that were tested, five showed an improved breakdown voltage from the best results given in Table I for large area (10×15 mil^2) MIS test structures on MBE HgCdTe(001) films. The other two small area gates in Fig. 6 probably contained a hillock that reduced the breakdown voltage to 0.2 V.

The breakdown voltage for the HgCdTe(001) film in Fig. 6 improved to a near typical value for our MBE HgCdTe(112)Te layers (which is given as 0.5 V for a standard 10×15 mil^2 gate area in Table I) when the pyramidal hillock was removed from the MIS gate. This twofold im-

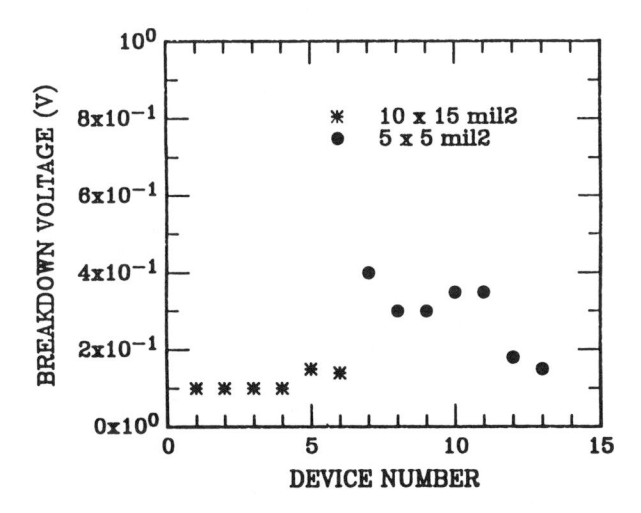

FIG. 6. The breakdown voltage for a MWIR HgCdTe(001) film is plotted for two different metal gate areas. Since most of the smaller gates avoided pyramids within the depletion volume, the average MIS breakdown voltage increased.

provement in the breakdown voltage of our MBE MWIR HgCdTe(001) film still falls short of high quality bulk MWIR HgCdTe which shows breakdown voltages up to 1.5 V.

Before considering the next possible cause for the high tunnel currents measured in our MBE HgCdTe layers, we will first describe the dominant mechanism for compound twin formation in MBE HgCdTe as well as our method for reducing their density. These compound twins are bounded by inclined (111) planes and produce a high dislocation density within the twin volume as evidenced by chemical defect etches. The local growth rate is also perturbed for HgCdTe(001) growth, although surface hillocks due to compound twins is not found on our HgCdTe(112)Te films

if the Hg flux applied is 20% above the minimum value necessary for single crystalline growth.[3]

Figure 7 shows the rings or clusters of pyramids that form during MBE HgCdTe(001) growth when a Hg pressure burst is detected with a remote ion gauge. The lateral dimension of these pyramids indicates a nucleation depth[4] that agrees well with the time during growth when the Hg pressure burst was recorded. The partially filled ring of pyramids probably occurs after a Hg droplet spreads on the HgCdTe growth front. This suggests that compound twins form as a result of Hg rich clusters on the HgCdTe growth front.

We find a 40-fold reduction in pyramidal hillock density for HgCdTe(001) growth as shown in Fig. 8 after passivating the crucible wall in our Hg effusion cell. In fact, a $100\times$ reduction of the compound twin density in MBE GaAs(001) layers was found by switching from a pyrolytic boron nitride (PBN) to a sapphire Ga crucible.[7] The formation of a volatile Ga_2O species within the Ga crucible, its subsequent transport to the GaAs substrate and then Ga agglomeration about the adsorbed oxide is thought to produce most of the compound twin density observed in MBE GaAs(001) growth.[8] It is likely that adsorption of residual gas (such as H2O and CO2) on the PBN crucible wall and subsequent reaction with the Ga vapor leads to the Ga2O formation.

In a similar fashion, we believe that the formation of HgO and its subsequent transport to the sample surface during MBE HgCdTe(001) growth is inhibited by reducing the amount of residual gas on the Hg crucible wall. The pyramid density of 2×10^3 cm^{-2} found in Fig. 8 is about a $10\times$ reduction from previously reported values,[9,10] although GaAs(001) layers now show compound twin densities near 100 cm^{-2}.[7]

Since line and plane defects that are revealed by chemical etching do not seem responsible for the higher tunnel currents in our MBE HgCdTe films, we now turn to the possibility of point defects. A Te antisite appears to be kinetically

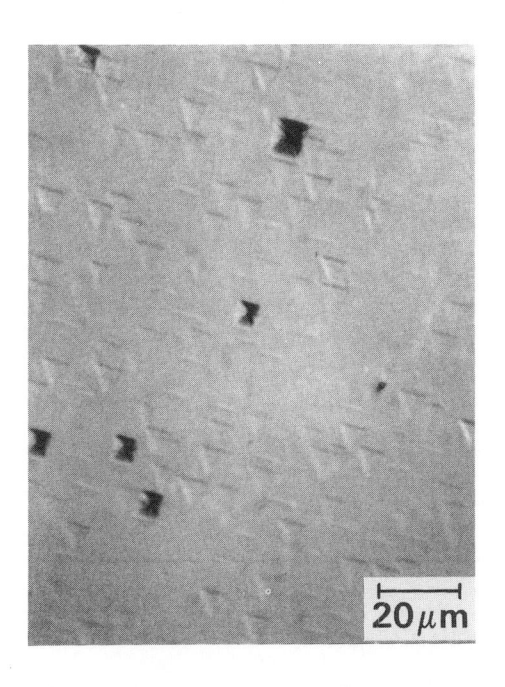

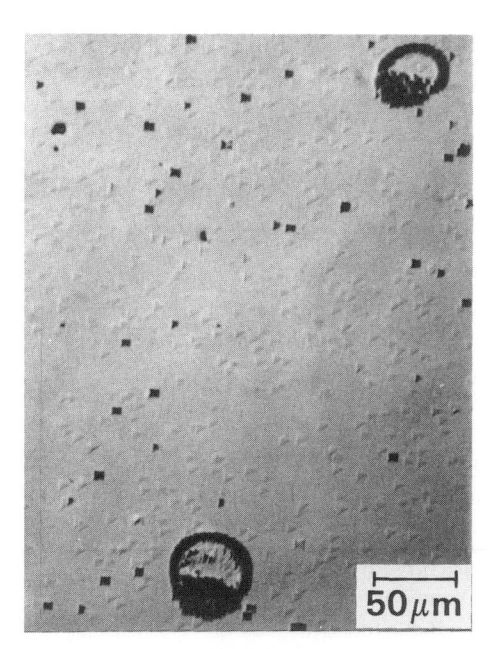

FIG. 7. Evidence is given for pyramidal hillock formation on MBE HgCdTe(001) films due to Hg clusters on the growth front.

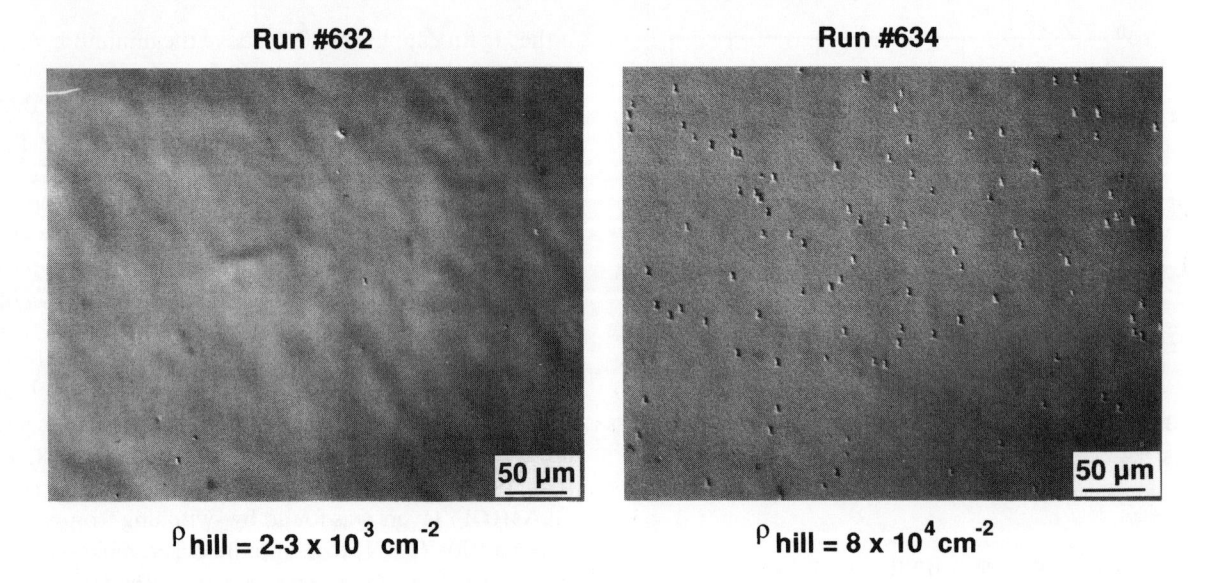

$\rho_{hill} = 2\text{-}3 \times 10^3 \text{ cm}^{-2}$ $\rho_{hill} = 8 \times 10^4 \text{ cm}^{-2}$

FIG. 8. A 40-fold reduction in the pyramid density is illustrated for MBE HgCdTe(001) films. This is accomplished by passivating the internal walls of the Hg source.

trapped in MBE HgCdTe layers[12] due to the low growth temperature; this antisite can lead to more mid-gap states that enhance trap-assisted tunneling at the moderate electric fields applied during MIS detector operation. In addition, a donor micro-inhomogeneity can lead to premature breakdown in MWIR MBE HgCdTe due to the high electric fields associated with the n^+ regions.

The metal vacancy concentration expected for MBE HgCdTe is approximately $1 \times 10^{16} \text{ cm}^{-3}$; this is the equilibrium vacancy concentration at 200 °C under Te-saturated conditions.[13] Since the metal vacancy acts as a doubly ionized acceptor at 80 K, the as-grown MBE films should show a *low* 10^{16} cm^{-3} acceptor density which, in fact, is observed for most MBE HgCdTe(112)Te layers. However, we find that MBE HgCdTe(001) films are n-type as-grown and do not change in Hall concentration after a standard 200 °C postgrowth anneal for 24 h under Hg saturated conditions (to remove any metal vacancies). This suggests a metal vacancy concentration $< 1 \times 10^{15} \text{ cm}^{-3}$ for our as-grown MBE HgCdTe(001) layers.

Figure 9 plots the MIS donor density in MBE HgCdTe(001) layers as a function of the Hg flux applied during growth. At the minimum equivalent beam pressure of 2×10^{-4} Torr, the HgCdTe(001) films have a low line and plane defect density, but the donor density is measured to be near $1 \times 10^{16} \text{ cm}^{-3}$. To reach the minimum donor density of $1\text{-}2 \times 10^{15} \text{ cm}^{-3}$, the Hg flux applied during growth should be at least twice the value required for low line and plane defect density film growth.

The simplest explanation for the Hg flux dependence in Fig. 9 as well as the n-type behavior of as-grown HgCdTe layers is the formation of a Te antisite. In this model, adsorbed Te will fill any metal vacancies present during growth which causes the as-grown n-type behavior. The Te species will compete less successfully for available metal sites as the Hg flux is increased which can explain the dependence found in Figure 9. In addition, group V dopants (As and Sb) are found to interact with the metal sublattice to cause n-type doping.[14]

The fact that MBE HgCdTe(001) layers depart from the thermodynamically predicted metal vacancy concentration is not very surprising since the MBE growth process does not occur near equilibrium. The low growth temperature allows kinetic effects to sometimes dominate the film properties and metastable structures can be produced. However, it appears that the Te antisites probably present in MBE HgCdTe(001) can easily cause the increased tunnel currents observed in our MIS test structures. A possible solution to this difficulty is found in higher temperature growth; we are now attempting to grow HgCdTe layers at or above 225 °C by MBE. The Te desorption rate is significantly higher in this regime and should reduce the Te antisite formation

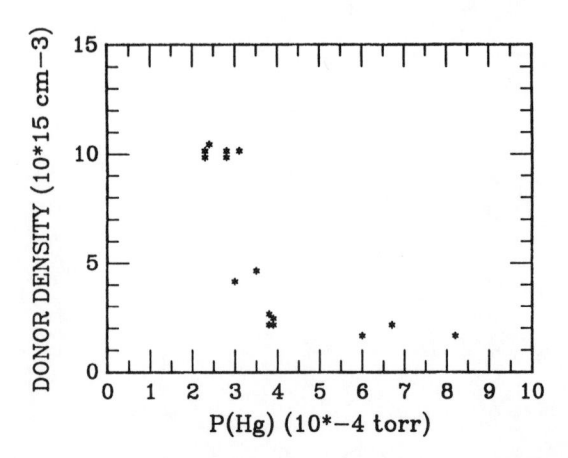

FIG. 9. The MIS donor density in MBE HgCdTe(001) films falls with increasing Hg flux applied during growth.

rate considerably. In fact, substrate temperatures over 200 °C are found to produce as-grown MWIR HgCdTe(001) layers by MBE that are *p* type.[15]

The presence of donor micro-inhomogeneities can also lead to higher tunnel currents in our MIS test structures. A mechanism for their formation is suggested from Fig. 4. The development of microfacets on the HgCdTe growth front is shown to result in locally varying rates of HgTe growth; however, the incorporation rate of unintentional dopants (that produce the low residual donor density in our HgCdTe(112)Te films) should also vary with the local crystallographic orientation. The (111)Te orientation, on the other hand, should provide the smoothest growth front morphology since it has the lowest surface free energy due to its high in-plane coordination. Since this orientation is still susceptible to in-plane twinning,[11] further development work appears necessary before it can be exploited.

IV. SUMMARY

Recent MBE grown HgCdTe(112)Te heterostructures show a $4\times$ increase in available charge capacity and a $2\times$ decrease in net donor density relative to our MBE HgCdTe(001) layers. To obtain full advantage of the MIS heterostructure however, another $2.5\times$ increase in the MWIR breakdown field (0.8 V/μm–2.0 V/μm) and a $2\times$ decrease in the MWIR donor density (1×10^{15} cm^{-3}–5×10^{14} cm^{-3}) is necessary. Since the MWIR breakdown field does not appear to be limited by compositional micro-inhomogeneities or by compound twin defects in the MBE HgCdTe layers, we suspect either Te antisites or donor micro-inhomogeneities are responsible for the significant tunnel currents measured in our MWIR MBE HgCdTe. At present, we are growing HgCdTe at a higher temperature (>225 °C) to increase the Te desorption rate from the HgCdTe surface and in turn to reduce the Te antisite formation rate.

ACKNOWLEDGMENTS

The authors gratefully acknowledge the technical assistance of Don Todd and Larry Presley.

[1] M. W. Goodwin, M. A. Kinch, and R. J. Koestner, J. Vac. Sci. Technol. A **8**, 1226 (1990).
[2] J. M. Arias, S. H. Shin, J. G. Pasko, R. E. DeWames, and E. R. Gertner, J. Appl. Phys. **65**, 1747 (1989).
[3] R. J. Koestner and H. F. Schaake, J. Vac. Sci. Technol. A **6**, 2834 (1988).
[4] A. Million, L. DiCioccio, J. P. Gailliard, and Piaguet, J. Vac. Sci. Technol. A **6**, 2813 (1988).
[5] D. A. Stevenson and M.-F. S. Tang, J. Vac. Sci. Technol. B **9**, XXXX (1991).
[6] T. H. Myers, R. W. Yanka, K. A. Harris, A. R. Reisinger, J. Han, S. Hwang, Z. Yang, N. C. Giles, J. W. Cook, Jr., J. F. Schetzina, R. W. Green, and S. McDevitt, J. Vac. Sci. Technol. A **7**, 300 (1989).
[7] D. G. Schlom, W. S. Lee, T. Ma, and J. S. Harris, Jr., J. Vac. Sci. Technol. B **7**, 296 (1989).
[8] S.-L. Weng, J. Vac. Sci. Technol. B **5**, 725 (1987).
[9] K. Harris, T. H. Myers, R. W. Yanka, L. M. Mohnkern, R. W. Green, and N. Otsuka, J. Vac. Sci. Technol. A **8**, 1013 (1990).
[10] However, a pyramid-free MBE HgCdTe(001) layer is reported elsewhere in these proceedings; see Z. Yang, Z. Yu, Y. Lansari, J. W. Cook, Jr. and J. F. Schetzina, J. Vac. Sci. Technol. B **9**, 1805 (1991).
[11] J. M. Arias, S. H. Shin, J. G. Pasko and E. R. Gertner, Appl. Phys. Lett. **52**, 39 (1988).
[12] J. P. Faurie, HgCdTe MBE Miniworkshop, DARPA-CNVEO, Washington 1987 (unpublished results).
[13] H. F. Schaake, J. Electron Mater. **14**, 513 (1985).
[14] M. Boukerche, P. S. Wijewarnasuriya, S. Sivananthan, I. K. Sou, Y. J. Kim, K. K. Mahavadi, and J. P. Faurie, J. Vac. Sci. Technol. A **6**, 2830 (1988).
[15] J. M. Arias, S. H. Shin, J. T. Cheung, J. S. Chen, S. Sivananthan, J. Reno, and J. P. Faurie, J. Vac. Sci. Technol. A **5**, 3133 (1987).

Review of the status of computational solid-state physics

A. Sher, M. van Schilfgaarde, and M. A. Berding
SRI International, Menlo Park, California 94025

(Received 5 November 1990; accepted 18 December 1990)

The current status of *ab initio* methods to study the properties of solids is reviewed. During the past ten years, a wealth of calculations have established that many properties of solids can be accurately calculated from first principles. Recent advances in methods now make it possible to treat a number of practical problems in materials science within tolerable times and costs.

Practically every solid-state physics text begins by pointing out the difficulty of solving the Schrödinger equation for $\sim 10^{22}$ particles, and that numerous approximations are required to make progress. Twenty-five years ago, Hohenberg and Kohn[1] established the framework for modern first-principles calculations by proving that the total energy of a system was a functional only of the electron density. This functional, however, is unknown. A subsequent "local" approximation[2] made it possible to cast the problem as a set of independent particles moving in an effective "one-electron" potential for which explicit expressions could be obtained. This approach, now known as density functional theory (DFT) and the local density approximation (LDA), has evolved into a parameter-free theory capable of predicting structural and electronic properties with reasonably high precision.[3] However, until recently, the computational methods required to actually solve these equations were so slow that even on large machines calculations have been restricted to simple systems, e.g., silicon and GaAs.[4] There has been a significant advance in computational methods in the last few years, making it possible now to solve problems much more efficiently than previously.[5,6] This constitutes an impressive advance in science, but, perhaps more important, a revolution in technology. Now it is practical in systems as complicated as HgTe, CdTe, and soon the alloys $Hg_{1-x}Cd_xTe$, to compute, for example, native defect energies,[7,8] impurity substitution energies,[7,8] impurity interstitial energies,[9] and diffusion coefficients,[9] as well as the simpler bond lengths, cohesive energies, and elastic constants.[10] Accurate calculations of defect electrical activity and transport properties based on these parameter-free theories are within reach. This capability should have a dramatic impact on device processing and performance design.

The purpose of this paper is to review the status of this advance and to provide the reader with some insight into its prospects. The nature of the analysis is illustrated schematically in Fig. 1. It begins with a trial density n_{in}, often a superposition of free-atomic densities. Using the LDA (see Fig. 1), this density generates an effective one-electron potential V_{in} and a Hamiltonian, which in turn, generates an electron density n_{out}. n_{out} is mixed with n_{in} and the cycle is repeated until self-consistency is reached, i.e., $n_{out} = n_{in}$. Coulomb and magnetic interactions as well as relativistic terms are included. The Schrödinger equation must be solved in terms of a basis set. Two leading, generally applicable basis sets are the linearized augmented plane wave (LAPW)[11] and linear muffin-tin orbital (LMTO)[3] methods. For any basis set, the computational time for a solution using conventional techniques is proportional to N^3, where N is the number of basis states per unit cell. The output is the wave functions $\Psi(r)$, and the corresponding electron density $n_{out}(r)$.

The exact functional that generates V_{in} from n_{in} is not known, and in practice it is possible to do this only within the LDA, for which explicit expressions can be calculated. The LDA has been shown to produce good valence band states[3] in semiconductors and good structural properties in most materials, e.g., atomic volume, cohesive energies, and elastic constants.[10] The band gap and conduction band states are less accurately predicted in LDA. However, it has recently been shown that these properties can be accurately calculated using a technique called the "GW" approximation.[12] GW is the first term in a perturbation expansion of the exact many-body Schrödinger equation. Very good optical properties have been obtained in semiconductors when the LDA is used as a starting point to calculate GW, the Green's function G and screen Coulomb interaction W.

Among the basis sets used to solve the Schrödinger equation, the pseudopotential method is the simplest because it uses plane waves. Plane waves (PW) are suitable only for smooth potentials as they are eigenfunctions of free electrons, but they are unsuitable for systems with d electrons, such as CdTe and HgTe. Even in a simple semiconductor such as Si, 100 to 1000 PW are needed per atom, depending on the precision sought. Because the computation time varies as N^3, the direct application of this method is limited to problems with relatively few atoms per unit cell. Car and Parrinello[5] devised an improvement to this method in which the Hamiltonian itself is allowed to evolve while iterating for the eigenstates. The computation time for one iteration in their method is proportional to NM^2, where M is the number of occupied states. This advance permits much larger systems to be treated, since $M \ll N$ in that method.

A second important method is LMTO. It employs a much more efficient basis set with only ~ 20 orbitals/atom required to reach convergence. The LMTO method, until re-

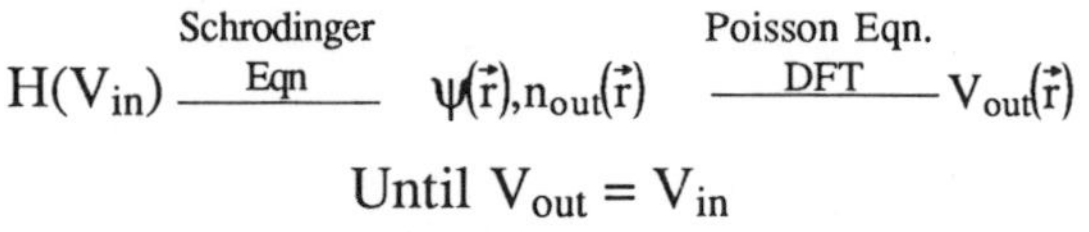

$$H(V_{in}) \xrightarrow{\text{Schrodinger Eqn}} \psi(\vec{r}), n_{out}(\vec{r}) \xrightarrow{\text{Poisson Eqn.}}_{\text{DFT}} V_{out}(\vec{r})$$

$$\text{Until } V_{out} = V_{in}$$

FIG. 1. Self-consistency loop.

cently, employed the atomic spheres approximation (ASA) in which within each iteration of the self-consistency loop the potential is approximated by its spherical average about the ions and some interstitial positions (empty spheres).[3] This approximation limited the applicability of the method to problems of high symmetry; it excluded problems like substitutional impurity atoms, or surfaces from those for which the precision was limited by LDA. Two years ago, a full-potential LMTO (FP-LMTO) was devised[6] that removed the ASA, thereby opening a whole new range of applications. The Car and Parrinello technique still works in FP-LMTO but does not yield much speed enhancement because the basis set is already small. There is another advance of value, however: the Harris-Foulks functional.[13,14] This is a technique for selecting a density functional that is insensitive to details of the electron density and obviates the need for the self-consistency loop. The technique increases the speed of calculations with precision still limited by LDA.

A concrete example of the speed of current calculations is presented in Table I. The example selected is a calculation of the tellurium antisite defect in CdTe using a supercell of 64 atoms. The quantities calculated are the substitution energy, the impurity levels, and strain distributions around the relaxed antisite.

In this approach, the crystal is treated as a periodic arrangement of supercells with one antisite in each cell. When the cell size is large enough so the antisites no longer interact with one another, then the calculated properties converge to those of an isolated antisite. We find that a cell size of 32 atoms is often sufficient but the numbers in Table I are for 64 atom cells.

The estimated computation times required to solve this problem on an Apollo DN10,000 computer, using a single CPU are listed in Table I. This machine is small enough to sit in a closet, costs < $100,000, and is run by our professional staff. Its clock time is roughly 1/6 that of a Cray, but because the Cray is a vector processor, it runs this kind of problem about 30 times faster than the Apollo. Quoted times and precisions are approximate, and four iterations to self-consistency are assumed.

As you can see from the table, conventional pseudopotential methods are out of the question for a problem with 64 atoms per unit cell. However, the Car-Parrinello (for smaller problems) and LMTO methods are quite practical. A time of 100 h, or about four days, for an ASA-LMTO solution, on a machine that sits in a closet operating in the background, is perfectly acceptable. Our machine actually has three CPUs and can accommodate four, so if one person is doing a production run it does not seriously affect the rest of the group. If the Harris-Foulks functional is used, a factor of four is gained. A time of 500 h for FP-LMTO requires some patience, but even this is acceptable. Using an optimally vectorized and parallelized code can greatly reduce this time:[15] indeed, we are now testing an experimental version of the ASA program that does this calculation in about 10 h, using the three CPUs in parallel. We wish to emphasize that problems involving 64 atoms per unit cell are well outside the capabilities of earlier methods and remain time consuming even using FP-LMTO. However, a problem dealing with host material properties with only two atoms per cell, e.g., the band structure or cohesive energy $r_{in}(64/2)^3 = 3 \times 10^4$ times faster, so instead of 500 h the results are obtained in about one minute.

In the past, solid-state theory was able only to calculate general trends of phenomena from first principles, or with more accuracy predict interrelations between various observables, i.e., parametrize potentials. Parametrized potentials generally can be trusted only in circumstances where the local atom arrangement deviates little from the one for which the parameters were chosen. Hence, for example, if parameters are chosen to fit bulk properties, quantitative predictions of surface reconstruction may not be trustworthy, although the symmetry may be properly given. The new *ab initio* methods reviewed here do not suffer from this kind of uncertainty and, as a consequence, can be trusted to within a given precision. This ability should be an invaluable help in interpreting observations, devising means to circumvent deleterious effects, and designing manufacturing processes.

Acknowledgments: We thank ONR (Contract Nos. N00014-89-K-0132 and N00014-88-C0096), AFOSR (Contract No. F4920-88-0009), and NASA (Contract No. NAS1-18226) for support of our program.

TABLE I. Computational effort to calculate an antisite defect in CdTe using a supercell approach (64 atoms/cell). Time is for an Apollo DN10,000, single CPU. Times and precisions are approximate. Four iterations to self-consistency are assumed.

Method	Orbitals/atom	Time (h)	Precision (eV)
Pseudopotential[a]	90	10 000	0.2
	300	< 10^5	LDA[b]
LAPW	50	5 000	0.2
	100	40 000	LDA[b]
Car-Parrinello[a]	300	1 000	LDA[b]
LMTO-ASA[c]	13	100	0.5
LMTO-FP	22	500	LDA[b]

[a] Pseudopotential form unsuitable for *d*-band materials.
[b] LDA error is unknown but is expected to be about 0.1 eV.
[c] Unsuitable for calculating lattice relaxations.

[1] P. Hohenberg and W. Kohn, Phys. Rev. **136**, B864 (1964); W. Kohn and L. J. Sham, *ibid*. **140**, A1133 (1965).
[2] W. Kohn and L. J. Sham, Phys. Rev. **140**, A1133 (1965).
[3] O. K. Anderson, O. Jepsen, and D. Glötzel, in *Highlights of Condensed Matter Theory*, edited by F. Bassani, P. Tosi, and P. Fimi (North Holland, Amsterdam, 1985).
[4] G. A. Baraff, M. Schlüter, and G. Allen, Phys. Rev. Lett. **50**, 739 (1988).
[5] R. Car and M. Parrinello, Phys. Rev. **55**, 2471 (1985).
[6] M. Methfessel and M. van Schilfgaarde, Phys. Rev. (to be submitted).
[7] M. Berding, M. van Schilfgaarde, A. T. Paxton, and A. Sher, J. Vac. Sci. Technol. A **8**, 1103 (1990).
[8] M. J. Caldas, J. Dabrowski, A. Fazzio, and M. Scheffler, Phys. Rev. Lett. **65**, 2046 (1990).
[9] Y. Bar-Yam and J. D. Joannopoulos, Phys. Rev. Lett. **52**, 1129 (1984).
[10] M. van Schilfgaarde, M. Methfessel, and A. T. Paxton (in preparation).
[11] S. H. Wei and H. Krakauer, Phys. Rev. Lett. **55**, 1200 (1985), and references therein.
[12] M. S. Hybertsen and S. G. Louis, Phys. Rev. Lett. **58**, 1551 (1987).
[13] J. Harris, Phys. Rev. B **31**, 1770 (1985).
[14] M. Foulkes and R. Haydock, Phys. Rev. B **39**, 12520 (1989).
[15] M. van Schilfgaarde and A. T. Paxton (in preparation).

Mercury cadmium telluride junctions grown by liquid phase epitaxy

C. C. Wang
Fermionics Corporation, Simi Valley, California 93063

(Received 2 January 1991; accepted 12 February 1991)

This article reports the advancement of HgCdTe epitaxial technology. Specifically, we will discuss the grown *p–n* junctions using liquid phase epitaxy (LPE). In our LPE approach, CdTe is used as the substrate. In the LPE technology studied, an HgCdTe active layer is first grown with a desired composition. After growth is terminated, the furnace is cooled down to room temperature. The HgCdTe layer is then characterized before reinserting into the furnace for the second layer growth. Both *P*-on-*n* and *N*-on-*p* type of heterostructures are obtained by controlling dopants in each layer properly. For the *P*-on-*n* type, the active layer is either un-doped or doped with In and the top layer is doped with column V species. For the *N*-on-*p* type, the active layer is undoped and the top layer is doped with In together with Cu. In either case, the location of the *p–n* junction can be well controlled.

I. INTRODUCTION

The commonly used methods for fabricating photovoltaic diodes in mercury cadmium telluride (HgCdTe) have been via ion implantation and diffusion. Abundant results of *n* on *p* diodes were reported.[1-8] However, for most applications at 77 K the low junction impedance and high leakage current of unknown origins of these diodes preclude their use in current electro-optic systems.

In the past decade, HgCdTe liquid phase epitaxy (LPE) has emerged as an important technology to produce device-grade materials.[9-12] Most efforts, however, are concentrated on being able to produce reproducible high quality single epitaxial layers. The goal is to produce *n*-type epilayers for photoconductive mode of operation and *p*-type epilayers as the base material for ion-implanted or diffused junctions. Recently, multilayers of HgCdTe using LPE became successful and theoretical analyses of heterojunctions have since been proposed.[13,14] In this paper, we will discuss multilayer HgCdTe heterostructure diodes, grown by LPE, in both *P*-on-*n* and *N*-on-*p* modes. *P*-on-*n* diodes exhibit better than 60% quantum efficiency and also appear to be more reproducible and have higher R_0A parameters than those of HgCdTe diodes fabricated by ion implantation and diffusion.

II. EPILAYER PREPARATION AND JUNCTION FORMATION

In our LPE approach, both $\langle 111 \rangle$ CdTe and $\langle 111 \rangle$ ZnCdTe lattice matched to HgCdTe have been used as the substrate. In this paper, only results on the CdTe substrate grown at Fermionics will be discussed. After standard cleaning procedures, the wafer is loaded in a horizontal tipping/sliding system and a HgCdTe epilayer ~25 μm thick with a desired composition is grown on top of it in a saturated Te solution. A predetermined amount of excess Hg is added in the melt to account for the bulk loss of Hg during growth. A secondary Hg source at a lower temperature zone is used to establish a partial equilibrium condition over the melt. After growth is terminated, the furnace is cooled down to room temperature. The HgCdTe layer is then measured for its infrared (IR) transmission characteristics (such as percent of transmission and cut on λ). It is then etched before reinserting into the furnace for the second layer growth. The second layer with a higher Cd composition 1 μm thick is grown on top of the active layer in Hg solution. That layer contains some Zn to adjust its lattice parameter as close as possible to that of the HgCdTe active layer. The function is to reduce defect-induced dislocations at the interface because of dangling bonds. At the end of growth cycle the furnace is cooled down to room temperature at a fast rate to minimize interdiffusion of Hg and Cd across the heterostructure interface.

The movement of Hg and Cd across the interface influences the relative band gap on either side of the heterostructure. That is to say that the band gap of the multilayer structure probably changes continuously from the top layer to the active layer and no discontinuity exists between the layers. What makes this structure behave as a *p–n* junction is the result of doping in the top layer. The basic and only requirement for this to happen is to have a slow diffusant in the top layer.

For the *P*-on-*n* type, the active layer is either undoped or doped with In to a level of $1 \times 10^{15}/\text{cm}^3$. The as-grown layer is *p* type, with a carrier concentration on the order of $5 \times 10^{17}/\text{cm}^3$ and a Hall mobility about 300 cm^2/V s at 77 K. The top wide band gap layer is doped with column V species such as As or Sb to a level of $10^{16}/\text{cm}^3$. After growth, the multilayer is annealed in a saturated Hg vapor environment. The active layer was then changed to *n* type with a carrier concentration on the order of $5-10 \times 10^{14}/\text{cm}^3$ and a Hall mobility of 4×10^4 cm^2/V s. The top layer stayed as *p* type as a result of doping. Since column V elements replace Te sites and move very slowly in HgCdTe, we have created a *P*-on-*n* heterojunction not easily achievable by the ion implantation or diffusion process.

For the *N*-on-*p* type, the active layer is undoped and the top layer is doped with In and Cu. The In level is to provide an *n*-type carrier concentration of high $10^{16}/\text{cm}^3$ and Cu low *P* type of $10^{16}/\text{cm}^3$. Since Cu diffuses quickly in HgCdTe and In moves very slowly at the growth temperature, we have created an *N*-on-*p* heterostructure *in situ*. Same Hg vapor

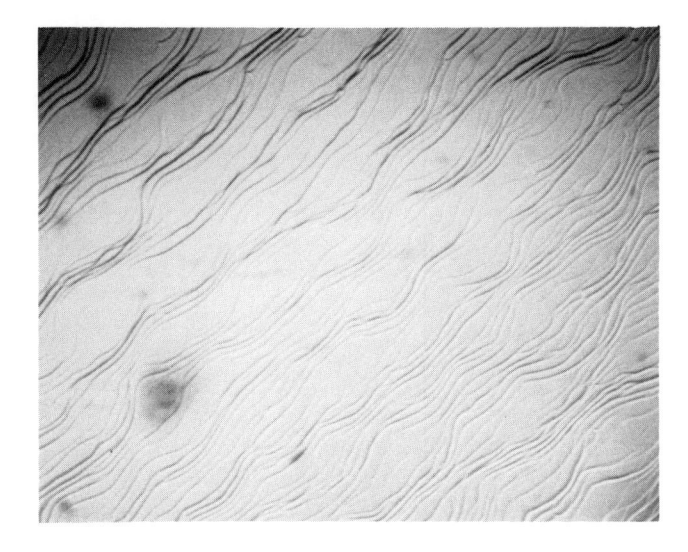

FIG. 1. The surface feature of a HgCdTe double-layer grown by LPE usually exhibits fine lines following the terracing lines of the first layer. The exact reason for this behavior is not quite clear.

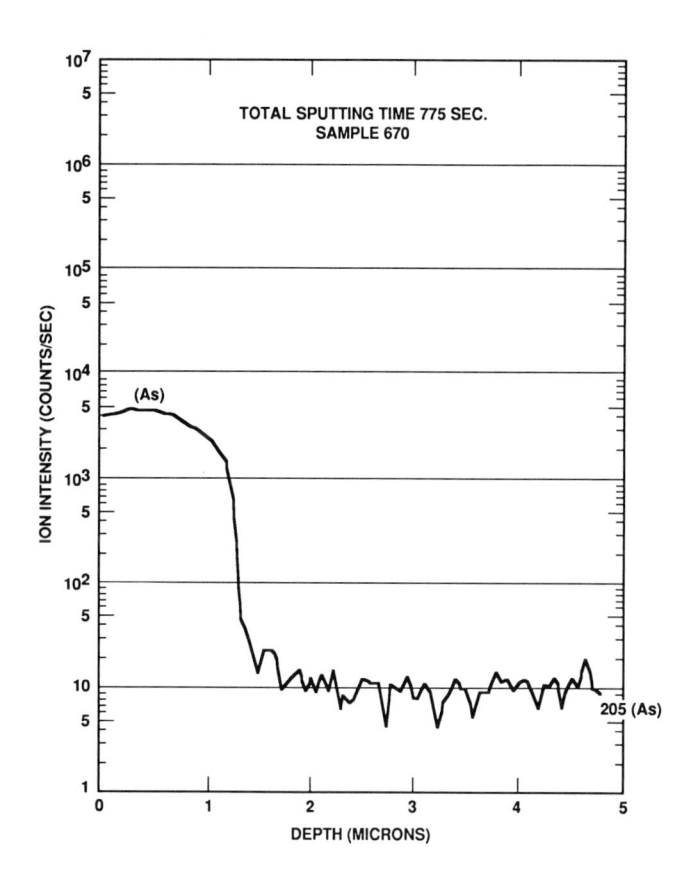

FIG. 2. The SIMS profile of As in a HgCdTe double-layer indicates that As is a slow moving species in HgCdTe. The location of rapid transition occurs at the metallurgical interface of the first and second layers.

annealing is followed to homogenize the defect structure in the multilayer.

III. MULTILAYER CHARACTERIZATION

The surface feature of single layer is quite common, containing some defects and the terracing structure characteristic of LPE growth. The surface of multilayer is slightly different. In addition to the usual features, the multilayer surface exhibit fine closely spaced lines along the terracing lines as shown in Fig. 1. The exact reason for this surface texture is not quite clear but it is suspected that these fine lines are the result of meltback prior to the top layer growth in the Hg solution. Figures 2 and 3 are, respectively, secondary ion mass spectroscopy (SIMS) data of As and Sb profile in the multilayer after the junction formation. Notice the relatively rapid transition of As and Sb concentration at the interface of the top layer and the active layer. It appears that no appreciable amount of As nor Sb was detected in the active layer and the substrate. Figure 4 shows SIMS data for an N-on-p heterostructure. As shown in Fig. 4, In does not move into the active layer while Cu distributes uniformly in the double layer. As described before, the basic and only requirement for either P-on-n or N-on-p type to exist in HgCdTe is to have a slow diffusant in the top layer.

Typical Auger line scans for Hg, Cd, and Te distributions in the multilayer structure are shown in Fig. 5. Similar line scans have been used routinely to evaluate the reproducibility of heterostructure growths. In the example here, the top layer is about 0.9 μm thick. The features shown in Fig. 5 at 0.9 μm from the surface are variations of composition at the metallurgical interface. These variations are caused by interdiffusion of all species during the period of heterostructure formation. As a result of interdiffusion, the compositions change continuously from the active layer to the surface indicating that the band gap changes continuously. The maximum bandgap in the top layer is usually designed to be $x = 0.3$–0.35. The lowering of Cd composition toward the surface is the result of depletion of Cd concentration in the melt. Since the top layer is both P-on-n and N-on-p types has a higher carrier concentration, the narrowing of bandgap near the surface has practically no negative effect. The position of p–n junction depends on the relative doping concentration across the junction. However, in practice it is not clear where the junction takes place. From our diode studies, it has been estimated that the electrical junction is about 0.5 μm from the metallurgical interface and resides in the active layer.

IV. DEVICE FABRICATION

In fabricating large diodes ($200 \times 200\ \mu$m), Au is used as the ohmic contact for p-type side and In for the n-type side. After p-side contact metallization, standard photolithographic techniques are used to delineate mesa diodes. The chip is then mounted on a supporting fixture to facilitate the backside-illuminated scheme, i.e., the radiation is incident from the substrate side and absorbed in the HgCdTe active layer. A thermal compression bonding technique is used to connect 1 mil gold wire from the mesa tops to the header pins before measurements are made. These large diodes are used to screen and characterize multilayers. The parameters examined are spectral response, capacitance-voltage data, di-

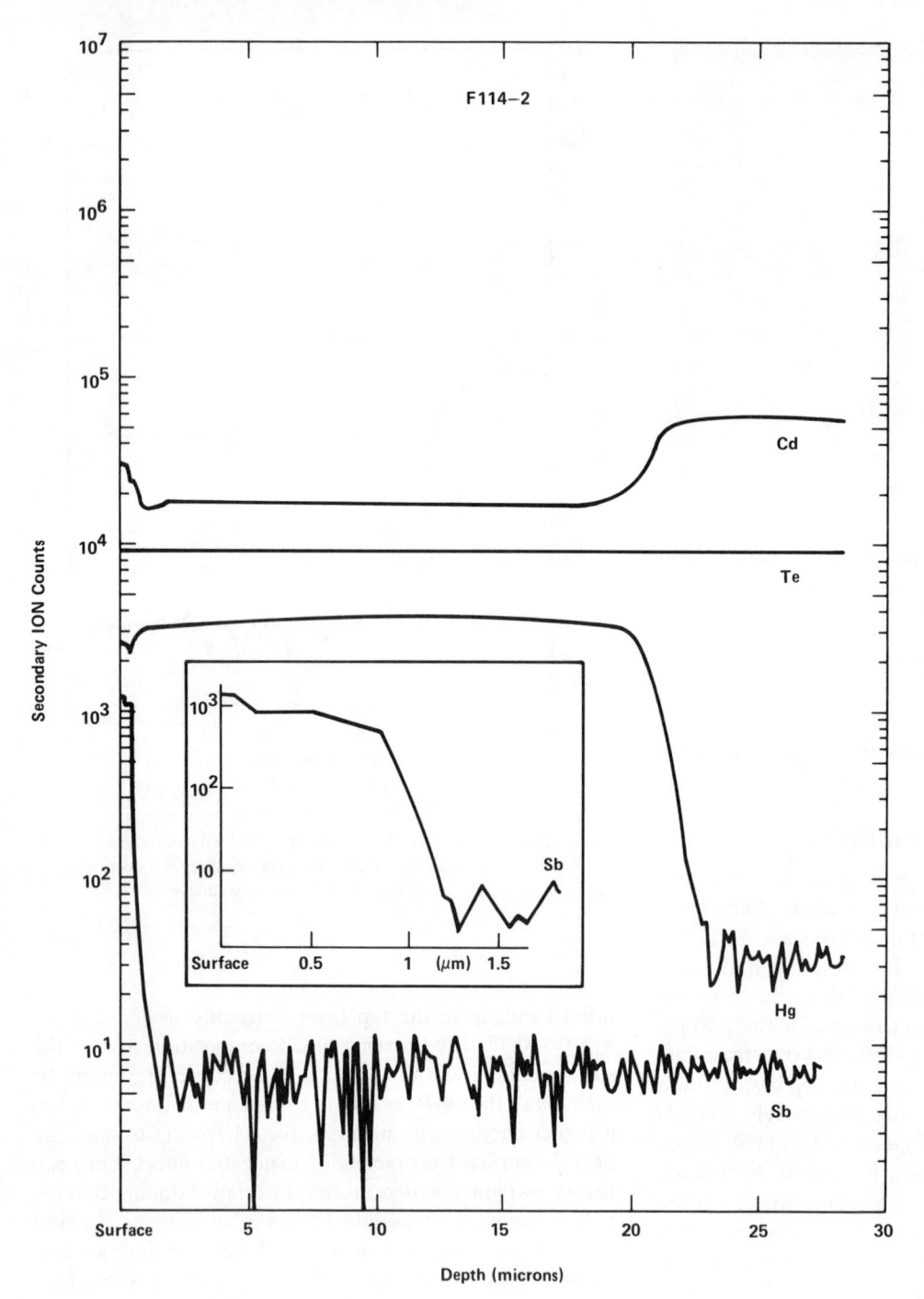

FIG. 3. Similar to the behavior of As, Sb is also found to be a slow diffusant in HgCdTe as shown here in the SIMS profile.

ode characteristics, surface treatment experiments, etc. In the work reported here all diodes have no surface passivation layer applied.

V. DETECTOR PERFORMANCE AND DISCUSSIONS

Figure 6 shows the I–V curves of several 11 μm P-on-n diodes at 77 K. The I–V curves are shifted vertically per division for each diode to illustrate the relative uniformity of detectors. The diodes are shown to have an open circuit voltage (V_{oc}) of ~45 mV and a short circuit current (I_{sc}) of ~25 μA. Using $I_{sc} = \eta Q_{bg} q A$, where η is the quantum efficiency, Q_{bg} is the background flux, q is the electric charge, and A is the detector optical area, the quantum efficiency is calculated to be ~60%.

For diodes with a longer wavelength cutoff, the open circuit voltage usually becomes smaller because of a narrower band gap in the active layer. For example, the open circuit voltage for a 13 μm cutoff diode is reduced to about 20 mV.

The mesa surface critically effects the diode performance. The problem becomes progressively less manageable as the detector cutoff wavelength moves to longer wavelengths. To a certain degree, this is true for all semiconductor p–n junctions. By comparing HgCdTe P-on-n and N-on-p type junctions, the surface of the former is found to be more manageable. This is probably because the surface of p-type HgCdTe is

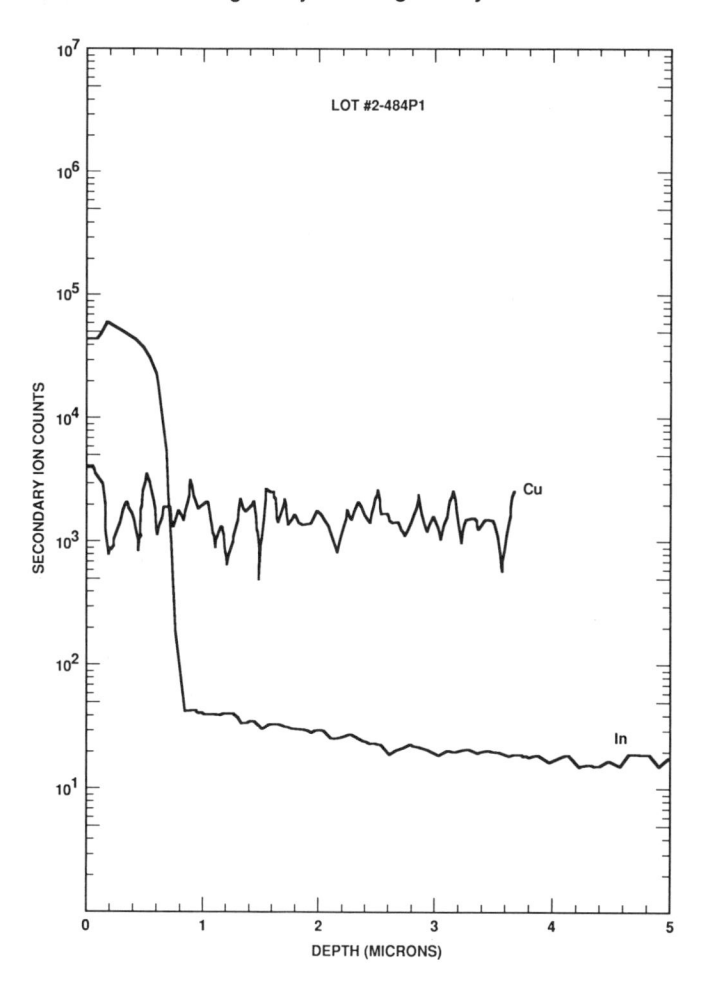

FIG. 4. In is believed to move in HgCdTe substitutionally. As a result, it is also a slow dopant. Its behavior in HgCdTe depends critically on Hg partial pressure. In the N-on-p device design reported here, the active layer is undoped and the top layer is doped with In and Cu. The SIMS profile reveals that In stays in the top layer while Cu permeates throughout HgCdTe during growth.

easily inverted while that of n-type HgCdTe is not. As a result, a less shunted conducting surface is present in the P-on-n type junction where the n-type HgCdTe has a lower carrier concentration.

Figure 7 shows the relative spectral response measured per watt of two representative P-on-n diodes. In our measurements, a reduced field of view (FOV) was used to cut down the background photon flux and a thermocouple detector was used as the reference. The 50% point in the relative spectral response is used to determine the spectral cutoff wavelength of detectors. The cutoffs of detectors at various places of a wafer are used to evaluate its lateral uniformity. From our development so far, a spread of $\lambda_0 \pm 0.5\,\mu m$ over a wafer size of 2.5×2.5 cm, where $12\,\mu m \geqslant \lambda_0 \geqslant 10\,\mu m$, can easily be achieved. With a better control, consequently a lower yield, a spread of $\pm 0.25\,\mu m$ has been possible.

The C–V curves for these diodes are routinely measured, from which the carrier concentration distribution in the active layer can be calculated. For a P-on-n structure the calculated carrier concentration distribution is mostly on the n-type side. The depletion width at zero bias is typically 0.4–0.5 μm and the junction appears to be graded. The carrier concentration at zero bias is about mid-$10^{14}/cm^3$ and usually is lower than that obtained from the Hall data. This is probably caused by compensation which does take place at the p–n junction interface. The carrier concentration calculated at > 1 V reverse bias usually matches well with the Hall measurement. A value of low $10^{15}/cm^3$ is generally obtained.

For an N-on-p structure, similar diode behavior is observed. The carrier concentration calculated at zero bias is typically mid-$10^{15}/cm^3$ and low $10^{16}/cm^3$ at > 1 V bias. The depletion width is typically 0.2–0.3 μm.

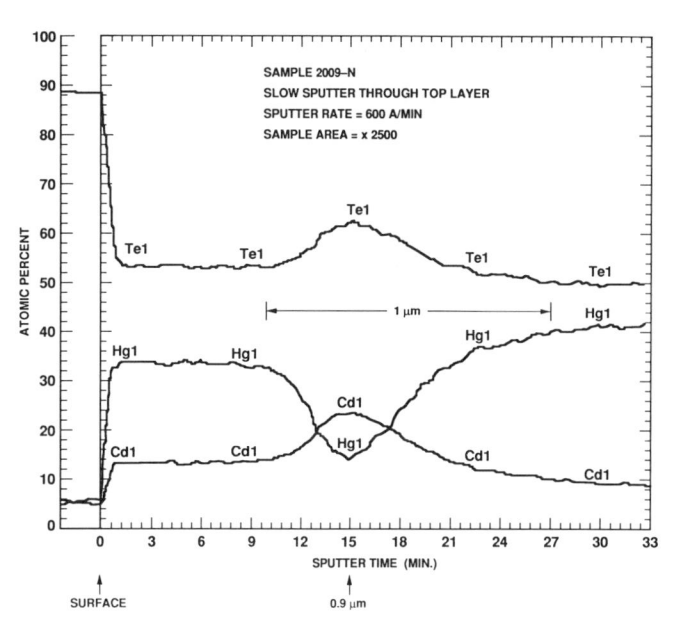

FIG. 5. The Auger profile of Hg, Cd, and Te in HgCdTe double-layer shows composition variations across the interface. In the example here, the top layer is about 0.9 μm thick. The compositions change continuously indicating that cross-diffusion has taken place during growth.

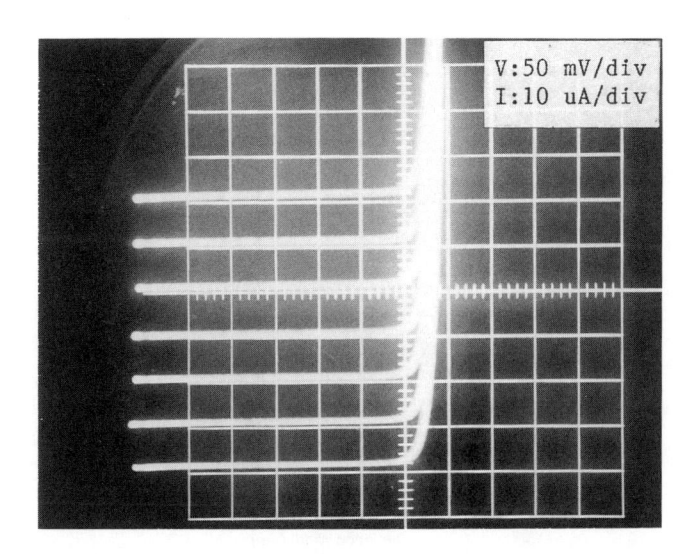

FIG. 6. The I–V curves shown here are for several 11 μm HgCdTe P-on-n diodes. These curves are shifted vertically per division for each diode to illustrate relative uniformity.

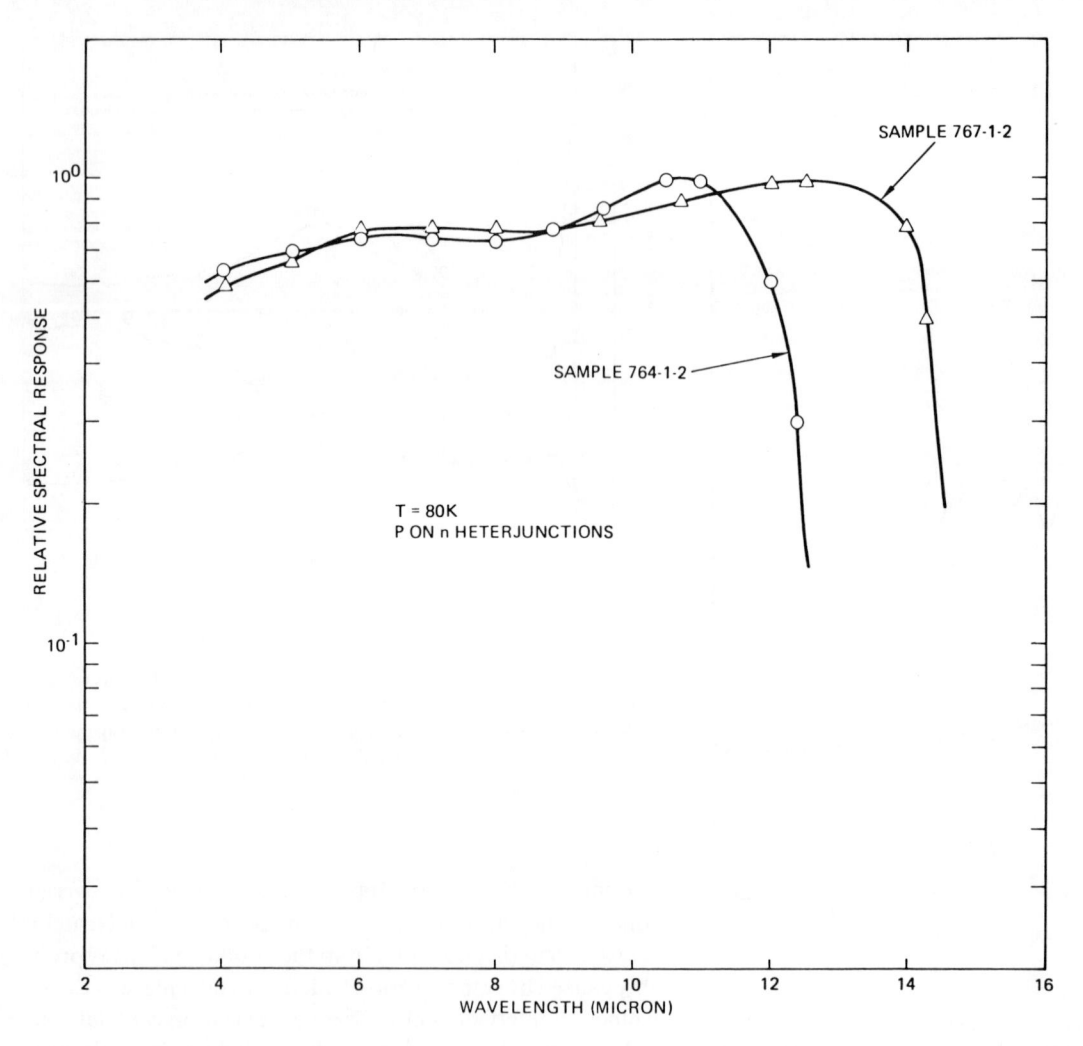

FIG. 7. The relative spectral response is measured per watt of incident power. The 50% point is used to determine the detector spectral cutoff wavelength.

Figure 8 illustrates the junction profile from C–V measurements at 77 K of both P-on-n and N-on-p diodes. The breakdown voltage for 10.5 μm P-on-n diodes (mesa area = 4×10^{-4} cm^2) is usually ≈ 1 V, with better ones being $\geqslant 1.5$ Volt. On the other hand, N-on-p diodes with similar

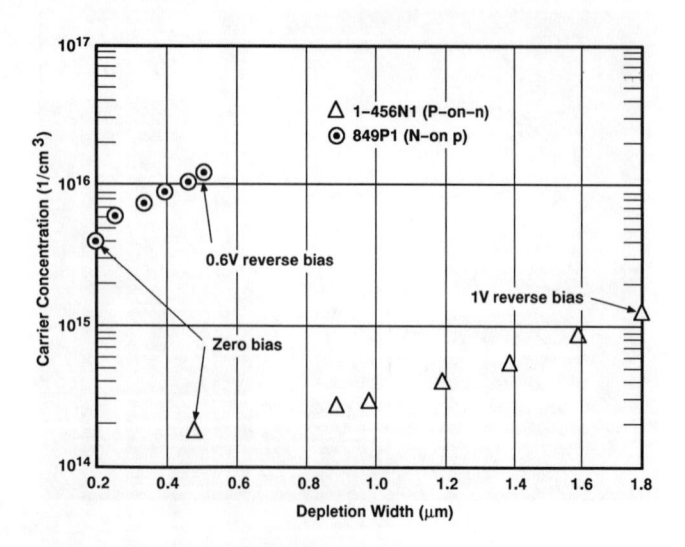

FIG. 8. The junction profile obtained from C–V data usually indicates that the diode has a graded junction. Data shown here are typical for P-on-n and N-on-p diodes.

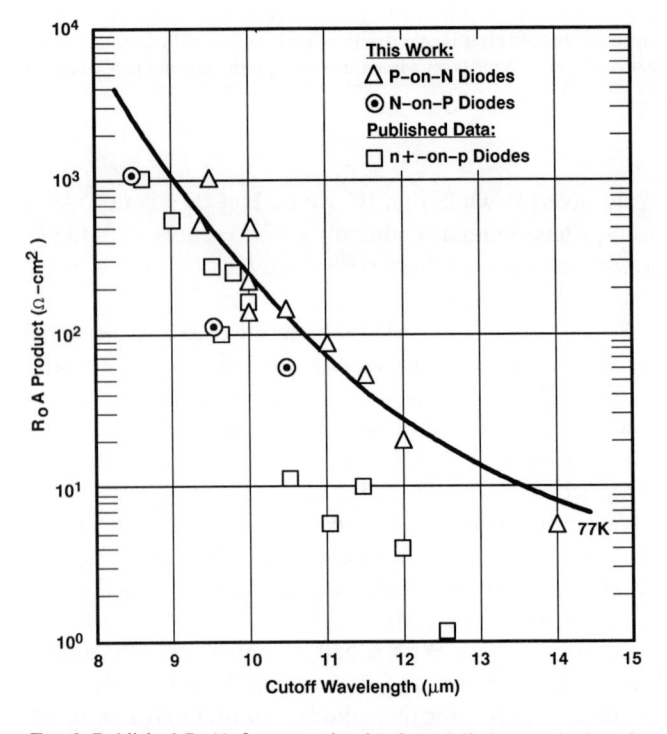

FIG. 9. Published $R_0 A$'s for n-on-p ion-implanted diodes are obtained from Refs. 6–8. The better diodes were selected. Our N-on-p grown junction diodes compare favorably with the reported data. However, our P-on-n diodes are clearly superior. The solid line is calculated based on simple diffusion model.

spectral cutoffs usually exhibit softer reverse characteristics. In both diode types, the longer the wavelength is, the smaller the breakdown voltage.

In the initial phase of this technology development, potential barrier has been observed at 77 K. The simple diode behavior of this type would be: large breakdown voltage, high $R_0 A$ product and low quantum efficiency. The approach to avoid potential barrier is to adjust, in concert, the carrier concentration of the top layer, its bandgap and the composition gradient at the interface. The diodes discussed here do not appear to have the potential barrier problem and their quantum efficiencies generally are in the range of 40%–60%.

In comparing performance of P-on-n and N-on-p diodes, we have observed that both exhibit similar $R_0 A$ products at 77K for $\lambda_c < 10 \mu m$. For $\lambda_c > 10 \mu m$, the $R_0 A$ products of N-on-p diodes are generally inferior; their quantum efficiency is lower and the forward impedance is higher. These problems are probably caused by the relative immaturity in fabricating N-on-p heterojunctions, especially in the surface control on the p-type side for diodes with a longer wavelength cutoff. In addition, in our N-on-p diode design, the p-type conduction is controlled by Cu doping which may not be an optimum choice for $> 10 \mu m$ long wavelength diode applications. Figure 9 shows a preliminary comparison of P-on-n and N-on-p diodes.

Also shown in Fig. 9 are published $R_0 A$ products of ion-implanted n-on-p diodes. Our N-on-p grown junction diodes compare favorably with the reported data. On the other hand, our P-on-n diodes clearly are superior, especially for $\lambda_c > 10 \mu m$. The solid curve shown in Fig. 9 is calculated using simple diffusion model[5] for a P-on-n structure with typical diode parameters: $n \approx 5 \times 10^{14} /cm^3$, $\tau_h \approx 1 \mu s$ and the empirical equation for Eg (x,T).[15]

VI. SUMMARY

We have demonstrated that long wavelength HgCdTe diodes in both P-on-n and N-on-p types can be achieved by LPE. The key feature in the realization of these types of devices is to use a slow moving conduction species in the top layer. For the P-on-n type, the dopant for p-type is column V species, either As or Sb. For the N-on-p type, the dopant for n type is In. From our limited study, P-on-n diodes appear to be better in $R_0 A$ and more controllable than N-on-p diodes. This behavior is probably the result of a combination of several factors: the relative immaturity of N-on-p heterojunction technology, the nature of surface control of low carrier concentration p-type HgCdTe and the inherent design of N-on-p diodes.

ACKNOWLEDGMENTS

The author is grateful to personnel at Fermionics for their skillful work in crystal growth and device processing. The work reported here was partially supported by Night Vision and Electro Optics Laboratory, Fort Belvoir, Virginia 22060-5677.

[1] A. Kolodny and I. Kidron, IEEE Trans. Electron Devices **ED-27**, 37 (1980).
[2] H. Ryssel et al., IEEE Trans. Electron Devices, **ED-27**, 58 (1980).
[3] C. C. Wang et al., IEEE Trans. Electron Devices, **ED-27**, 154 (1980).
[4] M. Lanir and K. J. Riley, IEEE Trans. Electron Devices, **ED-29** (1982).
[5] A. Rogalski, Infrared Phys. **28**, 139 (1988).
[6] M. B. Reine, A. K. Sood, and T. J. Treadwell, *Semiconductors and Semimetals*, edited by R. K. Willardson and A. C. Beer (Academic, New York, 1981), Vol 18, p. 201.
[7] G. L. Destefamis, G. Guernet, B. Pelliciari, G. Tartavel, J. L. Teszner, and J. L. Tissot, 3rd International Conference on Advanced Infrared Detectors and Systems, London, Conf. Publ. No. 263, p. 44 (1986).
[8] M. Lanir and K. J. Riley, IEEE Trans. Electron Dev. **ED-29**, 274 (1982).
[9] C. C. Wang, S. H. Shin, M. Chu, M. Lanir, and A. H. B. Vanderwyck, J. Electron. Chem. Soc. **127**, 175 (1980).
[10] J. E. Bowers, J. L. Schmit, C. J. Speerschneider, and R. B. Maciolek, IEEE Trans. Electron Devices, **ED-27**, 14 (1980).
[11] T. C. Harman, J. Electron. Mater. **8**, 191 (1979).
[12] S. L. Bell and S. Sen, J. Vac. Sci. Technol. A **3**, 112 (1985).
[13] P. R. Bratt, J. Vac. Sci Technol. A **1**, 1687 (1983).
[14] P. R. Bratt and T. N. Casselman, J. Vac Sci Technol. A **3**, 238 (1985).
[15] J. Chu, S. Xu, and D. Tang, Appl. Phys. Lett. **43**, 1064 (1983).

HgZnTe for very long wavelength infrared applications

E. A. Patten, M. H. Kalisher, G. R. Chapman, J. M. Fulton, C. Y. Huang, P. R. Norton, M. Ray, and S. Sen
Santa Barbara Research Center, Goleta, California 93117

(Received 2 October 1990; accepted 9 January 1991)

HgZnTe has been of considerable interest in the last few years as an alternative infrared (IR) detector material to HgCdTe because of HgZnTe's greater mechanical hardness, higher Hg vacancy formation energies, and other desirable properties which may lead to greater producibility (yield) and reliability (operation life). We are developing very long wavelength IR (VLWIR) HgZnTe detectors (17 μm at ≥65K) for longterm Earth Observing System NASA missions. In recent months we have been able to achieve, nearly routinely, outstanding quality HgZnTe on lattice-matched substrates ($Cd_{0.84}Zn_{0.16}Te$). By careful screening of substrate quality and refinement of the growth system and growth parameters, we overcame previous melt retention problems observed when using these high Zn% substrates. X-ray rocking curve widths (full width at half-maximum) as narrow as 26 arc-sec (for a 1×8 mm spot) for these epilayers have been achieved which is comparable to that for the best HgCdTe ever reported. In addition, we have, for the first time, fabricated and demonstrated high performance VLWIR ($\lambda_{co} > 17$ μm at 80K) photoconductors made from liquid phase epitaxy HgZnTe which exhibit near background limited performance (peak D^* up to 10^{11} cm$\sqrt{Hz}$/W) at high background. From these device results we see a clear correlation between the layer/substrate lattice match and the resulting device performance.

I. INTRODUCTION

HgZnTe has been of considerable interest in recent years as an infrared (IR) detector material since it should yield the same performance and wavelength tunability as HgCdTe, but potentially offers detectors with greater producibility and reliability. The introduction of Cd into HgTe tends to weaken an already weak HgTe bond whereas the substitution of the smaller atom Zn (versus Cd) results in a tighter lattice which is more resistant to dislocation formation as predicted by Arden Sher.[1] Many advantageous properties of HgZnTe have been predicted and/or experimentally observed including: increased mechanical hardness,[2,3] slower Hg diffusion,[2,4] and higher Hg vacancy formation energies.[5] No detailed predictions can be made at this time concerning the impact these material properties of HgZnTe will have on detector thermal stability and hence operating life. The relative stability of HgZnTe versus HgCdTe devices will need to be directly assessed by fabricating high performance HgZnTe detectors and determining their bake stability characteristics as compared with equivalent HgCdTe devices.

Researchers at Santa Barbara Research Center (SBRC) in conjunction with those at Stanford University and SRI International began investigating the properties of HgZnTe in 1986 because of NASA's concern about the longterm stability of satellite IR focal plane arrays (FPA) using HgCdTe technology.[1,4,5] In 1989, SBRC began developing the growth and processing technology specifically of very long wavelength IR (VLWIR) HgZnTe (target cutoff: $\lambda_{co} \approx 17$ μm at 65K) in support of detector development for the EOS instruments.

One important issue regarding the development of FPAs using VLWIR HgZnTe is the relative level of difficulty in controlling bandgap uniformity across a wafer versus that for HgCdTe with the same narrow band gap. One needs to add less Zn versus Cd to the semimetal HgTe to achieve a given bandgap (ZnTe band gap = 2.37 eV versus 1.59 eV for CdTe) and this results in a steeper E_g versus x slope for HgZnTe. This steeper dependence on x-value makes it more difficult to achieve a uniform bandgap across a HgZnTe wafer (versus HgCdTe) for a given x-value variation. This issue might have been troublesome particularly at low x values (VLWIR cutoffs) if it were not for the fact that the HgZnTe bandgap shows a larger deviation of the E_g versus x curve from the straight-line average or "bowing" than does HgCdTe.[6] The large bowing for HgZnTe results from the relatively large bond length mismatch (between the constituent binaries) occurring in this alloy, compared to HgCdTe with its nearly lattice-matched constituents. This large bowing results in a reduction of the slope of dE_g/dx for HgZnTe to 2.1 eV for small band gaps (near 0.1 eV) which is only about 10% higher than the value for HgCdTe (1.9 eV) with the same bandgap. Therefore, the required control of the Zn composition is not much greater for HgZnTe than for HgCdTe for long wavelength material. In addition, there is some evidence that the large binary bond length mismatch (6%) between HgTe and ZnTe has an additional benefit (besides dE_g/dx bowing). The strain energy associated with variations in the local ratio of the two binary constituents may act to suppress local composition variations. This hypothesis may be supported by both the sharp exciton lines measured for HgZnTe (evidence of well defined composition)[2] and higher degree of longitudinal composition uniformity[7] in bulk HgZnTe vs HgCdTe. Both of these factors may tend to result in nearly equivalent bandgap uniformity for VLWIR HgZnTe and HgCdTe.

To develop VLWIR HgZnTe, we first determined suitable

growth parameters for these longer cutoff wavelengths and tackled the problem of Te melt retention on the surface of our liquid phase epitaxial HgZnTe layers. Last year at the HgCdTe Workshop[8] we presented materials data on VLWIR layers with excellent surface morphologies grown on lattice-mismatched substrates ("low Zn%" or $Cd_{0.96}Zn_{0.04}Te$). At that time we were not able to obtain similar surfaces on layers grown on higher Zn% substrates (nominally $Cd_{0.84}Zn_{0.16}Te$). The first HgZnTe photoconductor arrays were processed therefore using layers grown on mismatched substrates. The results were very encouraging and will be discussed in brief in this paper. The main focus of this paper, however, will be on material and device results for layers grown more recently on lattice-matched substrates.

In the last six months, by improving the growth system and procedures (one objective was to reduce residual oxygen contamination in the growth system) and careful selection of high quality substrates, layers with excellent surface morphology have been grown on lattice-matched or "high Zn%" substrates substrates. A second lot of HgZnTe photoconductors (PCs) were fabricated using some of these layers (along with a control wafer grown on a mismatched substrate). The performance of the detectors on these lattice-matched wafers exhibited much higher performance levels than those on mismatched wafers (from either process lot). Blackbody responsivity, noise, spectral response and photoconductive lifetime data on these devices will be discussed. The importance of the substrate lattice constant and structural quality to these results will be emphasized.

II. EXPERIMENTAL

Epitaxial $Hg_{1-x}Zn_xTe$ layers were grown by the horizontal liquid phase epitaxy (HLPE) technique using a two-zone open-tube furnace system with a graphite slider boat. Small ZnTe-saturated Te-rich melts were used in conjunction with a separate Hg source heated in a lower temperature furnace (~ 300 °C) to set the Hg overpressure in the system and therefore the Hg incorporation in the layer. The layers were grown at approximately 460 °C with a cooling rate of 0.1 °C/min which resulted in growth rates near 20 μm/h. The substrates used were $Cd_{1-y}Zn_yTe$ ($0 \leq y \leq 0.20$), grown by the vertical modified-Bridgman technique. The substrates were oriented to (111)B $\pm$ 0.1°, polished, and cut into 1-in. squares for use in the slider boat. The structural quality of the substrates was typically screened using x-ray diffraction rocking curve analysis.

Epitaxial layers were routinely characterized using Fourier transform IR spectroscopy, variable temperature Van der Pauw Hall measurements, and x-ray diffraction. A few selected layers also were characterized using energy dispersive x-ray analysis (EDX) and photoconductive (PC) decay lifetime measurements.

Based upon the results of these routine characterization measurements, layers were selected for detector processing. In general the criteria were: good surface morphologies (low or negligible melt retention), 300 K transmission cutoffs from 8–9 μm, thickness between 10 and 20 μm, narrow x-ray rocking curves (< 50 arc-sec) and Hall characteristics (for annealed wafers) exhibiting low n-type carrier concentrations at 80 K (below 10^{15} cm^{-3}). The selected wafers were processed into linear arrays of photoconductors with 50×50 μm active areas, CdTe passivation and ZnS antireflection coating. The processing steps used were essentially those developed previously for HgCdTe.

These processed wafers contained arrays with three levels of contact expansion for diagnostic purposes. Contact expansion in photoconductors is commonly used to increase the effective lifetime of the minority carriers as well as the device resistance which will both lead to increased responsivity.[9] A schematic of a photoconductor with contacts which are extended away from the optically active area is shown in Fig. 1. The greater the contact expansion, the longer the electrical length L of the device and so its resistance. The electrical length of detectors on the three types of arrays are 6.35×10^{-2} cm ("no" expansion), 1.27×10^{-2} cm ("medium" expansion) and 1.78×10^{-2} cm ("full" expansion), while the optical area for all these devices is the same. After processing, selected arrays were diced, mounted on chip carriers and wire bonded for dewar testing.

Measurements of current–voltage characteristics, blackbody responsivity, noise, spectral response and PC decay lifetime were performed on selected devices at 80 K. All radiometric testing was performed using an 800 K blackbody, a signal chopping frequency of 1 KHz and a noise bandwidth of 10 Hz. Noise measurements were performed as a function of frequency on selected devices. 10 KHz noise was measured and used to calculate D^*. A low noise preamplifier was used which has a root-mean-square noise of 1.15 nV/$\sqrt{Hz}$ although for low detector bias even this preamplifier noise was greater than the detector noise. Spectral response measurements were obtained using a double-beam Perkin Elmer 983G grating spectrophotometer. The two background levels used for these measurements were 1.1×10^{17} and 1.5×10^{16} ph/cm^2/s.

III. RESULTS AND DISCUSSION

A. Material characterization

The growth of high quality HgZnTe layers is critically dependent on the use of closely lattice-matched, high quality CdZnTe substrates. Until recently, however, we were not

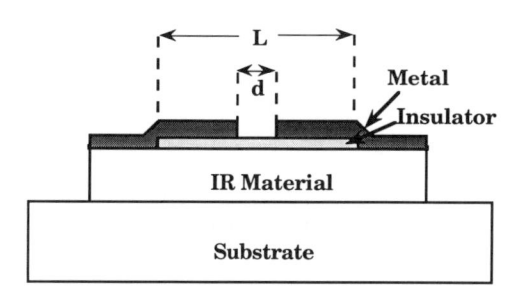

FIG. 1. Schematic of contact overlap structure in a photoconductor for extending the contacts to the IR material away from the active optical area. The electrical length of the device L is longer than that associated with the optical absorbing region (d) which results in an increased effective lifetime of the minority carriers and so increased gain of the device.

able to routinely achieve excellent surface morphologies on layers grown on these high Zn% substrates because of increased melt retention on these layers versus those on lower Zn% substrates. Last year at this workshop we reported on the excellent surface morphology, good structural quality and electrical properties of VLWIR HgZnTe layers grown on lattice-mismatched substrates ($Cd_{0.96}Zn_{0.04}Te$).[8]

In the last six months, we have succeeded in improving the growth of HgZnTe layers on lattice-matched substrates such that films with excellent surface morphology, structural quality and electro-optical properties are now nearly routinely achieved. This high yield of device quality material has resulted in part because of the high structural quality of the $Cd_{0.84}Zn_{0.16}Te$ substrates used for these growths as indicated by the narrowness of their x-ray diffraction rocking curves. Our criterion for the substrate is that the x-ray rocking curve (using a 1×8 mm x-ray beam size) full width at half-maximum (FWHM) be less than approximately 20 arc-sec and reasonably uniform across the wafer.

Shown in Figs. 2(a) and 2(b) are examples of x-ray diffraction rocking curves taken on a high Zn% substrate and for the HgZnTe layer (HLPE 289) grown on this substrate. The substrate has an average Zn composition of $y = 0.16$ based upon EDX while the layer has an average $x = 0.13$ near the substrate/layer interface which results in a 0.05% mismatch. This substrate exhibited uniformly narrow rocking curves with an average value of the FWHM of 13 arc-sec. Very narrow rocking curves were also measured uniformly across the layer HLPE 289. An average FWHM of 30 arc-sec was measured for this layer which is comparable to that for the best HgCdTe ever reported. These curves are so narrow that the compositional grading through the layer is evident by the curve's asymmetry.

By comparison, the rocking curves for a layer (HLPE 291) grown on a mismatched substrate are much broader, averaging 222 arc-sec. Example rocking curves for this layer and its substrate are shown in Figs. 3(a) and 3(b). From EDX, the x value of the layer near the substrate is $x = 0.14$ whereas the substrate has a Zn composition of $y = 0.03$ which results in a 0.9% mismatch.

In the past year we have observed that rocking curve widths for HgZnTe layers grown on lattice-mismatched substrates average 5–10 times wider than for the lattice-matched case. The electrical and optical properties were similar, however, for layers grown on low and high Zn% substrates. No photoconductive lifetime data has been taken on mismatched layers although it is anticipated that lifetimes should be shorter for these more defective layers.

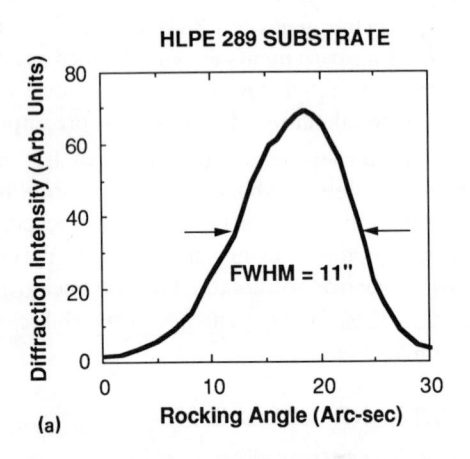

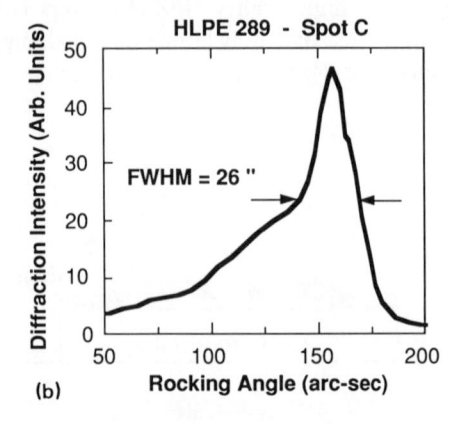

FIG. 2. (a) X-ray rocking curve for high Zn% substrate used for layer HLPE 289. The curves taken across this substrate are uniformly narrow with an average value of 13 arc-sec. (b) Rocking curve for HLPE 289 is so narrow that the vertical compositional grading in the layer is evident by the curve's asymmetrical shape. A full width at half maximum of 26 arc-sec is comparable to the best HgCdTe ever reported.

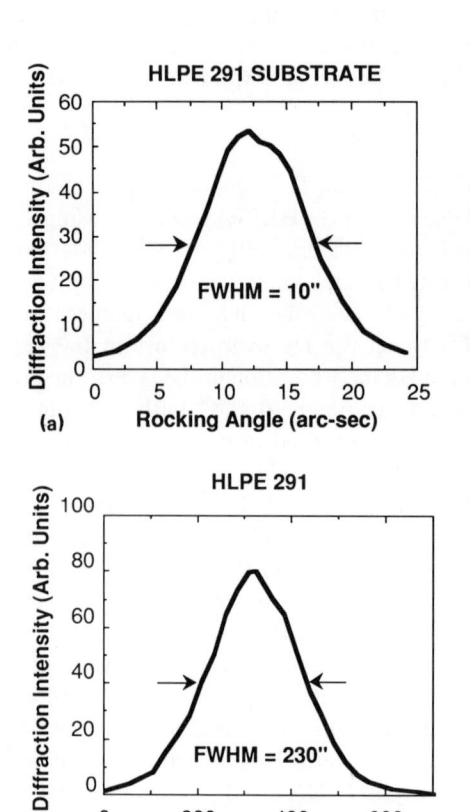

FIG. 3. (a) X-ray rocking curve for the low Zn% CdZnTe substrate used to grow layer HLPE 291. The uniformly narrow curves are typical of low Zn% substrates. (b) Wide x-ray rocking curves (average FWHM = 222 arc-sec) measured for HLPE 291 result from the lattice-mismatch with its substrate.

Besides substrate quality, another potential reason for the recent improvement in surface morphology (reduced melt retention) for layers grown on high Zn substrates was reduced oxygen in the HLPE growth system. The oxygen was reduced by replacing a possibly leaky hydrogen purifier (hydrogen flows through the system during growth) and by several procedural and hardware changes which were intended to reduce the influx and residence time of oxygen. We believe, based upon past experience with HgCdTe, that the presence of oxygen in the growth system can aggravate melt retention on the epitaxial layer.

B. Device testing

In the past year, the first two lots of HgZnTe photoconductors were fabricated. The target cutoff wavelength for the starting material was from 8–9 μm at 300 K to yield detector cutoffs from 16–20 μm at 80 K.

Lot 1 devices consisted of HgZnTe layers grown on low Zn% or lattice-mismatched substrates ($Cd_{0.96}Zn_{0.04}Te$). These detectors exhibited good performance (peak responsivity of 10^4 V/W and peak D^* near 10^{10} cm$\sqrt{Hz}$/W) with cutoff wavelengths near 19 μm at 80 K.

Lot 2 consisted of three wafers, two of which had lattice-matched substrates (HLPE 288, HLPE 289) and one wafer with a mismatched substrate (HLPE 291). The mismatched wafer devices performed at similar levels for both lots. By comparison, the detectors on the two lattice-matched wafers exhibited far superior performance levels with peak D^* at or near BLIP ($>10^{11}$ cm$\sqrt{Hz}$/W at 80 K) and very high responsivities ($>10^6$ V/W at 80 K) with cutoff wavelengths from 17 to 18.5 μm at 80 K. These results are discussed in more detail in the following paragraphs.

The relative response per watt versus wavelength for a typical device on wafer HLPE 288 is shown in Fig. 4 revealing an 80 K cutoff wavelength of 18.5 μm (average cutoff $\approx 18 \mu$m). The 300 K transmission cutoff (100% point) for this wafer was measured to be 9.2 μm. The peak responsivity of HLPE 288 (calculated using a blackbody-to-peak conversion factor of 2) versus applied electric field at a background of 1.1×10^{17} ph/cm^2/s (300 K background, f/1.5 field-of-view) is shown in Fig. 5. At an applied electric field of 45 V/cm, the responsivity begins to saturate near 5×10^5 V/W. Peak D^* for this device was determined to be 3.2×10^{10} cm$\sqrt{Hz}$/W which is 30% BLIP for this background. BLIP D^* was calculated using a peak wavelength of 14 μm, a quantum efficiency of 0.8 and blackbody-to-peak conversion factor of 2. The factor which limits D^* at 80 K appears to be thermally generated noise, based upon some preliminary measurements at 30 K where the noise is greatly reduced from that seen at 80 K. The noise value used to determine D^* is that measured at 10 KHz which is well above the 1/f noise regime for these devices. The 10 KHz noise is also much greater than the calculated Johnson noise for these devices (≈ 0.7 nV/$\sqrt{Hz}$ for 100 Ω device at 80 K and 10 Hz noise bandwidth). In view of the long cutoff wavelengths of these detectors, it is not surprising that they are thermally limited at 80 K.

Wafer HLPE 289 exhibited slightly shorter 80 K cutoff wavelengths as expected from its room temperature trans-

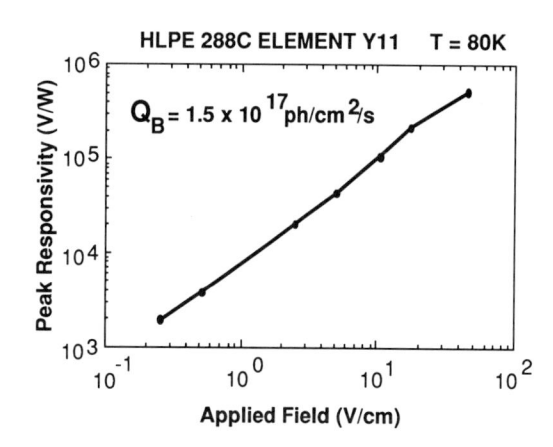

FIG. 5. Peak responsivity vs applied electric field at 80 K for HLPE 288C el. Y11 (full expansion). Peak responsivity reaches at 5×10^5 V/W at 45 V/cm assuming a blackbody-to-peak conversion factor of 2.

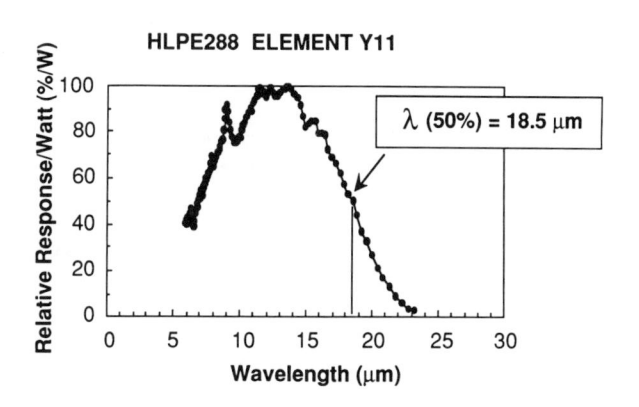

FIG. 4. Relative response per watt for HLPE 288 element Y11 at 80 K. The 50% response point or cutoff averaged 18 μm for devices measured on this wafer.

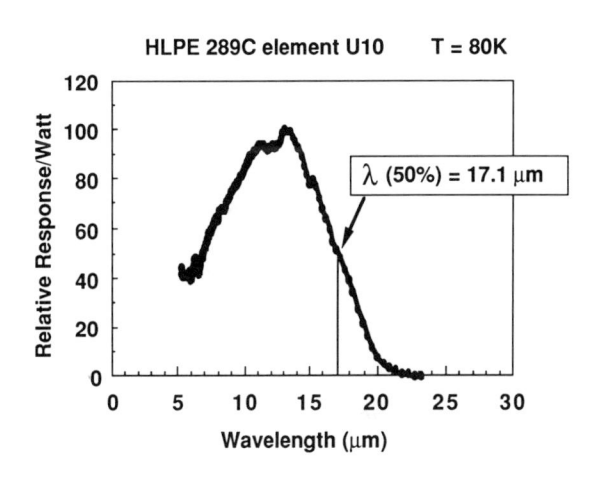

FIG. 6. Relative response per watt at 80 K for element U10 on wafer HLPE 289. The peak response of this device is at 13.5 μm and the 50% cutoff is seen to be 17.1 μm.

mission characteristics (100% point at 8.6 μm). The relative response per watt versus wavelength for element U10 (Fig. 6) has a peak of 13.5 μm and a cutoff of 17.1 μm. Spectral response curves for other devices on this array show a gradual trend from shorter (≈ 15 μm) to longer (≈ 17 μm) cutoffs across the array. Significant lateral compositional and thickness variations are not unexpected in our HgZnTe due to the temperature variations across the growth zone of the furnace. These HgZnTe wafers were not grown under specific conditions to optimize or reduce the compositional grading through the layer and so grading may be contributing to the softness of the spectral cutoff. No definitive measurement of the compositional grading through these layers has yet been made.

Because of the shorter cutoffs of the devices on HLPE 289, their resistances are somewhat higher (than for HLPE 288) and this is partly responsible for the higher responsivities seen for these detectors. The peak responsivities at 80 K under two different background levels (1.5×10^{16} and 1.1×10^{17} ph/cm^2/s) for six devices on HLPE 289 are plotted in Fig. 7. Two devices were characterized from each of the three types of array (no contact expansion, "medium" contact expansion and "full" contact expansion). Peak responsivities for the full expansion devices average 2×10^6 V/W which is comparable to the best VLWIR HgCdTe photoconductors reported at these background levels.

The corresponding peak D^* data for these photoconductors are shown in Fig. 8. At the higher background, D^* reaches 8×10^{10} cm$\sqrt{\text{Hz}}$/W which is 84% BLIP (at 1.1×10^{17} ph/cm^2/s background, BLIP $D^* = 9.5 \times 10^{10}$ cm$\sqrt{\text{Hz}}$/W with a peak at 14 μm). It is important to note that BLIP D^* here was calculated, for simplicity, using a single peak wavelength (14 μm) for all devices. This estimate of the peak wavelength is definitely too long for some of these devices and will result in overestimating BLIP D^* (proportional to peak wavelength). Just changing the peak wavelength to 13 μm in the calculation will result in a lower BLIP D^* of 8.8×10^{10} cm$\sqrt{\text{Hz}}$/W which raises the measured D^* to greater than 90% of BLIP. At the lower background, the peak D^* of 10^{11} cm$\sqrt{\text{Hz}}$/W is at 38% of BLIP (BLIP D^* is 2.6×10^{11} cm$\sqrt{\text{Hz}}$/W). With a cutoff near 17 μm, it appears that performance is thermally limited below a background of 10^{17} ph/cm^2/s.

Photoconductive decay lifetime data at 80 K for two devices on HLPE 289 are plotted in Fig. 9. The top data set shows the longer lifetimes measured for a device with contact expansion versus shorter lifetimes measured for a nonexpanded device whose data is at the bottom. The closed circles are for the "preferred" voltage polarity which yields longer lifetimes for the expanded devices because this polarity affords the longer transit time of the minority carriers. The open data points are for the reverse polarity. Negligible polarity effect is observed for the nonexpanded device as expected. These lifetimes are longer than the bulk limit predicted for 17 μm cutoff material and indicate that some trap-

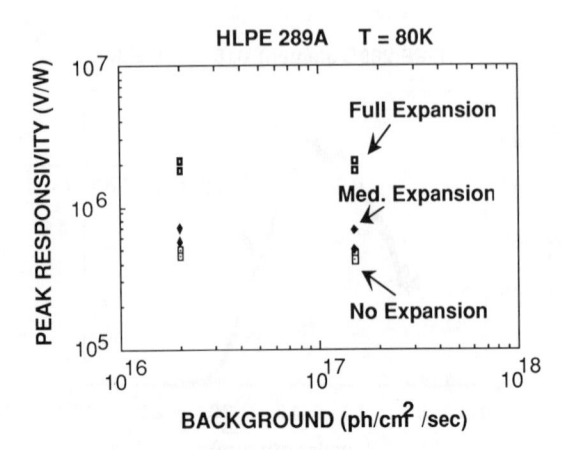

FIG. 7. 80 K peak responsivity at two background for six devices on HLPE 289. (Two detectors from each of the three types of arrays: no, medium and full contact expansion.) Responsivity is seen to track with the amount of contact expansion and reaches 2×10^6 V/W for the full expansion devices.

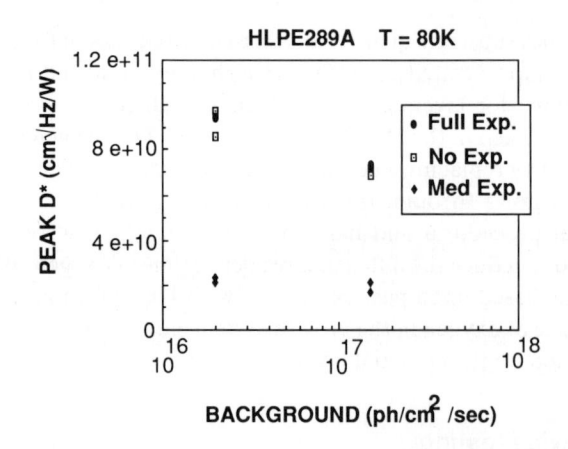

FIG. 8. Peak D^* at 80 K vs background for the detectors from Fig. 7. At the lower background, D^* reaches 10^{11} cm$\sqrt{\text{Hz}}$/W which is 38% of BLIP while at higher background, D^* peaks at 8×10^{10} cm$\sqrt{\text{Hz}}$/W which is 84% BLIP (9.5×10^{10} cm$\sqrt{\text{Hz}}$/W if a peak wavelength of 14 μm is used).

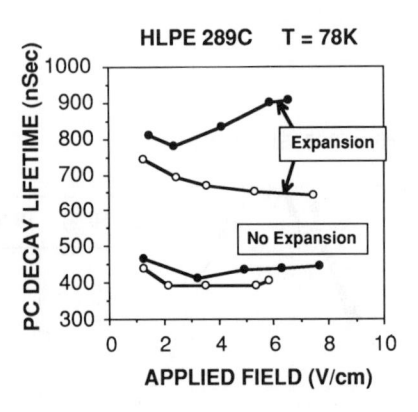

FIG. 9. 80 K PC decay lifetime vs applied field for two devices on HLPE 289, one with contact expansion and the other unexpanded. The closed data points correspond to the polarity which increases the minority carrier path length to the attractive contact and therefore its lifetime. The open circles are for reverse polarity. The expanded detector shows a significant dependence of the lifetime on polarity as expected.

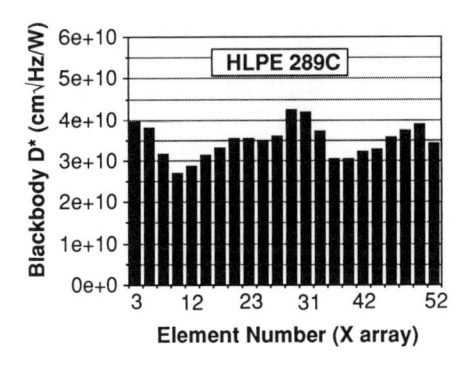

FIG. 10. Plot of blackbody D^* at 80 K for every other element in an medium expansion array segment (52 devices long) on HLPE 289. All devices were biased at 2 mA which does not necessarily result in the maximum D^* for a particular device. The variations in D^* across the array probably result from compositional and thickness variations in the starting material.

ping phenomena is likely present.

As mentioned earlier, spectral response data for HLPE 289 indicated nonuniformity in the composition across the wafer. Lateral variations in composition and thickness are not uncommon with Te melt-grown HgCdTe and it is known that the temperature uniformity is key to controlling these unwanted fluctuations. The temperature profile in the growth region of the HLPE furnace had variations which were present during the growth of the layers reported here. We therefore were not surprised to see the performance variations shown in Fig. 10 which is a plot of blackbody D^* (without correction for peak wavelength) for an array segment with 52 elements from HLPE 289. Resistance, responsivity, and noise were measured for nearly every other element in this array segment, each held at 2 mA bias current. The undulations in D^* across the array probably result from composition and thickness variations which lead to resistance and responsivity variations with the same lateral profile.

In comparison with the excellent performance of the lattice-matched wafers, detectors from the mismatched wafer in lot 2 (HLPE 291) exhibited characteristics similar to those of the mismatched wafer from lot 1 with a similar cutoff. The peak responsivity of HLPE 291 full expansion PCs at 80 K averaged 1.5×10^4 V/W while the peak D^* was 1.4×10^{10} cm$\sqrt{\text{Hz}}$/W at a background of 1.5×10^{16} ph/cm^2/s (BLIP $D^* = 2.6 \times 10^{11}$ cm$\sqrt{\text{Hz}}$/W). These results highlight the importance of lattice-matched substrates to achieving the highest performing photodetectors.

IV. SUMMARY

High performance VLWIR HgZnTe photoconductors have been fabricated in our first two fabrication lots using processing originally developed for HgCdTe. 80 K D^* near BLIP (peak $D^* = 8 \times 10^{10}$ cm$\sqrt{\text{Hz}}$/W at a background of 1.1×10^{17} ph/cm^2/s) have been obtained for the best wafers which were grown on lattice-matched Cd$_{0.84}$Zn$_{0.16}$Te substrates. These devices have cutoffs as long as 18.5 μm at 80 K. The responsivities observed ($>10^6$ V/W) are comparable with the best VLWIR HgCdTe photoconductors measured under these background conditions. Device performance on lattice-matched wafers was far superior to that on mismatched wafers, demonstrating the importance to detector performance of a close lattice match between layer and substrate.

These results are very encouraging considering these are our first efforts to fabricate HgZnTe photoconductors and demonstrate that HgCdTe processing is largely compatible with HgZnTe. The assessment of whether HgZnTe-based devices have greater thermal stability and therefore potentially greater operation life still needs to be made based upon bake stability studies which include equivalent HgCdTe devices.

ACKNOWLEDGMENTS

The authors wish to thank C. B. Tacelli for performing some of the spectral response measurements, and J. A. Kiele and V. L. Liguori for preparation of the CdZnTe substrates. We also wish to acknowledge helpful technical discussions with M. Berding and Arden Sher of SRI. This work was supported by NASA/Langley on NASA Contract No. NAS1-18232 with William E. Miller as the technical monitor.

[1] A. Sher, A. B. Chen, W. E. Spicer, and C. K. Shih, J. Vac. Sci. Technol. A **3**, 105 (1985).
[2] R. Triboulet, J. Cryst. Growth **86**, 79 (1988).
[3] S. Fang, L. J. Farthing, M.-F. S. Tang, and D. A. Stevenson, J. Vac, Sci, Technol. A **8**, 1120 (1990).
[4] E. J. Smith, T. Tung, S. Sen, W. H. Konkel, J. B. James, V. B. Harper, B. F. Zuck, and R. A. Cole, J. Vac. Sci. Technol. A **5**, 3043 (1987).
[5] M. A. Berding, A. Sher, and A. B. Chen, J. Vac. Sci. Technol. A **5**, 3009 (1987).
[6] M. A. Berding, S. Krishnamurthy, and A. Sher, J. Vac. Sci. Technol. A **5**, 3014 (1987).
[7] J. Ameurlaine, A. Rousseau, T. Nguyen-Duy, and R. Triboulet, SPIE **865**, 30 (1987).
[8] M. H. Kalisher, E. A. Patten, and S. Sen, 1989 U.S. Workshop on the Physics and Chemistry of Mercury Cadmium Telluride and Related II–VI Compounds (extended abstracts), p. I-51.
[9] M. A. Kinch, S. R. Borrello, and A. Simmons, Infrared Phys. **17**, 127 (1977).

A high quantum efficiency *in situ* doped mid-wavelength infrared *p*-on-*n* homojunction superlattice detector grown by photoassisted molecular-beam epitaxy

K. A. Harris, T. H. Myers, R. W. Yanka, and L. M. Mohnkern
Electronics Laboratory, General Electric Company, Syracuse, New York 13221

N. Otsuka
School of Materials Engineering, Purdue University, West Lafayette, Indiana 47907

(Received 2 October 1990; accepted 11 December 1990)

HgTe/CdTe superlattices in infrared (IR) detector structures have been theoretically shown to allow for better control over cutoff wavelength, minimize diffusion currents, and greatly reduce band-to-band tunneling currents as compared with the corresponding HgCdTe alloy. However, the few HgTe/CdTe superlattice detectors that have been fabricated exhibit little or no quantum efficiency. In this paper, we report the first high quantum efficiency mid-wavelength infrared (MWIR) detectors based on HgTe/CdTe superlattices. This result is significant because it represents the first experimental verification that IR detectors with useful characteristics can in fact be fabricated from HgTe/CdTe superlattices. The MWIR detectors were fabricated from an *in situ* doped *p*-on-*n* MWIR homojunction superlattice epilayer grown by photoassisted molecular-beam epitaxy (PAMBE). This growth technique produces low defect growth of superlattice material, as is described in this paper. Our development of an extrinsic doping technology using indium and arsenic as the *n*-type and *p*-type dopants, respectively, led to the successful doping of the superlattice and is also discussed.

I. INTRODUCTION

Since their initial proposal as an infrared (IR) detector, HgTe/CdTe superlattices have been predicted to possess inherent advantages over the corresponding HgCdTe (MCT) alloy for the fabrication of high performance long-wavelength infrared (LWIR) and very long-wavelength infrared (VLWIR) detectors.[1] These advantages include lower tunneling currents, better control of the band gap, and higher temperature operation. HgTe/CdTe superlattices grown by the molecular-beam epitaxy (MBE) technique[2,3] exhibit properties in agreement with the theoretical predictions. Until now, however, fabrication of IR detectors using a HgTe/CdTe superlattice as the IR sensing material has not been realized, casting doubt on the feasibility for the practical use of this material system.

In this paper, we report the first successful fabrication of mid-wavelength infrared (MWIR) superlattice (SL) detectors exhibiting high quantum efficiencies. This accomplishment was based on obtaining low defect superlattice material as well as the development of an extrinsic doping technology. In this paper, we first describe basic materials growth. We then report the successful *in situ* extrinsic doping suitable for HgTe/CdTe superlattice layers. Indium is used as the *n*-type dopant and arsenic is used as the *p*-type dopant. The combination of these technologies allowed for the growth of *in situ* doped homojunction superlattice material which resulted in the fabrication of the MWIR SL detectors discussed here.

II. EXPERIMENTAL

The growth of Hg-based epilayers was carried out in two custom MBE machines modified for photoassisted molecular-beam epitaxy (PAMBE) growth.[4] An argon ion laser equipped with broad band optics was used as the illumination source. Epilayers were grown without buffer layers directly on (211)B CdTe substrates obtained from II–VI, Inc. General Electric's (GE) proprietary substrate preparation, which includes a combination of standard wet etching techniques and thermal processing prior to growth, has been shown to produce high quality substrate–epilayer interfaces without the added complexity of a buffer layer.[5] Epilayers were grown at temperatures ranging from 150 to 180 °C during this study. A Varian beam flux monitor was used to provide a precise determination of the flux from the thermal sources, thus enabling reproducible growth conditions.

Defect etching on annealed samples was accomplished through the use of standard etches.[6] Transmission electron microscopy (TEM) characterizations using iodine thinning techniques were performed at Purdue University. Secondary ion mass spectrometry (SIMS) analysis was performed by Charles Evans and Associates.

Detectors were fabricated at GE's Electronics Laboratory's MCT processing facility. A mesa diode structure was used for superlattice processing. Upon completion of processing, the wafers were tested in a 78 K cold-probe station to evaluate *I–V* characteristics. Selected arrays from the wafers were packaged and mounted in a variable temperature cryocooler for more extensive testing, including spectral response, quantum efficiency, and response uniformity.

III. MATERIALS GROWTH

For the past three years GE has been investigating PAMBE for the growth of device quality Hg-based materials. While enhancement of growth kinetics due to laser illumination has led to device quality epilayers,[4,7,8] twinning related defects and line dislocation densities continued to limit yield.

In order to increase yields, a new approach to the growth of HgCdTe by PAMBE was developed at GE—compositionally modulated structures (CMS).[5] The CMS technique involves the growth of a multilayer structure or SL consisting of multiple repeats of MCT (or HgTe) thin layers (up to 1500 Å thick) and thin CdTe layers (20–80 Å thick). The film is then subjected to a post-growth interdiffusion anneal under a Hg ambient in a sealed quartz ampule producing a uniform MCT epilayer. This method is different from that of the interdiffused multilayer process used in metalorganic chemical vapor deposition which uses thick CdTe layers and *in situ* interdiffusion.[1,6] Growth by the CMS method has led to a reduction of microstructural defects and the propagation of line dislocations and has increased yields of twin-free (111)B HgCdTe epilayers.[5]

The reduction of microstructural defects is believed to be due to strain induced in the growing film by the thin CdTe layers, analogous to the use of lattice mismatched layers in III–V materials to block threading dislocations. Evidence of this effect is observed for HgTe/CdTe superlattices as shown by the TEM cross section of a superlattice grown on CdTe in Fig. 1. Dislocations due to the slight lattice mismatch can be seen at the interface. Threading dislocations originating from the substrate also appear. As the CMS(SL) structure continues to grow, however, the dislocation structure quickly dies out due to line dislocation bending.

Further evidence of threading dislocation blocking is illustrated in Fig. 2, which shows two MCT epilayers grown by PAMBE on (211)B CdTe substrates. The layer grown without CMS has the high etch pit density of 5×10^6 cm^{-2} [Fig. 2(a)] consistent with lattice mismatch. The much lower etch pit density, 1×10^5 cm^{-2} [Fig. 1(b)], observed for the layer grown by CMS, however, indicates that these dislocations have been suppressed during layer growth. Threading dislocation suppression is also stable with respect to low temperature annealing. This result indicates that the CMS technique allows for the use of CdTe substrates for high quality film growth. The ability of the CMS technique for blocking threading dislocations is thus attractive for the use of substrates with even greater lattice mismatch, such as GaAs.

The CMS(SL) technique is useful in reducing planar twinning on the (111)B orientation, but yields of twin free epilayers greater than 15 um thick are limited, continuing to restrict the utility of this orientation. The (100) orientation also has limited utility due to pyramidal hillock densities exceeding that required for reasonable device yields.[5] As previously reported,[4] the (211)B orientation does not exhibit either of these problems, resulting in an intrinsically higher material yield. GE has thus concentrated on the (211)B orientation for material growth.

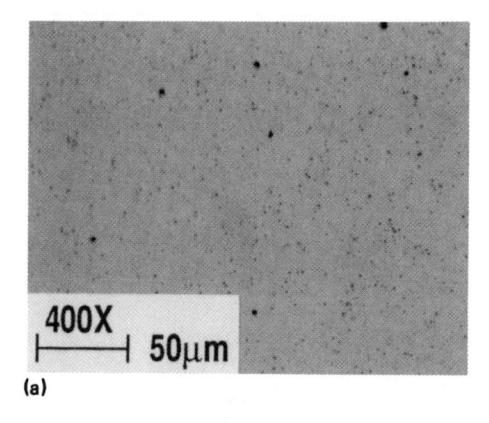

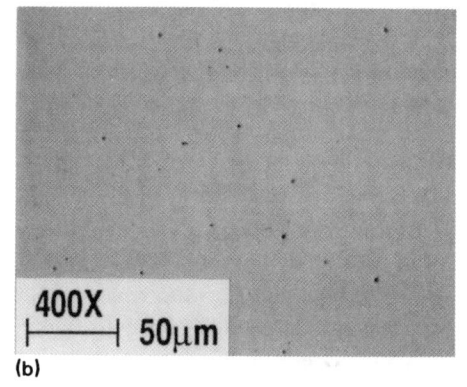

FIG. 2. Defect etch micrographs comparing (a) nonCMS(SL) growth and (b) CMS(SL) growth. The defect density of the CMS(SL) layer is lower, indicative of suppression of the propagation of threading dislocations.

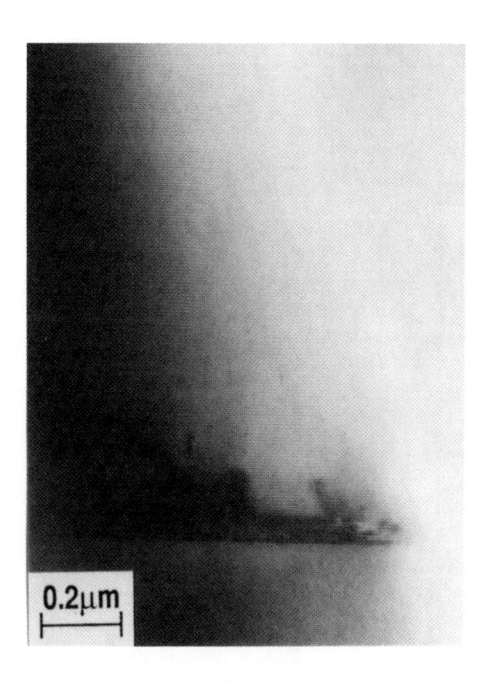

FIG. 1. TEM micrograph of a HgTe/CdTe superlattice grown on a (211)B CdTe substrate illustrating suppression of the line dislocation structure propagating from the substrate/epilayer.

IV. n-TYPE DOPING

Indium is typically the *n*-type dopant of choice for MCT since it is easily activated, has a reasonable diffusion coeffi-

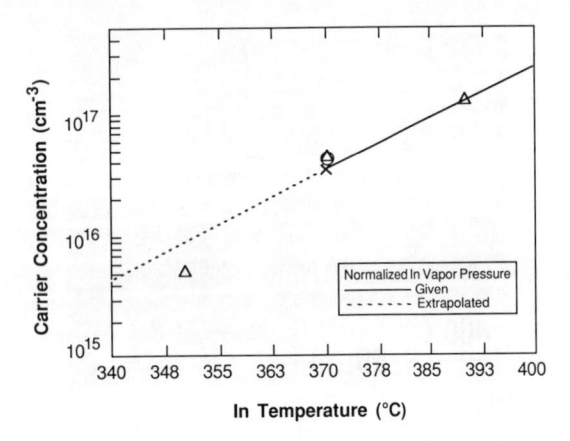

FIG. 3. Indium carrier concentration as measured by Hall effect vs indium oven temperature. The variation tracks the indium vapor pressure.

cient, and can be incorporated by almost all growth techniques. All of the MCT films and HgTe/CdTe superlattices grown at GE by PAMBE and doped with indium exhibit nearly 100% dopant activation similar to that reported by others.[9] The absolute indium concentration is easily controlled by the MBE source temperature. The variation of indium incorporation with source temperature tracks the indium vapor pressure curve as shown in Fig. 3, and is reproducible to within 20% from one oven reloading to the next.

In the case of CMS or SL structures, the layer is doped continuously for indium: the indium shutter is opened at the beginning of n-type layer growth and closed at completion. Dopant grading is accomplished by opening and closing the indium shutter at preselected time intervals, thus allowing for precise indium doping control. This is illustrated in Fig. 4 by the SIMS measurement obtained from an indium step-doped CMS structure. The abrupt onset and termination of indium incorporation with the opening and closing of the indium shutter is also indicative of the lack of a persistent memory effect when using indium, in contrast to earlier reports.[10]

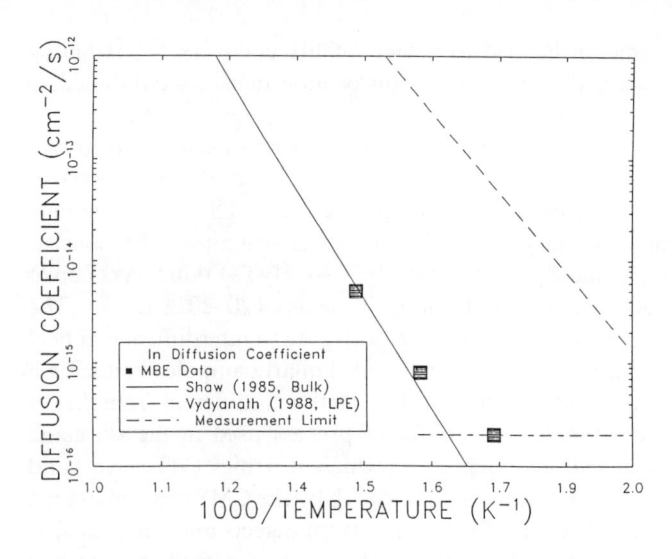

FIG. 5. Arrhenius plot of indium diffusion coefficient for the step-doped sample of the previous figure. The MBE data points correspond to three pieces of the wafer, each piece having undergone a 400, 350, and 300 °C anneal for 5 h, respectively. The figure also shows indium diffusion coefficient data obtained for bulk MCT and for LPE MCT.

The diffusion coefficient for indium has been recently reported to be relatively high as determined from indium doped liquid phase epitaxy (LPE) MCT films.[11] The thermal stability of indium was assessed by cutting the step-doped epilayer of Fig. 4 into four pieces and subjecting three of these pieces to 300, 350, and 400 °C anneals, respectively, under isothermal Hg overpressure conditions for 5 h. Figure 4 also shows SIMS data obtained from the sample having undergone the 300 °C anneal. No significant change in the indium profile was observed for the 300 °C annealed sample to the resolution of the SIMS. This lack of indium redistribution is indicative of the stability of indium in high quality material.

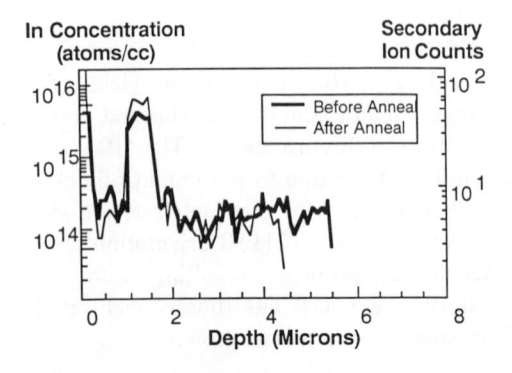

FIG. 4. SIMS depth profile obtained from an indium step-doped CMS epilayer grown on a (211)B CdTe substrate. One scan (thick line) was obtained from an as-grown piece of the wafer and the other scan was obtained from a piece annealed at 300 °C for 5 h.

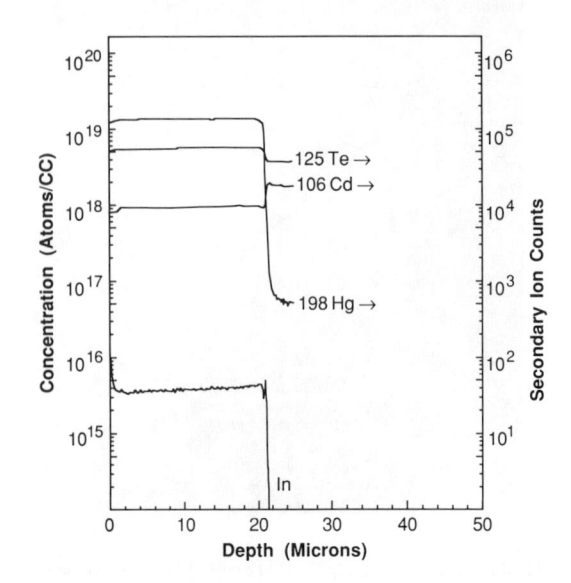

FIG. 6. SIMS depth profile obtained from a 22 μm thick MCT CMS layer.

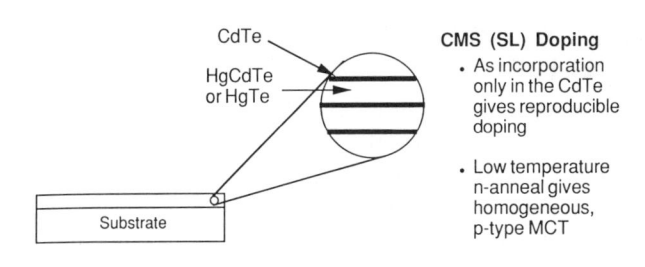

FIG. 7. Illustration of arsenic doping in CMS and superlattice structures.

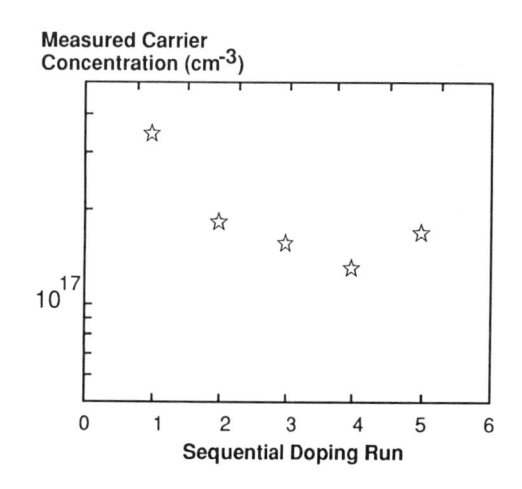

FIG. 9. Measured p-type carrier concentration at 78 K for five sequentially grown arsenic-doped epilayers.

Indium diffusion was observed at 350 and 400 °C which allowed for a direct measurement of the diffusion coefficient. Figure 5 shows a plot of the measured indium diffusion coefficient along with that measured for LPE MCT[11] and for bulk MCT[12] versus inverse temperature. The indium diffusion coefficient data measured for the step-doped epilayer is an order of magnitude lower than that observed for LPE MCT and is in agreement with the data obtained from bulk MCT.

Uniform indium doping of thick epilayers is also easily achieved. Figure 6 shows a SIMS atomic profile obtained from a 22 μm thick MCT CMS layer. The Hg, Cd, and Te profiles are constant, indicative of material homogeneous in depth, and the indium concentration is uniform throughout the sample illustrating long term source stability. This type of layer is currently being used at GE to investigate Hg-diffused diode methods for producing LWIR detectors.

V. p-TYPE DOPING

Several p-type dopants, Li, Ag, Sb, and As, for MCT have been investigated.[9] Of these potential p-type dopants, arsenic is the only technologically feasible p-type dopant for MCT and HgTe/CdTe layered structures, because it is both highly activated and stable with respect to redistribution. Using PAMBE, GE reported the first p-type As-doped MCT grown by MBE in 1987.[13] However, doping level control is hampered by the intrinsic Hg-deficient nature of MBE growth as Hg vacancies promote antisite doping, and thus compensation, when using p-type dopants that substitute on the Te site.

To overcome this problem we have utilized the fact that CdTe can be doped p-type with arsenic.[14] For superlattice growth, arsenic is incorporated only into the CdTe layers,[8,15,16] as shown in Fig. 7, effectively decoupling dopant incorporation from Hg vacancies and reducing the probability of antisite incorporation. The As-doped superlattice structure can be used as is for superlattice detector fabrication or interdiffused in a standard n-type anneal (the CMS technique) to remove residual Hg vacancies from the layer growth, resulting in p-type homogeneous MCT.

Indeed, this technique has proven to be a viable approach to controlled and reproducible arsenic doping of MCT and HgTe/CdTe superlattice epilayers. Figure 8 shows SIMS data obtained from an arsenic-doped annealed CMS layer. The profile illustrates a uniform arsenic dopant distribution

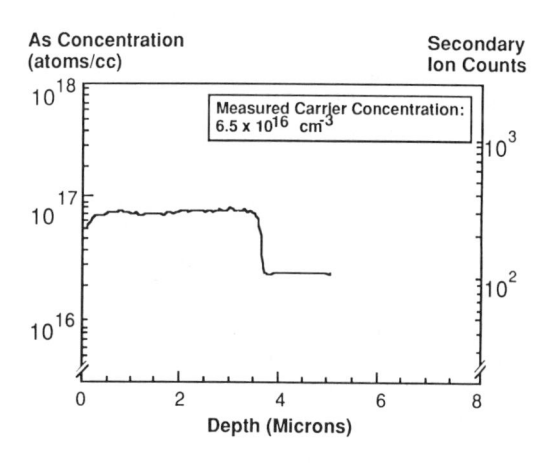

FIG. 8. SIMS depth profile obtained from an arsenic doped CMS(SL) epilayer. Measured Hall data at 78 K on this sample indicated nearly 100% arsenic activation.

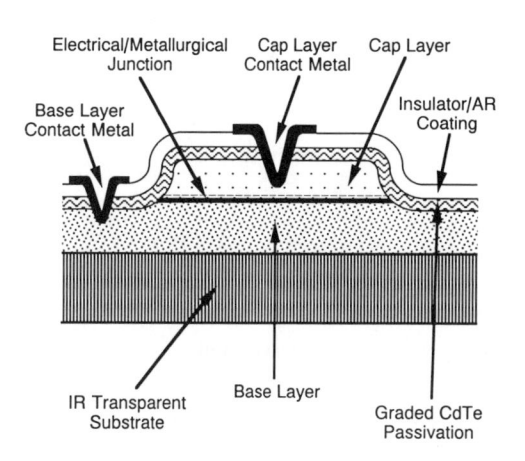

FIG. 10. Schematic of a processed mesa structure.

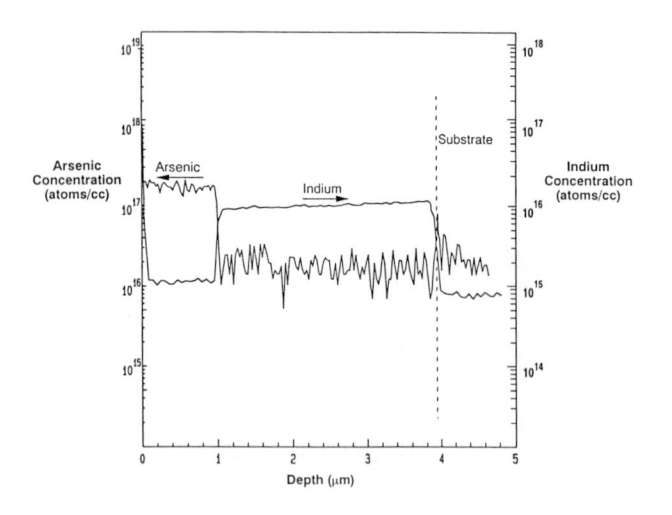

FIG. 11. Indium and arsenic SIMS depth profiles obtained from a section of the p-on-n homojunction superlattice wafer.

indicative of source stability and constant arsenic incorporation achievable with the CMS technique. Hall measurements obtained from this sample at 78 K correlated with the SIMS data indicated nearly 100% arsenic activation in this layer. Figure 9 shows the measured Hall carrier concentration at 78 K for five sequentially grown arsenic-doped layers illustrating reproducible arsenic doping levels from run to run.

Arsenic doping levels were found to be quite dependent on the substrate temperature during growth. A 30 °C reduction of the substrate temperature was found to result in an order of magnitude increase in the arsenic dopant incorporation for a constant arsenic flux. This result indicates that the arsenic sticking coefficient is strongly dependent on the substrate growth temperature in the measured temperature range.

Arsenic doping of multilayer structures is thus critically dependent on the simultaneous control of both the substrate temperature and the As source oven temperature. Arsenic doping levels can be reproducibly controlled to levels as high as low to mid 10^{18} cm^{-3} through appropriate adjustment of these two parameters.

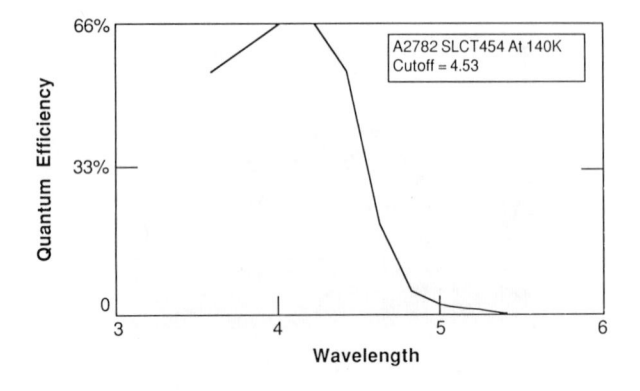

FIG. 12. Representative spectral response of a superlattice detector with $\lambda_c = 4.53$ μm at 140 K illustrating a sharp profile. The peak response corresponds to 66% peak quantum efficiency.

VI. MWIR HOMOJUNCTION SUPERLATTICE DETECTOR

As was detailed in the previous sections, the low temperature technique of PAMBE has allowed for the growth of multilayers exhibiting a high degree of crystalline perfection and routine *in situ* n-type and p-type extrinsic doping. GE is also developing a proprietary low temperature mesa diode fabrication process specifically engineered for multilayer structures susceptible to high temperature degradation. Using this combination of technologies, we have successfully fabricated MWIR superlattice detectors exhibiting high quantum efficiencies and uniform response.

The basic structure of the IR homojunction superlattice devices grown in this study is illustrated by the schematic cross section of a processed mesa structure in Fig. 10. The starting epilayer was grown on a (211)B CdTe substrate and consisted of a 3 μm n-type base layer followed by the growth of a 1 μm thick p-type cap layer. Dopant incorporation was verified by SIMS analysis on a small piece of the grown layer. The resultant SIMS depth profile of this homojunction SL, shown in Fig. 11, confirmed that the dopant profile and doping levels matched that expected from growth conditions. Mesa structures were etched into the film using standard photolithographical processes. The mesas were then passivated using MBE CdTe followed by an insulating dielectric. The final step was metallization of the base and cap layer contacts.

The measured spectral response for one of these devices with a cutoff wavelength of about 4.5 μm at 140 K is shown in Fig. 12. Direct measurements yielded peak quantum efficiencies as high as 66% (at 140 K) at the peak wavelength and an average over the 3–5 μm waveband of 55%. The variation in cutoff wavelength was 0.28 meV/K, in agreement with the theoretically predicted value for a superlattice device.[17] At 78 K the measured peak quantum efficiency was found to decrease to 45%–50% with a cutoff wavelength of 4.9 μm. These devices represent the first HgTe/CdTe superlattice-based IR detectors to exhibit significant quantum efficiencies.

IR transmittance was used to confirm the superlattice structure of the material. The IR microbeam spectrum obtained from an as-grown piece of the SL [Fig. 13(a)] illustrates a sharp cutoff of about 3.7 μm (300 K). The IR microbeam spectrum obtained from the same piece after an interdiffusion anneal [Fig. 13(b)] is shifted to shorter cutoff wavelength. This shorter cutoff wavelength was later determined to be about 1.2 μm by the group at North Carolina State University. The longer cutoff wavelength of the as-grown layer is indicative of quantum size effects associated with a periodically layered structure, thus indicating that the layer is a superlattice.

Figure 14 shows a representative TEM micrograph of the MWIR superlattice. All observed areas of the superlattice exhibited a highly regular layer spacing with a large degree of interfacial sharpness and layer uniformity. Line dislocation counts were consistent with a value of less than 10^6 cm^{-2}, and represented the resolution of the TEM technique. From the TEM image, the layer thicknesses for this superlattice are 30 (HgTe) and 50 Å (CdTe).

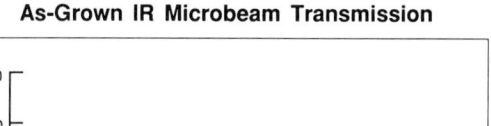

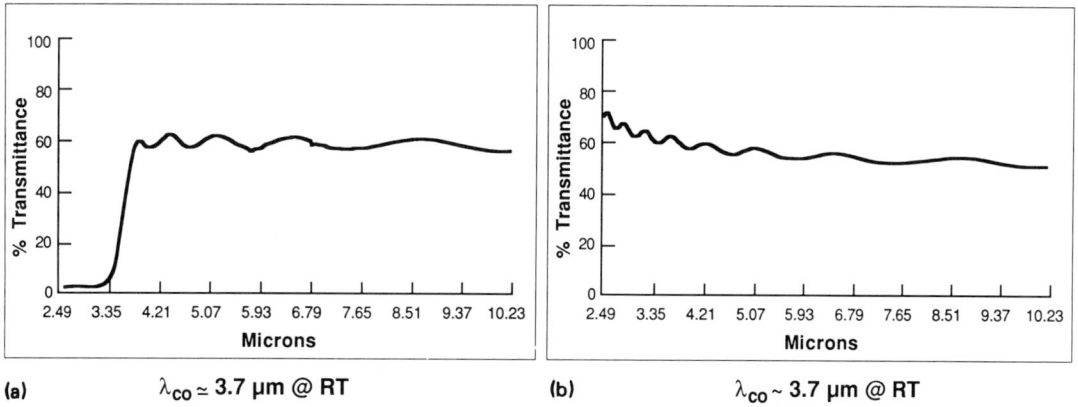

FIG. 13. IR microbeam transmission spectra obtained from two pieces of the SL wafer, one as-grown (a) and one having undergone an isothermal interdiffusion anneal (b). The post-anneal shift to shorter cutoff wavelength is indicative of quantum size effects associated with a superlattice structure.

The optically sensitive area was determined by IR spot scan as illustrated in Fig. 15. The response of the front-side illuminated device is uniform over the entire detector area except in the region of the contact. The optically active area is in good agreement with that of the physical area of the mesa device.

Figure 16(a) shows a representative I–V measurement obtained at 78 K of the MWIR SL detectors. The p-on-n nature of the homojunction is clearly shown as the I–V curve appears reversed from that obtained from the more conventional n-on-p diode. The variation of $R_0 A$ versus inverse temperature, as shown in Fig. 16(b), indicates that tunneling processes begin to dominate diffusion limited behavior at low temperatures. The low temperature behavior of $R_0 A$ may be due to both a bulk tunneling phenomenon as well as surface leakage currents. The I–V curve of a gated superlattice test device is shown in Fig. 17. The guard-ring evaluation as a function of guard bias indicates that the diode dark current is reduced by reverse biasing the guard gate. This result is indicative of a surface-limiting dark current process, and may indicate improper passivation of the surface. In spite of this, the $R_0 A$ values for these SL detectors were typically 5×10^5 Ω cm^2 which is comparable to that which has been achieved in the corresponding alloy.

The fabrication of high quantum efficiency MWIR p-on-n superlattice homojunction detectors is a significant milestone towards the ultimate development of this material for an LWIR or VLWIR device technology. The results presented here clearly demonstrate the feasibility of the use of this quantum material in IR detector structures.

VII. SUMMARY

The growth of thin, strain-inducing CdTe layers in a CMS (SL) configuration was shown to significantly reduce the propagation of line dislocations in MCT and HgTe/CdTe superlattice epilayers. The technique, when combined with the use of (211)B CdTe substrates, resulted in the growth of HgTe/CdTe superlattices of high structural perfection.

An n-type extrinsic doping technology using indium as the dopant was presented. Dopant incorporation was shown to be easily controlled by the indium oven temperature, yielding nearly 100% indium activation in the epilayer. Indium diffusion coefficient data obtained from annealing experiments performed on a step-doped epilayer indicated low diffusion coefficients consistent with high quality MCT crystal growth.

A p-type doping technology using arsenic was reported. The doping was accomplished by incorporating arsenic in the CdTe layers of the CMS(SL) structure. Arsenic atomic concentrations were reproducible and, when compared with corresponding Hall effect data, showed nearly 100% dopant

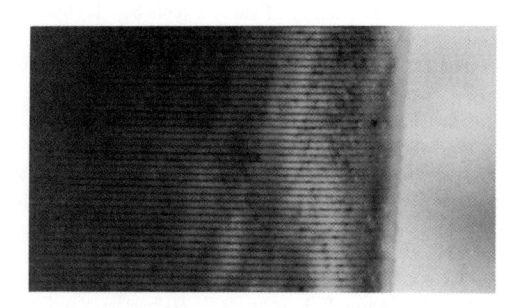

FIG. 14. TEM cross-section micrograph obtained from a small section of the p-on-n homojunction superlattice. The layer uniformity and interface sharpness demonstrates the deposition control provided by MBE.

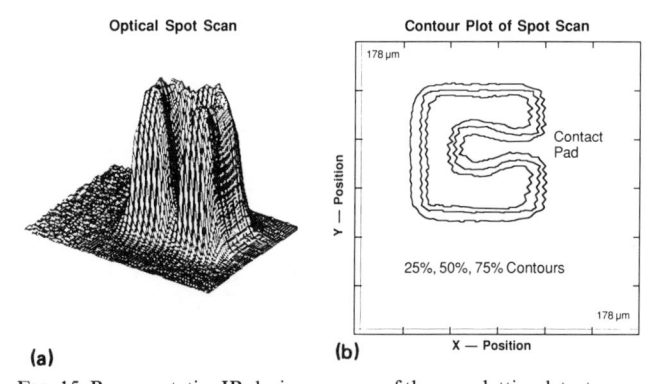

FIG. 15. Representative IR device response of the superlattice detectors are shown as (a) a detailed three-dimensional 3D map of photoresponse and (b) a spot scan contour plot of the 25%, 50%, and 75% isoresponse lines. IR photoresponse is uniform across the detector except over the region masked by the contact.

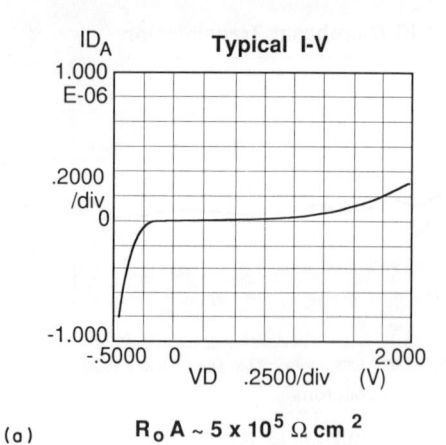

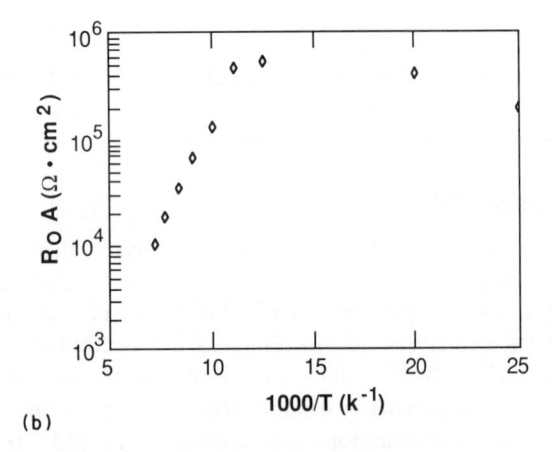

FIG. 16. Electrical behavior of the superlattice devices is illustrated by (a) the I–V curve of a representative SL and (b) a plot of R_0A vs inverse temperature. The I–V curve is reversed from that obtained from the more conventional n-on-p diode. Measured R_0A values are 5×10^5 Ω cm^2 at 78 K with low temperature characteristics indicative of nonoptimal surface passivation.

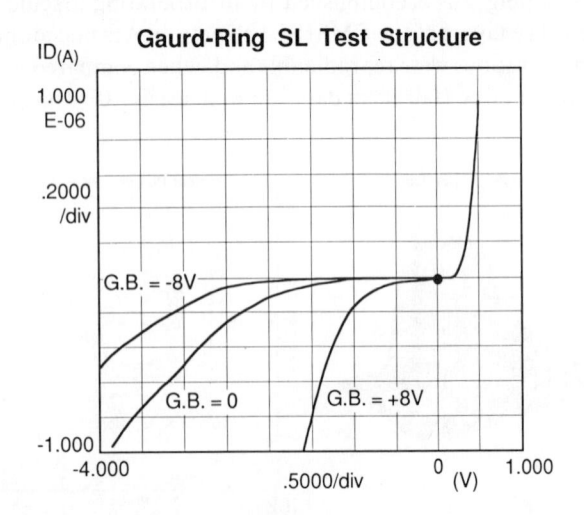

FIG. 17. I–V trace of a representative superlattice diode (gated test device) illustrating breakdown voltages comparable to those typically observed for devices fabricated from the corresponding alloy. The diode dark current is drastically reduced by reverse biasing the guard gate, which is indicative of surface limited passivation.

activation. Control of arsenic doping levels was found to depend on simultaneous control of the As oven temperature and the substrate temperature.

MWIR HgTe/CdTe superlattice mesa diode detectors were fabricated using GE's mesa diode process from PAMBE-grown SL material. SIMS depth profiling indicated a high degree of dopant profile control. TEM cross-section observations revealed sharp metallurgical interfaces and highly uniform layer thicknesses.

The detectors exhibited quantum efficiencies as high as 66% (140 K) at the peak wavelength and an average over the 3–5 μm waveband of 55%. Measured quantum efficiency was lower at 78 K, with a peak of 45%–50% and cutoff wavelength of 4.9 μm. Spot scans showed uniform response across the superlattice detector area. Resistances and breakdown voltages were reasonable for this wavelength and comparable to that which has been achieved for the corresponding alloy. These device results demonstrate for the first time the feasibility of using superlattice material in IR detector structures.

ACKNOWLEDGMENTS

The authors would like to acknowledge D. W. Dietz for device packaging and K. M. Girouard and S. C. H. Wang for device testing. The authors would like to thank Professor J. F. Schetzina and his group at North Carolina State University for kindly performing the near IR transmission analysis of the interdiffused superlattice. The work presented in this paper was supported by GE Internal Research and Development funds.

[1] J. N. Schulman and T. C. McGill, Appl. Phys. Lett. 34, 663 (1979).
[2] K. A. Harris, S. Hwang, D. K. Blanks, J. W. Cook, Jr., J. F. Schetzina, N. Otsuka, J. P. Baukus, and A. T. Hunter, Appl. Phys. Lett. 48, 396 (1986).
[3] J. P. Baukus, A. T. Hunter, O. J. Marsh, C. E. Jones, G. Y. Wu, S. R. Hetzler, T. C. McGill, and J. P. Faurie, J. Vac. Sci. Technol. A 4, 2110 (1986).
[4] T. H. Myers, R. W. Yanka, K. A. Harris, A. R. Reisinger, J. Han, S. Hwang, Z. Yang, N. C. Giles, J. W. Cook, Jr., J. F. Schetzina, R. W. Green, and S. McDevitt, J. Vac. Sci. Technol. A 7, 300 (1989).
[5] K. A. Harris, T. H. Myers, R. W. Yanka, L. M. Mohnkern, R. W. Green, and N. Otsuka, J. Vac. Sci. Technol. A 8, 1013 (1990).
[6] H. F. Schaake, J. H. Tregilgas, A. J. Lewis, and P. M. Everett, J. Vac. Sci. Technol. A 1, 1625 (1983).
[7] R. J. Koestner and H. F. Schaake, J. Vac. Sci. Technol. A 6, 2834 (1988).
[8] J. M. Arias, S. H. Shin, D. E. Cooper, M. Zandian, J. G. Pasko, E. R. Gertner, and J. Singh, J. Vac. Sci. Technol. A 8, 1025 (1990).
[9] M. Boukerche, P. S. Wijewamasuniya, S. Sivananthan, I. K. Sou, Y. J. Kim, K. K. Mahavadi, and J. P. Faurie, J. Vac. Sci. Technol. A 6, 2830 (1988).
[10] J. P. Faurie, DARPA II–VI Materials and Processing Conference, McLean, VA, April 7–9, 1987 (unpublished).
[11] H. R. Vydyanath, Aerojet Electrosystems Company, Azusa, CA.
[12] D. Shaw, Phys. Status Solidi (a) 89, 173 (1985).
[13] T. H. Myers, presented at the DARPA-NVEOL(CCNVEO) Sponsored Workshop on MBE of MCT, Nov. 1987, Washington, D. C. (unpublished).
[14] R. L. Harper, Jr., S. Hwang, N. C. Giles, J. F. Schetzina, D. L. Dreifus, and T. H. Myers, Appl. Phys. Lett. 54, 170 (1989).
[15] Jeong W. Han, S. Hwang, Y. Lansari, R. L. Harper, Z. Yang, N. C. Giles, J. W. Cook, Jr., and J. F. Schetzina, J. Vac. Sci. Technol. A 7, 305 (1989).
[16] P. Capper, P. A. C. Whiffin, B. C. Easton, C. D. Maxey, and I. Kenworthy, Mat. Lett. 6, 356 (1988).
[17] Y. Guldner, G. Bastard, and M. Voos, J. Appl. Phys. 57, 1403 (1985).

Subband spectroscopy and steady-state measurement of dark current in metal–insulator–semiconductor devices

R. A. Schiebel
Texas Instruments Incorporated, Dallas, Texas 75265

(Received 23 October 1990; accepted 3 February 1991)

We demonstrate a steady-state measurement technique of measuring dark current in HgCdTe metal–insulator–semiconductor (MIS) devices, implementing a gated diode. We show that this technique has several advantages over conventional MIS dark current measurement methods. Using this technique we are able to observe subbands due to inversion layer quantization, hence demonstrating a new technique of subband spectroscopy. The tunnel current into these bands is monitored as a function of temperature to deduce the nature of the tunneling transition.

I. INTRODUCTION

The metal–insulator–semiconductor (MIS) structure is widely used for HgCdTe materials characterization and study of inversion layer properties. Its high impedance frustrates direct measurement of material properties through steady state measurements (such as current–voltage) and necessitates pulsed or alternating current (ac) techniques. Stray capacitance and resistance and speed limitations of pulsed techniques limit analysis to relatively large area devices with low dark currents. Low field dark current measurements are limited by signal/noise limitations of pulsed techniques. These factors limit the sensitivity of MIS devices to material nonuniformities, and restrict characterization to relatively low-doped material, particularly for narrow band gap HgCdTe.

In this work we demonstrate a novel technique that circumvents these limitations and allows steady state current–voltage measurements on the MIS device in strong inversion. We apply the technique to the topic of subband spectroscopy, demonstrating direct observation of tunneling into subbands formed as a result of inversion layer quantization, and comparing these measurements to other techniques. We monitor the tunneling current into these subbands as a function of temperature in order to better understand the nature of the tunneling current.

II. EXPERIMENTAL

The particular devices used in this study were fabricated on *p*-type HgCdTe grown by liquid phase epitaxy on (111)B-oriented CdZnTe, with the intrinsic acceptor concentration adjusted by post annealing. Approximately 1900 Å ZnS provided insulation between the HgCdTe and the gate level. Metallurgical diodes were formed by donor-like damage induced by ion implantation of 100 keV boron at a 10^{13} cm^{-2} dose. The gated diode structures were realized with MIS field effect transistors (MISFETs), which in essence are gated diode structures with diodes on either side of the gate. The areas of the metallurgical diodes varied with the size of the gate, but were typically in the range 500–1500 μm^2.

III. STEADY STATE DARK CURRENT MEASUREMENT

To measure dark currents in MIS devices we use the familiar gated diode structure shown in Fig. 1. The gate is biased to inversion; the adjacent metallurgical diode provides direct, ohmic contact to the inversion layer and is used to monitor the current through the induced junction under the MIS gate. To understand how this device is used we consider separately the effect of gate bias V_G and diode bias V_D on its operation.

Figure 2 shows a measurement of the diode current versus gate bias. Three regimes are noted:

$V_{FB} > V_G$: At biases below flatband the surface under the MIS gate is accumulated, producing a substantial tunnel current at the periphery of the metallurgical diode.

$V_{FB} < V_G < V_T$: At biases between flatband and the threshold of strong inversion, the region under the MIS gate is depleted, and the measured current is dominated by the leakage current of the metallurgical diode.

$V_T < V_G$: At biases greater than threshold, the region under the MIS gate is inverted and the current from the induced junction, by virtue of its larger area or tunneling current, predominates. This is the main regime of interest to this work.

Previously,[1-3] gated diodes were used to measure surface generation-recombination current in the depletion regime to deduce properties of the semiconductor–insulator interface. No such currents were observed in the devices used in this study, due to the low test temperatures and the predominance of metallurgical diode current in the depletion regime.

In contrast, other gated diode studies have generally ignored the inversion regime. This is undoubtedly because the tunneling current in the induced junction, by virtue of the larger band gaps in these studies, was negligible compared to measurement capabilities or other dark current sources. Narrow band gap HgCdTe, however, is ideal for observing these currents.

To understand the role of diode bias, we consider the band diagram for the region under the gate, illustrated in Fig. 3.

In Fig. 3(a) the gate is biased to inversion and the diode is unbiased. The device is at equilibrium; no net current flows through the induced junction.

In Fig. 3(b) the diode is reverse biased, creating a potential difference between the inversion layer and the bulk. Just as in a *pn* junction, the applied bias manifests itself as a splitting of the quasi-Fermi levels, and the depletion region widens. To satisfy charge conservation, the inversion layer charge decreases in depth and density. A steady-state current flows across the induced junction and to the *n* side of the metallurgical diode.

At a greater reverse bias, shown in Fig. 3(c), the depletion region has grown at the expense of the inversion layer to the extent that there is no inversion layer. The absence of an inversion layer prevents the metallurgical diode from contacting the field induced junction. The diode bias must be decreased, or the gate voltage increased, for an inversion layer to exist.

These effects may be observed in the device's current versus diode voltage curves, shown in Fig. 4. At low gate biases, no inversion layer exists, and the dark current is simply that of the metallurgical diode. At high gate biases, the dark current is dominated by that of the field-induced junction. At intermediate gate biases, the current saturates as the diode

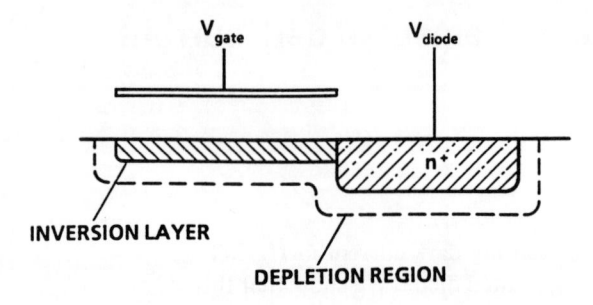

FIG. 1. Cross section of gated diode structure used for this study.

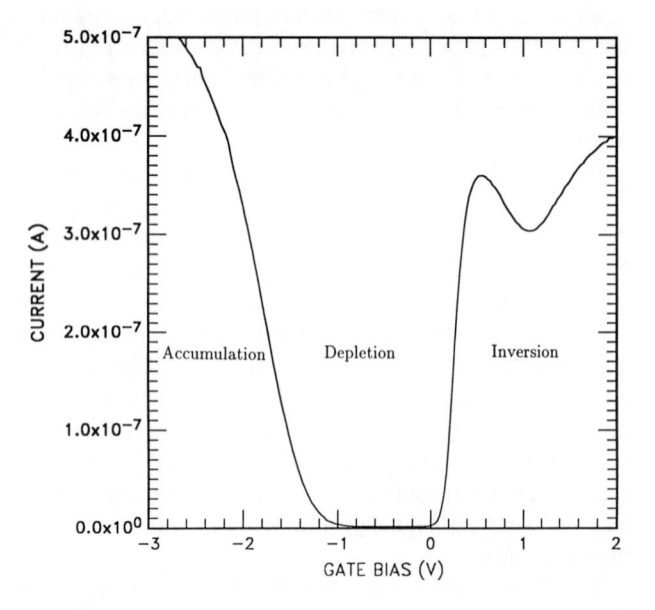

FIG. 2. Diode current vs gate bias showing accumulation, depletion, and inversion regimes. $T = 40$ K. $V_D = 2$ mV. $\lambda_C = 9.3$ μm at 77 K. MIS area $= 2.6 \times 10^{-5}$ cm^2.

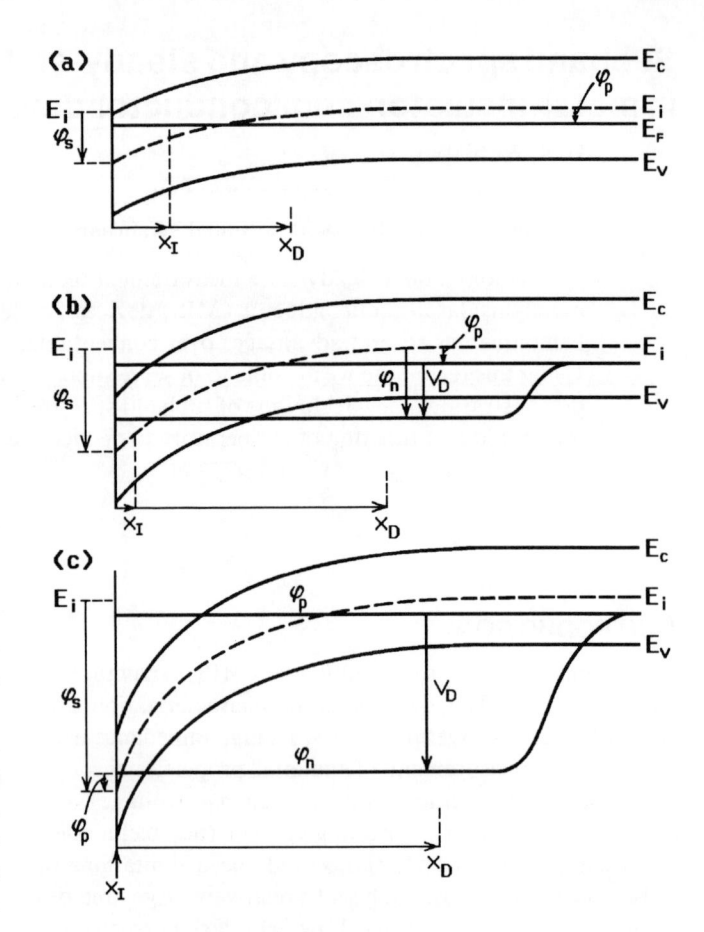

FIG. 3. Band diagrams (explained in the text) for the region under the gate for various diode biases in the gated diode measurement, or at various times in a transient measurement.

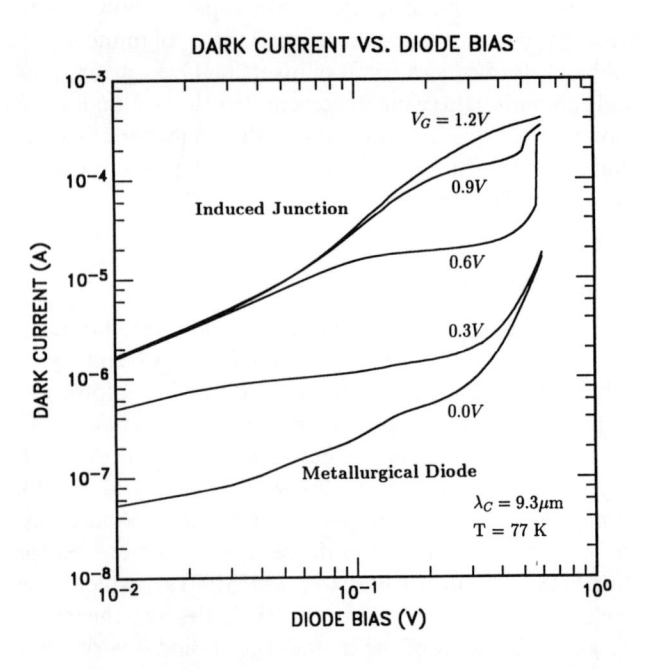

FIG. 4. Current–voltage characteristics for the metallurgical junction and the induced junction. $T = 77$ K. Same device as Fig. 2.

bias increases (just as in a MISFET) and the inversion layer is pinched off.

The band diagrams of Fig. 3 are identical to those for a pulsed MIS device as a function of time. Figure 3(c) shows the device in deep depletion, immediately after being pulsed. Figure 3(b) shows the device with the well partially full.

Figure 3(a) shows the device with a full well, at $t = \tau_{\text{storage}}$. Thus, a particular diode bias in the gated diode technique corresponds to a particular time in the pulsed technique.

Pulsed measurements,[4,5] where capacitance or voltage are measured as a function of time, are in principle able to make similar measurements to the gated diode technique but in

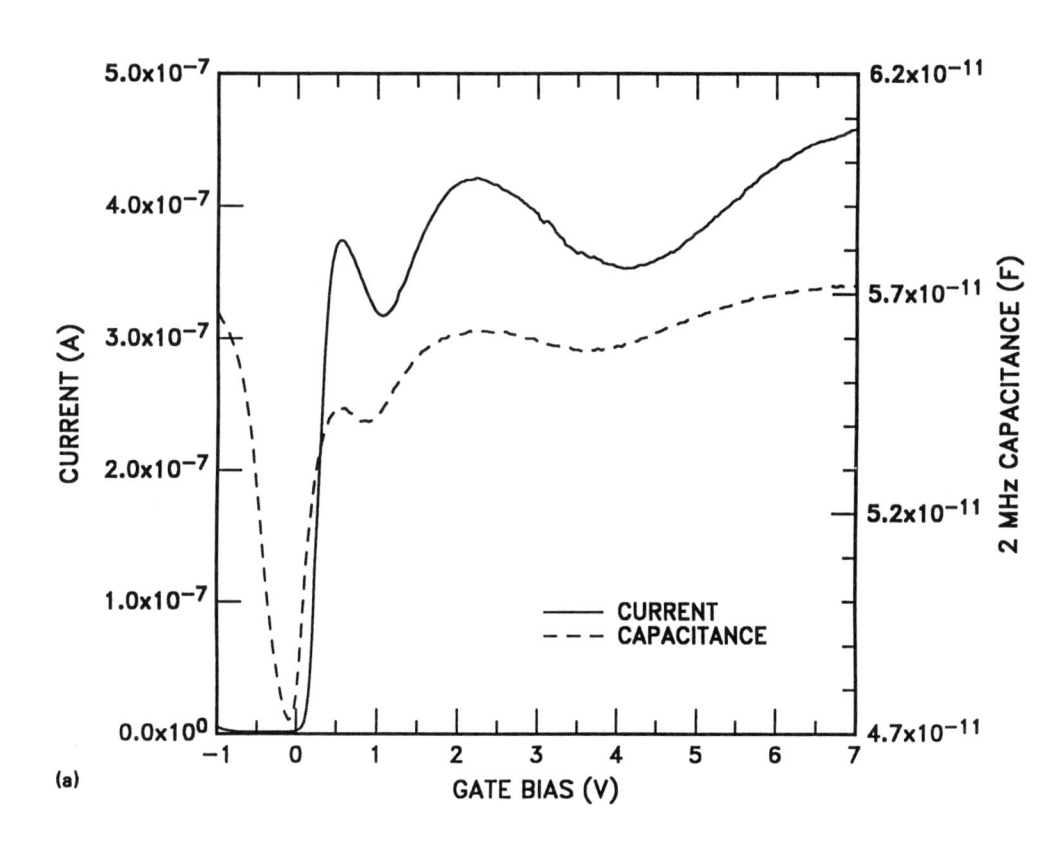

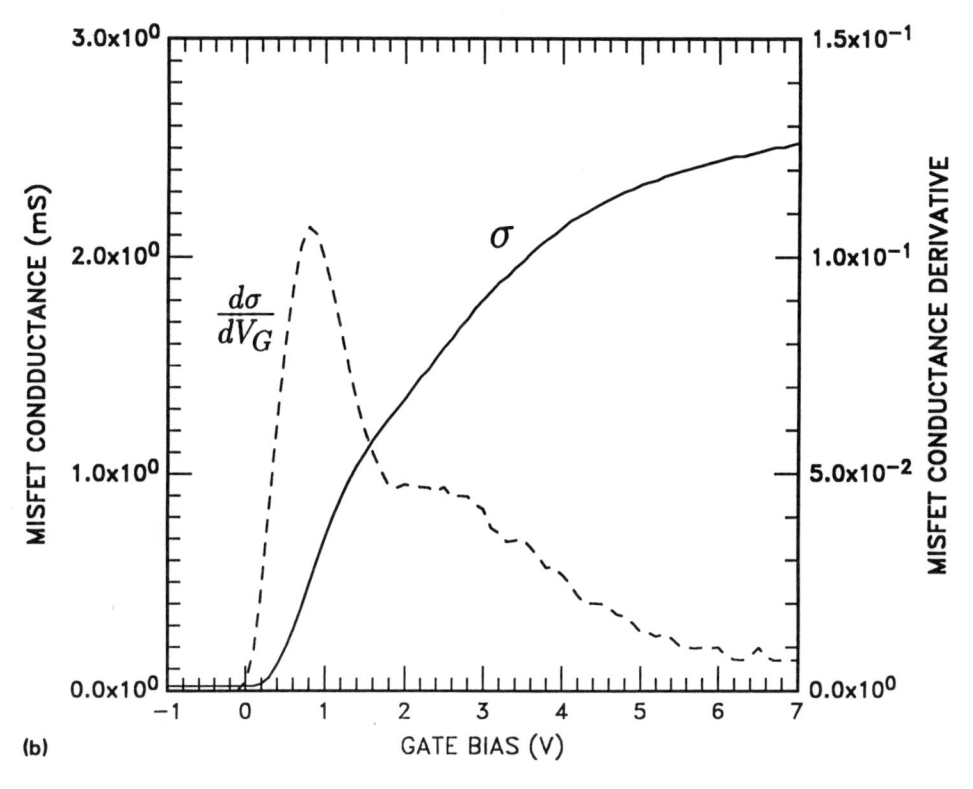

FIG. 5. Observation of subbands: (a) Current–voltage and capacitance–voltage curves showing oscillations due to inversion layer quantization, with each peak corresponding to a different subband. Capacitance data is from a 610 by 203 μm MIS device; the current from a 51 by 51 μm device. $V_D = 2$ mV. (b) MISFET channel conductance and its derivative. $V_{\text{DS}} = 5$ mV. $f = 1$ MHz. $T = 40$ K. $\lambda_C = 9.3\,\mu$m at 77 K.

practice have several limitations. Signal/noise limitations of these techniques limit low field measurements. Speed limitations frustrate measurement of samples with high dark current levels, preventing measurements on devices with higher dopings, or a complete analysis of poor devices.

Capacitance measurements require accurate measurement of the complex impedance of a device, increasing the vulnerability of the measurement to stray capacitance and resistance. Measurements are consequently limited to temperatures above freeze-out (to avoid stray resistance) and large area devices (to minimize stray capacitance effects). The sensitivity of the MIS device to material nonuniformities is thereby limited.

In contrast, the gated diode technique requires simple direct-current (dc) measurement of current versus voltage. We have successfully made current versus gate and diode voltage measurements on small (25 by 25 μm) MIS devices, with diode biases (quasi-Fermi level splittings) as low as 2 mV, at doping levels in excess of 10^{16} cm^{-3}. These are all beyond the limits of conventional MIS measurement techniques.

IV. SUBBAND SPECTROSCOPY

Charge confinement in the narrow potential well in which the inversion layer charge resides leads to one-dimensional quantization and the creation of subbands in the inversion layer (omitted for clarity in Fig. 3). The detection of these subbands in various semiconductors, including HgCdTe, has been the subject of several studies since the birth of metal-oxide semiconductor (MOS) technology. This topic was previously reviewed for semiconductors in general[6] and for HgCdTe in particular.[7]

In HgCdTe and silicon, capacitance or conductance measurements (perpendicular to the semiconductor-insulator interface), have been used to observe these levels. In these measurements, oscillations in capacitance/conductance are noted as the Fermi level passes through each subband.[8-10]

Current–voltage measurements have not heretofore been used to observe these levels in HgCdTe. On wider band gap semiconductors, tunnel junctions—either metallurgical,[11] or MIS[12]—were required to get appreciable current levels. In HgCdTe the field-induced junction, by virtue of the narrow band gap, is a tunnel junction, so no special device features are necessary. It is thus easy to observe these levels, manifested as oscillations in measured current versus gate voltage in a gated diode device, shown in Fig. 5(a). Three peaks in dark current are observed, corresponding to the first three subbands.

For comparison, the capacitance measured on a MIS device on the same film is also shown in Fig. 5(a). Excellent agreement is noted. However, the capacitance-voltage measurements, taken at 40 K, could not be taken at lower temperature because of the increasing series resistance due to freeze-out. Oscillations in the gated diode current were observed and easily measured from temperatures below 5 up to about 100 K.

Conductance parallel to the semiconductor-insulator interface may also be used to observe the subbands. Conductance steps should be observed as the gate voltage is in-creased and higher lying subbands are occupied. In practice, the effect is weak, and structure is best observed in the conductance derivative. In HgCdTe, structure in accumulation layer conductance was noted[13] but the step-like behavior was not observed, probably due to the gate bias-dependent mobility, lateral nonuniformities, and other broadening processes.

Figure 5(b) shows inversion layer conductance and its derivative measured on a MISFET on the same film as the devices of Fig. 5(a). The conductance exhibits three distinct regions of slope, made clearer by the derivative plot, that correspond to the three subbands shown in Fig. 5(a). While the agreement with the capacitance- and current-voltage measurements is good, it does not elucidate the subbands as well as those measurements. In particular, the conductance derivative at high gate biases approaches the level due to stray resistances external to the device.

By measuring the gated diode's current–voltage characteristics versus temperature we may identify dark current mechanisms and the nature of the tunneling transition to the quantized levels. Figure 6 shows an example of such a measurement, illustrating the temperature dependence of the current through the first subband of the gated diode device in Fig. 5. At far left, a sharp rise is noted in the diffusion limited regime; tunneling is responsible for the dark current measured at lower temperatures.

Usually such tunneling is ascribed to trap-assisted tunneling either through a tunnel–tunnel[10] or thermally assisted[9] process. The temperature dependence of the dark current at low temperatures in Fig. 6 may be explained by freeze-out, since it follow the temperature dependence of the Fermi level through $\exp - (E_F - E_V)/kT$. (The Fermi level was calculated assuming charge neutrality in the bulk, a 15 meV acceptor level, and accounting for nonparabolicity, degener-

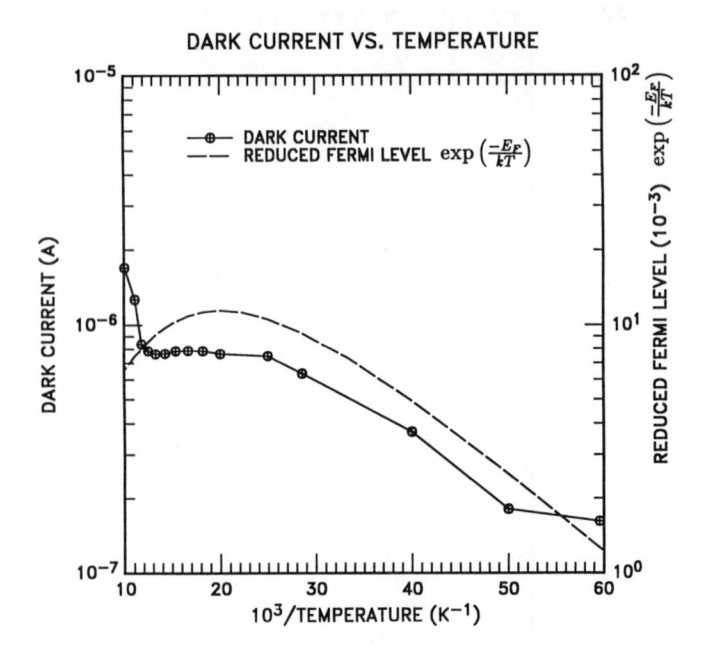

FIG. 6. Dark current vs inverse temperature for the first subband of the gated diode device in Fig. 5. $V_D = 5$ mV.

acy, and partially ionized impurities.[14]) This eliminates the possibility of thermally assisted tunneling in favor of the tunnel–tunnel process. The relatively high (1×10^{16} cm^{-3}) doping leaves open the possibility of band-to-band tunneling as well.

V. CONCLUSIONS

Using a gated diode structure we have demonstrated a technique that extends the measurement capabilities of HgCdTe MIS devices beyond that presently available with transient and/or capacitance measurements. We have used this technique to observe structure due to subbands in the inversion layer and demonstrated its superiority to other subband spectroscopic techniques. Through variable temperature measurements of the gated diode current we are able to gain a better understanding of the tunneling transition to the subbands.

[1] A. S. Grove and D. J. Fitzgerald, Solid State Electron. **9**, 783 (1966).
[2] A. S. Grove, *Physics and Technology of Semiconductor Devices* (Wiley, New York, 1967), Chap. 10.
[3] J. P. Rosbeck and E. R. Blazejewski, J. Vac. Sci. Technol. A **3**, 280 (1985).
[4] T. Yamamoto, Y. Miyamoto, and K. Tanikawa, J. Cryst. Growth **72**, 270 (1985).
[5] D. K. Blanks, J. D. Beck, M. A. Kinch, and L. Columbo, J. Vac. Sci. Technol. A **6**, 2790 (1988).
[6] F. Koch, *Solid State Sciences 53: Two Dimensional Systems, Heterostructures, and Superlattices*, edited by G. Bauer, F. Kuchar, and H. Heinrich (Springer, Berlin, 1984), p. 20.
[7] F. Koch, *Interfaces, Quantum Wells, and Superlattices*, edited by C. R. Leavens and R. Taylor (Plenum, New York, 1988).
[8] M. Kaplit and J. N. Zemel, Phys. Rev. Lett. **21**, 212 (1968).
[9] M. A. Kinch, J. D. Beck, and W. T. Zwirble, IEDM Technical Digest, 508 (1980).
[10] M. J. Yang, C. H. Yang, and J. D. Beck, Semicond. Sci. Technol. **5**, S118 (1990).
[11] U. Kunze, J. Phys. C **17**, 5677 (1984).
[12] D. C. Tsui, Phys. Rev. Lett. **24**, 303 (1970).
[13] W. Zhao, F. Koch, J. Ziegler, and H. Maier, Phys. Rev. B **31**, 241 (1985).
[14] I. Bloom and Y. Nemirovsky, Solid State Electron. **31**, 17 (1988).

Dark current processes in thinned p-type HgCdTe

D. K. Blanks

Texas Instruments Incorporated, Dallas, Texas 75265

(Received 16 October 1990; accepted 20 December 1990)

Previous studies of p-type $Hg_{1-x}Cd_x$Te indicate that diffusion is the dominant dark current mechanism for material with a 10 μm spectral cutoff ($x = 0.225$) at 77 K [J. P. Omaggio, IEEE Trans. Electron Dev. **37**, 141 (1990)]. This work demonstrates a 10X reduction of diffusion current in p-type HgCdTe upon thinning to a 10 μm thickness. Charge transient measurements applied to metal–insulator semiconductor (MIS) structures were used to determine dark currents over a temperature range of 40–110 K and electric fields of 0.5–1.4 V/μm for thick/thin pairs of p-type HgCdTe slices. Emperical fits to dark current measured in thick and thinned sister slices are compared to theoretical expressions for diffusion and tunnel current mechanisms.

I. INTRODUCTION

Dark current is a major limitation of sensitivity and dynamic range for $Hg_{1-x}Cd_x$Te infrared detectors with a 10 μm spectral cutoff ($x = 0.225$). Consequently, reducing dark current has been a major goal in the development of HgCdTe as an infrared detector material. At 77 K, the primary dark current processes include (a) tunnel current across the band gap via band-to-band or Shockley–Read–Hall (SRH) trap tunneling, (b) depletion current consisting of carriers thermally generated within the depletion region, and (c) diffusion current arising from carriers generated outside the depletion region followed by diffusion to the depletion region. Figure 1 illustrates these processes for a metal–insulator semiconductor (MIS) structure fabricated on p-type HgCdTe. Because of the high electron mobility in narrow-gap HgCdTe, diffusion lengths in p-type material can be as large as 250 μm. Since the depletion region in the detector is on the order of 1 μm or less, diffusion current is collected from a much larger volume than the active detector region and is the dominant dark current source for p-type HgCdTe. Thinning p-type HgCdTe has been proposed as a means of minimizing the diffusion current by eliminating most of the diffusion volume.[1] This effect is demonstrated here by comparing dark current measurements for thick p-type HgCdTe and a sister slice thinned to 10 μm. The dark current was measured using a charge transient technique performed on MIS structures over the 40–110 K temperature range. The electric field and temperature dependence of the dark current was used to determine the nature of dark current processes in the thick and thinned material.

II. EXPERIMENTAL TECHNIQUE

The p-type HgCdTe was grown by solid-state recrystalization (SSR) The net dopant concentration was measured using the maximum–minimum capacitance technique to be in the $2-3\times 10^{15}$ cm^2 range. Since the HgCdTe was annealed at 270 °C to produce an estimated background vacancy concentration of $5-7\times 10^{13}$, the majority of the p-type doping is dominated by copper introduced during the growth process. A 200×800 mils slice of this material was sawed in half to make two 200×400 mils slices. The thick control slice was epoxied to a ceramic carrier while the sister slice was waxed to another ceramic carrier. Each slice was passivated with a proprietary Texas Instruments process known as ASP-II, and insulated with approximately 2000 Å of ZnS. The sister slice was then epoxied to a ceramic carrier with the ZnS-coated surface face down and diamond turned to 40 μm. A bromine methanol polish was used to polish the slice to 10 μm. The top surface was passivated and coated with ZnS in the same manner as the back surface. Aluminum gates 10×15 mils in area were fabricated on both sister slices. Spectral measurements indicated a 10.0 μm cutoff which corresponds to a Cd concentration of $x = 0.225$.

The dark current was measured using a charge transient technique as illustrated in Fig. 2. As the MIS device is pulsed into deep depletion, a current-integrating op-amp connected to the gate outputs a voltage V_{out} proportional to the charge on the gate

$$Q(t) = C_f V_{out}, \quad (1)$$

where C_f is the feedback capacitance of the charge transient circuit. A typical charge transient is shown in Fig. 3 for an MIS device on p-type HgCdTe. Since the charge transient

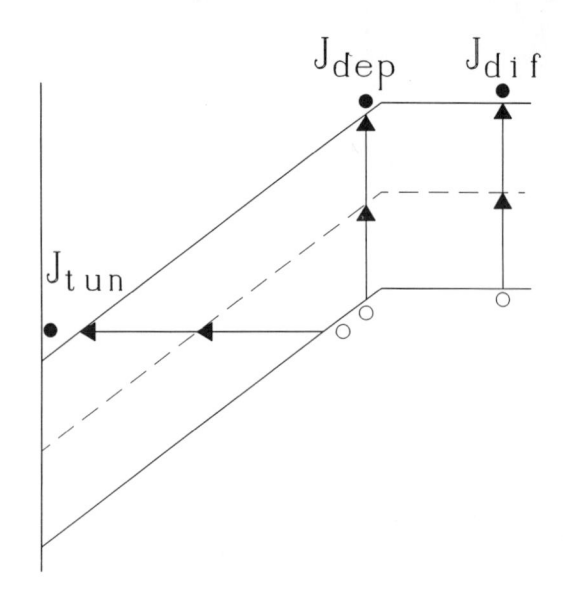

FIG. 1. Dark current processes in p-type HgCdTe, including tunnel, depletion, and diffusion currents.

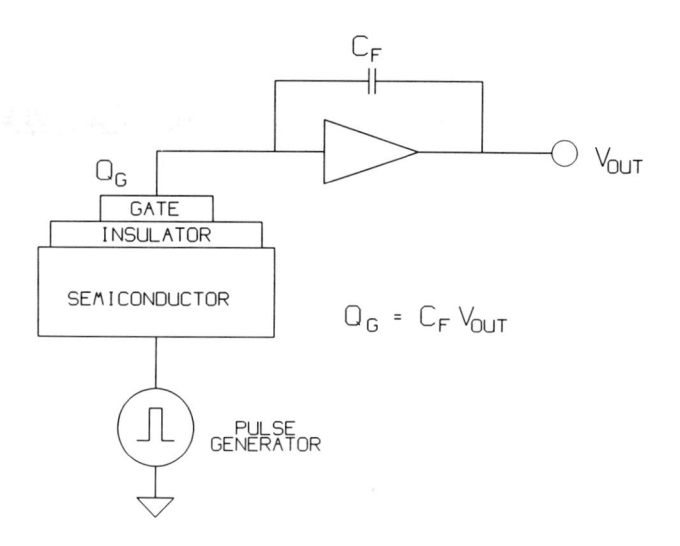

FIG. 2. Apparatus used to perform charge transient measurements on MIS devices.

circuit only measures relative changes in charge rather than absolute charge, the charge at equilibrium is used as a reference level such that $Q(t \to \infty) \to 0$. An estimate of material quality can be derived from the time τ_s required for the device to come to equilibrium. The time τ_s is referred to as the storage time of the MIS device since it represents the maximum time available for collecting and storing the photon-generated charge during device operation. After the device comes to equilibrium, the minority-carrier charge stored in the depletion region is given approximately by $C_i \Delta V$ where C_i is the insulator capacitance and ΔV is the applied bias with respect to the threshold of strong inversion. The storage time is related to the dark current by the relation

$$\tau_s \cong \frac{C_i \Delta V}{\langle J_d \rangle A}, \qquad (2)$$

where $\langle J_d \rangle$ is the average dark current during time τ_s. The storage time increases with ΔV until the onset of tunnel current. The tunnel current increases so rapidly with bias that τ_s stops increasing with ΔV. This condition is referred to as the tunnel breakdown of the MIS detector. Thus, the optimum

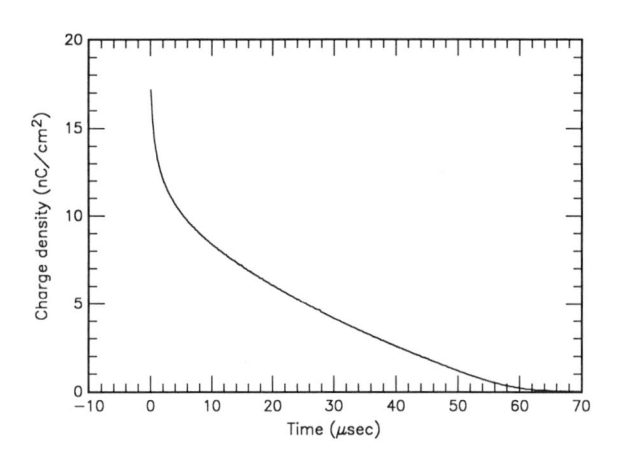

FIG. 3. Charge transient measured for an MIS device fabricated on p-type $Hg_{0.775}Cd_{0.225}Te$.

τ_s is a measure of the average dark current for a ΔV corresponding to the onset of tunnel breakdown. The net dopant concentration is usually the main parameter controlling τ_s in p-type HgCdTe. A low dopant concentration is desirable to reduce the curvature of the energy bands for a given ΔV, thus allowing a larger ΔV for the same tunnel breakdown field and resulting in a larger storage time. For good quality p-type HgCdTe in the 10 μm cutoff range with $N_b = 5 \times 10^{14}$ cm^{-3}, storage times at 77 K are on the order of 50 μs for $C_i/A \cong 5.2 \times 10^{-8}$ F/cm^2 and $\Delta V \cong 0.5$ V, indicating an average dark current of 0.65 mA/cm^2.

While the storage time is a general measure of device quality, further analysis can yield detailed information about the dark current. From basic circuit theory, the time dependence of the surface potential $\Psi_s(t)$ is directly related to the charge transient by the relation

$$\Psi_s(t) = \frac{-Q(t)}{C_i} + \Psi_{sbi}. \qquad (3)$$

As the device comes to equilibrium, $\Psi_s(t)$ approaches the built-in surface potential given by the depletion approximation

$$\Psi_{sbi} = \frac{2kT}{q} \ln\left(\frac{N_b}{n_i}\right). \qquad (4)$$

Another useful quantity is the depletion capacitance given by

$$C_d(t) = \left(\frac{qN_b \epsilon_s A^2}{2|\Psi_s|}\right)^{1/2}, \qquad (5)$$

which then yields the dark current as a function of time,

$$J_d(t) = \frac{1}{A}\left(1 + \frac{C_d(t)}{C_i}\right)\frac{dQ}{dt}. \qquad (6)$$

The values of C_i and N_b are determined from the capacitance–voltage measurements and E_g from spectral cutoff measurements. Equations from Ref. 2 were used to obtain n_i. The dielectric constant ϵ_s was extrapolated from data in Ref. 3. The dQ/dt derivative was calculated numerically using a quadratic derivative convolution technique.[4,5]

The electric field at the insulator–semiconductor interface can also be obtained from

$$E(t) = \frac{qN_b A}{C_d(t)}. \qquad (7)$$

The values of $J_d(t)$ and $E(t)$ for any time t can be paired to construct a J_d versus E plot. Thus, the charge transient technique automatically records dark current information as the MIS device sweeps from deep depletion to equilibrium. The J_d versus E plots presented later were constructed in this manner.

III. EXPERIMENTAL RESULTS

A. Thick p-type HgCdTe

To analyze the nature of the dark current sources, dark current scans were performed over a 40–95 K range for the thick slice and 40–110 K for the thinned slice. A semilog plot of J_d versus $1/E$ for the best thick MIS device illustrates the different extremes of dark current behavior (Fig. 4). At high

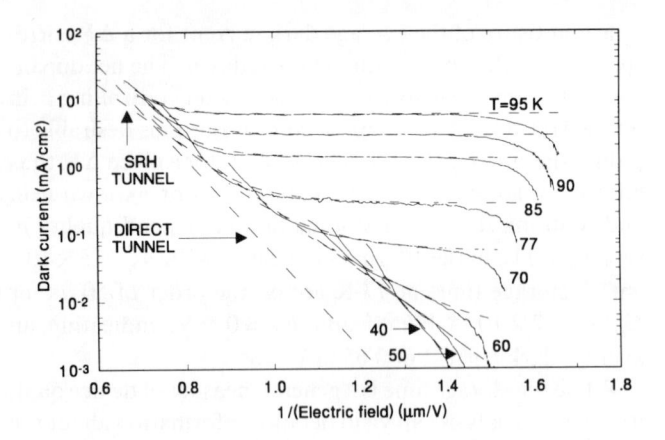

FIG. 4. Dark current scans performed over the 40–95 K temperature range for an MIS device fabricated on thick p-type HgCdTe. The solid curves are experimental data and the dashed lines are semiempirical fits. The SRH tunnel component was extrapolated using the fit to the 60 K dark current scan.

temperature and low electric field, the dark current is dominated by diffusion current. According to standard theory, the diffusion current given by

$$J_{\text{dif}} = \left(\frac{qn_i^2}{p_0}\right)\left(\frac{kT\mu_n}{q\tau_n}\right)^{1/2}, \quad (8)$$

is expected to be independent of electric field and vary as $\exp(-E_g/kT)$ due to the n_i^2 factor. Assuming constant values of $\mu = 1.5 \times 10^5$ cm^2/V s and $\tau_n = 10^{-7}$ s, Eq. (8) yields

$$J_{\text{dif}} \simeq 3 \times 10^7 \exp\left(\frac{-E_g}{kT}\right) \text{ mA/cm}^2. \quad (9)$$

Fitting the high-temperature portion of the dark current data yields

$$J_{\text{dif}}^{\exp} = 3.1 \times 10^6 \exp\left(\frac{-E_g'}{kT}\right) \text{ mA/cm}^2, \quad (10)$$

where $E_g' = 0.108$ eV. The disagreement between the prefactors is not surprising considering the rough estimate of the parameters in Eq. (9). In addition, it is difficult to measure the prefactor accurately in a semilog plot. The exponent can be determined more accurately and is simpler to estimate theoretically as E_g and m^* are the only adjustable parameters. The discrepancy with the measured $E_g = 0.124$ eV band gap may indicate some temperature dependence in the mobility or lifetime as well as experimental error.

In the low-temperature limit, thermal processes freeze out and tunneling is the major dark current source. The tunnel currents form the high-field low-temperature boundaries on the J_d versus $1/E$ plots in Fig. 4. There are two definite slopes on the boundary which are identified as the direct and trap-assisted tunnel currents. Anderson derived an expression for direct tunneling[6]

$$J_{\text{tun}}(E) = \left(\frac{3}{8}\right)^{1/2} \frac{(\alpha q)^3 \Psi_s E}{\hbar P(4+\theta)} \exp(-\theta), \quad (11)$$

where

$$\alpha = \frac{1}{\pi}\left(1 - \frac{E_g}{2q\Psi_s}\right)^{1/2} \quad (12)$$

and

$$\theta = \left(\frac{3}{32}\right)^{1/2}\left(\frac{E_g^2}{\alpha PqE}\right), \quad (13)$$

and P is the Kane interband matrix element. For trap-assisted tunneling through a midgap SRH center,[7,8]

$$J_{\text{SRH}}(E) = \frac{\pi^2 q^2 m^* E^2 M^2}{h^3 E_g}\left(\frac{N_r}{N_b}\right)\exp\left[\frac{-\pi(m^*/2)^{1/2}E_g^{3/2}}{4qE\hbar}\right] \quad (14)$$

The matrix element M associated with the trap potential can be estimated from the equation[9]

$$M = \frac{2\sqrt{2\pi}}{m_0}\hbar^2\left(\frac{2m_0}{\hbar^2}\right)^{1/4}\frac{E_g}{(0.5E_g)^{3/4}}. \quad (15)$$

The prefactor in the direct tunnel current term can be estimated by assuming $E \sim 1$ V/μm. The interband matrix element is approximately $P = 8.4 \times 10^{-8}$ eV cm over most of the x-value range for Hg$_{1-x}$Cd$_x$Te. This yields

$$J_{\text{tun}}(E) \sim 9.8 \times 10^7 \exp\left(\frac{-1.2 \times 10^7 E_g^2}{E}\right) \text{ mA/cm}^2, \quad (16)$$

where E_g is measured in eV and E in V/cm. The experimental fit gives

$$J_{\text{tun}}^{\exp}(E) = 6.5 \times 10^5 \exp\left(\frac{-1.1 \times 10^7 E_g^2}{E}\right) \text{ mA/cm}^2, \quad (17)$$

For the SRH tunnel component we estimate $E \sim 1.0$ V/μm and $(N_r/N_b) \sim 0.01$ to obtain

$$J_{\text{SRH}}(E) \sim 6 \times 10^5 \exp\left(\frac{-5.3 \times 10^6 E_g^2}{E}\right) \text{ mA/cm}^2. \quad (18)$$

From the experimental data,

$$J_{\text{SRH}}^{\exp}(E) = 1.1 \times 10^4 \exp\left(\frac{-7.0 \times 10^6 E_g^2}{E}\right) \text{ mA/cm}^2. \quad (19)$$

The exponential factors are in reasonable agreement with theory for the direct and trap-assisted tunnel currents. The measured prefactors in front of the exponent are much smaller than predicted. As discussed earlier with connection with the diffusion current, the disagreement in the prefactor is reasonable considering the unknowns and approximations involved.

The diffusion, band-to-band, and trap-assisted tunnel currents give reasonable agreement for the dark current scans above 70 K. However, these equations do not fit the scans below 70 K with very good accuracy. The details of fitting the dark current in this regime will be discussed after analysis of the thinned p-type dark current behavior.

B. Thin p-type HgCdTe

The first clear evidence of reduced diffusion current for the thinned p-type HgCdTe is reflected by an increase in MIS storage times. Figure 5 plots the storage at 77 K for each MIS device as a function of dopant concentration. Even

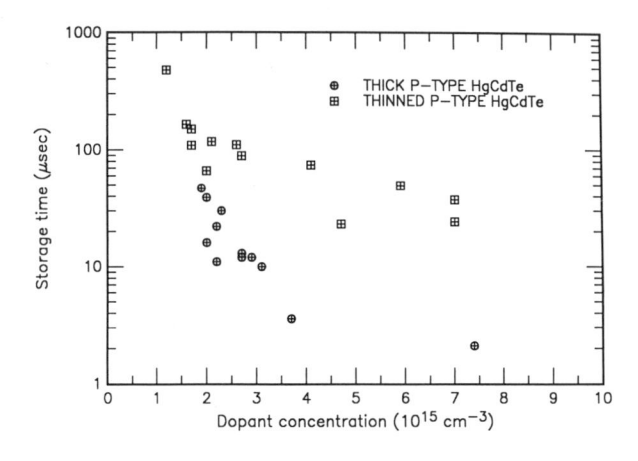

FIG. 5. Storage times vs net acceptor dopant concentration measured for MIS devices fabricated on thick and thinned p-type HgCdTe.

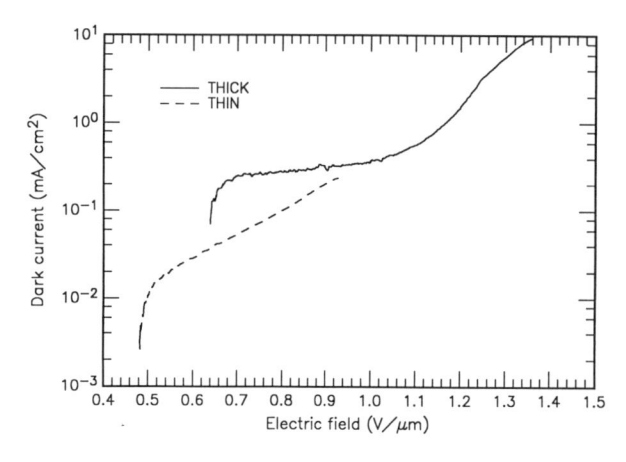

FIG. 7. Dark current scans of the best devices on the thick and thinned slices (#14 and #16 in Fig. 6, respectively) illustrating the reduction in diffusion current upon thinning.

though the storage and doping vary across each slice, the storage times are consistently higher for the thinned device at any given dopant concentration. The storage time varies from 2.1–46.9 μs on the thick slice and 24–477 μs on the thinned slice. At least 3/4 of the devices on the thinned slice have $\tau_s \geqslant 50$ μs, which is superior to storage times observed for high-quality thick p-type HgCdTe with 10 μm cutoff. To illustrate the increase in storage more clearly, the bar plot in Fig. 6 arranges the storage times in increasing order for each slice. The general trend is for a 3× increase in average storage time for the thinned slice, though the best thinned device with 477 μs storage is clearly above this trend. By comparison, the next best thin device has 164 μs of storage.

Figure 7 compares the dark current scans at 77 K for the best thick and thin devices. The dopant concentration determined using the maximum–minimum capacitance technique gave 2.5×10^{15} cm^2 for the thick device and 1.5×10^{15} cm^2 for the thinned device. Diffusion current dominates over the 0.7–1.0 V/μm electric field range in the thick device with tunnel breakdown occurring around 1.1 V/μm. The thinned device has a lower built-in field due to its lower doping. The dark current is strongly field dependent but smaller in magnitude than the thick counterpart over the common field range of 0.7–0.9 V/μm. To compare the effect of temperature, the dark current at a field of 0.8 V/μm is graphed in an Arrhenius plot in Fig. 8. The thin p-type HgCdTe demonstrates a 10× decrease in diffusion current for $T > 70$ K but a 10× increase in tunnel current for $T < 70$ K. The lower dopant concentration in the thinned device coupled with the increased tunnel current could indicate damage induced by the thinning procedure. It has been well established that structural damage of HgCdTe can introduce n-type donors into the material. In this case, the damage was moderate enough to prevent p-to-n-type conversion but increased the trap-assisted tunneling by introducing more SRH centers into the lattice. The reduction of dopant concentration must also enhance the storage as discussed earlier, though it is difficult to gauge the contribution to the device performance.

The dark current scans for the thinned device plotted vs inverse field for each temperature in Fig. 9. The diffusion current is fitted with the equation

$$J_{\text{dif}}^{\text{exp}} = 1.7 \times 10^6 \exp\left(\frac{-E_g'}{kT}\right) \text{ mA/cm}^2, \tag{20}$$

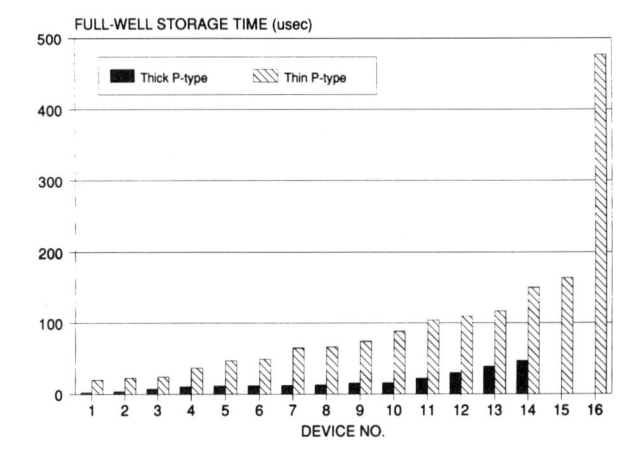

FIG. 6. Storage times for thick and thinned p-type HgCdTe MIS devices arranged in increasing order. The device numbers are for reference only.

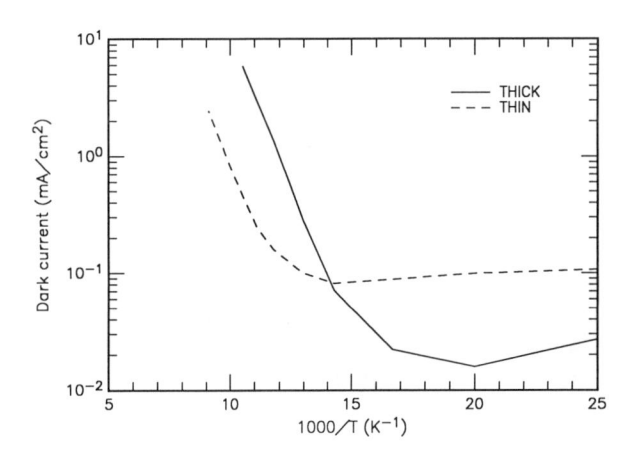

FIG. 8. Dark current vs reciprocal temperatures for the thick and thinned devices. The dark current level was measured at 0.8 V/μm.

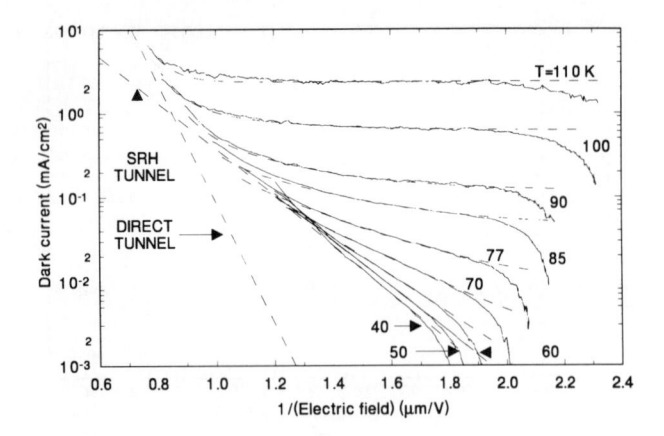

FIG. 9. Dark current scans performed over the 40–110 K temperature range for the thinned p-type HgCdTe MIS device. The SRH tunnel component was extrapolated from the 50 K dark current scan.

with $E'_g = 0.128$ eV. The high-field tunnel currents are fitted with the expressions

$$J^{\text{exp}}_{\text{tun}}(E) = 1.3 \times 10^6 \exp\left(\frac{-1.1 \times 10^7 E_g^2}{E}\right) \text{ mA/cm}^2, \quad (21)$$

$$J^{\text{exp}}_{\text{SRH}}(E) = 2 \times 10^2 \exp\left(\frac{-4.3 \times 10^6 E_g^2}{E}\right) \text{ mA/cm}^2. \quad (22)$$

The dark current characteristics have some similar features to the thick device in that the dark current is diffusion-limited at high temperatures (albeit with a lower amount of diffusion current) and limited by band-to-band tunnel current in the high-field regime. In particular, the direct tunnel current is almost identical for both devices. This is expected since direct band-to-band tunneling should only depend on the energy gap and effective electron mass. However, the dark current scans from 40–77 K are quite different compared to the same scans in the thick device. In addition to the lower diffusion current, the constant slope over a wide field range indicate that the scans consist of tunnel current. At the same time, the temperature dependence of the dark current rules out simple band-to-band processes and suggests a thermally assisted tunnel current mechanism. This concept will be discussed in more detail in the next section.

IV. DISCUSSION

A thermally assisted tunnel mechanism proposed by previous workers[10] considers an electron thermally generated from the valence band to an SRH trap followed by tunneling to the conduction band. Thermal transitions to and from the conduction band are neglected due to the small density of conduction band states compared to the heavy-hole valence band. In fact, thermally assisted tunneling was assumed to be the only tunnel current mechanism in p-type HgCdTe.[10] The difficulty with this view is that thermal transitions should freeze out at low temperatures and result in a rapid decrease in dark current. This disagrees with experimental observation demonstrated in Fig. 8 which shows that dark current remains constant or increases slightly with decreasing temperature below the onset of tunneling. Similar behavior has been observed in other examples of p-type HgCdTe as well.[11]

We have seen that dark current in the low-temperature high-field limit can be modeled with direct and trap-assisted tunnel processes without any thermal transitions. At the same time, a thermally activated tunnel current appears for low temperatures and field below the band-to-band tunnel limit. Therefore, an explanation consistent with the facts is that three different tunnel processes occur in the p-type HgCdTe. These processes illustrated in Fig. 10 include direct tunneling at high electric fields and trap-assisted tunneling at lower fields. The trap-assisted tunneling occurs via a tunneling or thermal transition from the valence band. The thermal transition will dominate if the thermal transition probability from the heavy-hole valence band exceeds the probability of tunneling from the light-hole band. Therefore, thermally assisted tunneling is observed at the higher temperatures and lower fields. The band-to-band thermal-assisted transitions would not necessarily go through the same traps. Since the total transition probability is limited by the slowest transition, the trap-assisted tunneling is dominated by the trap with equal probability of transitions to either band. As the temperature increases, the thermal transition should move higher in the band gap to adjust for the faster transition rate. This is reflected in the dark current scan in the decreased slope of the tunnel current scans with increasing temperature. The shallower slope indicates a smaller tunnel distance as the thermal transition goes to higher states in the band gap.

An empirical fit can be made to the thermally assisted tunnel current by including a thermal dependence in the exponent and prefactor of the standard tunnel current equation:

$$J_{\text{SRH}}(E,T) = \rho(T) \exp\left[\frac{-\varepsilon(T)}{E}\right]. \quad (23)$$

The empirical fits for $\rho(T)$ and $\varepsilon(T)$ are shown in Fig. 11 for the thick and thinned devices. The thermally assisted dark current is the major dark current source for the thinned device over the 40–77 K temperature range and for electric fields below the band-to-band direct and trap-assisted re-

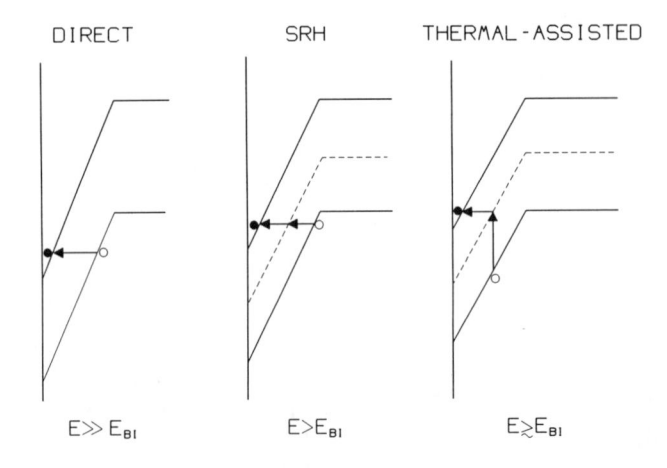

FIG. 10. Proposed tunnel current mechanisms in p-type HgCdTe include direct, SRH, and thermal-assisted tunnel current processes.

thinning process. The dark current was fitted using diffusion and the various tunnel current processes using a semiempirical approach. The experimental results indicate the presence of direct tunnel, band-to-band SRH trap tunneling, and thermally assisted SRH trap tunneling processes in p-type HgCdTe.

NOMENCLATURE

ΔV	breakdown voltage with respect to flatband,
μ_n	electron mobility,
$\Psi_s(t)$	surface potential at insulator–semiconductor interface,
Ψ_{sbi}	built-in surface potential at equilibrium,
τ_n	electron lifetime,
A	MIS gate area,
C_d	depletion capacitance,
C_f	feedback capacitance in charge transient circuit,
C_i	insulator capacitance,
E_g	energy gap,
E'_g	energy gap calculated using thermal dependence of diffusion current,
J_d	minority-carrier dark current,
J_{dif}	theoretical diffusion current,
J_{dif}^{exp}	measured band-to-band tunnel current,
J_{SRH}	theoretical trap-assisted tunnel current,
J_{SRH}^{exp}	measured trap-assisted tunnel current,
J_{tun}	theoretical band-to-band tunnel current,
J_{tun}^{exp}	measured band-to-band tunnel current,
M	tunnel transition matrix element,
m_0	electron rest mass,
m^*	electron effective mass,
N_b	net dopant concentration $(N_A - N_D)$,
N_r	Shockley–Read–Hall trap density,
n_i	intrinsic carrier concentration,
p_0	hole concentration,
$Q(t)$	charge transient as a function of time,
V_{out}	voltage outputted from charge transient circuit.

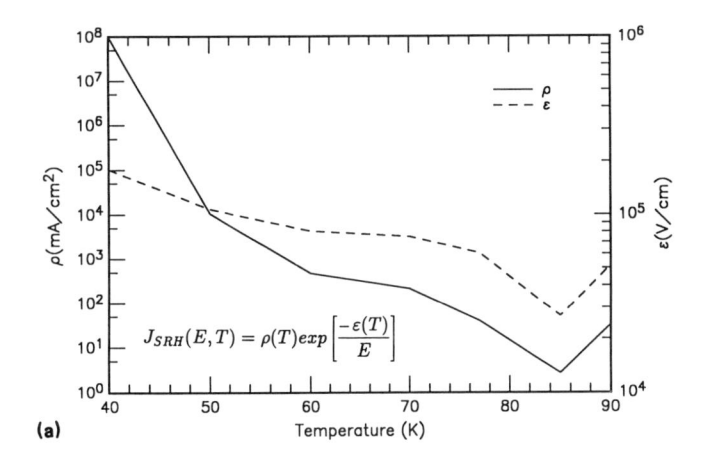

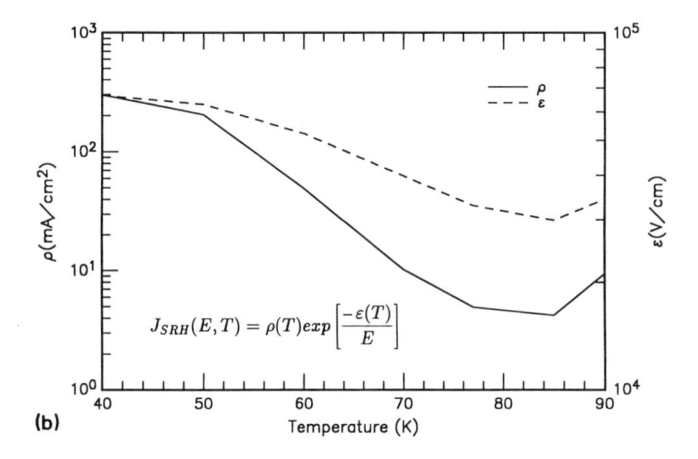

FIG. 11. Measured dependence of semiempirical parameters $\rho(T)$ and $\varepsilon(T)$ for the (a) thick and (b) thinned p-type HgCdTe MIS devices.

gime. This is also the case for the thick device over the 40–60 K temperature range. While the thermally assisted tunnel current is present in both thick and thinned devices, it is observed at higher temperatures in the thinned device because of the reduced diffusion current. This has important implications for device applications using thinned p-type HgCdTe. The diffusion-limited behavior commonly observed at 77 K no longer applies once the p-type HgCdTe is thinned below the diffusion length. Trap-assisted tunneling becomes the chief dark current source and will depend on the SRH trap density, energy level, and cross section within the band gap.

V. CONCLUSIONS

A study of p-type HgCdTe thinned to 10 μm has demonstrated a reduction in diffusion current associated with the

[1] J. P. Omaggio, IEEE Trans. Electron Dev. 37, 141 (1990).
[2] G. L. Hansen and J. L. Schmit, J. Appl. Phys. 54, 1639 (1983).
[3] J. C. Brice, EMIS Datareview RN = 15390, 29 (1987).
[4] A. Savitzky and M. J. E. Golay, Anal. Chem. 36, 1627 (1964). Note that corrected versions of some tables are given in Ref. 6.
[5] J. Steinier, Y. Termonia, and J. Deltour, Anal. Chem. 44, 1906 (1972).
[6] W. W. Anderson, Infrared Phys. 17, 147 (1977).
[7] M. A. Kinch, J. D. Beck, and W. T. Zwirble, Proc. IEDM 1980, 508.
[8] J. D. Beck, M. A. Kinch, and M. W. Goodwin, in Proceedings of the IRIS Detector Speciality Group, Boulder, CO, 1983 (unpublished).
[9] W. W. Anderson and H. J. Hoffman, J. Appl. Phys. 53, 9130, (1982).
[10] M. A. Kinch, J. Vac. Sci. Technol. 21, 215 (1982).
[11] D. K. Blanks, J. D. Beck, M. A. Kinch, and L. Colombo, J. Vac. Sci. Technol. A 6, 2790 (1988).

Properties of Schottky diodes on n-type Hg$_{1-x}$Cd$_x$Te

Patrick W. Leech and Martyn H. Kibel
Telecom Australia Research Laboratories, Clayton, 3168, Victoria, Australia

(Received 2 October; accepted 5 December)

The electrical properties of Schottky diodes formed by Ag, Au, Cu, Pd, Pt, Sb, and Ti on n-Hg$_{1-x}$Cd$_x$Te ($x = 0.6$–0.7) have been measured using current-voltage and capacitance-voltage techniques. Both LPE grown (111) Hg$_{0.3}$Cd$_{0.7}$Te and metal-organic chemical-vapor deposition grown (100) Hg$_{1-x}$Cd$_x$Te epitaxial layers were used with surface preparation either by chemical etching or by air exposure. The current–voltage characteristics of contacts fabricated on chemically etched surfaces have been described by a thermionic emission-recombination model. For these etched surfaces, the barrier heights produced by the metals Ag, Au, Cu, and Sb on (111) Hg$_{0.3}$Cd$_{0.7}$Te were in the range $\phi'_b = 0.74$ V to 0.79 V while for Pd and Pt, $\phi'_b = 0.69 \pm 0.01$ V. In close agreement, Pt contacts on etched (100) surfaces exhibited a barrier height of $\phi'_b = 0.70 \pm 0.01$ V for stoichiometries of $x = 0.6$–0.68. The effects of air exposure on diode characteristics were most significant for the Au contacts due to an inability of the metal to reduce interfacial TeO$_2$ or to react appreciably with the Hg$_{1-x}$Cd$_x$Te. The interfacial reaction of Au and of Pt layers with Hg$_{0.3}$Cd$_{0.7}$Te has been studied by Auger depth profiling.

I. INTRODUCTION

Hg$_{1-x}$Cd$_x$Te is becoming of increasing interest for application in optical communications devices operating in the near infrared spectral region from 1.0 to 2.5 μm. Discrete devices such as the high performance avalanche and p-i-n photodiodes and photodetector have been successfully fabricated on bulk Hg$_{1-x}$Cd$_x$Te[1,2] and metalorganic chemical-vapor deposition (MOCVD) grown layers.[3,4] The potential also exists for the development of integrated optical circuits in Hg$_{1-x}$Cd$_x$Te.[5] Of particular interest are integrated optoelectronic receivers comprising a metal–semiconductor field effect transistor (MESFET) and metal–semiconductor–metal (MSM) Schottky barrier photodetector which have recently been demonstrated in InGaAs.[6] Photoreceivers based on the MSM detector have shown high bandwidth and a simpler integration process compared with the nonplanar p-i-n photodiode. But the realization of these device structures in Hg$_{1-x}$Cd$_x$Te requires the ability to fabricate rectifying contacts both as the gate of the MESFET and the interdigitated electrodes of the MSM detector. Currently, little published data exists on the electrical properties of metal Schottky barrier contacts on n-Hg$_{1-x}$Cd$_x$Te.

The formation of Schottky barriers at metal contacts on n-Hg$_{1-x}$Cd$_x$Te has been theoretically predicted by Spicer et al. for x values greater than 0.4–0.5.[7] This prediction was given by each of the examined models for Schottky barrier pinning (the metal-induced gap states, the effective work function model, and the defect model).[7] Consistent with the above analysis, rectifying electrical characteristics have been reported for metal contacts on n-Hg$_{0.4}$Cd$_{0.6}$Te (Ref. 8) and n-CdTe ($x = 1.0$).[9] In the case of CdTe, the diode parameters were dependent on the degree of reactivity between metal and semiconductor[9] and on surface pretreatment.[10] Metal/Hg$_{1-x}$Cd$_x$Te interfaces have been characterized by a greater degree of reaction and more extensive intermixing than for CdTe, resulting in the possible doping of the Hg$_{1-x}$Cd$_x$Te by movement of the metal.[11] The extent of interaction between metal overlayer and Hg$_{1-x}$Cd$_x$Te has been shown by photoemission spectroscopy[12] to range from Au and Sb (classified as unreactive with respect to Hg$_{1-x}$Cd$_x$Te) with heats of formation of metal telluride ($\Delta H_f = -2.2$ kcal/mol-Te for AuTe$_2$ and $\Delta H_f = -4.5$ kcal/mol-Te for Sb$_2$Te$_3$) less in magnitude than for HgTe to the ultrareactive Ti with heats of telluride formation ($\Delta H_f = -25.4$ to -28.9 kcal/mol-Te) greater than for both HgTe and CdTe. The metals Ag and Cu with $\Delta H_f = -8.6$ kcal/mol-Te (Ag$_2$Te) and $\Delta H_f = -10.0$ kcal/mol-Te (Cu$_2$Te) have been defined as intermediate in reactivity.[12] For the metals Pt and Pd, the high heats of cation alloying ($\Delta H_f = -18.6$ kcal/mol-Te for PtTe and $\Delta H_f = -24.4$ kcal/mol-Te for PdTe) indicate the formation of a reactive interface.[13] In the present work, we report on the electrical characteristics of metal/n-Hg$_{1-x}$Cd$_x$Te junctions ($x = 0.6$–0.7) as a function of the metal type and surface pretreatment. Included in the study were metals classified[12] as ultrareactive (Ti), reactive (Pt, Pd), intermediate in reactivity (Ag, Cu) and unreactive (Au, Sb) with respect to Hg$_{1-x}$Cd$_x$Te; allowing evaluation of the effects of chemical reactivity on contact properties. In addition, the metals Ag, Au, Cu, and Sb are known acceptors on sublattice sites.[14] The pre-etching procedures (Br:HBr or I:KI:HBr) were chosen in order to obtain differing initial ratios of surface Cd/Hg[15] while air exposure generated an interfacial layer of oxide. The stoichiometric range $x = 0.6$–0.7 correspond to the near infrared spectral region.

II. EXPERIMENTAL DETAILS

The n-type Hg$_{1-x}$Cd$_x$Te used in these experiments was grown either by liquid phase epitaxy (LPE) or MOCVD. The LPE wafers were obtained from Fermionics Corp. and were comprised of Hg$_{0.3}$Cd$_{0.7}$Te on substrates of (111)CdTe; with In doping of the Hg$_{0.3}$Cd$_{0.7}$Te used to achieve a carrier concentration of 5×10^{16} cm^{-3}. Substrates for the MOCVD growth consisted of Si-doped (2° off 100)

GaAs on which the $Hg_{1-x}Cd_xTe$ layers were formed by an interdiffused multilayer process (IMP) with in-situ annealing at $\simeq 360$ °C.[16] The stoichiometric x ratio of these undoped layers grown by MOCVD was varied from $x = 0.60$ to 0.68 but with a constant carrier concentration of 1.5×10^{16} cm^{-3}. After definition of the photoresist pattern, the samples were either etched in 0.1 w/w% Br:HBr or I:KI:HBr solutions or air exposed to laboratory environment for 12 h, then rinsed in DI water and immediately loaded into the evaporation chamber. Metal layers of Ag, Au, Cu, Pd, Pt, Sb, or Ti/(30 nm) followed by Au/(200 nm) were deposited by electron beam at a base pressure of 1×10^{-6} mb. The evaporation rate of 2 A s^{-1} generated temperatures at the sample surface of ⩽45 °C. Following metal deposition, the excess metal was lifted off in acetone, defining arrays of circular contacts 0.008 cm^2 in diameter.

Current–voltage (I–V) and capacitance–voltage (C–V) (1 MHz) characteristics were measured on more than 20 contacts on each of a series of wafers at room temperature. In addition, the EBIC technique was used to measure minority carrier diffusion length L_e and minority carrier lifetime T_r using a procedure described previously for $Hg_{1-x}Cd_xTe$.[17]

Auger depth profiles and x-ray photoelectron spectroscopy (XPS) data were obtained using a Vacuum Generators HB-100 electron spectrometer which was operated at a base pressure of $\simeq 10^{-10}$ mb. A VG AG-60 gun was used to produce argon ions for removal of surface contaminants in the case of XPS and for sputter etching in the determination of AES depth profiles. For this latter application, a flat bottomed crater of approximately 0.75 mm × 1.0 mm was etched by the rastering of a finely focussed beam of ions, thus minimizing errors from any misalignment of the ion and electron beams. The ion gun was operated at 5 kV and the ion beam current density at the sample was measured as 40 μA cm^{-2}. A field emission source was employed to provide the electron beam for Auger analysis. This source was operated at 10 kV and provided a sample current of 20 nA into a spot of approximately 50 nm in diameter. The focussed beam was rastered over an area of 45 μm × 30 μm, resulting in an electron beam current density of 1.5 mA cm^{-2}, which is below the critical limit for electron beam induced damage to II–VI materials.[16,18] X rays were generated by a dual-anode (Mg/Al) source, utilizing the AlKα line (1486.6 eV) which was obtained at 15 kV and 32 mA. For both AES and XPS, the resulting electrons were analyzed via a VG CLAM-100 150° spherical sector analyzer, which was operated in the constant retarding ratio (CRR) mode for AES (analyzer energy of 10 eV), and in constant analyzer energy (CAE) mode for XPS (pass energy of 50 eV).

III. RESULTS

A. Electrical measurements

The I–V measurements have been considered initially in terms of the thermionic emission model[19] of current transport and subsequently in relation to a thermionic emission-recombination model. Typical ln J versus V characteristics for Pt/n-$Hg_{0.3}Cd_{0.7}Te$ and Au/$Hg_{0.3}Cd_{0.7}Te$ diodes at 293 K are shown in Fig. 1. Diodes formed on the same crystal were nearly identical in ln J-V properties with the exception of air exposed Au/n-$Hg_{0.3}Cd_{0.7}Te$ contacts for which some variability was evident. In Fig. 1, the deviation of the curves from the ideal linear characteristics at voltages above 0.2 V is due to the series resistance effect of the diodes. From the forward bias ln J-V data, the value of ideality factor n was measured from the slope of the ln J-V curve according to the thermionic emission model of current transport.[19] Calculated values of n for the Pt/n-$Hg_{0.3}Cd_{0.7}Te$ junctions were $n = 1.25 \pm 0.03$ for the etched surfaces to $n = 1.30 \pm 0.03$ on air-exposed surfaces. In comparison, Au/n-$Hg_{0.3}Cd_{0.7}Te$ contacts exhibited a stronger dependence of ideality factor on surface preparation with $n = 1.62 \pm 0.1$, 1.67 ± 0.1, and 3.5 ± 1.0 for diodes on Br:HBr and I:KI:HBr etched surfaces and on air exposed, respectively. Intermediate in behaviour was Sb/n-$Hg_{0.3}Cd_{0.7}Te$ with $n = 1.49 \pm 0.02$ for chemically etched and $n = 1.42 \pm 0.1$ for air exposed surfaces. The large deviation in the ideality factor of these diodes from unity indicates the operation of a parallel or concurrent mechanism of current transport additional to that of thermionic emission. Further indication of a parallel process was the magnitude of current in forward bias which exceeded that due to ideal thermionic emission by a factor of between 2 and 10.

At doping concentrations approaching 1×10^{17} cm^{-3}, the process of electron tunneling may contribute significantly to diode current, resulting in the presence of a regime of thermionic field emission. The relative importance of thermionic emission to thermionic field emission for a given semiconductor has been defined in the model by Padovani and Stratton[20] in terms of the tunneling parameter E_{00}:

$$E_{00} = hq/2(N_d/me_s)^{1/2}, \quad (1)$$

where $m = 0.08\, m_0$ is the effective mass of electrons and $e_s = 12.4$ is the dielectric constant for $Hg_{0.3}Cd_{0.7}Te$ at 300 K.[14] h, q, and N_d are the reduced Planck constant, the elementary charge, and carrier concentration, respectively. The range of applicability of thermionic emission is then defined by $1.0 \ll E_{00}/kT$. In the LPE grown $Hg_{0.3}Cd_{0.7}Te$ with $N_d = 5 \times 10^{16}$ cm^{-3}, $E_{00} = 4$ meV giving the ratio $E_{00}/kT = 0.15$. The effect on barrier lowering is then $\Delta\phi_b = (3E_{00}/2q)^{2/3} V_d^{1/3}$ with V_d the diffusion potential. The resulting value of $\Delta\phi_b = 0.03$ V is considered as a minor effect.

Alternatively, the current due to recombination in the space-charge region of a diode has been identified as an important cause of departure from ideal behavior in Schottky diodes on moderately doped semiconductors.[19] The theory of current due to recombination at localized centers within the depletion zone was described initially by Sah et al.[21] and later developed in a model by McLean et al.[22] Based on the later model, the total forward current I comprised parallel contributions from thermionic emission and recombination mechanisms:

$$I = I_s\{\exp[q/kT(V - IR)] - 1\} \\ + I_r\{\exp[q/kT(V - IR)/2] - 1\}, \quad (2)$$

where I_s is saturation current, R is series resistance of the

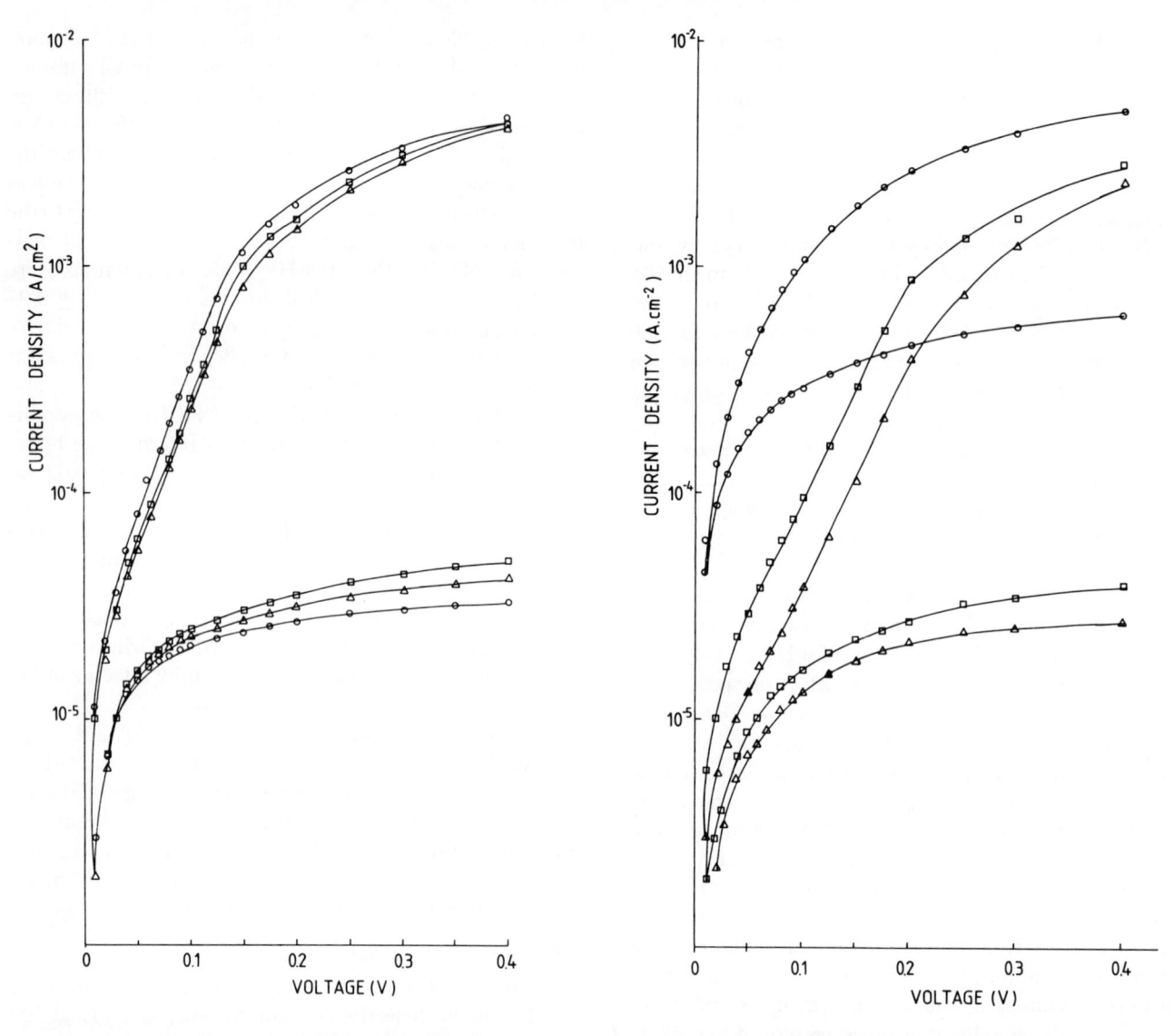

FIG. 1. Typical ln J vs V curves at 293 K for a) Pt/nHg$_{0.3}$Cd$_{0.7}$Te and b) Au/n-Hg$_{0.3}$Cd$_{0.7}$Te diodes formed on (111) surfaces: ○ Air exposed, □ I:KI:HBr etched, △ Br:HBr etched.

diode, and kT/q is the thermal energy. The recombination current, I_r, is given as

$$I_r = Sqn_i W/2T_r, \qquad (3)$$

in which $n_i = 1 \times 10^{11}$ cm^{-3} is the intrinsic carrier concentration of Hg$_{0.3}$Cd$_{0.7}$Te at 293 K,[21] $T_r = 35$ ns^{-1} is the carrier lifetime in the depletion region determined from EBIC measurements and W is the depletion width. S is diode area.

Comparison of the ln J–ln V curves for the Pd, Pt, Au, and Sb/n-Hg$_{0.3}$Cd$_{0.7}$Te systems with the predictions of the recombination model shows good agreement as in Fig. 2. Experimental plots for the other metals, Fig. 2, were also well described by the recombination model using measurements of series resistance R from the ln J–V curves. Included in this figure is the theoretical curve calculated on the basis of the p-n junction diffusion theory[24] with the diffusion constant $D_e = 4.90$ cm^2 s^{-1} and the diffusion length $L_e = 4.14$ μm based on the EBIC measurements. This theory predicts current densities which were a factor of $\approx 10^4$ less than the experimentally observed results indicating strongly against the formation of a p-n junction. Table I summarizes the J_s, n, and barrier height data for the metal/n-Hg$_{0.3}$Cd$_{0.7}$Te junctions. ϕ'_b from I–V measurements was obtained using a recombination correction to the barrier height[22] $\phi'_b = \phi_b + 2.303[(n-c)/m]kT/q$ where ϕ_b was determined from ideal thermionic emission.[19] $c = 0.91$ and $m = 0.34$ were constants empirically determined from the relationship $I_s = I'_s 10^{(c-n)/m}$ which gives the increase in n and in I'_s/I_s (ratio of measured saturation current/saturation current for ideal diode) with increasing magnitude of recombination current.[22] c and m thereby allow determination of the recombination correction to ϕ_b when $I_s = I'_s$ in Eq. 2. Also included in Table I were barrier heights, ϕ_b^{CV}, determined by the differential capacitance–voltage method.[24] ϕ_b^{CV} was calculated as the sum of the intercept V_i of the extrapolated $1/C^2$ versus V curve with the voltage axis, δ the difference between the conduction band edge and Fermi level and the thermal energy kT/q. Typically, plots of $1/C^2$ versus V were found to lie on a straight line as illustrated in Fig. 3.

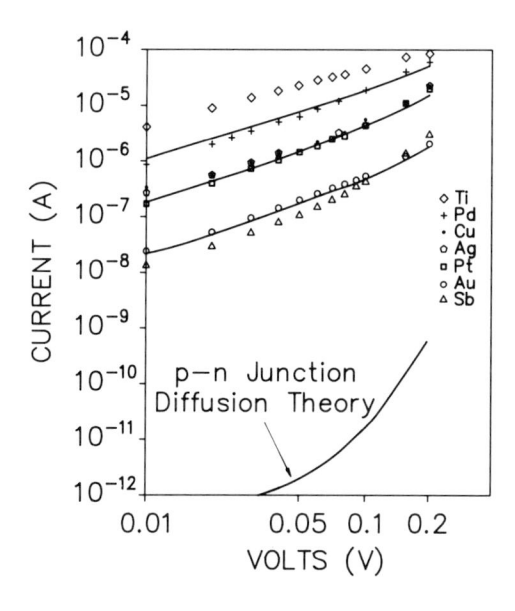

FIG. 2. Comparison of ln J vs ln V curves for various metal/n-Hg$_{0.3}$Cd$_{0.7}$Te junctions. The points were experimentally determined while the lines represent best fit with Eq. (2). The predicted curve for p-n diode diffusion theory for Au/Hg$_{0.3}$Cd$_{0.7}$Te is also shown.

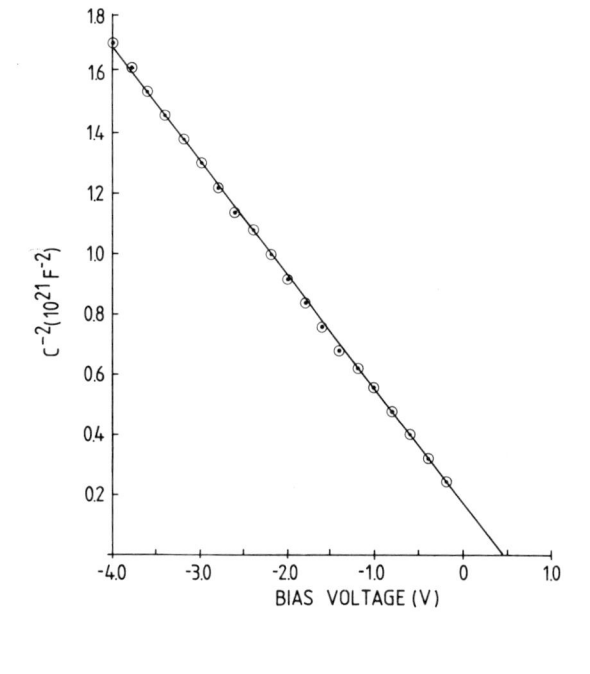

FIG. 3. Plot of $1/C^2$ vs reverse bias voltage for Pt/Hg$_{0.3}$Cd$_{0.7}$Te contact.

From Table I, equivalent values of ϕ'_b and ϕ_b^{CV} were obtained for the contact metals Pd, Pt, and Cu while C–V measurements providing higher barrier heights for the Ag and Au/Hg$_{0.3}$Cd$_{0.7}$Te systems. Larger barrier heights measured by C–V rather than I–V for Ag and Au may be associated with the sensitivity of the C–V technique to the presence of insulating layers.[19] For the contacts fabricated on I:KI:HBr etched surfaces, two broad groupings of barrier height were evident; Pd and Pt/n–Hg$_{0.3}$Cd$_{0.7}$Te contacts exhibited ϕ'_b of 0.69 V while Ag, Au, Cu, and Sb produced significantly higher values of ϕ'_b in the range 0.74–0.79 eV. Also, the effects of surface pretreatment on ϕ'_b were dependent on metal type. For Pd and Pt/n-Hg$_{0.3}$Cd$_{0.7}$Te diodes, the barrier height was essentially uneffected by pretreatment while Sb/Hg$_{0.3}$Cd$_{0.7}$Te contacts showed slightly lower ϕ'_b for air exposed than etched surfaces.

In contrast, Au/n-Hg$_{0.3}$Cd$_{0.7}$Te contacts on air exposed surfaces showed erratic and low ϕ'_b. The values of n (Table I) for diodes formed on I:KI:HBr etched surfaces ranged from Pt with $n = 1.25$ to Ag with $n = 2.23$ and Cu with $n = 2.45$; while the saturation current density J_s was lowest for Au and Sb contacts, increasing in order of Ag, Cu, Pd and Ti.

For MOCVD grown layers, Pt and Au contacts prepared on etched surfaces exhibited rectifying properties. For these contacts, the current density in forward bias was also well described by the thermionic emission-recombination model,[10] with the ideality factor n decreasing from $n = 2.0$ at $x = 0.60$ to $n = 1.38$ at $x = 0.68$ as shown in Fig. 4(a). Included in Fig. 4(a) is the value of $n = 1.25$ for the Pt contacts formed on LPE Hg$_{0.3}$Cd$_{0.7}$Te; agreement between this result and for the MOCVD layers is evident despite the differences in carrier concentration. Within the range $x = 0.60$–$x = 0.68$, the barrier height was determined from the ln J-V plots as $\phi'_b = 0.70 \pm 0.01$ eV, with the recombination correction decreasing from $\Delta\phi_{rec} = 0.16$ V at $x = 0.6$

TABLE I. Diode characteristics for metal/n-Hg$_{0.3}$Cd$_{0.7}$Te (111) contacts prepared on Br:HBr, I:KI:HBr etched or air-exposed surfaces.

Metal	ϕ'_b (V) Br:HBr	ϕ'_b (V) I:KI:HBr	ϕ'_b (V) Air exp	n I:KI:HBr	ϕ_b^{CV} (V) I:KI:HBr	J_s (A cm^{-2}) I:KI:HBr	$\phi m(V)$
Au	0.79	0.79	0.50	1.67	0.893	7.0×10^{-6}	5.1
Sb	0.74	0.74	0.72	1.49	...	5.0×10^{-6}	4.7
Pt	0.69	0.69	0.70	1.25	0.693	2.0×10^{-5}	5.7
Pd	0.69	0.695	0.69	1.85	0.706	3.0×10^{-4}	5.12
Ag	...	0.74	...	2.23	0.787	2.0×10^{-4}	4.26
Cu	...	0.74	...	2.45	0.737	2.0×10^{-4}	4.65
Ti	...	Near ohmic	...	3.10	...	2.0×10^{-3}	4.33

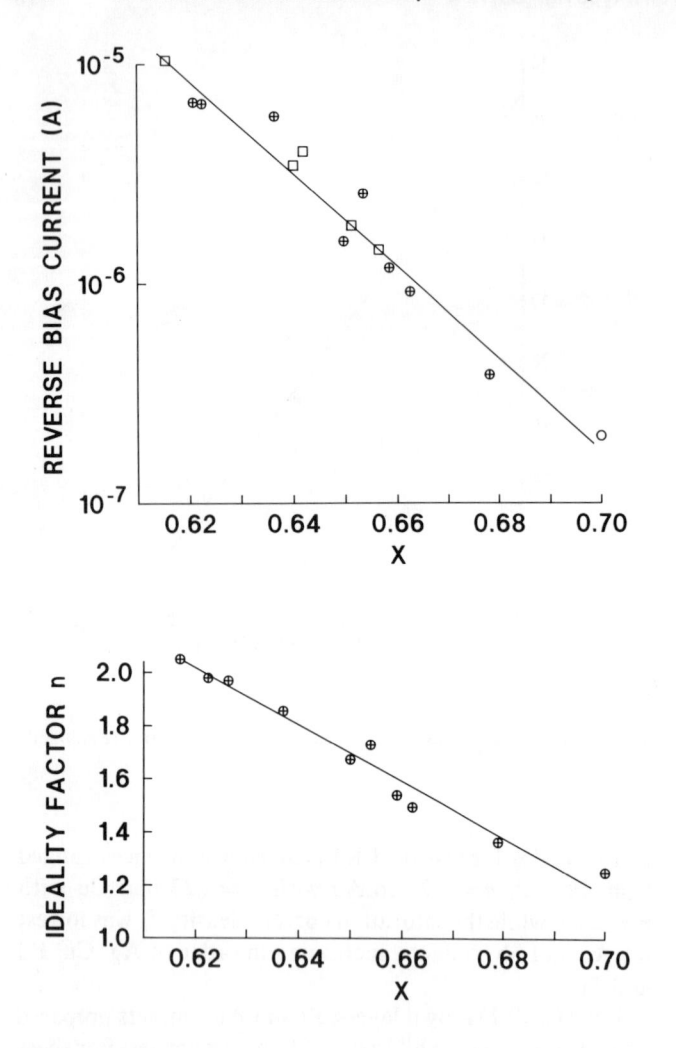

FIG. 4. Reverse leakage current and ideality factor n as a function of x ratio for ○ Pt and □ Au diodes on MOCVD $Hg_{1-x}Cd_xTe$ (2° off 100).

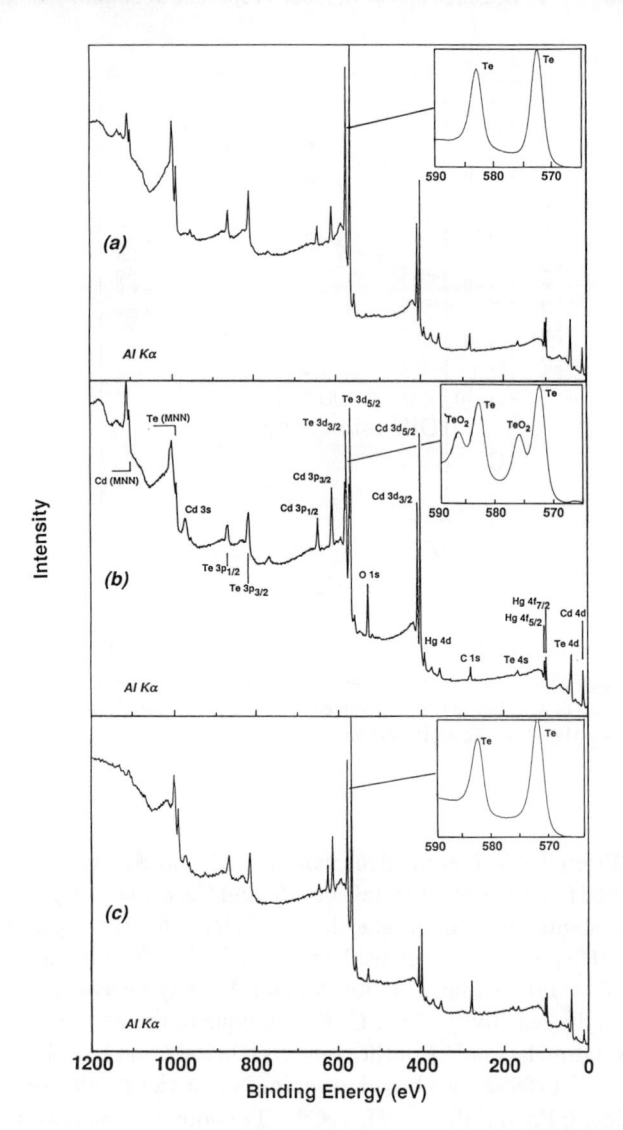

FIG. 5. XPS Spectra of (111) $Hg_{0.3}Cd_{0.7}Te$ surfaces after (a) argon ion bombardment, (b) air exposure for 12 h, and (c) etch in I:KI:HBr solution.

to $\Delta\phi_{rec} = 0.08$ V at $x = 0.68$. Increasing x value also correlated with a reduction in reverse saturation current density, J_0, as shown in Fig. 4(b). A similar reduction in J_0 with x was found for diodes formed by both Au and Pt contacts on n-$Hg_{1-x}Cd_xTe$ with the results also consistent with J_0 measured for the Pt contacts on the LPE(n-$Hg_{0.3}Cd_{0.7}Te$).

B. XPS analysis and Auger depth profiles

XPS has been used in this work to provide a semi-quantitative assessment of the surfaces studied. Relative sensitivity factors were obtained from the literature and verified by the use of standard samples. Figure 5 shows XPS Spectra of (111) $Hg_{0.3}Cd_{0.7}Te$ surfaces prepared by argon-ion bombardment (to remove surface contaminants), air exposure and chemical etching. The insert presented in each spectrum is an expansion of the Te 3d region, which is a sensitive indication of surface oxidation. A chemical shift is experienced by Te on oxidation due to its altered environment, and this is reflected in a movement to a higher binding energy of the Te peaks as seen in Fig. 5(b). Spectrum 5(c) shows the total removal of this oxide by I:KI:HBr as well as a substantial modification in surface stoichiometry. Note that a similar effect was obtained using Br:HBr. The surfaces prepared by these chemical etchants show a depletion in both Cd and Hg, with Hg:Cd:Te ratios of 0.2:0.4:1.0 for Br:HBr and 0.2:0.3:1.0 for I:KI:HBr.

For the Auger depth profiles, an NBS Ta_2O_5/Ta standard was used in conjunction with Tables of relative elemental etch rates to determine the sputter etch rates of the materials investigated. Relative Auger sensitivity factors were also determined for the individual elements from examination of standard materials. Due to the overlap of the Auger peaks corresponding to Pt, Au, and Hg, (peaks centered at approximately 64, 69, and 76 eV, respectively), the calculated profiles may indicate slightly more Pt or Au present deep into the material than is actually the case. This overlap is much more prevalent in the case of Au/Hg, although the shape of the profile should not be unduly affected. Figure 6(a) shows an Auger sputter profile of the as-deposited interface of 20 Å Pt/$Hg_{0.3}Cd_{0.7}Te$. Both Te and Cd were present within the deposited layer but without evidence of O, C or other contaminants. At the interfacial region (Pt etch

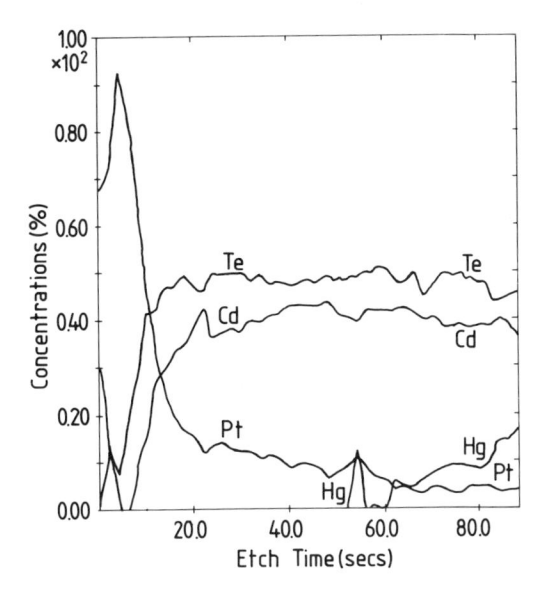

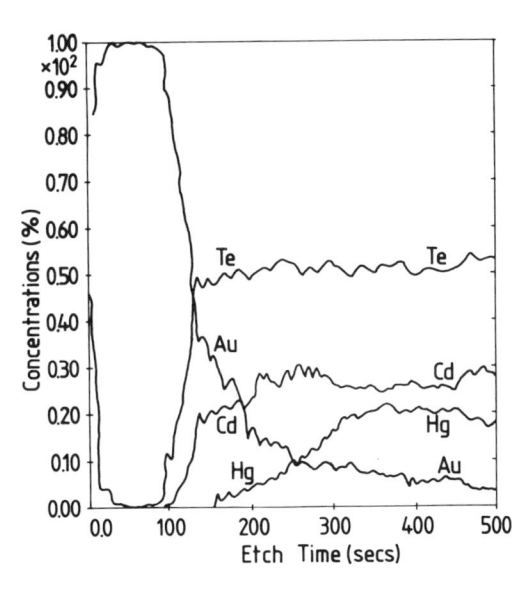

FIG. 6. Auger depth profiles of (a) Pt/Hg$_{0.3}$Cd$_{0.7}$Te and (b) Au/Hg$_{0.3}$Cd$_{0.7}$Te interfaces.

rate = 1.85 Å s^{-1}, interface width 28 Å), a substantial intermixing of Pt with the Hg$_{0.3}$Cd$_{0.7}$Te was indicated, with some associated loss of Cd. All of the profiles examined also showed a significant Hg loss in the region immediately adjacent to the interface as shown in Fig. 6(a). These features such as the Hg loss and Te movement into the overlayer are consistent with the Pt cation alloying observed in p-Hg$_{0.7}$Cd$_{0.3}$Te.[13] An Auger sputter profile which was typical of the Au/Hg$_{0.3}$Cd$_{0.7}$Te interface is shown in Fig. 6(b) for 200 Å thickness. No significant amounts of Cd and Te were evident within the Au overlayer while the presence of Te at the free surface was indicated. For Au/Hg$_{0.3}$Cd$_{0.7}$Te, the profiles showed the formation of a broader interface (Au etch rate = 2.8 Å s^{-1}, interface width 140 Å) than for Pt; characterized by the movement of Au to a significant depth into the Hg$_{0.3}$Cd$_{0.7}$Te. The loss of Hg from the interfacial region was comparatively minor. Again, these features were in general agreement with previous studies performed by photoemission spectroscopy of the Au/Hg$_{0.8}$Cd$_{0.2}$Te interface.[25]

IV. DISCUSSION

The J–V characteristics of the metal/n-Hg$_{1-x}$Cd$_x$Te ($x = 0.6$–0.7) contacts prepared on chemically etched surfaces have been modelled by thermionic emission theory modified for the effects of recombination in the depletion region. The junctions with characteristics nearest to ideal thermionic emission were formed by Pt, Sb, and Au with $n = 1.25$, 1.49, and 1.67, respectively. Sb and Au contacts also exhibited the highest diode resistance and J_s values correlating with the more abrupt interfaces formed with Hg$_{1-x}$Cd$_x$Te by these unreactive metals. The low value of n for Pt contacts may result from the formation of a strongly reacted but narrow interface, as indicated by the Auger depth profiles. Friedman et al.[13] have suggested that the reaction at the Pt/Hg$_{1-x}$Cd$_x$Te interface reaches completion at low metal coverages due to the formation of a layer of Pt–Cd compound which prevents further inward movement of the deposited Pt. Within the other metals, an increasing reactivity (Cu, Ag, Pd, and Ti) with Hg$_{1-x}$Cd$_x$Te correlated with lower diode resistance, and higher J_s and n values. These results suggest a higher interface recombination during current transport with increasing extent of interfacial disorder. Stoichiometric x ratio also exerted a strong influence on n value attributed to an increasing recombination contribution to current with decreasing x. The change in the value of intrinsic carrier concentration from $n_i = 1 \times 10^{11}$ cm^{-3} at $x = 0.6$ to $n_i = 3 \times 10^{12}$ cm^{-3} for $x = 0.7$ at 293 K (Ref. 27) can account for the increase in both the recombination current from Eq. (3) and in reverse saturation current from the equation $I_s = Sq\, n_i\, w/2T_r$ (Ref. 24) with variation in these parameters from $x = 0.6$ to 0.7.

The measured barrier heights for the metal/Hg$_{1-x}$Cd$_x$Te junctions on etched surfaces were essentially independent of metal work function (Table I). The grouping of Ag, Au, Cu, and Sb with $\phi'_b = 0.74$–0.79 V was comprised of the unreactive and intermediate reactivity metals. For these metals, only a partial or incomplete reaction with the etch-induced layer (Cd:Te rich and imperfect in crystallinity) may occur. The barrier height $\phi'_b = 0.74$–0.79 V may then be attributed to Fermi level pinning associated with surface states created by the etch-induced layers. A pinning of barrier height in the range $\phi'_b = 0.70$–0.74 V was previously reported for contacts of low reactivity metals on chemically etched n-CdTe.[9] But in Hg$_{1-x}$Cd$_x$Te, the role of acceptor doping of the interfacial region on electrical contact properties requires further investigation. Differences in the Cd:Te ratio due to the two etchants had little effect on ϕ'_b, but the more stoichiometric interfaces formed by Br:HBr resulted in slightly lower values of diode resistance and J_s than for I:KI:HBr. In contrast, reactive metals may effectively consume the etch induced layer but create other defects as a result of the interaction during deposition. Spicer et al. have proposed that the native defect dominant in determining E_{fi} for both Hg$_{1-x}$Cd$_x$Te and CdTe was the Te$_{Cd}$ antisite.[7] Based on a movement of the valence band maximum of 0.35 eV from $E_{fi} = 0.75$ eV for

TABLE II. Heats of formation of metal oxides (Ref. 29)

Oxide	ΔH_f (kcal/mol)
Ag_2O	-7.42
Ag_2O_2	-5.80
Au_2O_3	-0.77
Cu_2O	-37.6
CuO	-40.3
PdO	-20.4
Pt_3O_4	-39.0
PtO_2	-41.0
Sb_2O_4	-216.9
Sb_2O_5	-232.3
TiO	-124.2
Ti_2O_3	-363.5
TeO_2	-72.3

CdTe to $Hg_{1-x}Cd_xTe$,[7] by extrapolation, the magnitude of E_{fi} for the Te_{Cd} antisite at $x = 0.7$ equals $\simeq 0.66$ eV.

For diodes prepared on air-exposed surfaces, the electrical properties were determined both by stability of the metal oxide and by the reactivity of the metal/$Hg_{1-x}Cd_xTe$ interface. For Sb, with large heat of oxide formation (Table II) but unreactive with respect to $Hg_{1-x}Cd_xTe$, J–V characteristics of Sb diodes prepared on air oxidized surfaces were closer to unity ($n = 1.42$) than diodes on etched surfaces ($n = 1.49$), consistent with a strong reduction of TeO_2 without disruption of the interface. In comparison, the inability of Au either to reduce TeO_2 (Table II) or to react appreciably with $Hg_{1-x}Cd_xTe$ correlates with a strong effect of air oxidation in both increasing the ideality factor ($n = 3.5$) and reducing barrier height ($\phi'_b = 0.5$ V). Similar effects of a thin oxide layer at Au/p-$Hg_{1-x}Cd_xTe$ junctions have been previously attributed to a reduction in the density of interface states in the presence of the layer; with the oxide acting as a diffusion barrier for Au.[26] Sensitivity to the effects of interfacial oxide may also be expected for Ag and Cu contacts with low heat of oxide formation (Table II) and intermediate reactivity with $Hg_{1-x}Cd_xTe$. Pt also possesses a low heat of oxide formation (Table II) but forms a reactive interface with $Hg_{1-x}Cd_xTe$. The equivalence of the J–V characteristics for diodes formed on etched and air-exposed surfaces indicates that the reactivity of Pt at the interface was sufficiently large to disrupt any TeO_2 films.

In conclusion, the formation of Schottky diodes on n-$Hg_{1-x}Cd_xTe$ $x \geqslant 0.4$ predicted in theory has been experimentally demonstrated. The I–V characteristics of the contacts prepared on chemically etched surfaces have been described by a thermionic emission model modified by the effects of recombination in the depletion region. For contacts on etched surfaces, I–V and C–V measurements have shown that the barrier height produced by the group of metals Ag, Au, Cu, and Sb was in the range $\phi'_b = 0.74$–0.79 V while for the metals Pd and Pt, $\phi'_b = 0.69$ V. Diode resistance also decreased with chemical reactivity of the metal. The lowest n values associated with the metals Pt, Sb, and Au and with increasing x ratio were attributed to a minimum interface recombination during current transport. For contacts on air-exposed surfaces, the diode characteristics were dependent on the ability of the metal overlayer to reduce interfacial TeO_2 or to disrupt the interface by reaction with the $Hg_{1-x}Cd_xTe$. These results enable the fabrication of devices including MESFETs and MSM photodetectors in $Hg_{1-x}Cd_xTe$ for the realization of optoelectronic integrated circuits. Other results[30] have shown that the formation of Schottky barriers on $Hg_{1-x}Cd_xTe$ only occurred at stoichiometries of $x \geqslant 0.5$

ACKNOWLEDGMENTS

The authors wish to thank T. Rogers for performing the EBIC measurements and S. Li for assistance in device modelling. The MOCVD grown material was provided by Dr. G. Pain. The permission of the Executive General Manager, Research, of Telecom Australia to publish this paper is acknowledged.

[1] R. Tribolet, T. Nguyen Duy, and A. Durand, J. Vac. Sci. Technol. A 3, 95, (1985).
[2] B. Orsal, R. Alabedra, M. Valenza, G. Lecoy, J. Meslage, and C. Y. Boisrobert, IEEE Trans. Electron Devices ED-35, 101, (1988).
[3] J. Thompson, P. Mackett, G. T. Jenkin, T. Nguyen Duy, and P. Gori, J. Cryst. Growth 86, 917, (1988).
[4] L. M. Smith, J. Thompson, G. T. Jenkin, T. Nguyen Duy, and P. Gori, SPIE 866, 163 (1987).
[5] R. A. Schiebel, J. Dodge, and R. Gooch, Electron. Lett. 25, 530, (1989).
[6] L. Yang, A. Sudbo, W. T. Tsang, P. A. Garbinski, and R. M. Camarda, IEEE Photon. Technol. Lett. 2, 59 (1990).
[7] W. E. Spicer, D. J. Friedman, and G. P. Carey, J. Vac. Sci. Technol. A 6, 2746 (1988).
[8] P. W. Leech, J. Appl. Phys. 68, 907 (1990).
[9] I. M. Dharmadasa, A. B. McLean, M. H. Patterson, and R. H. Williams, Semicond. Sci. Technol. 2, 404 (1987).
[10] I. M. Dharmadasa, W. G. Herrenden, and R. H. Williams, Appl. Phys. Lett. 48, 1802 (1986).
[11] W. E. Spicer, J. Vac. Sci. Technol. A 8, 1174 (1990).
[12] G. D. Davis, J. Vac. Sci. Technol. A 6, 1939 (1988).
[13] D. J. Friedman, G. P. Carey, I. Lindau, and W. E. Spicer, Phys. Rev. B 35, 1188 (1987).
[14] J. Brice and P. Capper, *Properties of Mercury Cadmium Telluride* (INSPEC, The Institution of Electrical Engineers, London, 1987), p. 79.
[15] P. W. Leech, M. H. Kibel, and P. J. Gwynn, J. Electrochem. Soc. 137, 705 (1990).
[16] G. N. Pain et al., J. Vac. Sci. Technol. A 8, 1067 (1990).
[17] D. L. Polla and A. K. Sood, J. Appl. Phys. 51, 4908 (1990).
[18] M. H. Kibel, Proceedings of the II–VI Semiconductors Conference, Adelaide, 1989 (unpublished).
[19] E. H. Rhoderick, *Metal–Semiconductor Contacts* (Clarendon, Oxford, 1978), p. 84.
[20] F. A. Padovani and R. Stratton, Solid State Electron. 9, 695 (1966).
[21] C. T. Sah, R. N. Noyce, and Shockley, Proc. IRE 45, 1228 (1957).
[22] A. B. McLean, I. M. Dharmadasa, and R. H. Williams, Semicond. Sci. Technol. 1, 137, (1986).
[23] F. L. Madarasz, F. Szmulowicz, and J. R. McBath, J. Appl. Phys. 58, 361 (1985).
[24] E. H. Rhoderick, *Metal–Semiconductor Contacts* (Clarendon, Oxford, 1978), p. 131.
[25] G. D. Davis, W. A. Beck, M. K. Kelly, N. Tache, and G. Margaritondo, J. Appl. Phys. 60, 3157 (1986).
[26] V. Krishnamurthy, A. Simmons, and C. R. Helms, Appl. Phys. Lett. 56, (1990).
[27] F. L. Madarasz, F. Szmulowicz, and J. R. McBath, J. Appl. Phys. 58, 361 (1985).
[28] J. Werthen, J.-P. Haring, A. L. Fahrenbruch, and R. H. Bube, J. Appl. Phys. 54, 5982 (1983).
[29] *Handbook of Physics and Chemistry*, 69th ed. (Chemical Rubber, Boca Raton, FL, 1988).
[30] C. G. Kelly and P. W. Leech (to be published).

Uncooled 10.6 μm mercury manganese telluride photoelectromagnetic infrared detectors

P. Becla
Massachusetts Institute of Technology Cambridge, Massachusetts 02139

N. Grudzien[a] and J. Piotrowski[a]
Boston Optrocics Corporation Brookline, Massachusetts 02146

(Received 2 October 1990; accepted 18 December 1990)

Photoelectromagnetic PEM detectors for 10.6 μm utilizing $Hg_{1-x}Mn_xTe$ alloys have been designed and fabricated. Numerical and experimental analysis of the PEM effect has been performed. The normalized responsivity R_v detectivity D^* and time constant τ were calculated using Lile's generalized theory of the PEM effect. The responsivity ($R_v = 0.1$ V/W) and detectivity ($D^* = 1.2 \times 10^7$ cm Hz$^{1/2}$ W^{-1}) were achieved with p-type of $Hg_{1-x}Mn_xTe$ having the composition $x \sim 0.08$ and acceptor concentration N_D of about 2×10^{17} cm^{-3}. The experimental values of R_r and D^* are a factor of 3 below the calculated values.

I. INTRODUCTION

$Hg_{1-x}Cd_xTe$ (MCT) has long been the material of choice for fabrication of various infrared devices. Currently, $Hg_{1-x}Mn_xTe$ (MMT) with comparable electronic properties, may be a potential competitor. In MCT alloys, Cd destabilizes weak Hg–Te bonds.[1,2] This leads to low mechanical strength[3] and longterm changes in transport properties.[4] Manganese as well as zinc is reported to strengthen the CdTe[5,6] and HgTe[2] lattices. Thus, Mn added to Hg–Te may stabilize the bonds, resulting in improved structural and mechanical properties. Furthermore, at room temperature (300 K), the semiconducting properties of MMT are expected to be similar to those of $Hg_{1-x}Cd_xTe$.[7]

The $Hg_{1-x}Mn_xTe$ and $Hg_{1-x}Cd_xMn_yTe$ alloys have been used for fabrication of photovoltaic detectors,[8,9] light emitting diodes (LEDs) and laser heterostructure.[10,11] It has been reported[8] that both the current–voltage characteristics and R_0A products are superior in MMT detectors to those of MCT detectors with similar bandgaps. A strong electroluminescence observed from MMT p–n junctions[10,11] suggested that this material may be better for some infrared applications.

In this article we report on the preparation and performance of the uncool photoelectromagnetic (PEM) detectors for 10.6 μm wavelength range utilizing bulk $Hg_{1-x}Mn_xTe$ (0.08 > x > 0.9) single crystals grown by the traveling heat method (THM).[12] Our work consists of theoretical and experimental parts. A generalized model was used to calculate the voltage responsivity R_v, resistivity R_0, and response time τ_r of the MMT/PEM detectors.[13] Detectivity was calculated from the standard photoelectric equation[14] assuming that the noise (V_N) of the PEM detector is limited by the thermal Johnson–Nyquist noise.

The experimental part focuses primarily on detector preparation and various (electrical, optical, and photoelectrical) characterizations. The experimental and modeling results are anticipated to establish the maximum performance of the MMT/PEM detectors, and provide a rational basis for comparison with similar detectors made of $Hg_{1-x}Cd_xTe$.

II. THEORETICAL DESIGN

Figure 1 illustrates the PEM effect in semiconductors. The PEM effect is caused by an in-depth diffusion of photogenerated carriers whose trajectories are deflected in a magnetic field. The driving force for diffusion is the gradient of carrier concentration. Typically, the gradient is caused by the nonuniform absorption of radiation in the semiconductor. In the narrow gap semiconductors, which are suitable for room temperature long wavelength photodetectors, the ambipolar diffusion length is small ($L_d \sim 2$ μm) while the absorption depth of 10.6 μm radiation is much larger ($1/\alpha \sim 10$ μm). In such a case, the radiation is almost uniformly absorbed within the diffusion length. There are several ways to achieve good photoresponse in such a case. One of them is to use a device with a low recombination velocity at the front side and a high recombination velocity at the back side.

Lile[13] developed a generalized model of PEM effect and derived analytical expressions for the PEM photovoltage. The Lile model is based on assumptions of the homogeneity of the semiconductor, steady state conditions, weak optical excitation, nondegenerate statistics, negligibility of interference and edge effects, independence of material properties on magnetic field and exponential absorption of radiation ($r_2 = 0$). The voltage responsivity of the PEM detector can be expressed by[13]

$$R_v = \frac{\lambda}{hc}\frac{B}{wd}\frac{\alpha z(b+1)}{n_i(b+z^2)}\frac{Z(1-r_1)}{Y(a^2-\alpha^2)}, \quad (1)$$

where

$$Z = [(\alpha - s_2 a^2\tau) + (\alpha + s_1 a^2\tau)e^{-\alpha d}]$$
$$\times ch(ad) - a[(1 - s_2\alpha\tau) - (1 + s_1\alpha\tau)e^{-\alpha d}]$$
$$\times sh(ad) - [(\alpha + s_1 a^2\tau) + (\alpha - s_2 a^2\tau)e^{-\alpha d}], \quad (2)$$

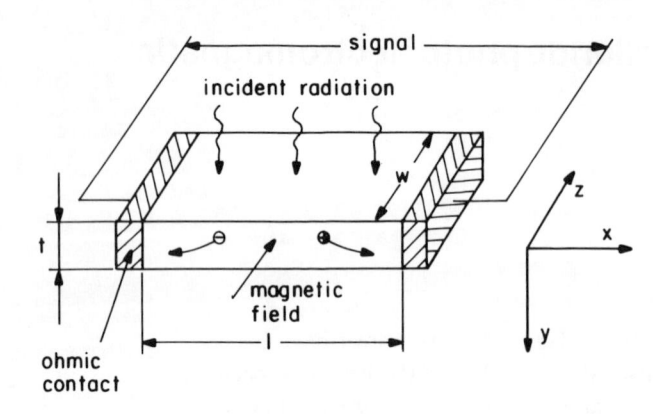

$$Y = a[a\tau(s_1 + s_2)ch(ad) + (1 + s_1s_2a^2\tau^2)sh(ad)]$$

$$- \frac{b(b+1)z^2\mu_h^2 B^2}{a^2(a^2-\alpha^2)L_e^2 d(1+a^2)(b+z^2)n_i^2}$$

$$\times \{(a^2-\alpha^2)n_i^2 2 - [2 - a^2 d\tau(s_1+s_2)ch(ad)]$$

$$- a[\tau(s_1+s_2) - d(1+s_1s_2a^2\tau^2)sh(ad)]\}. \quad (3)$$

a is the reciprocal diffusion length in the magnetic field.

$$a = \left[\frac{(1+\mu_n^2 B^2)z^2 + b(1+\mu_h^2 B^2)}{L_e^2(z^2+1)}\right]^{1/2}. \quad (4)$$

The remaining symbols have the usual meaning: λ is the wavelength, B is the magnetic field, r_1 is the frontside reflection coefficient, h is the Planck constant, c is light velocity, b is the electron to hole mobility ratio, z is the hole to intrinsic concentration ratio, τ is the recombination time, α is the absorption coefficient, μ_e and μ_h are the electron and hole mobilities, and L_e is the electron diffusion length.

The resistivity of the PEM detector is

$$R = \frac{zl}{qn_i\mu_h(b+z^2)wd}\left[1 - \frac{b(b+1)^2 z^2 \mu_h^2 B^2}{a^2 L_e^2(1+z^2)(b+z^2)}\right]^{-1}. \quad (5)$$

If the backside reflection coefficient, $r_2 > 0$, multiple reflections occur and the resulting voltage responsivity can be calculated by summing the contributions from the consecutive radiation passes.

As no bias voltage is applied, the noise of PEM detectors is essentially the thermal Johnson–Nyquist noise.

$$V_J = (4kTR\Delta f)^{1/2}, \quad (6)$$

where k is the Boltzmann constant and Δf is the frequency band. The detectivity of the PEM detectors can be calculated from the definition as

$$D^* = \frac{R_v(A\Delta f)^{1/2}}{V_j}. \quad (7)$$

As in the case of a Dember detector, the response time τ_r of a PEM detector is the time required to establish the equilibrium concentration and is shorter than the bulk recombination time τ. For devices with high recombination velocity at the back surface, it can be expressed[15] as

$$\tau_r = \frac{\tau d^2}{(2D\tau + d^2)} \quad (8)$$

and

$$\tau = \frac{n_0\tau_0 + p_0\tau_p}{n_0 + p_0}, \quad (9)$$

where D is the ambipolar diffusion coefficient.

For calculations of performance of the MMT PEM detectors, we used the energy gap-composition dependence:[17]

$$E_g[eV] = -0.303 + 3.903x + 10^{-4} \times (5.17 - 14.803x) \quad (10)$$

and for the intrinsic concentration:[18]

$$n_i[m^{-3}] = (2.325 + 42.69x + 0.007698T$$
$$- 0.07346xT - 102.6x^2)$$
$$\times 10^{20} E_g^{3/4} T^{3/2} \exp(-E_g/2kT). \quad (11)$$

The absorption coefficient for optical carrier generation can be calculated according to Anderson expressions,[19] which were derived within the Kane model, including the Moss–Burstein shift. The Auger mechanism is a fundamental band-to-band generation-recombination process which determines the recombination lifetime in narrow gap MMT at room temperature. The Auger recombination lifetime has been calculated according to Casselman and Petersen,[20,21] taking the overlap integrals equal to 0.2.

Figure 2 presents calculated data for R_0, R_v, D^*, and τ_r of an uncooled 10.6 μm MMT/PEM detector as a function of the acceptor concentration N_D. Of interest is the region where detectivity D^* reaches a maximum (i.e., the value of about 2.3×10^7 cm Hz$^{1/2}$ W^{-1}). This value is achievable at the acceptor concentration of about 2.2×10^{17} cm^{-3}. At this point the responsivity, resistivity and response time are 0.27 V/W, 90 Ω, and 60 ns, respectively. Further increase in the acceptor concentration slightly increases responsivity, but decreases detectivity due to an increase at the resistivity and the noise.

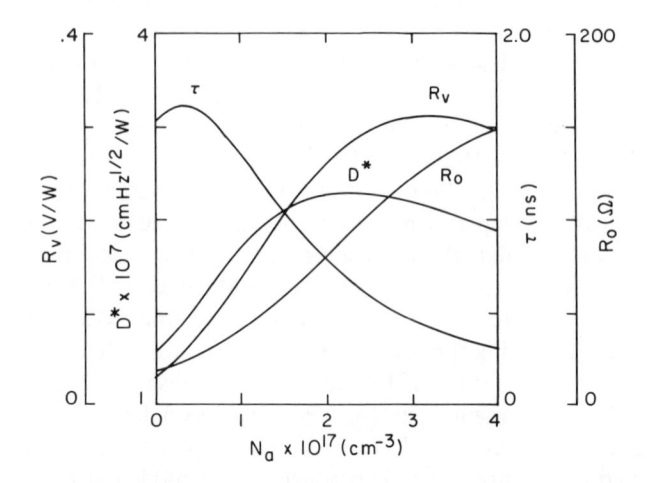

FIG. 2. Normalized resistivity, responsivity, detectivity, and response time of uncooled 10.6 μm MMT/PEM detectors as a function of acceptor concentration.

III. FABRICATION AND MEASURED PERFORMANCE

The preparation of sensitive elements of MMT/PEM detectors was similar to the manner described in Refs. 14–16.

The bulk crystals of $Hg_{0.92}Mn_{0.08}Te$ have been grown by the THM method. The native acceptor concentration has been adjusted at a level of about $2-3\times10^{17}$ cm^{-3} by the post-grown annealing of the wafers in a mercury saturated atmosphere. The wafers have been epoxied to silicon substrates, thinned to final thickness of about 5 μm and cut into bars of about 1 mm $\times$ 2 mm. Electrical contacts were made by gold electroplating to the end of the HgMnTe bars to which the gold wires were attached. The typical size of active areas was 1 mm^2. The sensitive elements were then mounted in the narrow slot of the miniature permanent magnet having a field strength of about 1.5 T. The devices have been characterized by the measurement of the resistivity R_0, voltage responsivity R_v, and the time constant τ_r. The measurements of R_v and τ_r were performed typically with a chopped CO$_2$ laser beam. In a few cases, spectral response, $R_v = f(h\nu)$, has also been measured. Detectivity has been determined by taking the measured responsivity. The PEM detector noise has been measured at zero bias voltage and was found to be equal to the value determined from Eq. (6).

Figure 3 shows the experimental spectral responses, R_v and D^*, of the MMT PEM detectors. Due to a high level of doping, the spectral characteristics are very selective (the spectral half width is about 6 μm). The best performance, i.e., the highest responsivity and detectivity at 10.6 μm wavelength, was achieved using $Hg_{1-x}Mn_xTe$ with composition x of about 0.08–0.09. At that composition range, the peak sensitivity of the detectors was in the region of 7–8 μm. In both cases, the reduction or increase of the composition of $Hg_{1-x}Mn_xTe$ lead to a reduction of the responsivity and detectivity at the 10.6 μm wavelength region. See for example R_v' and R_v'' on Fig. 3.

Typical experimental values of responsivity, resistivity, detectivity, and response time of the MMT PEM detectors optimized for 10.6 μm wavelength are in the range of 0.1 V/W, 70 Ω, 1.2 cm Hz$^{1/2}$ W^{-1} and 60 ns, respectively.

Experimental values of responsivity and detectivity are lower about 3.0 times from the theoretical values. The reason for that is the imperfection in the composition x uniformity of doping and the construction of the detectors.

The MMT PEM detector has been tested using CW and pulse radiation. The maximum signal voltages for low frequency (10 Hz–1 kHz) square chopped CO$_2$ radiation exceed 30 mV per 1mm detector length. When very short pulses were used, the signal voltage up to 1 V per 1 mm length of detector was obtained. The damage threshold of the detector and its longterm stability are under investigation.

IV. SUMMARY

Single crystals of $Hg_{1-x}Mn_xTe$ solid solution of well controlled composition, purity and morphology were obtained by the THM method. Uncooled PEM detectors optimized for 10.6 μm were designed and constructed. Lile's theoretical model was used to optimize the spectral characteristics (R_v and D^*), the resistivity R_0, and response time τ_r. The measured photoelectric characteristics confirmed the expectations of Lile's theoretical model. Both the theoretical and the measured performance of the PEM detectors are inferior to that of PC detectors. PEM detectors, however, have additional important advantages which make them useful in many applications. In contrast to photoconductors, they do not require electrical biasing, the noise is very low, and the frequency characteristics are flat in a wide frequency region, starting from cw to 1 GHz.

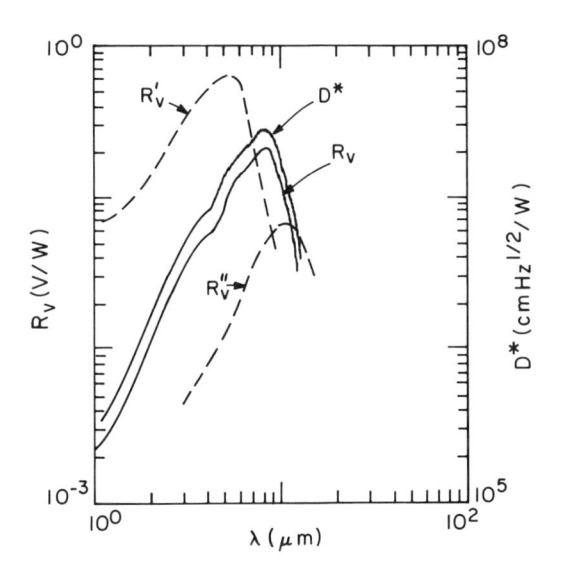

FIG. 3. Measured spectral responsivity and detectivity of uncooled 10.6 μm MMT/PEM detectors; R_v ($x = 0.08$); R_v' ($x = 0.1$); R_v'' ($x = 0.07$).

ACKNOWLEDGMENTS

This work was supported by the National Science Foundation through Contract No. CTS890956.

[1] A. Wall, C. Caprile, A. Franciosi, R. Reifenberg, and U. Debska, J. Vac. Sci. Technol. A **4**, 818 (1986).
[2] A. Shar, A. B. Chen, W. E. Spicer, and C. K. Shih, J. Vac. Sci. Technol. A **3**, 105 (1985).
[3] W. E. Spicer, J. A. Silberman, I. Lindan, A. B. Chen, A. Sher, and J. A. Wilson, J. Vac. Sci. Technol. A **1**, 1735 (1983).
[4] G. Nimc, B. Schicht, and R. Dornhaus, Appl. Phys. Lett. **34**, 490 (1979).
[5] S. B. Qadri, E. F. Skelton, A. W. Webb, and S. Kennedy, Appl. Phys. Lett. **46**, 257 (1985).
[6] S. B. Qadri, E. F. Skelton, A. W. Webb, and S. Kennedy, J. Vac. Sci. Technol. A **4**, 1971 (1986).
[7] J. K. Furdyna, J. Vac. Sci. Technol. **21**, 220 (1982).
[8] P. Becla, J. Vac. Sci. Technol. A **4**, 2014 (1986).
[9] S. Wong and P. Becla, J. Vac. Sci. Technol. A **4**, 2019 (1986).
[10] P. Becla, J. Vac. Sci. Technol. A **6**, 2725 (1988).
[11] P. Becla, U. S. Patent, Patent and Trademark Office, September 1987, Serial No. 100119.
[12] P. Becla (to be published).
[13] D. L. Lile, Phys. Rev. B **8**, 4708 (1973).
[14] D. Genzow, M. Grudzien, and J. Piotrowski, Infrared Phys. V **20**, 133 (1980).

[15] J. Piotrowski, W. Galus, and M. Grudzien, Infrared Phys. (in press).
[16] D. Genzow, A. Jozwikowska, K. Jozwikowski, T. Niedziela, and J. Piotrowski, Infrared Phys. **24**, 21 (1984).
[17] N. D. Gavaleshko, P. N. Gorley and V. A. Shendirovski, *Narrow Band Gap Semiconductors: Technology and Physical Properties*, edited by Naukova Dumka (Naukova, Dumka, Kijev, 1984).
[18] A. Rogalski and J. Rutkowski, Infrared Physics **29**, 887 (1989).
[19] W. W. Anderson, Infrared Phys. **20**, 363 (1980).
[20] T. N. Casselman, J. Appl. Phys. **52**, 848 (1981).
[21] P. E. Peterson, Semiconductors Semimetals **18**, 121 (1981).

Novel device concept for silicon based infrared detectors

G. Scott, D. E. Mercer,[a] and C. R. Helms
Stanford Electronics Laboratories, Department of Electrical Engineering, Stanford University, Stanford, California 94305

(Received 3 October 1990; accepted 18 December 1990)

We propose a novel device concept for infrared detectors which utilizes a thin film of a narrow gap semiconductor deposited on a p-Si substrate. Such a device would retain the advantages of using a Si-based technology, while operating with higher quantum efficiency. The operation of the device would be similar to a Schottky barrier detector, in that carriers would be excited from the overlayer into the substrate. However, the absence of undesired states near the Fermi level should lead to more efficient generation and transport of photoexcited carriers. Calculations using a recently developed diffusion model for the behavior of Schottky barrier detectors indicate the potential for an order of magnitude improvement in quantum efficiency over PtSi/p-Si detectors at a wavelength of 4 μm.

I. INTRODUCTION

The use of Si Schottky barriers for photodetection, especially in the infrared, is well established. Although quantum efficiencies are typically quite low (a few percent) extremely uniform detector arrays using Si technology are possible leading to superior performance for cases which are limited by so-called fixed pattern noise. Of course, overall system performance will still benefit from gains in individual pixel quantum efficiency and considerable recent effort has been spent to increase the quantum efficiency of PtSi/p-Si detectors which have an effective threshold of ~5 μm.

There are four factors which limit the quantum efficiency of these types of detectors. First, optimum performance is achieved for PtSi thicknesses < 50 Å even though the absorption depth ($1/\alpha$) is approximately 300 Å. Thus, typically 80% of the available photons are lost. Second, even for photons that produce excitations in the PtSi a substantial fraction produces carriers below threshold which cannot be collected. The third and fourth factors lead to the requirement of very thin active layers. They are associated with inelastic mean free path of the excited carriers and the small transmission probability for carriers over the barrier. This limits the thickness of the active layer to on the order of only the product of the mean free path and the transmission probability which has been shown to be on the order of tens of angstroms. For thicker films more photons can be absorbed but the overall efficiency is reduced due to inelastic scattering.

An ideal detector similar to the Schottky structure could be fabricated by replacing the Schottky contact with a narrow-gap semiconductor (NGS) leading to the heterostructure shown in Fig. 1. In this idealization the n-type doping of the low bandgap semiconductor is assumed large enough so that any band bending would be negligible. In addition, as the diagram indicates the band lineups would need to provide for the appropriate Si barrier and the ideal bandgap would be in a range from near threshold to 1/2 that value to minimize inelastic processes. A similar structure was proposed by Spratt et al.,[1] but our device would utilize a bandgap less than threshold to maximize absorption in the overlayer. Comparing the performance of this structure to the metal gate structure just discussed the potential performance gains are clear. First, as will be discussed in more detail below, expected transmission probabilities and inelastic mean free paths will be considerably higher leading to better efficiency. This also provides for optimum performance for much thicker films, larger overall absorbance, and in addition, subthreshold excitations will be reduced due to the "gate" bandgap.

To this point, one might consider this all to be an intellectual exercise since it is unlikely that an 0.1–0.2 eV bandgap semiconductor with the proper doping and band lineups could be grown epitaxially on Si. However if we consider the point of comparison, polycrystalline PtSi, it seems likely that even relatively poor quality polycrystalline films of a suitable NGS may offer performance gains over metal films. The purpose of this paper is to develop these ideas from a more rigorous point of view, and suggest candidate systems for the gate.

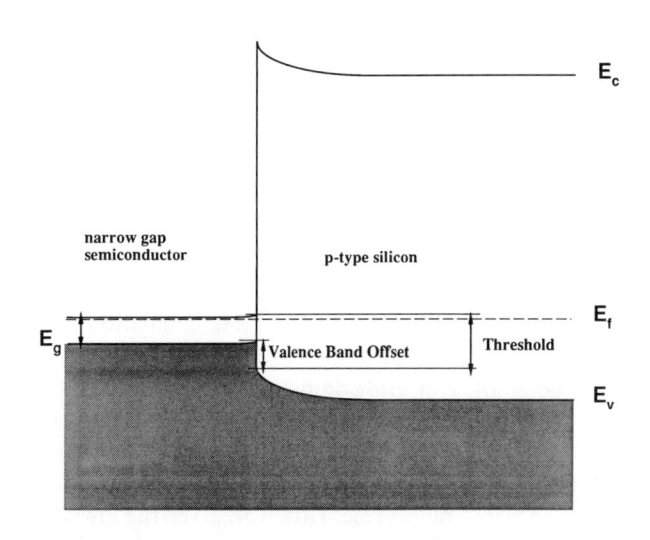

FIG. 1. Band diagram of NGS/Si heterostructure. Photoresponse threshold ϕ is the energy difference between the valence band maximum of the Si and the conduction band minimum of the NGS.

II. ANALYSIS AND MODELING OF DETECTION PROCESS

A more quantitative picture of the heterostructure detector performance can be obtained through an analysis analogous to that employed for normal Schottky photodiodes. For this purpose we will apply the theory developed by Mercer and Helms[2] for PtSi/p-Si photodetectors. The process of photodetection in a Schottky diode may be divided into three distinct steps. In the absorption phase, photons incident on the detector interact with the metal film to produce electron-hole pair excitations, a concept normally reserved for semiconductors but applicable to metals in this case. After absorption, the relevant carriers, generally holes in infrared Schottky detectors, move from the point of excitation to the actual Schottky barrier during the transport phase; after which they are either reflected back into the metal film or transmitted over the barrier into the semiconductor material where they can be collected. While fairly straightforward, this process is inherently inefficient. The transmission of the Schottky barrier is quite small for carriers having energies of excitation near the threshold of the detector as a result of poor k-vector matching between the metal and the semiconductor. Tunnelling can generally be neglected for the low barriers and doping levels used for infrared detection. Given this weak emission, a reasonably long hot carrier lifetime would be desirable. However, the average lifetime of carriers at energies exceeding the Schottky barrier height is extremely short in metals due to hole-hole scattering. The large density of cold carriers at the Fermi level and the similarly large density of available final states make hole-hole scattering an extremely efficient inelastic process. While these shortcomings sharply limit the performance of normal Schottky photodiodes, many are reduced or virtually eliminated in the heterostructure device. We will use the three-step detection process to compare the performance of the heterostructure to a Schottky barrier detector.

In the first step, absorption, a much smaller amount of the absorbed radiation is lost in the production of subthreshold excitations. Carriers cannot be excited into the energy gap, and the density of available states near the band edge is quite low. For a given level of absorption, the low bandgap semiconductor will produce more potentially detectable carriers than a metal or silicide. Of course, the lower density of states also means that the absorption length in the semiconductor is greater than in a metal, so we will require a thicker film for the same amount of absorption.

The second phase of the detection process involves transport of hot holes within the overlayer, which is limited by different mechanisms than transport of thermal holes (those within a few kT of the band edge). Recombination mechanisms would be too slow to be the limiting inelastic process. While hole-hole scattering dominates in metals, there would be very few holes near the valence band maximum of an NGS doped n-type, and so hole-hole scattering may be neglected. For hole energies above 1.0–1.5 E_g, impact ionization will start to play a role. However, given the low density of final states available near the band edges, this will probably not be a major factor until much higher energies. In most cases, we believe that optical phonon emission will be the mechanism which most rapidly thermalizes the hot holes.

The third phase of the detection process, transmission, may be analyzed using the Fowler method.[3] This method assumes that any carrier with sufficient energy directed perpendicularly to the barrier will be able to cross it. The simplest transmission function for our system involves the semiconductors with a spherical parabolic band with maximum at Γ. In nearly all cases, the NGS has hole effective masses which are less than Si. This means that the Fowler analysis is straightforward, with the condition for transmission simply being $h^2 k_\perp^2 / 2m_h > \phi_b$. As the excited carrier energy is relatively close to the band edge, we cannot make the simplifying assumption of a constant density of states. We must work with ratios of volumes and surface areas of the actual spheres (ellipsoids) in k-space. In the simple case described above, the transmission function is

$$T = \frac{1}{2}\left(1 - \frac{\sqrt{\phi_b - E_g}}{\sqrt{h\nu - E_g}}\right). \quad (1)$$

Once the barrier transmission function has been calculated, a barrier escape velocity may be found and used in the diffusion model described below to determine photoyield.

The simple Fowler analysis does not take into account quantum mechanical reflections for carriers with energies slightly above threshold. While we will not make any quantitative calculations of the phenomena in this paper, a qualitative observation is in order. Using semiconductors with the valence band maximum at Γ, we would obtain better $k_\perp$ matching than for the case of a Schottky barrier, which would reduce reflections near threshold.

The variation of the photoyield per absorbed photon with the thickness of the low bandgap semiconductor layer for a detector characterized by these parameters may be examined using the diffusion model of Mercer and Helms.[2] For the case of a normal infrared Schottky photodetector, it was determined that the yield per absorbed photon (Y) may be expressed in the simplified form

$$Y \sim \left(\frac{h\nu - \phi}{h\nu}\right)\frac{1}{1 + d/C\tau_r} \quad Cd/D \ll 1, \quad (2)$$

where ϕ is the Schottky barrier height, d is the metal (silicide) film thickness, C is the barrier escape velocity, τ_r is the mean hot carrier lifetime, and D is the hot carrier diffusion constant. Simple kinetic arguments give $C \sim vT/2$ and $D \sim vL_s/3$, where v represents the mean carrier velocity and L_s the mean free path for elastic scattering, which implies that the condition $L_s/d \gg T$ should be satisfied for the rigorous application of the result. The term in parentheses, which corrects for the presence of subthreshold excitations, must be modified to account for the presence of the bandgap and the fact that the density of states cannot be considered to be constant over the range of excitation, as can be done in the case of a metal. In addition, the existence of interface traps cannot be overlooked for this structure. If the traps at the NGS/silicon interface are characterized by a filling velocity C_{t1} and those at the air(cavity)/NGS interface by C_{t2}, the presence of the traps can be included in the diffusion formulation in the same manner as the escape velocity C. It will be assumed that carriers emitted from traps are thermalized. For details of the calculation, see Ref. 2. With the incorporation of interface traps and the assumption of a parabolic density of states function, Eq. (2) can be revised for the case of the heterostructure detector as

$$Y \sim \left[\frac{(h\nu - E_g)^{3/2} - (\phi - E_g)^{3/2}}{(h\nu - E_g)^{3/2}} \right]$$
$$\times \frac{1}{1 + d/C\tau_r + C_{t1}/C + C_{t2}/C},$$
$$Cd/D, C_{t1}d/D, C_{t2}d/D \ll 1 \quad (3)$$

From this expression, it is clear that interface traps will significantly impact the photoyield only if the trap filling velocity is comparable to the barrier escape velocity. The complete expression for the quantum efficiency of the heterostructure detector is obtained by multiplying Eq. (3) by the absorption.

III. CALCULATIONS FOR A SPECIFIC SYSTEM

We will now apply the modified diffusion model to a particular heterostructure, lead-tin-telluride (LTT)/p-Si. The fraction of absorption, the first phase in the detection process, is shown in Fig. 2 for 4 μm radiation as a function of film thickness for InSb and Pb$_{0.8}$Sn$_{0.2}$Te. Each device is assumed to be equipped with a tuned resonant optical cavity of the sort commonly used to enhance the absorption in Schottky photodetectors.[4] The cavity is tuned to maximize absorption in a 1000 Å Pb$_{0.8}$Sn$_{0.2}$Te film. The optical constants for InSb were taken from Palik,[5] while the data for Pb$_{0.8}$Sn$_{0.2}$Te (bandgap ~0.1 eV) were estimated from the work of Palik, Korn,[6] and Opyd.[7] The absorption length $1/\alpha$ is approximately 6000 Å for the (LTT) alloy, much longer than the value of ~300 Å obtained for platinum silicide.[8] However, with increases in the carrier lifetime and barrier transmission, relatively thick films on the order of 1000 Å may be used without sacrificing overall quantum efficiency.

In the transport phase, a variety of recombination and scattering mechanisms will affect the hot hole. We analyze these in an attempt to determine the process which will limit the lifetime of hot holes in LTT. Recombination mechanisms dominate the lifetimes of thermal carriers. Radiative recombination, normally the limiting step for direct gap materials, only seems to dominate for extremely high quality films of LTT at low temperatures.[9] Auger recombination, which limits thermal hole lifetime in heavily doped LTT, would act at a similar rate for hot holes. The decrease in interaction matrix element would be balanced by an increase in final density of states for the excited hole, and so lifetime should remain in the range of to 10^{-9} s.[10] In epitaxial layers of PbSnTe, lifetimes of 10^{-8} s are typically measured in photoconductivity decay measurements[9,11] In poor quality films, where defect concentrations are typically 2 orders of magnitude higher, we estimate that the worst case Shockley-Read-Hall limited recombination lifetime would be on the order of 10^{-10} s. As mentioned above, we will probably be able to neglect impact ionization in most cases.

Hot carrier transport in the NGS overlayer will probably be limited by optical phonon emission, which would scatter the hole to an energy below threshold. Holes at threshold would have a kinetic energy of $\phi_b - E_g$, which would vary from 0.05 to 0.2 eV, depending on the choice of bandgap for the overlayer. This is considerably higher than the energy of an optical phonon ($h\nu_{LO} = 0.013$ eV for LTT),[12] and so we expect to have a high rate of phonon emission. As LTT is a polar material, the dominant process should be polar optical phonon scattering. We calculate a Frölich coupling constant of $\alpha = 0.12$ for LTT with $x = 0.2$,[13] which leads to a single scattering lifetime of $t = 2 \times 10^{-12}$ s, assuming a hot hole energy of 0.2 eV.[14] For holes with energy greater than .013 eV above threshold, multiple phonon scatterings would be required to bring the hole energy below threshold. This is a Poisson process, but we may make a crude estimate of the carrier lifetime above threshold by multiplying the number of scattering events by the scattering time. As the polar optical phonon interaction is electrostatic, there is a slight decrease in the interaction (increase in scattering time) with carrier energy. The energy dependence of the scattering time is roughly $E^{1/2}$.

The model parameter L_s may be calculated to be ~60 Å for thermal carriers using the Hall mobility, which we will estimate at approximately 1000 cm^2/V s for thermally evaporated films operating at 77 K.[15,16] L_s, which includes the effects of structural imperfections as well as acoustic phonon scattering, should vary slowly with energy and will be assumed to be constant over the range of photon energies considered here. The parameter values obtained above indicate that for LTT films up to 1000 Å, the $L_s/d \gg T$ condition will probably not hold over the entire 3–5 μm range; however, experience has shown that Eq. (2) remains a good approximation to the more rigorous solution given by the full expression of the diffusion model[2] until $L_s/d \ll T$.

Evaluation of Eq. (3) using the parameters obtained for LTT indicate that the yield per absorbed photon should be

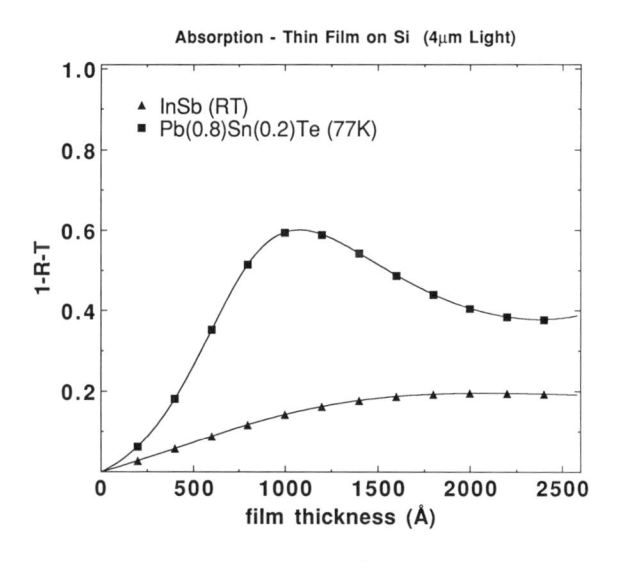

FIG. 2. The absorption fraction as a function of LTT film thickness for $\lambda = 4$ μm. An optical cavity tuned for an LTT film thickness of approximately 1000 Å is assumed to be employed in each case. The absorption for InSb using the same optical cavity is shown for comparison.

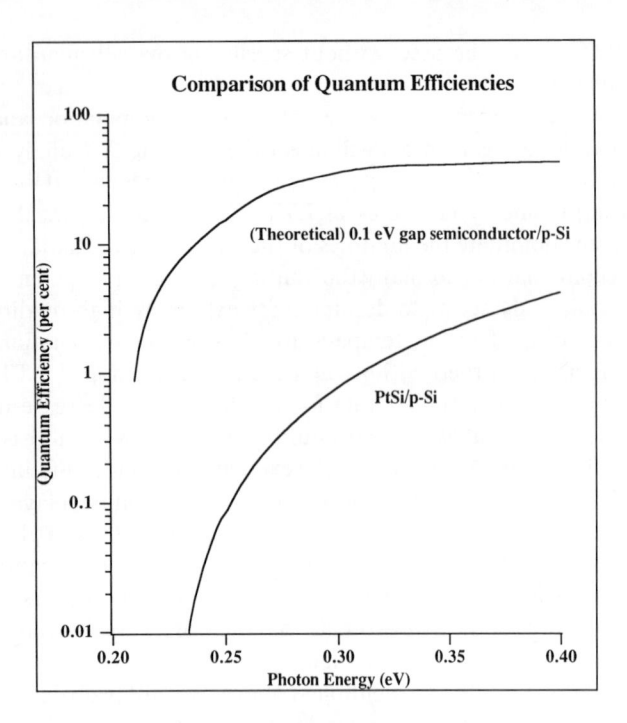

FIG. 3. Theoretical quantum efficiency for a 1000 Å heterostructure detector equipped with an optical cavity tuned for operation at 4 μm as a function of incident photon energy. The device is assumed to have 0.2 eV threshold and 0.1 eV bandgap in the overlayer. The quantum efficiency of a PtSi/p-Si detector is shown for comparison.

virtually constant for film thicknesses up to 1000 Å at $\lambda = 4$ μm. An overestimate of an order of magnitude in either the barrier transmission or the carrier lifetime would still result in a degradation of less than 25% at 1000 Å.

The theoretical quantum efficiency versus incident photon energy for 1000 Å of LTT on Si is shown in Fig. 3. We assume that the overlayer has a 0.1 eV bandgap, and that the system has a 0.2 eV threshold. We calculate an efficiency of approximately 37% at a photon energy of 0.31 eV (4 μm wavelength), compared to a benchmark standard of 1% for PtSi devices. Near threshold, the ratio of the absorption efficiency of the heterostructure detector to the Schottky detector goes as $h\nu/(h\nu - \phi)$. Thus, the quantum efficiency of the heterostructure detector increases greatly relative to the Schottky detector as the incident photon energy approaches threshold.

The major controllable parameter in our device design is the bandgap of the overlayer material. There are tradeoffs to be considered in selecting a bandgap. A narrow bandgap leads to more absorption of incident light in the overlayer, but there will also be more subthreshold excitations, depending on whether the valence band offset increases. A change in overlayer bandgap may also affect the threshold, depending on the band lineup.

IV. CONCLUSIONS

The novel device structure described here can result in a substantial increase in quantum efficiency over other Si based IR detectors through improved absorption efficiency and hot carrier lifetime. If an overlayer material with the proper bandgap and valence band offset can be found, such a device may be used for imaging arrays in the 3–5 μm window.

ACKNOWLEDGMENTS

This work was supported by Air Force Contract No. F19628-86-K-0023 and DARPA through ONR Contract No. N00014-88-K-0674.

[a] Current address: Texas Instruments Inc., Dallas, Texas 75265.

[1] J. P. Spratt, R. F. Schwartz, and V. M. Cheng, IEEE Trans. Electron Devices **ED-24**, 1117 (1977).
[2] D. E. Mercer and C. R. Helms, J. Appl. Phys. **65**, 5035 (1989).
[3] R. H. Fowler, Phys. Rev. **38**, 45 (1931).
[4] M. Kimata, M. Denda, N. Yutani, S. Iwade, N. Tsubouchi, M. Daido, H. Furukawa, R. Tsunoda, and T. Kanno, Opt. Eng. **26**, 209 (1987).
[5] E. D. Palik, *Handbook of Optical Constants of Solids* (Academic, Orlando, 1985).
[6] D. M. Korn, PhD dissertation, University of California, Los Angeles, 1971.
[7] W. G. Opyd, MS thesis, Naval Postgraduate School, 1973. Defense Documentation Center No. 767667.
[8] J. M. Mooney, J. Appl. Phys. **64**, 4664 (1988).
[9] K. Weiser, E. Ribak, A. Klein, and M. Ainhorn, Infrared Phys. **21**, 149 (1981).
[10] R. Rosman and A. Katzir, IEEE J. Quantum Electron. **QE-18**, 814 (1982).
[11] K. H. Herrmann, Solid-State Electron. **21**, 1487 (1978).
[12] *Landolt-Bornstein Numerical Data and Functional Relationships in Science and Technology*, III/17f, edited by K. H. Heliwege and O. Madelung (Springer, Berlin, 1983).
[13] *Polarons in Ionic Crystals and Polar Semiconductors*, edited by J. T. Devreese (North Holland, Amsterdam 1972).
[14] *Hot-Electron Transport in Semiconductors*, edited by L. Reggiani (Springer, Berlin, 1985).
[15] G. E. Folmanis, E. V. Dorofeev, V. P. Zlomanov, E. V. Kul'bachevskaya, and O. I. Tananaeva, Inorg. Mater. **11**, 299 (1975).
[16] H. Holloway, E. M. Logothetis, and E. Wilkes, J. Appl. Phys. **41**, 3543 (1970).

Characterization of PbTe/p-Si and SnTe/p-Si heterostructures

G. Scott and C. R. Helms
Stanford Electronics Laboratories, Department of Electrical Engineering, Stanford University, Stanford, California 94305

(Received 3 October 1990; accepted 18 December 1990)

Heterostructures which utilize a thin film of narrow-gap semiconductor deposited on a p-Si substrate have the potential to function as high-efficiency, easily fabricated infrared detectors. We have characterized two such systems, namely PbTe and SnTe. The structures were prepared by congruently evaporating films of PbTe and SnTe from single sources of the compounds onto ⟨100⟩ p-Si wafers which were chemically cleaned prior to deposition. We have analyzed the composition and electrical properties of the deposited films, as well as the electrical and photoresponse properties of the heterostructures. Current–voltage measurements indicated rectifying behavior with good ideality factor and low leakage down to 77 K for the case of SnTe/Si. High series resistance of the PbTe/Si structures made electrical characterization difficult, though they also showed rectifying behavior and low reverse leakage. Photoresponse measurements indicate a threshold of 0.36–0.4 eV for SnTe/p-Si and 0.3 eV for PbTe/p-Si. These results indicate that the PbSnTe alloy on p-Si will have a lower energy threshold than either of the binary compounds.

I. INTRODUCTION

Heterostructures which utilize a thin film of narrow-gap semiconductor (NGS) deposited on a p-Si substrate have the potential to function as high-efficiency, easily fabricated infrared detectors.[1] The low density of unwanted states in the overlayer means that the optical absorption and hot carrier transport processes may occur more efficiently than in a Schottky barrier detector. A typical device (see Ref. 1 for a band diagram) would have a photoresponse threshold roughly equal to the valence band offset between the NGS and the Si plus the bandgap of the NGS. The low threshold desired for infrared detectors therefore requires a low valence band offset.

There has been a great deal of interest over the years in NGS/Si systems, such as PbS/Si,[2–5] PbSe/Si,[6] Pb(S,Se)/Si,[7] and PbTe/Si.[8] However, most of these studies have involved depositing films several microns thick on n-type Si substrates, with the exception of a preliminary study on PbSnTe/p-Si.[9] The major objective of previous work has been to maximize light absorption in the overlayer, and to create devices with threshold equal to the overlayer bandgap. The devices discussed in this paper utilize films with thickness on the order of 1000 Å, chosen to maximize photoresponse of the heterostructure itself.

II. EXPERIMENT

The PbTe/Si and SnTe/Si devices we investigated were prepared by evaporation from single sources of the overlayer material onto unheated p-type ⟨100⟩ Si wafers (typical doping 10^{15} cm^{-3}) in a diffusion-pumped vacuum chamber (base pressure 10^{-6} Torr). Prior to deposition, the wafers were cleaned using the RCA wet chemical process with a final HF-dip step. A shadow mask defined the pattern of the devices and a large area front-side contact, which we used as an "ohmic" contact. Film thickness and deposition rate were monitored with a Maxtek TM-100 thickness monitor, and the results confirmed with stylus probe step height measurements. Typical film thicknesses were in the range of 300–1000 Å. We analyzed the composition of our films by Auger sputter profiling, performing Auger spectroscopy with a 0.3% resolution CMA with integral electron gun (1 μA e^- beam) while sputtering the films with a Xe$^+$ ion beam (current density 12 μA/cm^2).

Current–voltage (I–V) measurements were done using an HP 4140B picoAmmeter. The sample was mounted in an MMR Low-temperature microprobe station, and I–V measurements were done over a range of temperatures from 80 to 300 K. At first, we tried backside electrical contacts, mounting the device on a copper sheet with conductive epoxy. However, we found that we could obtain more consistent results with two front-side contacts, one of which was several orders of magnitude larger in area than the other. We used two sets of small diodes, areas 1.9×10^{-4} and 2×10^{-3} cm^2, and the large area contact was typically 0.5 cm^2.

The samples prepared for the photoresponse measurements were similar to the I–V structures, except that the device area was larger (0.04 cm^2), to generate more signal. The light source was an Oriel globar with a Kratos grating monochromator. Interference filters with cut-on wavelengths of 1.6 and 2.3 μm removed second harmonic light from the monochromator. The beam was chopped at 80 Hz. and short-circuit current measured with a Keithley 721 current amplifier, with the resulting ac voltage signal measured by a lock-in amplifier. Different overlayers exhibited a wide variation in dynamic resistance, and so photocurrent measurement gave a more consistent basis of comparison than did open circuit voltage. Unless noted otherwise, the samples were cooled to 80 K during measurement.

III. RESULTS AND DISCUSSION

PbTe evaporates congruently in molecular form, while heated SnTe produces both SnTe and Te$_2$ molecules in the

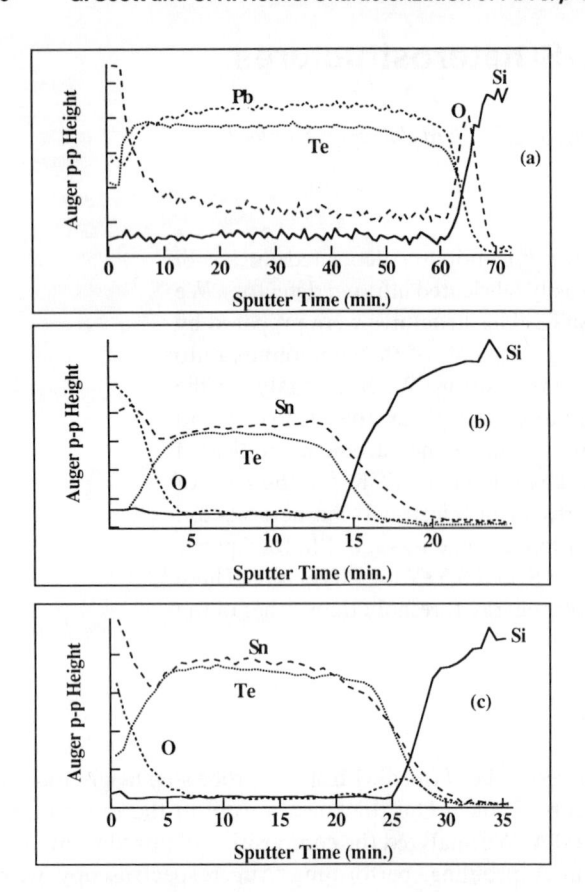

FIG. 1. Auger sputter profiles of PbTe and SnTe films deposited on Si. (a) 1000 Å PbTe/Si system. Note the uniform composition of the film and the abrupt interface. (b) 300 Å SnTe/Si system. While most of this film has a uniform composition, the interface is rich in Sn, as is the oxidized surface. (c) Same as b, but with 10 Å of Te deposited before SnTe evaporation. This appears to give a more stoichiometric SnTe layer at the interface.

Overlayer resistivity was measured by a four-point probe. SnTe had a sheet resistance of 29.6 Ω/square for a 1000-Å-thick film, which gave a resistivity of 2.96×10^{-4} Ω cm, corresponding to heavily doped material. The slight deviation from stoichiometry that we observed in the SnTe film may explain the heavy doping, as these materials tend to be mainly vacancy doped. The PbTe film showed a considerably higher resistivity (0.46 Ω cm), which may explain the high series resistance in the diode measurements.

I–V measurements of SnTe/p-Si diodes showed rectifying characteristics, but high leakage below 120 K. Addition of a thin Te layer at the interface eliminated most leakage, but reduced the current by an order of magnitude. Figure 2 shows a plot of $\log(I_0/T^2)$ versus $1/kT$, where I_0 is the reverse saturation current of the diode. This activation energy plot makes it possible to measure thermionic emission barriers in devices where the active area and precise transmission characteristics at the interface are unknown.[11] The slope of the plot gives the thermionic emission barrier, while the intercept gives a measure of the active area and transmission probability at the interface. In Fig. 2 we see two thermally activated processes, with activation energies of 0.1 and 0.33 eV, which would be consistent with two parallel diodes. The difference between these two energies is roughly equal to the bandgap of SnTe at 80 K, which is somewhere between 0.2 and 0.3 eV. The Burstein shift makes it difficult to assign a precise bandgap to SnTe at that temperature.[12] Therefore, the low temperature data may be due to hole injection over a gas phase.[10] We believed that both compounds would produce stoichiometric films when evaporated onto Si. An Auger sputter profile of a 1000 Å film of PbTe on Si [Fig. 1(a)] shows the chemical composition of the film with depth. As can be seen from the figure, the PbTe surface has a thin native oxide layer, and in this particular sample, there was a large oxide peak at the PbTe/Si interface. We observed good stoichiometry in the film itself within the quantitative limits of the technique, low oxygen content, and an abrupt interface. The apparently large oxygen signal in the film is actually a Te secondary peak. There was some difficulty in interpreting the data at the interface of the PbTe/Si system due to the overlap of the Si LVV peak and Pb NVV peak at 92 eV. We attempted to analyze another Pb NVV peak at 236 eV, but noise prevented the recording of meaningful data. The SnTe films also exhibited good stoichiometry in the bulk of the film and low oxygen content. However, a Sn-rich layer was observed at the surface and Si interface of all samples [Fig. 1(b)]. There is no stable mixed phase of Sn and Si, and so a chemical reaction seems unlikely. A thin layer of Te (10 Å) deposited on the substrate before evaporating SnTe led to a more stoichiometric composition at the interface [Fig. 1(c)].

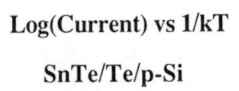

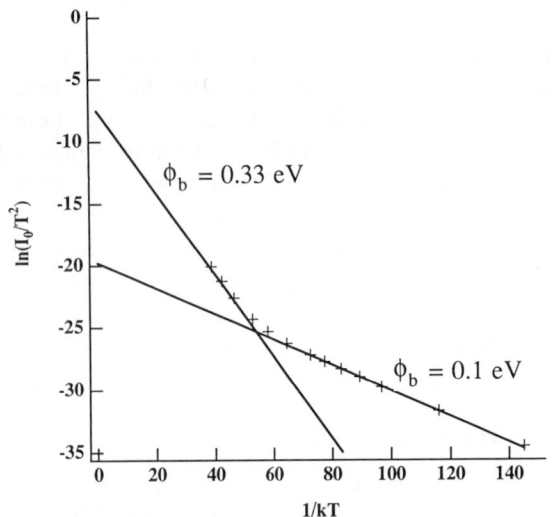

FIG. 2. Temperature-dependent current-voltage characteristics of SnTe/p-Si with thin Te layer at the interface, shown as an activation energy plot. Two thermally activated processes are present. From 80 to 180° K, a 0.1 eV barrier dominates, while at 225° K and higher, a 0.33 eV process supplies most of the current. This higher energy barrier is roughly the same as the photoresponse threshold. We believe that the 0.1 eV barrier is the valence band offset, while the 0.33 eV barrier corresponds to holes generated by thermal excitation across the SnTe band gap, with sufficient energy to pass into the Si.

valence band offset of 0.1 eV, while the high temperature data may be caused by thermal emission over the SnTe bandgap, followed by injection of the hot hole into the Si. The latter process would have a threshold energy equal to the valence band offset plus SnTe bandgap, and so the assumption that band offsets are acting as thermionic emission barriers is consistent with the observed activation energies.

For photoresponse measurements on both SnTe/p-Si and PbTe/p-Si, we applied the Fowler theory,[13] normally used for analyzing metal-semiconductor contacts, to our structures. The theory should be approximately correct for energies well above threshold, where the final density of states varies less rapidly. Fig. 3(a) shows a Fowler plot, square root of normalized photoyield versus photon energy, for the SnTe/p-Si system. The slope of the plot appears to be fairly uniform from photon energy 50 meV above threshold to the Si bandgap. This indicated that the photoresponse was produced by interband absorption in overlayer followed by hot hole emission into Si valence band, and that photoexcitation of holes from the valence band of the SnTe into the Si valence band could be neglected. Otherwise, there would have been a dramatic change in slope of the Fowler plot about 0.2 eV above threshold. There was a change in slope much closer to threshold [Fig. 3(b)]. This might have been due to a nonuniform region at the SnTe/Si interface, resulting in two parallel diodes. The nonuniformity may be due to native oxide growth on the Si substrate or clustering in the initial SnTe film growth. Another possibility is that the semiconducting overlayer has a rapidly varying density of states near threshold, and so the simple Fowler theory would not hold in that case. The lower threshold (0.34–0.35 eV) would then be the actual photothreshold, with the increased density of final states giving a more Fowler-like plot above 0.4 eV photon energy. Note that the lower photothreshold is consistent with the high-temperature barrier height determined from the activation energy measurement. This indicates that the barriers observed in the temperature dependent I–V data are consistent with the band offsets. PbTe/p-Si photoresponse gave a threshold of about 0.3 eV. Responsivity was considerably less than in the case of SnTe.

IV. CONCLUSIONS

We have investigated films of PbTe and SnTe on Si substrates. PbTe films had a high resistivity, while SnTe had an extremely low resistivity, possibly due to slight variations in the film composition. The devices exhibited good rectification, ideality factor near unity, and low reverse leakage. Activation energy measurements showed contributions from both the conduction and valence bands of the overlayer to the diode current. Internal photoemission measurements showed that the SnTe/p-Si structure had a 0.33 eV photoresponse threshold, while PbTe/p-Si had a 0.3 eV threshold. The threshold response is a Schottky-like process, with interband absorption in the overlayer followed by emission of a hot hole into the Si substrate. These results indicate that the valence band offset is roughly equal for PbTe/Si and SnTe/Si. The ternary alloy $Pb_{1-x}Sn_xTe$, which has a narrower bandgap than either of the binary compounds, may thus have a lower photoresponse threshold when deposited on p-Si.

ACKNOWLEDGMENTS

The authors wish to thank Phil McKernan and Stan Wanner for their help in preparing the samples. This work was supported in part by DARPA through ONR Contract No. N00014-88-K-0674 and gift funds from Texas Instruments.

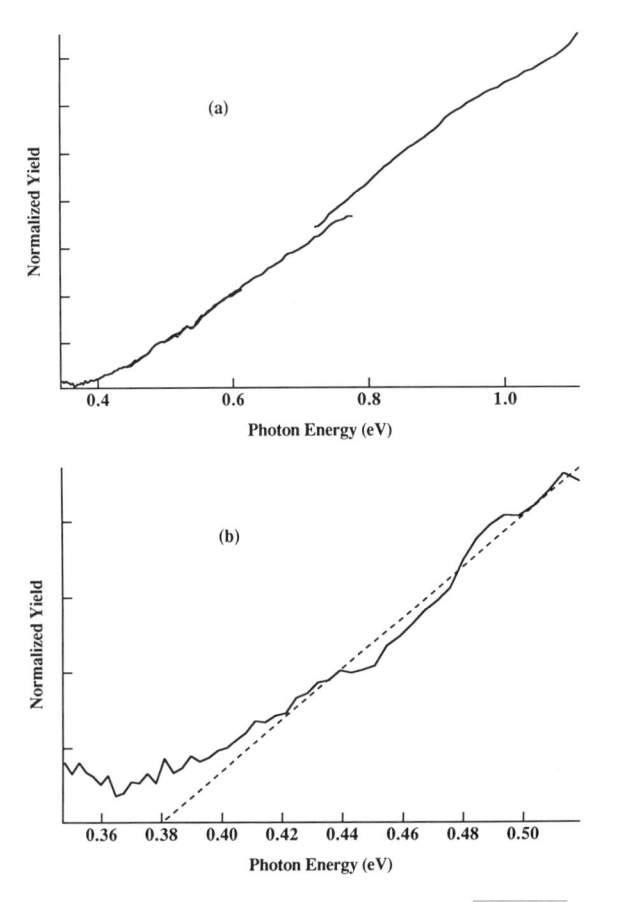

FIG. 3. Photoresponse data presented as a Fowler plot, $\sqrt{Yield \times h\nu}$ vs $h\nu$. Two different monochromator gratings were used to cover the photon energy range from 0.3 to 1.1 eV. The straight line characteristic indicates that a single process dominates the photoresponse, namely interband absorption in the SnTe followed by hot hole injection into the Si. Deviation from the Fowler characteristic near threshold may be due to a decrease in the density of states near the SnTe band edges. Depending on where the data is extrapolated to determine threshold, we get a threshold ranging from 0.36 to 0.4 eV.

[1] G. Scott, D. E. Mercer, and C. R. Helms, J. Vac. Sci. Technol. B **9**, 1781 (1991).
[2] H. Sigmund and K. Berchtold, Phys. Status Solidi **20**, 255 (1967).
[3] H. Elabd and A. J. Steckl, J. Appl. Phys. **51**, 726 (1980).
[4] A. J. Steckl, H. Elabd, K. Y. Tam, S. P. Sheu, and M. E. Motamedi, IEEE Trans. Electron. Devices **ED-27**, 126 (1980).
[5] Y. S. Sarma, H. N. Acharya, and N. K. Misra, J. Mater. Sci. **21**, 137 (1986).
[6] D. A. Kiewit, Phys. Status Solidi A **1**, 725 (1970).
[7] Y. S. Sarma, H. N. Acharya, and N. K. Misra, Infrared Phys. **24**, 425 (1984).
[8] P. R. Vaya, J. Majhi, B. S. V. Gopalam, and C. Dattatreyan, Phys. Status Solidi A **93**, 353 (1986).

[9] G. E. Folmanis, E. V. Dorofeev, V. P. Zlomanov, E. V. Kul'bachevskaya, and O. I. Tananaeva, Inorg. Mater. **11**, 299 (1975).

[10] K. C. Mills, *Thermodynamic Data for Inorganic Sulphides, Selenides and Tellurides* (Butterworths, London, 1974).

[11] E. H. Rhoderick and R. H. Williams, *Metal-Semiconductor Contacts*, 2nd ed. (Clarendon Oxford, 1988).

[12] *Landolt-Bornstein Numerical Data and Functional Relationships in Science and Technology*, III/17f, edited by K. H. Heliwege and O. Madelung (Springer, Berlin, 1983).

[13] R. H. Fowler, Phys. Rev. **38**, 45 (1931).

Long-wavelength infrared detection in a photovoltaic-type superlattice structure

Byungsung O, J.-W. Choe, M. H. Francombe, K. M. S. V. Bandara, E. Sorar, and D. D. Coon
Microtronics Associates, University Technology Development Center, Pittsburgh, Pennsylvania 15213

Y. F. Lin and W. J. Takei
Westinghouse Science and Technology Center, Pittsburgh, Pennsylvania 15235

(Received 2 October 1990; accepted 5 December 1990)

The first successful demonstration of long wavelength infrared (LWIR) detection with a photovoltaic-type AlGaAs/GaAs superlattice structure is reported. The experimental response band of the detector is centered near 10 μm in good agreement with the theoretical response band provided that electron–electron interactions are taken into account. The detector operates at significantly lower bias voltage than photoconductive multiple quantum well LWIR detectors. This could lead to important advantages in applications to photovoltaic detector arrays. Optimization through the use of termination layers at the ends of superlattices is also discussed. The response at 83 K is about 50% of the response at 24 K which is 5 mA/W. Optimization of the response, operating temperature or bias voltage has not been carried out.

I. INTRODUCTION

Recent research on optical excitation of intersubband transitions in semiconductor quantum well structures[1–7] has demonstrated that by careful engineering of quantum well and superlattice parameters, tailored infrared (IR) response at chosen wavelength can be achieved over a very wide spectral range extending to the long wavelength IR (LWIR) band and beyond. As a result of the use of intraband processes rather than interband processes, it is possible to use wide band gap materials such as GaAs, which have fewer growth and processing problems than mercury cadmium telluride, and to implement detector structures in a monolithic format. Research has proceeded to the point where new types of infrared detectors, capable of greatly improved performance,[1–3,5] are now feasible. Quantum well detectors with response at various wavelengths from 5 to 12 μm have now been demonstrated (for review see Ref. 5). Progress in this new area of GaAs quantum well infrared detection has been incredibly rapid, due to the previous development of an advanced technology base in GaAs molecular beam epitaxy. Successful fabrication of quantum well infrared detectors in AlGaAs/GaAs has been carried out at Bell Labs,[1–3,8] Westinghouse[5,6] and Bell Communication Research (BellCore).[4] Much of the above work has addressed photoconductive multiple quantum well (MQW) structures designed for peak IR absorption at 8–12 μm. Detector structures showing high performance ($D^* > 3 \times 10^{13}$ cm Hz$^{1/2}$ W^{-1} at 4 K at about 10 μm), e.g., have been reported[1] and we have published results indicating IR response with excellent detector uniformity at wavelengths out to 12 μm.[5]

Although the photoconductive mode of operation is capable of high current responsivity, this mode involves high dark current and high levels of operating power. A low-power type of structure has been described by Kastalsky et al.[4] at BellCore, which involves photovoltaic IR detection in the optical range 3.6–6.2 μm. In the present work we present experimental observations and related theoretical results for photovoltaic-type detectors with absorption peaks in the LWIR spectral range (8–12 μm).

The use of superlattice minibands provides a short wavelength as well as a long wavelength cutoff. The short wavelength cutoff is associated with suppression of thermal excitation of carriers. The reduction in thermionic dark current should permit operation at higher temperatures. This approach to achieving higher temperature operation in conjunction with long wavelength detection is natural from the point of view of device fabrication because barriers can be of a convenient height, comparable to the barrier height already employed in shorter wavelength heterostructure IR detectors. In the photoconductive approach, excited state subbands are placed near the classical threshold for maximum response.[7] This gives rise to issues regarding the tolerances required for composition in the barrier regions, and provides additional motivation for exploration of the photovoltaic alternative. For photovoltaic detectors, spectral band widths are mainly attributed to the tunneling between wells which can be controlled by changing the barrier thickness of the superlattice, i.e., thin and high barriers allow broad band detection with lower dark current.

Optimization through the use of specially tailored termination layers at the ends of superlattices is also discussed. Such layers are analogous to coatings on optical elements.[9] In the case of optical coatings, one does not have to compute the composite properties of the optical element together with various coatings in order to deduce the specifications of an optimal coating because one has formulas which directly specify the properties of an optimal coating. Similarly we deduce formulas which specify the properties of optimal superlattice termination layers without requiring that one perform quantum mechanical calculations for the composite structure consisting of the superlattice with the attached termination layer.

II. EXPERIMENTAL RESULTS

The structure, which was grown by molecular beam epitaxy (MBE), consists of 50 periods of 73 Å GaAs quantum wells (Si doped at 3.0×10^{17} cm^{-3}) and 65 Å undoped Al$_{0.27}$Ga$_{0.73}$As barriers (see Fig. 1). A 1000 Å Al$_x$Ga$_{1-x}$As layer graded from $x=0$ to $x=0.18$ was grown adjacent to the superlattice and the combination was sandwiched between doped ($n=1.0\times10^{18}$ cm^{-3}) GaAs contact layers. With the parameters shown, there are two minibands below the top of the barrier, with the energy of the graded barrier at $x=0.18$ being lower than the bottom of the first excited miniband, but much higher than the top of the ground state miniband to block the flow of the electrons (dark current) to the collector contact layer.[10] This idea of employing such a blocking layer in a quantum well IR detector first appeared in work of Coon et al.[11] Detector arrays, fabricated in the geometry of 4-mil square mesas making ohmic contact to each side of the device, were illuminated from the backside via an array of etched grooves on the GaAs surface. Such structures permit the attainment of relatively high quantum efficiencies without having to resort to edge-illumination conditions, of the type used by Kastalsky.[4]

In our present studies the spectral photoresponse was measured with a globar, a circular-variable-filter and a chopper–lock-in-amplifier system. The sample was mounted in a flatpack and cooled on the coldhead inside the closed-cycle-refrigerator.

Figure 2 shows the diode-like dark current–voltage characteristics at 24 K. Figure 3 shows the bias dependent photoresponse at 24 K and indicates that no response was observed at zero bias. This is in contrast to the results of Kastalsky,[4] and may be attributable to a lower effective quantum efficiency than is obtainable with edge illumination. The peak response is found to decrease by a factor of 2 on increasing the temperature from 24 K to 83 K. Figure 4 shows the spectral photoresponse under a bias voltage of -0.58 V at 24 K, corresponding to the voltage for peak response in Fig. 3. We suggest that the rapid fall-off in response at bias values above 0.58 V can probably be attributed to progressive decoupling of the miniband structure depicted in Fig. 1. The measured dark current under -0.58 V bias at 24 K was approximately 10^{-8} A. From the photoresponse and the noise measurements, the detectivity at 24 K was calculated to be about 2×10^8 cm$\sqrt{\text{Hz}}$/W.

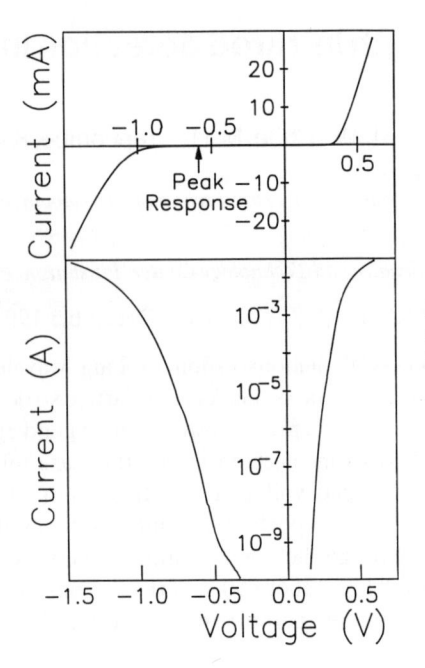

FIG. 2. Current–voltage characteristics at $T=24$ K.

III. MODELING APPROACH

We use the Kronig–Penney model generalized to include position-dependent effective masses and Batey and Wright's parameters[12] to understand the observed spectra. Without inclusion of the electron–electron interactions, the ground state miniband is found to be located at 43 meV above the bottom of the well and the first excited state miniband to be located at 161 meV. The widths of the minibands are found to be 2.2 and 21.4 meV, respectively. The absorption coefficient is calculated from the standard dipole approximation. We accommodate line broadening effects by assuming that interminiband transitions for levels with a particular miniband momentum have a Lorentzian probability distribution

$$f(E') = \frac{1}{\pi} \frac{\Gamma}{(E'-E_{kn})^2 + \Gamma^2} \qquad (1)$$

with a peak at $E'=E_{kn}$, the transition energy in the absence of such broadening and the electron–electron interactions. Here Γ is the line halfwidth. The exchange interaction effect is included in leading order:[6,13]

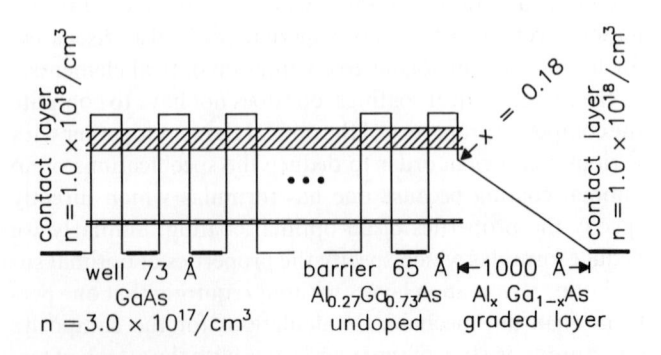

FIG. 1. Structure of a multiple quantum well photovoltaic device.

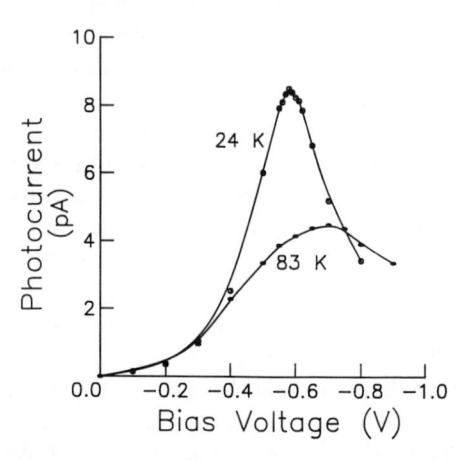

FIG. 3. Photocurrent vs bias voltage at $T=24$ and 83 K with a broad spectrum source (8–14 μm).

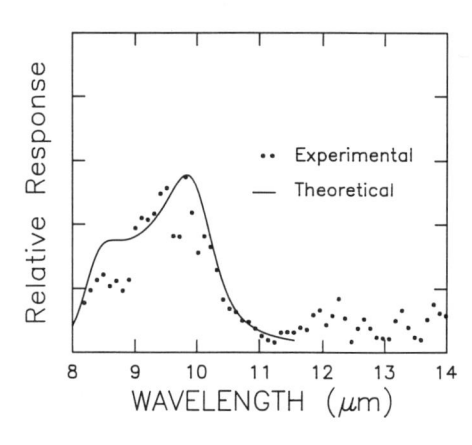

FIG. 4. Experimental and estimated spectral photoresponse at $T = 24$ K under a bias voltage of -0.58 V. The double bump structure is related to the density of states near the top and bottom of the excited state miniband.

$$E_{\text{exch}}(k) \simeq \frac{-e^2 k_F}{4\pi\epsilon} \left[\frac{2\mathbf{E}(k/k_F)}{\pi} \right] \quad (2)$$

which is independent of the detailed shape of the unperturbed wave function. Here k is the magnitude of the two-dimensional (2D) lateral wave vector and k_F is the corresponding 2D Fermi wave vector, ϵ is the permittivity of media, and $\mathbf{E}$ is a complete elliptic integral.[14] Over the range from $k = 0$ to $k = k_F$, the exchange energy varies from 14 to 9 meV. Integrating over the photon energy now yields the absorption coefficient α as

$$\alpha = \left(\frac{\pi e^2 \hbar^3}{m_e^2 c \epsilon_0} \right) \frac{n_e \cos^2 \theta}{\hbar \omega} \frac{1}{N} \sum_{\substack{\text{miniband} \\ \text{states}}} |T_z|^2 Q, \quad (3)$$

where m_e is the free electron mass, ϵ_0 is the permittivity of the vacuum, n_e is the average electron density, θ is the angle between the polarization vector and the normal to the plane of the quantum well, and N is the number of cells in the superlattice. The photoexcitation matrix element T_z is given as

$$|T_z| = \left| \frac{1}{f_b \sqrt{n_b}} \int_{-s}^{0} u_k^*(z) \frac{\partial u_n(z)}{\partial z} dz \right.$$
$$\left. + \frac{1}{f_w \sqrt{n_w}} \int_{0}^{t} u_k^*(z) \frac{\partial u_n(z)}{\partial z} dz \right|, \quad (4)$$

where f_i's are the ratios of effective electron masses to the free electron mass, n_i's are refractive indices of the medium, s is the barrier thickness, t is the well width, $u_i(z)$'s are the miniband cell periodic functions, and b and w refer to the barrier and well, respectively. The quantity Q is given as

$$Q = \int_{0}^{k_F} f\left[E_{\text{exch}}(k) + \hbar\omega \right] \frac{g(k)dk}{N_e}, \quad (5)$$

where $g(k)$ is the 2D density of states, N_e is the total number of electrons available in a given level, and f is the aforementioned Lorentzian line shape. Also there are other effects we did not include in the above calculation, i.e., the depolarization field effect (~ 2 meV),[15,16] the intersubband exchange interaction energy (~ 1 meV),[6] the correlation energy ($\sim$ a few millielectron volts),[17] etc. We accommodate these effects (~ 5 meV) by a shift of the peak calculated in the above, and the result is shown in Fig. 4. We note that the agreement between the experimental spectral response band and the theoretical spectral response band is very good.

IV. ENHANCEMENT OF PERFORMANCE BY TERMINATION LAYERS

Now we discuss the enhancement of photoresponse by controlling transmission and reflection of miniband carriers at the ends of the superlattice through the use of tailored termination layers[18] (see Fig. 5). We find that a key formula for design of termination layer involves a non-linear mapping called a Möbius transformation.[19] The mapping acts on a complex plane. In our case Möbius transformations represent the mathematical relationship between reflected Bloch wave amplitudes R' and reflected plane wave amplitudes R, which have the form

$$R' = \frac{-cR + d}{d^*R - c^*}, \quad (6)$$

where c and d are complex parameters which depend on the carrier energy as well as the parameters of the superlattice, and $|c| > |d|$. The nonlinearity of the transformation and the importance of phases make it difficult to substitute intuition for mathematics.

The above transformations in Eq. (6) have the additional geometric properties: (A) that they map the unit circle $|R| = 1$ onto the unit circle $|R'| = 1$ and (B) that they map the interior of the unit circle $|R| < 1$ into the interior of the unit circle $|R'| < 1$. These properties are illustrated in Fig. 6 for which both the barrier thickness (s) and the well width (t) are 90 Å and the barrier height (V_0) is 145 meV. These

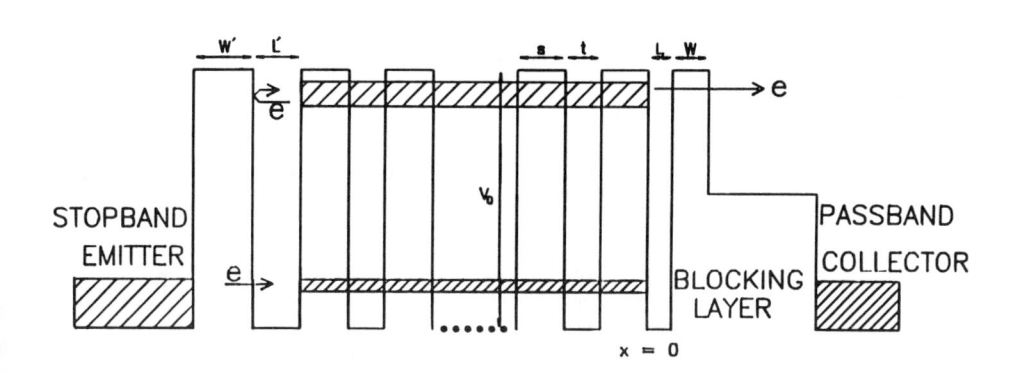

FIG. 5. Illustration of termination layers in conjunction with a superlattice photovoltaic infrared detector of the Kastalsky type. Arrows at the ends of the upper miniband indicate transmission and reflection which are achievable by choosing appropriate termination layer parameters L, W, L', W'.

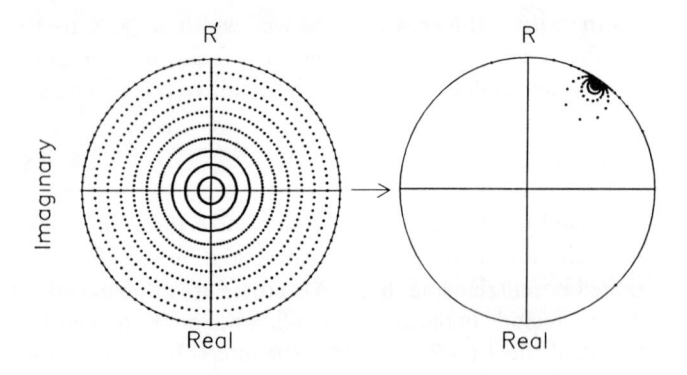

FIG. 6. Example of a Möbius transformation mapping of the unit circle of reflected Bloch wave amplitudes R' onto the unit circle of reflected plane wave amplitudes R. This conformal mapping is used for optimal design of superlattice termination layers. For a reflective stopband layer $|R'| = 1$. For an antireflective passband (antiblocking) layer $|R'| = 0$.

properties are related to the fact that normalized reflected wave amplitudes should lie on or within the unit circle whether they involve Bloch waves or plane waves. Physically, this corresponds to the fact that the reflected probability current cannot exceed the incident probability current.

We next determine the parameters of the Möbius transformation. Inside the superlattice we use a linear combination of Bloch waves

$$\Psi(x) = A'\Psi_K(x) + B'\Psi_{-K}(x). \tag{7}$$

The miniband Bloch waves for the superlattice can be written as

$$\Psi_K(x) = \{[ce^{ik(x-n\Lambda)} + de^{-ik(x-n\Lambda)}]e^{-iK(x-n\Lambda)}\}e^{iKx} \tag{8}$$

with

$$\Psi_K^*(x) = \Psi_{-K}(x) \tag{9}$$

for x lying in the $(n+1)$th well of the superlattice. In the above K is the Bloch wave vector, $\Lambda = s + t$ and c,d are complex parameters which depend on the carrier energy as well as the parameters of the superlattice.[20] Outside the superlattice we use a linear combination of plane waves

$$\Psi(x) = Ae^{ikx} + Be^{-ikx} \tag{10}$$

with

$$k = \frac{\sqrt{2mE}}{\hbar} \tag{11}$$

to represent a carrier with energy E. The Bloch wave reflection amplitude $R' = B'/A'$ is related to the plane wave reflection amplitude $R = B/A$ by Eq. (6).

We find that the incident, reflected and transmitted probability currents associated with a miniband carrier are given by

$$J_{\text{incident}} = \frac{\hbar k}{m}|A'|^2[|c|^2 - |d|^2] > 0, \tag{12}$$

$$J_{\text{reflected}} = \frac{\hbar k}{m}|B'|^2[|d|^2 - |c|^2] < 0, \tag{13}$$

$$J_{\text{transmitted}} = \frac{\hbar k}{m}|T|^2 > 0, \tag{14}$$

where $|T|^2$ is the transmission coefficient. From Eq. (6) it is obvious we get zero Bloch wave reflectivity, if

$$R = \frac{d}{c}. \tag{15}$$

The magnitude and phase required by Eq. (15) can be implemented by specifying the width and position of a barrier adjacent to the end of the superlattice. By adjusting the barrier width, the magnitude of R can be continuously varied between one and zero. Therefore we can always find a barrier width such that $|R| = |d/c|$. The required phase can be obtained by adjusting the distance between superlattice and the barrier. The dependence of the phase of R on the distance L between the end of the superlattice $x = 0$ and the adjacent barrier at $x = L$ is

$$R = R(L) = e^{2ikL}R(0). \tag{16}$$

In the case of a square barrier, $R(0)$ would be given by the usual expression for reflection from a square barrier with the barrier width adjusted to give the required value of

$$|R| = |R(0)| = |d/c| \quad \text{passband barrier width rule.} \tag{17}$$

One can then determine L from the equation

$$e^{2ikL} = \frac{d}{cR(0)} \quad \text{passband barrier positioning rule.} \tag{18}$$

The same phase adjustment and magnitude adjustment formulas apply to any barrier shape. Numerically computed passband and stopband reflectivities are shown in Fig. 7 for a superlattice with $s = t = 90$ Å, $V_0 = 145$ meV in the case of a square barrier termination layer. Stopband blocking layer results are deduced from similar arguments:

$$|R| = |R(0)|$$

$$= \frac{|cd\,||1-\rho|}{c^*c - d^*d\rho} \quad \text{stopband barrier width rule,} \tag{19}$$

$$e^{2ikL} = \frac{c^*d(1-\rho)}{[c^*c - d^*d\rho]R(0)}$$

stopband barrier positioning rule, (20)

where ρ is a real parameter which represents the degree of blocking, and is larger for thicker barriers.

V. DISCUSSION

We have demonstrated LWIR detection with a photovoltaic-type AlGaAs/GaAs superlattice structure in good

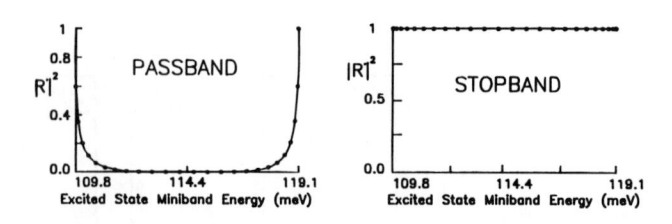

FIG. 7. Typical reflectivities for a square barrier passband and a stopband termination layer obtained from the quantitative design rules. The energy range corresponds to the excited state miniband of a superlattice, measured relative to the bottom of the wells of the superlattice.

agreement with the theoretical response band provided that electron–electron interactions are taken into account. This detector operates at significantly lower bias voltage than photoconductive MQW LWIR detectors and could lead to important advantages in applications to photovoltaic detector arrays. The signal current limitation can be improved through the use of specially tailored termination layers at the ends of superlattices to optimize the transmission or reflection of miniband carriers. With modifications to the blocking layer design and reduction of carrier concentration in the wells as a means of reducing thermionic emission, one can expect D^* values approaching 10^{11} cm$\sqrt{\text{Hz}}$/W at 40 K in the LWIR range.

The full range of device parameters should now be explored for the purpose of optimizing performance, since the device reported here represents only a preliminary design with $D^* = 1.6 \times 10^8$ cm$\sqrt{\text{Hz}}$/W and responsivity 5 mA/W. The success of the theoretical modeling and the use of termination layers described here indicates that development of the new class of LWIR detectors need not be primarily empirical.

ACKNOWLEDGMENTS

This work was supported in part by NASA under Contract No. NAS7-1105 and by the U.S. Air Force under Contract No. F33615-85-C-5171.

[1] B. F. Levine, C. G. Bethea, G. Hasnain, V. O. Shen, E. Pelve, R. R. Abbott, and S. J. Hsieh, Appl. Phys. Lett. 56, 851 (1990).
[2] B. F. Levine, G. Hasnain, C. G. Bethea, and Naresh Chand, Appl. Phys. Lett. 54, 2704 (1989).
[3] B. F. Levine, C. G. Bethea, G. Hasnain, J. Walker, and R. J. Malik, Appl. Phys. Lett. 53, 296 (1988).
[4] A. Kastalsky, T. Duffield, S. J. Allen, and J. Harbison, Appl. Phys. Lett. 52, 1320 (1988).
[5] D. D. Coon, J. Vac. Sci. Technol. A 8, 2950 (1990).
[6] K. M. S. V. Bandara, D. D. Coon, Byungsung O, Y. F. Lin, and M. H. Francombe, Appl. Phys. Lett. 53, 1931 (1988).
[7] D. D. Coon and R. P. G. Karunasiri, Appl. Phys. Lett. 45, 649 (1984).
[8] B. F. Levine, C. G. Bethea, K. K. Choi, J. Walker, and R. J. Malik, J. Appl. Phys. 64, 1591 (1988).
[9] A. Thelen, *Design of Optical Interference Coatings* (McGraw–Hill, New York, 1989).
[10] Byungsung O, J.-W. Choe, M. H. Francombe, K. M. S. V. Bandara, D. D. Coon, Y. F. Lin, and W. J. Takei, Appl. Phys. Lett. 57, 503 (1990).
[11] D. D. Coon, R. P. G. Karunasiri, and H. C. Liu, J. Appl. Phys. 60, 2636 (1986).
[12] J. Batey and S. L. Wright, J. Appl. Phys. 59, 200 (1986).
[13] J.-W. Choe, Byungsung O, K. M. S. V. Bandara, and D. D. Coon, Appl. Phys. Lett. 56, 1679 (1990).
[14] Bateman Manuscript Project, *Higher Transcendental Functions* (McGraw–Hill, New York, 1953), Vol. II.
[15] W. P. Chen, Y. J. Chen, and E. Burstein, Surf. Sci. 58, 263 (1976).
[16] S. J. Allen, Jr., D. C. Tsui, and B. Vinter, Solid State Commun. 20, 425 (1976).
[17] B. Vinter, Phys. Rev. Lett. 35, 598 (1975).
[18] D. D. Coon and E. Sorar, Appl. Phys. Lett. 56, 1790 (1990).
[19] J. D. Paliouras, *Complex Variables for Scientists and Engineers* (Macmillan, New York, 1990).
[20] A. Yariv, *Theory and Applications of Quantum Mechanics* (Wiley, New York, 1982).

High-efficiency infrared light emitting diodes made in liquid phase epitaxy and molecular beam epitaxy HgCdTe layers

P. Bouchut, G. Destefanis, J. P. Chamonal, A. Million, B. Pelliciari, and J. Piaguet
LETI/DOPT/SLIR, CENG-85 X-38041, Grenoble, France

(Received 2 October 1990; accepted 10 December 1990)

We report in this article, for the first time the performance of light emitting diodes (LEDs) made in HgCdTe (MCT) epilayers on CdZnTe (CZT) lattice-matched substrates grown either by liquid phase epitaxy (LPE) or molecular beam epitaxy (MBE). Diodes are n/p homojunctions made by ion implantation. The MCT composition was chosen in order to get the emission wavelength between 3 and 5 μm. Electroluminescence spectra were recorded between 20 K and 300 K. A comparison of the two materials is made, demonstrating that they exhibit similar performances. Internal quantum efficiencies as high as 25% (λ peak = 4.1 μm at 77 K) and 6% (λ peak = 3.5 μm at 300 K) were obtained on both LPE and MBE materials and they reach a maximum for a temperature of about 90 K. For a material with λ peak = 5.5 μ at 77 K, internal quantum efficiency remains as high as 5%. The external quantum efficiency is considerably increased using backside emission through the CZT substrate and optical coupling with a CZT lens: in such conditions 20% of the emitted light could be extracted from the structure. These results show that MCT can already be used for LEDs in the 3–5 μm region.

I. INTRODUCTION

Since the pioneering work of Verie and Granger in 1965[1] that demonstrated that HgCdTe (MCT) can be used to make infrared light emitting diodes (LEDs), only very little effort has been devoted to study the emission properties of this material.

During the last 25 years, a huge effort has been made in order to improve MCT growth techniques for infrared detectors, both by liquid phase epitaxy (LPE) and vapor phase epitaxy (VPE) techniques.

Some studies of the emission properties were carried out by optical excitation either for photoluminescence measurements to check the quality of the material,[2] or to prove that MCT can generate stimulated emission.[3–6]

More recently, improvements in MCT growth by molecular beam epitaxy (MBE)[7] and metalorganic chemical vapor deposition (MOCVD)[8] have led to the observation of laser action by optical pumping.

Recent papers have been published on LEDs in MCT material (bulk[9,10] and LPE on sapphire[11]), however up to now no injection stimulated emission has been reported.

In this paper we present a study on LEDs made in MCT epilayers grown either by LPE or MBE. The composition of MCT was chosen in order to give an emission wavelength ranging from 3 to 5 μm.

II. EXPERIMENTAL PROCEDURE

The vacancy doped p-type MCT layers were either grown by (LPE)[12] or (MBE)[13] on a CZT lattice-matched substrate. The layers were 8 μm thick and their p-type carrier concentration after technological processing was about 10^{16} atoms/cm^3.

Diodes were made using our standard implanted planar n/p technology whose details have been previously reported.[14] The implanted area is 100×100 μm^2 and the cutoff wavelength at 77 K is in the 4–5.5 μm range. Series resistances were about 50 Ω at 77 K. Devices are 5×5 diode arrays. Each array was interconnected via indium bumps to an injection circuit and the backside emission (through the substrate) of each diode was characterized (Fig. 1).

Arrays were integrated in a variable temperature cryostat providing a 20–300 K temperature range. The junction of each diode is forward biased with a square wave signal from a current generator. The diode was imaged on a detector using a ZnSe lens. The detector output was fed into a lock-in amplifier, with the signal from the current generator used as a reference signal. The detector (1.2×1.2 mm^2) is a cooled MCT photovoltaïc diode with a cutoff wavelength of 5.7 μm at 77 K.

Emission spectra were recorded with a 320 mm focal length Czerny Turner type monochromator.

The external quantum efficiency η_{ext} was calculated considering the transmission of the lens and the field of view defined by the optical geometry of the experiment.

The extraction yield Y was estimated assuming that the light emission inside the material was isotropic and photons emitted outside the escape angle were lost

$$Y = T \frac{\Omega_e}{4\pi},$$

where T is the resulting transmission at normal incidence of the various interfaces and Ω_e the useful emission solid angle.

Then the internal quantum efficiency is equal to $\eta_{\text{int}} = \eta_{\text{ext}}/Y$.

No attempt was made to correct the extraction yield for self absorption.

It should be noted that the backside transmission (0.78) is slightly enhanced over the frontside transmission (0.71) due to the lower refractive index of CZT (2.8) over MCT (3.3). No antireflection coating was added to further improve the transmission coefficient.

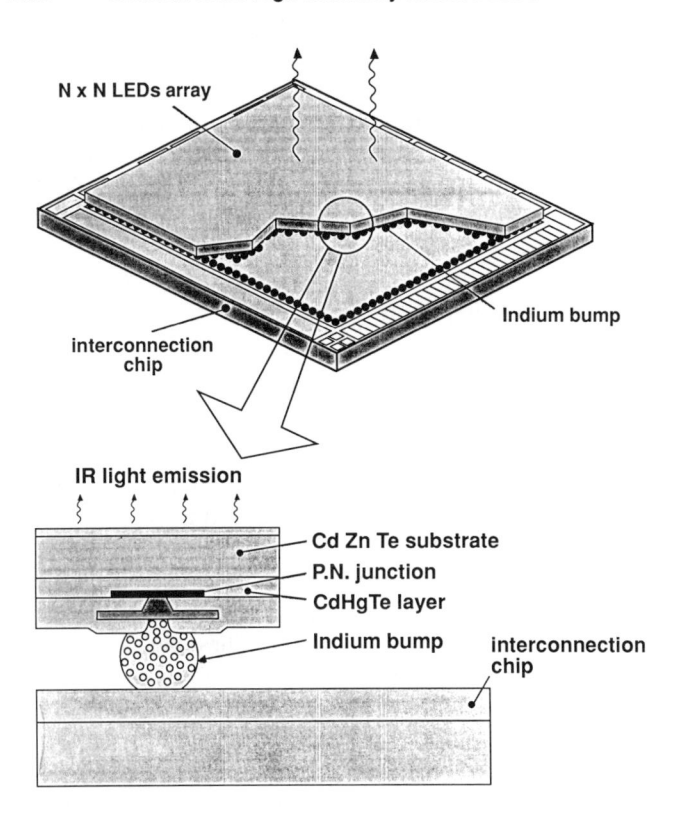

FIG. 1. Schematic view of the device used for the electroluminescence experiment.

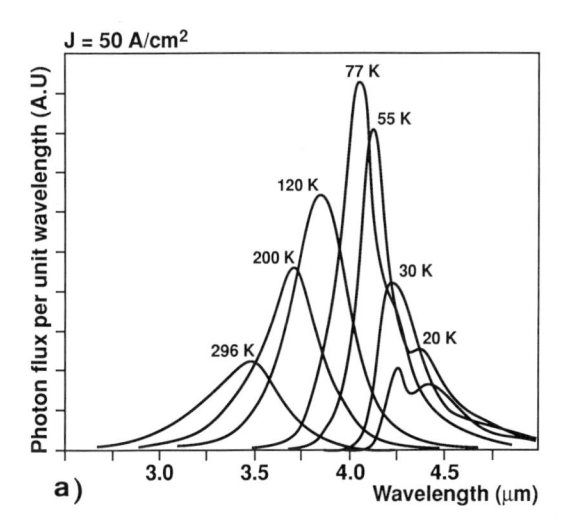

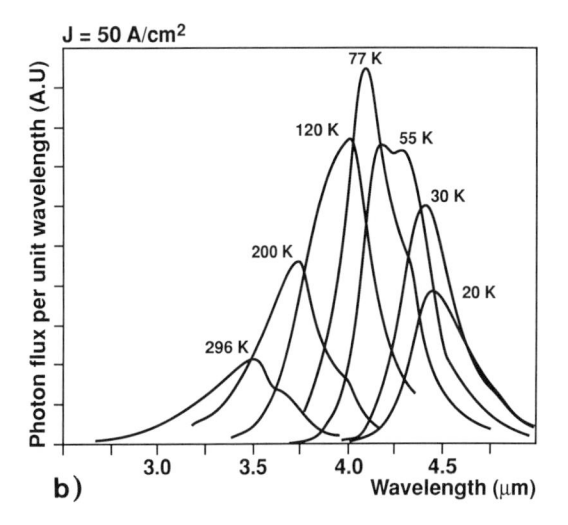

FIG. 2. Electroluminescence spectra of 4.1 μm HgCdTe diode at several temperatures. (a) LPE material, (b) MBE material.

III. RESULTS

Figures 2(a) and 2(b) show typical electroluminescence intensity spectra taken at several temperatures for two samples of similar emission wavelengths made respectively by LPE and MBE.

The spectra exhibit the same overall behavior as was also observed in the previous work of Zucca et al.[11] For both samples, the peak intensity increases when the temperature decreases, reaches a maximum at about 90 K and then decreases with the occurrence of a second peak at lower energy. This peak is clearly seen in the LPE sample spectrum [Fig. 2(a)]. By analogy, the low energy shoulder seen in the MBE spectrum can be attributed to a second peak which occurs even at higher temperature.

The internal quantum efficiency of the diode versus temperature (Fig. 3) shows the same variation as that seen for the peak intensity in the previous spectra. The maximum occurs at 90 K and values as high as 25% and 20% are obtained respectively for MBE and LPE samples. The emission peak wavelength is 4 μm in each case. Below 60 K the internal quantum efficiency decreases exponentially with $1/T$ with different activation energies for the two samples.

The external quantum efficiency of this backside emission structure is very low due to the high value of the refractive index of these materials. In order to improve the extraction yield of the structure an immersion lens was made. This consisted of a CZT half-sphere glued to the CZT substrate and centered on the array (Fig. 4).

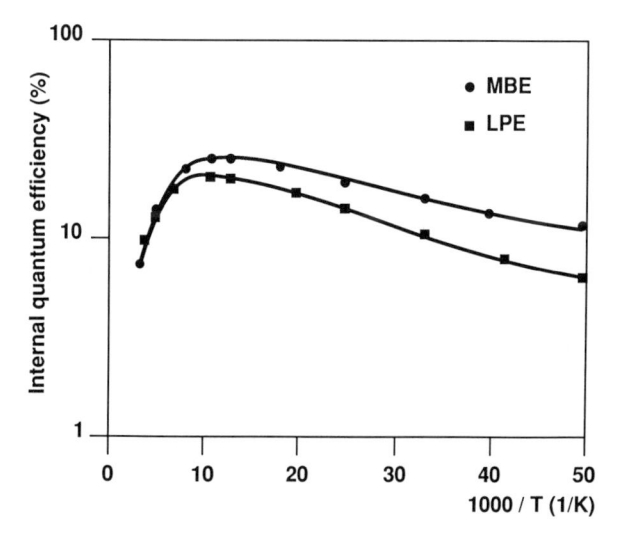

FIG. 3. Internal quantum efficiency vs temperature for a 4.1 μm MCT diode for LPE and MBE material.

Longer wavelength diodes exhibited the same qualitative emission spectra with a maximum of the electroluminescence moving toward lower temperatures and the internal quantum efficiency decreasing with the peak wavelength (Fig. 5). In the 3–5 μm range, LPE and MBE samples with same emission wavelength have similar internal quantum efficiencies.

IV. DISCUSSION

Below 100 K, the main peak energy (Fig. 6) of the electroluminescence for LPE and MBE samples significantly differs from the law deduced from[15] for the energy gap. This shift is attributed[16] to the localization energy of the exciton. Following from this explanation the main peak results from the recombination of excitons localized in potential fluctuations related to alloy disorder at low temperatures and the recombination of free excitons at high temperatures.

The linewidth increases linearly with temperature with a coefficient of 0.15 meV/K (Fig. 7). These data are very consistent with results obtained by Ravid[6] in photoluminescence study of shorter wavelength CMT alloy ($x = 0.5$). First the slope ($1.75\, k_B T$) of the linewidth variation with temperature is the same although wavelengths of interest are merely shifted. This value is consistent with a k conservation model for electron–hole recombination.

Second the discrepancy between their data of peak's energy variation with temperature and this model can be accounted by the localization energy of exciton at low temperature. Only at high temperature > 100 K when free exciton exists can we find a theoretical slope of $0.5\, k_B T$. In their data, as in our data, we can observe a two slope temperature dependence of the peak's energy. When corrected by the gap's law variation the high temperature slope is more or less equal to the theoretical $0.5\, k_B T$. So, only at high temperature are those results consistent with a k conservation model for electron–hole band to band recombination.

At temperatures less than 120 K the MBE sample linewidth shows a slight increase due to the contribution of the unresolved second peak. In the LPE sample the peak is

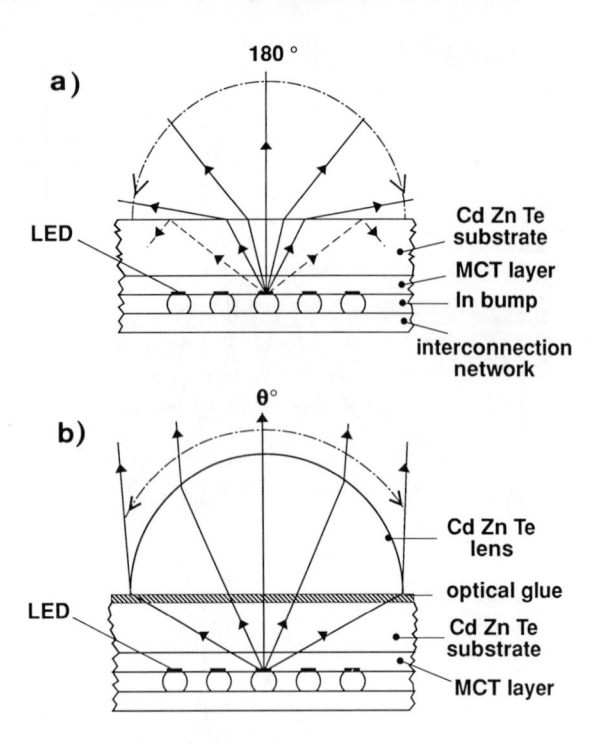

FIG. 4. Schematic of the backside emission (a) without coupling lens, (b) with an optically coupled immersion lens.

The 2 mm diam lens opens up the escape angle of photons from the 18° Bragg limited angle to an impressive 68° limited by geometry. This results in a theoretical increase in the extraction yield by a factor of 13.4. For such a device a tenfold increase of the external quantum efficiency was achieved with a value as high as 5% at a wavelength of 4.1 μm and 77 K. This shows that the emission is isotropic inside the material as was assumed for the evaluation of the internal quantum yield. Moreover, we observe that the divergence of the source is reduced to ± 9° [Fig. 4(b)].

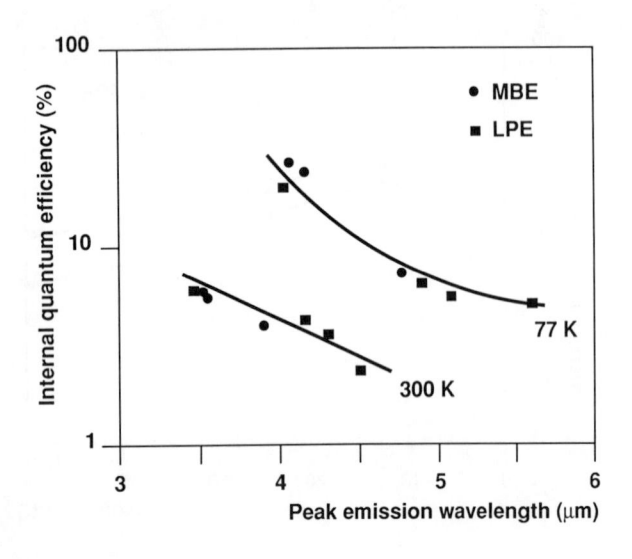

FIG. 5. Internal quantum yield vs peak emission wavelength at 300 and 77 K.

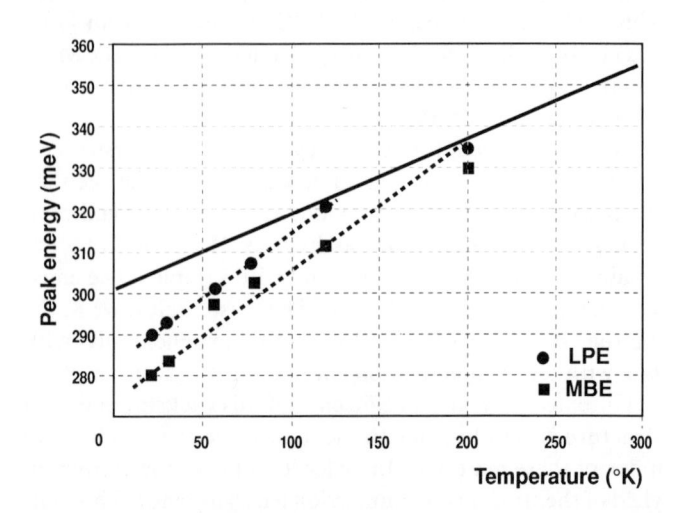

FIG. 6. Energy of main peak vs operating temperature for LPE and MBE sample. Line is the gap energy law from Ref. 11.

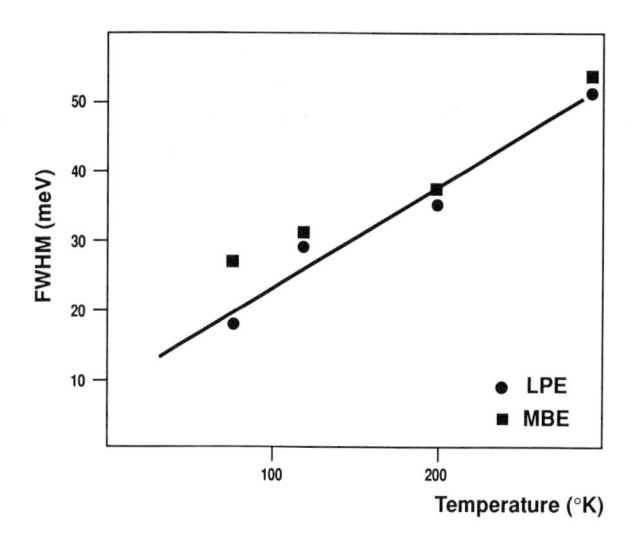

FIG. 7. Linewidth of the main peak vs operating temperature for LPE and MBE sample.

resolved and separated by 13 ± 1 meV from the main peak. This second peak is generally attributed to impurity centers.[2,16,17]

As a 12 meV acceptor binding energy is found in both our samples (LPE and MBE) by Hall measurements, we consider that this second peak is due to a band to acceptor, or donor to acceptor recombination.

The relative intensity of this second peak is higher in the MBE sample than in the LPE sample and contributes to the asymmetry of the main peak, even at high temperature.

As no compositional shift was observed even at the array scale, we are induced to think that this shoulder is not an artefact of the energy gap variation. The p-type carrier concentration of the samples being the same, we attribute the difference in intensity at low temperature to more donor to acceptor recombination in the MBE sample. MBE material, when converted to n-type still results in a greater doping level than LPE material $|N_D - N_A| \sim 10^{16}$ atoms/cm^3 compared to $|N_D - N_A| < 5 \times 10^{14}$ atoms/cm^3 in our LPE material.

As the donor level is completely ionized even at very low temperature, transitions from this level are not distinguishable from conduction band transitions, but could enhance the internal quantum efficiency of electroluminescence.

At high temperature (300 K) there is 1% of nonionized acceptors which can still contribute to donor–acceptor transitions. This ratio seems low but could produce the observable low energy shoulder in high donor concentration materials.

The internal quantum efficiency η_{int} is equal to τ/τ_{rad} where τ is the lifetime of minority carriers and τ_{rad} the radiative lifetime. When considering only Auger recombination as the nonradiative recombination mechanism,[18] η_{int} should have a different behavior than that observed (Fig. 3). The maximum should appear around 200 K, and below 80 K, due to carrier freezout, η_{int} is radiatively dominated and increases toward unity.

Other limiting factors must be introduced to account for the observed variation of η_{int}.

Some authors[19-21] attributed the observed values and variation of lifetime in photodecay or photoabsorption experiments to Schockley–Read recombination. We have no experimental evidence to prove that this type of recombination can account for the observed variation in our samples.

Nevertheless, Zucca et al.[11] have proposed a model to fit the decrease of η_{int} at low temperature. They suggested that there is a current crowding effect on forward bias planar diodes with a high resistivity p-type layer. Such a current effect quenches the internal yield at the periphery of the diode. With such a model, they found that the internal yield was inversely proportional to the resistivity of the p-type layer. In our case, the activation energy of the series resistances is 2.3 meV and we find that the experimental decrease of the internal yield has an activating energy of 2 and 2.9 meV, respectively, for MBE and LPE samples. This appears to be in good agreement with their theory. This crowding effect could constitute the main obstacle of planar technology to achieve good internal yields in more resistive materials for lower wavelength emission.

The 25% maximum value of the internal quantum efficiency obtained in the MBE sample at 4.1 μm peak emission, and 77 K, is the highest value reported at such wavelengths in MCT epilayers. However the decrease with emission wavelength is very sharp and leads to a value of 6% at 5 μm and 77 K.

The 300 K values decrease from 6% at a 3.5 μm peak emission wavelength to 2.5% at 4.5 μm. An empirical law could be fitted as

$$\eta_{int} = \exp(-E_A/E),$$

where $E_A \sim 1$ eV and E(eV) is the energy corresponding to the peak wavelength.

Only at this latest temperature could the Auger limited lifetime account for the variation of the internal quantum efficiency with the peak wavelength.[22]

V. CONCLUSIONS

This electroluminescence study of MCT LEDs shows that even for wavelengths above 4 μm, internal quantum efficiencies as high as 25% can be obtained (at 77 K) in both LPE and MBE epilayers grown on CZT lattice-matched substrates.

Room temperature operation is also possible, since the internal quantum efficiency remains sufficiently high (6% for $\lambda = 3.5$ μm).

An improvement of the external quantum efficiency was obtained using immersion lenses made in CZT and backside-coupled to the chosen hybrid structure.

As far as electroluminescence properties are concerned LPE and MBE materials exhibit comparable performance. This opens up a very encouraging perspective for future injection devices that will require more sophisticated structures which only MBE or MOCVD techniques will be able to shape.

More work is required to understand the internal quantum efficiency limitations which are not only due to Auger recombination mechanism.

As a conclusion MCT appears attractive for infrared LEDs in the 3–5 μm wavelength range.

ACKNOWLEDGMENTS

We would like to thank J. C. Gay for providing us with CZT immersion lenses and M. Ravetto for mounting facilities. This work was supported by DRET (French Ministry of Defense).

[1] C. Verie and R. Granger, C. R. Acad. Sci., Paris **261**, 3349 (1965).
[2] A. T. Hunter and T. C. Mc Gill, J. Vac. Sci. Technol. **21**, 205 (1982).
[3] T. C. Harman, J. Electron Mater. **8**, 191 (1979).
[4] I. Melngailis and A. J. Strauss, Appl. Phys. Lett. **8**, 179 (1966).
[5] V. I. Ivanov Omskii, A. SH. Mekhtiev, R. B. Rutamov, and V. A. Smirnov, Phys. Status Solidi B **130**, K43 (1985).
[6] A. Ravid and A. Zussman, J. Appl. Phys. **67**, 4260 (1990).
[7] K. K. Mahavadi, S. Sivananthan, M. D. Lange, X. Chu, J. Bleuse, and J. P. Faurie, J. Vac. Sci. Technol. A **8**, 1210 (1990).
[8] A. Ravid, A. Zussman, G. Cinader, and A. Oron, Appl. Phys. Lett. **55**, 2704 (1989).
[9] H. A. Tarry, Electron. Lett. **22**, 416 (1989).
[10] J. C. Flachet, M. Royer, Y. Carpentier, and G. Pichard, SPIE **587**, 149 (1985).
[11] R. Zucca, J. Bajaj, and E. R. Blazejewski, J. Vac. Sci. Technol. A **6**, 2725 (1988).
[12] B. Pelliciari, J. Cryst. Growth **86**, 146 (1988).
[13] A. Million, L. Di Cioccio, J. P. Gaillard, and J. Piaguet, J. Vac. Sci. Technol. A **6**, 2813 (1988).
[14] G. Destefanis, J. Cryst. Growth **86**, 700 (1988).
[15] J. Chu, S. Xu, and D. Tang, Appl. Phys. Lett. **43**, 1064 (1983).
[16] A. Lusson, F. Fuchs, and Y. Marfaing, J. Cryst. Growth **101**, 673 (1990).
[17] B. L. Gelmont, V. I. Ivanov-Omskii, V. A. Mal'tseva, and V. A. Smirnov, Sov. Phys. Semicond. **15**, 638 (1981).
[18] S. E. Schacham and E. Finkman, J. Appl. Phys **57**, 2001 (1985).
[19] D. L. Polla, S. P. Tobin, M. B. Reine, and A. K. Sood, J. Appl. Phys. **52**, 5182 (1981).
[20] D. L. Polla and R. L. Aggarwal, Appl. Phys. Lett. **43**, 941 (1983).
[21] J. S. Chen, J. Bajaj, W. E. Tennant, D. S. Lo, M. Brown, and G. Bostrup Mat. Res. Soc. Symp. Proc. **90**, 287 (1987).
[22] G. C. Osbourn and D. L. Smith, Phys. Rev. B **20**, 20 (1979).

Electrical properties of modulation-doped HgTe–CdTe superlattices

S. Hwang, Y. Lansari, Z. Yang, J. W. Cook, Jr., and J. F. Schetzina

Physics Department, North Carolina State University, Raleigh, North Carolina 27695-8202

(Received 3 October 1990; accepted 17 December 1990)

Growth of modulation-doped HgTe–CdTe superlattices (SLs) at very low temperatures (140 °C) by photoassisted molecular beam epitaxy is reported. SL layer thicknesses were intentionally chosen such that most of the SLs studied are inverted-band semimetals or inverted-band semiconductors. Both p- and n-type samples were successfully prepared and studied. The doped superlattices exhibit excellent electrical properties. Lack of carrier freeze-out at low temperatures provides convincing evidence that modulation-doping has been achieved.

I. INTRODUCTION

The potential use of HgTe–CdTe superlattices (SLs) in infrared (IR) detector applications, particularly at very long wavelengths ($>12\,\mu$m), has stimulated interest in these new quantum structures. Theoretical and experimental studies, focused primarily on determining the band gap, band structure, and valence band offset, have revealed a number of unique SL properties that are of potential importance in device applications.[1-4] For example, highly anisotropic electron and hole effective masses, with large mass components in the superlattice growth direction, could lead to smaller tunneling currents in superlattice-based p-on-n detector structures than in corresponding HgCdTe alloy detectors.

However, the use of SLs in infrared detector and other optoelectronic applications generally requires precise control of the SL electrical properties through substitutional doping. In order to address the issue of doping, we have employed photoassisted molecular beam epitaxy (MBE), along with modulation doping techniques, to produce both n- and p-type superlattices. Low substrate growth temperatures (140 °C) were employed to minimize SL layer interdiffusion effects.

Modulation doping involves the transfer of carriers from a substitutionally doped barrier layer (modifier layer) to an adjacent matrix material having a smaller band gap.[5] It has been shown to enhance carrier transport in planar III–V heterostructures. More recently, this technique has also been employed to achieve doping of HgCdTe alloys[6-8] and HgTe–CdTe superlattices.[9,10]

In the present study, $Hg_{0.15}Cd_{0.85}Te$ barrier layers of thickness L_b, doped with either In or As, and HgTe matrix layers of thickness L_z were alternately deposited by means of photoassisted MBE to produce modulation-doped SLs. The electrical properties of these SLs were studied in detail by recording and analyzing magnetic-field-dependent Hall data over the temperature range 4.2–300 K. Both n-type and p-type samples exhibited clear evidence of modulation doping, with doping levels controlled by the MBE dopant oven temperature employed during growth.

SL layer thicknesses were intentionally chosen such that most of the SLs studied are *inverted-band* semimetals or *inverted-band* semiconductors—a new SL semiconducting regime only recently proposed theoretically.[11] The three principal regimes of the HgTe–CdTe SL are illustrated qualitatively by the band structure diagrams shown in Fig. 1. The diagram at the left in Fig. 1 illustrates the normal semiconducting regime. In this case, the SL miniband $E1$ is the lowest-energy conduction band, $H1$ is the lowest-energy heavy hole miniband, $H2$ is the second heavy hole miniband, and $L1$ is the lowest-energy light hole miniband, respectively. This semiconducting regime occurs for *small* HgTe layer thicknesses. For example, in a (100) SL having barrier layers of thickness $L_b \approx 50$ Å, the normal semiconducting regime occurs for HgTe layer thicknesses $L_z \approx 75$ Å or less. In this regime, the SL band gap is determined by the energy difference $E1-H1$ at the center of the SL Brillouin zone ($k=0$). In this *normal semiconducting* regime, the band gap decreases as L_z increases until the SL band gap becomes zero. This signals the onset of the *semimetallic* regime, illustrated by the center diagram of Fig. 1. With any further increase in L_z, the $E1$ and $H1$ bands invert and overlap to produce the *inverted-band semimetallic* regime. In this regime, the zero-band-gap point moves from $k=0$ towards the edge of the Brillouin zone at $k_\perp = \pi/d$ ($d=L_z+L_b$) as L_z increases (assuming negligible dispersion for $H1$ along $k_\perp$, corresponding to thick barrier layers). With still further increase in L_z, the $E1$ band falls below the $H1$ band and the *inverted-band semiconducting* regime occurs. Note that in this regime, the band gap is determined by the energy differ-

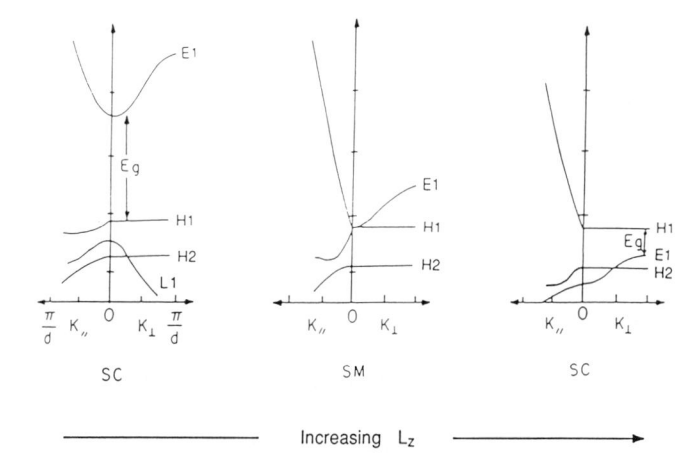

FIG. 1. The band structure of HgTe–CdTe SLs shown qualitatively as a function of well thickness L_z. With increasing L_z, the SL evolves from the normal semiconducting regime, through the semimetallic regime, and into the inverted-band semimetallic regime.

ence $H1 - E1$ at $k_\perp = \pi/d$, the edge of the Brillouin zone. This is a real, *positive* band gap as manifested by an optical absorption edge in the IR. Note also that in this inverted-band semiconducting regime, the $H1$ band serves as an electron conduction band while $E1$ becomes a hole valence band. Ultimately, with further increase in L_z, the $E1$ band falls below $H2$. In this case, the SL "thermal-activation" band gap is determined by the energy difference $H1 - H2$, while the "optical" band gap is determined by the energy difference $H1 - E1$, since the optical transition $H2 - H1$ is forbidden.

II. EXPERIMENTAL DETAILS AND DATA ANALYSIS

The modulation-doped superlattice samples were grown in a Hg-compatible MBE system designed and built at NCSU. A description of this MBE system together with the techniques that have been developed to grow HgCdTe by both conventional and photoassisted MBE are given in earlier publications.[7,12] In the present work, modulation-doped superlattices were grown by alternately depositing $Hg_{0.15}Cd_{0.85}Te$ barrier layers that were heavily doped with either indium or arsenic with undoped HgTe layers. In this way, superlattices consisting of 200 double layers were prepared. The thickness of the HgCdTe doping layers was about 50 Å in each of the superlattices, with the dopant uniformly distributed. The temperatures of the In and As ovens were varied to achieve different doping levels. The HgTe layers in the various SLs have thicknesses ranging from about 23–104 Å. The individual layer thicknesses for each SL were determined by using DEKTAK total thickness measurements, the angular separation of x-ray satellite peaks relative to the main diffraction peak, and the growth rates of HgTe and $Hg_{0.15}Cd_{0.85}Te$ determined using calibration layers grown under the same conditions. We estimate that the individual layer thicknesses obtained by this method are accurate to about one monolayer (± 3.2 Å). This is supported by direct layer thickness measurements on selected SLs by means of vertical cross-section transmission electron microscopy. A nominal x value of 0.85 was obtained for the SL barrier layers by determining the band gap of thicker calibration layers grown under the same conditions using low temperature photoluminescence measurements. Although the x value of the actual SL layers, which are much thinner than the calibration layers, may be somewhat different due to shuttering effects, etc., the precise value of the barrier layer x value is not required for accurate band structure calculations. For example, changing the barrier layer x value from 0.85 to 0.7 changes the calculated band gap of SLs in the semiconducting regime by ~ 2 meV, or about the band gap change that occurs when layer thicknesses are changed by one monolayer.

The SL layer thicknesses were intentionally chosen such that most of the SLs studied are inverted-band semimetals or inverted-band semiconductors. To study their electrical properties, the SL samples were mounted in a Janis Super-Varitemp dewar equipped with a 9 T superconducting magnet and cooled to liquid helium temperatures. Standard van der Pauw Hall effect measurements were completed as a function of applied magnetic field and temperature. The magnetic-field-dependent transverse components $\sigma_{xx}(B)$ and $\sigma_{xy}(B)$ of the conductivity tensor were calculated from the Hall coefficients and resistivities for the data analysis described below.

Meyer et al.[14] first reported the presence of multiple carrier species in HgTe–CdTe SLs. They used a least-square fitting procedure to determine the number and type of carrier species as well as the corresponding densities and mobilities from magnetic-field-dependent Hall data.[15] In the present study, we use two different but complementary techniques to extract the information. The first is the mobility spectrum analysis developed by Beck and Anderson,[16] and the second consists of a combined linear and non-linear least-square algorithm that was developed at NCSU.

The mobility spectrum analysis is a mathematical mapping technique that determines the maximum conductivity density function $s(\mu)$ as a function of mobility μ (or the mobility spectrum) from σ_{xx} and σ_{xy}. Carrier species with different mobilities appear as peaks in the spectrum, and the corresponding carrier concentration can be obtained from the peak value. The mobility spectrum analysis does not require an *a priori* assumption about the number of carrier species and has been shown to be very useful in predicting the number and type of carrier species in the sample.[17,18] However, we have noticed that, in some cases, peak values and positions in the mobility spectrum change when complementary set of conductivity data taken for the same sample at similar but different magnetic fields are employed in the analysis. In order to obtain a more quantitative result, we have developed a least-square algorithm to extract the carrier concentrations and mobilities. In the case of mixed-conduction involving m species of carriers, σ_{xy} can be expressed as[15,16]

$$\sigma_{xy}(B) = \sum_{j=1}^{m} S_j \frac{n_j e \mu_j^2 B}{1 + \mu_j^2 B^2}, \qquad (1)$$

where n_j and μ_j are the concentration and mobility, respectively, of the jth species, and S_j is $+1$ for holes and -1 for electrons. We note that σ_{xy} is linear (in the least-square sense) with respect to n_j but nonlinear with respect to μ_j. If we know the exact values for the μ_j's, the corresponding n_j's can be obtained with a linear least-square technique in one step. On the other hand, μ_j can be obtained with a nonlinear least-square technique in an iterative fashion, provided that n_j's are known and initial estimates for the μ_j's are available. In our analysis, we start from the number and type of carrier species, as well as their mobilities, that were determined from the mobility spectrum analysis. We then perform a linear least-square fit to Eq. (1) to obtain an estimate for the n_j's. This is followed by a nonlinear least-square fit to the same equation to improve the values for the μ_j's. This fitting process is repeated self-consistently until the chi-square is minimized. We have employed the singular value decomposition (SVD) technique and the Levenberg-Marquardt method for the linear and nonlinear least-square fit, respectively.[19] This approach has the advantage of being free from the problems associated with matrix singularity that could undermine the fitting process. The goodness-of-fit was deter-

TABLE I. Sample characteristics.

			n-type superlattices				
Sample	L_z (Å)	L_b (Å)	E_g (meV)	T_{In} (°C)	$n^{4.2\,K}$ (cm^{-3})	$\mu_n^{4.2\,K}$ (cm^2/V s)	
A66A	78	52	0[a]	400	$\sim 1.0 \times 10^{16}$	$\sim 2.8 \times 10^4$	
A67B	81	52	0[a]	425	$\sim 1.3 \times 10^{16}$	$\sim 4.9 \times 10^4$	
A68A	78	49	0[a]	450	$\sim 1.8 \times 10^{16}$	$\sim 3.9 \times 10^4$	
A69A	58	49	37.9[b]	450	1.4×10^{17}	2.6×10^4	
A70A	104	49	24.4[c]	450	8.2×10^{16}	2.6×10^4	
A71A	87	49	9.3[d]	450	8.2×10^{16}	2.1×10^4	

			p-type superlattices				
Sample	L_z (Å)	L_b (Å)	E_g (meV)	T_{As} (°C)	$p^{4.2\,K}$ (cm^{-3})	$\mu_p^{4.2\,K}$ (cm^2/V s)	
A72A	84	58	8.8[d]	200	5.0×10^{15}	1.8×10^3	
A73	68	49	15.5[b]	220	2.0×10^{16}	9.3×10^2	
A74	32	30	157.3[b]	220	1.4×10^{17}	2.2×10^2	
A75	84	49	0[a]	220	2.1×10^{16}	6.1×10^3	
A76	23	26	267.9[b]	220	8.0×10^{16}	2.4×10^2	

[a] $E_g = 0$ due to $E1 - H1$ anticrossing (inverted-band semimetallic).
[b] $E_g = E1 - H1$ (normal semiconducting).
[c] $E_g = H1 - H2$ (inverted-band semiconducting).
[d] $E_g = H1 - E1$ (inverted-band semiconducting).

mined by the chi-square and the incomplete gamma function (IGF). Applying this algorithm to both simulated and real experimental data, we found that the number of different carriers, carrier type, and values for both carrier concentration and mobility are consistent with those determined by the mobility spectrum analysis.

Using the mixed-conduction analysis described above, we have investigated the magnetotransport properties of a series of ten modulation-doped SL samples in the temperature range 4.2–300 K. The SL layer thicknesses, band gaps, dopant oven temperature, and low temperature electrical data are summarized in Table I. The SL band gaps listed were calculated using a six-band $\mathbf{k} \cdot \mathbf{p}$ model.[20] Because neither the mobility spectrum analysis nor Eq. (1) is valid at high magnetic fields, all of the magnetotransport data analyzed in the present work were taken using magnetic fields of 2 T or less.

III. RESULTS AND DISCUSSION

A. Indium-doped superlattices

Previous investigations have shown that undoped HgTe–CdTe SLs often exhibit p-type conductivity at low temperatures.[14,15] In contrast, all of the SL samples that were modulation-doped with indium exhibited n-type conductivity over the entire temperature range. In the present study, the indium dopant oven temperature was varied between 400 and 475 °C. For lightly doped samples (A66A, A67B—Table I), the electron concentration exhibits an activated behavior as the temperature increases from 4.2 to 300 K. This is as expected, since the doping level is low and most of the carriers are intrinsic. As the indium oven temperature was raised to 450–475 °C, the effect of modulation doping sets in and the electron concentration reaches to a plateau at low temperatures. This is illustrated by the carrier concentration versus temperature data sets shown in Fig. 2 for SL samples A68A, A69A, A70A, and A71A, respectively. In these SL samples, substantial numbers of donor electrons transfer from the Hg$_{0.15}$Cd$_{0.85}$Te modifier layers to the HgTe matrix layers, the donor level being more than 1 eV higher than the bottom of the conduction band. At very low temperatures these transferred electrons are distributed among the available states according to Fermi–Dirac statistics. Since essentially all the available states are the delocalized superlattice states, the electron concentration remains constant in this temperature range. That no electron freeze-out is observed even at 4.2 K indicates that essentially all the indium donors reside in the modifier layer. Consequently, no substantial reduc-

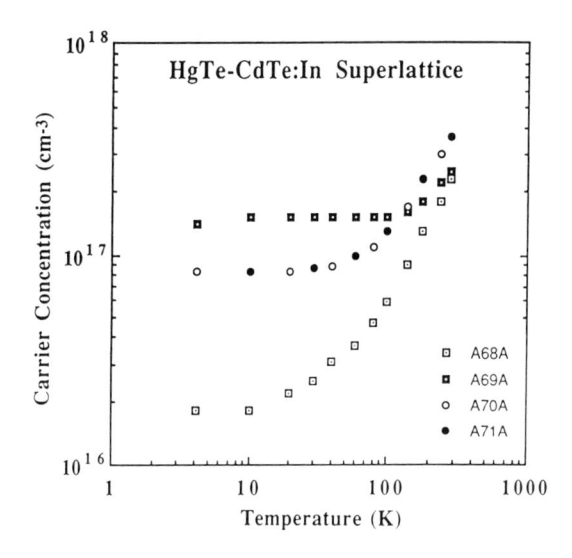

FIG. 2. Carrier concentration vs temperature for A68A, A69A, A70A, and A71A, whose characteristics are summarized in Table I. The curves for A70A and A71A essentially overlap each other.

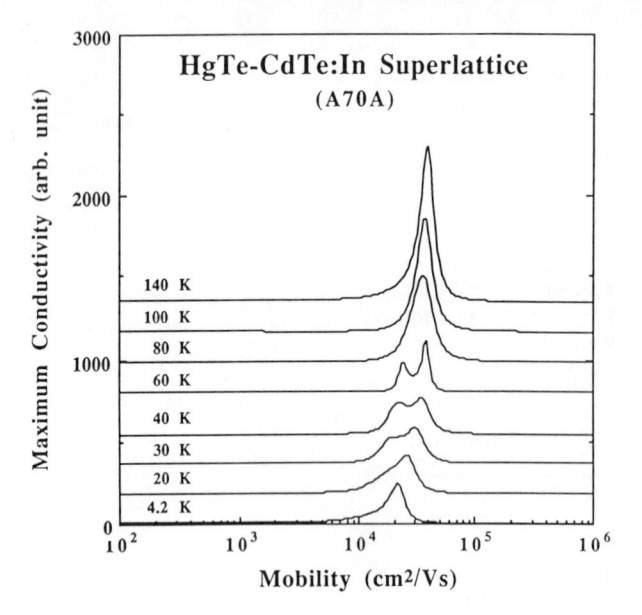

FIG. 3. The low temperature mobility spectra for A70A. The spectra were calculated using magnetic fields of 0, 0.2, 0.4, 0.6, 0.8, and 1 T and a relative error of 10⁻⁴.

tion in electron mobility due to impurity scattering was observed as the doping level was raised. We expect the mobility to be further improved when undoped spacer regions are introduced in the modifier layers, providing additional spatial separation between the transferred electrons and the ionized indium impurities. The modulation doping technique is very reproducible, as is demonstrated by the electron concentration curves for samples A70A and A71A, which are essentially overlapping.

Theory predicts that a band broadening effect exists in semimetallic HgTe–CdTe superlattices as a result of the anticrossing of the $E1$ and $H1$ bands in the $k_\perp$ direction.[15] Our mobility spectrum analysis for samples A70A and A71A indicates that the band broadening effect exists in these two samples, although it is more pronounced in A70A, which has thicker wells. The band broadening effect appears as a broadened peak in the mobility spectrum for A71A and as two overlapped peaks in that for A70A. Figure 3 shows the mobility spectrum for A70A between 4.2 and 140 K. The spectrum exhibit a single, sharp peak at high temperatures.

As the temperature decreases to below 100 K, however, the peak broadens and develops into two overlapped peaks. In addition, least-square analysis shows that attempts to fit the field-dependent Hall data in this temperature range with one type of electron leads to poor agreement between model and data, as indicated by the large chi-square and small IGF. However, substantial improvement in goodness-of-fit (chi-square decreases by at least an order of magnitude) can be achieved by assuming two or more species of electrons. Therefore, both techniques suggest the existence of multiple carrier species in these samples. Figure 3 also indicates a reduced mass broadening effect at very low temperatures. To further study the matter, we performed least-square fit at each temperature. In the analysis, two types of electrons were assumed, for the improvement in goodness-of-fit is not large enough to justify an assumption of three. Figure 4(a) shows the mobility data obtained for A70A. We note that at low temperatures there exist two types of electrons, which exhibit similar mobility behavior as temperature is lowered. Since scattering mechanisms are expected to be nearly identical, the two electron mobilities are most likely due to two different electron effective masses associated with different regions of the overlapped $E1$ and $H1$ bands. The corresponding concentration data for SL A70A are shown in Fig. 4(b). In the temperature range between 4.2 and 60 K, in which the total electron concentration is a constant, the electrons appear to be redistributing themselves among the available states since, as the temperature decreases, the population of the high mobility electrons gradually increases at the expense of low mobility electrons due, perhaps, to the temperature dependence of the $E1 - H1$ inverted-band crossing effect.

Note, from Table I, that n-type modulation doping has been successfully demonstrated for both regular and inverted-band semiconducting SLs, as well as for inverted-band semimetallic SLs.

B. Arsenic-doped superlattices

Although undoped HgTe–CdTe superlattices usually exhibit p-type characteristics at low temperatures, p-type doping control is by no means straightforward, since the introduction of external dopant atoms may be accompanied by

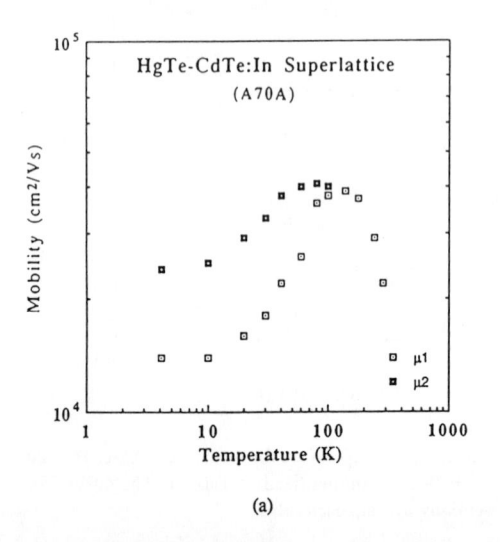

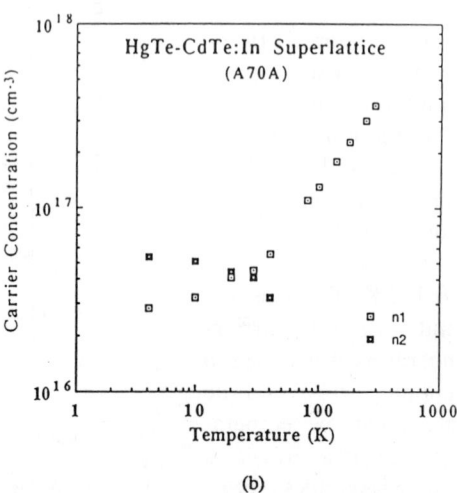

FIG. 4. Mobility (a) and carrier concentration (b) vs temperature for A70A. n1 and n2 correspond to the carrier concentrations of the high mobility (μ1) and low mobility (μ2) electrons, respectively.

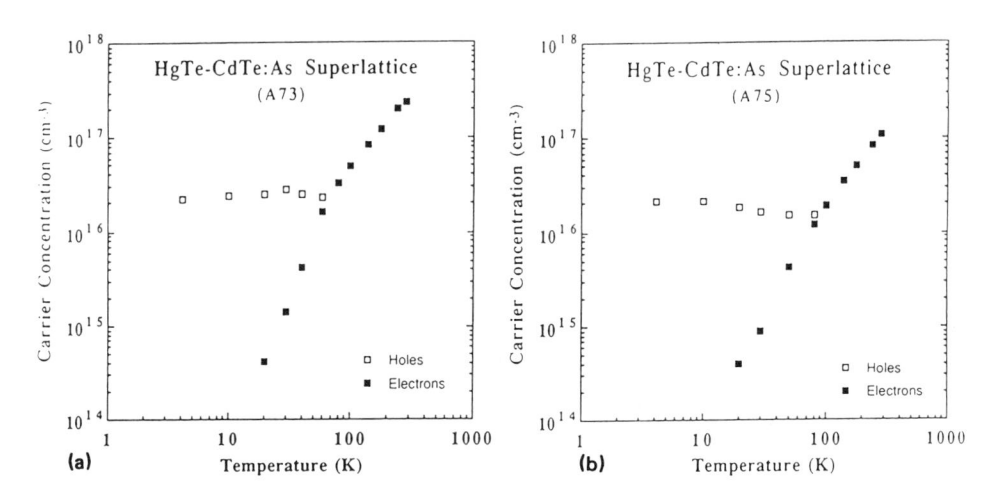

FIG. 5. Carrier concentration vs temperature for A73 (a) and A75 (b).

compensating defects. Our previous studies have shown that arsenic atoms can be successfully incorporated into modifier layers to produce stable p-type modulation-doped HgCdTe quantum alloys.[6-8] In the present work, arsenic oven temperatures were varied from 200 to 220 °C to produce p-type HgTe–CdTe superlattices in the semiconducting and semimetallic regions. The mixed conduction analysis is particularly valuable in this case, because the electrical properties are often overwhelmed by the presence of intrinsic electrons. For sample A72A the arsenic oven temperature of 200 °C is too low to produce a sizable hole concentration. The sample exhibits n-type conductivity even at 4.2 K, but the mobility spectrum revealed a hole concentration somewhat higher than the electron concentration, as listed in Table I. Higher arsenic oven temperature increases the doping level. Figure 5 shows hole concentration versus temperature data obtained by the least-square fitting method for samples A73 and A75. Again, the effect of modulation doping is clearly demonstrated by the constant hole concentration at temperatures between 4.2 and 80 K. At higher temperatures, the transport properties are dominated by electrons, and the least-square fitting process becomes insensitive to the holes. These data for SLs A73 and A75 demonstrate that reproducibility is also achieved for p-type modulation doping.

In addition to the mass broadening effect for electrons, the $E1 - H1$ anticrossing also predicts a broadening of the hole effective mass. Although the hole effective mass at the zone center increases rapidly as the Γ-point gap becomes more negative ($L_z > 20$ monolayers), the effective mass near the anticrossing point remains small, and holes with a range of different effective masses could coexist at the same time. The 4.2 K mobility spectrum for A75, an inverted-band semimetallic SL (Table I), is shown in Fig. 6. The multiple peaks in the spectrum are interpreted as preliminary evidence of mass-broadening for holes. However, 90% of the holes reside in the states with the lowest mobility, and the population of high mobility holes decreases as temperature is raised. At 60 K, only low mobility holes can be resolved in the mobility spectrum. Also shown in Fig. 6 is the mobility spectrum for A73, which exhibits a less pronounced broadening effect. According to our calculations, A73 is in the normal semiconducting regime, with $E1$ being the conduction band (Table I). Thus, the mass broadening observed in this sample is most likely due to band nonparabolicity.

Note, from Table I, that p-type modulation doping has been successfully demonstrated for both regular and inverted-band semiconducting SLs, as well as for inverted-band semimetallic SLs.

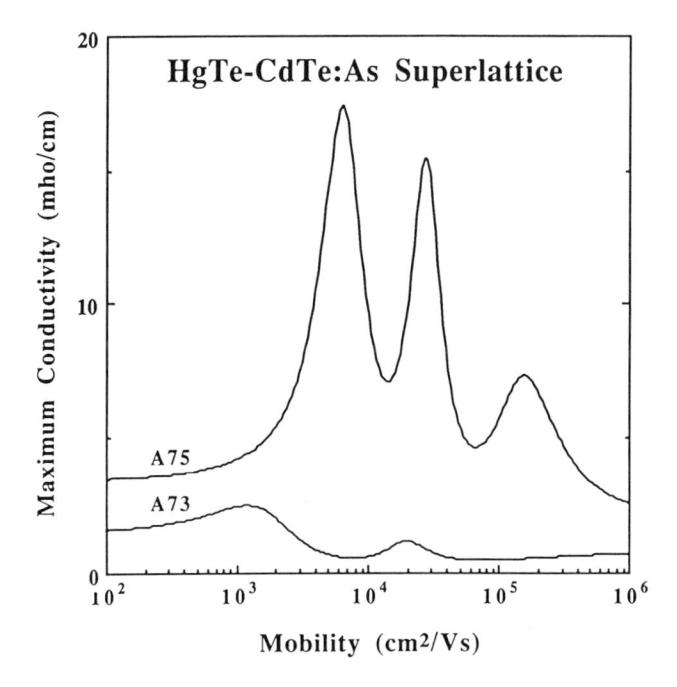

FIG. 6. The 4.2 K mobility spectra for A73 and A75. The spectra were calculated using magnetic fields of 0, 0.2, 0.4, 0.6, 0.8, and 1 T and a relative error of 10^{-4}.

IV. SUMMARY

Modulation-doped HgTe–CdTe superlattices with band gaps in the semimetallic region have been grown at low temperatures (140 °C) by means of photoassisted MBE. Indium and arsenic were used as n- and p-type dopants, respectively. The electrical properties of the SL samples were investigated using a mobility spectrum analysis, combined with a linear and nonlinear least-square fitting technique. Clear evidence of modulation doping has been demonstrated, and the doping results are highly reproducible. Carrier mass broadening effects have also been observed in both *n*- and *p*-type superlattices.

ACKNOWLEDGMENTS

The authors wish to acknowledge J. Matthews, M. Bennett, and A. Mohan for their assistance with substrate preparation, and P. K. Baumann for his assistance with Hall data acquisition. One of the authors (S. H.) also wishes to thank Dr. W. A. Beck for valuable discussions and assistance in coding the mobility spectrum analysis program, and Dr. J. R. Meyer for a useful discussion on the least-square technique employed at the Naval Research Laboratory (NRL). This work was supported by NRL Contract No. N00014-89-J-2024.

[1] J. N. Schulman and T. C. McGill, Appl. Phys. Lett. **34**, 663 (1985).
[2] D. L. Smith, T. C. McGill, and J. N. Schulman, Appl. Phys. Lett. **43**, 180 (1983).
[3] T. C. McGill, G. Y. Wu, and S. R. Hetzler, J. Vac. Sci. Technol. A **4**, 2091 (1986).
[4] J. R. Meyer, C. A. Hoffman, and F. J. Bartoli, Semicond. Sci. Technol. **5**, S90 (1990).
[5] R. Dingle, H. L. Stormer, A. C. Gossard, and W. Wiegmann, Appl. Phys. Lett. **37**, 805 (1978).
[6] Jeong W. Han, S. Hwang, Y. Lansari, R. L. Harper, Z. Yang, N. C. Giles, J. W. Cook, Jr., J. F. Schetzina, and S. Sen, J. Vac. Sci. Technol. A **7**, 305 (1989).
[7] J. F. Schetzina, Jeong W. Han, S. Hwang, Y. Lansari, R. L. Harper, Z. Yang, N. C. Giles, J. W. Cook, Jr., and S. Sen, in *Proceedings of the International Conference on Beam-Solid Interactions (Ankara, 1989)*, edited by R. Ellialtioglu and S. Ellialtioglu [DOGA-Tr. J. Phys. **14**, 65 (1990)].
[8] J. F. Schetzina, J. W. Han, Y. Lansari, N. C. Giles, Z. Yang, S. Hwang, J. W. Cook, Jr., and N. Otsuka, J. Cryst. Growth **101**, 23 (1990).
[9] Jeong W. Han, S. Hwang, Y. Lansari, Z. Yang, J. W. Cook, Jr., and J. F. Schetzina, J. Vac. Sci. Technol. B **8**, 205 (1990).
[10] S. Hwang, Z. Yang, Y. Lansari, J. W. Han, J. W. Cook, Jr., N. C. Giles, and J. F. Schetzina, Mat. Res. Soc. Symp. Proc. **161**, 263 (1990).
[11] P. M. Hui, H. Ehrenreich, and N. F. Johnson, J. Vac. Sci. Technol. A **7**, 424 (1989).
[12] K. A. Harris, S. Hwang, D. K. Blanks, J. W. Cook, Jr., N. Otsuka, and J. F. Schetzina, J. Vac. Sci. Technol. A **4**, 2061 (1986).
[13] A. C. Beer, *Galvanomagnetic Effects in Semiconductors* (Academic, New York, 1963).
[14] J. R. Meyer, C. A. Hoffman, F. J. Bartoli, J. W. Han, J. W. Cook, Jr., J. F. Schetzina, X. Chu, J. P. Faurie, and J. N. Schulman, Phys. Rev. B **38**, 2204 (1988).
[15] C. A. Hoffman, J. R. Meyer, F. J. Bartoli, J. W. Han, J. W. Cook, Jr., J. F. Schetzina, and J. N. Schulman, Phys. Rev. B **39**, 5208 (1989).
[16] W. A. Beck and J. R. Anderson, J. Appl. Phys. **62**, 541 (1987).
[17] W. A. Beck, R. A. Wilson, and A. C. Goldberg, J. Cryst. Growth **81**, 136 (1987).
[18] W. A. Beck, F. Crowne, J. R. Anderson, M. Gorska, and Z. Dziuba, J. Vac. Sci. Technol. A **6**, 2772 (1988).
[19] W. H. Press, B. P. Flannery, S. A. Teukolsky, and W. T. Vetterling, *Numerical Recipes: The Art of Scientific Computing* (Cambridge University, Cambridge, 1989), Chap. 14.
[20] Z. Yang, J. F. Schetzina, and J. K. Furdyna, J. Vac. Sci. Technol. A **7**, 360 (1989).

Optical and magneto-optic properties of HgTe/CdTe superlattices in the inverted-band semiconducting regime

Z. Yang, Z. Yu, Y. Lansari, J. W. Cook, Jr., and J. F. Schetzina

Department of Physics, North Carolina State University, Raleigh, North Carolina 27695

(Received 3 October 1990; accepted 16 November 1990)

Low temperature infrared transmission and far-infrared magneto-optic transmission experiments were completed for a series of HgTe/CdTe superlattices (SLs). The SLs studied were grown with layer thicknesses intentionally chosen to make the samples inverted-band semimetals or inverted-band semiconductors, a new regime of the HgTe/CdTe SL which only recently has been predicted by theory. Cyclotron resonance of electrons in the first conduction subband $H1$ was observed in the far-infrared magneto-transmission experiments. From these measurements electron effective masses were calculated. The optical transition from the second heavy hole subband $H2$ to the second conduction subband $E2$ was observed in the infrared transmission experiments. The $H2$–$E2$ transition energy was observed to *decrease* with *increasing* SL well width L_Z. In addition, the electron effective mass was found to *increase* as L_Z increases. Both of these observations indicate that all of the SL samples are, indeed, *inverted-band* SLs. This in turn implies that the valence band offset between HgTe and CdTe must be large, ~400 meV at 4.5 K.

I. INTRODUCTION

In HgTe/CdTe superlattices (SLs) having a fixed barrier layer thickness L_B, the band gap, determined by the energy difference between the first conduction subband $E1$ and the first heavy hole valence subband $H1$, decreases as the HgTe well width L_Z increases until the band gap becomes zero. In this *normal semiconducting* regime, the in-plane effective masses of the electrons and holes in the SL layer plane also decrease with increasing L_Z and became very small when the SL band gap approaches zero. With a further increase in L_Z, the $E1$ and $H1$ bands invert and the SL becomes *semimetallic* with overlapping bands.

Recently, however, Hui, Ehrenreich, and Johnson[1] showed theoretically that as L_Z increases still further, the semimetallic regime ends and a new *inverted-band semiconducting* regime occurs. This unexpected result successfully explains the valence-band offset controversy in HgTe/CdTe SLs in favor of a large offset value (>300 meV). Recent magnetotransport measurements[2,3] support this theoretical model. The band structure for an inverted-band semiconducting HgTe/CdTe superlattice is shown qualitatively in Fig. 1. For a (100) superlattice at 4.2 K having $L_B \approx 50$ Å, the inverted-band semiconducting regime occurs for $L_Z \approx 80$ Å or greater. Note that in this regime, the SL band gap is determined by the energy difference $H1 - E1$ at $k_\perp = \pi/d (d = L_Z + L_B)$, the edge of the Brillouin zone. This is a real, *positive* band gap as manifested by an optical absorption edge in the infrared (IR). Note also that in the inverted-band semiconducting regime, the $H1$ band serves as an electron conduction band while $E1$ becomes a hole valence band. Ultimately, with further increase in L_Z, the $E1$ band falls below $H2$. In this case, the SL "thermal-activation" band gap is determined by the energy difference $H1 - H2$, while the "optical" band gap is determined by the energy difference $H1 - E1$, since the optical transition $H2 - H1$ is forbidden. Far-infrared (IR) magneto-transmission experiments are ideal for observing cyclotron resonance of free carriers from which effective masses can be calculated. In addition, IR transmission spectra of HgTe/CdTe SLs reveal optical transitions between the subbands (such as $H1 - E1$ or $E1 - E2$ in Fig. 1), the transition energies being directly related to SL well and barrier widths and the valence band offset V_p.

In this article we report, for the first time, the optical and magnetooptic properties of a set of SLs specifically grown with layer thicknesses corresponding to the inverted-band SL regime, with most of the samples being inverted-band semiconductors at 4.5 K. Low temperature electron cyclotron resonance was observed for all SL samples in the far-IR magnetooptic transmission experiments when the magnetic field was perpendicular to the SL layers (Faraday geometry). The $H2 - E2$ optical transition in the same samples was observed in low temperature IR transmission experiments. It was found that the $H2 - E2$ transition energy *decreased* as the well width L_Z *increased*, indicating that there is less and less quantum confinement with increasing L_Z. At the same time, it was found that the HgTe/CdTe SL in-plane

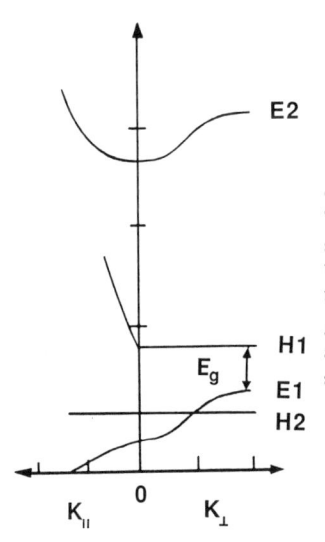

FIG. 1. Subband structure of an inverted-band HgTe/CdTe superlattice. The lowest conduction subband is $H1$ subband and the highest valence subband is $E1$. The energy difference between these bands is the SL band gap. Higher energy conduction ($E2$) and valence ($H2$) bands are also shown.

electron effective mass *increased* with *increasing* L_Z. To our knowledge, this is the first direct experimental evidence that such an unusual dependence of the electron mass on L_Z exists. Our results demonstrate unambiguously that all the SLs studied are, in fact, inverted-band SLs. The results also provide additional direct experimental evidence for a large valence band offset for the HgTe/CdTe SL.

II. EXPERIMENTAL DETAILS

Nine SLs were used in the IR transmission and far-IR magneto-optic transmission experiments. The SL samples were grown in a Hg-compatible molecular beam epitaxy (MBE) system designed and built at NCSU. All of the samples were grown at 140 °C by photoassisted molecular beam epitaxy and consisted of 200 double layers. The well material in each SL was HgTe and the barrier material was $Cd_{0.85}Hg_{0.15}Te$. The SLs were all n-type at low temperatures with carrier concentrations ranging from the low 10^{16} cm^{-3} to low 10^{17} cm^{-3}, obtained by conventional Hall measurements. The well and barrier thicknesses of each SL were obtained by double-crystal x-ray diffraction, a knowledge of the layer growth rates, and a measurement of the total SL thickness. The HgTe well thickness L_Z and $Hg_{0.15}Cd_{0.85}Te$ barrier thickness L_B for each SL are listed in Table I.

The low temperature IR transmission spectra of the SLs in the spectral region from 2–14 μm were measured at 4.5 K. In these experiments, the SLs were mounted in an optical IR cryostat and cooled using liquid He. A glowbar light source was used to pass broad-band light through the SL sample, and the transmitted light was collected and analyzed using a Nicolet model 60-SXR Fourier transform infrared spectrometer (FTIR). From these measurements, SL IR absorption coefficients were obtained.

Far-IR magnetooptic transmission spectra were also measured at 4.5 K, in the spectral range from 96.5 to 495 μm, using magnetic fields up to 7 T. In these experiments, the SL samples were placed in a Janis model 12-CNDT split-coil superconducting magnet cryostat equipped with far-IR optical windows. A far-infrared laser (Apollo Laser model 122) was used to illuminate the SLs at normal incidence with the laser beam parallel to the applied magnetic field (Faraday geometry). The incident laser beam was circularly polarized so that either electron cyclotron-resonance-active (CRA) polarization or cyclotron-resonance-inactive (CRI) polarization could be obtained. The far-IR laser light transmitted through the SL was detected with a Si bolometer. A portion of the incident far-IR laser beam was monitored with a second Si bolometer in order to eliminate the effect of laser intensity fluctuations on the measured SL transmittance.

III. RESULTS AND ANALYSIS

Figure 2 shows IR absorption spectra for three SLs at 4.5 K. It is noted that a large absorption edge occurs near photon energies at 310, 237, and 125 meV for samples A67B, A58B, and A27A, respectively. Note that this absorption edge shifts to lower photon energy as the well width increases. The sharp absorption peak at 125 meV for sample A27A indicates that the sample was of excellent optical quality. Similar sharp peaks are observed in the absorption spectra of the other SLs used in this study, attesting to their high quality. This is the first time that sharp edge peaks have been observed in the IR absorption of HgTe/CdTe SLs. The absorption is non-zero over the entire spectral region studied, indicating that the optical band gaps of these SLs are, indeed, very small (< 90 meV).

The experimental results were analyzed using a six-band **K·P** model.[4] Both the $E1$–$E2$ and the $H2$–$E2$ transitions are strong transitions. The $H2$–$E2$ transition is expected to be about 3 times stronger than the $E1$–$E2$ transition. We believe the sharp absorption edge peak observed for each of the three SLs of Fig. 2 (labeled $H2$–$E2$ in the figure) is due to the combined contribution of these two quantum transitions.

TABLE I. The list of the well and barrier thicknesses of the superlattice samples used in this work. The numbers inside the parentheses are the corresponding number of monolayers for each layer thickness.

Sample no.	Well width L_Z (Å) (monolayers)	Barrier width L_B (Å) (monolayers)
A68B	77.5 (24)	48.5 (15)
A67A	80.8 (25)	51.7 (16)
A57B	84.0 (26)	103.4 (32)
A71A	87.2 (27)	48.5 (15)
A58B	90.5 (28)	103.4 (32)
A28A	103.4 (32)	51.7 (16)
A59B	116.3 (36)	103.4 (32)
A27A	138.9 (43)	54.9 (17)
A60B	158.3 (49)	103.4 (32)

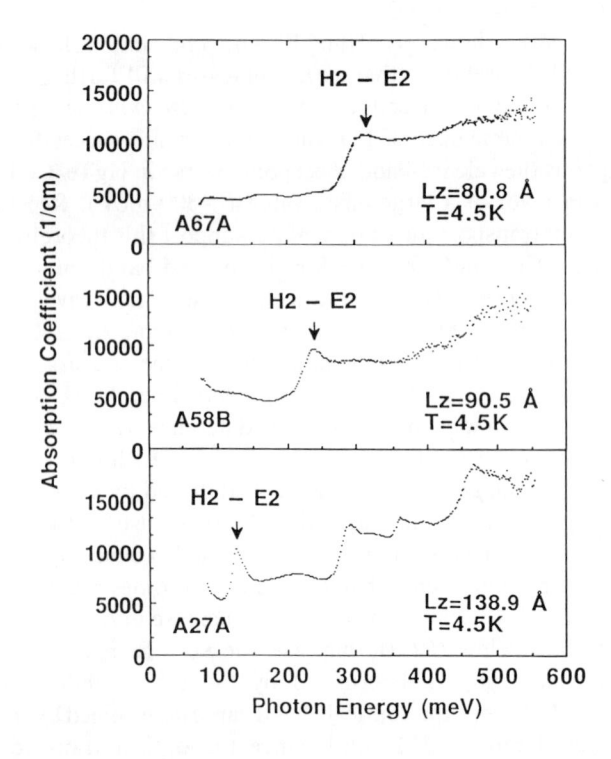

FIG. 2. Infrared transmission spectra of three inverted-band superlattices at $T = 4.5$ K. The absorption peak due to the $H2$–$E2$ transition is shown. The well widths for the three samples are indicated.

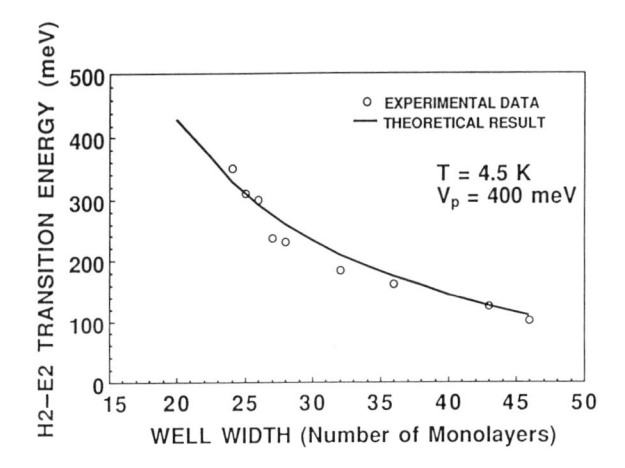

FIG. 3. The experimental $H2$–$E2$ absorption peak position as a function of the well width (open circles). The solid curve is the theoretical absorption peak position as a function of the well width, using 400 meV as the valence band offset between HgTe and CdTe.

The observed sharp transition edge structure is *not* due to excitons. Rather, the nonparabolicity of the $E1$ and $H2$ subbands and the k_x-dependent transition probability of these two transitions give rise to the sharp peak structure (k_x is the in-plane component of the electron wave vector). All of the SL absorption edge structures shift to higher energy with increasing temperature, and the sharp-peaked edge structures reduces to steplike edges at room temperature.

Figure 3 shows the comparison between the experimental peak energy (open circles) and the theoretical ones (solid line). The well widths L_Z listed in Table I were used in the calculation. The valence band offset at 4.5 K between HgTe and CdTe was taken to be 400 meV; other bulk band parameters used in the calculation are given in Ref. 4. It is seen that the theoretical results agree well with the experimental data.

Figure 4 shows the far-infrared magneto-transmission spectra of the same samples shown in Fig. 3, obtained at 4.5 K and at a fixed far-IR wavelength of 118.9 μm. Two strong absorption lines, labeled A and B respectively, are clearly observable in the CRA polarization in most samples. In other samples, one strong absorption line was observed. No lines were observed in the opposite sense of polarization. This is consistent with the Landau level scheme of the lowest conduction subband $H1$ in the SLs shown in Fig. 5. The numbers labeled on each level are the Landau level numbers and the + or − sign indicates the spin state of each level. With given free carrier concentrations at 4.5 K and the period $L_B + L_Z$ of each SL, it is deduced that at magnetic fields above 1 T, either one or two of the lowest Landau levels are partially occupied with electrons. The optical transitions follow the selection rules that the Landau level number changes by either + 1 or − 1 depending on whether the polarization is CRA or CRI, and the spin states remain unchanged. The allowed transitions from the two lowest Landau levels to the higher levels are indicated by the two arrows in the figure. The magnetic field values of the two absorption lines are consistent with the assumption that line A is due to the − 2 to − 1 transition and line B is due to the − 1 to 0 transition.

The in-plane electron effective mass is directly proportional to the magnetic field value of the cyclotron resonance absorption line at a given photon energy. The fact that there are two absorption lines indicates that the conduction subband $H1$ is strongly nonparabolic. In Fig. 6 each dot represents the magnetic field position of line A at a fixed far-infrared wavelength of 118.9 μm of a set of SL samples plotted as a function of well width L_Z. It is seen that the magnetic field position *increases* as the well width *increases*. It is thus evident that the in-plane electron effective mass

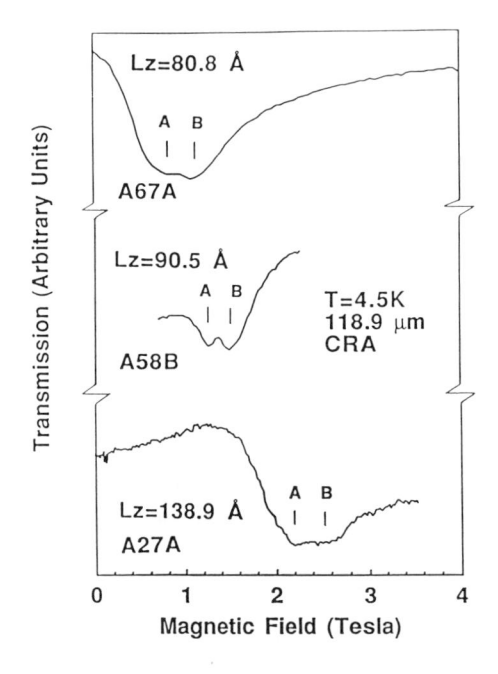

FIG. 4. Far-IR magneto-transmission spectra ($T=4.5$ K) for the same superlattices as in Fig. 3 at a fixed wavelength of 118.9 μm. The far-infrared laser beam is circularly polarized in the electron cyclotron resonance active sense, and both the magnetic field and the laser beam are perpendicular to the superlattice layer plane. The two absorption lines are labeled "A" and "B."

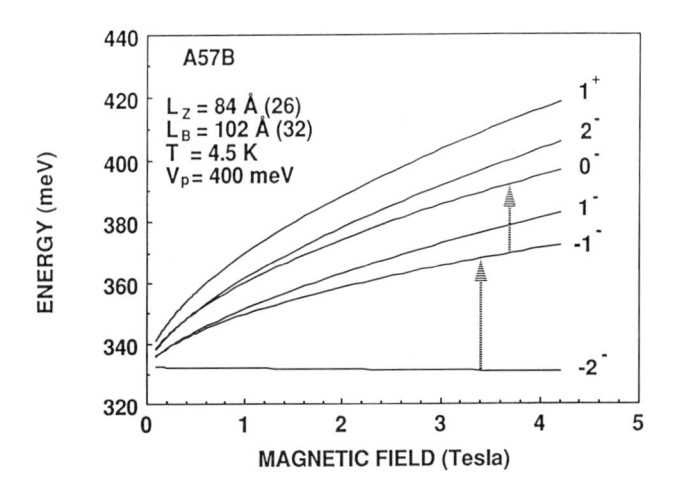

FIG. 5. The Landau levels of the $H1$ subband. The numbers labeled on each level are the Landau quantum number and the + or − sign denotes the spin state of the level. The allowed CRA optical transitions are indicated by the two arrows.

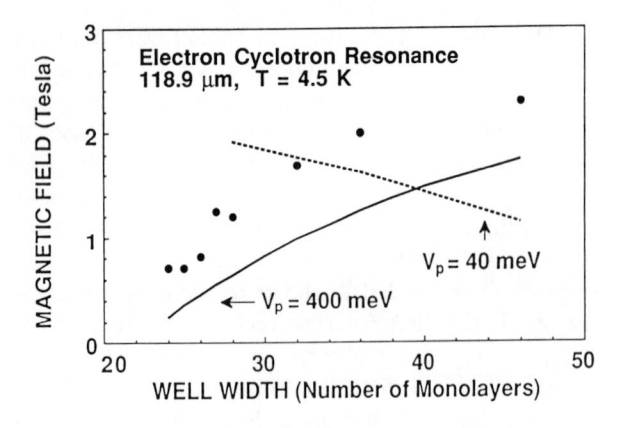

FIG. 6. The magnetic field position of line A of each sample at 118.9 μm as a function of SL well width (dots). The solid curve is the theoretical result using 400 meV as the valence band offset between HgTe and CdTe. The dashed line indicates the theoretical result obtained using a valence band offset of 40 meV.

also *increases* with *increasing* well width, and the SLs are indeed inverted-band SLs.

As was pointed out earlier in this paper, the theoretical model used in the analysis predicts that all the SLs used in this study are inverted-band SLs, if the valence band offset between HgTe and CdTe is taken to be ~400 meV. The solid line in Fig. 6 is the calculated magnetic field position of the optical transition from the lowest Landau level in the CRA polarization at a wavelength of 118.9 μm as a function of the well width L_z. It is noted that the theory correctly predicts the trend of the change of the effective mass as a function of L_z, i.e. the effective mass increases as the well width increases, when a valence band offset of 400 meV is used in the calculation. It is not clear at this point as to why the predicted cyclotron resonance magnetic fields for various well widths are smaller than the experimental ones. It is possible that the actual valence band offset at 4.5 K may be even larger than 400 meV. Another possibility is that the estimated carrier concentrations of the SLs used in the calculation are incorrect (too small). As is pointed out in Ref. 5 because mixed conductivity generally occurs in small band gap SLs, more sophisticated analysis of the Hall experiment is needed to obtain the correct free carrier concentrations and their mobilities.

The experimental results obtained in this work, however, are completely inconsistent with a small valence band offset for the HgTe–CdTe SL. The dashed line in Fig. 6 shows the theoretical dependence of the cyclotron resonance magnetic field versus SL well thickness that was obtained using 40 meV as the valence band offset. It is seen that the theory predicts the wrong trend of the change of the effective mass with SL well width. In fact, using $V_p = 40$ meV, the theory predicts that the SLs used in this work are all in the *normal* semiconducting regime, so that the cyclotron resonance magnetic field (or electron effective mass) is predicted theoretically to *decrease* with increasing well width, as shown in the figure. We must therefore conclude that the HgTe/CdTe valence band offset at 4.5 K is large; of the order of 400 meV.

IV. SUMMARY AND CONCLUSIONS

Inverted-band HgTe/CdTe superlattices are studied systematically for the first time by means of low temperature IR transmission and far-IR magneto-transmission experiments. Cyclotron resonance of the conduction band electrons has been observed and the effective mass of the electrons on the inverted first conduction subband $H1$ in the direction parallel to the SL layer plane has been measured directly and systematically for the first time. The electron effective mass *increases* with *increasing* SL well width L_z, indicating that the SLs studied in this work are, indeed, inverted-band SLs. The $H2$ to $E2$ transition energy obtained from the infrared transmission spectra decreases as the well width increases, in agreement with the fact that there is less and less quantum confinement as the well width increases. The theoretical results obtained using 400 meV for the valence band offset agree well with the $H2/E2$ transition energies obtained from the low temperature IR transmission experiments. The theory also agrees qualitatively with the in-plane electron effective masses obtained from the far IR magneto-transmission experiment. Theoretical results obtained using 40 meV as the HgTe–CdTe valence band offset are qualitatively inconsistent with the experimental far IR magnetooptic results. We thus conclude that the valence band offset between HgTe and CdTe must be large, of the order of 400 meV at 4.5 K.

ACKNOWLEDGMENTS

This work was supported by Naval Research Laboratory Contract No. N00014-89-J-2024 and by the National Science Foundation through the computer resource allocation at the Cornell University IBM Supercomputer Center.

[1] P. M. Hui, H. Enrenreich, and N. F. Johnson, J. Vac. Sci. Technol. A **7**, 424 (1989).
[2] J. R. Meyer, C. A. Hoffman, F. J. Bartoli, J. W. Han, J. W. Cook, Jr., J. F. Schetzina, X. Chu, J. P. Faurie, and J. N. Schulman, Phys. Rev. B **38**, 2204 (1988).
[3] C. A. Hoffman, J. R. Meyer, F. J. Bartoli, J. W. Han, J. W. Cook, Jr., J. F. Schetzina, and J. N. Schulman, Phys. Rev. B **39**, 5208 (1989).
[4] Z. Yang, J. F. Schetzina, and J. K. Furdyna, J. Vac. Sci. Technol. A **7**, 360 (1989).
[5] S. Hwang, Y. Lansari, Z. Yang, J. W. Cook, Jr., and J. F. Schetzina, J. Vac. Sci. Technol. B **9**, 1799 (1991).

Magneto-optical transitions between subbands with different quantum numbers in narrow gap HgTe–CdTe superlattices

H. Luo, G. L. Yang, and J. K. Furdyna
Department of Physics, University of Notre Dame, Notre Dame, Indiana 46556

L. R. Ram-Mohan
Department of Physics, Worcester Polytechnic Institute, Worcester, Massachusetts 01609

(Received 2 October 1990; accepted 15 January 1991)

Magneto-optical transitions induced by the coupling between the conduction and the valence bands through the momentum matrix element, and by the coupling terms between light and heavy holes resulting from an applied magnetic field are studied theoretically in narrow gap HgTe–CdTe superlattices. Selection rules and transition probabilities for the above transitions are presented and compared with allowed transitions. The numerical results for the transition probabilities show that some of the interband transitions with $\Delta N = \pm 1$ are significant and have to be considered in the studies of interband magneto-optical spectra of narrow gap superlattices.

I. INTRODUCTION

The growth and the physical properties of HgTe–CdTe superlattices have been studied extensively because of their possible applications as infrared optical devices. The energy gap of such systems can vary from zero to nearly the value of the CdTe energy gap, corresponding to a wide range of the spectrum in the infrared and far infrared regions. Transport, optical, and magneto-optical measurements have been performed on such superlattices with a variety of dimensions.[1–5] With constant improvements in the growth process, the band structure and optical properties can be explored to increasingly more detail.

It is well known that in symmetric quantum wells, and superlattices, the allowed optical transitions satisfy the selection rule $\Delta N = 0$, where N is the subband quantum number.[6] However, it was pointed out that the mixing of the heavy-hole and the light-hole relaxes the above selection rule, resulting in additional transitions with $\Delta N \neq 0$.[7,8] Such transitions have been studied in GaAs–GaAlAs systems, which could be enhanced by applied electric fields.

Studies of magneto-optical spectra in HgTe–CdTe superlattices so far have focused on transitions involving $\Delta N = 0$. In narrow gap materials, however, the conduction and the valence bands are strongly coupled through the momentum matrix element. In the presence of an applied magnetic field, there is also coupling between the light hole and the heavy hole states. It has been shown that such couplings will also introduce subband mixing,[9] leading to transitions having $\Delta N \neq 0$. It is the purpose of this paper to analyze these transitions in narrow-gap superlattices of HgTe–CdTe, which should be considered in interpretations of magneto-optical spectra.

Throughout the paper, we limit ourselves only to the case of the magnetic field parallel to the growth direction, which is the most common configuration used in both theoretical and experimental studies. In the following section, we will briefly discuss the selection rules. Results of numerical calculations on superlattices of HgTe–CdTe will then be presented in Sec. III.

II. SELECTION RULES

In a superlattice consisting of semiconductors with the zinc-blende crystal structure, the Hamiltonian in the **k·p** approximation can be written as

$$H(z) = \begin{cases} H_1, & z \text{ in material 1} \\ H_2, & z \text{ in material 2,} \end{cases} \quad (1)$$

where H_1 and H_2 are the Hamiltonians for (bulk) material 1 and 2, respectively, and explicitly include contributions from the Γ_6, Γ_7, and Γ_8 bands. Both H_1 and H_2 have the following form[10]:

$$H_i = \begin{bmatrix} H_{a,i} & H_{c,i} \\ H_{c,i}^\dagger & H_{b,i} \end{bmatrix}, \quad i = 1,2, \quad (2)$$

where $H_{a,i}$ and $H_{b,i}$ are 4×4 matrices describing the two spin states at $k_z = 0$ (Ref. 11) in the i th material. Also in Eq. (2), $H_{c,i}$ is the 4×4 coupling matrix between the two spin states, and is proportional to the momentum matrix element P and k_z if no magnetic field is present (we set $k_x = k_y = 0$). The matrix elements in $H_{c,i}$ involving P couple the conduction band and the valence band with opposite spins. Since these terms involve the interaction between the conduction band and the valence band, their effects are stronger in materials with smaller energy gaps.

When a magnetic field is applied, there will also be couplings between the Γ_8 bands with different spins, namely, between the light and the heavy hole bands in different spin states. Although the coupling between the valence bands are the effect of higher bands (bands other than the Γ_6, Γ_7, and Γ_8 bands), they are significant because the light hole and the heavy hole states are very close to each other, which leads to a large mixing. This coupling is proportional to the magnetic field and is therefore expected to have a greater effect at higher fields.

The difference between the Hamiltonian of a superlattice and that of a bulk crystal is, among others, that in the superlattice case k_z does not commute with the Hamiltonian, whereas it does in the bulk case. Thus for superlattices k_z has to be replaced by $-i\partial/\partial z$ and cannot be set to zero, unlike the case of bulk crystal, where calculations at $k_z = 0$ are generally sufficient.

It has been shown that coupling terms in H_c both between the conduction band and the valence bands, and between the light and the heavy holes, cause the mixing of a given subband with other subbands that have the opposite parity.[9] Since the subband ladder in a one-band single quantum well follows the pattern of subbands with alternating parities, the mixing will come from subbands with ΔN equal to 1, 3, 5 and so on, among which $\Delta N = \pm 1$ will have the dominant effect. This is schematically illustrated in Fig. 1. Such mixings result in transitions involving $\Delta N \neq 0$, and in particular $\Delta N = \pm 1$ transitions. By considering the parities of the wave functions in the presence of an applied magnetic field parallel to the growth direction, the selection rules for the allowed ($\Delta N = 0$) transitions are[9]

CRA (or σ_L): $a_n \to a_{n+1}$, $b_n \to b_{n+1}$,
CRI (or σ_R): $a_n \to a_{n-1}$, $b_n \to b_{n-1}$,
$\mathbf{E} \| \mathbf{B}$ (or π): $a_n \to b_n$, $b_n \to a_n$,   (3)

where a and b represent the two spin states, and n is the Landau level quantum number. For $\Delta N = \pm 1$, we have

CRA: $a_n \to b_{n+1}$, $b_n \to a_{n+1}$,
CRI: $a_n \to b_{n-1}$, $b_n \to a_{n-1}$,
$\mathbf{E} \| \mathbf{B}$: $a_n \to a_n$, $b_n \to b_n$.   (4)

We note that the quantum numbers n for Landau levels belonging to the a set and the b set are defined in the same way as in Ref. 11, which is different from the convention commonly used for bulk materials.[12] In the current definition, $a_{n+1} \to b_n$ represents the so called spin–flip transition. In the $\mathbf{E} \| \mathbf{B}$ configuration, the electric field $\mathbf{E}$ of the incident electromagnetic wave is parallel to the growth direction (since $\mathbf{B}$ lies in that direction). Since the wave vector $\mathbf{q}$ of the electromagnetic wave is perpendicular to $\mathbf{E}$, $\mathbf{q}$ is then necessarily in the superlattice planes. This situation can be achieved by the strip line technique.[13]

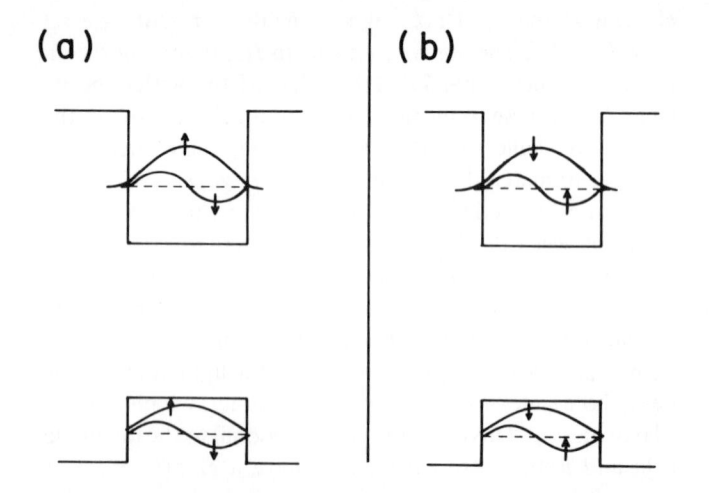

FIG. 1. Schematic illustration of the mixing of the $N = 1$ subbands with subbands having different parities and different spin states. The symmetric functions are the original wave functions, and the asymmetric functions are the admixture. (a) The original subbands without the mixing are in the spin–up state; (b) the original subbands are in the spin–down state before mixing.

As can be seen in Eq. (4), $\Delta N = \pm 1$ transitions allowed in the Faraday (CRA or CRI) configuration involve a change of the spin state. If any of the transitions involving ΔN equal to ± 3, ± 5, ± 7 and so on are measurable, they will follow the same selection rules as those in Eq. (4).

It has been shown that, when dealing with transitions in narrow gap HgTe–CdTe superlattices, the entire superlattice Brillouin zone has to be considered.[5] The selection rules discussed above can be used anywhere in the Brillouin zone, even though the actual transition probabilities are different for different values of the superlattice wave vector. Furthermore, the selection rules can be used for all (100), (110), and (111) superlattices. In the next section, we will focus on the case of (100) superlattices.

It must be pointed out that in deriving the selection rules no further perturbation theory is used other than the original $\mathbf{k} \cdot \mathbf{p}$ perturbation. The existence of $\Delta N = \pm 1$ transitions is only the result of considering all the matrix elements in the original Hamiltonian given in Eq. (2) in the $\mathbf{k} \cdot \mathbf{p}$ approximation. This is important because, as will be shown in the next section, some of the $\Delta N = \pm 1$ transitions turn out to be of the same order of magnitude as the $\Delta N = 0$ transitions. In fact, there exist $\Delta N = \pm 1$ transitions which are stronger than certain $\Delta N = 0$ transitions, including the cyclotron resonance in the conduction band!

III. NUMERICAL CALCULATIONS

In this section, we will present the numerical calculations of the transition probabilities for $\Delta N \neq 0$ transitions discussed in the previous section, and their dependence on the applied magnetic field. We will also consider the signifigance of subband crossing. A value of 300 meV was used for the valence band offset in the calculations such that both the electrons and the holes are confined in the HgTe layers.

Numerical calculations of the transition probabilities were carried out in the $\mathbf{k} \cdot \mathbf{p}$ approximation for (100) HgTe–CdTe superlattices in a magnetic field parallel to the growth direction, using a scheme described in Ref. 14. The strain effect is not included in this study. The effects of warping and inversion asymmetry are also neglected.

Some of the Landau levels belonging to different subbands involved in the transitions of interest are shown in Fig. 2 for a HgTe–CdTe (60–40 Å) superlattice, where the solid lines and dashed lines (labeled by primes) correspond to the two different spin states a and b, respectively.

First of all, the calculated probabilities of the $\Delta N = \pm 1$ transitions are in agreement with the selection rules presented in the last section. The ratio of the calculated probabilities of $\Delta N = \pm 1$ transitions (at $B = 2$ T) to the probability of the cyclotron resonance in the first conduction subband, namely, the $0(E1) \to 1(E1)$ transition, are listed in Table I. This is done for the Landau levels in the first few subbands, including subbands $E1$, $E2$, $LH1$, $HH1$ and $HH2$. In the above cyclotron resonance transition, $0(E1)$ is the Landau level labeled by 0 (see Fig. 2) in the $E1$ subband, and similarly for the other Landau levels. The results listed in Table I are for a HgTe–CdTe (60–35 Å) superlattice. Table I shows that the $\Delta N = \pm 1$ transitions, which have been considered forbidden, are in fact comparable in transition probability to

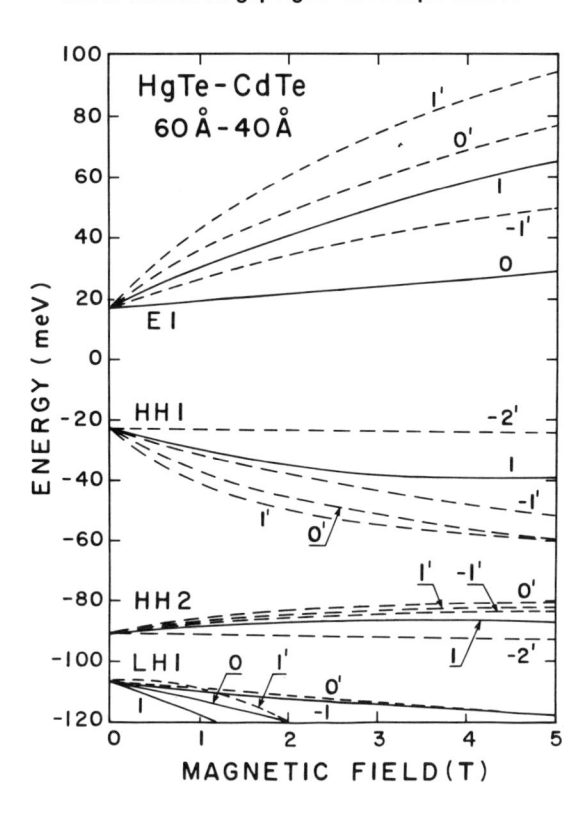

FIG. 2. Landau levels at $q_z = 0$ (the wave vector along the growth direction) as a function of magnetic field. The solid lines and the dashed lines (the latter labeled with primes) represent the two spin states.

the cyclotron resonance in the conduction subband. In fact, the transition $-1(HH1) \to 0'(E2)$ has even a greater probability than that of the cyclotron resonance in this superlattice.

In Fig. 2, the light hole $LH1$ subband lies slightly below the heavy hole $HH2$ subband. According to the discussion in the last section, the coupling between the two spin states in these two subbands will become stronger as the energy difference between them is reduced. When we reduce the barrier width, or increase the well width, both the $LH1$ and the $HH2$ subbands will move up, with the $LH1$ subband moving faster. At certain well and barrier widths, the $LH1$ subband will eventually move above the $HH2$ subband. In the process of moving up, the $LH1$ subband will cross the $HH2$ subband,

TABLE I. The probabilities of the $\Delta N = \pm 1$ transitions divided by the probability of the cyclotron resonance in the first conduction subband (relative probabilities), and the polarizations needed to observe the transitions. There are also other $\Delta N = \pm 1$ transitions than those listed in the table, but their probabilities are too small to be observable. The superlattice used here is HgTe–CdTe (60–35 Å).

Transition	Relative probability	Polarization
$-1'(HH2) \to 0(E1)$	0.35	CRA/σ_L
$0'(HH2) \to 1(E1)$	0.23	CRA/σ_L
$0(LH1) \to 1'(E2)$	0.79	CRA/σ_L
$1'(LH1) \to 0(E2)$	0.45	CRI/σ_R
$-1(LH1) \to 0'(E2)$	1.05	CRA/σ_L

at which point the mixing of these two subband should reach a maximum in a resonant manner. Correspondingly, there will be a maximum in the probabilities of the $\Delta N = \pm 1$ transitions involving these two subbands.

We will consider the case of changing the barrier width, namely, the CdTe layer width, to achieve this subband crossing. We fix the HgTe layer width to 60 Å and change the width of the CdTe layers d_{CdTe} from 15 to 60 Å. Numerical results show that the crossing occurs near $d_{CdTe} = 30$ Å.

Probabilities were calculated for superlattices with 60 Å HgTe layers and with CdTe layer width d_{CdTe} in the range from 15 to 60 Å (the superlattices remain open gap with such CdTe layer thicknesses), in the presence of a magnetic field of 2 T. The calculated ratio,

$$\rho = \frac{W_{0'(HH2) \to 1(E1)}}{W_{0(LH1) \to 1(E1)}}, \quad (5)$$

where $W_{i \to f}$ is the probability of the transition from state i to state f, is plotted in Fig. 3, as a function of the CdTe layer thickness. The reason for using the allowed transition $0(LH1) \to 1(E1)$ [appearing in the denominator in Eq. (5)] as the basis of comparison is that it has a similar transition energy as that of the $0'(HH2) \to 1(E1)$ transition, making the process of numerical calculations simpler. In a HgTe–CdTe (60–35 Å) superlattice, its probability is about 1/5 of that of the cyclotron resonance transition.

Similar dependence of the transition probabilities on the movement of different subbands can be studied when the HgTe layer width is changed while the width of the CdTe layers is fixed. One can even keep the width of both the HgTe layers and the barriers fixed, and replace the CdTe layers with CdHgTe layers, and observe the same effects.

As mentioned before, the coupling terms between the light hole and the heavy hole in different spin states are induced by the applied magnetic field. It is natural to expect that these $\Delta N = \pm 1$ transitions will be stronger at higher magnetic fields. Such a behavior was found numerically. As an example, we will discuss numerical results for a HgTe–CdTe (60–40 Å) superlattice. To illustrate the magnetic field de-

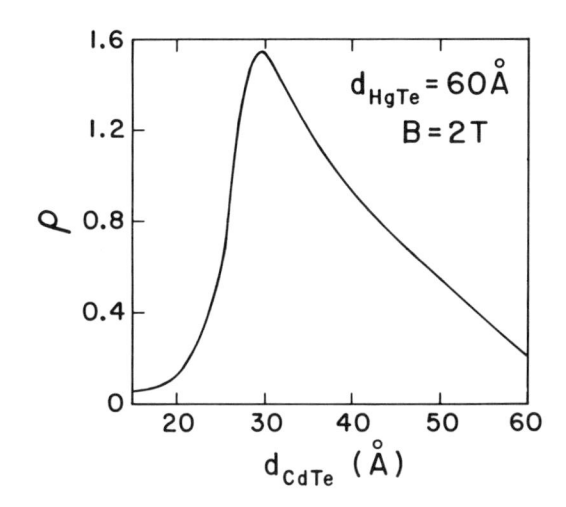

FIG. 3. The ratio $\rho = W_{HH2(0') \to E1(1)} / W_{LH1(0) \to E1(1)}$ calculated at $B = 2$ T as a function of the width of CdTe layers. A resonant effect occurs near $d_{CdTe} = 30$ Å.

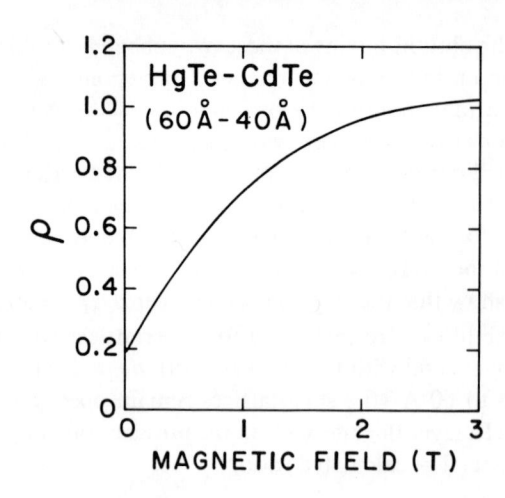

FIG. 4. The magnetic field dependence of the ratio $\rho = W_{0'(HH2)\to 1(E1)}/W_{0(LH1)\to 1(E1)}$. The value of ρ becomes larger as the magnetic field increases.

pendence of the transitions due to the above coupling process, we will again use the ratio ρ given in Eq. (5). Figure 4 shows the calculated values of ρ as a function of the magnetic field. As anticipated, one finds that ρ increases as the magnetic field becomes higher.

With the numerical results presented above, we find that the $\Delta N = \pm 1$ transitions may play an important role in magneto-optical studies of HgTe–CdTe superlattices, especially those where subbands with $\Delta N = \pm 1$ are close to each other. The same is true for HgTe–CdTe single quantum wells, except that the exact transition probabilities will then be different.

IV. CONCLUSIONS

We have demonstrated that in narrow gap HgTe–CdTe superlattices the coupling matrix elements between the conduction band and the valence band, as well as those between the light hole and the heavy hole, lead to significant $\Delta N = \pm 1$ transitions, which were previously considered forbidden. We find that the probabilities of such transitions are very sensitive to the distance in energy between subbands with $\Delta N = \pm 1$, and a resonant behavior is observed as this distance goes to zero. We also studied the probabilities of the $\Delta N = \pm 1$ transitions as functions of an applied magnetic field parallel to the growth direction. The calculated probabilities show a strong dependence on the magnetic field, becoming larger as the field increases.

ACKNOWLEDGMENT

The research at the University of Notre Dame was supported by the NSF Grant No. DMR-8904802.

[1] Z. Yang, M. Dobrowolska, H. Luo, J. K. Furdyna, and J. T. Cheung, Phys. Rev. B **38**, 3407 (1988).
[2] J. M. Perez, R. J. Wagner, J. R. Meyer, J. W. Han, J. W. Cook, Jr, and J. F. Schetzina, Phys. Rev. Lett. **61**, 2261 (1988).
[3] M. Dobrowolska, T. Wojtowicz, H. Luo, J. K. Furdyna, O. K. Wu, J. N. Schulman, J. R. Meyer, C. A. Hoffman, and F. J. Bartoli, Phys. Rev. B **41**, 5084 (1990).
[4] J. M. Berroir, Y. Guldner, J. P. Vieren, M. Voos, X. Chu, and J. P. Faurie, Phys. Rev. Lett. **62**, 2024 (1989).
[5] J. R. Meyer, R. J. Wagner, F. J. Bartoli, C. A. Hoffman, M. Dobrowolska, T. Wojtowicz, and J. K. Furdyna, Phys. Rev. B (in press).
[6] R. Dingle, *Festkörperprobleme XV (Advances in Solid State Physics)*, edited by H. J. Queisser (Pergamon, New York, 1975), p. 21.
[7] Y. C. Chang and J. N. Schulman, Phys. Rev. B **31**, 2069 (1985).
[8] G. D. Sanders and K. K. Bajaj, Phys. Rev. B **35**, 2308 (1987).
[9] H. Luo and J. K. Furdyna, Phys. Rev. B **41**, 5188 (1990).
[10] W. Leung and L. Liu, Phys. Rev. B **8**, 3811 (1973).
[11] Even though the eigenstates of the Hamiltonian in Eq. (2) are not the eigenstates of the spin operator, they behave the same way in optical transitions as the spin-up and spin-down states. We will, therefore, refer to them as spin states throughout the paper for the convenience of discussion.
[12] M. H. Weiler, R. L. Aggarwal, and B. Lax, Phys Rev. B **17**, 3269 (1978).
[13] M. von Ortenberg, Infrared Phys. **18**, 735 (1978).
[14] L. R. Ram-Mohan, K. H. Yoo, and R. L. Aggarwal, Phys. Rev. B **38**, 6151 (1988).

Shubnikov–de Haas oscillations and quantum Hall effect in modulation-doped HgTe–CdTe superlattices

C. A. Hoffman, J. R. Meyer, D. J. Arnold,[a] and F. J. Bartoli
Naval Research Laboratory, Washington, D. C. 20375

Y. Lansari, J. W. Cook, Jr., and J. F. Schetzina
North Carolina State University, Raleigh, North Carolina 27695

(Received 2 October 1990; accepted 29 October 1990)

We have investigated quantum oscillations in the magneto-transport properties of HgTe–CdTe superlattices grown by molecular-beam epitaxy. Modulation doping was achieved by incorporating either indium donors or arsenic acceptors into the CdTe barriers. In a *p*-type sample, quantized plateaus were observed in the Hall conductivity down to $i = 3$ conduction channels. Since the structure contained 200 periods, this implies that the quantized holes populated only a small fraction of the total superlattice volume. A mixed conduction analysis of the nonoscillating component of magneto-transport data provided confirming evidence for the presence of a two-dimensional hole gas with the appropriate density in addition to the superlattice holes. Previous reports of the quantum Hall effect in HgTe–CdTe also yielded i far less than the total number of superlattice wells. In contrast, an *n*-type sample from the present study displayed a single quantum Hall plateau at $i \approx 140$, indicating that in this case most of the 200 superlattice periods contributed to the conduction. We argue that this represents the first observation of the quantum Hall effect associated with carriers distributed throughout the interior of a HgTe–CdTe superlattice.

I. INTRODUCTION

We report here an experimental investigation of quantum oscillations in the magneto-conductance of *n*-type and *p*-type modulation-doped HgTe–CdTe superlattices. Several previous investigators[1-3] have observed Shubnikov–de Haas oscillations in this system at magnetic fields as low as a few kG. However, derivative techniques were required to enhance the small amplitudes of the signals because the condition $\mu B \gg 1$ (μ is mobility and B is field), which assures that the Landau level separation is larger than the lifetime broadening of the levels, was generally not satisfied at fields for which the oscillations were observed. Increasing the field to improve μB did not necessarily help, since for typical doping levels the carriers should all occupy the lowest Landau level[4] and there should be no further oscillations whenever B exceeds ~1 T. Nonetheless, other investigators[5,6] have observed strong magneto-conductance oscillations and the quantum Hall effect at much higher magnetic fields, ranging up to 29 T. As long as the free carriers are taken to be uniformly distributed throughout the superlattice, it is difficult to understand why oscillations should be observed at such high magnetic fields.

In the present work, we similarly find that in one case the Shubnikov–de Haas and quantum Hall effects are observed at fields far above those at which the extreme quantum limit should have been reached. However, it will be seen that the observed behavior is related to the presence of a quasi-two-dimensional (2D) carrier gas which does not occupy the interior of the superlattice. On the other hand, we also report data showing a well-resolved quantum Hall plateau at $B < 1$ T, which we believe to represent the first observation of the quantum Hall effect in HgTe–CdTe due to carriers occupying the interior of the superlattice.

II. EXPERIMENT

Table I lists the characteristics of six 200-period HgTe–$Hg_{0.15}Cd_{0.85}$Te superlattices, which were grown by molecular-beam epitaxy (MBE) in a system which has been described previously.[7] Deposition was directly onto lattice-matched [100] $Cd_{1-x}Zn_x$Te substrates with no buffer layers. The well and barrier thicknesses listed in the table were accurately determined from x-ray satellite peaks in conjunction with growth-rate data.[8] Energy gaps were obtained from the temperature dependence of the intrinsic carrier density.[8]

Modulation doping was achieved by incorporating indium donors or arsenic acceptors into the CdTe barriers of the superlattices. Note from Table I that in *n*-type samples the low-temperature electron mobilities remained above 4×10^4 cm^2/Vs for net donor densities up to 8×10^{15} cm^{-3}, and μ_n did not show any obvious dependence on $N_D - N_A$ in this range. However, the hole mobilities in *p*-type samples with $N_A - N_D$ up to 6×10^{16} cm^{-3} were found to depend more strongly on doping level. These preliminary results are consistent with the previous observation that for CdTe grown by MBE, controlled doping with high majority-carrier mobilities is more reproducibly achieved for *n*-type than for *p*-type.

Diagonal (ρ_{xx}) and Hall (ρ_{xy}) resistivities were measured as a function of magnetic field (0–7 T) and temperature (2–300 K) by the Van der Pauw technique. A mixed-conduction analysis was then employed to extract temperature-dependent densities and mobilities for the various electron and hole species contributing to the transport. In cases where Shubnikov–de Haas oscillations were significant at low temperatures, the nonoscillating components of the conductivities were used in the analysis. The data for samples 3, 4, and 5 showed evidence for an additional hole

TABLE I. Sample characteristics. Well and barrier thicknesses, energy gap, dopant, net doping level, and low-temperature majority carrier mobility for the six superlattices.

No.	d_W (Å)	d_B (Å)	E_g (meV)	Dopant	n_0 (cm^{-3})	p_0 (cm^{-3})	μ (cm^2/Vs)	p_s (cm^{-2})	μ_s (cm^2/Vs)
1	26	36	>100	As		6.5×10^{16}	1.4×10^{2}		
2	84	49	0	As		1.7×10^{16}	0.65–7.3×10^{4}		
3	23	26	150	As		4.6×10^{15}	4.6×10^{3}	1.3×10^{11}	2.4×10^{4}
4	68	49	0	As	1.5×10^{14}		1×10^{5}	3.4×10^{12}	2.3×10^{3}
5	78	52	0	In	2.3×10^{15}		4.4×10^{4}	1.6×10^{12}	5.3×10^{3}
6	81	52	0	In	8.0×10^{15}		4.2×10^{4}		

species besides the superlattice majority and minority carriers. The densities and mobilities of these additional carriers were relatively independent of temperature. Such carriers have frequently been observed in HgTe–CdTe heterostructures, and they are believed to correspond to a quasi-2D "charge transfer" population which resides within the superlattice near the interface with the substrate. A previous investigation demonstrated a correlation between the magneto-transport and magneto-optical properties of charge-transfer carriers, whose mobilities were found to scale with superlattice well thickness.[9] The densities (in cm^{-2}) and mobilities of the additional carrier species are listed as p_s and μ_s in Table I. Distinct from this was the observation of two hole species in sample 2, whose densities both increased with temperature. These data are consistent with the "mass-broadening" effect which has been reported previously for semimetallic HgTe–CdTe superlattices.[8]

Only samples 3 and 6 displayed unambiguous Shubnikov–de Haas oscillations in the dc magneto-conductivity at low temperatures. In all other cases (except perhaps sample 2, which should have been marginal), the measured mobilities and majority carrier concentrations are such that $\mu B > 1$ only at fields beyond which the density of states in the lowest Landau level is large enough to accommodate the entire carrier population, i.e., no further oscillations are expected.

III. RESULTS

The low-temperature resistivity of sample 3, which was p-type, displayed strong Shubnikov–de Haas oscillations at high magnetic fields. Figure 1 illustrates the ratio of the oscillating and nonoscillating components of the magneto-conductivity as a function of inverse field, where $\Delta\sigma_{xx} \equiv \sigma_{xx} - \sigma_{xx}^0$ and the conductivity tensor is related to the resistivity tensor as follows:

$$\sigma_{xx} = \frac{\rho_{xx}}{\rho_{xx}^2 + \rho_{xy}^2}, \qquad (1)$$

$$\sigma_{xy} = \frac{\rho_{xy}}{\rho_{xx}^2 + \rho_{xy}^2}. \qquad (2)$$

The nonoscillating background (σ_{xx}^0) has been approximated using a numerical procedure in which the net conductivity (σ_{xx}) is recursively fit to the superposition of a slowly varying polynomial and a rapidly varying oscillatory component. At $T = 2$ K (solid curve), up to 5 oscillations are observed and the magnitude of the signal is quite high, becoming as great as 70% of σ_{xx}^0 at $B = 7$ T. Comparison with the data at 10 K (dashed curve) shows that with increasing T, the amplitude of the oscillations decreases rapidly. A fit to the temperature decay[10] yields an effective mass of $\approx 0.12 m_0$. This is a factor of 3 larger than the theoretical value[11] for the superlattice in-plane hole mass calculated when the well and barrier thicknesses are taken to have the appropriate values.

A conventional analysis of the Shubnikov–de Haas data in Fig. 1 does not appear to be strictly appropriate. This is partly because a theory is not yet available which properly includes the details of the HgTe–CdTe band structure. Beyond this, the data indicate an increase of the frequency of the oscillations with decreasing B^{-1}, possibly implying a variation of the hole density with magnetic field. If the spacing between the first two minima is used to estimate the areal carrier concentration one obtains 1.7×10^{11} cm^{-2} whereas a value 40% smaller is obtained from the spacing between the second and third minima. Either value is more than a factor of 2 smaller than the concentration for the superlattice hole

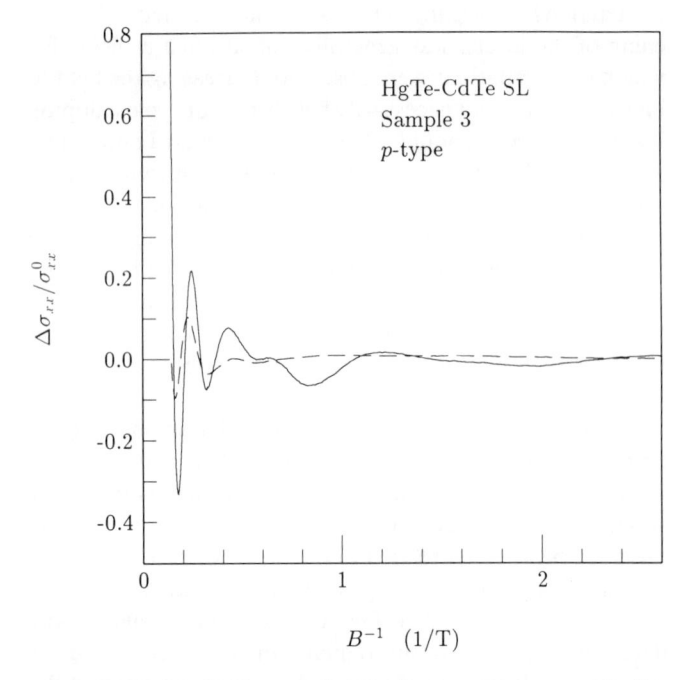

FIG. 1. Ratio of the oscillatory and nonoscillatory components of the diagonal conductivity vs inverse magnetic field, for sample 3. Two temperatures are shown, 2 K (solid curve) and 10 K (dashed curve).

obtained from the mixed conduction analysis, but is in relatively good agreement with p_s listed in Table I (as discussed in the previous section, p_s is taken to correspond to a quasi-2D hole located within the superlattice but near the interface with the substrate). The mobility obtained from the decay of the amplitude with B^{-1} is 7.2×10^3 cm^2/Vs, which is between the values for μ_p and μ_s in Table I. However, if the hole density varies with B, one should not expect this mobility determination to be highly accurate.

When the diagonal and Hall resistivities at T = 2 K are plotted as a function of magnetic field, as in Fig. 2, we see that this sample displays the quantum Hall effect. Each of the plateaus in ρ_{xy} is seen to be accompanied by a minimum in ρ_{xx}. Parallel conduction by the second hole species is probably responsible for the failure of ρ_{xx} to closely approach zero resistivity. The various plateaus are more easily identified if we replot the data as a Hall conductivity in units of $e^2/2\pi\hbar$. This is done in Fig. 3, which shows resolved plateaus for $i = 3$ through 6, and further structure corresponding to 7 and 8. Since the index i corresponds to the total number of conduction channels (a product of the number of quantum wells contributing and the number of occupied Landau levels in each well), observation of the $i = 3$ plateau implies that at most 3 of the 200 periods in the superlattice contribute to the quantized Hall conductivity. The obvious conclusion is that it is the quasi-2D population, p_s, which produces the effect rather than the holes residing in the interior of the superlattice. The same conclusion inescapably follows from the observation of Shubnikov–de Haas oscillations at fields up to 7 T. For a uniformly distributed hole population with the measured concentration, only the lowest Landau level should have been occupied at any magnetic field above 0.2 T. However, the conductivity of a quasi-2D hole gas with the density p_s would continue to oscillate in a manner entirely consistent with the Shubnikov–de Haas and

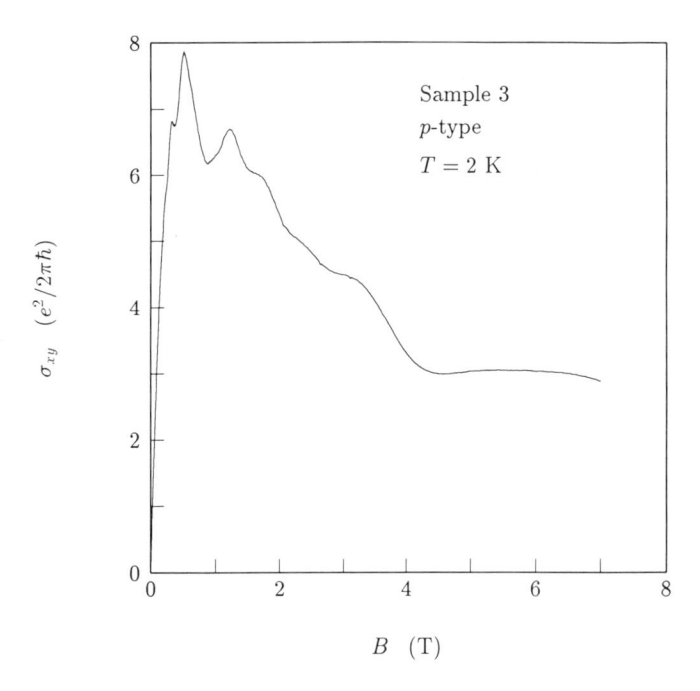

FIG. 3. Hall conductivity vs magnetic field at $T = 2$ K, for sample 3. Lower-order quantum Hall plateaus are well resolved.

quantum Hall data. This would also account for the observation of an effective mass larger than the value expected for superlattice holes.

We point out that these data are similar to previous results showing the quantum Hall effect in HgTe–CdTe superlattices at high magnetic fields.[5,6] At B between 3 and 29 T, Ong et al. reported quantum Hall plateaus corresponding to $i = 2$ through 12 in an n-type superlattice ($d_W = 90$ Å, $d_B = 40$ Å) with 12 periods. Those authors suggested that i less than the number of wells may have been observable because the superlattice miniband for each Landau level was split into 12 nondegenerate sublevels [the conductance per channel then becomes $(e^2/2\pi\hbar)/12$ since the density of states per sublevel is only 1/12 as great as the total density of states per Landau level]. For 40 Å barriers, the miniband width of E1 (the conduction band in semiconducting HgTe–CdTe) is ≈ 30 meV. However, in a semimetallic superlattice such as that studied by Ong et al., the electrons populate HH1 (or $-2'$ when a magnetic field is applied[12]), whose miniband width nearly vanishes. Thus, it is unreasonable to expect the formation of sublevels with finite gaps containing localized states in between, as required for observation of the quantum Hall effect. Even for E1, there has thus far been no experimental evidence (e.g., from magneto-optics) that the miniband splits into resolvable sublevels. Ong et al. note that both their magneto-transport and their magneto-optical data show evidence for at least two carrier species. As a simple explanation for their results, we suggest that it is not the superlattice electron but the second species (presumably quasi-2D) which contributes the high-field conductivity with quantum Hall plateaus. Taking the net areal electron density of 1.38×10^{12} cm^{-2} and assuming a uniform distribution among the 12 wells, the density of states in the lowest Landau level should have been large enough to contain all of the electrons whenever $B > 4.8$ T. It

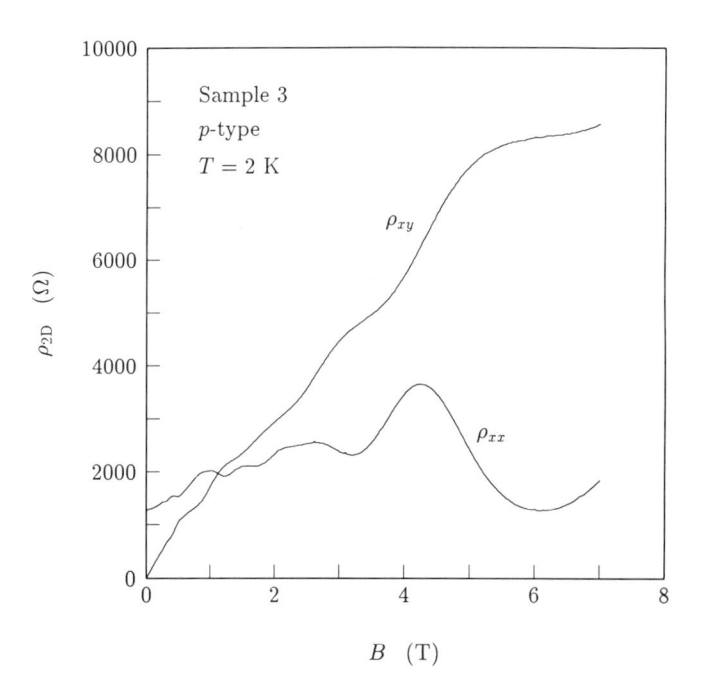

FIG. 2. Diagonal and Hall resistivities vs magnetic field at $T = 2$ K, for sample 3.

is much easier to understand why quantum oscillations continue to be observed up to much higher fields if the same number of electrons are assumed to populate only one or two wells.

In a similar study of a p-type $Hg_{0.92}Cd_{0.08}Te$–CdTe superlattice with 100 periods ($d_W = 70$ Å, $d_B = 40$ Å), Woo et al. observed quantum Hall plateaus corresponding to $i = 9$ and $i = 18$ at $B = 5.5$ and 11 T. The implication is clearly that only 9 of the 100 wells contributed to the quantized Hall conduction. While those authors discussed the result in terms of only 9 layers being "contacted," it seems more likely that band bending near the top surface or near the interface of the superlattice with the substrate made it energetically favorable for the holes to populate only 9 wells.

We believe that in all of the cases considered above (sample 3 as well as the previous results of Ong et al. and Woo et al.), magneto-conductivities displaying the quantum Hall effect were due to carriers occupying only a small fraction of the total superlattice volume. This is because (a) the plateaus had indices $i = \sigma_{xy}/(e^2/2\pi\hbar)$ which were much smaller than the number of superlattice periods, and (b) the steps occurred at magnetic fields much higher than those at which the extreme quantum limit should have been reached for a uniform carrier distribution. We now show that in sharp contrast, neither (a) nor (b) applies to the quantum Hall plateau observed in the low-temperature magneto-transport for sample 6.

For this n-type superlattice at $T = 2$ K, Fig. 4 shows the diagonal and Hall resistivities as a function of magnetic field. Beginning at only 0.7 T and extending to 1.5 T is a quantum Hall plateau in ρ_{xy}, accompanied by a deep minimum in ρ_{xx}.[13] For the electron concentration of 8.0×10^{15} cm^{-3}, one calculates that the density of states should become large enough to contain all of the electrons in one Landau level whenever $B > 0.44$ T. Although this is slightly lower than the experimental value for the onset of the plateau, it will be seen below that the dispersion in k_z causes the lowest and first excited Landau levels to overlap in energy until $B \approx 0.7$ T.

The plot of σ_{xy} versus B in Fig. 5 shows that in units of $e^2/2\pi\hbar$, the index of the plateau is $i \approx 140$. That is, most of the 200 periods in the superlattice contribute to the quantized conduction of sample 6. We argue that this represents the first observation of the quantum Hall effect associated with carriers populating the interior of a HgTe–CdTe superlattice. That quantum oscillations are observed in the dc resitivity at fields as low as 0.4 T demonstrates the success of this first attempt to achieve controlled, modulation doping in Hg-based superlattices. The broad quantum Hall plateau would not have been observable had the structure not been of extremely high quality and the doping uniform over most of the layers.

IV. DISCUSSION

In principle, one should observe the quantum Hall effect only in two-dimensional systems, since dispersion in the third dimension prevents the formation of energy gaps between successive Landau levels. The vanishing of σ_{xx} (if there are no parallel conduction paths) and plateaus in σ_{xy} occur whenever the Fermi energy lies in or near one of the gaps, where only localized tail states are available for conduction. A superlattice is generally considered to be an anisotropic 3D system, since interactions between the quantum states in neighboring wells lead to dispersion along the growth axis. However, Störmer et al.[14] have pointed out that as long as the miniband width is smaller than the Landau level spacing, gaps will be present between the Landau levels and the quantum Hall effect may still be observable. They demonstrated this experimentally for a 60-period GaAs–$Ga_{1-x}Al_xAs$ superlattice, which yielded plateaus in ρ_{xy} corresponding to $i = 48$ and 96. They attributed the difference

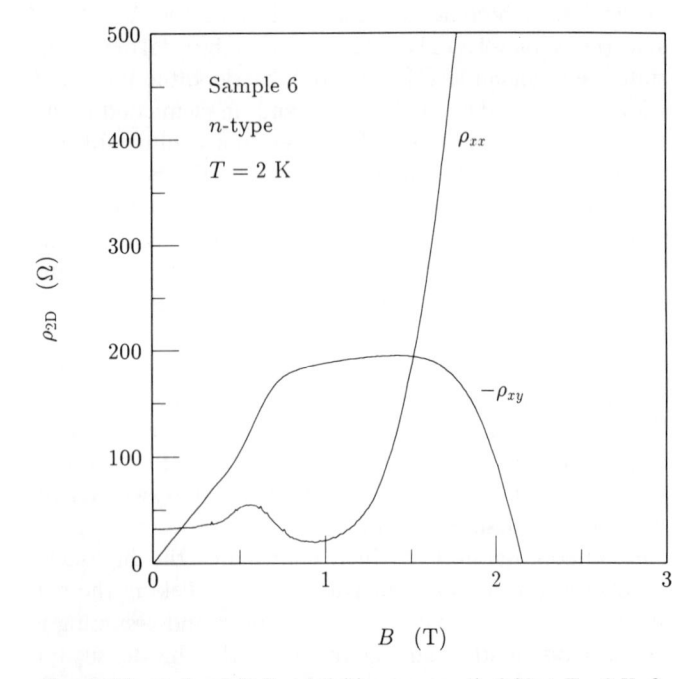

FIG. 4. Diagonal and Hall resistivities vs magnetic field at $T = 2$ K, for sample 6.

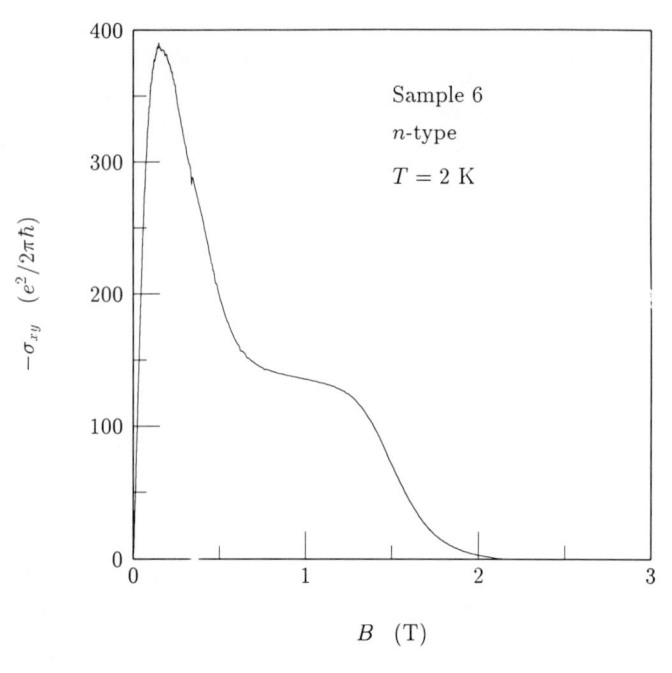

FIG. 5. Hall conducitivity vs magnetic field at $T = 2$ K, for sample 6. For this 200-period superlattice, the quantum Hall plateau occurs at $i \approx 140$.

between 60 periods and 48 conducting wells to depletion of part of the sample.

Band structure calculations[11,12] verify that despite the significant k_z dispersion predicted for Sample 6, there should be an energy gap between the lowest electron Landau level at $k_z = \pi/d$ and the first excited level at $k_z = 0$ whenever $B > 0.7$ T. This is in excellent agreement with the field at which the quantum Hall plateau in Fig. 5 begins. As in the GaAs–Ga$_{1-x}$Al$_x$As superlattice discussed above, the difference between 200 total periods and 140 channels contributing to the quantized conduction is probably due to some type of nonuniformity of the electron concentration. Nonetheless, the majority of the superlattice volume may be considered "active" in contributing to the quantum Hall effect.

This conclusion is far different from that which must be reached concerning the results for sample 3, as well as for the previous experiments of Ong et al. and Woo et al. In each of those cases, not only did the quantum oscillations continue up to very high magnetic fields, but the finding of a small number of conduction channels i suggests that the relevant carriers occupied only a few wells at the top or bottom of the superlattice. Furthermore, both sample 3 and the sample studied by Ong et al. showed evidence for conduction by multiple carrier species, providing independent confirmation that the net conduction was not by a single carrier occupying the interior of the superlattice.

ACKNOWLEDGMENTS

We thank L. R. Ram-Mohan for the use of his superlattice band structure software. This research was partially supported by SDIO/IST and managed by NRL.

[a] Present address: IBM, T. J. Watson Research Center, Yorktown Heights, New York 10598.

[1] M. W. Goodwin, M. A. Kinch, R. J. Koestner, M. C. Chen, D. G. Seiler, and R. J. Justice, J. Vac. Sci. Technol. A **5**, 3110 (1987).

[2] D. G. Seiler, G. B. Ward, R. J. Justice, R. J. Koestner, M. W. Goodwin, M. A. Kinch, and J. R. Meyer, J. Appl. Phys. **66**, 303 (1989).

[3] L. Ghenim, R. G. Mani, J. R. Anderson, J. T. Cheung, Phys. Rev. B **39**, 1419 (1989).

[4] Whereas in wide-gap heterostructures the two spin states associated with a given Landau level are often difficult to resolve until the magnetic field becomes quite high, in narrow-gap superlattices the spin splitting and Landau splitting are nearly equal. Therefore, when we refer to a given Landau level in this paper, we mean a spin-split level for which no factor of 2 degeneracy is assumed.

[5] N. P. Ong, J. K. Moyle, J. Bajaj, and J. T. Cheung, J. Vac. Sci. Technol. A **5**, 3079 (1987); J. T. Cheung, G. Nizawa, J. Moyle, N. P. Ong, B. M. Paine, and T. Vreeland, Jr., *ibid.* A **4**, 2086 (1986).

[6] K. C. Woo, S. Rafol, and J. P. Faurie, Phys. Rev. B **34**, 5996 (1986); J. Vac. Sci. Technol. A **5**, 3093 (1987).

[7] K. A. Harris, S. Hwang, Y. Lansari, J. W. Cook, Jr., J. F. Schetzina, and M. Chu, J. Vac. Sci. Technol. A **5**, 3085 (1987).

[8] C. A. Hoffman, J. R. Meyer, F. J. Bartoli, J. W. Han, J. W. Cook, Jr., J. F. Schetzina, and J. N. Schulman, Phys. Rev. B **39**, 5208 (1989).

[9] C. A. Hoffman, J. R. Meyer, R. J. Wagner, F. J. Bartoli, X. Chu, J. P. Faurie, L. R. Ram-Mohan, and H. Xie, J. Vac. Sci. Technol. A **8**, 1200 (1990).

[10] L. M. Roth and P. N. Argyres, in *Semiconductors and Semimetals*, edited by R. K. Willardson and A. C. Beer (Academic, New York, 1966), Vol. 1, p. 159.

[11] L. R. Ram-Mohan, K. H. Yoo, and R. L. Aggarwal, Phys. Rev. B **38**, 6151 (1988).

[12] J. R. Meyer, R. J. Wagner, F. J. Bartoli, C. A. Hoffman, and L. R. Ram-Mohan, Phys. Rev. B **40**, 1388 (1989).

[13] At higher fields, ρ_{xy} converts to p-type and the transport appears to be dominated by a hole with high density and very low mobility.

[14] H. L. Störmer, J. P. Eisenstein, A. C. Gossard, W. Wiegmann, and K. Baldwin, Phys. Rev. Lett. **56**, 85 (1986).

Theory for electron and hole transport in HgTe–CdTe superlattices

J. R. Meyer, D. J. Arnold,[a] C. A. Hoffman, and F. J. Bartoli
Naval Research Laboratory, Washington, D. C. 20375

L. R. Ram-Mohan
Worcester Polytechnic Institute, Worcester, Massachusetts 01609

(Received 2 October 1990; accepted 29 October 1990)

We present results of the first detailed theory for electron and hole transport in HgTe-CdTe superlattices. The calculation incorporates the superlattice band structure in full generality, and also treats multi-well scattering and screening processes which have been ignored in previous theories. It is predicted that whereas the electron and hole mobilities should be nearly equal at low temperatures, the hole mobility falls far below the electron value at somewhat higher temperatures due to the extreme nonparabolicity of the valence band. This prediction is entirely consistent with experimental results reported previously. Excellent quantitative agreement with the data over a broad temperature range is achieved if interface roughness scattering is considered in addition to ionized impurity scattering, acoustic and optical phonon scattering, and electron–hole scattering. It is pointed out that low-temperature electron mobilities for a number of thin-well HgTe–CdTe superlattices follow the d_W^6 dependence expected for the interface roughness mechanism.

I. INTRODUCTION

The mobilities of electrons and holes in HgTe–CdTe superlattices have been studied in a number of recent experimental investigations.[1-9] However, despite the extensive data base which has become available, there has been no previous attempt to treat the problem of free carrier transport in the HgTe–CdTe system theoretically. This is primarily because there was no formalism which could account for the highly unusual narrow-gap superlattice band structure. In this paper we present the results of a comprehensive new theory for electron and hole mobilities in HgTe–CdTe heterostructures. Not only is the band structure built into the calculation, but ionized impurity scattering is treated more generally than in past theories, since scattering by impurities in neighboring wells and screening by carriers in other wells are fully accounted for.

Due to space limitations, we defer a complete exposition of the formalism to a separate work. Following a brief summary of the theory, we discuss its application to temperature-dependent mobilities in a specific superlattice for which experimental data are available. The conclusions for this example will then be broadened into a more general consideration of transport in the HgTe–CdTe system. In particular, evidence will be given that interface roughness scattering represents a dominant mechanism limiting the electron mobility in structures with thin wells.

II. THEORY

Figure 1 illustrates the theoretical band structure for a typical positive-gap HgTe–CdTe superlattice at $T = 4.2$ K. We have employed an 8-band transfer-matrix algorithm,[10] using a valence band offset of 350 meV,[11] including strain, and assuming the barriers to contain 15% HgTe. Note that whereas the conduction band dispersion in the plane (k_x) is well behaved, the valence band is exceptionally nonparabolic. While the hole mass at the top of the band is just slightly larger than that of the electrons, it becomes infinite and then negative (electron like) at only slightly lower energies. This feature should be expected to dominate the temperature dependence of the hole mobility, since at low T the holes occupy light-mass states near the valence band maximum whereas at higher T they are thermally activated to the larger-mass region. Note also that the conduction and valence bands are extremely anisotropic, i.e., the mass in the growth direction (k_z) is much larger than that in the plane. Any realistic modeling of the transport problem should take into account these essential aspects of the band structure. We have therefore employed a formalism which is fully general with respect to the free carrier dispersion relations, and band struc-

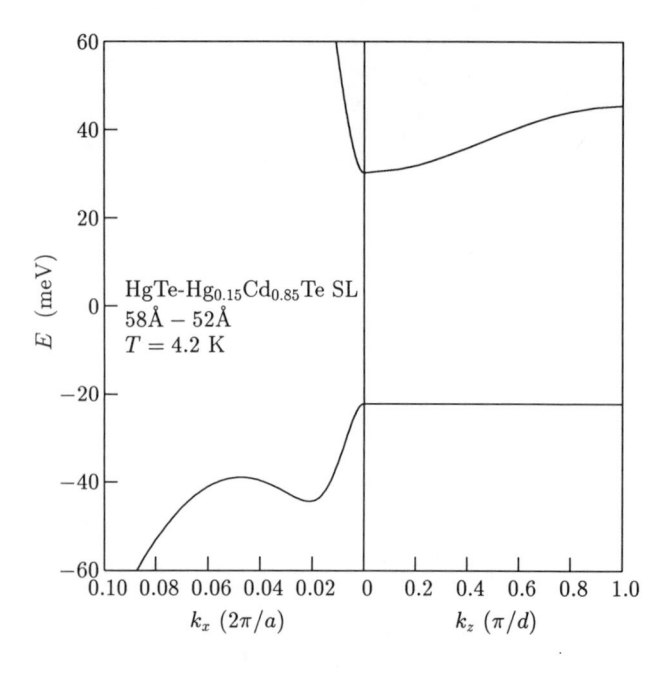

FIG. 1. [100] HgTe–CdTe superlattice band structure for input to the mobility calculations. The well thickness (58 Å) has been increased slightly from the nominal experimental value (52 Å) in order to make the calculated energy gap agree with experiment.

tures obtained for each temperature between 4.2 and 300 K are incorporated into the calculation in numerical form.

Those scattering mechanisms which dominate the electron and hole transport in HgTe are also expected to be significant in Hg-based heterostructures. We have treated ionized impurity scattering using a generalized approach which fully accounts for multi-well interactions in the superlattice. Although the case of transport by quasi-two-dimensional (2D) electrons at a heterojunction has been considered in extensive detail,[12,13] previous formalisms for treating multiple quantum wells and superlattices are surprisingly incomplete. Whereas past calculations have usually considered only the screening of impurities by other carriers in the same well as the test electron,[14,15] we have derived a generalized scattering potential which is three-dimensional (3D) yet anisotropic, due to the asymmetry of screening in the plane as compared to screening along the growth axis (where the screening is less effective because the charge must be located only in the wells). An iterative approach has been used to solve Poisson's equation self-consistently. We also consider the distinction between quasi-2D transport and anisotropic 3D transport. Even if the mass anisotropy of a 3D system is infinite this does not imply equivalency to the 2D limit, since the density of states into which a carrier can scatter depends on whether k_z is a good quantum number. Which of the 2D or 3D formalisms is appropriate depends on the relation between the scattering time and the rate of tunneling between neighboring wells.

Electron–hole scattering is somewhat similar to ionized impurity scattering in that both occur via the screened Coulomb potential. However, the mobility of the center of mass in a carrier–carrier process requires that one account for the dynamic nature of the screening, whereas the screening of an ionized impurity is static. This effect has been approximated using limiting forms discussed in Ref. 16. Since the hole-to-electron effective mass ratio is large in most situations where electron–hole scattering is important, we ignore the effect of this mechanism on the hole mobility. The electron scattering rate is then calculated in the limit of an infinite mass ratio. Acoustic, polar optical, and nonpolar optical phonon scattering have been treated using expressions discussed by Price.[17,18] Phonon parameters are taken from values which have previously given good agreement with experiment for electron and hole mobilities in bulk HgTe.[19]

We have also considered a further mechanism which does not arise in bulk semiconductors: interface roughness scattering.[20,21] Whereas most previous investigators have estimated the variation of the energy levels with well thickness from a simple expression involving the bulk effective mass, in narrow-gap superlattices this can lead to significant errors. We have therefore used the band structure formalism to explicitly calculate $\partial E/\partial d_W$ as a function of d_W for the $E1$ and $HH1$ bands.

Wavevector-dependent scattering rates have been calculated for each of the mechanisms discussed. Net electron and hole mobilities are then derived for comparison with experiment by combining these within the relaxation time approximation.

III. COMPARISON WITH EXPERIMENT

In this work, we compare results of the transport calculation with data[3] for a p-type HgTe–CdTe superlattice of well and barrier thicknesses $d_W = 52$ Å and $d_B = 52$ Å. The 200-period superlattice was grown by molecular-beam epitaxy at 175 °C, the substrate was [100]CdTe, and a 2-μm-thick CdTe buffer layer separated the substrate from the superlattice. Low-temperature Hall measurements determined the net acceptor concentration to be 3×10^{14} cm^{-3}. The filled and open circles in Fig. 2 represent the experimental electron and hole mobilities as a function of temperature. For T between 10 and 40 K, the minority electron mobility was determined from photo-Hall measurements as a function of CO_2 laser excitation intensity, while for $T \geq 70$ K the intrinsic electron concentration was large enough to allow a determination of μ_n from the magneto-transport data without excitation.

The curves represent theoretical electron and hole mobilities due to ionized impurity scattering alone, assuming the superlattice to be uncompensated. The three sets of curves correspond to different limiting cases for the dimensionality (2D or 3D) and the scattering potential (a simplified isotropic screened Coulomb potential[22] or the more general anisotropic potential which was discussed in the previous section). The dash-dot curves represent the 2D limit (initial and final wavevectors must lie in the plane), for scattering by the simplified potential. Scattering by impurities in neighboring wells is included, and is found to be quite important

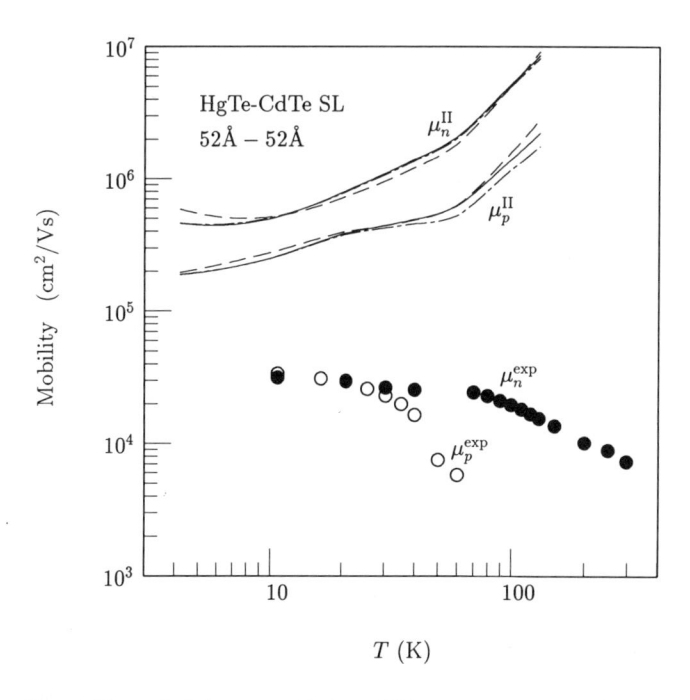

FIG. 2. Theoretical electron and hole mobilities vs temperature for ionized impurity scattering only. The dash-dot curve is for 2D transport and an isotropic screened Coulomb potential, the dashed curve is for 3D transport with the isotropic potential, while the solid curve is for 2D transport and the more general anisotropic potential. No compensation is assumed ($N_D + N_A = 3 \times 10^{14}$ cm^{-3}). Experimental (Ref. 3) mobilities are shown for comparison.

since the screening length (λ_s) at low temperatures is more than 3 times the superlattice period (d). The solid curves represent 2D transport, but for scattering by the more general anisotropic potential. The correction is modest in this example because of the relation between λ_s and d. For larger carrier concentrations where they become comparable, the restriction of the screening charge to the wells will have a much more significant effect on the scattering potential and hence on the mobility. The dashed curves were obtained assuming the simplified scattering potential, but for 3D transport (scattering out of the plane is allowed). We find that for this mechanism and these superlattice parameters, the 2D and 3D mobilities agree to within 30%.[23] Since the distinction between 2D and 3D does not appear to be crucial in this example, we will employ the 2D results (solid curves) in the analyses below without making further comparisons between the different limits.

Considering the strong nonparabolicity of the valence band, it may seem surprising that the ratio μ_n^{II}/μ_p^{II} is relatively independent of temperature. Although there is a slight increase near $T = 30$ K, this effect is small because the mobility associated with the ionized impurity scattering mechanism has a weak dependence on effective mass (it goes roughly as $m^{-1/2}$). The sharp upturn of the mobilities above 70 K is due to increased screening, which results from the thermal activation of intrinsic carriers. However, we will see below that while ionized impurity scattering becomes less effective in this region, there is eventually a net decrease in the electron mobility due to enhanced electron-hole scattering.

Clearly, the magnitudes of the theoretical ionized impurity scattering mobilities in Fig. 2 are far higher than the experimental values, probably indicating that the assumption of zero compensation is inappropriate. This should not be surprising, since the low-temperature hole concentration for this sample ($p_0 \approx 3 \times 10^{14}$ cm^{-3}) is somewhat below the range 10^{15}–10^{16} cm^{-3} which is more typical of net unintentional doping levels in HgTe-CdTe superlattices. Figure 3 illustrates the effect of increasing the net impurity concentration to $N_D + N_A = 4 \times 10^{15}$ cm^{-3}, i.e., 86% compensation. In this case electron-hole scattering, acoustic phonon scattering, and scattering by polar and nonpolar optical modes have also been included in the calculation. We note first that the strong downturn of the experimental hole mobility at temperatures above 30 K is now accurately reproduced by the theory. This is because phonon processes yield a strong dependence of the mobility on effective mass (e.g., $\mu \propto m^{-5/2}$ for acoustic phonons). Since the electron and hole deformation potentials are taken to be equal, the divergence of μ_n and μ_p with increasing T is due solely to the strong increase of the mass ratio, which comes about as holes are thermally activated into states with higher k_x (see Fig. 1).

Although the level of compensation has been adjusted to bring the calculated electron mobility in Fig. 3 into good agreement with the data at $T = 10$ K, the theoretical and experimental temperature dependences remain significantly different. While μ_n^{exp} gradually decreases as T is varied between 10 and 60 K, the theoretical mobility increases by a factor of 4 over this range. Not until T exceeds 60 K does μ_n begin to turn down due to phonon scattering.

The insensitivity of the ionized impurity scattering mobilities in Fig. 2 to the details of the scattering potential or to the 2D versus 3D character of the transport make it appear unlikely that further refinements to the impurity scattering calculation would significantly improve the incorrect temperature dependence. Comparison with the data instead suggests that another scattering mechanism may be affecting the transport, specifically a mechanism which has more effect on the electrons than on the holes. This suggests that we examine the possible effects of interface roughness scattering, which has been demonstrated to limit the electron mobility in GaAs–Ga$_{1-x}$Al$_x$As single and multiple quantum wells with relatively thin d_W.[21,24] Phenomenologically, it is apparent that roughness in the well boundaries represents a spatial variation in d_W, which in turn leads to fluctuations in the quantization energy for free carriers in the wells. This fluctuating energy may be thought of as a scattering potential, which yields an inverse relaxation time of the form[20,21]

$$\tau_{\text{IR}}^{-1} \propto g(\Lambda)\Delta^2\Lambda^2(\partial E_0/\partial d_W)^2, \qquad (1)$$

where Δ is the magnitude of the fluctuations and Λ is their correlation length. The function $g(\Lambda)$, which includes free carrier screening, becomes small whenever the spatial extent Λ is long compared to the Fermi wavevector, i.e., there is little scattering since the wells are then locally smooth. From the particle-in-a-box model with infinite barriers ($E_0 \propto d_W^{-2}$), one estimates that in lowest order: $\partial E_0/\partial d_W \approx d_W^{-3}$. Although we will employ a more exact dependence obtained from the band structure calculation

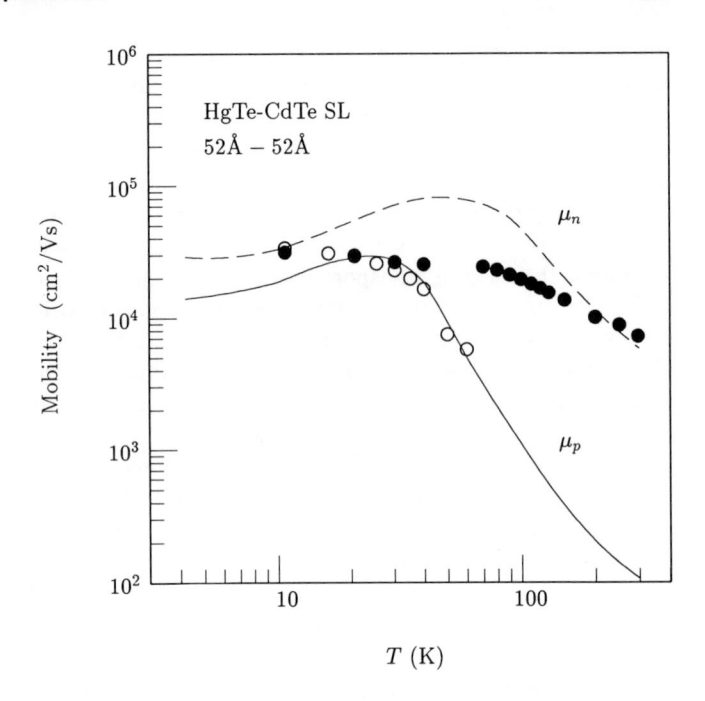

FIG. 3. Experimental (Ref. 3) (points) and theoretical (curves) electron and hole mobilities vs temperature. The calculation assumes $N_D + N_A = 4 \times 10^{15}$ cm^{-3} and no interface roughness scattering.

(see the previous section), these simple considerations imply that μ_{IR} should scale roughly as d_W^6. This is the dependence observed experimentally in GaAs–Ga$_{1-x}$Al$_x$As.[21,24]

For a number of HgTe–CdTe superlattices from various experimental investigations,[3,5,7–9] low-temperature electron mobilities are plotted as a function of superlattice well thickness in Fig. 4 (the sample considered above in Figs. 2 and 3 is represented by the open circle). The two lines in Fig. 4 correspond to $\mu_n \propto d_W^6$. That superlattices from two different series (the triangles[7] and the filled circles[5]) appear to follow this dependence strongly suggests that interface roughness scattering is in fact a dominant mechanism governing electron mobilities in HgTe–CdTe superlattices with relatively thin wells. Hole mobilities are expected to be affected far less by this mechanism, since $\partial E_0/\partial d_W$ is roughly a factor of 5 smaller for the HH1 valence band than it is for the E1 conduction band. However, we also note that for wells thick enough that the superlattice becomes semimetallic, both the conduction band minimum and the valence band maximum reside in the HH1 band.[5] Interface roughness scattering should then become weak for both types of carriers, which is why the curves in Fig. 4 have been terminated at $d_W \approx 65$ Å.

The detailed calculation gives agreement with the data from Ref. 7 (triangles) if we take Δ to be 1 monolayer (the usual assumption for GaAs–Ga$_{1-x}$Al$_x$As) and $\Lambda \approx 60$ Å (GaAs–Ga$_{1-x}$Al$_x$As values[21,24,25] span the range 40–200 Å).[26] Using $\Lambda = 60$ Å and $N_D + N_A = 2.5 \times 10^{15}$ cm^{-3}, the results of including interface roughness scattering in the present mobility calculation are given in Fig. 5. Since $\partial E_0/\partial d_W$ for the HH1 band is small, the hole mobility is relatively unchanged from Fig. 3. However, the calculated electron mobilities are now in quite good agreement with the data. The increase with temperature is now slight, and the discrepancy between theory and experiment is never greater than 40%. We emphasize that since λ is taken from a fit to other samples and the phonon parameters are HgTe values, the only parameter adjusted in calculating both $\mu_n(T)$ and $\mu_p(T)$ is the degree of compensation.

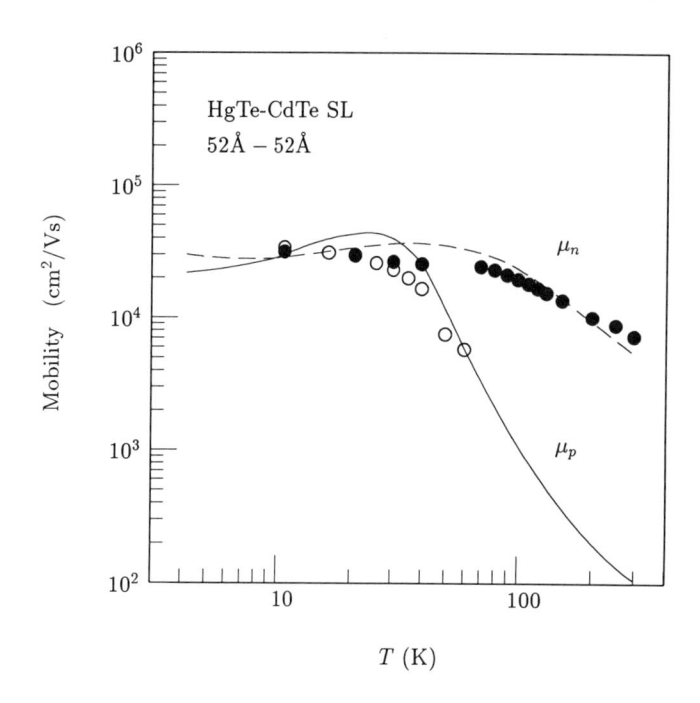

FIG. 5. Experimental (Ref. 3) (points) and theoretical (curves) electron and hole mobilities vs temperature. The calculation assumes $N_D + N_A = 2.5 \times 10^{15}$ cm^{-3} and one monolayer interface roughness with a correlation length $\Lambda = 60$ Å.

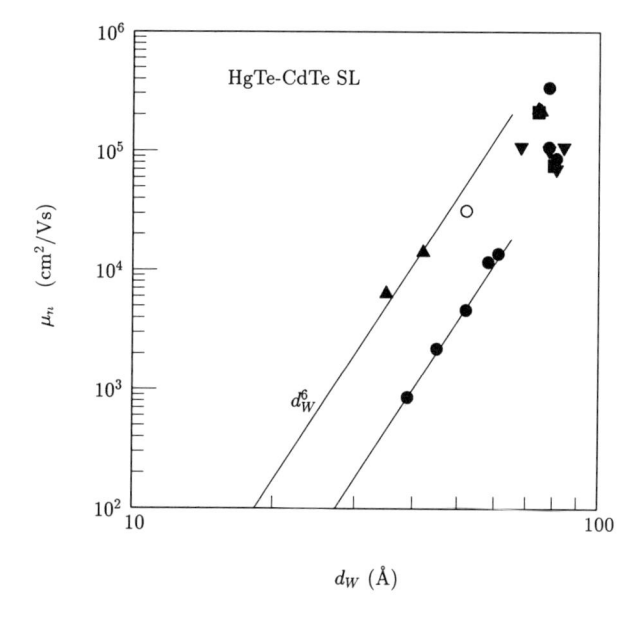

FIG. 4. Experimental low-temperature electron mobilities vs superlattice well thickness, compared to curves representing $\mu_n \propto d_W^6$. The filled circles are from Ref. 5, triangles from Ref. 7, boxes from Ref. 9, inverted triangles from Ref. 8, and the open circle (Ref. 3) is the same superlattice considered in Figs. 1–3.

IV. CONCLUSIONS

We have carried out the first detailed calculation of electron and hole mobilities in HgTe–CdTe superlattices, using a formalism which fully incorporates the complex band structure. We have also employed a generalized anisotropic scattering potential for ionized impurity scattering and have treated both the 2D and 3D limits. When ionized impurity scattering (with compensation), electron–hole scattering, and scattering by acoustic, polar optical, and nonpolar optical phonon modes are taken into account, the theoretical hole mobilities are found to be in excellent agreement with experiment. A sharp drop of the hole mobility with increasing temperature, which has frequently been observed experimentally, is accurately reproduced by the theory and shown to be due to the extreme nonparabolicity of the superlattice valence band. However, the calculated temperature variation for the electron mobility agrees poorly with the experimental dependence unless interface roughness scattering is also included. We show that when superlattice mobilities from previous experimental studies are plotted as a function of well thickness, two different series of samples follow the d_W^6 dependence which is characteristic of the interface roughness mechanism. From a fit to data for thin-well superlattices with the highest mobilities, a value of $\Lambda = 60$ Å is estimated for the roughness correlation length. Using this value, the inclusion of interface roughness scattering in the calculation leads to excellent quantitative agreement between theoretical and experimental mobilities for both electrons and holes.

ACKNOWLEDGMENTS

This research was supported in part by the Office of Naval Research and in part by SDIO/IST.

[a] Present address: IBM, T. J. Watson Research Center, Yorktown Heights, New York 10598
[1] J. P. Faurie, M. Boukerche, S. Sivananthan, J. Reno, and C. Hsu, Superlatt. Microstruct. **1**, 237 (1985).
[2] M. W. Goodwin, M. A. Kinch, R. J. Koestner, M. C. Chen, D. G. Seiler, and R. J. Justice, J. Vac. Sci. Technol. A **5**, 3110 (1987).
[3] C. A. Hoffman, J. R. Meyer, E. R. Youngdale, J. R. Lindle, F. J. Bartoli, K. A. Harris, J. W. Cook, Jr., and J. F. Schetzina, Phys. Rev. B **37**, 6933 (1988).
[4] J. R. Meyer, F. J. Bartoli, C. A. Hoffman, and J. N. Schulman, Phys. Rev. B **38**, 12457 (1988).
[5] C. A. Hoffman, J. R. Meyer, F. J. Bartoli, J. W. Han, J. W. Cook, Jr., J. F. Schetzina, and J. N. Schulman, Phys. Rev. B **39**, 5208 (1989).
[6] C. A. Hoffman, J. R. Meyer, F. J. Bartoli, J. W. Han, J. W. Cook, Jr., and J. F. Schetzina, Phys. Rev. B **40**, 3867 (1989).
[7] C. A. Hoffman, J. R. Meyer, R. J. Wagner, F. J. Bartoli, X. Chu, J. P. Faurie, L. R. Ram-Mohan, and H. Xie, J. Vac. Sci. Technol. A **8**, 1200 (1990).
[8] C. A. Hoffman, J. R. Meyer, D. J. Arnold, F. J. Bartoli, Y. Lansari, J. W. Cook, Jr., and J. F. Schetzina, J. Vac. Sci. Technol. B **9**, 1808 (1991).
[9] M. Dobrowolska, T. Wojtowicz, H. Luo, J. K. Furdyna, O. K. Wu, J. N. Schulman, J. R. Meyer, C. A. Hoffman, and F. J. Bartoli, Phys. Rev. B **41**, 5084 (1990).
[10] L. R. Ram-Mohan, K. H. Yoo, and R. L. Aggarwal, Phys. Rev. B **38**, 6151 (1988).
[11] J. R. Meyer, C. A. Hoffman, and F. J. Bartoli, Semicond. Sci. Technol. **5**, S90 (1990).
[12] T. Ando, A. B. Fowler, and F. Stern, Rev. Mod. Phys. **54**, 437 (1982).
[13] A. Gold, Phys. Rev. B **41**, 8537 (1990).
[14] K. Hess, Appl. Phys. Lett. **35**, 484 (1979).
[15] S. Mori and T. Ando, J. Phys. Soc. Japan **48**, 865 (1980).
[16] J. R. Meyer and F. J. Bartoli, Phys. Rev. B **28**, 915 (1983).
[17] P. J. Price, Ann. Phys. **133**, 217 (1981).
[18] P. J. Price, Surf. Sci. **113**, 199 (1982).
[19] J. R. Meyer, C. A. Hoffman, F. J. Bartoli, J. M. Perez, J. E. Furneaux, R. J. Wagner, R. J. Koestner, and M. W. Goodwin, J. Vac. Sci. Technol. A **6**, 2775 (1988).
[20] A. Gold, Phys. Rev. B **35**, 723 (1987).
[21] H. Sakaki, T. Noda, K. Hirakawa, M. Tanaka, and T. Matsusue, Appl. Phys. Lett. **51**, 1934 (1987).
[22] For the simplified potential, the screening length is taken to have the conventional bulk form corresponding to a carrier distribution which is uniform throughout the volume of the superlattice.
[23] The difference can in principle be much larger, e.g., nearly a factor of 10 for electron scattering by acoustic phonons at low temperatures. However, this is not of great practical importance since phonon scattering is insignificant in this limit.
[24] R. Gottinger, A. Gold, G. Abstreiter, G. Weimann, and W. Schlapp, Europhys. Lett. **6**, 183 (1988).
[25] M. Tanaka and H. Sakaki, J. Cryst. Growth **81**, 153 (1987).
[26] $\Lambda \approx 200$ Å is required to fit the filled circles.

Minority carrier lifetimes of metalorganic chemical vapor deposition long-wavelength infrared HgCdTe on GaAs

R. Zucca, D. D. Edwall, J. S. Chen and S. L. Johnston
Rockwell International Science Center, Thousand Oaks, California 91360

C. R. Younger
Rockwell International/Electro-Optical Center, Anaheim, California 92803

(Received 3 October 1990; accepted 26 December 1990)

Metalorganic chemical vapor deposition (MOCVD) growth of HgCdTe on GaAs is a promising technique that overcomes the size and crystal quality limitations of CdTe substrates. An important material parameter is the minority carrier lifetime, which determines the ultimate zero bias impedance and quantum efficiency of a photodiode. We present the first systematic study of the temperature and carrier concentration dependence of minority carrier lifetimes on n-type and p-type layers of MOCVD long-wavelength infrared HgCdTe grown on GaAs substrates. The temperature dependencies of the lifetime are compared with theoretical predictions based on Auger, radiative, and Shockley–Read recombination. Excellent fits are obtained over a broad temperature range, from 20 K to room temperature. The experimental lifetimes of n-type material reach the theoretical limit imposed by Auger + radiative recombination for carrier concentrations higher than 2×10^{15} cm^{-3}. For lower carrier concentrations, the measured lifetimes are shorter than those predicted from Auger + radiative recombination, and Shockley–Read recombination must be added to the calculations. The lifetimes of arsenic-doped and vacancy-doped p-type material are Shockley-Read limited. They are one order of magnitude longer than those previously observed on vacancy-doped liquid phase epitaxy material.

I. INTRODUCTION

HgCdTe is the preferred material for detectors and focal plane arrays over a broad range of the infrared spectrum. Great interest is focused on the long wavelength infrared (LWIR) spectral region, between 8 and 12 μm. Among several crystal growth techniques that can be used to fabricate epitaxial material with the required narrow band gap, metalorganic chemical vapor deposition (MOCVD) is particularly attractive.[1] This technique allows for changes of doping and composition within an uninterrupted growth sequence. Therefore in principle, heterostructure photodiodes can be built with tailored composition and doping.[2] Furthermore, the MOCVD technique is being used successfully to grow HgCdTe on GaAs substrates,[3] thus eliminating the constraints imposed by the small size of CdTe substrates. Thus, MOCVD growth of HgCdTe on GaAs has the potential of producing sophisticated high-performance photovoltaic focal plane arrays on large area substrates.[4]

Since MOCVD HgCdTe on GaAs is a relatively young material, it must be fully characterized. Minority carrier lifetimes are important parameters, because they determine minority carrier diffusion lengths, which determine the ultimate dynamic resistance and quantum efficiency of photodiodes.[5] While crystallographic, optical, and electrical transport properties have been investigated extensively,[3] data on minority carrier lifetimes of this material have not been reported.

We present the results of the first systematic study of minority carrier lifetimes on LWIR HgCdTe grown by MOCVD on GaAs substrates. Samples of both n-type and p-type material with a broad range of carrier concentrations have been studied. Minority carrier lifetimes were measured by the photoconductivity transient technique,[6] over a broad temperature range, from 20 K to room temperature. The temperature dependences of the lifetimes were compared with theoretical calculations. In Sec. II, we discuss material growth and sample preparation. Measurement techniques and data analysis techniques are discussed in Secs. III and IV, respectively. In Sec. V, we present minority carrier lifetime data for n-type material and we interpret the results in terms of possible recombination mechanisms. In Sec. VI, we analyze systematic trends with carrier concentration. In Sec. VII, we present minority carrier lifetime data for p-type material; and, in Sec. VIII, we study systematic trends with composition.

II. MATERIAL GROWTH AND SAMPLE PREPARATION

The substrates were bulk semi-insulating GaAs with (100) orientation. CdTe buffer layers with thickness ranging between 0.2 and 4 μm were grown at 425 °C with a growth rate of 3–4 μm/h prior to HgCdTe growth. The HgCdTe material was grown by the interdiffused method by which ternary HgCdTe is formed via interdiffusion of multiple binary HgTe and CdTe layers.[7] The primary organometallic alkyls were dimethylcadmium (DMCd), dimethyltelluride (DMTe) and di-isopropyltelluride (DipTe). DMTe and DMCd were used for CdTe growth, while DIPTe was used for HgTe growth. Elemental Hg served as the Hg source and maintained alloy stability. Growth temperatures ranged between 370 and 400 °C with growth rates of 3–4 and 10–12 μm/h for CdTe and HgTe, respectively. The CdTe–HgTe period thickness was typically 1000–2000 Å. Extrinsic doping was accomplished by injecting indium and arsenic alkyls during the CdTe growth cycles of the interdiffused process.[7] More detailed descriptions of material growth techniques can be found in Ref. 3.

The composition of the material was calculated from the room temperature optical transmission following the procedure of Finkman and Schacham.[8] The carrier type, concentration and mobility were determined from Van der Pauw and Hall measurements at 78 K.

The samples for minority carrier lifetime measurements were approximately 0.5×0.5 cm squares. They were lightly etched with Br–Methanol, mounted on TO-5 headers and contacted on two sides, using indium on n-type material and gold on p-type material. The samples remained exposed to air for 30–60 min after the etch, while they were being prepared and mounted in a dewar.

III. MEASUREMENT TECHNIQUES

The photoconductivity transient technique was used for the minority carrier lifetime measurements.[6-9] Minority carriers are created by a short subband gap laser pulse, and the corresponding photocurrent pulse is recorded. The minority carrier lifetime is equal to the time constant of the exponential decay of the photocurrent.

The excitation pulses with a length of approximately 3 ns were produced by a GaAs laser, highly attenuated to avoid nonlinear responses.[9] Some measurements were done with a Tektronix OT502 Electrical/Optical converter driven by a Tektronix PG501 pulse generator as a laser source operating at 0.846 μm with peak power of 1 mW. The photoresponse signal was collected and averaged with a Tektronix 7854 waveform processing oscilloscope. The averaged decay pulses were retrieved by a computer for analysis. Figure 1 shows, as an example, a transient as retrieved by the computer. The signal is also displayed on a logarithmic scale with the result of a linear least-squares fit performed by the computer.

The laser beam was normally incident on the front surface of the sample. Occasional measurements with back side illumination agreed with the front side illumination measurements. The sample headers were covered with a cap having a 2-mm-diam sapphire window in the center. This cold cap keeps the background radiation low and assures that no laser light reaches the contact region of the sample.

Misleading results can be obtained if the material is not homogeneous. Anomalous layers with a built-in p/n junction yield transients that correspond to the RC time constant of the junctions rather than the minority carrier lifetime. We have avoided this pitfall by measuring the temperature dependence of the lifetime for almost all the samples. Anomalous layers can be identified because their photoconductive decay time constant is exceptionally large, and it does not vary through the extrinsic temperature region.

IV. DATA ANALYSIS

The temperature dependence of the lifetime was compared with curves calculated with theoretical expressions of the Auger and radiative lifetime. These curves are based on fundamental material properties and do not allow for parameter adjustments. When the experimental lifetimes fell short of the calculated ones, Shockley–Read (SR) recombination was added, with some parameter adjustment.

The minority carrier lifetime τ was calculated as

$$1/\tau = 1/\tau_A + 1/\tau_R + 1/\tau_{SR}, \qquad (1)$$

where τ_A, τ_R and τ_{SR} represent the Auger, radiative, and Shockley–Read lifetimes, respectively.

The Auger lifetime of n-type material was calculated with the well-known equations for the e–e transition (Auger 1)[10]

$$\tau_A = 2n_i^2 \tau_{Ai}/(n_0 + p_0)n_0, \qquad (2)$$

where low-level excitation is assumed. n_i is the intrinsic carrier concentration, n_0 and $p_0 = n_i^2/n_0$ are the electron and hole concentrations, and τ_{Ai} is the intrinsic Auger lifetime (in seconds) given by

$$\tau_{Ai} = 3.8 \times 10^{18} \epsilon_\infty (1+\mu) \exp[(1+2\mu)E_g/$$
$$(1+\mu)kT]/(m_e/m_0)|F_1 F_2|^2 (kT/E_g)^{3/2}. \qquad (3)$$

ϵ_∞ is the optical dielectric constant, μ is the electron/hole effective mass ratio, E_g is the energy gap, k is Boltzmann's constant, T is the absolute temperature, m_e is the electron effective mass, m_0 is the electron mass, and $|F_1 F_2|$ is the overlap integral.

All the parameters in Eqs. (2) and (3) can be calculated with standard equations available in the literature (E_g from Ref. 11, n_i from Ref. 12, and m_e from Ref. 13), once the composition x and the doping density are known. n_0 in Eq. (2) is calculated from

$$n_0 = N_d/2 + [(N_d/2)^2 + n_i^2]^{1/2}, \qquad (4)$$

derived from the charge neutrality condition. Only the overlap integral in Eq. (3) is not known precisely. We have adopted $|F_1 F_2| = 0.2$ for all our calculations.

We have assumed that Auger 7 is the dominant Auger process for p-type material,[14] and we have calculated the intrinsic Auger lifetime, τ_{Ai7}, using Casselman's approximation,[15]

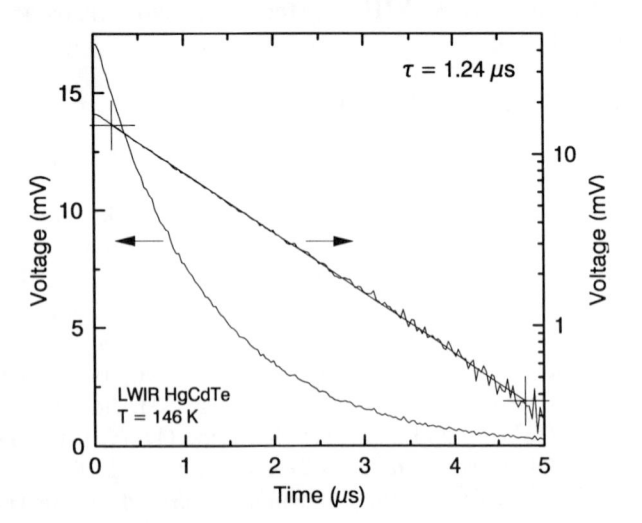

FIG. 1. Example of minority carrier lifetime measurement by the photoconductive transient technique. The pulse decay is shown as recorded by the oscilloscope and transmitted to the computer. The conversion to a logarithmic scale and the least-squares fit, both done by the computer, are also shown.

$$\tau_{Ai7} = \gamma \tau_{Ai}. \qquad (5)$$

Here

$$\gamma = 6(1 - 1.25\, kT/E_T)/(1 + 1.5\, kT/E_T), \qquad (6)$$

where E_g was used for the threshold energy, E_T. The remaining equations for p-type material are obtained by reversing the role of electrons and holes in Eqs. (2) and (4). The validity of this calculation is discussed in Sec. VII.

Radiative lifetimes were calculated with the well-known equation

$$\tau_R = 1/B(n_0 + p_0), \qquad (7)$$

valid in the low excitation level case. B is calculated from the following expression, assuming parabolic, nondegenerate bands,[16]

$$B = 5.8 \times 10^{13} \epsilon_\infty^{1/2}$$
$$\times [m_0/(m_e + m_h)]^{3/2} (1 + m_0/m_e + m_0/m_h)$$
$$\times (300/T)^{3/2} (E_g^2 + 3\, kTE_g + 3.75\, k^2 T^2), \qquad (8)$$

where m_h is the hole effective mass, kT and E_g are in eV and B in cm^3/s.

The Shockley–Read lifetime was calculated assuming one discrete trap level, with equation

$$\tau_{SR} = [\tau_{n0}(p_0 + p_1) + \tau_{p0}(n_0 + n_1)]/(p_0 + n_0), \qquad (9)$$

valid for low excitation levels.[10] n_1 and p_1 are the electron and hole densities when the Fermi level, E_f, is at the trap level, E_t. We assumed that the minimum lifetimes for electrons and holes were equal, so that $\tau_{n0} = \tau_{p0} = \tau_0$. With this simplification, and after simple manipulation of the equations using Boltzmann statistics, Eq. (9) transforms into

$$\tau_{SR} = \tau_0 \{1 + 2\cosh[(E_i - E_t)/kT]/(n_0/n_i + n_i/n_0)\}, \qquad (10)$$

where E_i is the position of the intrinsic Fermi level. This equation is valid for both n-type and p-type material, since it is invariant with respect to a substitution of n_0 for p_0. We derive E_i from n_i and the hole density as

$$E_i - E_v = kT \ln(N_v/n_i), \qquad (11)$$

where E_v is the valence band energy, and N_v is the effective valence band density of states, which is calculated as in Ref. 17.

When Shockley–Read recombination was taken into account, the minimum lifetime τ_0 and the trap position as a fraction of the energy gap $(E_t - E_v)/E_g$ were used as adjustable parameters.

V. n-TYPE MATERIAL RESULTS

In Fig. 2, we present as an example the measured minority carrier lifetime as a function of inverse temperature from room temperature to 14 K for an indium-doped n-type LWIR sample with a carrier concentration of 2.7×10^{15} cm^{-3} at 77 K. The data show normal behavior, with nearly constant lifetime at low temperature, raising gradually with temperature, and then dropping rapidly near room temperature, when the semiconductor becomes intrinsic.

Superimposed are the calculated radiative and Auger lifetimes, and the combined lifetime. The agreement between experimental and calculated data is remarkable, considering that no adjustable parameter was used. Note that radiative recombination plays only a minor role for this LWIR material. Although we considered it in all our calculations, it represents only a small correction to the dominant Auger lifetime. The slight discrepancy between slopes in the intrinsic region suggests that the composition determined from an optical transmission measurement was not accurate, or that the composition of the sample used for optical transmission was slightly different.

In Fig. 3, we present a different example. This LWIR sample has lighter doping than the previous one, 6.6×10^{14} cm^{-3} at 78 K. The minority carrier lifetime between room temperature and 20 K is also well behaved, but an attempt to compare the experimental data with radiative and Auger recombination does not show the good agreement reached in Fig. 2. Well in the intrinsic region, above 200 K, the experi-

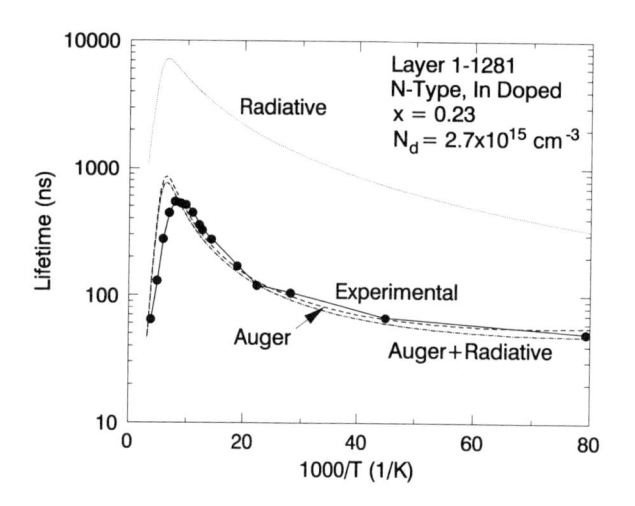

FIG. 2. Example of temperature dependence of the minority carrier lifetime of n-type LWIR MOCVD HgCdTe on GaAs. Superimposed is the result of a calculation taking into account Auger and radiative recombination, matching the experimental data without parameter adjustment.

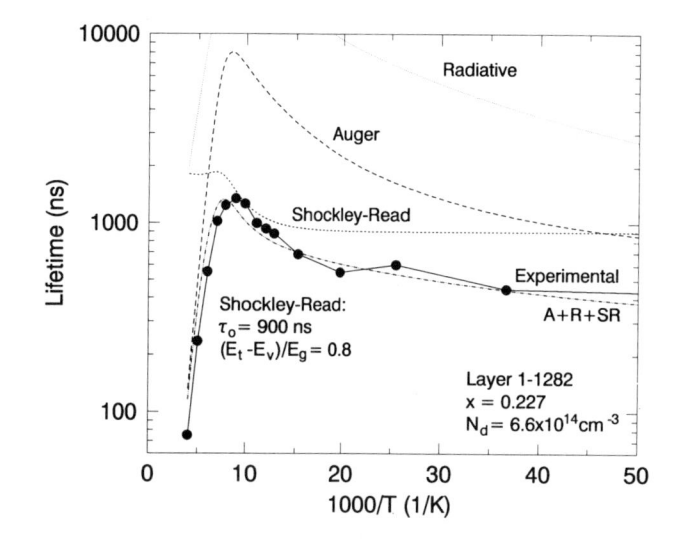

FIG. 3. Another example of temperature dependence of the minority carrier lifetime of n-type LWIR MOCVD n-type HgCdTe on GaAs. Shockley–Read recombination was added to Auger and radiative recombination to fit the experimental data.

mental curve is close to the Auger lifetime curve, but at lower temperature the experimental lifetimes are shorter than those predicted by Auger + radiative recombination. We were able to obtain a good fit by adding Shockley–Read recombination with a trap level at $0.8\ E_g$ above the valence band. This energy level is in the same range where Pratt et al.[18] have estimated a single trap in n-type doped LWIR bulk HgCdTe.

VI. n-TYPE MATERIAL DISCUSSION

We summarize our data on n-type material in Fig. 4, where we display minority carrier lifetimes at 78 K as a function of carrier concentration. Since lifetimes are sensitive to composition, and the LWIR layers measured did not have exactly the same composition, we avoided the complexity of a three-dimensional plot by scaling the measured lifetimes to a composition of 0.225. The scaling factor was determined from the ratio of theoretical Auger + radiative lifetimes at $x = 0.225$ and at the sample composition.

The dashed line in Fig. 4 represents the theoretical lifetime at 78 K for $x = 0.225$, when only Auger and radiative recombination are taken into account. At high carrier concentration, the measured lifetimes are very close to this line. At carrier concentrations lower than 2×10^{15} cm^{-3}, the measured lifetimes are shorter than those predicted by Auger + radiative recombination, and are better described by the solid line, which is obtained by adding Shockley–Read recombination.

The Shockley–Read parameters were chosen so that the theoretical curve fits the data with the longest lifetimes (Fig. 4). The minimum lifetime, τ_0, of 1 μs corresponds to a trap density of 1.8×10^{13} cm^{-3}, if we assume a capture cross section for electrons of 10^{-15} cm^2 and a thermal velocity of 5.7×10^7 cm/s. The trap level, $0.8\ E_g$ above the valence band, is in agreement with the observations in Ref. 18. Although this single-level Shockley–Read model provides a good fit to the temperature dependence of the minority carrier lifetimes for the samples with the longest lifetimes in Fig. 4, other values of the parameters must be used to fit the temperature dependence of the lifetime for the samples which are below the theoretical curve in Fig. 4. This is an indication of a more complex trap configuration for the lower lifetime samples.

VII. p-TYPE MATERIAL RESULTS

In Fig. 5, we present, as an example, the measured minority carrier lifetime as a function of inverse temperature from room temperature to 25 K for an arsenic-doped p-type LWIR sample with a carrier concentration of 4×10^{15} cm^{-3}. Superimposed are the calculated radiative, Auger, and Shockley–Read lifetimes, and the total lifetimes obtained by combining either radiative and Auger lifetimes or radiative and Shockley–Read lifetimes. Good agreement is observed between experimental and calculated data in both cases. Since minority carrier lifetimes much longer than those predicted with the current theory for the Auger 7 process have been observed,[16,19] the interpretation of the minority carrier lifetime in Fig. 5 in terms of radiative + Shockley–Read recombination is more appropriate than the interpretation in terms of radiative + Auger recombination. However, the radiative + Shockley–Read fitting in Fig. 5, excellent in the extrinsic region ($1000/T > 8$), predicts too long lifetimes in the intrinsic region. This is an indication that Auger recombination is needed to fit the data completely.

In Fig. 6, we present a case where the experimental lifetime is shorter than that predicted by Auger + radiative recombination, and the lifetime in the extrinsic region is unquestionably determined by Shockley–Read recombination. However, Auger recombination must be added to account for the value of the lifetime in the intrinsic region. These observations suggest that Auger recombination cannot be neglected completely in p-type material, although the theory

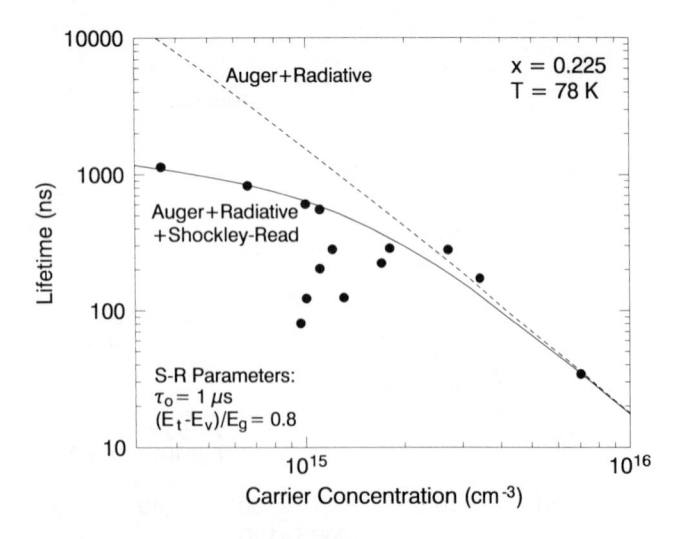

FIG. 4. Carrier concentration dependence of the minority carrier lifetimes at 77 K for LWIR MOCVD n-type HgCdTe on GaAs. Theoretical curves for Auger and radiative recombination with and without Shockley–Read recombination added to Auger and radiative recombination are shown with solid and dashed lines, respectively.

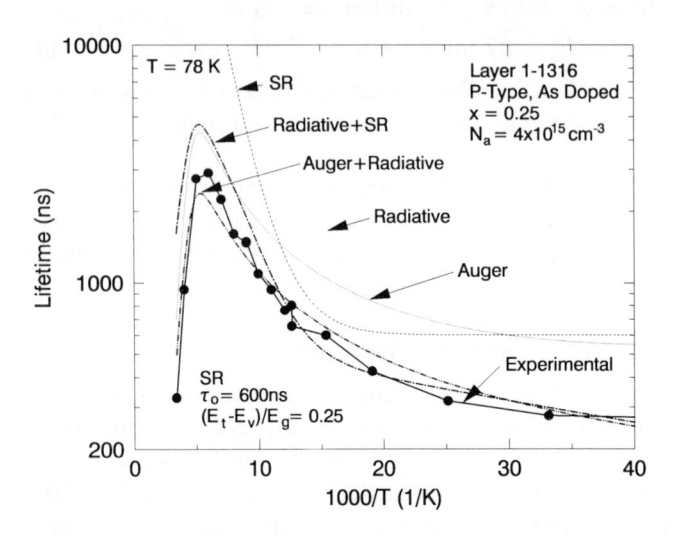

FIG. 5. Example of temperature dependence of the minority carrier lifetime of p-type LWIR MOCVD HgCdTe on GaAs. Superimposed are theoretical curves for radiative, Auger, and Shockley–Read recombination, as well as the results of combining radiative + Auger and radiative + Shockley–Read recombination.

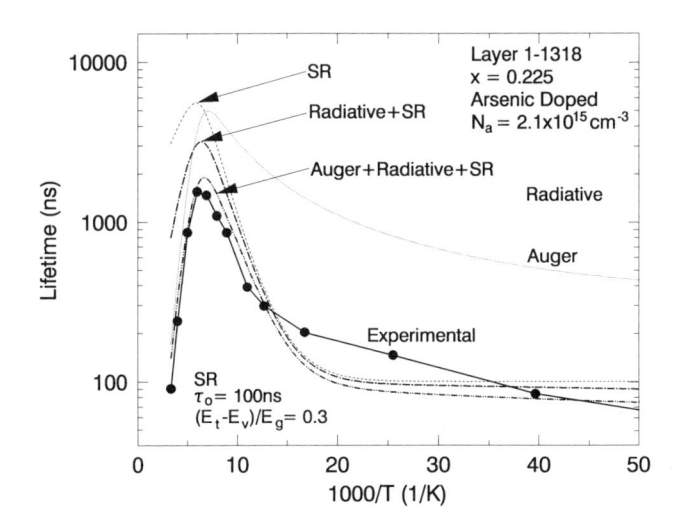

FIG. 6. Another example of temperature dependence of the minority carrier lifetime of p-type LWIR MOCVD HgCdTe on GaAs. Superimposed are theoretical curves for radiative, Auger, and Shockley–Read recombination, as well as the results of combining radiative + Auger and radiative + Shockley–Read recombination.

may need to be revised.[20] The position of the trap level at 0.25–0.3 E_g above the valence band is consistent with reports of a trap level in the lower part of the band gap for p-type HgCdTe.[21,22] Some lack of accuracy of the fit obtained in Fig. 6 suggests that the trap structure in this material is more complex than the single trap assumed.

VIII. p-TYPE MATERIAL DISCUSSION

We summarize our data on p-type material in Fig. 7, where we display minority carrier lifetimes at 78 K as a function of carrier concentration. Arsenic-doped and vacancy-doped samples are identified. As with the n-type data in Fig. 4, we have adjusted for variations in composition by scaling the measured lifetimes to a composition of 0.225. Published data on minority carrier lifetimes in p-type material liquid phase epitaxy (LPE) grown and vacancy doped,[23] LPE grown and arsenic doped,[19] and Bridgman grown and both impurity and vacancy doped[24] are indicated with dashed lines for comparison. The lifetimes of the MOCVD material are similar to those observed on the Bridgman material, higher than those of the LPE vacancy-doped material, and lower than those of the LPE arsenic-doped material.

The solid line in Fig. 7 represents the theoretical lifetime at 78 K for $x = 0.225$, calculated from Auger + Shockley–Read recombination. We assumed that the recombination centers were at 0.3 E_g above the valence band and that their concentration was proportional to the carrier concentration, resulting in a minimum lifetime inversely proportional to the carrier concentration. A value of $\tau_0 = 4 \times 10^8$ s cm^{-3}/N_a was used in Fig. 7. The assumption that the concentration of Shockley–Read recombination centers is proportional to the carrier concentration implies an association between recombination centers and arsenic acceptors. This assumption is rather simplistic because it does not involve the defect structure of the material, which must be considered to explain the differences with arsenic-doped LPE material. However, it succeeds in explaining the strong carrier concentration dependence of the minority carrier lifetime for MOCVD HgCdTe on GaAs.

Differences between the minority carrier lifetimes of impurity- and vacancy-doped p-type HgCdTe have been observed and discussed.[25] Such differences do not appear clearly in the MOCVD material; the lifetimes of the MOCVD vacancy-doped samples, represented by squares in Fig. 7, seem to be in line with the lifetimes of the arsenic-doped samples. Unfortunately, the doping ranges of the arsenic-doped and vacancy-doped samples did not overlap in Fig. 7, so that this observation could be better tested.

IX. SUMMARY AND CONCLUSIONS

We have presented a systematic study of the temperature and carrier concentration dependence of minority carrier lifetimes of n- and p-type MOCVD LWIR HgCdTe grown on GaAs substrates. We have compared the temperature dependence of the lifetimes with theoretical predictions based on Auger, radiative, and Shockley–Read recombination, obtaining excellent fits over a broad temperature range, from 20 K to room temperature.

The experimental lifetimes of n-type material can be described by Auger + radiative recombination for carrier concentrations above 2×10^{15} cm^{-3}. At lower carrier concentrations, Shockley–Read recombination must be added to account for the carrier concentration dependence of the minority carrier lifetimes.

The lifetimes of arsenic-doped and vacancy-doped p-type material are one order of magnitude longer than those previously observed on vacancy-doped LPE material, but still lower than those observed on arsenic-doped LPE material. The lifetimes are limited by Shockley–Read recombination. The concentration of recombination centers was assumed to be proportional to the doping density to explain the strong carrier concentration dependence of the lifetimes.

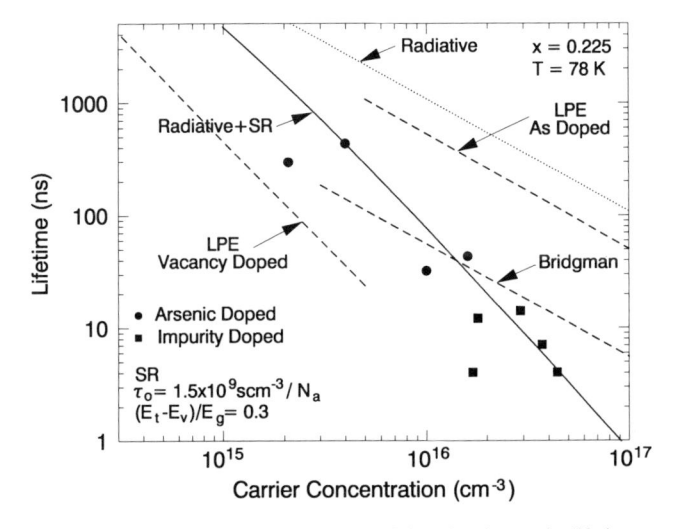

FIG. 7. Carrier concentration dependence of the minority carrier lifetimes at 78 K for LWIR MOCVD p-type HgCdTe on GaAs. The dashed lines represent LPE vacancy-doped data, Bridgman data, and LPE arsenic-doped data from Refs. 23, 24, and 19, respectively. The solid line represents a theoretical calculation of radiative and Shockley–Read recombination. The minimum S-R lifetime was assumed to vary with the doping density.

ACKNOWLEDGMENTS

We are grateful to J. Bajaj for his advice on minority carrier lifetime measurements, to R. E. DeWames for discussions on the minority carrier lifetimes of p-type material, and to D. L. McConnell for valuable assistance with sample preparation. Work funded by SDIO/TNS, sponsored by AFSTC/SWS, managed by WRDC/MLPO, Contract No. F33615-89-C-5557.

[1] J. B. Mullin and S. J. C. Irvine, J. Phys. D **14**, L149 (1981).
[2] J. S. Chen, D. D. Edwall, D. S. Lo, L. O. Bubulac, and R. Zucca, Proceedings of the IRIS Specialty Group on Infrared Materials, Gaithersburg, Maryland, Aug. 1990.
[3] D. D. Edwall, J. Bajaj, and E. R. Gertner, Proceedings of the 1989 IRIS Specialty Group on IR Materials, Monterey, California, pp. 41–56; D. D. Edwall, J. S. Chen, J. Bajaj, and E. R. Gertner, Semicond. Sci. Technol. **5**, S221 (1990).
[4] L. J. Kozlowski, W. E. Tennant, L. O. Bubulac, and E. R. Gertner, Proceedings of the IRIS Specialty Group on Infrared Detectors, Gaithersburg, Maryland, Aug. 1990.
[5] M. B. Reine, A. K. Sood, and T. J. Tredwell, in *Semiconductors and Semimetals*, edited by R. K. Willardson and A. C. Beer (Academic, New York, 1981), Vol. 18, pp. 201–311.
[6] R. Fastow and Y. Nemirovsky, J. Appl. Phys. **66**, 1705 (1989).
[7] S. J. C. Irvine, J. B. Mullin, J. Giess, J. S. Gough, and A. Royle, J. Cryst. Growth **93**, 732 (1988).
[8] E. Finkman and S. E. Schacham, J. Appl. Phys. **56**, 2896 (1984).
[9] J. Bajaj, S. H. Shin, J. G. Pasko, and M. Khoshnevisan, J. Vac. Sci. Technol. A **1** (1983).
[10] J. S. Blakemore, *Semiconductor Statistics* (Dover, New York, 1987).
[11] G. L. Hansen, J. L. Schmit, and T. N. Casselman, J. Appl. Phys. **53**, 7099 (1982).
[12] G. L. Hansen and J. L. Schmit, J. Appl. Phys. **54**, 1639 (1983).
[13] M. H. Weiler, in Ref. 5, Vol. 16, pp. 119–191.
[14] T. N. Casselman, J. Appl. Phys. **52**, 848 (1981).
[15] T. N. Casselman and P. E. Petersen, Solid State Commun. **33**, 615 (1980).
[16] S. E. Schacham and E. Finkman, J. Appl. Phys. **57**, 2001 (1985).
[17] P. Migliorato and A. M. White, Sol. St. Electr. **26**, 65 (1983).
[18] R. G. Pratt, J. Hewett, and P. Capper, J. Appl. Phys. **60**, 2377 (1986).
[19] T. Tung, M. H. Kalisher, A. P. Stevens, and P. E. Herning, Proc. Mat. Res. Soc. Symp. **90**, 321 (1986).
[20] T. N. Casselman's comments at the 1990 U.S. Workshop on the Physics and Chemistry of Mercury Cadmium Telluride and Novel IR Detector Material, San Francisco, California, October 2–4, 1990 (unpublished).
[21] D. L. Polla, S. P. Tobin, M. B. Reine, and A. K. Sood, J. Appl. Phys. **52**, 5182 (1981).
[22] C. E. Jones, K. James, J. Merz, R. Braunstein, M. Burd, M. Eetemadi, S. Hutton, and J. Drumheller, J. Vac. Sci. Technol. A **3**, 131 (1985).
[23] J. S. Chen, J. Bajaj, W. E. Tennant, D. S. Lo, M. Brown, and G. Bostrup, Proc. Mat. Res. Soc. Symp. **90**, 287 (1986).
[24] D. E. Lacklison and P. Capper, Semicond. Sci. Technol. **2**, 33 (1987).
[25] R. Fastow and Y. Nemirovsky, J. Vac. Sci. Technol. A **8**, 1245 (1990).

Trapping effects in HgCdTe

Y. Nemirovsky, R. Fastow, M. Meyassed, and A. Unikovsky
Kidron Microelectronics Research Center, Department of Electrical Engineering, Technion-Israel Institute of Technology, Haifa 32000, Israel

(Received 2 October 1990; accepted 4 December 1990)

Carrier trapping influences the performance of HgCdTe infrared detectors in the 8–12 μm range by enhancing tunneling currents, reducing excess carrier lifetimes, and increasing g–r and $1/f$ noise. In this work, the effects of carrier trapping on the tunneling currents in n^+p diodes are calculated, and the dependence of the tunneling current on temperature, bias, doping level, and trap characteristics is illustrated. It is shown that by assuming the dominant trap energy to be at the Fermi level, the calculated tunneling currents exhibit many of the peculiar features which have been observed experimentally. The related effects that minority carrier traps have on the excess carrier lifetimes are also discussed, and a simple method of estimating many of the relevant trapping characteristics from lifetime measurements is presented.

I. INTRODUCTION

Understanding the dynamics of trapping and the properties of traps in HgCdTe is required for characterizing material grown by various methods and obtained from different sources. The presence of traps affects the performance of all HgCdTe detectors regardless of their type or configuration.

In the past two decades, different techniques have been applied to identify traps and to determine their energy distribution and density, such as deep level transient spectroscopy (DLTS),[1,2] diode pulse recovery,[3,4] optical modulation absorption (OMA), and more recently the two-photon magneto-optical (TPMO) technique.[5] In addition, a number of theoretical approaches have been applied to estimate the properties of defects and traps in HgCdTe.[6-9]

This paper focuses on trapping phenomena that is directly related and extracted from current HgCdTe detectors [i.e. photoconductors and metal-insulator semiconductors (MIS) as well as junction photodiodes].

Three topics are considered:

(a) Trap-assisted tunneling in MIS and junction photodiodes.

(b) Trapping and recombination processes, as related to the excess carrier lifetime in steady state and transient measurements.

(c) Trapping as related to the noise current spectral density of devices.

A. Trap-assisted tunneling

At operating temperatures ~77 K, trap-assisted tunneling is the dominant dark current mechanism over a wide range of biases in long-wavelength infrared LWIR HgCdTe MIS diodes[10-12] as well as in junction photodiodes.[13-22] Trap-assisted tunneling exhibits a distinct temperature dependence that is remarkably different from band-to-band tunneling or thermally limited mechanisms.

In this study, we model the bias dependence as well as the temperature dependence of trap-assisted tunneling. The model is based on two assumptions: that the dominant energy level in trap-assisted tunneling coincides with the Fermi level, and that the tunneling proceeds via thermally excited bulk Shockley-Read centers. Hence, we term this mechanism as thermal-trap-assisted tunneling. The model presented in this study qualitatively fits previously measured data for MIS diodes and junction photodiodes fabricated on p-type material. It should be noted that the process of surface recombination is not included.

B. Trapping and the excess carrier lifetime

Trapping is responsible for the large variation in reported "lifetimes" in p-type material. In particular, in material dominated by Shockley-Read recombination, trapping is manifested by the large differences between steady state "lifetimes" and transient "lifetimes" measured on the same samples.[23-26] A simple approach is presented which illuminates the role of trapping in lifetime experiments and correlates the transient and steady state "lifetimes" with the properties of the trapping centers.

C. Trapping and noise

Traps contribute two types of excess noise in devices: $1/f$ noise which is correlated with the dark current associated with trap-assisted tunneling, and generation-recombination noise.[27] The noise current spectral density can be modeled with the two types of tunneling currents. However, this topic is beyond the scope of the current paper and it will be reported elsewhere.

II. THERMAL-TRAP-ASSISTED TUNNELING IN HgCdTe MIS AND JUNCTION PHOTODIODES

Thermal-trap-assisted tunneling is the dominant dark current mechanism for HgCdTe MIS and junction photodiodes, over a wide range of operating temperatures and voltages. By modeling thermal-trap-assisted tunneling, trapping effects in HgCdTe can be quantitatively analyzed.

Tunneling is often a dominant current mechanism in HgCdTe photodiodes due to the narrow band gap of $x \cong 0.22$ material (~0.1 eV at 77 K), the small effective mass of electrons in HgCdTe ($m_e^*/M_0 = 7 \times 10^{-2} E_g$), and the cryogenic temperatures of its operation.

De Wames et al.,[13–15] Kinch et al.,[10–12] Wong,[20] Anderson et al.,[21,22] and more recently Nemirovsky et al.,[16–18] demonstrated the existence of two types of tunneling: trap-assisted and band-to-band, with distinctively different temperature dependencies.

Thermal-trap-assisted tunneling, its temperature and voltage dependencies, is modeled in this study on the basis of the following assumptions:

(1) The transition of electrons takes place by thermal excitation from the valence band to a Shockley-Read recombination center, and then via tunneling from the trapping center to the conduction band (see Appendix A).

(2) The dominant trap energy in the thermal-trap-assisted tunneling process, coincides with the Fermi level.

The dominant trap energy is calculated with

$$E_t = E_F = \frac{E_g}{2} + \frac{kT}{q} \ln\left(\frac{m_h^*}{m_e^*}\right)^{3/4} - \frac{kT}{q} \ln\left(\frac{N_A}{n_i}\right), \quad (1)$$

where N_A is the effective doping in the depletion region, E_g is the bandgap energy, and (m_h^*/m_e^*) is the ratio of the effective masses of holes and electrons. The Fermi energy is measured relative to the valence band.

The physical picture underlying the second assumption of this model is that there is a uniform distribution of trapping centers, throughout the bandgap and that the barrier for tunneling is lowest at the uppermost center that is still occupied. Hence, the occupied trapping center that coincides with the Fermi level, has the highest transition probability and plays the most dominant role in the thermal-trap-assisted tunneling process.

This assumption is responsible for the unique features of the model and explains the different temperature and bias dependencies of thermal-trap-assisted tunneling compared to band-to-band tunneling: tunneling currents that increase with increasing temperature and decreasing doping level and which exhibit ohmic-like regions and soft breakdown.

For modeling, we use the expression for the current density derived in Appendix A. The thermal-trap-assisted tunneling current density J_{TAT} is given by

$$J_{TAT} = qN_t \left(\frac{c_p p_1 w N_c}{c_p p_1 + w N_c}\right) x_D \, (\text{A/cm}^2), \quad (2)$$

where N_t and c_p are the trap density and hole capture coefficient, respectively and x_D is the depletion region width.

Since the model assumes that $E_t = E_F$,

$$p_1 = N_v \exp(-E_F/kT), \quad (3)$$

and the tunneling rate is

$$wN_c = \frac{6 \times 10^5 E}{(E_g - E_F)} \times \exp\left[-\frac{1.7 \times 10^7 E_g^{1/2} (E_g - E_F)^{3/2}}{E}\right] \, [\text{s}^{-1}], \quad (4)$$

where E is the electric field associated with the tunneling barrier (see Appendix A).

The dynamic resistance-area product due to thermal-trap-assisted tunneling is given by

$$(RA)_{TAT} = \left(\frac{\partial J_{TAT}}{\partial V}\right)^{-1}, \quad (5)$$

and an analytic expression is derived in Appendix B. The following two limiting cases emerge.

(a) For $c_p p_1 \ll w N_c$, Eq. (2) reduces to

$$J_{TAT} \cong q(N_t c_p) p_1 x_D. \quad (6)$$

In this case, the tunneling rate is larger than the thermal transition rate, and the dark current is limited by the thermal transition. It should be noted that the thermal-assisted dark current in this limiting case of thermal-trap-assisted tunneling, can be significantly larger than the thermally limited generation current from the depletion region obtained by the normal Shockley-Read model, $J = q(n_i/2\tau_0) x_d$ as $p_1 \gg n_i$.

(b) For $c_p p_1 \gg w N_c$, Eq. (2) reduces to

$$J_{TAT} \cong qN_t (wN_c) x_D. \quad (7)$$

In this case, the tunneling rate is smaller than the thermal transition rate, and it determines the magnitude of the dark current.

In both cases, the dark current density is directly proportional to the density of trapping centers N_t. In the first case, limited by the thermal transitions, the dark current is directly proportional to the product of the density of trapping centers and the capture coefficient for holes, i.e. $N_t c_p$.

According to the model presented in this study, the tunneling rate wN_c exhibits the expected tunneling dependence for composition x and electric field E. It is easily seen from Eq. (4) that wN_c increases with increasing electric field and reduced composition and bandgap. However, wN_c exhibits an unexpected temperature and doping level dependence. With increasing temperature and decreasing doping level, the Fermi level and hence the dominant trap level is shifted to higher energies. Thus, the barrier for tunneling is reduced and wN_c increases, in contrast to direct band-to-band tunneling.

The magnitude of the tunneling rate wN_c as compared to the hole emission rate $c_p p_1$, determines the significance of the two limiting cases. At higher temperatures and electric fields, the limiting case (a) is obtained and the dark current can be approximated with Eq. (6). At lower temperatures and electric fields the limiting case (b) is obtained and the dark current can be approximated with Eq. (7).

To illustrate the unique features of this model thermal-trap-assisted tunneling is compared to direct band-to-band tunneling. Direct band-to-band tunneling is modeled with the simple approach of a triangular barrier determined by the bandgap.[10,28] The direct band-to-band tunneling current density J_{BTB} is given by[10]

$$J_{BTB} = 10^{-2} N_A^{1/2} V_t^{3/2} \exp\left[-\frac{3 \times 10^{10} E_g^2}{(N_A V_t)^{1/2}}\right] (\text{A/cm}^2). \quad (8)$$

The RA product associated with direct band-to-band tunneling is given by

$$(RA)_{BTB}^{-1} = 10^{-2} N_A^{1/2} \exp\left(\frac{-4 \times 10^{10} E_g^2}{(N_A V_t)^{1/2}}\right)$$
$$\times \left(\frac{3}{2} V_t^{1/2} + \frac{3 \times 10^{10} E_g^2}{2 N_A^{1/2}}\right). \quad (9)$$

The total dark tunneling current is calculated and modeled with

$$J = J_{BTB} + J_{TAT}. \quad (10)$$

The total dynamic resistance-area product is calculated and modeled with

$$\left(\frac{1}{RA}\right)_{TOT} = \left(\frac{1}{RA}\right)_{BTB} + \left(\frac{1}{RA}\right)_{TAT}. \quad (11)$$

The temperature dependence of the thermal-trap-assisted tunneling current density J_{TAT} is exhibited in Fig. 1. The unique features of thermal-trap-assisted tunneling become apparent by comparing it with the temperature dependence of direct band-to-band tunneling.

Figure 1(a) exhibits the remarkably different temperature dependencies of the two types of tunneling. Whereas direct band-to-band tunneling is reduced as temperature and hence bandgap increase, thermal-trap-assisted tunneling increases. This behavior results directly from the assumption $E_t \simeq E_F$.

Figure 1(b) exhibits the remarkably different dependencies on doping level of the two types of tunneling. Whereas direct band-to-band tunneling increases with doping level due to reduced barrier width, thermal-trap-assisted tunneling decreases with doping level. Again, this distinctive behavior results directly from the assumption $E_t \simeq E_F$.

Figure 1(c) exhibits the temperature dependence of the two tunneling mechanisms with electric field as a parameter. Again, the remarkably different behavior of the two types of tunneling processes, become apparent. Direct band-to-band tunneling is very sensitive to the electric field whereas the dark current associated with thermal-trap-assisted tunneling is not sensitive to the higher electric fields. This stems from the fact that at high fields $wN_c \gg c_p p_1$ and the limiting case (a) prevails.

Figure 1(d) exhibits the effect of the trapping center density upon the dark current associated with thermal-trap-assisted tunneling. The dark current is directly proportional to the density of trapping centers. In the thermal-limited case, it is also directly proportional to the hole capture coefficient.

The bias dependence of the thermal-trap-assisted tunneling current density is exhibited in Fig. 2. This type of tunneling is characterized by current-voltage characteristics that at medium and high bias voltage exhibit nearly "Ohmic regions". The "Ohmic regions" occur for the thermally limited case (a) where the tunneling rate is larger than the capture rate. As we change the parameters that increase wN_c, the "Ohmic regions" are obtained at lower bias voltages. In Fig. 2(a) the "Ohmic region" is obtained at lower bias voltages as the temperature and hence wN_c is increased. Figure 2(b) and 2(c) exhibit the effect of trapping density and capture coefficient.

Figure 3 exhibits the transition from hard breakdown to soft breakdown as thermal-trap-assisted tunneling is includ-

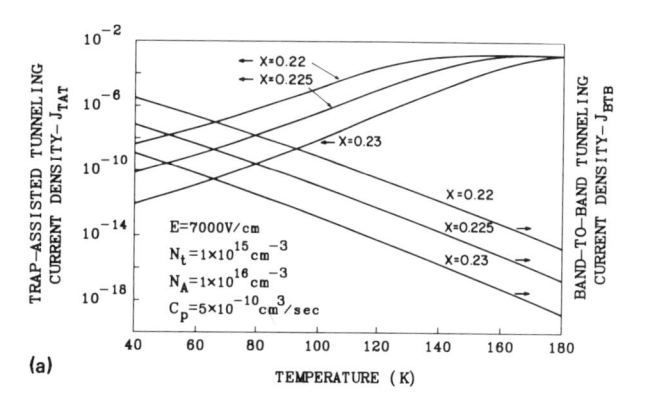

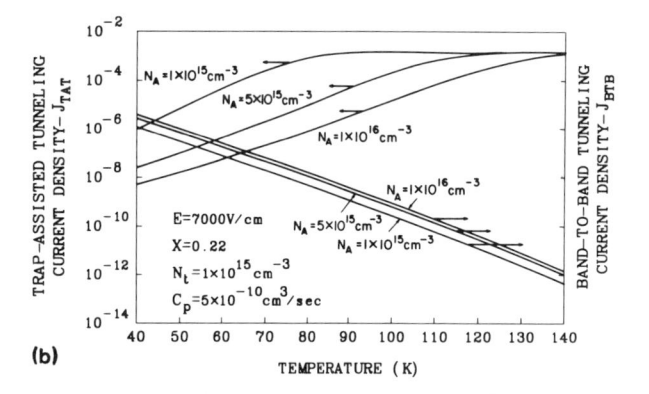

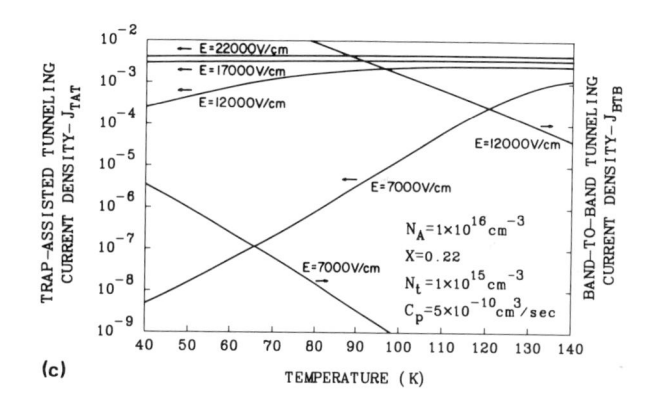

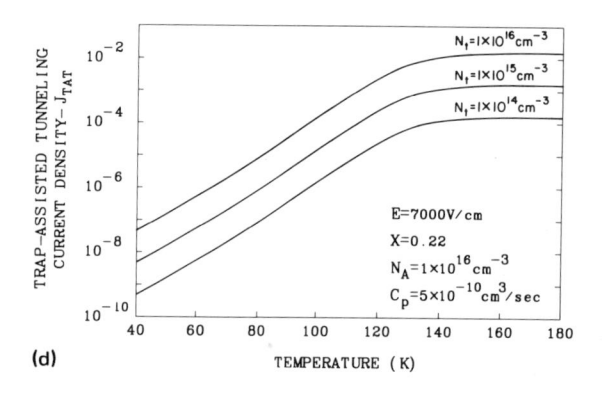

FIG. 1. Calculated thermal-trap-assisted tunneling current density J_{TAT} [Amp/cm^2] and band-to-band tunneling current density J_{BTB} [Amp/cm^2] versus temperature for Hg$_{1-x}$Cd$_x$Te. (a) with composition x as a parameter, (b) with doping level N_a [cm^{-3}] as a parameter, (c) with electric field E as a parameter, and (d) with density of trapping centers N_t [cm^{-3}] as a parameter. Note: The current scale is logarithmic.

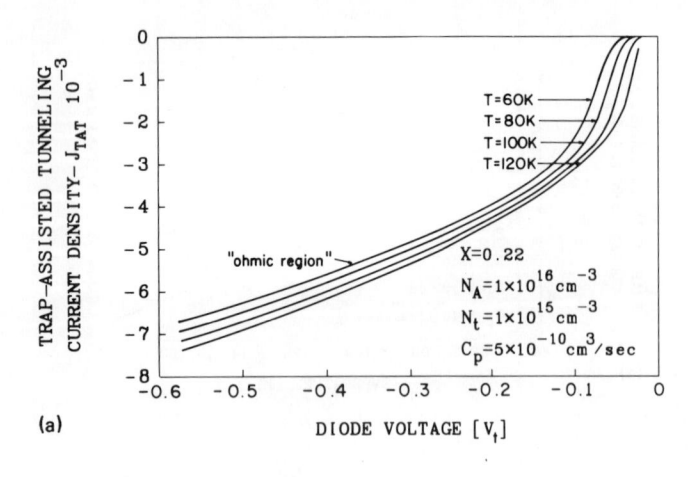

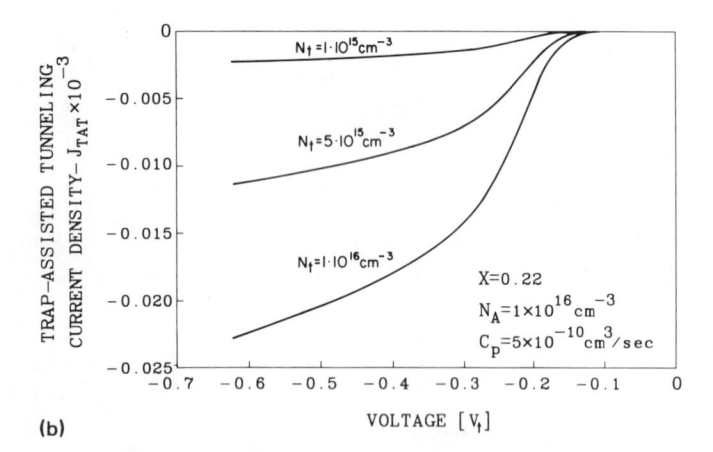

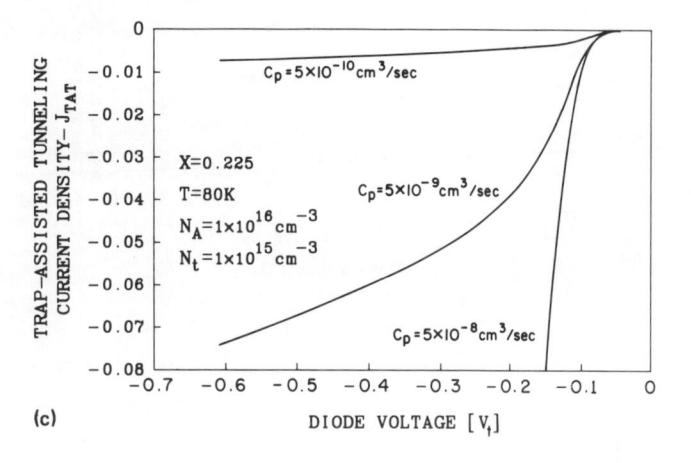

FIG. 2. Calculated thermal-trap-assisted tunneling current density J_{TAT} [Amp/cm^2] versus diode voltage for Hg$_{1-x}$Cd$_x$Te. (a) with temperature T as a parameter, (b) with density of trapping centers N_t [cm^{-3}] as a parameter, and (c) with capture coefficient of holes c_p [cm^{-3}] as a parameter. Note: The current scale is linear.

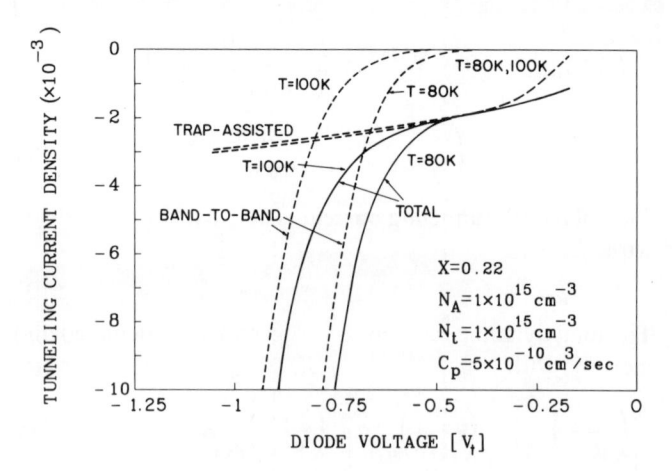

FIG. 3. A comparison of the calculated current-voltage characteristics of Hg$_{1-x}$Cd$_x$Te diodes with $x = 0.22$ for trap-assisted tunneling and band-to-band tunneling (dashed lines). The total (combined) tunneling current-voltage characteristics are also shown (solid lines).

ed in the model. Direct band-to-band tunneling is characterized with hard breakdown behavior and the breakdown voltage is shifted to higher bias voltages as temperature increases. The "Ohmic region" exhibited by thermal-trap-assisted tunneling tends to soften the breakdown. This behavior has been observed empirically in numerous experiments (see for instance Fig. 5 in Ref. 16 and Fig. 3 in Ref. 13).

The calculated dependence of the dynamic resistance-area (RA) product of HgCdTe photodiodes as a function of reciprocal temperature, with the diode reverse-bias voltage as a parameter, is shown in Fig. 4. The total (RA)$_{TOT}$ product of Fig. 4(a) is calculated with Eq. (11) and the contribution of both types of tunneling—direct band-to-band and thermal-trap-assisted tunneling, is taken into account. The total (RA)$_{TOT}$ product exhibits a maximum around 60 K. Above approximately 60 K, as temperature increases, (RA)$_{TOT}$ decreases significantly. Below ~60 K, as temperature decreases, (RA)$_{TOT}$ decreases again but less sharply. Figure 4(b) exhibits the relative contribution of the two tunneling processes. The thermal-trap-assisted tunneling limited (RA)$_{TAT}$ product increases as temperature decreases whereas the direct band-to-band tunneling limited (RA)$_{BTB}$ product decreases as temperature is reduced. Hence, the resulting total (RA)$_{TOT}$ product exhibits a maximum. At higher temperatures for lower and medium reverse bias voltages, the total (RA)$_{TOT}$ product is limited by thermal-trap-assisted tunneling and at lower temperatures and higher reverse bias voltages, the total (RA)$_{TOT}$ product is limited by direct band-to-band tunneling.

The model presented in this study explains, for the first time, the observed maxima of the dynamic resistance (RA)$_{TOT}$ as a function of reciprocal temperature, as well as the overall measured temperature dependence of the (RA) product that was previously reported (for example, Fig. 7 of Ref. 16).

The relative contribution of the two tunneling processes determine the measured temperature dependence of the dynamic resistance. At low and medium reverse-bias regions, thermal-trap-assisted tunneling dominates, and the measured dynamic resistance (RA)$_{TOT}$ product increases as temperature is reduced. At lower temperatures and higher reverse-bias voltages, the contribution of band-to-band tunneling increases, and the associated (RA)$_{TOT}$ product de-

culated values of the $(R_0A)_{TAT}$ product correspond to the typically measured data (e.g., Fig. 2 of Ref. 16).

It was previously observed that LWIR HgCdTe photodiodes exhibit (R_0A) product that become temperature insensitive with decreased temperature of operation and the values can vary by several orders of magnitude.[13-15] The values of the (R_0A) product of Fig. 4(c) as predicted by the model, are very sensitive to the density of traps, N_t and the capture coefficient c_p. Apparently, the trapping properties of state-of-the-art arrays fabricated on current materials, vary significantly.

The calculated dependence of the dynamic resistance-area $(RA)_{TOT}$ product as a function of diode reverse bias, is shown in Fig. 5. The thermal-trap-assisted tunneling limited $(RA)_{TAT}$ product is characterized by a nearly constant value at medium and high bias voltage. This behavior stems from the "Ohmic region" exhibited by the current–voltage characteristics. The direct band-to-band tunneling limited $(RA)_{BTB}$ product is characterized by a sharp fall as diode reverse-bias voltage increases. Typically, below 60 K the $(RA)_{TOT}$ product is limited by direct band-to-band tunnel-

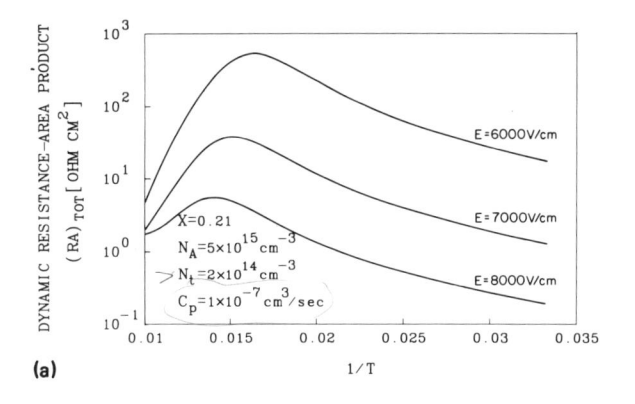

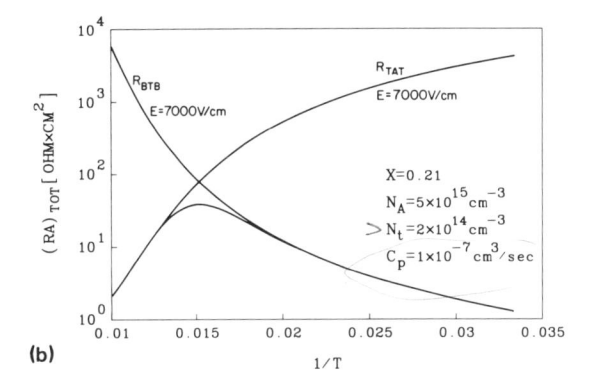

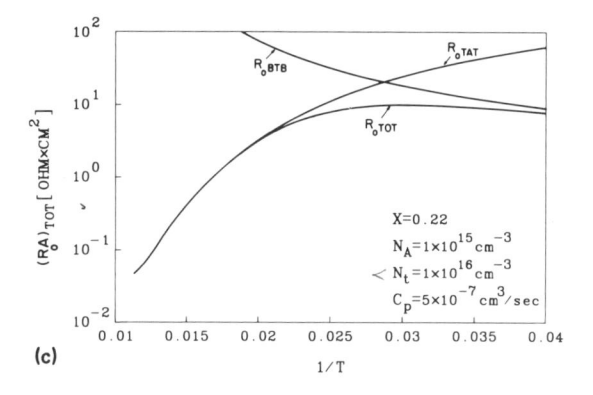

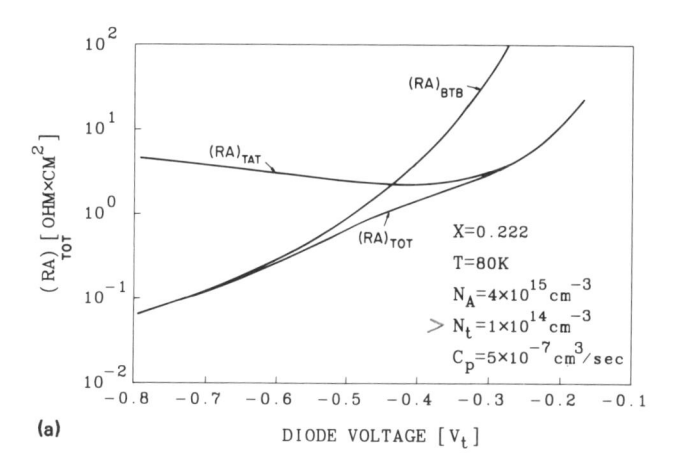

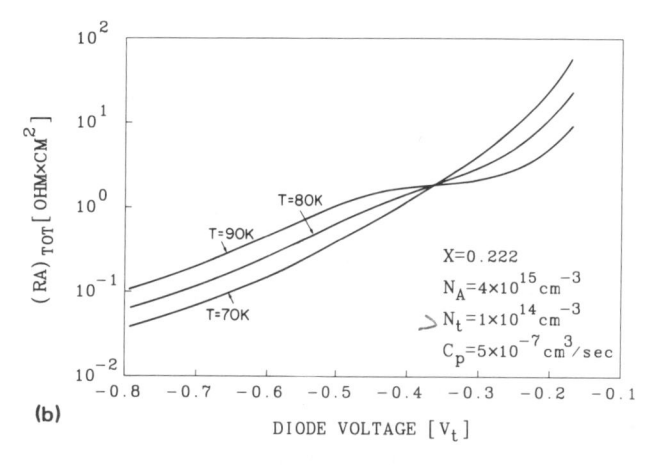

FIG. 4. (a) Dependence of the calculated total dynamic resistance of $Hg_{1-x}Cd_xTe$ $N+P$ diodes with $x = 0.21$ upon reciprocal temperature with the maximum electric field at the junction as a parameter. (b) Dependence of the calculated trap-assisted tunneling and band-to-band dynamic resistance (dashed lines) and combined (total) dynamic resistance (solid line) upon reciprocal temperature. (c) Calculated dependence of the R_0A product upon reciprocal temperature.

creases. Hence, the measured $(RA)_{TOT}$ versus $1/T$, characteristics exhibit a maximum and either a plateau that is almost independent of the temperature or a decrease in the $(RA)_{TOT}$ at higher biases and lower temperature.

The model also accounts for the somewhat unexpected temperature dependence of the measured (R_0A) product at zero bias, at low temperatures. The measured diodes show a plateau in the R_0A value and a gradual increase toward lower temperatures.[13-22] Figure 4(c) exhibits the calculated thermal-trap-assisted tunneling limited resistance $(RA)_{TAT}$ at zero bias. The temperature dependence as well as the cal-

FIG. 5. Dependence of the calculated trap-assisted tunneling and band-to-band dynamic resistance (dashed lines) and combined (total) dynamic resistance (solid line) upon diode voltage. (a) At $T = 80$ K, (b) Dependence of the total tunneling dynamic resistance upon diode voltage with the temperature T as a parameter.

ing whereas above 100 K the $(RA)_{TOT}$ product is limited by thermal-trap-assisted tunneling. At normal operation temperature of approximately 80 K, thermal-trap-assisted tunneling limits the $(RA)_{TOT}$ product at medium reverse-bias voltage and as the diode reverse bias voltage increases, a transition occurs and band-to-band tunneling limits the $(RA)_{TOT}$ product.

Figures 4 and 5 illustrate the predictions of the model. However, the exact shape and the (RA) values are strongly dependent on the parameters used for modeling. The relative contribution of the two tunneling processes depends strongly on the physical structure of the junction (n^+p^-p or n^+p for ion implanted junctions and n^-p for diffused junctions), vacancy or impurity doping, the doping level, and obviously the density and nature of the trapping centers. At this stage, we lack an independent direct method for characterizing the traps in HgCdTe and the trap parameters are determined by the best fit to the measured data.

III. TRAPPING AND THE EXCESS CARRIER LIFETIME

The excess carrier lifetime is, in general, an extremely sensitive indicator of material quality. This is especially true for p-type $Hg_{1-x}Cd_xTe$ ($x \sim 0.22$) at 77 K, whose lifetime is limited by Shockley–Read recombination.[23–26] Although it has not yet been established that material with higher excess carrier lifetimes yield "superior" diodes, a number of experiments have shown that the connection between the excess carrier lifetime and the dark current is accurately described by the diffusion model, for n^+/p diodes at zero bias and at temperatures $\geqslant 77$ K.[13,16,19] According to this model, the dark current produced by the diffusion of thermally generated minority carriers in the neutral p-region is proportional to $(\tau_n)^{1/2}$, so that a larger excess carrier lifetime results in a higher R_0A product.

The simple relationship between the lifetime and the dark current, however, ceases to be valid under reverse bias conditions, when the tunneling processes described in Sec. II are significant. In this case, the lifetime is not explicitly contained in the expression for the tunneling current, but is an implicit parameter which is determined by the characteristics of the trapping centers. The relationship between the lifetime and the dark current is further complicated by the different tunneling mechanisms, and is dependent on whether band-to-band or thermal trap-assisted tunneling dominates and on the rate limiting step for the thermal trap-assisted process. One consequence of this is that, unlike the case of zero bias, the excess carrier lifetime is not a fundamental parameter in determining reverse bias characteristics. Measurements of the excess carrier lifetime, however, remain a valuable technique for characterizing the trapping centers which cause the tunneling current.

The concentration and capture rates of the trapping centers, along with bandgap, temperature, and electric field in the depletion region, determine the magnitude of the dark current due to thermal trap-assisted tunneling. As the excess carrier lifetime is limited by Shockley–Read recombination, measurements of the lifetime can be used to estimate many of the relevant parameters of the trapping centers. At this preliminary stage in relating lifetime measurements to thermal trap assisted tunneling, two questions are particularly important; the first is whether lifetime measurements alone can be used to estimate the concentration and capture rates of the trapping centers; and the second is whether the trapping centers which are characterized by lifetime measurements are the same as those which cause the thermal trap-assisted tunneling current.

In interpreting lifetime data, it is important to make a distinction between the three commonly measured lifetimes; the excess minority carrier lifetime, τ_n, the excess majority carrier lifetime, τ_p, the transient lifetime, τ_t. Each of these time constants is associated with the recombination of electrons and holes, however, each refers to a different aspect of this process and, therefore, provides different information. In p-type $Hg_{0.78}Cd_{0.22}Te$ the transient and the majority carrier lifetimes may be as much as 25 times greater than the minority carrier lifetime due to the presence of trapping centers.[23–26]

The meanings of the relevant lifetimes are briefly reviewed in Appendix C. The first step in the recombination process is the capture of the minority carrier by a S-R center. This process occurs within a characteristic time of $(N_t c_n)^{-1}$, where N_t is the density of trapping centers, and c_n is the minority carrier capture rate of the center. This time constant is referred to as the excess minority carrier lifetime, τ_n. Ignoring carrier re-emission, the second step in the recombination process is the capture of a majority carrier by the filled center. This process occurs within a characteristic time of $(p_0 c_p)^{-1}$, where p_0 is the carrier concentration, and c_p is the majority carrier capture rate. The majority carrier lifetime is simply the total time that the hole remains in the valence band, and is given by $\tau_p = (N_t c_n)^{-1} + (p_0 c_p)^{-1}$. Thus it is seen that the excess majority carrier lifetime is always greater than or equal to the minority carrier lifetime. When trapping is significant (i.e. $\tau_{\text{majority}} \gg \tau_{\text{minority}}$), as is often found in p-type HgCdTe with $x \sim 0.22$ and $p_0 \sim 10^{16}$ cm^3, the transient lifetime, τ_t, can be shown to be approximately equal to the majority carrier lifetime. The mathematical expressions for the minority, majority, and transient lifetimes are given in Appendix C, along with useful approximations for p-type $Hg_{0.78}Cd_{0.22}Te$.

Traps can essentially be described by the Shockley-Read (SR) recombination statistics. Trapping differs from recombination only in the relative values of the capture coefficients. If the capture coefficient for the minority carrier is many times larger than that for the majority carrier, then the SR center is called a "trap" rather than a "recombination center". In this case, the SR centers are not only "stepping stones" between the valence and conduction bands, but also trap electrons and the density of excess trapped electrons is significant.

Several experimental methods have been used for measuring the "lifetimes" of p-type HgCdTe. These techniques can be divided into transient and steady-state methods. A large difference between steady-state and transient lifetimes is an indication of trapping effects (see Appendix C).

The steady-state methods (including steady-state photoconductivity, zero bias resistance and diffusion length) yield the excess minority carrier lifetime which to a good approximation is given by:

$$\tau_n \approx \frac{1}{c_n N_t}. \quad (12)$$

The transient methods (including photoconductive decay and diode reverse recovery) yield to a good approximation the majority carrier lifetime

$$\tau_t \approx \tau_p \approx \frac{1}{c_n N_t} + \frac{1}{c_p P_0}. \quad (13)$$

Independent measurement of the minority and majority carrier lifetimes (i.e. the steady-state and transient lifetimes) can be used to determine trap properties. However, Eqs. (12) and (13) include three unknown parameters: the density of traps, N_t, and the capture coefficients for minority and majority carriers c_n, c_p. Hence, additional information obtained by deep-level transient spectroscopy (DLTS)[1,3,4] is required.

Since the large difference between the steady-state minority carrier lifetime and the transient lifetime should occur due to the highly asymmetric capture coefficients, we assume

$$\frac{c_n}{c_p} \approx 10\text{--}10^2. \quad (14)$$

Trap properties determined by DLTS indicate that the capture coefficients are indeed highly asymmetric with the electron capture coefficient approximately two orders of magnitude larger than the hole capture coefficient.

Table I contains measured lifetimes (steady state and transient) for vacancy doped (undoped) material as well as impurity (Au or As) doped material. The corresponding traps properties are calculated with Eqs. (12)–(13).

$$c_p \approx \frac{1}{(\tau_t - \tau_n) p_0}, \quad (15)$$

$$c_n \approx 50 c_p, \quad (16)$$

and

$$N_t \approx \frac{1}{c_n \tau_n} \cong \left(\frac{c_p}{c_n}\right)\left(\frac{\tau_t - \tau_n}{\tau_n}\right) p_0. \quad (17)$$

Table I summarizes the results of 15 samples grown and doped by several methods, yet the derived capture coefficient for holes is almost the same for all the measured samples and is approximately 3×10^{-9} cm^3/s which is a reasonable value for the majority carrier capture coefficient and compares well to the DLTS data. The density of traps varies between 1×10^{14} cm^3 in the higher quality HgCdTe to $\sim 6 \times 10^{15}$ cm^3 at the higher doped HgCdTe. This range of values is similar to that obtained by fitting the dc properties of measured photodiodes with the model described in Sec II.

Given the values of the majority carrier capture rates determined by lifetime measurements ($\sim 10^{-10}$–10^{-9} cm^3/s), it is somewhat surprising that those used in the tunneling model in order to yield reasonable dark currents are at least two orders of magnitude larger ($\sim 10^{-7}$ cm^3/s). Since the majority carrier capture rates determined by DLTS tend to be in the range of 10^{-11}–10^{-10} cm^3/s, the discrepancy between DLTS measurements and the tunneling model is even greater. Besides the discrepancies in the absolute values of the capture rates, the type of trapping or recombination center is also in question. The large differences observed in the transient and minority carrier lifetimes indicate the recombination occurs at a donor-like trap. DLTS measurements also show that the traps are donor-like, for the material which has been characterized. However, the large value of the majority carrier capture rate needed in the tunneling model suggests that an acceptor-like center is responsible for the trap-assisted tunneling current. Other trap-assisted tunneling models have used similarly large values of the majority carrier capture rate in order to fit the data.[12]

At the present time, the discrepancy between the lifetime and DLTS data, and the trap-assisted tunneling parameters has not been resolved. One possible explanation is that the

TABLE I. Calculated traps properties from measured "lifetimes" in undoped and doped p-type HgCdTe from several sources.

Sample no.	Growth method	Dopant	Doping level at 77 K p_0(cm^{-3})	τ_n (ns)	τ_t (ns)	$c_p \times 10^{-9}$ [cm^3/s]	$c_n \times 10^{-7}$ [cm^3/s]	Trap density N_t [cm^{-3}]
SAT86	THM	None	3.5×10^{15}	28	100	4	2	1.4×10^{14}
N628	SSR	None	6.8×10^{15}	11	72	2.4	1.2	7.5×10^{14}
862-7	SSR	None	8.9×10^{15}	2.7	35	3.5	1.7	2.1×10^{14}
P1637	Slush	None	1.5×10^{16}	1.4	23	3.1	1.5	4.7×10^{15}
THM6-36	THM	None	4.4×10^{15}	10	23	3.1	1.5	4.7×10^{14}
855-1	SSR	None	9.0×10^{15}	8	36	4	2	6×10^{15}
861-3	SSR	As	1.2×10^{16}	62	90	3	1.5	1×10^{14}
850-3	SSR	As	1.6×10^{16}	11	40	2	1	8.4×10^{14}
ND74-A1	SSR	As	5.6×10^{15}	47	115	2.6	1.3	1.6×10^{14}
860-1	SSR	As	1.1×10^{16}	2.5	38	2.5	1.3	3.1×10^{15}
867-1	SSR	As	6.1×10^{15}	8	90	2	1	1.2×10^{15}
809-1	SSR	Au	6.2×10^{15}	49	220	0.9	0.5	4.3×10^{14}
716-3-6	SSR	Au	7.0×10^{15}	10	80	2	1	1×10^{15}
818-1	SSR	Au	8.3×10^{15}	1.6	50	2.5	1.2	5×10^{15}
835-3	SSR	Au	8.3×10^{15}	6.0	62	2.8	1.4	5×10^{15}

centers which limit the lifetime are different than those which contribute to the tunneling current. In addition, it should be noted that the measured lifetimes of Table II characterize the traps in the neutral p-type region of the substrate, while thermal-trap-assisted tunneling dark currents are generated at the depletion region. Hence, the relevant traps for this mechanism physically reside at the junction and it is highly probable that the nature and the properties of these traps are determined by the junction formation technology. A second possible explanation is that the value used for the matrix element of the trap potential (see Appendix A) is too small, resulting in an underestimation of the tunneling current.

IV. SUMMARY

Comprehensive experimental studies of HgCdTe ($x \cong 0.22$) MIS and junction photodiodes, fabricated by alternative junction formation technologies, utilizing different substrates and different passivation techniques, have one common feature; they indicate that traps and trap-assisted processes play a dominant role in limiting device performance and contributing large nonuniformities along arrays.

Trap-assisted tunneling has previously been proposed as a dominant mechanism contributing a significant component to dark currents and noise currents, over a wide range of operating temperatures and bias voltages.[10-22] The mechanism associated with trap-assisted tunneling was identified and characterized with a positive temperature coefficient. In addition, it was observed that the physical processes associated with trap-assisted tunneling contributed an Ohmic current component in the so-called reverse bias saturation region at relatively high temperatures with an empirical temperature dependence that follows $\exp \gamma T$. The same mechanism also limited the zero bias dynamic resistance R_0 at lower temperatures.[13-15]

However, the observed temperature and voltage dependence and behavior of trap-assisted tunneling remained unexplained, and the role of traps in limiting device performance could not be taken into account quantitatively.

In this paper, we model the trap-to-band process that limits the performance of LWIR HgCdTe diodes (junction as well as MIS structures). We analyze thermal-trap-assisted tunneling where the transition of electrons takes place by thermal excitation from the valence band to a Shockley–Read recombination center, and then via tunneling from the trapping center to the conduction band.

This combined mechanism was considered and studied in the past.[10-12] However, it was previously assumed that even though there are probably a large number of SR levels through the band gap, the trap-assisted tunnel current should be dominated by traps located somewhere near the middle of the gap where tunnel transitions from the valence and conduction bands are approximately equal.

In our study, we propose a significantly different assumption, i.e., that the dominant trap energy coincides with the Fermi level. Indeed, there is most probably a distribution of S-R levels throughout the band gap. However, the traps adjacent to the Fermi level are characterized by a high probability of occupation and a high transition probability, since the tunneling barrier is relatively low.

The combined process of thermal-trap-assisted tunneling together with the assumption that the dominant trap coincides with the Fermi level, are modeled in this paper to yield the correct temperature and bias dependence (i.e. the positive temperature coefficient, the Ohmic behavior and the soft breakdown that have been experimentally observed). The experimental observations and the associated models emphasize that state-of-the-art photodiodes are limited by thermal-trap-assisted tunneling at low and medium reverse bias regions at higher temperatures of operation, and a combination of band-to-band tunneling and thermal-trap-assisted tunneling at lower temperatures and higher reverse-bias voltages. These observations illuminate the highly significant role of material properties in the overall performance of detectors.

The experiments and modeling indicate that control of traps is a key problem in the program for high-performance infrared focal plane arrays. Control of doping, electrical field profiles and barrier heights associated with control of band gap, trap density, and capture coefficients, become inter-related and stringent because the tunneling probability factors (band to band as well as thermal trap assisted) contain these parameters in an exponential.

Progress in the elucidation and characterization of traps and trapping effects in Hg CdTe is required, if we are to set to improve the performance of state-of-the-art arrays. The role of traps and trapping effects should be addressed by the modern and advanced epitaxial growth methods.

APPENDIX A

The major mechanism of thermal-trap-assisted tunneling (TAT) in MIS and n^+p HgCdTe photodiodes is assumed to include thermal excitation from the valence band to a Shockley–Read recombination and trapping center N_t, and then tunneling from the center to the conduction band.

The net capture rate of electrons into N_t centers is

$$U_n = c_n n(N_t - n_t) - c_n n_1 n_t - n_t w(N_c - n_c) + (N_t - n_t)wn_c, \quad (A1)$$

where N_t is the density of trapping centers, n_t is the density of N_t centers occupied by electrons, n_c is the density of electrons in the conduction band at the tunneling energy, c_n is the capture coefficient for electrons, N_c is the density of states in the conduction band, w is the tunneling probability from the N_t center to the conduction band, and $n_1 = N_c \exp[-(E_g - E_t)/kT]$, and E_t is the trap energy measured from the valence band.

The first two terms in Eq. (A1) describe the normal Shockley–Read thermal generation-recombination processes and the third and fourth terms describe the tunneling transitions between the trapping center and the conduction band.

The net capture rate of holes is given by the Shockley-Read theory

$$U_p = c_p p n_t - c_p p_1 (N_t - n_t), \quad (A2)$$

where c_p is the capture coefficient for holes, $p_1 = N_v \exp[-E_t/kT]$, and N_v is the density of states in the valence band.

In steady-state, $U_n = U_p$, giving the function of occupied centers

$$\frac{n_t}{N_t} = \frac{c_n n + c_p p_1 + w n_c}{c_p(p+p_1) + c_n(n+n_1) + w N_c}. \quad (A3)$$

By combining Eqs. (A1)–(A3), the net combination rate which is determined by the density of trapping centers, their capture properties, tunneling rate, and doping level, is obtained:

$$U = U_n = U_p$$
$$= N_t \frac{c_p c_n (np - n_i^2) + c_p w [n_c p - p_1 (N_c - n_c)]}{c_p(p+p_1) + c_n(n+n_1) + w N_c}. \quad (A4)$$

For depletion regions where $n \approx 0, p \approx 0$,

$$U = N_t \frac{-c_p c_n n_i^2 - c_p p_1 (N_c - n_c) w}{c_p p_1 + c_n n_1 + w N_c}. \quad (A5)$$

Equation (A5) is further simplified by assuming that the dominant transition between the trapping centers and the conduction band is via tunneling and $c_n n_1 \ll w N_c$. In addition $N_c \gg n_c$ and $c_p p_1 w N_c \gg c_p c_n n_i^2$. Hence,

$$U = -N_t \frac{c_p p_1 w N_c}{c_p p_1 + w N_c}. \quad (A6)$$

The expression for the dark current density due to trap-assisted tunneling is

$$J_{TAT} = q N_t \left[\frac{c_p p_1 w N_c}{c_p p_1 + w N_c}\right] x_d \quad [\text{A/cm}^2], \quad (A7)$$

where x_d is the width of the depletion layer.

The thermal-trap-assisted tunneling current density of Eq. (A7) describes thermal transitions of electrons from the valence band into trapping centers and subsequently tunneling transitions of electrons from the trapping centers into the conduction band. Thermal transitions from the trapping centers into the conduction band are neglected as well as the tunneling transitions into the trapping centers from the valence band.

The thermal transition rate (e_n) from mid-band-gap traps into the conduction band is much lower than the corresponding thermal transition rate to the valence band (e_p) due to the low density of states associated with the conduction band ($e_n = c_n n_1 \sim c_n N_c$)

The tunneling rate of electrons from the valence band into the traps is much lower than band-to-band tunneling from the valence band into the conduction band, since the barrier for tunneling is nearly the same (E_g) and the density of states in the conduction band is higher than the density of traps.

Following Sah[29] and Kinch,[10] the tunneling rate is given by

$$w N_c = \frac{\pi^2 q m_e^* E M^2}{h^3 (E_g - E_t)} \exp\left[-\frac{4(2m_e^*)^{1/2}(E_g - E_t)^{3/2}}{3hqE}\right], \quad (A8)$$

where M is the matrix element of trap potential, m_e^* is the effective mass of electrons, and E is the electric field associated with the tunneling barrier.

The experimentally determined value of $[M^2(m_e^*/m)]$ for silicon is 10^{-23} V cm^3 and is adopted for the HgCdTe calculations. Hence,

$$w N_c = \frac{6 \times 10^5 E}{(E_g - E_t)} \exp\left[-\frac{1.7 \times 10^7 E_g^{1/2}(E_g - E_t)^{3/2}}{E}\right], \quad (A9)$$

where E_g and $(E_g - E_t)$ are in volts, and E in V/cm.

APPENDIX B

The dynamic resistance-area product due to thermal-trap-assisted tunneling is derived with:

$$(RA)_{TAT}^{-1} = \frac{\partial J_{TAT}}{\partial w N_c} \cdot \frac{\partial w N_c}{\partial E} \cdot \frac{\partial E}{\partial V} \quad (B1)$$

where J_{TAT}, $w N_c$ are given by Eqs. (2) and (3), respectively. E is the electric field and is given by $E = (qN_A/\epsilon_0 \epsilon_s)^{1/2}(2V_t)^{1/2}$. Hence,

$$\frac{\partial J_{TAT}}{\partial w N_c} = \frac{(c_p P_1 + w N_c) \cdot q N_t c_p P_1 - q N_c c_p P_1 w N_c}{(c_p P_1 + w N_c)^2}$$
$$= \frac{q N_t \cdot (c_p P_1)^2}{(c_p P_1 + w N_c)^2}, \quad (B2)$$

$$\frac{\partial w N_c}{\partial E} = \frac{6 \times 10^5}{(E_g - E_t)} \exp\left(-\frac{1.7 \times 10^7 \cdot E_g^{1/2}(E_g - E_t)^{3/2}}{E}\right)$$
$$\times \left(1 + \frac{1.7 \times 10^7 \cdot E_g^{1/2} \cdot (E_g - E_t)^{3/2}}{E}\right)$$
$$= \frac{w N_c}{E} \cdot \left(1 + \frac{1.7 \times 10^7 \cdot E_g^{1/2} \cdot (E_g - E_t)^{3/2}}{E}\right), \quad (B3)$$

$$\frac{\partial E}{\partial V} = \sqrt{q N_a / 2 \epsilon_0 \epsilon_s V_t} \quad (B4)$$

$$\frac{1}{(RA)_{TAT}} = \frac{q \cdot N_t (c_p P_1)^2}{(c_p P_1 + w N_c)^2} \cdot \frac{w N_c}{E}$$
$$\times \left(1 + \frac{1.7 \times 10^7 \cdot E_g^{1/2} \cdot (E_g - E_t)^{3/2}}{E}\right)$$
$$\times \left(\frac{q N_A}{2 \epsilon_s \epsilon_0 V}\right)^{1/2}. \quad (B5)$$

By combining Eqs. (B1)–(B5) we obtain

$$\frac{1}{(RA)_{TAT}} = \left\{ qN_t(C_pP_1)^2 \cdot \frac{6\times10^5}{(E_g-E_t)} \exp\left[-\frac{1.7\times10^7 \cdot E_g^{1/2} \cdot (E_g-E_t)^{3/2}(\epsilon_s\epsilon_0)^{1/2}}{(2qN_AV)^{1/2}}\right]\right\}$$
$$\left(C_pP_1 + \left\{\frac{6\times10^5(2qN_AV)^{1/2}}{(E_g-E_t)(\epsilon_s\epsilon_0)^{1/2}} \exp\left[\frac{1.7\times10^7 \cdot E_g^{1/2}\cdot(E_g-E_t)^{3/2}(\epsilon_s\epsilon_0)^{1/2}}{(2qN_AV)^{1/2}}\right]\right\}\right)$$
$$\times \left[\frac{1.7\times10^7\cdot E_g^{1/2}\cdot(E_g-E_t)^{3/2}(\epsilon_s\epsilon_0)^{1/2}}{(2qN_AV)^{1/2}}\right]\left(\frac{qN_A}{2\epsilon_s\epsilon_0 V}\right)^{1/2}. \tag{B6}$$

APPENDIX C

The three lifetimes which are relevant to the characterization of p-type $Hg_{1-x}Cd_xTe$ are the excess minority carrier lifetime, τ_n, the excess majority carrier lifetime, τ_p, and the transient lifetime, τ_t. The particular lifetime which is measured depends on the experimental technique. The mathematical expressions for these lifetimes are given below, and differ depending on the amount of minority carrier trapping. According to the Shockley–Read theory, the steady state excess minority and majority carrier lifetimes are given by

$$\tau_n = \frac{\tau_{p0}(n_0+n_1) + \tau_{n0}\{p_0+p_1+N_t[1+(p_0/p_1)]^{-1}\}}{n_0+p_0+N_t[1+(p_0/p_1)]^{-1}[1+(p_1/p_0)]^{-1}} \tag{C1}$$

and,

$$\tau_p = \frac{\tau_{n0}(p_0+p_1) + \tau_{p0}\{n_0+n_1+N_t[1+(n_0/n_1)]^{-1}\}}{n_0+p_0+N_t[1+(n_0/n_1)]^{-1}[1+(n_1/n_0)]^{-1}}. \tag{C2}$$

According to Sandiford, the transient lifetime, which is the time constant describing the decay of excess carriers is given by:[30]

$$\tau_t = \frac{\tau_{n0}\{p_0+p_1+N_t[1+(p_0/p_1)]^{-1}\} + \tau_{p0}\{n_0+n_1+N_t[1+(n_0/n_1)]^{-1}\}}{n_0+p_0+N_t[1+(n_0/n_1)]^{-1}[1+(n_1/n_0)]^{-1}}, \tag{C3}$$

where $\tau_{n0} = (N_t \cdot c_n)^{-1}$, $\tau_{p0} = (N_t \cdot c_p)^{-1}$, N_t = density of trapping centers,

$$n_1 = N_c \cdot \exp(E_t - E_c)/kT,$$
$$p_1 = N_v \cdot \exp(E_v - E_t)/kT.$$

For p-type $Hg_{0.78}Cd_{0.22}Te$ at 77 K the relationship $p_0 \gg N_t \gg p_1$, $n_1 \gg n_0$, is usually valid, so we can simplify the above lifetime expressions to:

excess minority carrier lifetime:

$$\tau_n \approx \tau_{n0} = \frac{1}{N_t \cdot c_n}; \tag{C4}$$

excess majority carrier lifetime:

$$\tau_p \approx \frac{1}{N_t \cdot c_n} + \frac{1}{c_p \cdot p_0}; \tag{C5}$$

transient lifetime:

$$\tau_t \approx \frac{1}{N_t \cdot c_n} + \frac{1}{c_p \cdot p_0}. \tag{C6}$$

The following discussion briefly reviews the several time constants which characterize the recombination and trapping processes and are known as the "lifetimes."[26]

The excess electron lifetime, τ_n is defined by

$$U_n = \frac{\Delta n}{\tau_n} = c_n[n(N_t - n_t) - n_1 n_t], \tag{C7}$$

where U_n is the net recombination rate for electrons.

The first step in the SR recombination process is the capture of the minority carrier by a center. This process occurs within a characteristic time of $1/c_n N_t$ where N_t is the density of trapping centers, and c_n is the minority carrier capture coefficient of the center. This time constant is to a good approximation a measure of the excess electron lifetime. Thus,

$$\tau_n \approx \frac{1}{c_n N_t}. \tag{C8}$$

Accordingly, the excess hole lifetime, τ_p, is defined by

$$U_p = \frac{\Delta p}{\tau_p} = c_p[pn_1 - p_1(N_t - n_t)], \tag{C9}$$

where U_p is the net recombination rate for holes.
Under steady-state conditions

$$U_n = U_p, \tag{C10}$$

and hence,

$$\frac{\Delta n}{\tau_n} = \frac{\Delta p}{\tau_p}. \tag{C11}$$

It is usually assumed that $\Delta n \approx \Delta p$ and in this case, under steady-state conditions, the common recombination rate is related to the common excess carrier lifetime and

$$\tau_n = \tau_p \approx \frac{1}{c_n N_t}\left(1 + \frac{p_1}{p_0}\right). \tag{C12}$$

In the case of SR recombination involving trapping centers, space charge neutrality includes the trapped electrons and hence,

$$\Delta n + \Delta n_t = \Delta p, \tag{C13}$$

where Δn, Δn_t, and Δp are the excess electrons, trapped electrons and holes, respectively.

Thus, if trapping centers are present in the p-type material, $\Delta n < \Delta p$, and the electron and hole lifetimes will be different with $\tau_p > \tau_n$.

The excess trapped electrons are calculated with

$$\Delta n_t = N_t (f - f_0), \tag{C14}$$

where f and f_0 are the occupation factor of the centers in terms of electron and hole concentration, under steady-state and equilibrium, respectively. Hence,

$$\Delta n_t \approx \left(\frac{c_n N_t}{c_p p_0}\right) \Delta n. \tag{C15}$$

The density of trapped electrons and trapping effects become significant, if either $N_t \gg p_0$ or $c_n \gg c_p$.

With trapping, $\Delta n \neq \Delta p$ and the majority carrier lifetime becomes longer than the minority carrier lifetime. Equations (C11), (C13), and (C15) yield

$$\tau_p = \tau_n \left(\frac{\Delta p}{\Delta n}\right) = \tau_n \left(1 + \frac{\Delta n_t}{\Delta n}\right), \tag{C16}$$

and the majority carrier lifetime is

$$\tau_p \approx \frac{1}{c_n N_t} + \frac{1}{c_p p_0}. \tag{C17}$$

Thus, it is seen that the majority carrier lifetime is simply the time needed to capture a minority carrier ($1/c_n N_t$) plus the time needed for an occupied center to capture a majority carrier ($1/c_p p_0$).

In transient photoconductive decay, basically two time constants are measured and defined as the "lifetimes". These time constants are obtained by solving the three rate equations for electrons, holes and trapped electrons, with appropriate initial and final conditions.

Because of the electroneutrality condition, there are only two independent equations for the case of one trapping level, and hence, two time constants or "lifetimes" completely define the transient decay (τ_i, τ_t). To a good approximation,

$$\tau_i \approx \frac{1}{c_n N_t} \approx \tau_n, \tag{C18}$$

$$\tau_t \approx \frac{1}{c_n N_t} + \frac{1}{c_p p_0} \approx \tau_p. \tag{C19}$$

Thus, the two time constants characterizing the transient decay are approximately given by the steady-state minority and majority excess carrier lifetimes. These two lifetimes are different only if trapping effects are significant.

Experimental techniques that measure under steady-state conditions yield τ_n; transient techniques yield τ_t. A large difference between transient and steady-state lifetimes, is an indication of a considerable trapping effect.

[1] C. E. Jones, V. Noir, D. L. Polla, Appl. Phys. Lett. **39**, 248 (1981).
[2] D. Polla, S. G. Tobin, M. B. Reine, and A. K. Sood, J. Appl. Phys. **52**, 5182 (1981).
[3] C. E. Jones, V. Nair, J. Lindquestand and D. L. Polla, J. Vac. Sci. Technol. **21**, 1987 (1982).
[4] D. L. Polla, C. G. Jones, J. Appl. Phys. **52**, 5118 (1981).
[5] C. L. Littler, and D. G. Seiler, and M. R. Loloee, J. Vac. Sci. Technol. A **8**, 1133 (1990).
[6] D. T. Cheung, J. Vac. Sci. Technol. A **3**, 128 (1985).
[7] C. E. Jones et al., J. Vac. Sci. Technol. A **3**, 128 (1985).
[8] C. W. Myles, J. Vac. Sci. Technol. A **6**, 2675 (1988).
[9] S. Goetting and C. G. Morgan-Pond, J. Vac. Sci. Technol. A **6**, 2675 (1988).
[10] M. A. Kinch, in *Semiconductors and Semimetals*, edited by R. K. Willardson and A. C. Beer, (Academic, New York, 1981), Vol. 18, p. 336.
[11] M. A. Kinch, J. Vac. Sci. Technol. **21**, 172 (1982).
[12] D. K. Blanks, J. D. Beck, M. A. Kinch, and L. Colombi, J. Vac. Sci. Technol. A **6**, 2790 (1988).
[13] R. E. DeWames, J. G. Pasko, E. Siyao, A. H. B. Vanderwyck, and G. M. Williams, J. Vac. Sci. Technol. A **6**, 2655 (1988).
[14] R. E. DeWames, G. M. Williams, J. G. Pasko, and A. H. B. Vanderwyck, J. Cryst. Growth **86**, 849 (1988).
[15] R. E. DeWames, *Tunneling in Small Band Gap HgCdTe P-N Junction* (Publisher, city 1989), page no.
[16] Y. Nemirovsky, D. Rosenfeld, R. Adar, and A. Kornfeld, J. Vac. Sci. Technol. A **7**, 528 (1989).
[17] Y. Nemirovsky, R. Adar, A. Kornfeld, and I. Kidron, J. Vac. Sci. Technol. A **4**, 1986 (1986).
[18] Y. Nemirovsky and D. Rosenfeld, J. Vac. Sci. Technol. A **8**, 1159 (1990).
[19] M. Reine, A. K. Sood, and T. J. Fredwell, in Semi-Conductors and Semimetals edited by R. K. Willardson and A. C. Beer (Academic, New York, 1981) Vol. 18.
[20] J. Y. Wong, IEEE Trans. Electron Devices, **ED-27**, 48 (1980).
[21] W. W. Anderson and H. J. Hoffman, J. Appl. Phys. **53**, 9130 (1982).
[22] W. W. Anderson and H. J. Hoffman, J. Vac. Sci. Technol. A **1**, 1730 (1983).
[23] R. Fastow and Y. Nemirovsky, Appl. Phys. Lett. **55**, 1982 (1989).
[24] R. Fastow and Y. Nemirovsky, J. Appl. Phys. **66**, 1705 (1989).
[25] R. Fastow and Y. Nemirovsky, J. Vac. Sci. Technol. A **8**, 1245 (1990).
[26] R. Fastow, D. Goren, and Y. Nemirovsky, J. Appl. Phys. **68**, 3405 (1990).
[27] A. van der Ziel, *Noise in Solid State Devices and Circuits* (Wiley, New York, 1986).
[28] S. M. Sze, *Physics of Semiconductor Devices* (Wiley Interscience, New York, 1982).
[29] C. T. Sah, Phys. Rev. **123**, 1594 (1961).
[30] D. J. Sandiford, Phys. Rev. **105**, 524 (1957).

Correlation of HgCdTe epilayer defects with underlying substrate defects by synchrotron x-ray topography

B. E. Dean, C. J. Johnson, S. C. McDevitt, G. T. Neugebauer, and J. L. Sepich
II–VI Incorporated, Saxonburg, Pennsylvania, 16056

R. C. Dobbyn and M. Kuriyama
National Institute of Standards and Technology, Gaithersburg, Maryland, 20899

J. Ellsworth and H. R. Vydyanath
Aerojet Electrosystems, Azusa, California, 91702

J. J. Kennedy
US Army CECOM/CNVEO, Ft. Belvoir, Virginia, 22060

(Received 3 October 1990; accepted 12 February 1991)

Synchrotron x-ray topography studies have been conducted at the National Synchrotron Light Source at Brookhaven National Laboratory to correlate defects in HgCdTe epilayers with those in underlying CdTe family substrates. Infrared detectors have been fabricated on these epilayers to investigate the performance impact of specific defects. This paper describes synchrotron x-ray facilities and methods. Images of substrates and epilayers are discussed and mapping of epilayer/substrate defects, such as microtwins, subgrain boundaries and slip lines, is demonstrated. Efforts to map detector array performance to epilayer and substrate topographs are described.

I. INTRODUCTION

A major obstacle to the improvement of yields in the manufacture of infrared focal plane arrays is the presence of structural defects in HgCdTe epitaxial layers. Few systematic studies of the impact, role, and origins of specific structural defects have been conducted; thus our understanding of the correlation of structural defects with IR device yield is limited. This paper describes efforts we are making to correlate HgCdTe epilayer defects such as microtwins, subgrain boundaries, lattice distortion and slip lines to substrate defects via synchrotron x-ray topography. We will present preliminary results that show the affects of specific epilayer defects on devices by correlating the performance of a detector array with an epilayer topographical image.

II. SYNCHROTRON TOPOGRAPHY METHODS

X-ray topography has long been used to investigate the structural quality of various materials.[1–3] Laboratory topography procedures are currently utilized in many facilities performing electronic materials research. Synchrotron x-ray topography can have several advantages over typical laboratory methods, e.g., rotating anode generator. The NIST beamline X-23A3 at NSLS/Brookhaven is equipped with a large aperture monochromator that provides parallel (to 0.2 arc s) monochromatic light with tunability from 5 to 30 keV and a narrow energy bandpass ($\Delta E/E \approx 10^{-4}$ at 8 keV).[4–6] The large photon flux, which is characteristic of synchrotron radiation, coupled with the high resolution achievable with the NIST monochrometer have enabled us to resolve individual dislocation loops in epiready substrates using transmission topography.[7]

All substrates utilized were 2×3 cm^2 CdTe, Cd$_{1-x}$Zn$_x$Te, or CdTe$_{1-x}$Se$_x$. The CdTe and Cd$_{1-x}$Zn$_x$Te ($x = 0.045$) crystals were grown using the vertical Bridgman technique. The CdTe$_{1-x}$Se$_x$ ($x = 0.0396$) crystals were grown using the horizontal gradient freeze technique. In both techniques a separate cadmium reservoir with independent temperature control was used to suppress melt evaporation. The ingots were polycrystalline, although they were generally composed of one dominant grain.

The substrates were chemomechanically polished on their (111) A or B face for epitaxial growth. Prior to topography each substrate was subjected to the following characterization to complete its preparation: infrared (IR) transmission (2–40 μm), IR microscopy, etch pit density (EPD), Laue orientation, double crystal rocking curve (DCRC) and resistivity on companion material. We intentionally selected some substrates with known defects along with others that met normal commercial specifications and contained no normally identifiable defects such as twins or large precipitates. The details of growth and characterization methods are provided elsewhere.[8–10]

III. SUBSTRATE IMAGES

Reflection topographs of CdTe, CdZnTe, and CdTeSe substrates are shown in the upper portions of Figs. 1–3 respectively. The prominent defects identified in each substrate image are listed along with pretopograph characterization data. The upper portion of Fig. 1, a topograph of a CdTe substrate, shows significant lattice distortion. The dark areas of the figure correspond to the areas of the substrate that did not satisfy the Bragg condition. Despite significant strain, the single exposure coverage of the 2×3 cm^2 substrate was almost complete. A subgrain boundary, slip lines, and localized distortion are prominent.

The single exposure image of the CdZnTe substrate of Fig.

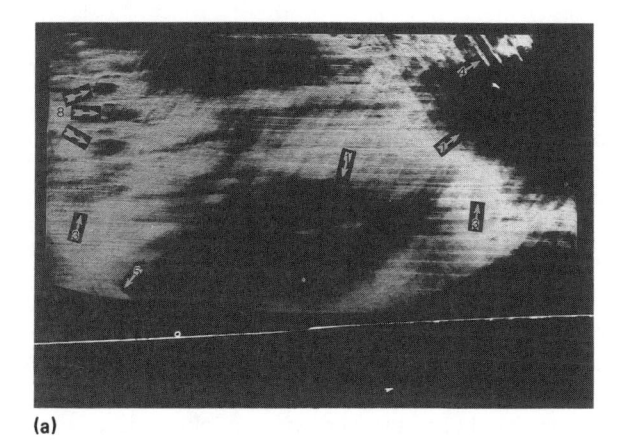

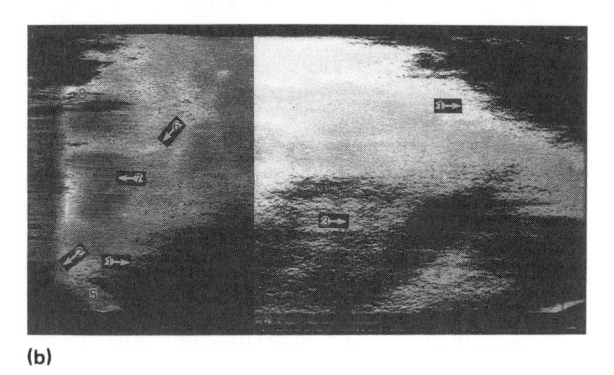

FIG. 1. (a) Synchrotron $(444)_S$ reflection topograph of a (111) A-face CdTe substrate 5277-17 (1 exposure). (b) Synchrotron $(444)_S$ reflection topograph of HgCdTe epitaxial layer 5277-17-734 (1 exposure). Defect Key, (1) lattice distortion, (2) microtwin, (3) $[0\bar{1}1](1\bar{1}\bar{1})$ slip, (4) $[10\bar{1}](\bar{1}1\bar{1})$ slip, (5) subgrain boundary, (6) Te precipitate, (7) cross-hatching, (8) localized distortion. Substrate characterization showed: IR transmission > 50% from 2.5–25 μm; EPD < 4×10^5 cm^{-2}; DCRC FWHM = 25 arc s; and one region of long linear precipitates.

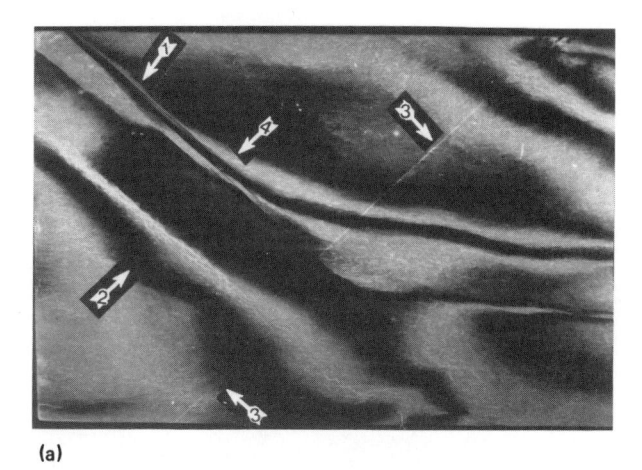

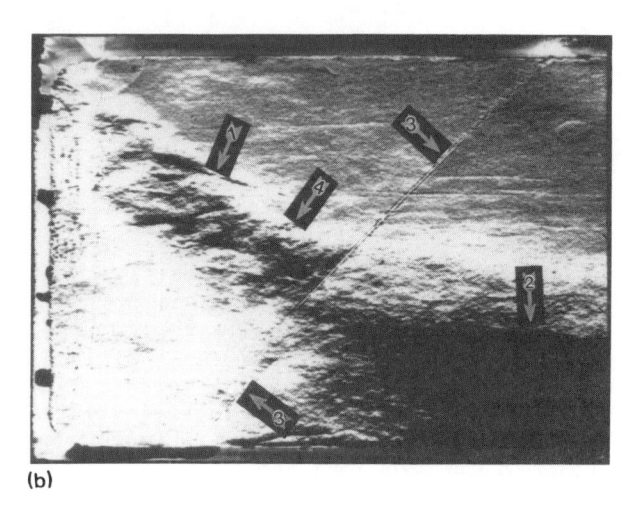

FIG. 2. (a) Synchrotron $(444)_S$ reflection topograph of a $(\bar{1}\bar{1}\bar{1})$ B-face CdZnTe substrate 4899-20 (8 exposures $\Delta\theta \approx 0.01°$). (b) Synchrotron $(444)_S$ reflection topograph of HgCdTe epitaxial layer 4899-20-676 (1 exposure). Defect Key, (1) subgrain boundary, (2) lattice distortion, (3) microtwins, (4) defect. Substrate characterization showed: IR transmission > 65% from 2.5–25 μm; EPD < 5×10^4 cm^{-2}; DCRC FWHM = 20 arc s; and typical precipitates size of 16 μm.

2 gave a very small percentage coverage (approximately 10%) of the total area of the sample. This indicates that lattice distortion was strong in the CdZnTe substrate. In order to obtain corner to corner coverage of this sample, we subjected the nuclear emulsion plate to a total of eight exposures stepped 0.01° apart. The total lattice distortion indicated was thus in the range of 0.07° or 250 arc s. A subgrain boundary is again prominent and the bright images twins are easily seen in Fig. 2. The local or small area crystalline quality of this sample was somewhat better than that of the CdTe sample shown in Fig. 1, however, as represented by the lower EPD and DCRC full width at half-maximum (FWHM) values.

The CdTeSe reflection image shown in the upper portion of Fig. 3 is typical of this material. The lattice distortion was significantly less and the single exposure coverage of the total area is impressive. The nearly featureless image indicates the absence of macroscopic defects. Note that the small area crystallinity was found to be far superior to that of the preceeding CdTe and CdZnTe, as both the EPD and DCRC FWHM were very low.

To date we have examined dozens of samples of each substrate material. The results described above are representative of CdTe family materials produced at II–VI Incorporated. It must be noted, however, that materials produced using different growth procedures could yield different results. This point is mentioned in order to emphasize the need for continued work in the growth of CdTe materials that are free of dislocations and defects.

IV. LIQUID PHASE EPITAXY GROWTH METHOD

The HgCdTe growth method we chose for this investigation is liquid phase epitaxy (LPE). This is the most mature method and the one that gives us the best chance of correlating device performance to substrate defects. The LPE process that we utilized at Aerojet Electrosystems involves bringing the CdTe or CdZnTe substrates in contact with a Te rich ternary solution of composition $(Hg_{1-x}Cd_x)_{1-z}Te_z$ at 500 °C and then cooling the solution at a rate of 0.1–0.2 °C/min to cause precipitation of solid $Hg_{1-x}Cd_x$Te on the substrate.[11]

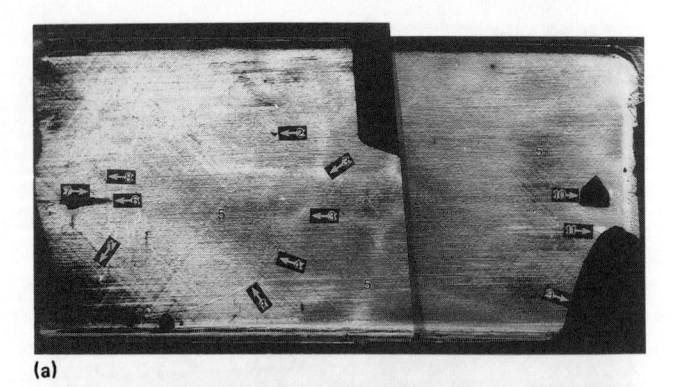

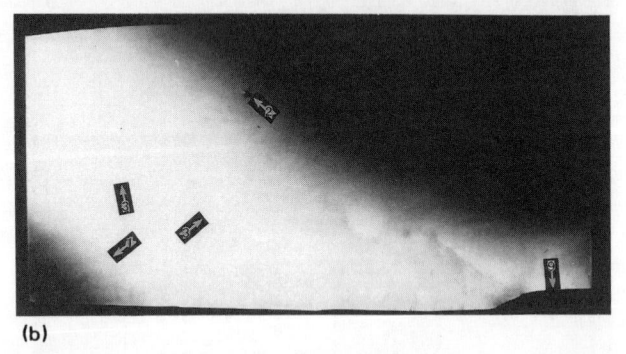

FIG. 3. (a) Synchrotron $(444)_S$ reflection topograph of a $(\bar{1}\bar{1}\bar{1})$ B-face CdTeSe substrate NSF4-A-7Y (1 exposure). (b) Synchrotron $(444)_S$ reflection topograph of HgCdTe epitaxial layer NSF4-A-7Y-740 (1 exposure). Defect key, (1) lattice distortion, (2) large defect, (3) mottled texture, (4) terrace boundary, (5) cross-hatching, (6) large defect, (7) small defects, (8) small defects, (9) subgrain boundary decorated by Te melt, (10) triangular column, (11) triangular column and Te melt. Substrate characterization showed: IR transmission $>66\%$ from 2.5–25 μm; EPD $< 8 \times 10^3$ cm^{-2}; DCRC FWHM = 10 arc s; and typical precipitates size < 10 μm.

V. EPILAYER IMAGES AND EPILAYER/SUBSTRATE DEFECT MAPPING

The epilayer to substrate defect mapping results of our investigations are shown in Figs. 1–3. Listed in each figure is the defect key for that epilayer/substrate pair and the substrate pre-topograph, characterization results.

A. HgCdTe/CdTe

The prominent CdTe substrate defects propagating into corresponding epilayers, as shown in Fig. 1 are: lattice distortion, slip lines, a star-shaped precipitate near the surface, and localized lattice distortions. The extent of overall lattice distortion in the substrates is limited to the range of 40–50 arc s and lattice distortion is reduced in the corresponding epilayers.

B. HgCdTe/CdZnTe

The prominent CdZnTe substrate defects propagating into corresponding epilayers, as shown in Fig. 2 are: lattice distortion, subgrain boundaries, microtwins, and cellular structures. The slip lines, prominent in the case of CdTe, were notably absent. This may be caused by dislocation tangles (Lomer–Cottrell locks); therefore, the Burgers vectors no longer lie in their slip plane.[7–9]

C. HgCdTe/CdTeSe

According to Fig. 3, the features mapping from a CdTeSe substrate to the corresponding epilayer were a large defect, a subgrain boundary, mottled texture, and a small amount of lattice distortion. The local or small area crystallinity of the CdTeSe substrate is again far superior to the CdTe and CdZnTe which we have investigated, as evidenced by the EPD and DCRC results listed in Fig. 3.

The cross-hatching, which is very prominent in the epilayer topograph but not observed in the substrate topograph, is attributed to inhomogeneous microstrain originating at the substrate/epilayer interface. Similar cross-hatching in epilayers has been observed with closely latticed matched (to within 0.0013) III–V materials. The microstrain in those cases disappeared when the epilayer was etched off.[15]

D. Defect mapping summary

A summary of the identification and defect mapping that we have demonstrated in our study is given in Table I.

VI. DEVICE FABRICATION METHODS

Test arrays of IR detectors have been fabricated on selected epilayers at Aerojet Electrosystems. The array pattern on a wafer, magnified photographs of two individual arrays and descriptive information on the devices are shown in Fig. 4.

The epilayers produced during this investigation were screened on the basis of crystalline quality and electrical quality as indicated by Hall data. Five partial epilayers were selected for device fabrication as listed in Table II. Substrate topographs exist for all five units and epilayer topographs were generated for three of the units. We note that mobilities vary greatly for these epilayers; from $-99\,000$ to $-173\,000$ cm^2/V s at 100 G and from $-81\,000$ to

TABLE I. Epilayer/substrate defect mapping summary. S refers to a defect confined to the substrate; E refers to a defect confined to the epilayer; and S & E refers to a defect located in the substrate and epilayer.

Structural defect	CdTe	CdTeSe	CdZnTe
Lattice bending	S & E	S & E	S & E
Subgrain boundary		S & E	S & E
$[01\bar{1}](1\bar{1}\bar{1})$ slip	S & E		
$[\bar{1}10](11\bar{1})$ slip	S & E		
$[\bar{1}01](\bar{1}1\bar{1})$ slip	S & E		
Cross-hatching	S & E	E	
Star-shaped Te precipitate	S & E		
Cellular pattern	S & E		S & E
Triangular column	E	E	
Microtwin	E		S & E
Linear defect	S		
Localized lattice distortion	S & E		
Mottled texture		S & E	
Terracing boundary		E	
Large defect		S & E	
Large defect		E	
Small defect		E	

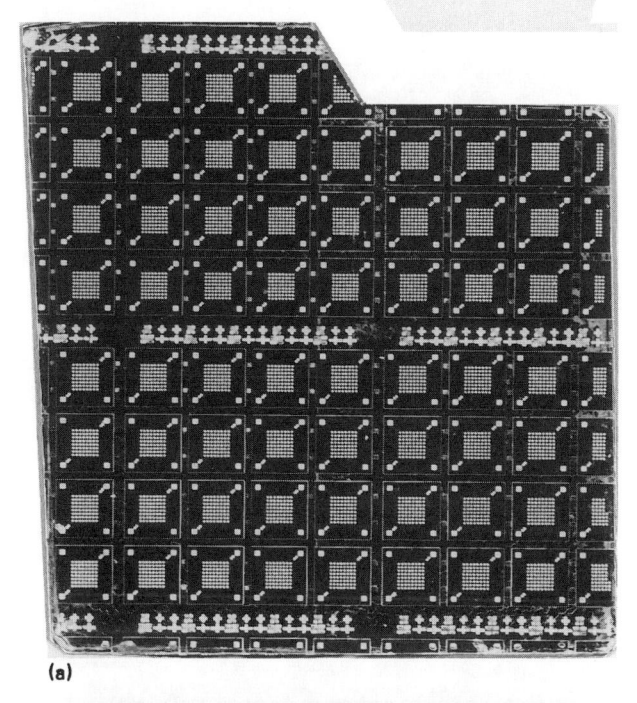

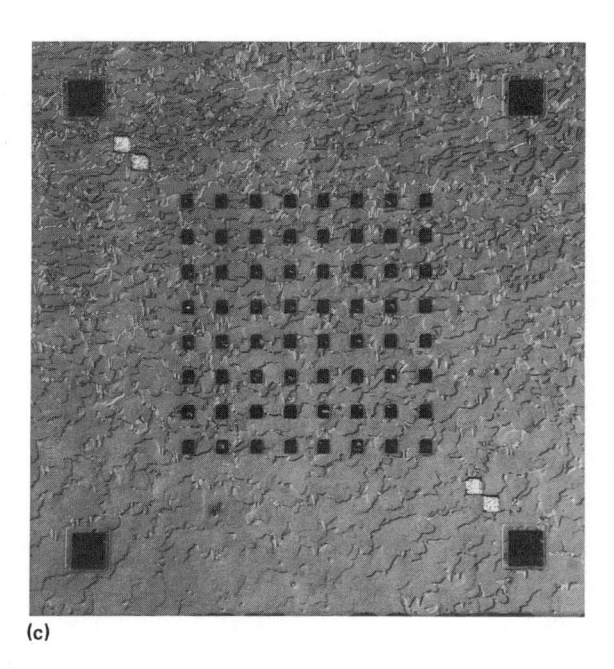

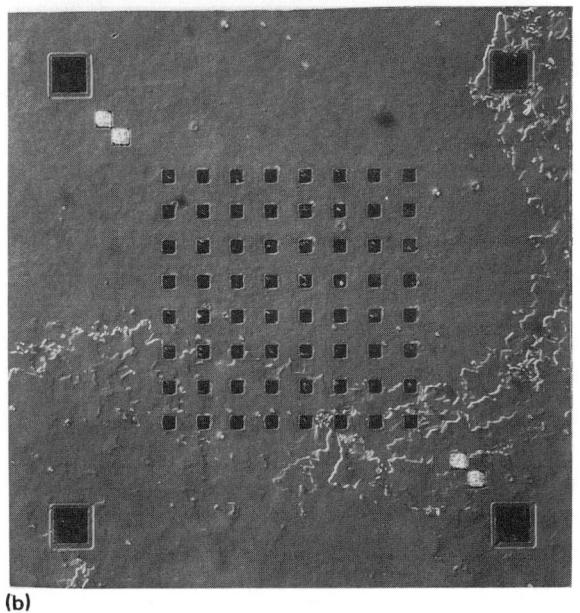

FIG. 4. (a) Device fabrication and testing. 8×8 array on HgCdTe 5277-8X-727 with 100 μm spacing, p on n configuration. Device performance is characterized by I vs V, R_0A, spectral response, resistivity carrier lifetime, DLTS, and CV measurements. (b) Close up of a device on a highly terraced region of the array shown in (a). Close up of a device on a relatively smooth region of the array shown in (a).

TABLE II. Data on epilayers selected for device fabrication. A topograph was obtained for each substrate prior to epitaxial growth. Topographs of the epilayers were obtained for 727, 728, and 734. The Hall data were measured at 77 K. Crystal quality was ascertained using the substrate topograph/the epitaxial topograph/and visual inspection of the epitaxial layer under a low power microscope.

Epi no.	Substrate	λ_{CO} RT	Hall data 100 G	Hall data 12 000 G	Res.	Electrical quality	Crystal quality
727 (B)	5277-8X CdTe	6.02	-4.5×10^{14} $-99\,000$	-5.5×10^{14} $-81\,000$	0.1373	Fair $\mu < 10^5$	Poor/fair/good
728 (B)	5277-7X CdTe	6.20	-1.8×10^{14} $-139\,000$	-2.9×10^{14} $-85\,000$	0.2481	Fair $\mu < 10^5$	Poor/fair/good
734 (A)	5277-17 CdTe	6.30	-3.0×10^{14} $-150\,000$	-4.1×10^{14} $-111\,000$	0.1372	Good $\mu > 10^5$	Fair/good/fair
762 (B)	5277-11X CdTe	6.42	-2.4×10^{14} $-173\,000$	-4.1×10^{14} $-105\,000$	0.1454	Good $\mu > 10^5$	Poor/···/good
763 (B)	5277-12X CdTe	6.53	-2.3×10^{14} $-165\,000$	-3.6×10^{14} $-109\,000$	0.1581	Good $\mu > 10^5$	Good/···/good

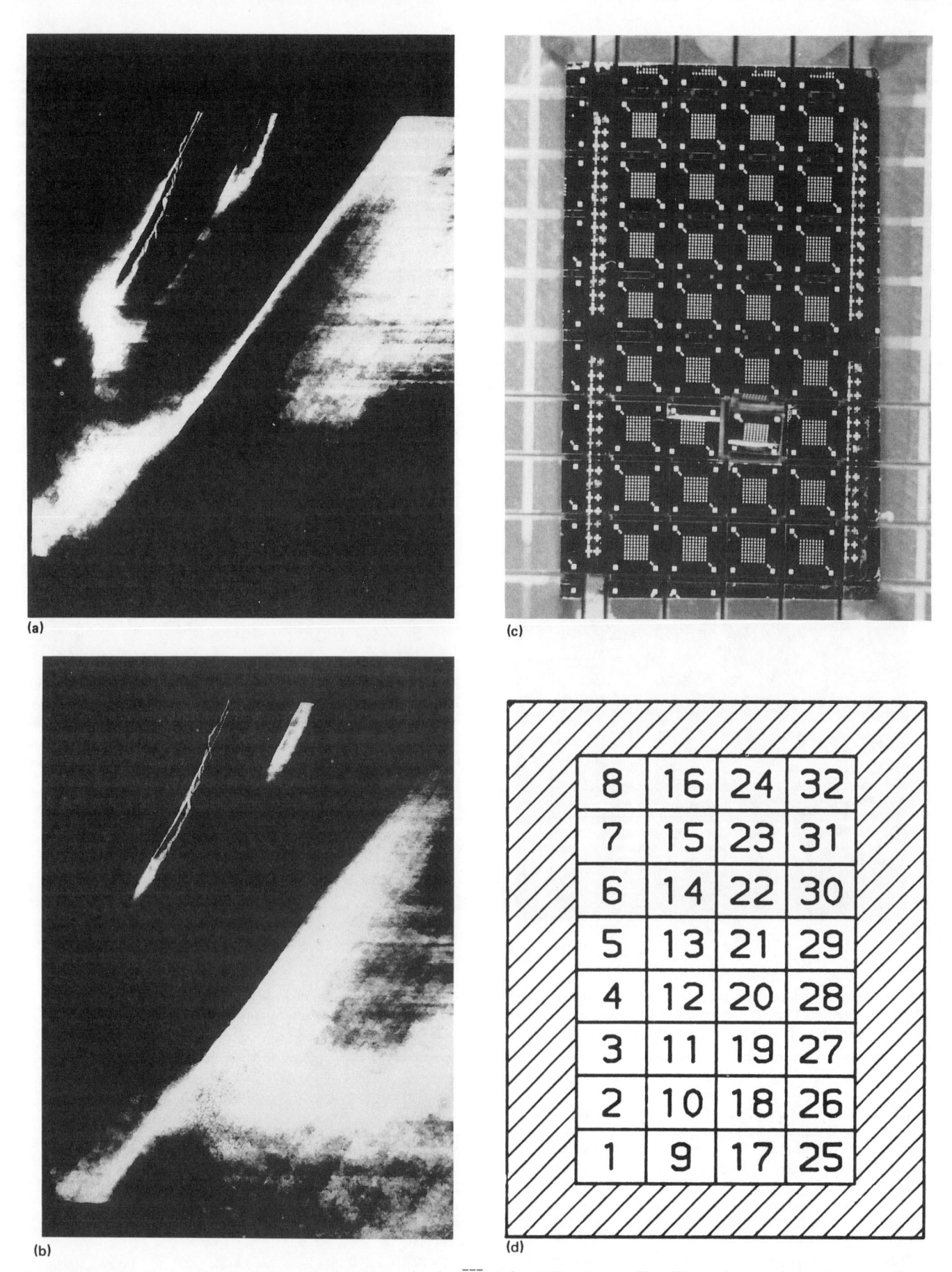

FIG. 5. (a) and (b). Two synchrotron $(444)_S$ reflection topographs of a $(\bar{1}\bar{1}\bar{1})$ B-face CdTe substrate. The difference in scattering angle between the two topographs was $\simeq 0.025°$ at 12 keV, which suggests that lattice distortion was moderate to minimal. (c) A photo of the IR detector arrays on 5277-11X-762. (d) A schematic showing the location of the IR detector arrays on 5277-11X-762. The orientation and scale of the schematic corresponds to topographs shown in (a) and (b). The devices for which performance data is available are 13, 17, 29, and 31.

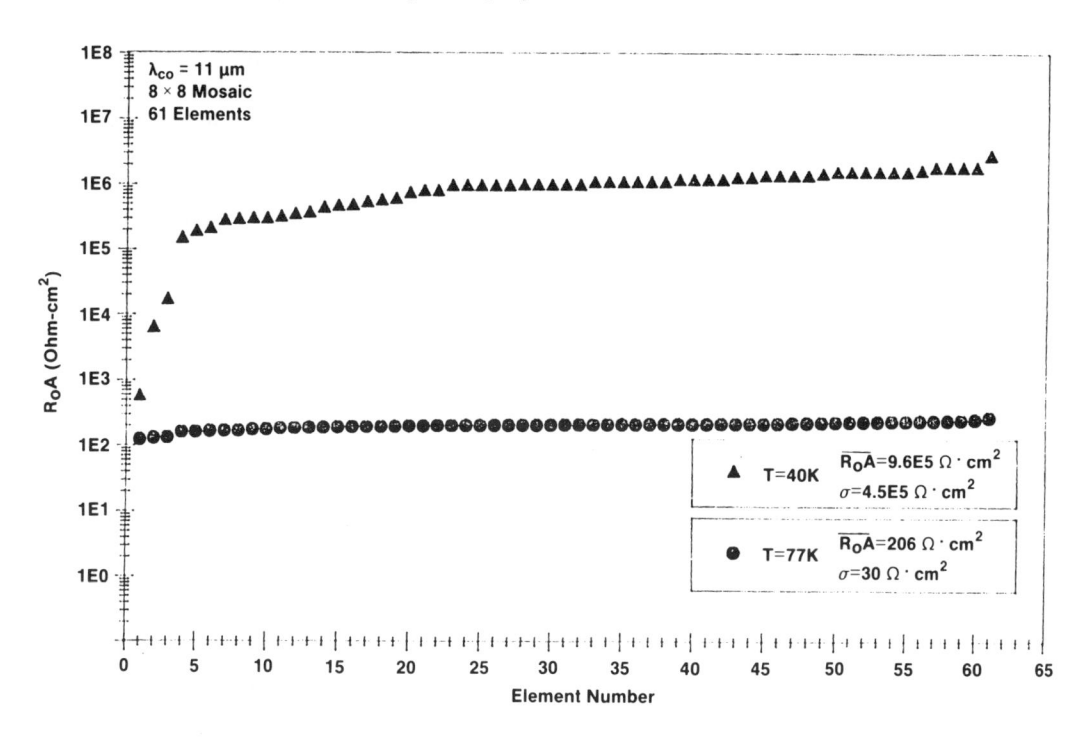

FIG. 6. R_0A ranking in order of performance for array 762-M2-29.

$-111\,000$ cm^2/V s at 12 000 G. All carrier concentrations for these epilayers were in the range -1.8×10^{14} to -5.5×10^{14} cm^{-3} for both magnetic field strengths.

VII. DEVICE TEST DATA

Our goal is to relate IR detector and test array performance to substrate and epilayer defects. Despite our best efforts, we have succeeded in obtaining only preliminary data for one the units listed in Table II: HgCdTe epilayer number 762 grown on CdTe 5277-11X. Figure 5 shows two reflection topographs of the substrate, a photo of the device array, and a schematic of the device numbering scheme. Defects such as lattice distortion, slip lines, and cellular structures were observed in the substrate topographs. No topographs of the epilayer were taken.

Four 8×8 arrays were tested from different regions of epitaxial layer 762. The R_0A performance data are shown for two of the arrays 762-M2-29 and 762-M2-31 in the form of R_0A ranking plots in Figs. 6 and 7. The area of the epitaxial layer in which array 762-M2-31 was fabricated included a sharp subgrain boundary, while array 762-M2-29 was fabricated in a region of the film with readily indentified strain fields (lattice distortion) and slip patterns. The R_0A perfor-

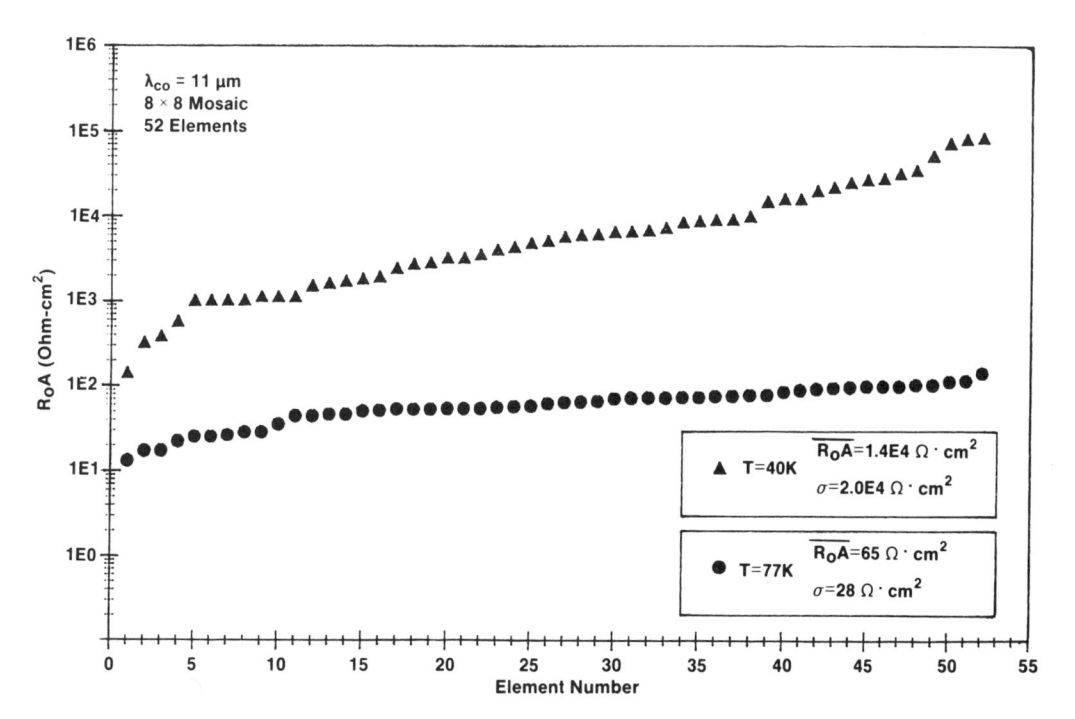

FIG. 7. R_0A ranking in order of performance for array 762-M2-31.

mance data shown in Figs. 6 and 7 indicate that array 762-M2-29 shows better performance than array 762-M2-31 at both 77 and 40 K, although substantially more pronounced at 40 K.

These results lead to the inference that sharply defined substrate subgrain boundaries seriously affects device performance particularly at lower temperatures, whereas the presence of strain fields and slip patterns do not seem to. The more pronounced effect on device performance at 40 K compared to 77 K may be due to the fact that the devices are diffusion current noise limited at 77 K whereas they are surface leakage/tunneling dark current limited at 40 K. The subgrain boundary under array 762-M2-31 probably plays a significant role in the defect assisted tunneling at 40 K and hence the reason for the poor performance of this array at 40 K compared to the array 762-M2-29.

Because of the small amount of device data available at this time, no general conclusions can be made regarding the effects that other specific substrate/epilayer defects have on device performance. However, based on devices made from one substrate/epilayer, it can be said that a small amount of lattice distortion in the substrate or an active slip system does not preclude good IR detector performance.

VIII. CONCLUSIONS

Our work to date has shown that synchrotron x-ray topography can non-destructively screen CdTe family substrates for lattice distortion, microtwins, subgrain boundaries, cellular structures, slip lines, and large precipitates prior to HgCdTe epitaxy. Also, growth by LPE reproduced all substrate defects in $Hg_{1-x}Cd_xTe$, $x = 0.2$, epilayers as observed by synchrotron topography, and substrate subgrain boundaries and large precipitates when near the substrate–epilayer interface, always degrade epilayer quality.

Lattice distortion is present in all substrates and epilayers, however, its magnitude is reduced in epilayers. Lattice distortion is more pronounced in CdZnTe substrates than in CdTe and CdTeSe. For several CdTe substrates, slip lines and cross-hatching were reduced in epilayer images as compared to substrate images. For the one CdTeSe epilayer/substrate case, cross-hatching was not observed in the substrate image but was pronounced in the epilayer image.

Device performance mapping to synchrotron topography identified defects is complicated by a myriad of epilayer growth and device fabrication process variables. Statistically significant conclusions will require mapping of many additional array patterns. We especially need to achieve array patterns of the highest device quality so that the impact of localized epilayer/substrate structural defects is not masked. To date, we have developed a methodology for tracking R_0A, quantum efficiency, and other pertinent device test results relative to location on the corresponding epilayer or substrate topographs. No statistically significant performance or mapping data have been generated.

ACKNOWLEDGMENTS

This work has been partially supported by the US Army CECOM/CNVEO under Small Business Innovation Research Contract No. DAAB07-87-C-FO88; J. Kennedy, CECOM/CNVEO at Ft. Belvoir, VA, monitor, SDIO under Contract No. F29601-88-C-0025; Capt. S. Stapp, US Air Force Weapons Laboratory at Kirkland AFB, NM, monitor, Aerojet Electrosystems, Azusa, CA, and II–VI Incorporated, Saxonburg, PA.

---

[1] M. Kuriyama, B. W. Steiner, and R. C. Dobbyn, Annu. Rev. Mater. Sci. **19**, 183 (1989).
[2] M. Kuriyama, B. Steiner, R. C. Dobbyn, U. Laor, D. Larson, and M. Brown, Phys. Rev. B **38**, 421 (1988).
[3] Y. C. Lu, R. S. Feigelson, R. K. Route, and Z. U. Rek, J. Vac. Sci. Technol. A **4**, 2190 (1986).
[4] R. Spal. R. C. Dobbyn, H. E. Burdette, G. G. Long, W. J. Boettinger, and M. Kuriyama, Nucl. Instrum. Methods Phys. Res. **222**, 189 (1984).
[5] B. Steiner, M. Kuriyama, and R. C. Dobbyn, Progr. Cryst. Growth Charact. **20**, 189 (1990).
[6] M. Kuriyama, R. C. Dobbyn, R. D. Spal, H. E. Burdette, and D. R. Black, J. Res. Natl. Inst. Stand. Technol. **95**, 559 (1990).
[7] Results to be published.
[8] K. Y. Lay, D. Nichols, S. McDevitt, B. E. Dean, and C. J. Johnson, J. Cryst. Growth **86**, 118 (1988).
[9] W. P. Allred, A. A. Khan, C. J. Johnson, N. C. Giles, and J. F. Schetzina, Mater. Res. Soc. Symp. Proc. **90**, 103 (1987).
[10] A. A. Khan, W. P. Allred, B. Dean, S. Hooper, J. E. Hawkey, and C. J. Johnson, J. Electron. Mater. **15**, 181 (1986).
[11] H. R. Vydyanath, J. A. Ellsworth, and C. M. Devaney, J. Electron. Mater. **16**, 13 (1987).
[12] J. P. Hirth and J. Lothe, *Theory of Dislocations* (McGraw-Hill, New York, 1968), p. 727.
[13] M. Kuriyama, J. G. Early, and H. E. Burdette, J. Appl. Cryst. **7**, 535 (1974).
[14] W. J. Boettinger, H. E. Burdette, and M. Kuriyama, Philos. Mag. **34**, 119 (1976).
[15] I. C. Bassignana, Bell Northern Research, Ltd., Ottawa, Canada (private communication, 1990).

Photoexcited hot electron relaxation processes in n-HgCdTe through impact ionization into traps

D. G. Seiler and J. R. Lowney
Semiconductor Electronics Division, National Institute of Standards & Technology, Gaithersburg, Maryland 20899

C. L. Littler, I. T. Yoon, and M. R. Loloee[a]
Department of Physics, University of North Texas, Denton, Texas 76203

(Received 2 October 1990; accepted 26 December 1990)

In this article we report on a new type of spectroscopy for impurity and/or defect levels in the energy gap of narrow-gap semiconductors using the near-band-gap photon energies from a laser. This spectroscopy is done under the conditions of intense laser photoexcitation and is associated with the Auger relaxation processes of hot electrons involving impact ionization of valence electrons into impurity or defect levels. Wavelength-independent structure in the photoconductive response versus magnetic field is observed at high intensities in samples of $Hg_{1-x}Cd_xTe$ with $x \approx 0.22$ and 0.24. This structure arises from hot electrons photoexcited high into the conduction band by sequential absorption of CO_2 laser radiation. The hot electrons lose their energy by impact ionizing valence electrons into impurity/defect levels in the gap. For the sample with $x \approx 0.22$ and an energy gap of 95 meV, three levels are found at 15, 45, and 59 meV above the valence band. A level at 61 meV is found for the sample with $x \approx 0.24$ and a gap of 122 meV.

I. INTRODUCTION

The semiconducting $Hg_{1-x}Cd_xTe$ alloys are used extensively as infrared detector materials in a wide range of both civilian and military systems. One of the important challenges in making substantial improvements in the quality and uniformity of the material is to develop better characterization techniques for the detection and identification of defects and impurities. Even though a large effort has been expended in characterizing these levels, most still remain rather poorly understood. Consequently, any new techniques or methods that can be developed to detect and characterize these levels should be exploited. Here we report a new method for observing and studying impurity/defect levels in narrow-gap semiconductors. This new method involves the combined use of intense laser radiation and the techniques of magneto-optical spectroscopy. Structure is observed in the photoconductive (PC) response as a function of magnetic field, the peak positions of which are independent of laser photon energy.

There have been a number of previously observed effects in semiconductors where resonance positions do not depend upon photon energy such as the magneto-impurity effect and the impurity-shifted magnetophonon effect.[1] Hot-electron conditions always seem to be necessary for observing these resonances, and they are achieved either by the non-ohmic conditions created by applying large enough electric fields or else by photoexcitation of carriers deep into the conduction band by optical excitation across the band gap. These effects are therefore distinct from intrinsic and extrinsic oscillatory photoconductivity effects which have been studied in many semiconductors at zero magnetic field. Magneto-impurity resonances (MIRs) have been seen in n-GaAs, n-InP, n- and p-Ge, and n-CdTe.[1] These resonances arise from inelastic scattering processes whereby a free carrier resonantly exchanges energy with a second carrier bound to a shallow acceptor or donor impurity in the presence of a magnetic field. Impurity-assisted magnetophonon resonances arise from a process whereby an electron emits a longitudinal optical (LO) phonon in falling from a Landau level into a bound state of an impurity.

In this work, the near-band-gap photon energies from a CO_2 laser are used to perform a new type of spectroscopy for characterizing impurity/defect levels that differs from the magneto-impurity and the impurity-shifted magnetophonon effects. This spectroscopy is done under the conditions of intense laser photoexcitation and is associated with the Auger relaxation processes of hot conduction band electrons relaxing by impact ionization of valence electrons into impurity levels. This work is important because it provides a new technique for studying impurity/defect levels in narrow-gap semiconductors in which these levels have been generally difficult to measure.

II. EXPERIMENTAL

The experiments reported here were carried out on two single-crystal, bulk grown n-type samples of $Hg_{1-x}Cd_xTe$ with x values of approximately 0.22 and 0.24. Both samples were lapped with alumina grit and then chem-mechanically polished using a 2% bromine–methanol solution. Electrical contacts were made to the samples using pure indium. The magneto-optical system used has been described elsewhere (See Fig. 1 of Ref. 2). Light from a grating tunable continuous wave (cw) CO_2 laser was focused onto a sample placed in a superconducting-magnet/variable temperature dewar system (0–12 T). To prevent lattice heating, the light was mechanically chopped at a low duty cycle of $\approx 1\%$. The propagation direction of the linearly polarized laser light was parallel to the magnetic field. The photoconductive response of the samples was obtained under constant current, ohmic conditions. The magneto-optical spectra (PC response versus magnetic field) were obtained using either ac magnetic field modulation and lock-in amplifier techniques or boxcar averager methods.

III. RESULTS AND DISCUSSION

The interaction of laser radiation with semiconductors is in general a complex phenomenon involving many different types of processes occurring before equilibrium is regained. Which particular processes dominate or are important depend upon such parameters as the laser photon energy, laser intensity, laser pulse width, sample temperature, and sample doping. Illumination of semiconductors with intense radiation also leads to carrier heating effects. Laser-based magneto-optical spectroscopy permits the study of this interaction of laser radiation and allows the observation of a number of separate features or resonances in the PC response versus magnetic field. These features must be understood and appreciated in order to be able to interpret our new high-intensity structure. As shown in Fig. 1 for a sample with $x \approx 0.24$ and a low-temperature band gap of $E_g = 122$ meV, four different features are observed in the spectra, depending upon which combination of laser wavelength, magnetic field range, laser intensity, and lattice temperature is used. First, a two-photon absorption structure is seen at long wavelengths, high intensities, and high magnetic fields. These two-photon magneto-optical spectra have been extensively studied by us in a number of samples of HgCdTe, and this structure is now well understood.[2] At shorter wavelengths, lower magnetic fields, and low laser intensities two sets of magneto-optical structure are observed and identified by upward pointing arrows: a one-photon magneto-absorption (OPMA) structure and a broad impurity magneto-absorption (IMA) peak just to the high field side of the largest OPMA peak. This IMA peak has been attributed to electron transitions from a shallow acceptor to the lowest conduction band Landau level.[2,3] These three sets of resonant structures (one-photon, two-photon, and impurity) can be understood by a Landau-level model (where peaks in the density-of-states occur at energies corresponding to the extremum of each Landau level) along with the corresponding optical transitions.[4] At these peaks there are enhanced optical transition rates or electron scattering rates. The resulting optically created electrons in the conduction band then cause an increase in the conductivity or photoconductive response of the samples.

Figure 2 shows that at low intensities only a one-photon absorption structure is present, while at higher intensities a series of structures develops on the low-field (or high-energy) side of the dominant one-photon peak at ≈ 4.2 T. The magnetic field positions of this new, high-intensity structure are independent of wavelength, as shown in Fig. 3. In the higher field regions of these spectra, one sees one-photon and impurity absorption structure which clearly has a dependence on wavelength or photon energy. In order to interpret better this wavelength-independent structure, ac magnetic-field modulation and lock-in amplifier techniques were employed to increase the resolution of the oscillations. An example of the improved resolution obtainable is seen in Fig. 4 where the second derivative of the PC response is plotted as a function of magnetic field.

Another sample with $x \approx 0.22$ exhibits similar wavelength-independent structure in the PC response, as shown in Fig. 5. It is strikingly clear that the two high-field peaks are much larger than the rest of the lower field series. Since the band gap energy is 95 meV for this sample, the largest one-photon peak has shifted out of our measurable field range; thus all of the structure can be attributed to the new mechanism. Figure 6 shows the second derivative of the PC

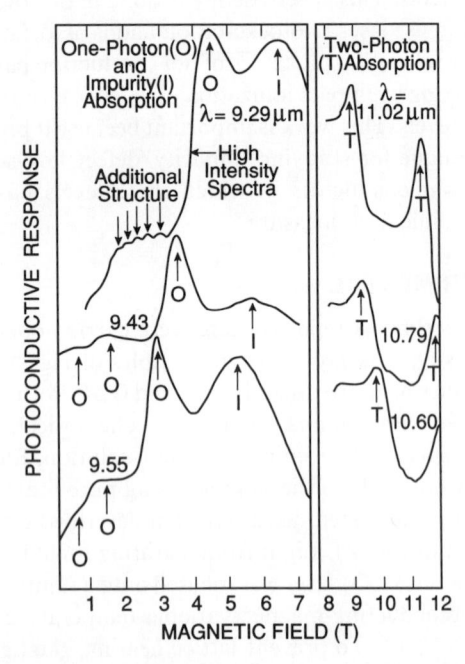

FIG. 1. Wavelength dependence of magneto-optical spectra for a sample with $x \approx 0.24$ at 7 K showing four different sets of structure related to one-photon absorption (O), two-photon absorption (T), impurity absorption (I), and the additional high-intensity structure, which is the subject of this paper. The spectra for $\lambda = 9.29\,\mu m$ have been obtained at much higher laser intensities. Boxcar averaging techniques have been used.

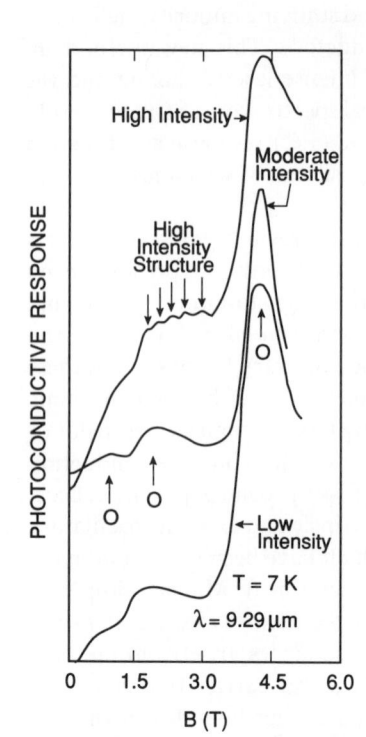

FIG. 2. Intensity dependence of the one-photon and the high-intensity structure for a sample with $x \approx 0.24$ using boxcar averager techniques.

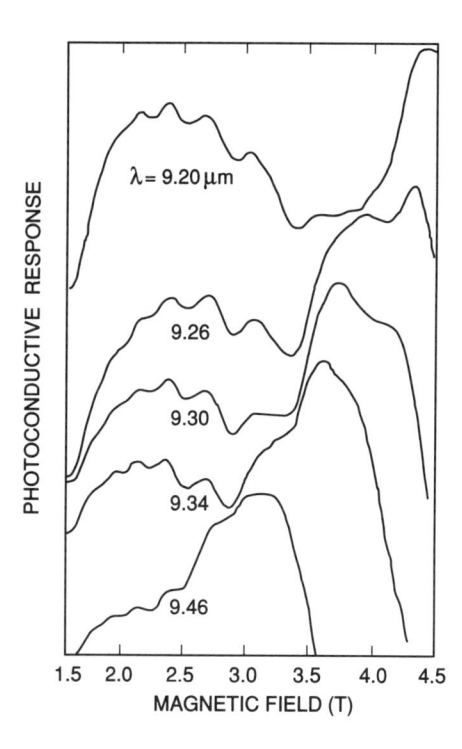

FIG. 3. Wavelength dependence of the high-intensity structure showing the independence of the peak positions on photon energy using boxcar averager techniques.

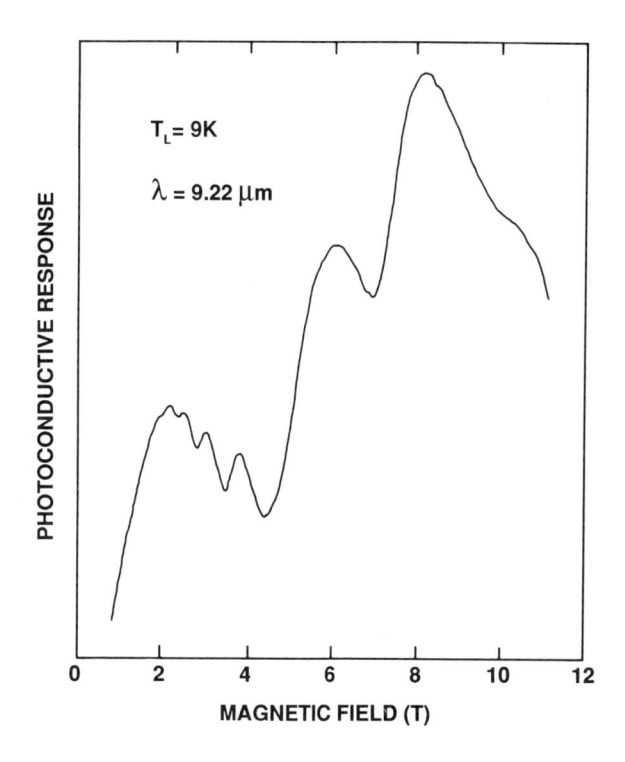

FIG. 5. Photoconductive response vs magnetic field showing the high-intensity structure obtained for a sample with $x \approx 0.22$ by using boxcar averager techniques.

response using the ac magnetic field and lock-in amplifier technique. The sensitivity of the derivative method is clearly superior at low fields, with many more oscillations observable below 2 T (2 T is the limit of observability of the structure seen in Fig. 5). Figure 6 clearly shows two dominant sets of structure: long period structure at lower fields and higher frequency structure above 1.5 T. The complexity of the data at high fields also indicates another possible set of structure.

The resonance conditions that lead to the oscillations in the photoconductive response are pictured in Fig. 7. Electrons are photoexcited across the energy gap and subsequently high into the conduction band by absorption of a second photon of energy $\hbar\omega$. Note that if the photon energy is less than the separation between the highest Landau level in the valence band and the lowest in the conduction band,

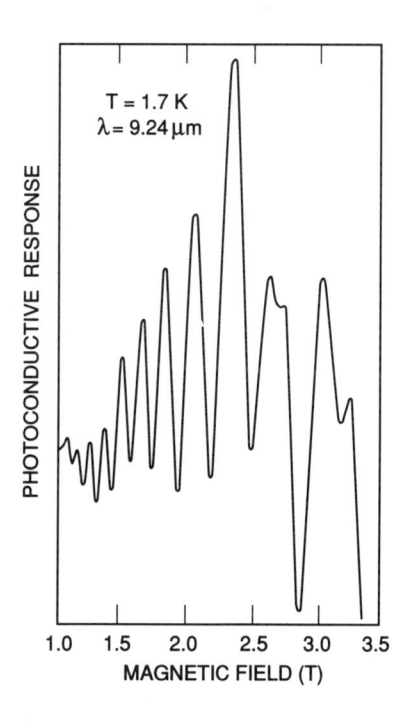

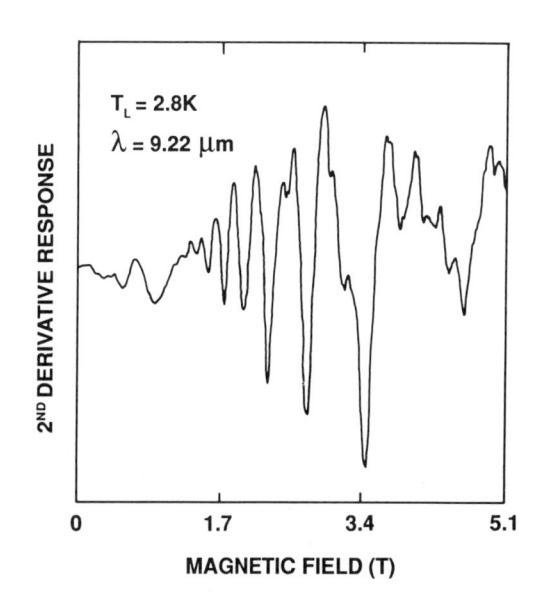

FIG. 4. High-resolution data obtained by using, in addition, ac magnetic field modulation and lock-in amplifier techniques. The lock-in amplifier output plotted on the y-axis represents the second derivative of the photoconductive response vs magnetic field.

FIG. 6. High-resolution data obtained by using, in addition, ac magnetic field modulation and lock-in amplifier techniques. This second derivative behavior shows important structure at low fields and other structure at high fields that cannot be seen with the boxcar technique.

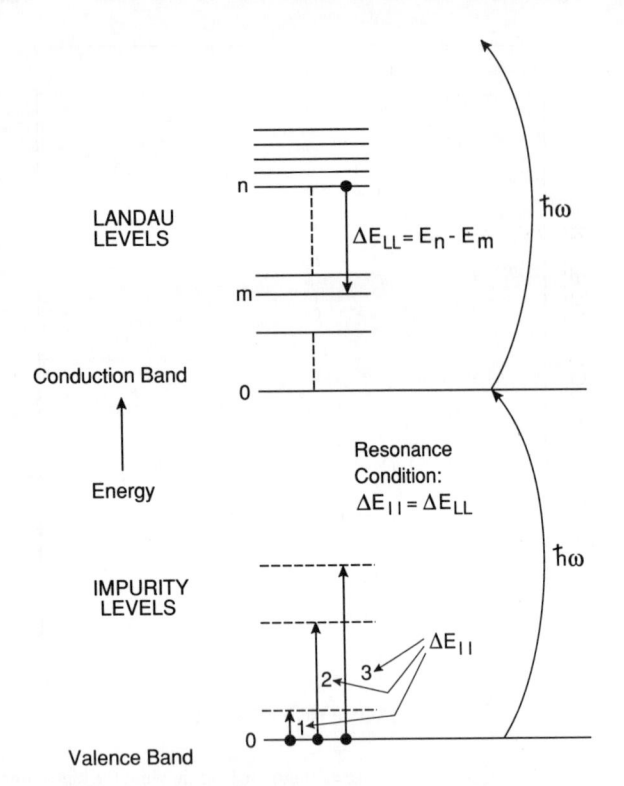

FIG. 7. A schematic of the electronic transitions that lead to the observed resonances. ΔE_{LL} is the energy difference between the initial and final conduction-band Landau-level energies. ΔE_{II} is the difference between the energy of the impurity level and the highest valence-band Landau level. Resonances occur when $\Delta E_{LL} = \Delta E_{II}$. Photons of energy $\hbar\omega$ excite electrons across the gap and subsequently well into the conduction band.

the sample is observed to be almost transparent and the signal goes to zero. For a resonance the transition energies ΔE_{LL} between conduction-band Landau levels must equal the transition energies ΔE_{II} between the highest valence-band Landau level and an impurity level in the gap. The arrows indicate the electron transitions from their initial to final states. Landau levels that are more than a photon energy above the lowest conduction-band Landau level are not populated and therefore do not contribute. An alternative possibility by which the electrons can be promoted high into the conduction band is through Auger processes. This would not affect the interpretation of our data.

A computer code was written to determine the transitions that correspond to the observed peaks. The series of conduction-band Landau levels and the upper valence-band Landau level were computed for the energy gaps of the two samples, 95 and 122 meV. A modified Pidgeon–Brown band model[5] was used with Weiler's set of band parameters[5]: $E_p = 19$ eV, $\Delta = 1$ eV, $\gamma_1 = 3.3$, $\gamma_2 = 0.1$, $\gamma_3 = 0.9$, $\kappa = -0.8$, $F = -0.8$, $q = 0.0$, and $N_1 = 0.0$. The sample orientation was assumed to be $\langle 111 \rangle$, and no exciton corrections were made since the structure is introduced not by the absorption process but by the relaxation process. Within the precision of our measurements, anisotropy effects are not expected to be important. The energy versus magnetic field dependence of the Landau levels is fitted with a parabolic spline. Note that the small gap of mercury cadmium telluride leads to a strongly nonparabolic band and thus to a nonlinear dependence of Landau level energies on magnetic field. A value is then given as input to the code for the energy of the impurity level above the valence band at zero magnetic field. The code searches for magnetic fields that satisfy the resonance condition. Successive values for the impurity-level energy are used until agreement is obtained between theory and experiment. The code furnishes as output the values of the magnetic fields, Landau level numbers for the transitions, and the actual energies of the impurity level above the upper valence-band Landau level. It is assumed that the impurity/defect level does not shift with magnetic field.

The results for the two samples are given in Tables I and II. Table I gives the values for the experimental and theoretical magnetic fields, B^{exp} and B^{th}, respectively, in tesla, the Landau level numbers for the transitions, and the resonance energies in milli electron volts for sample I with $x \approx 0.22$ and a gap of 95 meV. All of the resonances correspond to the series of Landau levels with spin state "a." This result implies that those states are preferentially pumped by the radiation and that spin is conserved as well in the downward transitions involving impact ionization. The data would show further splittings than are observed if the "b" spin state Landau levels were involved. Three distinct impurity levels were extracted from the data. The first is at very low field and corresponds to an impurity level of 15 meV at zero field. Only transitions to the lowest Landau level were observed ($1\to0$, $2\to0$, $3\to0$, $4\to0$). Those corresponding to the upper-level transitions were either too "smeared" at these low fields or quenched by phonon-assisted transitions between the levels. This energy corresponds to a commonly observed shallow acceptor level.[3] An alternate explanation would be that the low field structure may be associated with an impurity-shifted magnetophonon effect where the donor is nearly degenerate with the lowest conduction-band Landau level.

TABLE I. Experimental B^{exp} and theoretical B^{th} magnetic field positions for the peaks in the photoconductive response, along with the conduction band Landau level numbers associated with the transitions and the impact ionization energies E_{II} between the upper valence band Landau level and the impurity/defect level for the sample with $x \approx 0.22$.

B^{exp}(T)	B^{th}(T)	Transitions	E_{II} (meV)
0.263	0.276	$4\to0$	15.0
0.357	0.372	$3\to0$	15.1
0.584	0.568	$2\to0$	15.1
1.20	1.19	$1\to0$	15.2
1.25	1.23	$10\to4, 7\to2$	45.2
1.38	1.38	$8\to3, 5\to1$	45.2
1.54	1.50	$3\to0$	45.2
1.75	1.68	$9\to4, 6\to2$	45.2
2.01	2.00	$7\to3, 4\to1$	45.3
2.40	2.35	$8\to4, 2\to0$	45.3
2.48	2.55	$5\to2$	45.4
2.85	2.95	$6\to2, 4\to1$	59.4
3.13	3.13	$6\to3$	45.4
3.61	3.54	$3\to1$	45.5
3.78	3.77	$5\to2, 2\to0$	59.5
4.83	4.83	$4\to2$	45.7
6.01	5.75	$3\to1$	59.8
8.07	8.25	$1\to0$	60.0

TABLE II. Experimental B^{exp} and theoretical B^{th} magnetic field positions for the peaks in the photoconductive response, along with the conduction band Landau level numbers associated with the transitions and the impact ionization energies E_{II} between the upper valence band Landau level and the impurity/defect level for the sample with $x \approx 0.24$.

B^{exp}(T)	B^{th}(T)	Transitions	E_{II} (meV)
1.12	1.18	$9 \to 1, 7 \to 0$	61.2
1.21	1.19	$10 \to 2$	61.2
1.28	1.26	$8 \to 1, 6 \to 0$	61.2
1.39	1.41	$9 \to 2$	61.2
1.51	1.52	$7 \to 1, 5 \to 0$	61.2
1.64	1.69	$10 \to 3, 8 \to 2$	61.2
1.81	1.87	$6 \to 1$	61.3
2.02	1.98	$9 \to 3, 4 \to 0$	61.3
2.27	2.24	$10 \to 4, 7 \to 2$	61.3
2.55	2.54	$8 \to 3, 5 \to 1$	61.3
2.65	2.65	$3 \to 0$	61.4
2.91	...	...	...
3.11	3.03	$6 \to 2$	61.4

However, the determination of 15 meV ± 0.5 meV for this energy seems to preclude this assignment since the LO phonon energy is approximately 17 meV.

The second level identified is at 45 meV, which is near the middle of the gap. These mid-gap levels have been seen many times before (see review of impurity levels in Ref. 3). A third level was found at 59 meV above the valence band (or at 0.62 E_g), which is closer to the conduction band. This level is the origin of the four high-field peaks in the data of Fig. 5. The lower-field peaks correspond to the 45-meV level. Some of these peaks also appear in the doubly differentiated signal given in Fig. 6. It is interesting to examine the structure of the data given in Fig. 6. The structure at low field tend grow in amplitude and "stretch" with increasing field. This behavior is also seen in the structure for the other two sets of data. In fact, the calculations of the peak positions bear this out by showing how the peaks should become progressively broader and less frequent with increasing field. There is an interference between the second and third set of peaks in the vicinity of 3–4 T, and this interference is a sign that two different sets of resonances occur.

The data given in Table II are for the sample with $x \approx 0.24$ and $E_g = 122$ meV. Only one set of resonances was observed with an impurity-level energy of 61 meV above the valence band at zero field. This energy is at the middle of the gap. All of the peaks have been accounted for except one at 2.91 T. Presumably, this peak is associated with an impurity level closer to the conduction band as is true for the third set of Table I. However, the data do not go to a sufficiently high field to obtain a value for this third level. Likewise, the data did not extend to a field low enough to see the peaks corresponding to level one of Table I. This method has found mid-gap levels in both samples which are expected to exist in these materials. This result lends support to the validity of our interpretations.

Recent magneto-optical spectroscopy work on a p-type HgCdTe sample with $x = 0.216$ and $E_g = 91$ meV found deep levels at 0.49 and 0.66 E_g above the valence band.[6] This is in good agreement with our 0.47 and 0.62 E_g determination for the $x \approx 0.22$ sample, and 0.5 E_g for the $x \approx 0.24$ sample.

Finally, we point out that the magneto-impurity effect involving donor levels can be ruled out as an explanation for the origin of the observed high-intensity structure reported in this paper. Because of the extremely light mass of the electrons, the energy separation calculated using this model between ground and excited states of a donor and the observed periodicities gives values comparable to the energy gap of the samples.

IV. CONCLUSIONS

We have investigated structure in the photoconductive response of n-type $Hg_{1-x}Cd_xTe$ samples with $x \approx 0.22$ ($E_g = 95$ meV) and 0.24 ($E_g = 122$ meV) at 2–10 K in the presence of a magnetic field and under intense laser illumination. Oscillatory behavior as a function of magnetic field was observed in the photoconductive signal. The second derivative of this signal showed even more oscillations that were too small to be observed in the primary signal. The peaks of these oscillations were shown to correspond to resonances associated with Landau-level transitions. When the energy difference between any two conduction-band Landau levels is equal to the energy needed to impact-ionize a valence electron from the highest valence-band Landau level into a trap, there is a peak in the data. Similar resonances have been seen before, associated with impact ionization of shallow impurities from ground to excited states.[1] However, this is the first time that such an effect has been seen for impact ionization into deep traps. We have seen resonances in the sample with $x \approx 0.22$ that correspond to impact ionization from the highest valence-band Landau level into the shallow acceptor at about 15 meV, and into two deep levels, one at 45 meV and the other at 59 meV above the valence-band edge at zero field. We have also seen a level at 61 meV in the sample with $x \approx 0.24$. These mid-gap levels have been seen before in similar samples and support these results.[3] This technique should be useful for the investigation of deep-level impurities in narrow-gap semiconductors which are difficult to measure by conventional methods.

---

[a] Present address: Department of Physics, Michigan State University, Lansing, Michigan.

[1] See, for example, the review of the magneto-impurity effect, L. Eaves and J. C. Portal, J. Phys. C **12**, 2809 (1979).

[2] D. G. Seiler, C. L. Littler, M. R. Loloee, and S. A. Milazzo, J. Vac. Sci. Technol. A **7**, 370 (1989).

[3] C. L. Littler, D. G. Seiler, and M. R. Loloee, J. Vac. Sci. Technol. A **8**, 1133 (1990).

[4] See the discussions presented in Refs. 2 and 3 above.

[5] M. H. Weiler, in *Semiconductors and Semimetals*, edited by R. K. Willardson and A. C. Beer (Academic, New York, 1981), Vol. 16, p. 119.

[6] Z. Kucera, P. Hlidek, P. Hoschl, V. Koubbele, V. Prosser, and M. Zvara, Phys. Status Solidi B **158**, K173 (1990).

Dislocation density variations in HgCdTe films grown by dipping liquid phase epitaxy: Effects on metal–insulator–semiconductor properties

D. Chandra, J. H. Tregilgas, and M. W. Goodwin
Central Research Laboratories, Texas Instruments Inc., Dallas, Texas 75265

(Received 13 November 1990; accepted 26 December 1990)

The dislocation density variations in HgCdTe films grown by liquid phase epitaxy (LPE) from tellurium-rich melts were measured and were found to depend on film growth temperature, substrate dislocation density, epitaxial film thickness, and postgrowth annealing in Hg vapor. Thin films, less than about 30–40 μm in thickness, exhibit dislocation densities which generally follow the dislocation densities of the substrates. In thicker epitaxial films, however, dislocation multiplication can accompany Te precipitation and subsequent low temperature annealing in Hg vapor, thereby, raising the density of dislocations by as much as a factor of 3–10. Use of a high temperature preanneal in Hg vapor can be employed to eliminate dislocation multiplication during low temperature annealing. When this preanneal is used in conjunction with very thick film growth with thicknesses greater than about 80 μm, dislocation densities in the LPE films can fall to values below those in the substrate. The dependence of metal–insulator–semiconductor storage times on dislocation densities in these films has also been measured.

I. INTRODUCTION

Since dislocations in HgCdTe limit dark current in metal–insulator–semiconductor (MIS) capacitors and lead to a reduction in the charge storage time, minimizing the dislocation density of HgCdTe materials is important for improving MIS detector performance.[1,2] Dislocations can arise during growth, but as in the case of bulk HgCdTe, they can also multiply by climb upon cooling when Te precipitation is encountered, as well as during the dissolution of Te precipitates which are heterogeneously nucleated on dislocations.[3] In contrast to bulk HgCdTe, dislocations in HgCdTe liquid phase epitaxy (LPE) typically arise from threading dislocations from the substrate and from misfit dislocations which accommodate the change in lattice parameter across a compositional interface. Dislocation densities in some Hg annealed epitaxial films have been observed to follow defect densities in the substrate materials,[4] but other films have shown dislocation densities which can be significantly higher than those of the substrate.[5] It is apparent that other factors can influence the dislocation content of LPE films and control of these factors is necessary for optimizing MIS performance.

This paper is aimed at understanding some of the factors which can alter dislocation distribution of thick HgCdTe LPE films, in addition to relating this dislocation content to the behavior of MIS capacitors. Unlike previous investigations[4,6] which have been confined to relatively thin films below about 40 μm in thickness, this work will examine the dislocation variations encountered with epifilm thicknesses between about 40–150 μm. This thickness range has not received much attention previously, but facilitates comparisons with the behavior of bulk grown HgCdTe. As will be apparent, the behavior of thick LPE films is quite different from relatively thin films.

II. EXPERIMENTAL

LPE films of $Hg_{1-x}Cd_xTe$ with $x = 0.2$ were grown by dipping from a Te-rich melt. The details concerning film growth have been previously cited.[7] The substrates were typically lattice matched (111)B CdZnTe, with ZnTe mole fraction varying between 0.04 and 0.05, and CdTe was used in a few cases. Annealed films were angle-lapped at either about 0.5° or 2° from the backside of the grown film and defect etched to reveal dislocations.

MIS devices were fabricated on these films for material evaluation. The films were initially prepared using a bromine–methanol polish on a pellon pad and were passivated with an anodic oxide grown from a KOH solution. Following oxidation, deposition of 2000 Å of ZnS and 1200 Å of Al were applied. MIS gates with dimensions of 254×381 μm^2 were patterned in the Al film. Substrate leads were formed by etching vias through the ZnS to the sample. Connections to the gate metal and substrates were made by depositing 1.5 μm of In metal. Details of MIS device operation and device physics have been previously published.[8] Following the MIS evaluation, the MIS structures were carefully stripped and the underlying HgCdTe LPE was dislocation etched. Outlines of the original MIS gate could be observed, and dislocation densities of each MIS pixel were measured.

III. RESULTS AND DISCUSSION

The as-grown dislocation microstructure for two angle-lapped LPE films grown at widely different temperatures is shown in Figs. 1(a) and 1(b). The first film grown at a low temperature of about 440 °C, in Fig. 1(a), shows a uniform dislocation density which is about equal to the density of dislocations measured in the substrate. No separate misfit band is observed, due to an almost exact lattice matching

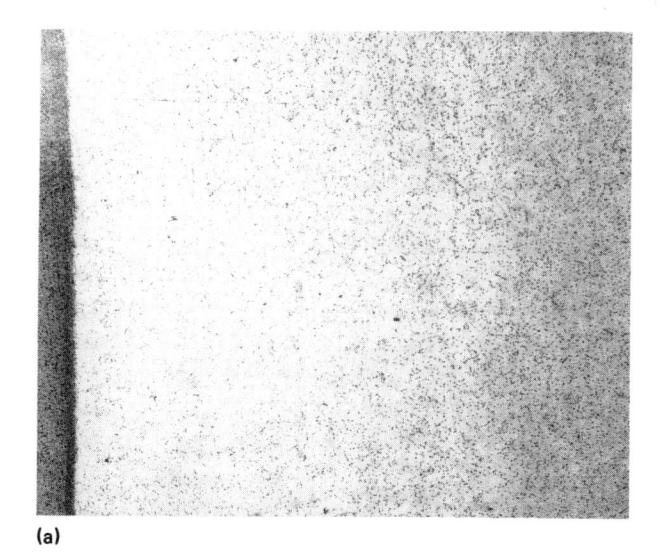

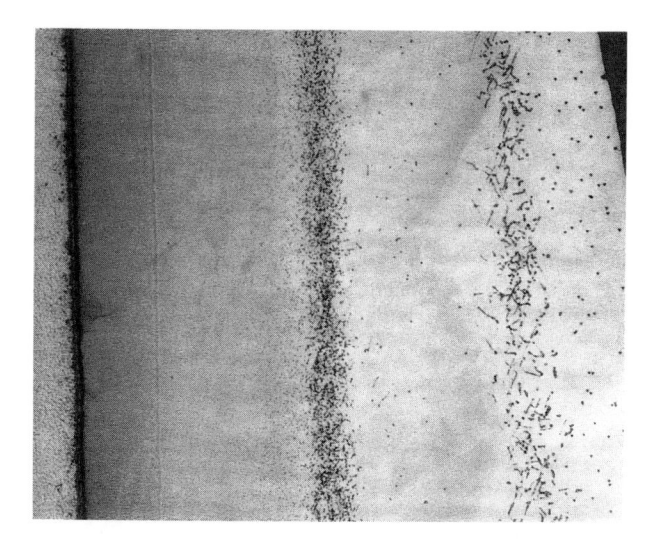

FIG. 1. As-grown HgCdTe LPE films angle-lapped 2° and defect etched. (a) LPE grown at about 440 °C showing a uniform dislocation density equal to that of the substrate (mag. 400×); (b) LPE grown at 550 °C showing substrate (far left), misfit band (center) and precipitation band (right) adjacent to LPE surface (mag. 400×).

between the substrate and the epitaxial layer. In the second layer grown at a high temperature of about 550 °C shown in Fig. 1(b), two bands of high dislocations are present. The first of these bands nearest the substrate is a misfit band, while the second band which is about 10 μm nearer to the film surface results from dislocation multiplication by climb[3,9] during Te precipitation when excess metal vacancies condense out on dislocations. The excess metal vacancy concentration for the film grown at growth temperature of 550 °C is 1.2×10^{18} cm^{-3} compared to about 5×10^{17} cm^{-3} at a 440 °C growth temperature.[10] Lower temperature LPE growth does not appear to result in dislocation multiplication during cooling. The apparent absence of dislocation multiplication near the surface of the high temperature grown film, in Fig. 1(b), is probably caused by Hg indiffusion once the sample has been removed from the Te-rich melt. Between the misfit dislocation band and the precipitation band, the absence of dislocation multiplication probably results from fewer metal vacancies in this region possibly due to precipitation on the misfit dislocations. Isolated Te precipitates within the matrix are generally not observed within about 10–15 μm of the misfit dislocation band, even in film grown from high temperatures, but matrix precipitation between dislocations, in Fig. 2, can be observed further from the misfit band of other films. The absence of matrix precipitation adjacent to the misfit band supports the view of a decreased metal vacancy concentration in this region which could occur if Te preferentially nucleated and precipitated on misfit dislocations. This view is analogous to denuding of precipitates by grain boundaries and subgrain boundaries in many other materials.

During stoichiometric adjustment annealing in Hg vapor at low temperatures in the range of about 270 °C, additional dislocation multiplication can occur in the portions of a film where Te has precipitated on dislocations. An example of the multiplication in the precipitation band from the sample in Fig. 1(b) both before and after low temperature annealing is shown in Figs. 3(a) and 3(b), respectively. In Fig. 3(a) before annealing, dislocations and Te precipitate clusters can be observed. After annealing, in Fig. 3(b), the dislocation density in the clusters has increased significantly after dissolution of the Te precipitates. The multiplication produced during annealing has been shown to result from dislocation climb and prismatic punching of dislocation loops in order to accommodate the volume expansion which occurs when Te precipitates react with metal interstitials to form a metal telluride.[3]

The dislocation profiling of LPE films which are more than about 30–40 μm in thickness typically shows regions of high and low dislocation density. An example of an intermediate thickness (62 μm) film grown at 500 °C and subsequently annealed at low temperatures in Hg vapor is shown in Fig. 4. Three distinct regions are observed: (i) a the misfit band, (ii) a low dislocation density (denuded) region, and

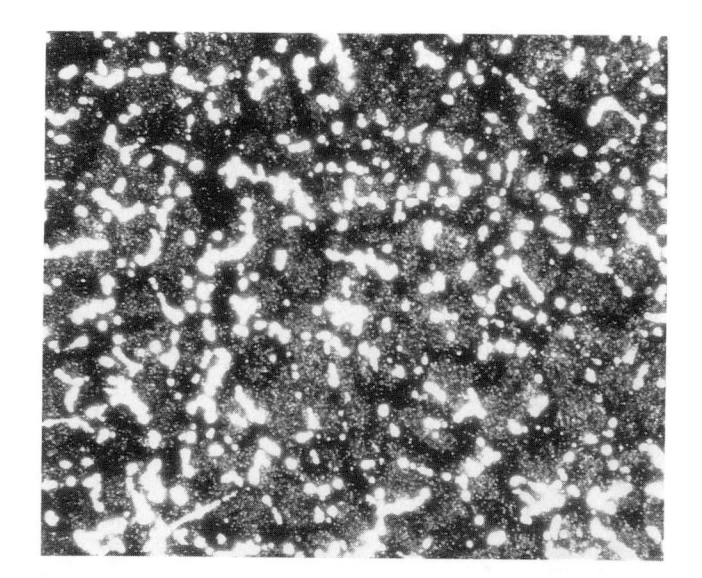

FIG. 2. Dark field micrograph of a high temperature LPE film after defect etching showing matrix Te precipitation (fine spots) and dislocations (large bright spot clusters) in a region removed from the LPE-substrate interface (mag. 400×).

(a)

(b)

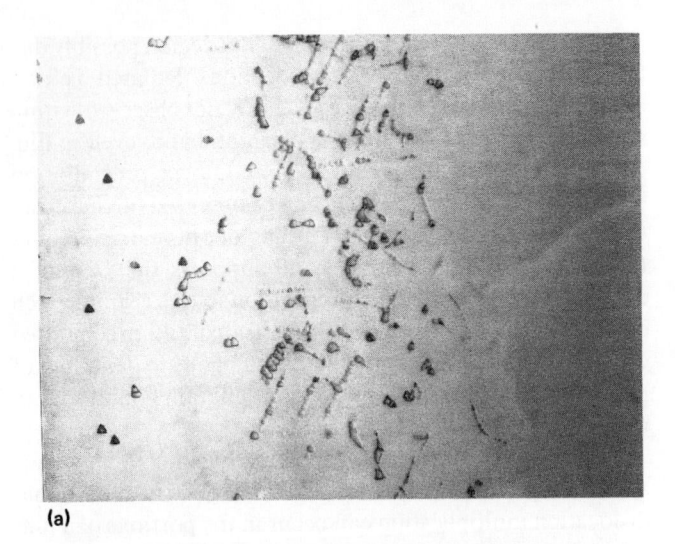

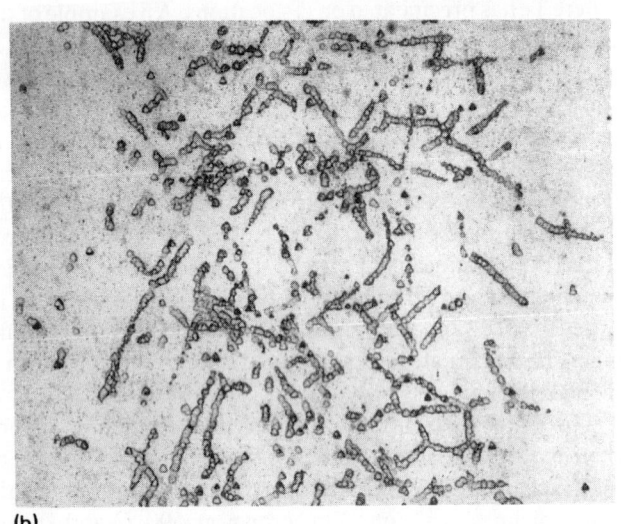

FIG. 3. Dislocation multiplication in the Te precipitation band of Fig. 1(b). (a) As-grown LPE showing shallow etch pits from Te precipitates and dislocations (mag. 400×); (b) Same region after low temperature annealing in Hg vapor showing an increased dislocation density (mag. 400×).

(iii) a dislocation multiplication region. These three regions are also evident in the dislocation profile for another epitaxial film shown in Fig. 5. Both samples show a low dislocation density region about 10–15 μm wide where the dislocations have about the same density as in the substrate. No dislocation multiplication is observed in this region, and threading

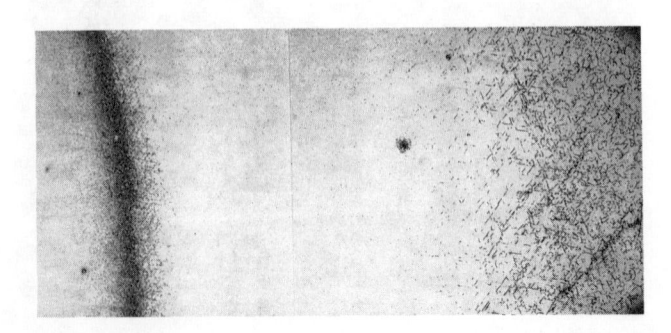

FIG. 4. Angle-lapped LPE film grown at 500 °C after low temperature annealing showing misfit band (left), low dislocation denuded region (center), and dislocation multiplication region (right).

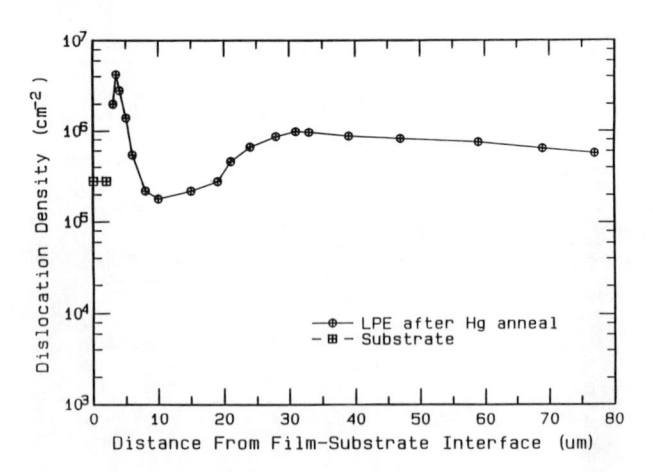

FIG. 5. Dislocation profile across an LPE film after dislocation multiplication showing a high density of misfit dislocations (far left), an adjacent low dislocation denuded region, and dislocation multiplication in the rest of the film.

dislocations from the substrate are the primary contribution to dislocation density. The absence of dislocation multiplication confirms that little or no precipitation of Te occurs on dislocations in region (ii), and that denuding of Te precipitates by the misfit band is a likely explanation. Further from the interface in region (iii), dislocation multiplication is observed out to the surface of the film. Dislocation densities in this region have been measured for both as-grown films and for portions of these films after low temperature annealing. This plot in Fig. 6 shows that dislocations in as-grown films

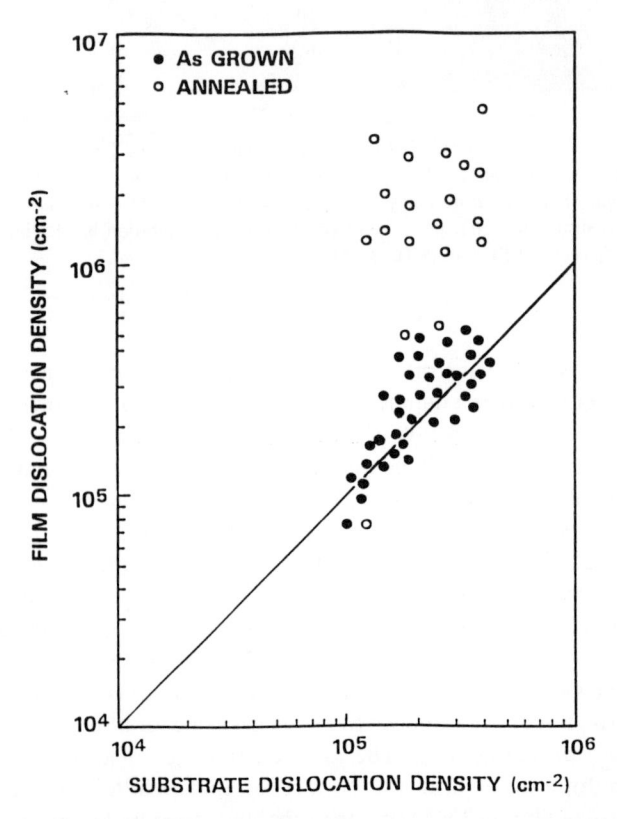

FIG. 6. Plot showing the dependence of LPE dislocation densities on substrate dislocation density for both as-grown films and after low temperature annealing. Dislocation multiplication during annealing can significantly increase dislocation density.

track the density of dislocations in the substrate, but that dislocation multiplication can increase the density of defects by a factor ranging from 3 to 10 after the low temperature anneal in Hg.

In order to circumvent dislocation multiplication upon low temperature stoichiometric adjustment annealing in Hg vapor, a high temperature preannealing process was performed on the epifilms prior to the low temperature anneal. This LPE process is essentially the same as a process[11] developed and applied to bulk HgCdTe,[9] with the exception that the preannealing temperature for epifilms was about the same as the film growth temperature. In contrast to the dislocation multiplication encountered in Fig. 6, use of the high temperature dislocation multiplication reduction (DMR) preanneal can suppress dislocation multiplication during low temperature annealing as the open triangles plotted in Fig. 7 show. Use of this preanneal on films grown at temperatures of less than about 500 °C allows uniform dislocation densities to be obtained in thicker LPE films which are characteristic of both the as-grown film and the dislocation density of the substrate. Consistent with earlier observations,[4] these dislocation densities in the epifilms are independent of the degree of lattice mismatch with the substrate, regardless of whether the films are grown on CdTe or lattice matched CdZnTe substrates.

In contrast to the growth of thin or intermediate thickness LPE films, very thick films with thicknesses greater than 100 μm have also been examined in conjunction with the DMR preanneal. The variation in dislocation density for one of these very thick samples is shown in Fig. 8. A gradual decrease in the density of dislocations is observed up to about 120 μm. Beyond about 120 μm from the interface, the dislocation density falls to a value below that of the substrate at a slightly faster rate, and then continues to decline at a lower rate toward the film surface where a dislocation density of about half that of the substrate is observed. Other LPE films have shown similar results with knees in the dislocation profiles at about 75 μm and with surface dislocation densities of about one third of those in the substrate. The decline in dislocation density with film thickness to densities below those in the substrate suggests that threading dislocations are being annihilated during growth. A binary dislocation recombination mechanism[12] has been proposed to explain dislocation gradients in III–V materials, and this mechanism may also be active in HgCdTe thin films during growth. The data for thick HgCdTe films needs to be examined in light of both this recombination model and possible compositional variations of the LPE films with depth.

First, the binary recombination mechanism[12] has been suggested to explain possible annihilation of threading dislocations. This model employs a "self-interaction" constant

$$\frac{dD}{dx} = -\lambda D^2,$$

where D is the density of threading dislocations, x is distance, and λ is a constant with units of length. Integration of this equation leads to

$$D_x = \frac{D_0}{1 + D_0 \lambda x}$$

which can be rewritten as

$$\frac{D_x D_0}{D_0 - D_x} = \frac{1}{\lambda x}.$$

The left-hand side of this equation can now be defined as a normalized dislocation density function. This expression can now permit an easy assessment of the applicability of the binary recombination law. Plotting the normalized dislocation density function against the film thickness logarithmically should yield a -1 slope if the law is obeyed. Replotted data from Fig. 8 is shown in Fig. 9. The slope of the least squares fit to the data is actually -1.53. Therefore, this simple binary recombination model alone cannot explain the dislocation density variation observed in very thick HgCdTe LPE films. It is important to emphasize, however, that a

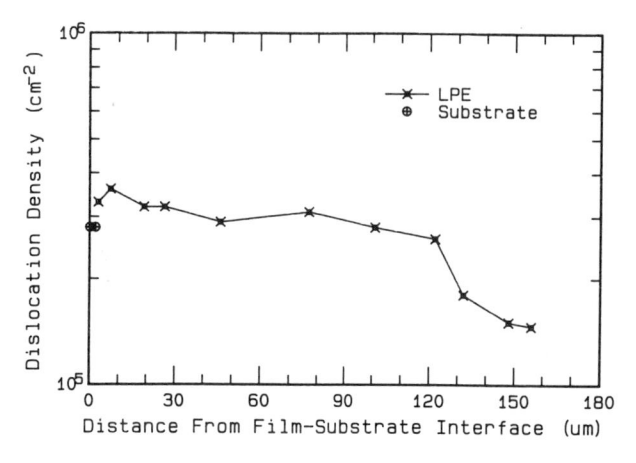

FIG. 8. Dislocation profile through a 150 μm LPE film after both DMR preannealing and low temperature anneal in Hg vapor showing dislocation densities lower than those of the substrate.

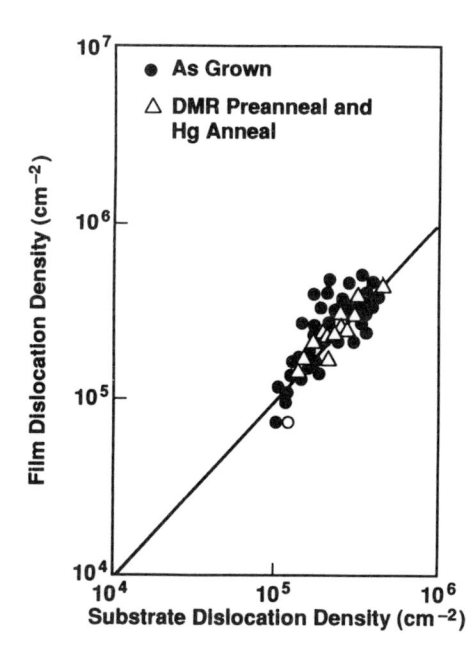

FIG. 7. Plot showing the dependence of LPE dislocation density for as-grown films and films given a high temperature preanneal in Hg followed by low temperature Hg anneal. DMR preannealing prevents dislocation multiplication.

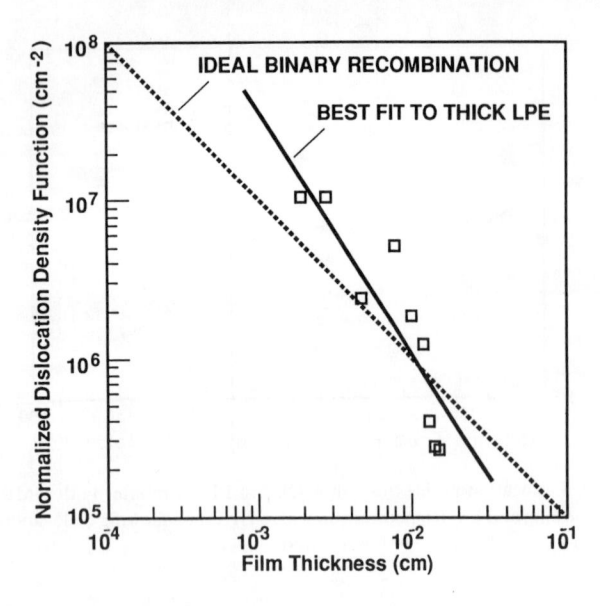

FIG. 9. Replotted data from Fig. 8 showing the dependence of the dislocation density function for ideal binary dislocation recombination.

mechanism for dislocation reduction is occurring, and that recombination still offers a qualitative explanation for our thick film dislocation gradient.

A second factor which may affect the dislocation density through a thick film is the compositional variation with thickness. Both the dislocation density and compositional gradient have been measured as a function of depth for a 150 μm LPE film as Figs. 10(a) and 10(b) show. In this film the initial high dislocation density band extends to about a thickness of 65 μm, but the gradient falls to a value below $3\times10^{-4} \mu m^{-1}$ after 7 μm, continuing to slowly decrease to less than $2\times10^{-4} \mu m^{-1}$ after 25 μm, and reaching a magnitude of about $7\times10^{-5} \mu m^{-1}$ after 45 μm. Szilagyi et al.[13] modeled the misfit dislocations density as a function of compositional gradient and showed good agreement with data for graded HgCdTe heterojunctions. Their work indicates a compositional gradient of about $8\times10^{-4} \mu m^{-1}$ is required to sustain a dislocation density of about 4×10^5 cm^{-2}, which is a factor of 3 or 4 higher than the observed compositional gradient in Fig. 10(b). Compositional variations, therefore, do not appear to be establishing the dislocation density in Fig. 10(a).

One possible factor which may influence dislocation density in LPE samples is the film growth rate. The increase in dislocation density observed in Fig. 10(a) adjacent to the episubstrate interface may be associated with a higher initial growth rate resulting from a higher degree of melt supersaturation compared to the previous film in Fig. 8. These growth rates, however, do not result in melt inclusions which are observed if constitutional supercooling is present. Further work to establish a dependence of dislocation density on growth rate needs to be performed.

Lastly, the MIS storage time has been measured as a function of the local dislocation density for a number of different LPE films with similar 77 K cutoffs in the 10.4–10.6 μm range. Figure 11 shows that variations in dislocation density between 4×10^4 cm^{-2} and 3×10^5 cm^{-2} have a pronounced impact on MIS storage time. This data is compared to the dependence measured on bulk recrystallized HgCdTe with a 77 K cutoff of about 10 μm which has been previously published.[14] The slightly lower measured storage times and the difference in slope for LPE films compared to bulk are attributed to the slightly higher cutoffs for the LPE.

IV. CONCLUSIONS

Dislocation densities of HgCdTe LPE films grown from Te-rich melts can be influenced by growth temperature, substrate dislocation density, film thickness, and postgrowth annealing in Hg vapor. Dislocation multiplication can result from the precipitation of Te which is promoted by high temperature growth, as well as by dislocation climb during the dissolution of Te precipitates which heterogeneously nucleate on dislocations. Denuding of excess metal vacancies by misfit dislocations over a depth of about 10–15 μm can limit Te precipitation in this region and explains why dislocation

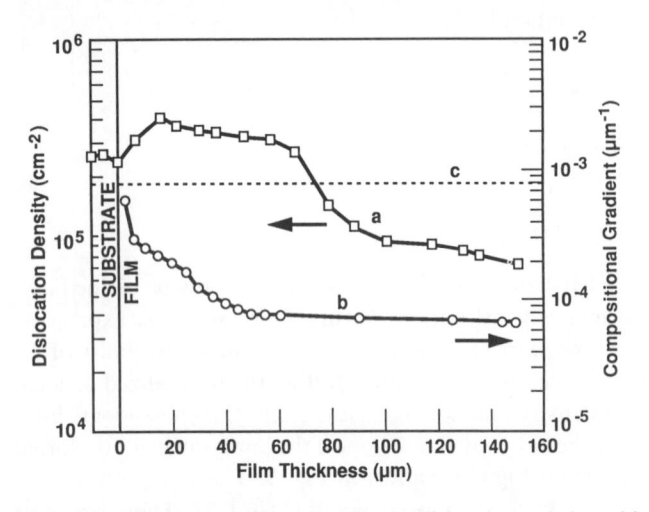

FIG. 10. Profiles of an LPE film showing the dislocation variation with depth in (a), the compositional gradient measured by microprobe in (b), and (c) compositional gradient required to sustain a dislocation density of about 4×10^5 cm^{-2}.

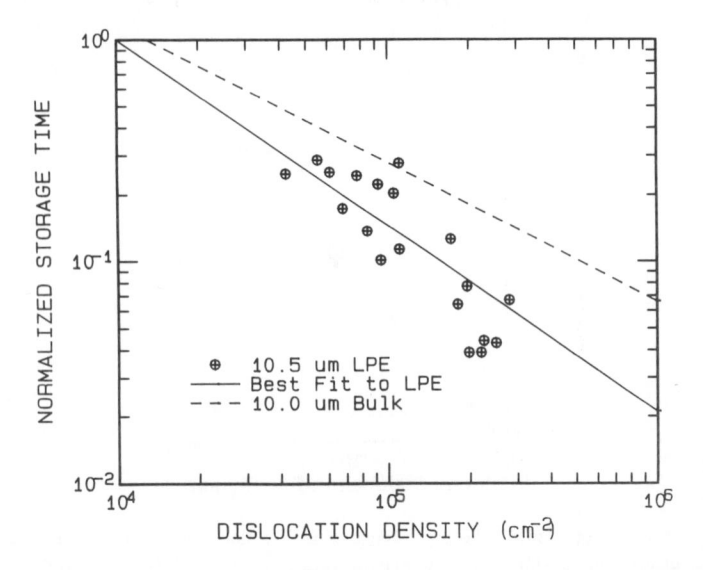

FIG. 11. Dependence of MIS storage time on dislocation density for LPE films compared to that in bulk HgCdTe with a slightly lower cutoff.

multiplication is not usually observed in very thin films. Both lower temperature growth and DMR preannealing can be employed to eliminate dislocation multiplication in epifilms which are greater than about 30–40 μm in thickness. Dislocation densities which are about equal to those in the substrate can be achieved by eliminating dislocation multiplication. Growth of very thick films in the range of about 80 μm can be employed with DMR preannealing to achieve dislocation densities which are below those in the substrate. Reducing dislocation density of an LPE film can significantly increase MIS storage time.

[1] A. J. Syllaios and L. Colombo, *IEDM Technical Digest* (IEEE, New York, 1982), p. 137.
[2] J. H. Tregilgas, T. L. Polgreen, M. C. Chen, J. Cryst. Growth **86**, 460 (1988).
[3] H. F. Schaake and J. H. Tregilgas, J. Electron. Mater. **12**, 931 (1983).
[4] H. Takigawa, M. Yoshikawa, and T. Maekawa, J. Cryst. Growth **86**, 446 (1988).
[5] D. Chandra and M. W. Goodwin, in *The 1990 Proceedings of IRIS Materials Specialty* (to be published).
[6] B. Pelliciari and G. Baret, J. Appl. Phys. **62**, 3986 (1987).
[7] D. Chandra and M. W. Goodwin, *Proc. IRIS Materials*, (ERIM, Ann Arbor, 1989), p. 215.
[8] M. A. Kinch, in *Semiconductors and Semimetals*, edited by R. K. Willardson and A. C. Beer (Academic, New York, 1981), Vol. 18, p. 313.
[9] J. H. Tregilgas and H. F. Schaake, 1989 Electronic Material Conference, Cambridge, Mass., June 21–23, 1989 (unpublished).
[10] H. F. Schaake, J. Electron. Mater. **14**, 513 (1985).
[11] H. F. Schaake and J. H. Tregilgas, U.S. Patent No. 4,481,044 (1984).
[12] H. Kroemer, T.-Y. Liu, P. M. Petroff, J. Cryst. Growth **95**, 96 (1989)
[13] A. Szilagyi and M. N. Grimbergen, J. Vac. Sci. Technol. A **4**, 2200 (1986)
[14] L. Colombo and A. J. Syllaios, *Proc. IRIS Materials* (ERIM, Ann Arbor, 1982), p. 264.

# Surface energies for molecular beam epitaxy growth of HgTe and CdTe

M. A. Berding, Srinivasan Krishnamurthy, and A. Sher
SRI International, Menlo Park, California 94025

(Received 5 November 1990; accepted 18 December 1990)

We present results for the surface binding energies for HgTe and CdTe that will serve as input for molecular beam epitaxy growth models. We have found that the surface binding energies are surface orientation dependent and are not simply proportional to the number of first-neighbor bonds being made to the underlying layer. Moreover, because of the possibility of charge transfer between cation and anion surface states, one may have large differences between the binding energy for the first and the last atom in a given layer, and these differences will be different for the narrow-gap, less ionic materials than for the wide-gap, ionic materials. We also find that the surface states associated with an isolated surface atom or vacancy are extended in materials with small gaps and small effective masses, and thus call into question the modeling of surface binding by simple pair interactions.

While considerable advances have been made in recent years in molecular beam epitaxy (MBE) growth of HgCdTe, modeling of the growth process is still primitive because of the complexity of the process. One important input common to all growth models is the binding energy of atoms to the particular growth surfaces. These energies are usually approximated as being equal to the number of first-neighbor bonds made to the surface times the energy per bond in the bulk. We have shown[1] that this is in fact a very poor approximation, and that the surface binding energies often differ considerably from those estimates, and are sensitive to the local surface stoichiometry. In this article, several implications of these results on the nature of the growth surface of HgTe and CdTe and their alloys are discussed.

We have calculated[1] the surface binding energies in two limits. The first is the *concentrated* limit in which the atom is added to complete the surface layer, resulting in an ideal surface. The second is the *dilute* limit in which an isolated atom is added to an underlying ideal surface; this limit corresponds to the initiation of a new layer. The surface binding energies were calculated using a tight-binding Green's function technique that has been presented elsewhere.[2,3] Surface binding energies of the constituents in these two limits for HgTe and CdTe were calculated for the (111)B and (100) surfaces. Results are summarized in Table I. Also shown for reference are the experimental bulk energies per bond.

The results in Table I demonstrate that surface binding energies are not proportional to the number of bonds made to the surface. Because no in-plane bonds are made on the (111) and (100) growth surfaces, this is evident by the fact that the binding energies in the dilute (E_d) and concentrated (E_c) limits are not in general equal, as they would be if the simple proportionality relationship held. Even when the dilute and concentrated energies are equal, as in the case of tellurium on the HgTe (111)B surface, these energies are not equal to three times the bulk energy per bond. Although the surface binding energies are not linearly proportional to the number of bonds being made to the surface, we do find that averaging over the dilute and concentrated cation and anion binding energies for a particular surface yields the cohesive energy per bulk layer, as it should.

Differences in the E_d and E_c can be attributed, in part, to charge transfer effects on the surface.[1,2] Briefly, because the anion surface states lie at lower energies than the cation surface states, surface cations will always transfer charge to surface anions, until all the surface anion states are fully occupied. Thus, for example, for cadmium on the (100) surface, differences in the binding energies result because in the dilute limit the cation added to the surface can transfer charge to the surface anions, thereby lowering the Cd surface binding energy, while in the concentrated limit, no surface anion states are available into which to transfer charge. In contrast, in the concentrated limit electrons must be promoted out of the anion surface states (which had been transferred there from other cations already present on the surface) when the cation completing the layer is added. Both of these result in a stronger binding of the cation in the dilute limit, as seen in Table I. This charge transfer effect will be largest in ionic semiconductors with large band gap because of the large difference in anion and cation surface state energies.

From Table I we note that on both the (111)B and (100) surfaces mercury is more weakly bound in the dilute limit than in the concentrated limit. As a consequence, the initiation of a new layer of mercury atoms will be more difficult than the completion of a mercury layer, and thus initiation of islands in layer-by-layer growth will be disfavored with respect to the completion of islands. The Te is strongly bound to both the (111)B and (100) surfaces, indicating that the tellurium-stabilized growth will be preferred.

For CdTe, we see that the nucleation of a new layer will not be as troublesome as in HgTe, as indicated by the large values for E_d in Table I, while the completion of the surface will be more difficult. This has an important consequence on the nature of the growing surface. Because $E_d < E_c$ for Cd in CdTe, the cations on the surface prefer to nucleate new layers, as opposed to completing layers. This can be interpreted as an effective repulsive surface interaction among the Cd atoms on the surface. This may have important consequences on the growth of this and other wide-gap ionic materials such as ZnTe. Because the surface atoms prefer not to sit adjacent to one another, there is the possibility of the atoms on the surface forming a superlattice with the surface

TABLE I. Surface binding energies for HgTe and CdTe (eV).

		(111)B Dilute	(111)B Concentrated	(100) Dilute	(100) Concentrated	Bulk
HgTe	Hg	~0.0	−0.3	~0.0	−1.0	−0.82
	Te	−2.8	−2.8	−2.4	−3.1	
CdTe	Cd	−2.2	−1.3	−2.1	−0.1	−1.10
	Te	−5.2	−2.1	−5.0	−2.6	

vacancies. This phenomenon has been predicted and observed on GaAs, where the same mechanism is responsible.[4,5] The nature of this surface superlattice will depend on the particular surface, the magnitude of the energy differences between E_d and E_c, and the temperature of the growing surface.

Because the surface binding energies are dependent on the surface stoichiometry in so far as it controls the states available for charge transfer, we expect the contributions to surface binding energies to change relatively abruptly at some specific surface coverage. As an example we consider the addition of Cd to the (100) surface of CdTe. As discussed above, when an isolated Cd atom is added to the surface, the binding energy is lowered because of the charge transfer to adjacent Te atoms, while when Cd is added a to nearly complete Cd surface, the binding energy does not benefit from such a charge transfer because there are no empty Te surface states. At 50% surface coverage, the number of Cd and Te surface states will be equal, with the lower-energy Te states being completely filled and the higher-lying Cd states being empty. Now we consider an arbitrary intermediate surface coverage to determine if the Cd surface binding energy will benefit from a charge transfer to adjacent Te. If the surface coverage is below 50%, empty Te states will still be available for charge transfer, while above 50% coverage, all Te states will be already full. Thus, near 50% we expect the value of the surface binding energies to change rather abruptly, with binding energies below 50% coverage being closer to E_d and binding energies for coverages above 50% being closer to E_c. The situation is similar for other polar surfaces. For example, for addition of Te to the (111)B surface the changeover between dilute-like and concentrated-like binding energies occurs at 75% coverage. This is because the density of surface states per cation on the surface is three times the density of surface states per anion. Thus, when the surface is 75% anions, the density of cation and anion surface states will be equal, and full charge transfer will occur. This is similar to the results found by Chadi[4] and Tong et al.[5] for GaAs.

Because the charge transfer depends on the band gap of the material, we expect the behavior in HgCdTe and HgZnTe with band gaps in the infrared to have effective attractive surface interactions. This is consistent with preliminary calculations using the supercells/slab method.[6] Thus, at sufficiently low temperatures, although higher than temperatures for which surface diffusion is too slow for surface equilibration, the surfaces will grow via the formation of smooth islands of like atoms. As the surface temperature is raised, entropy will dominate and the surface will be rough, with smooth islands being replaced by disordered, randomly arranged atoms.

The lateral extent of the surface states produced by an isolated surface atom or vacancy is found to differ for HgTe and CdTe. This has been demonstrated using the same tight-binding Hamiltonian in a slab calculation[6] and in a Green's function calculation. In the slab calculation, surfaces are modeled by constructing multiple layers of the semiconductor and vacuum, where the number of vacuum layers is large enough to completely decouple the two surfaces and the number of semiconductor layers is chosen large enough so that the center layer looks bulk like. In the Green's function method, a truly semiinfinite surface can be created. Although the slab method does not mimic a real surface as well as the Green's function calculation, it permits the calculation of the formation energy for a periodic array of defects. To examine the in-plane coupling of the vacancies, a regular array of Hg atoms was removed from the (111)A HgTe surface, where each surface vacancy created was completely surrounded by atoms. This removal energy per atom is expected to be equal to $-E_c$, if the wave functions are localized at the removal site. However, we find that the energy required (per atom from a unit cell of four atoms) to remove a regular array of Hg atoms from the (111) HgTe A surface differs substantially from $-E_c$. When the size of the supercell is increased to include nine atoms, the calculated energy (per atom) is larger than the previous one but still smaller than E_c. This is because the created surface vacancies are far from each other and consequently couple less when compared to the previous case. However, when these calculations were repeated for removal of cations from CdTe and ZnTe surfaces, the removal energies calculated with nine atoms per unit supercell agreed exactly with E_c. It suggests that the surface wave function is well extended on HgTe surfaces, but terminates near the third neighbor on CdTe and ZnTe surfaces. Whenever surface wave functions have a large spatial extent, coupling substantially to one another, the removal energies per atom will be affected.

The difference in spatial extent on HgTe and CdTe surfaces can be understood from the complex band structure of these materials. The complex band structure is the relationship between real energy and complex states[7] and connects the real bands through the forbidden gap in the complex wave vector plane. In the region of fundamental gap, the complex bands start from valence and conduction band edges and meet at a branch point in the complex plane. The extent of the wave function decreases exponentially with the magnitude of the imaginary part k_i of this branch point. When the hole mass is much larger than electron mass, it can be found that k_i is directly proportional to conduction band effective mass and band gap. Consequently, for HgTe in which both band gap and electron mass are very small, the surface wave functions are well extended. In the case of CdTe and ZnTe, with large gap and large effective mass, the calculated k_i is large and wave function is less extended.

We conclude by noting several features of the surface binding energies that must be incorporated into any MBE growth model. First, surface binding energies are surface orientation dependent and are not simply proportional to the number of first-neighbor bonds being made to the underlying layer. Moreover because of the possibility of charge transfer between cation and anion surface states, one may have $E_d < E_c$ or $E_d > E_c$, where the former will occur mostly for wide-gap ionic compounds and the latter for narrow- (or zero) gap, less ionic compounds. Finally, because the surface states associated with an isolated surface atom or vacancy are extended in materials with small gaps and small effective masses, the approximation of surface binding using simple pair interactions is highly suspect. We expect that the narrow-gap semiconductors HgCdTe and HgZnTe will behave more like HgTe than CdTe

Acknowledgments: The financial support provided for this work through ONR Contract No. N00014-88-C0096 and NASA Contract No. NAS1-18226 is gratefully acknowledged.

[1] S. Krishnamurthy, M. A. Berding, A. Sher, and A.-B. Chen, Phys. Rev. Lett. **64**, 2531, (1990).
[2] S. Krishnamurthy, M. A. Berding, and A. Sher, J. Appl. Phys. (in press).
[3] A.-B. Chen, Y.-M. Lai-Hsu, and W. Chen, Phys. Rev. B **39**, 923 (1989).
[4] D. J. Chadi, Phys. Rev. Lett. **52**, 1911 (1984).
[5] S. Y. Tong, G. Xu, and W. N. Mei, Phys. Rev. Lett. **52**, 1693 (1984).
[6] M. A. Berding, S. Krishnamurthy, and M. van Schilfgaarde (unpublished).
[7] C. Herring, Phys. Rev. **52**, 365 (1937); S. Krishnamurthy and J. A. Moriarty, Phys. Rev. B **32**, 1027 (1985).

Structural characterization of the (111) surfaces of CdZnTe and HgCdTe epilayers by x-ray photoelectron diffraction

M. Seelmann-Eggebert[a]
Department of Electrical Engineering, Stanford University, Stanford, California 94305

H. J. Richter
Fraunhofer-Institut für Angewandte Festkörperphysik, W-7800 Freiburg, Federal Republic of Germany

(Received 3 October 1990; accepted 17 December 1990)

HgCdTe epilayers and lattice matched CdZnTe (111) substrates have been investigated by x-ray photoelectron diffraction to characterize their surface structure in regard to termination, reconstruction and relaxation as well as in regard to degradations. Well-ordered CdZnTe surfaces of stoichiometric composition were realized by Ar ion sputtering, while sputter cleaning of the HgCdTe epilayers led to a significant increase of the CdTe mole fraction in the surface region. However, the crystallinity of the HgCdTe surfaces is found to be maintained in spite of their compositional degradation. Ordered and nondegraded HgCdTe(111) surfaces are obtained by electrochemical etching. The examined (111)A and (111)B surfaces can be easily differentiated since their photoelectron diffraction patterns indicates a surprisingly perfect termination. Evidence for any reconstruction or relaxation of the investigated (111) surfaces of CdZnTe and HgCdTe is not found.

I. INTRODUCTION

Photovoltaic infrared sensors are commonly based on HgCdTe material, grown as epilayers on CdTe- or on lattice-matched CdZnTe substrates. Substrates with (111) orientation are the preferential choice for the epitaxy owing to the problem that twins are usually present perpendicular to the (111) crystal axis. Epitaxial growth of HgCdTe on the Te terminated B face of the substrate results in layers of higher crystalline quality. Hence, structural properties of the substrate surface such as termination, relaxation, reconstruction or various imperfections, are very important for the epitaxy. The crystalline quality of the surface region of the epitaxial HgCdTe layers is also of technological importance, since it influences the electrical performance of the infrared detector devices. The purpose of this paper is to gain a better insight into the structural properties of these (111) surfaces of HgCdTe and CdZnTe, since they have not been systematically investigated as yet in spite of their technological importance.

For the study of the structure of the (111)A and B surfaces of CdZnTe and HgCdTe x-ray photoelectron diffraction (XPD) is used as the analytical tool and we first demonstrate that XPD has the potential for assessing the atomic structure of such surfaces. In the case of compound semiconductors XPD offers the particular advantage that the atomic species forming the compound and its surface structure can be separately analyzed via their characteristic photoemission. On the basis of simple geometric considerations we were able to interpret the arrangement of atoms at the surface of a crystal in a rather straightforward manner.

In a previous study we have published preliminary data on XPD obtained for the (111)B surfaces of CdZnTe and HgCdTe.[1] As exemplified by selected angular (polar and azimuthal) scans, those first measurements already revealed characteristic anisotropies in the angle-resolved photoemission originating from the Te $3d_{5/2}$, Cd $3d_{5/2}$, and Hg $4f_{7/2}$ core levels. Lately, similar results were also reported by other authors.[2] In this continued and more detailed study, we now present complete sets of XPD data taken for all directions of photoelectron escape with improved angular resolution, and we examine the (111)B as well as the (111)A surface of CdZnTe. Moreover, we study HgCdTe epilayers grown by liquid phase epitaxy on these CdZnTe substrates in order to characterize and correlate the surface structure of epilayer and substrate.

II. EXPERIMENTAL

The intensity of the core-level photoemission was measured with a concentric hemispheric analyzer of an ESCA-LAB MK II system (VG) using Al $K\alpha$ of an X-ray twin anode for the excitation of the photoemission. An adjustment of any pair of polar and azimuthal electron escape angles was achieved by rotating the sample and aligning the respective electron trajectory with the axis of the input lens of the analyzer. The polar angle of the photoelectron escape is defined to be zero for photoemission normal to the sample surface, i.e., for photoemission into the $\langle 111 \rangle$ direction, whereas the the azimuthal escape angle is referred to the $\langle 11-2 \rangle$ direction.

To obtain a complete set of XPD data (diffraction pattern) azimuthal scans were sequentially taken for polar angles varying from 88° (grazing emission) to 0° (normal emission). Aperture limited resolution was obtained with azimuthal and polar incremental steps of 2.5° and 2.2°, respectively, and mirror symmetries at $\varphi = 0°$, $\varphi = 60°$, and $\varphi = 120°$ were confirmed by rapid survey scans for all investigated samples. Data were taken for the Cd $3d_{5/2}$, Te $3d_{5/2}$, Hg $4f_{7/2}$, and Zn $2p_{3/2}$ photoelectrons, respectively. The zinc signals were weak and too noisy for detailed evaluation; however, they did not indicate any change of the stoichiometric composition near the CdZnTe surfaces.

Since minor details of the energy dispersion of the photoelectrons were of no importance in the context of this study the energy analyzer was operated with a relatively large pass energy corresponding to a resolution of 1.25 eV. The photoelectron counts integrated over this energy interval were taken for the known characteristic photoelectron energies (including charge-up) which yielded the peak count rate. To obtain reference data for (linear) background subtraction also the count rates at two further energies separated from the energy of the peak by ± 5 eV were recorded. The respective background data measured at 5 eV above the kinetic energies of the Cd $3d$, the Te $3d$ and the Hg $4f$ photoelectrons were not modulated as the XPD data but decreased proportional to the cosine of the polar escape angle (after correcting for the angle-dependent acceptance of the energy analyzer). The presented data are normalized with respect to this background level.

The CdZnTe single crystal samples were wafers cut from Bridgman-grown crystals oriented in the $\langle 111 \rangle$ direction using an x-ray diffractometer. Since the CdZnTe wavers were prepared as substrates for the epitaxy of HgCdTe (CdTe mole fraction $x = 0.2$) layers they contained about 4 mol % ZnTe to provide optimum lattice match. After the (111)B face of the CdZnTe substrates had been identified by means of a standard etching test,[3] the HgCdTe epilayers were grown on this surface by liquid phase epitaxy (vertical dipping technique). Mirror-like reflecting surfaces were realized by mechanical and chemical polishing (Br_2 in methanol) of the samples. In addition, the HgCdTe samples were electrochemically etched since only this technique proved to be appropriate for obtaining high-quality surfaces of stoichiometric composition. After transfer into the analysis chamber an additional ion bombardment (1 keV Ar^+ ions) was applied to some of the samples removing 5 to 10 monolayers. After ion bombardment the CdZnTe samples were annealed at temperatures up to 500 °C for about 1 h. However, the annealing procedure was later found to be of no consequence for the recorded XPD pattern. Before and after the XPD data were taken the sputtered surface was analyzed by XPS survey scans and proved to be free of oxygen and carbonaceous contaminants. During the analysis a base pressure of lower than 5×10^{-8} Pa was maintained.

III. THE EFFECT OF CRYSTALLINE ORDER ON ANGLE RESOLVED XPS

For a single-crystal surface the diffraction of core-level photoelectrons gives rise to a characteristic anisotropy in the angular distribution of the photoemission.[4-6] This anisotropy originates from coherent elastic scattering of the photoelectron wave at lattice atoms and from the interference between the unscattered wave and the waves thus scattered. Since the scattering cross section has a maximum[7] at zero scattering angle and since the phase shift upon scattering is small, constructive interference occurs giving rise to intensity maxima along internuclear axes (zero-order interference). Higher order interference may also occur at larger scattering angles but it tends to be suppressed due to the fact that the scattering cross section is strongly peaked in the forward direction. This "forward direction focusing"[4-6,8] along internuclear axes can also be understood as a focusing effect that the screened core potential of a scattering atoms exerts on the divergent electronbeam emitted from an atomic point source.[9] Since for each scatterer the directional changes the escaping electrons might undergo are restricted to the narrow angle cone for which the cross section is nonzero, the XPD structures are averaged out by integrating over a sufficiently large solid angle. We make use of this to separate the variations in the photoemission resulting from a compositional depth profile from those angular variations caused by the crystalline structure. For this purpose we average over the azimuthal angle and over a polar angle interval which decreases with increasing polar angle to maintain a constant depth resolution[10] ($\Delta 1/\cos \vartheta \approx 0.2$). This polar angle dependent average is then analyzed with respect to a possible compositional depth profile utilizing the linear relation between the escape depth and the reciprocal cosine of the polar escape angle.[11] Changes in the angular distribution of photoemission induced by differences in the atomic surface structure can be studied even in the presence of a compositional depth gradient by referring the intensity of the XPD structure to this angle-averaged background (relative intensities). An atom at the surface, if imaged by focusing a wave originating from a deeper lying emitter in a distance R, will lead to a relative intensity increase[10] $\chi_0 \exp(-R/\lambda)[1 - \exp - d_M/\lambda \cos \vartheta] \approx \chi_0 d_M/\lambda \cos \vartheta$, where the ratio $d_M/\lambda \cos \vartheta$ of the monolayer spacing and the escape depth is ≈ 0.2 for normal electron escape ($\vartheta = 0°$) in the considered case. The forward direction enhancement factor[6] χ_0 is in the order of one for Cd,Hg or Te scatterers[10] and about a factor of 2 smaller for lower Z elements like oxygen and carbon.[7]

IV. COMPARISON OF THE (111)A AND (111)B SURFACES OF CdZnTe

Figure 1 shows the angular distribution of the Cd $3d$ electrons (upper part) and the Te $3d$ electrons (lower part) as four hemispherical x-ray photoelectron diffraction patterns obtained for the two complementary (111) CdZnTe surfaces. The experimental results obtained for the Cd terminated A face are shown on the left hand side, whereas the right part of Fig. 1 refers to the Te terminated B face which is normally used as substrate for the HgCdTe epitaxy.

To cope with the sizeable amount of information contained in this set of XPD data, a hemispherical display format was chosen. For this purpose the recorded photoemission intensity is referred to a false color scale (in arbitrary units). The intensity at a pair of escape angles ϑ,φ can thus be displayed as a color on the unit sphere. A parallel projection of this unit sphere onto a plane parallel to the sample surface is convenient for a display of the data in two dimensions. The data were artificially extended to the entire hemisphere of angles utilizing the respective mirror symmetries.

This display format is convenient to separate the maxima caused by forward direction focusing (zero order interference) from those caused by higher order interference, since the latter appear as concentric fringes about the emitter-scatterer axes.[10] For an illustration of the phenomenon of forward direction focusing we assume that the emitter is lo-

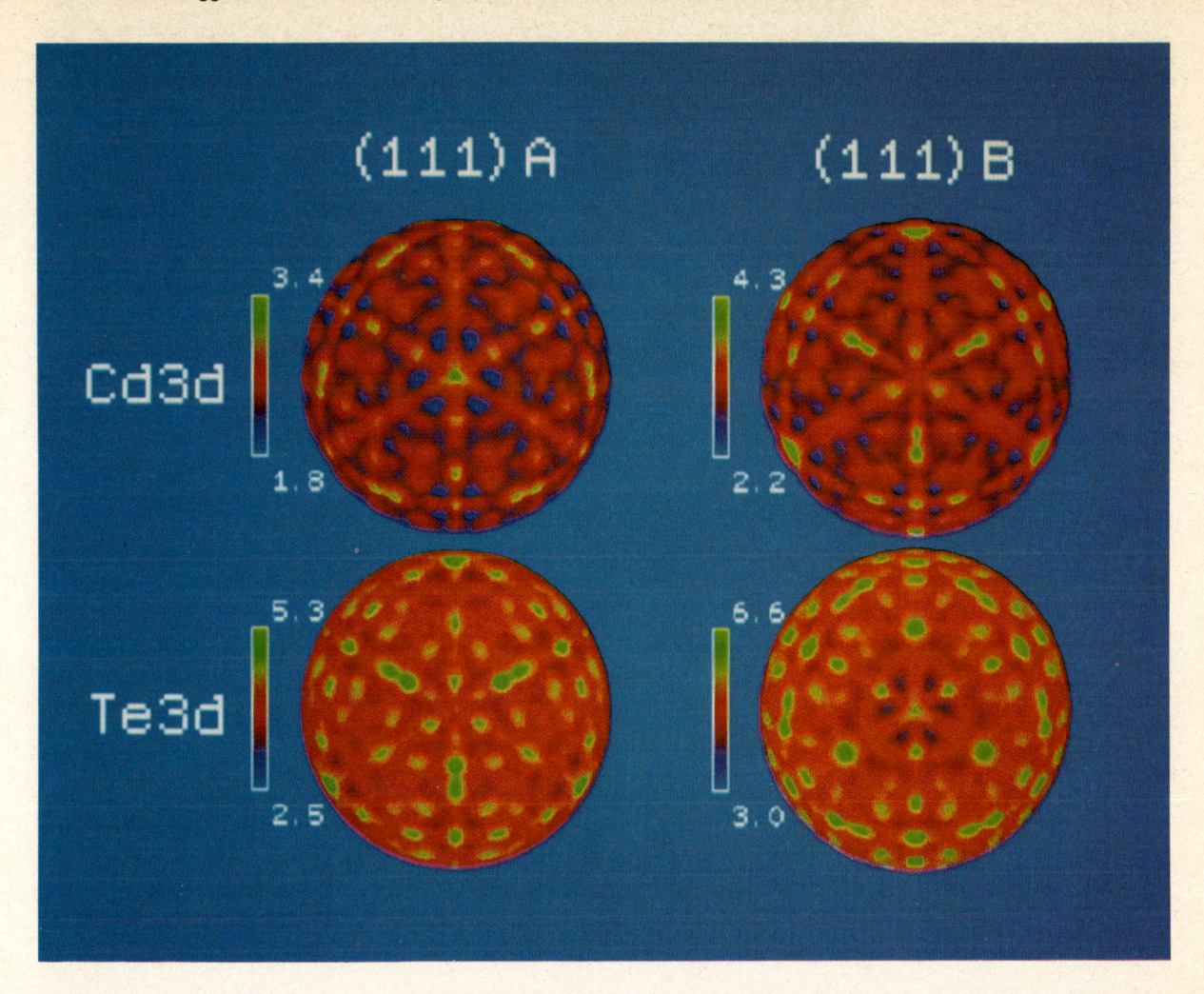

FIG. 1. X-ray photoelectron diffraction patterns of the (111) CdZnTe surfaces showing the angular resolved photoemission intensity as displayed on a sphere by false colors in a parallel projection onto a plane perpendicular to the surface normal for the Cd $3d_{5/2}$ electrons (upper part) and for the Te $3d_{5/2}$ electrons (lower part). The respective false color scales are indicated by the inserted boxes. The results for face A and B are shown on the left side and on the right side of the figure, respectively.

cated in the center of a ficticious sphere which surrounds it as a projection screen at a distance which is large compared to the inelastic mean free path of photoelectrons. The atoms of the lattice "illuminated" by the emitter point source can be considered to cast shadows (of reverse contrast) on the fictitious screen. Actually these contours result from the forward direction focusing effect and represent a directional increase in intensity. Eventually, the combined effect of photoemission from different nonequivalent sites can be interpreted as a superposition of the contours generated in the course of a "multiple exposure" by subsequently excited emitters.

Figure 1 exemplifies for the case of the (111) surfaces of CdZnTe, that the maxima revealed by XPD are clearly related to the directions of the main crystal axes and that the circular fringes of higher order interference are of minor importance.[10] For the (111)A face the predominant maxima are found in the directions $\langle 11-1 \rangle, \langle 010 \rangle, \langle 121 \rangle, \langle 12-1 \rangle$ for the Te $3d_{5/2}$ electrons, and in the directions $\langle 111 \rangle, \langle 13-1 \rangle, \langle 11-1 \rangle, \langle 151 \rangle$ for the Cd $3d_{5/2}$ electrons, respectively. On the other hand for the (111)B face, we find maxima at $\langle 111 \rangle, \langle 110 \rangle, \langle 010 \rangle, \langle 12-1 \rangle$ for the Te $3d_{5/2}$ signal and at $\langle 131 \rangle, \langle 11-1 \rangle, \langle 15-1 \rangle, \langle 010 \rangle$ for the Cd $3d_{5/2}$ signal. Please note, that the $\langle 111 \rangle$ axis vector is considered to point out of the crystal for each of both (111) faces. The different surface terminations of the A and the B face are immediately evident.

Table I summarizes the emission directions and emitter–scatterer distances for the two emitter–scatterer configurations related to the (111)A and the (111)B crystal orientation in the bulk of a zincblende crystal. All scatterers are listed up to a distance of five times the bond distance unit (1 bdu $\approx$ 2.8 Å). Indeed, all observed maxima occur under one of the directions given by Table I. The relative intensities of the measured maxima are also listed. The prevalent maxima have a nearest scatterer in a distance not larger than three bdu. Please note, that for emission from deep enough layers there is always a chain of scatterers for all of the listed directions. In the following we will refer to the configurations A and B, according to the crystal face for which the Te emitter has the corresponding scatterer configuration. If the Cd and the Te lattice atoms are scatterers of the same strength then

TABLE I. Polar and azimuthal escape angles ϑ, φ of forward direction scattering at near neighbors as a function of the distance R between emitter and scatterer for the two atomic configurations A and B of the (111) orientations of zinc blende (bdu = bond distance units). The maxima observed in the experimental data of Fig. 1 under the respective directions were qualitatively characterized by their relative intensities according to: $+++$ = very strong, $++$ = strong, $+$ = weak, $-$ = not observed.

Crystal direction	ϑ (deg)	φ (deg)	R (bdu)	Type	Intensity	Crystal direction	ϑ (deg)	φ (deg)	R (bdu)	Type	Intensity
Configuration A: (emitter of type B at face terminated by atoms of type A)						⟨111⟩	0.0	0.0	1.00	b	+++
⟨11−1⟩	70.5	0.0	1.00	a	+++	⟨110⟩	35.3	0.0	1.63	i	++
⟨110⟩	35.3	0.0	1.63	i	++	⟨13−1⟩	58.5	30.0	1.91	b	++
⟨131⟩	29.5	60.0	1.91	a	+++	⟨010⟩	54.7	60.0	2.31	i	+++
⟨−13−1⟩	80.0	60.0	1.91	a	+++	⟨331⟩	22.0	0.0	2.52	b	+
⟨010⟩	54.7	60.0	2.31	i	+++	⟨121⟩	19.5	60.0	2.83	i	+++
⟨33−1⟩	48.5	0.0	2.52	a	+	⟨12−1⟩	61.9	19.1	2.83	i	++
⟨13−3⟩	82.4	19.1	2.52	a	+	⟨151⟩	38.9	60.0	3.00	b	+++
⟨121⟩	19.5	60.0	2.83	i	+++	⟨11−1⟩	70.5	0.0	3.00	b	+++
⟨12−1⟩	61.9	19.1	2.83	i	++	⟨−15−1⟩	70.5	60.0	3.00	b	+++
⟨111⟩	0.0	0.0	3.00	a	++	⟨35−1⟩	46.9	19.1	3.42	b	++
⟨15−1⟩	56.3	40.9	3.00	a	++	⟨15−3⟩	73.0	30.0	3.42	b	+
⟨351⟩	28.6	30.0	3.42	a	+	⟨130⟩	43.1	40.9	3.65	i	++
⟨130⟩	43.1	40.9	3.65	i	++	⟨03−1⟩	68.6	46.1	3.65	i	++
⟨03−1⟩	68.6	46.1	3.65	i	++	⟨353⟩	14.4	60.0	3.79	b	++
⟨35−3⟩	63.9	13.9	3.79	a	+	⟨551⟩	27.2	0.0	4.12	b	−
⟨55−1⟩	43.3	0.0	4.12	a	−	⟨17−1⟩	55.5	46.1	4.12	b	+
⟨171⟩	43.3	60.0	4.12	a	+	⟨231⟩	22.2	30.0	4.32	i	+
⟨−17−1⟩	66.2	60.0	4.12	a	+	⟨23−1⟩	51.9	13.9	4.32	i	+
⟨231⟩	22.2	30.0	4.32	i	+	⟨13−2⟩	72.0	23.4	4.32	i	++
⟨23−1⟩	51.9	13.9	4.32	i	+	⟨371⟩	34.2	40.9	4.43	b	++
⟨13−2⟩	72.0	23.4	4.32	i	+	⟨55−3⟩	58.2	0.0	4.43	b	++
⟨553⟩	12.3	0.0	4.43	a	+++	⟨35−5⟩	77.0	10.9	4.43	b	++
⟨37−1⟩	47.4	30.0	4.43	a	++	⟨−17−3⟩	77.0	49.1	4.43	b	++
⟨17−3⟩	67.9	36.6	4.43	a	++	⟨37−3⟩	60.4	23.4	4.73	b	++
⟨373⟩	23.5	60.0	4.73	a	+	⟨141⟩	35.3	60.0	4.90	i	+
⟨141⟩	35.3	60.0	4.90	i	+	⟨14−1⟩	57.0	36.6	4.90	i	+
⟨14−1⟩	57.0	36.6	4.90	i	−	⟨−14−1⟩	74.2	60.0	4.90	i	+
⟨−14−1⟩	74.2	60.0	4.90	i	+	⟨57−1⟩	42.8	13.9	5.00	b	+
⟨571⟩	29.9	19.1	5.00	a	−	⟨17−5⟩	78.5	30.0	5.00	b	+
⟨120⟩	39.2	30.0	5.16	i	−	⟨120⟩	39.2	30.0	5.16	i	−
⟨02−1⟩	75.0	40.9	5.16	i	−	⟨02−1⟩	75.0	40.9	5.16	i	−

the exchange of emitter and surface polarity would not affect the resulting pattern. In fact, this symmetry is reflected in the respective patterns of Fig. 1.

For a quantitative assessment of the data of Fig. 1, the relative peak intensities of the Cd $3d_{5/2}$ and Te $3d_{5/2}$ photoelectrons obtained for the CdZnTe (111)A and (111)B surfaces were plotted in Fig. 2 for the low crystal index directions given in Table I. In order to compare the Cd $3d_{5/2}$ with the Te $3d_{5/2}$ signal in Fig. 2 we have introduced "relative intensities" by relating the intensities to a pattern internal background, established by averaging over an appropriate solid angle (see Sec. III). For both CdZnTe surfaces the dependence of the ratio of the Cd- and Te- background signal on the polar angle was only very weak and indicated a homogeneous and stoichiometric composition of the samples up to the top monolayers. With respect to the background the relative intensity increase induced by forward direction focusing was always observed to be less than 40%, i.e. only a fraction of the forward direction enhancement factor. This result is in agreement[10] with the fact, that for a particular crystal direction only emitter and scatterer in short chains with up to 2 or 3 scatterers contribute to the forward direction focusing peak[8] ("defocusing" due to multiple scattering[12,13]).

In Fig. 2 each direction for which the emitter has a nearest scatterer atom of the identical type has been labeled "i." In the respective chains along these directions the scatterer are equidistant. The directions for which the emitter and the nearest scatterer are of different type have been labeled either "a" or "b" according to the configuration discussed above (see Table I). In these directions the chains have scatterers in alternating shorter and longer distances, since they in addition contain scatterers of the emitter type. Please note, that if e.g. for type a a nearest scatterer occurs for configuration A in the distance R then, in the same crystal direction, there are also scatterers for configuration B, but the nearest scatterer is located in the larger distance $3R$. In the configuration for which the scatterer is nearer the next nearest scatterer is located in the distance $4R$ and is of the same type as the emitter.

The most striking features of Fig. 2 can be explained in terms of these scattering process types. In the case of type i the intensities are all the same for all patterns of Fig. 1 whereas in the cases of type a or b the relative intensity is large for one of the two configuration only, but clearly smaller for the other one, hence the peaks in these directions are well suited for an experimental distinction between the two configurations A and B. The configuration with the larger relative peak intensity is always observed to be that with the shorter emitter–scatterer distance (see Table I).

At the bottom of Fig. 2 the scattering distance R is indicated by shaded (configuration A) and solid (configuration B) bars. In addition to the directions with $R < 3$ bdu given by Table I in Fig. 2 we have included the low-index $\langle 120 \rangle$ direction which corresponds to a near neighbor direction in a simple cubic lattice, but for the zincblende lattice the nearest scatterer is at $R = 5.16$. The absence of any significant maximum under this direction at any of the XPD patterns shown in Fig. 1 proves that scatterers at such large distances are negligible (see also Table I). This conclusion is confirmed by the fact that for all directions shown in the right hand part of Fig. 2 the maximas are small or negligible for that configuration which does not match the scattering process type. By inspection of Table I it is found for these configurations that the scattering distances corresponding to the respective directions are larger than $R = 5$. On the other hand for the $\langle 111 \rangle$ and $\langle 11-1 \rangle$ direction a pronounced maximum is found even for the non matching configuration since the respective scattering distances are small ($R = 3.00$).

In Fig. 2 it is surprising that the maxima in the $\langle 33-1 \rangle$ and $\langle 331 \rangle$ directions ($R = 2.52$) are very weak for both configurations. The respective scatterers are located in a solid angle region around the $\langle 110 \rangle$ direction which is obstructed by the next nearest scatterer at $R = 1.63$ (see Fig. 1). By focusing the beam along the $\langle 110 \rangle$ axis this scatterer has removed intensity from off-axis directions and, consequently, all maxima located in this obstructed region are weak or absent (see also Table I). Off-axis multiple scattering and first order interference effects of this type can explain,[10] why there is no simple functional relation between the relative intensity and the scattering distance R.

Figure 2 also reveals differences in the scattering properties of the Te and Cd atoms. Considering a given configuration we find for most of the shown directions that the relative intensity is larger for the Te $3d_{5/2}$ electrons if the type is i and for the Cd $3d_{5/2}$ electrons if the type is a or b. By comparing the intensities in excess of the background (dashed line in Fig. 2) we conclude that the scattering amplitude at zero scattering angle is about 15%–20% larger for the Te atoms than for the Cd atoms.

Figure 3 exemplifies the potential of XPD for the depth sequential imaging of atomic layers by inspecting the experimental data for the Cd $3d_{5/2}$ signal of the Te-terminated (111)B face of CdZnTe. The angular positions of 49 Te scatterers forming a rhombic section of the top layer viewed from a Cd emitter site below are indicated by large circles. By considering a Cd emitter from increasingly deeper crystal layers labeled $N = 2, 4, 6, 8$, and 10 the shown sequence of projections patterns for the top layer is obtained. The minimum polar angle for which Te scatterers can be observed reduces from 71° ($N = 2$) to 9° ($N = 10$), while the corresponding distance between emitters and scatterer increases from $R = 1.0$ to $R = 5.74$ (2.8–16 Å). An assignment of the observed maxima to the "atomic shadows" is easy. The maximum at $\vartheta = 9.5°$, $\varphi = 60°$ is clearly resolved, which demonstrates that a depth region of at least 16 Å is probed. The Cd scatterers of layer 2 account for the remaining maxima (small circles) in the diffraction pattern of the Cd photoemission from the layers 4, 6, and 8, respectively. If there were a surface reconstruction involving a displacement of atoms in the order of some tenths of an angstrøm, additional or displaced maxima would be expected to occur.[10] Hence, within these limitations the observed pattern is indicative for an unreconstructed surface. In the case of an unreconstructed surface scatterers in the third or any deeper layer will not contribute any but redundant information if forward scattering is considered only.

V. CRYSTALLINITY AND STRUCTURE OF ETCHED HgCdTe SURFACES

Figure 4 shows the results of a XPD analysis for epitaxial HgCdTe ($x = 0.2$) layers grown on CdZnTe (111) substrates of orientation B. Prior to analysis these surfaces were subjected to the two different surface treatments described above in the experimental section. The results on the left hand side of Fig. 4 refer to a surface from which about 100 monolayers were removed by electrochemically etching, whereas the results on the right-hand side of Fig. 4 refer to a similarly prepared surface from which in addition about 10 monolayers were removed by sputtering with 1 keV Ar$^+$ ions.

After both surface treatments the angular distributions of the Te $3d_{5/2}$, Cd $3d_{5/2}$, and Hg $4f_{7/2}$ electrons exhibit distinct XPD patterns. Despite clear differences of these distributions with respect to the polar angle, the structure in these XPD patterns is in close agreement with the results for the (111)B surface of the CdZnTe substrate [Figs. 2(c) and (d)]. In particular, the patterns of the Hg $4f_{7/2}$ electrons and the Cd $3d_{5/2}$ electrons show the same structure indicating the presence of a (111)B surface also on the epitaxial layer. After electrochemically etching the Te $3d$ patterns of

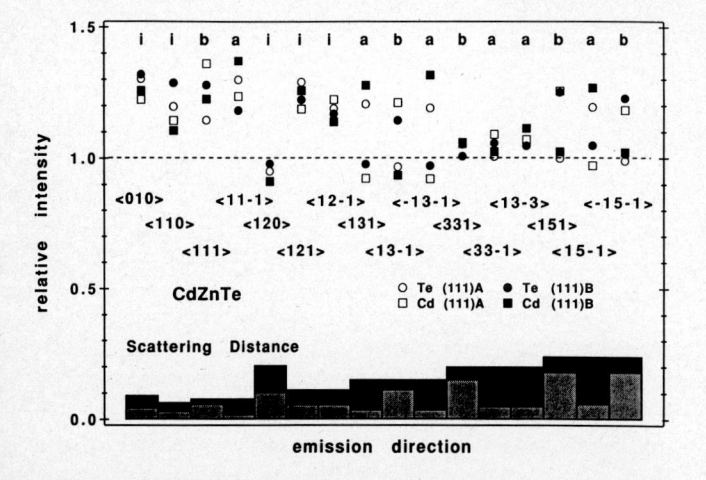

FIG. 2. Relation between low crystal index emission directions and the relative peak intensities (ratio of the peak countrate to the angle averaged background countrate) of the Te $3d_{5/2}$ and the Cd $3d_{5/2}$ photoelectrons emitted from the free CdZnTe (111)A and (111)B surfaces. For each crystal direction the distance between emitter and nearest scatterer is indicated at the bottom of the figure by the shaded and the solid bars for the configurations A and B, respectively. At the top of the figure it is indicated if emitter and scatterer are of identical type (i) or of differing type. For differing emitter and scatterer type a and b denotes the emitter–scatterer configuration, which has the shorter emitter-scatterer distance along the considered crystal axis.

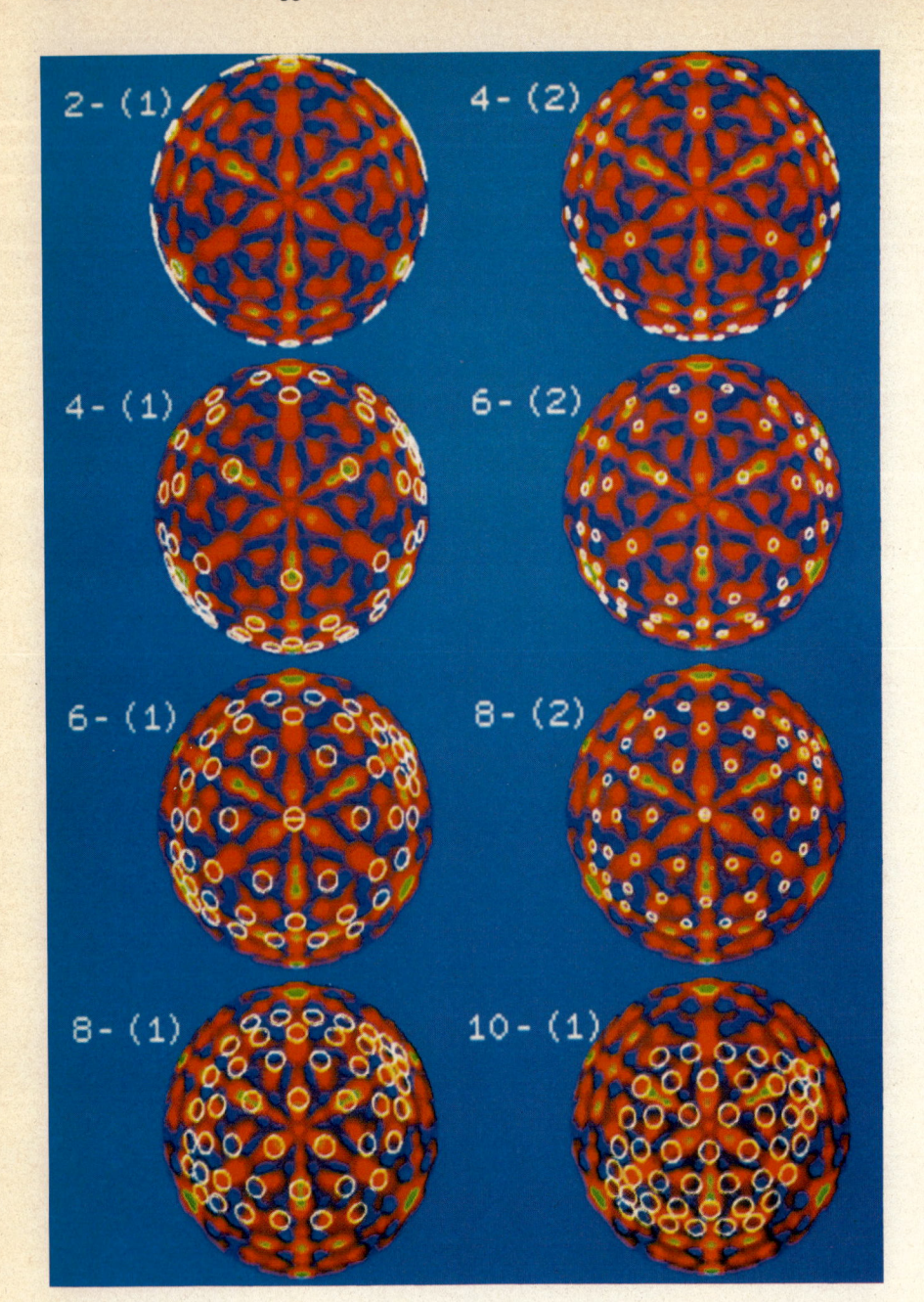

FIG. 3. XPD pattern of the Cd 3d photoemission from CdZnTe (111)B with overlays showing a central projection of Te atoms (large circles) and Cd atoms (small circles) of the outermost monolayer (atomic layer 1 and 2) with a Cd emitter of layer 2,4,6,8, and 10 being the center of projection, respectively. By the agreement of the observed peaks with the indicated scatterer positions the potential of XPD for depth-resolved atomic imaging up to a depth of 5 monolayers is exemplified.

the HgCdTe samples showed additional dim stripes in a sixfold symmetric arrangement. The nature of this structure is not understood as yet.

Although the surface of the electrochemically etched sample was lightly contaminated, the XPD patterns were basicly well developed even at grazing electron escape indicating a high crystalline quality of the epitaxial layer up to the top. The contamination layer does not give rise to additional structures in the XPD patterns but it causes the intensity of the Te $3d_{5/2}$, Cd $3d_{5/2}$, and Hg $4f_{7/2}$ photoemissions to decrease with increasing polar angle in a very similar way. A general structural agreement is observed with the XPD results recorded for the sputtered surface (right-hand side of Fig. 4); however, for the latter case there are different trends regarding the intensity distribution over the polar angle giving evidence for an inhomogeneity of the crystal composition near the surface. For the Hg $4f_{7/2}$ pattern the maxima occuring at smaller polar angles are pronounced, whereas in the Cd $3d_{5/2}$ pattern the maxima at large polar angle are emphasized, in particular the Cd $3d_{5/2}$ maximum at $\langle 11-1 \rangle$ appears to be enhanced if compared with the CdZnTe (111)B surface of Fig. 2(c). Since Hg and Cd atoms occupy equivalent lattice sites this result is indicative of an increase of the mole fraction of CdTe towards the surface.

The compositional depth profiles implied by the angle dependence of the angle averaged background were derived on the basis of the simple model of photoemission presented in

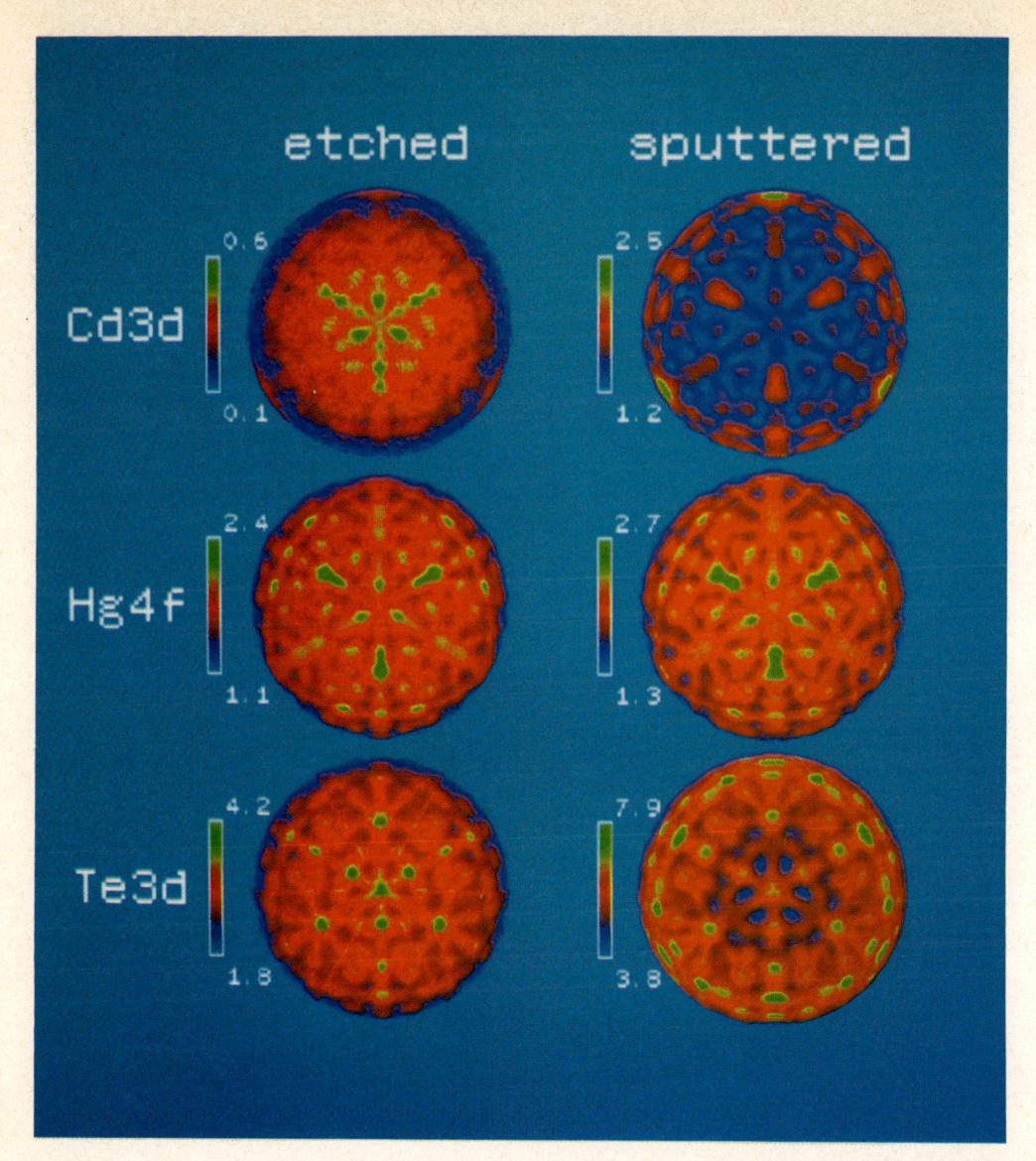

FIG. 4. Comparison of XPD patterns of the Cd $3d_{5/2}$, Hg $4f_{7/2}$, and Te $3d_{5/2}$ photoelectrons as recorded for an epitaxial HgCdTe($x = 0.2$) layer (grown on a (111)B CdZnTe substrate) after different surface treatments. "etched" refers to a surface from which about 100 monolayers were removed by electrochemical etching. "sputtered" refers to a surface from which about 10 monolayers were removed by sputtering with 1 keV Ar ions.

Ref. 14 (planar HgCdTe bulk covered with two homogeneous overlayers) and will be discussed elsewhere in detail.[10] The model suggested that the sputtered HgCdTe(0.2) sample has a depletion zone of HgTe ($0.51 \leqslant x \leqslant 0.56$) which extends into a depth $1.0 \leqslant d/\lambda \leqslant 1.5$. In addition an excess of tellurium (phase fraction 30%) is found in a top layer of at least $d = 0.14\lambda$ (≈ 3 Å). For the electrochemically etched samples we found that the contamination layer was uniform and had a thickness of about one monolayer. However, underneath this layer the probed sample volume was homogeneous and of the stoichiometric composition of the underlying epitaxial layer (HgCdTe with $x = 0.2$).

A better insight regarding the contribution of the contamination layer to the XPD patterns can be obtained by relating the intensity of each XPD maximum to the angle averaged background of the photoemission signal at the corresponding polar angle (see Sec. III). If the adatoms forming the contamination were ordered on the surface, there would be a high chance to detect their forward direction focusing image at least for grazing electron escape (see Sec. III). The fact that no additional features are observed could still be explained if the contamination atoms would occupy regular lattice sites. However, an additional signal contribution from the uppermost emitters of the substrate would possibly occur since they would find a corresponding scatterer counterpart resulting in a change of the forward direction focusing intensity with respect to the angle averaged signal.

For a quantitative analysis of such ordering effects and of possible structural changes induced by the different etch treatments we have plotted in Fig. 5 the relative intensities measured along the prevalent scattering axes as a function of the corresponding internuclear distances given in Table I and we use these data sets as "fingerprints." The data for the Te $3d_{5/2}$ electrons (configuration B) and for the Hg $4f_{7/2}$ and Cd $3d_{5/2}$ electrons (configuration A) are shown in Figs. 5(a) and 5(b), respectively, and besides the data for the sputtered and for the electrochemically etched HgCdTe surfaces the data of the free CdZnTe (111)B surface were included. In order to differentiate maxima which occur under directions with identical internuclear distance R the data are plotted with a small offset in the order given by Table I.

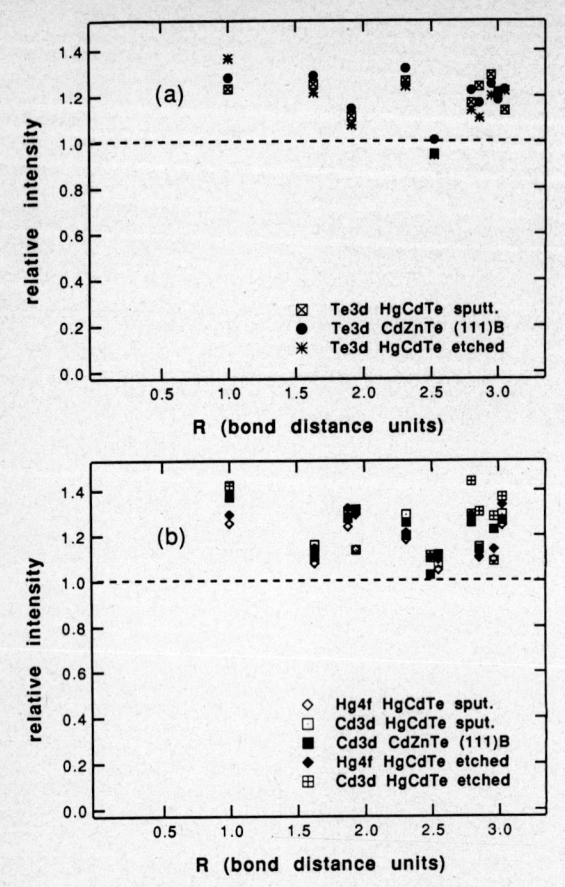

FIG. 5. Relative intensities of the prevalent XPD peaks as a function of the emitter–scatterer distance R as obtained (a) for the Te $3d_{5/2}$ electrons (configuration B) and (b) the Hg $4f_{7/2}$ and Cd $3d_{5/2}$ electrons (configuration A). The intensity data obtained for the electrochemically etched and the Ar ion sputtered HgCdTe are compared with the data of the free CdZnTe(111)B surface. In order to differentiate maxima with identical R the data are plotted with a small R offset in the order given by Table I.

After both surface treatments the relative intensity versus R data sets of the HgCdTe samples show very good agreement with the data sets obtained for the CdZnTe(111)B surface implying the same structure and a similarly high degree of order for these surfaces. Moreover the relative intensities for the Hg $4f_{7/2}$ electrons match closely the relative intensities for the Cd $3d_{5/2}$ electrons.

By comparison of the electrochemically etched HgCdTe surface with the uncontaminated CdZnTe surface we find a relative increase in the forward direction focusing intensity along the $\langle 111 \rangle$ axis for the Te signal and also along the $\langle 12-1 \rangle$ and $\langle 121 \rangle$ axes for the Cd signal. This result would be expected if contamination atoms were occupying the cation site on top of the bulk terminating Te layer in a regular manner. However, there is no clear evidence for such an order of the contamination layer, since an expected intensity increase is absent for other respective directions.

Finally we conclude from Fig. 5 that Hg and Cd atoms have very similar scattering properties at the considered high electron energies. Since those data points for the Te $3d_{5/2}$ and Cd $3d_{5/2}$ electrons which correspond to cation scattering tend to be of slightly higher intensity for the HgCdTe samples Hg may be a somewhat stronger scatterer than Cd.

VI. SUMMARY AND CONCLUSIONS

(111) surfaces of LPE-grown HgCdTe epilayers and of Bridgman-grown CdZnTe single crystals (used as substrate material for the epitaxial samples) were investigated by XPD to assess the crystalline structure to a depth of a few monolayers. The dependence on the direction of electron escape found for the recorded intensity of the photoemission from the Hg $4f$, Cd $3d$, and Te $3d$ core levels clearly revealed the phenomenon of forward direction focusing for all of the examined surfaces. Even carbonaceous contaminations on the order of one monolayer proved to be no obstacle for the XPD investigation of the underlying sample. In the hemispherical format the XPD pattern can be visualized as an image of the atomic neighbors which are centrally projected (in the real lattice space) along internuclear axes from the photoemitting atom (as the center of projection) onto a distant screen.

The results demonstrate that the XPD technique can be used as a tool for locating the atoms within the basis of the primitive unit cell of the lattice in a region of a few monolayers adjacent to the surface. The A and the B faces of the investigated (111) surfaces could be clearly differentiated by the uniqueness of their XPD patterns.

Deviations from the zinc blende structure were indicated neither by azimuthal nor by polar variations in the angular distribution of the photoemission. Since a surface relaxation in the context of a surface reconstruction is expected to distort the respective cubic symmetry, within the limits of this experimental method this result confirms[15] that there is no evidence for any reconstruction or relaxation of the (111) surfaces of CdZnTe or HgCdTe, respectively.

Electrochemical etching proved to be the superior method for the preparation of nondegraded HgCdTe surfaces since it led to well-ordered surfaces of stoichiometric composition without any etching residue. Dry etching by Ar ion sputtering caused a decrease of the mole fraction of HgTe in the HgCdTe surface region and an excess of Te in the outermost monolayer; however, the crystalline structure of the HgTe depleted region was maintained up to the first few monolayers in spite of the sputter treatment.

ACKNOWLEDGMENTS

Expert technical assistance by P. Meisen is gratefully acknowledged. The CdZnTe substrate crystals and the HgCdTe epilayers were grown and kindly provided by M. Bruder and J. Ziegler (AEG Heilbronn). This work was supported by the Bundesminister der Verteidigung, Federal Republic of Germany and in part from gift funds from Texas Instruments Inc.

[a] Permanent address: Fraunhofer-Institut für Angewandte Festkörperphysik, D-78 Freiburg, W. Germany.
[1] M. Seelmann-Eggebert and H.-J. Richter Proc. SPIE, 1106, 181 (1989).
[2] G. Granozzi, G. A. Rizzi, G. S. Herman, D. J. Friedman, C. S. Fadley, J. Osterwalder, and S. Bernardi, Phys. Scr. 41, 913 (1990).
[3] K. Nakagawa, K. Maeda, and S. Takeuchi, Appl. Phys. Lett. 34, 574 (1979).
[4] S. Kono, S. M. Goldberg, N. F. T. Hall, and C. S. Fadley, Phys. Rev. B 22, 6085 (1980).
[5] C. S. Fadley, Prog. Surf. Sci. 16, 275 (1984).
[6] H. C. Poon and S. Y. Tong, Phys. Rev. B 30, 6211 (1984).

[7] M. Fink and A. C. Yates, At. Data **1**, 385 (1970); M. Fink and J. Ingram, *ibid.* **4**, 129 (1972).
[8] W. F. Egelhoff, Phys. Rev. B **30**, 1052 (1984).
[9] W. F. Egelhoff, Crit. Rev. Solid State Mater. Sci. **16**, 213 (1990).
[10] M. Seelmann-Eggebert and H.-J. Richter, Phys Rev. B (submitted).
[11] M. Seelmann-Eggebert and H.-J. Richter, J. Electron. Spectrosc. Relat. Phenom. **52**, 273 (1990).
[12] S. Y. Tong, H. C. Poon, and D. R. Snider, Phys. Rev. B **32**, 2096 (1985).
[13] M. L. Xu, J. J. Barton, and M. A. Van Hove, Phys. Rev. B **39**, 8275 (1989).
[14] M. Seelmann-Eggebert and H.-J. Richter, J. Vac. Sci. Technol. A **6**, 2699 (1988).
[15] Y.-C. Lu, R. S. Feigelson, and R. K. Route, J. Appl. Phys. **67**, 2583 (1990).

X-ray photoelectron diffraction from the HgCdTe(111) surface

G. S. Herman, D. J. Friedman, T. T. Tran, and C. S. Fadley
Department of Chemistry, University of Hawaii, Honolulu, Hawaii 96822

G. Granozzi and G. A. Rizzi
Dipartimento di Chimica, Universita' di Padova, Padova, Italy

J. Osterwalder
Institut de Physique, Universite' de Fribourg, CH-1700 Fribourg, Switzerland

S. Bernardi
Centro Studi e Laboratori Telecomunicazioni Spa, Torino, Italy

(Received 2 October 1990; accepted 4 December 1990)

The surface polarity of a mercury cadmium telluride (MCT) (111) crystal surface has been determined by x-ray photoelectron diffraction (XPD). Emission from the core levels of Hg, Cd, and Te gave reproducible photoelectron diffraction patterns with considerable fine structure. Comparisons between experiment and single scattering cluster calculations via R factors very well distinguished the different kinds of lattice sites of Cd, Hg, and Te, and also permitted unambiguously assigning a cationic termination to the sample studied. This is thus a demonstration of the capability of XPD to study the type of termination involved at MCT and other compound semiconductor surfaces.

I. INTRODUCTION

Surface polarity determinations for (111) zinc-blende structures have been of much interest as a result of the sensitivity of chemical and physical properties to the nature of the surface termination.[1–7] Due to the lack of a center of symmetry in the zinc-blende structure, these (111) surfaces can have either cationic (here called type A) or anionic (type B) surfaces. A number of different methods have been used to try and determine the surface polarity of these (111) zinc-blende crystals.[1–7] X-ray photoelectron diffraction (XPD) has in previous studies been found to be a useful structural probe of surfaces and epitaxial overlayers,[6–11] and we have in this study applied it to HgCdTe(111). (A preliminary account of this work has appeared previously.[6]) In XPD, the emission direction of core level photoelectrons above a surface is varied, and intensity modulations produced by scattering and interference effects are measured. At high kinetic energies E_{kin} greater than about 500 eV, forward scattering peaks are the strongest features in these interference patterns, with these peaks being produced by scattering from nearest-neighbor atoms along low-index directions in a crystal. Additional fine structure is also seen in the diffraction patterns due to more complex higher-order interference effects. Since the core photoemission process is localized at the site of a specific emitting atom, structural information concerning a given site type can be obtained due to the differences in intensity modulations for atoms in different sites. It is the goal of this study to see how detailed such information can be for HgCdTe(111), including a consideration of all of the diffraction features produced. [Another XPD study of HgCdTe(111) concentrating only on the forward-scattering peaks can be found in these proceedings.[7]]

In analyzing our data, we have used single-scattering cluster (SSC) diffraction calculations of a type described in detail elsewhere.[8,12] These calculations thus automatically incorporate both the forward scattering peaks and all other interference features. This theoretical modeling also incorporates the correct spherical wave nature of the final state photoelectron waves, as well as the various angular momenta and interferences involved because of the dipole transition from some initial angular momentum l_i to the two allowed final momenta of $l_f = l_i \pm 1$.[12] In comparing experimental and theoretical curves of the variation of intensity with azimuthal angle for different structures, we have furthermore used R factors as a quantitative measure of the goodness of fit.[13] The same nonstructural parameters of inner potential (10 eV), inelastic electron attenuation length [$\Lambda_e(E_{kin} = 1386$ eV$) = 20.3$ Å, $\Lambda_e(E_{kin} = 1082$ eV$) = 16.8$ Å, and $\Lambda_e(E_{kin} = 914$ eV$) = 14.8$ Å],[14] and muffin-tin scattering phase shifts have been used for both surface terminations tested, and our structural conclusions were not found to be sensitive to these choices.

II. EXPERIMENTAL

The $Hg_{0.78}Cd_{0.22}Te$ sample was grown by slider liquid phase epitaxy (LPE) from Te-rich solutions using a two zone reactor with an internal source.[15] The sample was then annealed in a Hg atmosphere to reduce vacancies and lapped prior to analysis.

The XPD measurements were performed on a Hewlett–Packard 5950A spectrometer which was specially modified for automated angle scanning as described elsewhere.[8] All data were taken as azimuthal scans of intensity at a constant polar angle θ defined with respect to the surface. The full 360° range was scanned and the data were then threefold averaged into 120°. Such data should be less influenced by slight crystal misalignments, and the noise level is also reduced. Monochromatized Al $K\alpha$ radiation was used for excitation, and the angular acceptance cone was limited to $\pm 4.5°$. The photoelectron peaks studied were: Hg $4f_{7/2}$ ($E_{kin} = 1386$ eV, de Broglie wavelength $= 0.329$ Å), Cd $3d_{5/2}$ ($E_{kin} = 1082$ eV, de Broglie wavelength $= 0.373$ Å), and Te $3d_{5/2}$ ($E_{kin} = 914$ eV, de Broglie wavelength $= 0.406$ Å). The high-symmetry directions of the crystal were located to within ± 0.5° using strong forward-scattering XPD peaks along low-index directions. During all the runs, base pressures were in the 5.0×10^{-9} Torr range, although the specimen was not subjected to the bakeout required for normal operation at 10^{-10} to 10^{-11} Torr before analysis to avoid Hg depletion. The XPD scans of the air-exposed sample as inserted into the spectrometer and the same sample after mild Ar$^+$ bombardment (10^{-5} Torr Ar$^+$, 600 eV, 45° off normal incidence, 40 min) are very similar, even though a net depletion of Hg was observed at the surface after sputtering, with a net increase in the experimental intensity anisotropies [defined as $(I_{max} - I_{min})/I_{max} = \Delta I/I_{max}$] after the sputter-cleaning procedure. The stoichiometry as determined by XPS was $Hg_{0.79}Cd_{0.21}Te$ before Ar$^+$ bombardment, and $Hg_{0.63}Cd_{0.37}Te$ after.

III. RESULTS AND DISCUSSION

A set of experimental XPD patterns taken at polar angles of 19°, 35°, and 55° with respect to the surface is illustrated in Fig. 1. These three polar angles were chosen so that the scans passed through low-index directions of [11−1], [010], and [110], respectively, thus leading at some point in azimuth to strong forward scattering peaks for one or more of the core peaks studied (cf. Fig. 2). These data for emission from the Hg $4f_{7/2}$, Cd $3d_{5/2}$, and Te $3d_{5/2}$ core levels illustrate the strong similarity between the diffraction patterns of Hg and Cd emission and how they both differ from the diffraction patterns of Te. This is evident in both the relative intensities of the features as well as the location of the peaks.

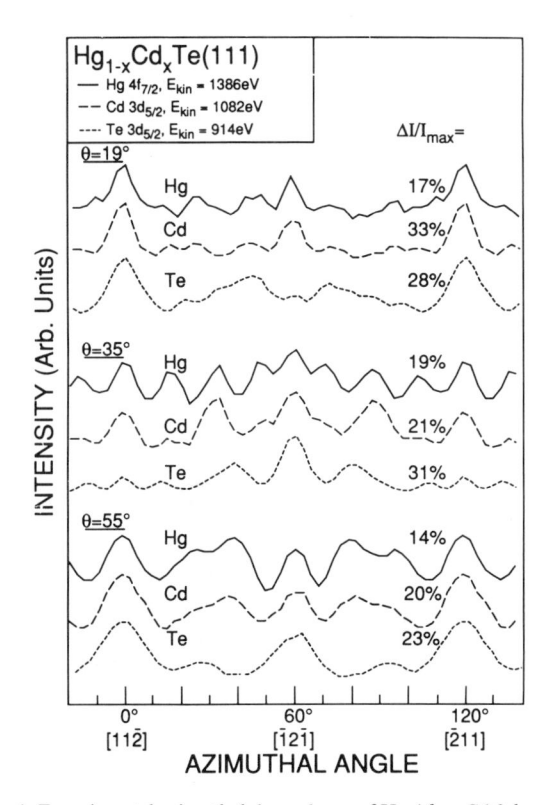

FIG. 1. Experimental azimuthal dependence of Hg $4f_{7/2}$, Cd $3d_{5/2}$, and Te $3d_{5/2}$ emission from $Hg_{0.78}Cd_{0.22}Te(111)$ at polar angles of 19°, 35°, and 55° with respect to the surface.

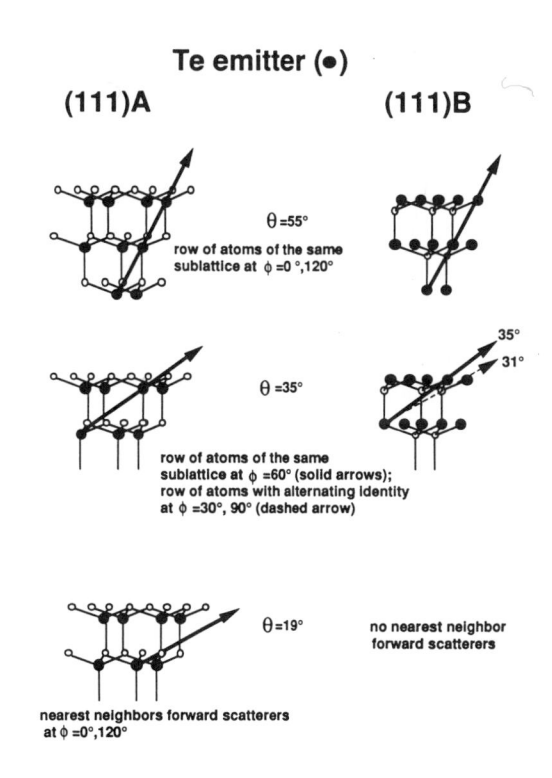

FIG. 2. Schematic illustration of the near-neighbors responsible for the strongest forward-scattering events in the two structural models discussed here (cationic termination = A and anionic termination = B) and at the three polar angles studied.

Qualitatively this is easily understood, since Hg and Cd are expected to occupy similar crystal sites, while Te occupies a different site with respect to the surface. The mirror plane at $\phi = 60°$ can be used as a measure of reproducibility of these experimental data. The data at the two higher polar angles have excellent mirror symmetry, while those at $\theta = 19°$ are slightly less symmetric along this mirror plane. The 19° data are also somewhat noisier due to a purely instrumental reduction in intensity as θ is reduced.

As noted earlier, the simplest approach for interpreting the strongest features present in XPD scans is based on the well known fact that the electron-atom scattering process yield final intensities that are strongly peaked in the forward direction. The neighbor atoms responsible for the strongest forward scattering events in these curves, assuming the Te atom as emitter, are schematically illustrated in Fig. 2 for the (111)A and (111)B surface polarities at the three different polar angles used. Other near-neighbors are present at polar angles slightly different from the nominal ones, and their corresponding forward scattering peaks will have an influence on some experimental features. Switching from Te to the cationic emitter (Cd or Hg) is equivalent to interchanging the A and B models. From this figure it can be understood why an azimuthal scan at $\theta = 19°$ can distinguish between the two surface polarities, since the type A surface for Te emission will give strong forward scattering peaks at $\phi = 0°$ (defined to be in the [11−2] azimuth) and 120°, while the type B surface for Te emission will not show these, and in general be more complex due to higher-order diffraction effects off the forward direction and/or weaker non-nearest neighbor forward scattering (cf. Fig. 2 and Te theoretical curves in Fig. 3). For $\theta = 35°$ similar forward-scattering events can occur for either model, but forward scattering at angles slightly off 35° can add extra features which may help in determining the structure. The diffraction patterns for $\theta = 55°$ are very similar for both surface terminations since there are similar rows of atoms of either the (Hg,Cd) or (Te) type along which forward scattering occurs. A further important concept in understanding photoelectron diffraction along high-density chains of atoms, as for $\theta = 55°$, is "defocusing" of the photoelectron flux due to multiple elastic scattering; for higher density chains this effect can reduce the forward scattering intensity from that expected on the basis of simple SSC calculations by a factor of about 1/2.[16-18]

In Fig. 3 we compare our experimental data at a polar angle of 19° with SSC calculations for the two different surface terminations. The uppermost set of curves is for emission from Hg $4f_{7/2}$. Visual inspection reveals that the A termination gives a much better fit to our experimental data than the B termination, with all peak positions and relative intensities agreeing excellently between experiment and theory. R factor analysis further confirms the visual inspection in that the R factor gives a much lower value for the A surface (0.08) than for the B surface (0.13). Similar conclusions are reached for Cd $3d_{5/2}$ emission in Fig. 3, with the A surface termination again giving a much better fit to experiment. Note also that, for the case of theory as well, the Hg $4f_{7/2}$ and Cd $3d_{5/2}$ curves are essentially identical, as expected. Considering last Te $3d_{5/2}$ emission, we find that both the A and B terminations give approximately the same fit to experiment. However, if we suppose that the intensities of the peaks at $\phi = 0°$ and 120° have been overpredicted in theory due to a lack of inclusion of defocusing, then the amplified structure in experiment from $\phi = 15°$ to 105° is still much better reproduced by termination A.

An analogous comparison for a polar angle of 35° is presented in Fig. 4. At this angle, visual inspection indicates a significantly better fit for all three core peaks for the A termination. The data for Te and Cd both give very good R factors for A termination, while that for Hg is only slightly better for A. Nonetheless, the prediction of peak positions and most relative intensities for Hg is much better for A, thus demonstrating a weakness in depending entirely on R factors in such analyses.

A final comparison for the highest takeoff angle of 55° shown in Fig. 5 is not as good visually unless we look carefully at the weaker structure from $\phi = 15°$ to 105° and suppose again that theory has overestimated the forward scattering peaks at $\phi = 0°$ and 120° due to defocusing effects. If the latter is done, we see a much better correlation of peak positions, and to a degree also relative intensities, for the A structure. The R factors for this angle are all very close for the two structures, but they are lower for all three core peaks with the A structure.

As a final qualitative comment, one would at first sight expect that, in a perfectly terminated lattice with no concentration gradients of Cd and Hg, the diffraction patterns of

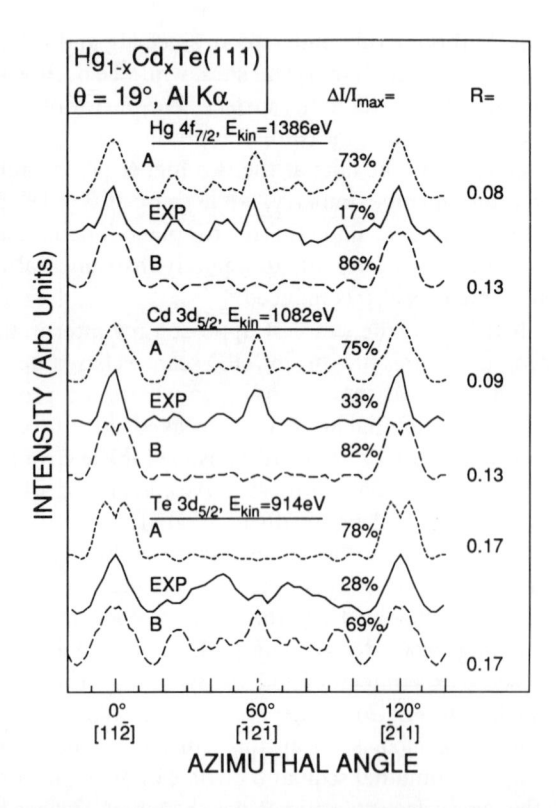

FIG. 3. Comparison between experimental scans taken at $\theta = 19°$ and SSC theoretical curves obtained for both the A and B surface terminations. The overall anisotropy as judged from $(I_{max} - I_{min})/I_{max}$ in % is also indicated for each curve. R factor values are also included to indicate the goodness of fit. It is typical in such analyses for theory to predict a significantly higher anisotropy than is observed, with reasons being the presence of surface disorder, the presence of an amorphous layer of surface impurities, and/or multiple scattering effects such as defocusing along low-index directions.

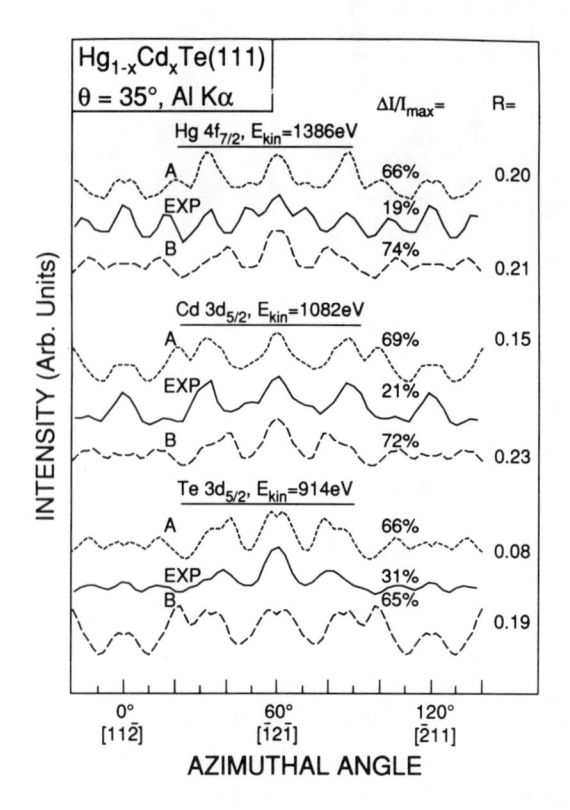

FIG. 4. As for Fig. 3, but for $\theta = 35°$.

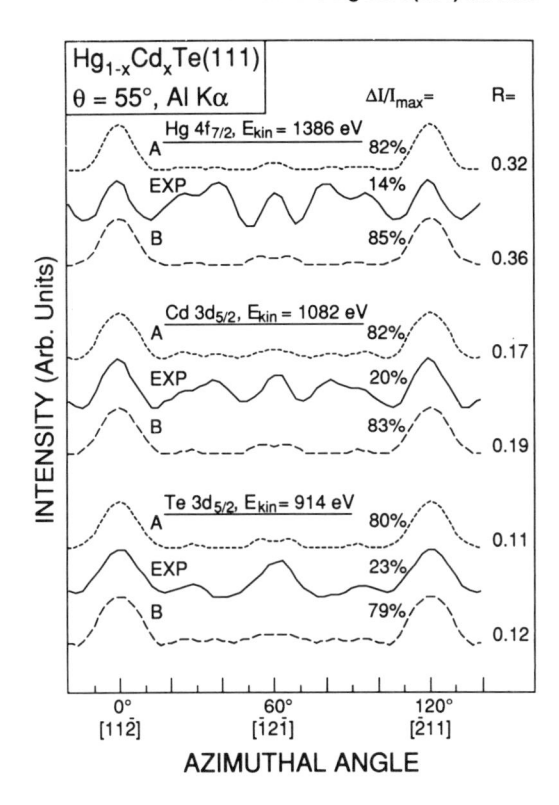

FIG. 5. As for Fig. 3, but for $\theta = 55°$.

Hg $4f_{7/2}$ and Cd $3d_{5/2}$ should be identical. This is indeed very nearly true for all of the data in Figs. 3–5, although some peaks are weaker in one pattern than the other. These deviations from this oversimplified expectation could be due to the presence of some surface disorder, a concentration gradient near the surface, the difference in kinetic energies of the two peaks (1386 versus 1082 eV, respectively) that will affect both the scattering strength and the de Broglie wavelength (0.32 versus 0.37 Å), and/or the known sensitivity of diffraction fine structure to the exact angular momenta present in the final state (cf. Ref. 12). Another simple structural expectation is that Te diffraction from an A or B surface should be nearly identical to Hg or Cd diffraction from a B or A surface, respectively, with the identity expected to be most applicable to Te and Cd which are closer in kinetic energy and have the same final-state angular momenta. This expectation is nicely borne out in all of Figs. 3–5 for Te and Cd, with greater deviations from it for the comparison to Hg due to the effects discussed above.

IV. CONCLUSIONS

The crystallographic polarity of a HgCdTe(111) surface has been determined using x-ray photoelectron diffraction measurements in combination with single scattering theory and R factor analysis. SSC calculations gave very good agreement when compared to experiment, and the strong dependence of the emitter site location on both experiment and theory clearly illustrate the strength of this technique for such studies. For all three takeoff angles measured and all three emitter types, the SSC calculations showed better agreement as judged both visually and by R factors for an A type or cationic surface termination. Future work will focus on better-defined epitaxial films studied under fully ultrahigh vacuum conditions. In the latter situation, the fine structure in the diffraction patterns should permit deriving more detailed structural information beyond the termination,[8,9] such as the depth distribution or the type of surface reconstruction present.

ACKNOWLEDGMENTS

This work has been supported by the Office of Naval Research (USA) under Contract No. N00014-90-K-0512, and by the New Energy Development Organization (NEDO) of Japan.

[1] A. W. Stevenson, S. W. Wilkins, M. S. Kwietniak, and G. N. Pain, J. Appl. Phys. **66**, 4198 (1989).
[2] A. C. Chami, E. Ligeon, R. Danielou, and J. Fontenille, Appl. Phys. Lett. **52**, 1502 (1988).
[3] E. A. Hewat, L. DiCioccio, A. Million, M. Dupuy, and J. P. Gailliard, J. Appl. Phys. **63**, 4929 (1988).
[4] Y.-C. Lu, C. M. Stahle, J. Morimoto, R. H. Bube, and R. S. Feigelson, J. Appl. Phys. **61**, 924 (1987).
[5] P. F. Fewster, S. Cole, A. F. W. Willoughby, and M. Brown, J. Appl. Phys. **52**, 4568 (1968).
[6] G. Granozzi, G. A. Rizzi, G. S. Herman, D. J. Friedman, C. S. Fadley, J. Osterwalder, and S. Bernardi, Phys. Scr. **41**, 913 (1990).
[7] M. Seelmann-Eggebert and H. J. Richter, J. Vac. Sci. Technol. B **9**, 1861 (1991).
[8] C. S. Fadley, Prog. Surf. Sci. **16**, 275 (1984).
[9] C. S. Fadley, Phys. Scr. **T17**, 39 (1987).
[10] D. A. Steigerwald and W. F. Egelhoff, Phys. Rev. Lett. **60**, 2558 (1988).
[11] P. Alnot, J. Olivier, and C. S. Fadley, J. Electron Spectrosc. Relat. Phenom. **49**, 159 (1989).
[12] D. J. Friedman and C. S. Fadley, J. Electron Spectrosc. Relat. Phenom. **51**, 689 (1990).
[13] M. A. van Hove, S. Y. Tong, and M. H. Elconin, Surf. Sci. **64**, 85 (1977).
[14] D. R. Penn, J. Electron Spectrosc. Relat. Phenom. **9**, 29 (1976).
[15] S. Bernardi, J. Cryst. Growth **87**, 365 (1988).
[16] S. Y. Tong, H. C. Poon, and D. R. Snider, Phys. Rev. B **32**, 2096 (1985).
[17] M.-L. Xu and M. A. van Hove, Surf. Sci. **207**, 215 (1989).
[18] A. P. Kaduwela, G. S. Herman, D. J. Friedman, C. S. Fadley, and J. J. Rehr, Phys. Scr. **41**, 948 (1990).

Electrochemical approaches to cleaning, reconstruction, passivation, and characterization of the HgCdTe surface

S. Menezes,[a] W. V. McLevige, E. R. Blazejewski, and W. E. Tennant
Rockwell International Science Center, Thousand Oaks, California 91360

J. P. Ziegler
Rockwell International Electro-Optical Center, Anaheim California 92803-3105

(Received 23 October 1990; accepted 18 January 1991)

A novel electrochemical approach based on voltammetric studies of reactions occurring at the HgCdTe/electrolyte interface has been devised to clean and reconstruct the HgCdTe surface prior to passivation. Sequential potential steps, applied to a HgCdTe specimen immersed in an electrolyte, oxidize the surface, remove Hg and leave a thin protective CdTe layer, reconstructed from the lattice atoms. A thick CdTe dielectric can be subsequently electrodeposited on this surface. The electrochemically processed interfaces were characterized by a low (10^{10} $cm^{-2} V^{-1}$) density of fast interface states, negligible concentration of slow traps and unusual stability to prolonged air exposure and to 100 °C air anneal. The electrochemical technique was further developed as a surface characterization tool to assess the quality of Br_2-etched HgCdTe surfaces. Monolayers of Te remaining on Br_2-etched surfaces convert sequentially to TeO_2 and then to $HgTeO_3$ on air exposure.

I. INTRODUCTION

The complexity of surface chemistry of the ternary HgCdTe compound is primarily responsible for the extreme sensitivity of infrared devices to processing parameters and the irreproducibility of device performance. Surface passivation has been a difficult task because the surface components are highly reactive, leading to its chemical and electrical instability. A survey of various passivants employed indicates that intrinsic or native films have been most effective in terminating the lattice with minimum distortion. These include anodic oxides (including tellurates), sulfides, selenides and fluorides of Hg, Cd and Te.[1–5] In each case, the native films formed on the ternary material are multicomponent. Thermodynamic considerations indicate that only the Cd compounds are stable with respect to HgCdTe.[6,7] The Hg and Te compounds react with the CdTe phase to generate undesirable reaction products, Hg, HgTe, and Te at the interface. Electrically, these effects are manifested as increased fixed charge and type conversion from p to n on annealing.

Electrochemical processing of HgCdTe allows rigorous control of the surface chemistry and provides an *in situ* monitor of surface reactions, including the insoluble reaction products remaining on the surface. Since electrochemical reactions occur at specific potentials, the undesirable species on the surface may be selectively eliminated and the surface reactions manipulated to reproducibly attain the desired stoichiometry. In this paper, we describe a new electrochemical method for *in situ* etching and passivation of the HgCdTe surface while retaining lattice continuity and stability. The method has been extended to assess the quality of chemically treated HgCdTe surfaces.

The (111B) oriented $Hg_{1-x} Cd_x Te$ material used in this study was grown by liquid phase epitaxy (LPE) at a temperature of 420 °C from Te melt using a horizontal semi-open tube apparatus.[8] The technique uses a specially prepared single crystal (0001) oriented sapphire substrate. A thin (5–7 μm) buffer layer of single crystal (111B) CdTe is grown on the sapphire by metal organic chemical vapor deposition. Finally, a ~14 μm thick layer of $x = 0.3$ $Hg_{1-x} Cd_x Te$ is grown by LPE.

II. SURFACE RECONSTRUCTION

Voltammetric techniques were used to monitor the species reacting on the HgCdTe surface in a given medium. The new surface reconstruction approach uses an acidic medium (pH 2) to permit the dissolution of one or more of the constituent ions which may be removed or incorporated into the surface layers of the HgCdTe lattice. The pH 2 electrolyte also provides an appropriate medium for electrodeposition of a thick CdTe layer after the electrochemical surface cleaning step.

Surface processes occurring at the HgCdTe/electrolyte interface have been investigated with a rotating ring-(HgCdTe)disk electrode, modulated photocurrent and surface analysis.[9] The potentials and the reaction sequence for the formation, dissolution and reduction of Hg, Cd, and Te based compounds and oxides were deduced from the results of these investigations in conjunction with thermodynamic data.

A typical current–voltage (I–V) curve, depicting the oxidation and reduction reactions occurring on the HgCdTe surface in a pH 2 electrolyte, is shown in Fig. 1. The unoxidized HgCdTe surface (dashed curve) was relatively inert in the potential region between -0.5 and 0.5 V vs a saturated calomel reference electrode (SCE). The anodic oxidation of the HgCdTe surface at potentials >0.5 V versus SCE produces a mixture of oxides comprising CdO, $CdTeO_3$, HgO, $HgTeO_3$, and TeO_2 that is partially soluble in the pH 2 electrolyte. The soluble surface reaction products were analyzed with a rotating ring-disk electrode.[9] The insoluble portion of the oxide components is reduced during the cathodic scan in

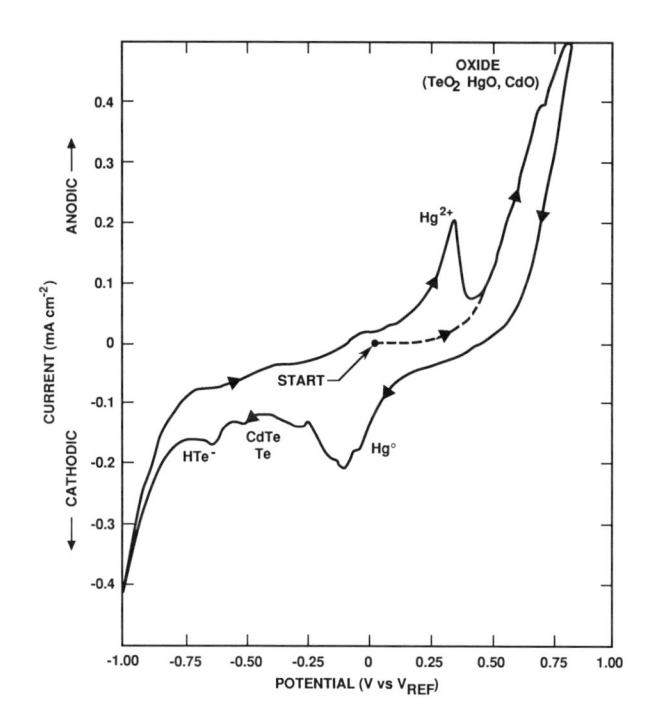

FIG. 1. Current–voltage curve of HgCdTe in 0.2 M K_2SO_4 showing oxidation, reduction and dissolution of surface species.

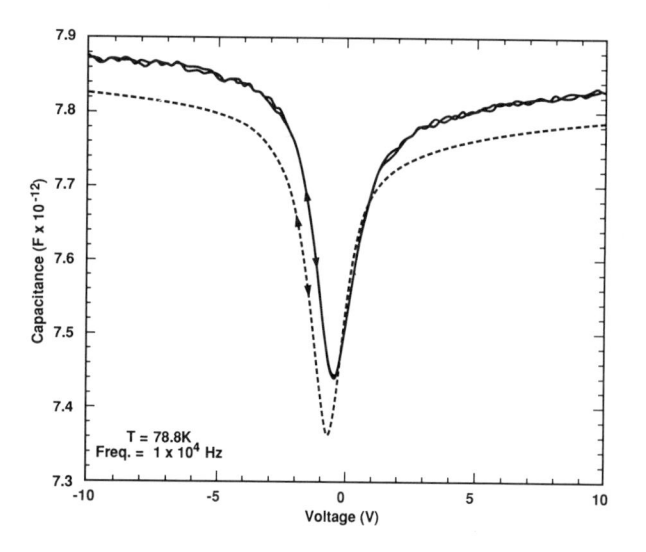

FIG. 2. C–V characteristics at 10 KHz for an electrochemically treated p-type MWIR (HgCd)Te/ZnS/Au test structure before (solid) and after (dashed) a 100 °C anneal.

the sequence: $HgO > HgTeO_3 > TeO_2 > CdO > CdTeO_3$. The preferentially reduced Hg oxides produce a Hg layer on the surface. This layer is dissolved off the surface by reoxidization as Hg_2^{2+}. The TeO_2 and $CdTeO_3$ components are subsequently reduced to Te and CdTe, respectively. Excess Te is cathodically dissolved as HTe^- or reacted with Cd^{2+} ions added to the electrolyte to form a thin CdTe overlayer. The presence of dissolved Cd^{2+}, $HTeO_2^+$, or Hg^{2+} in the electrolyte can alter the course of surface reactions at HgCdTe. Thus, an auxiliary electrode held at a constant potential was used to scavenge the dissolved species during surface processing.

Based on the insights provided by the voltammetric studies, the reactions the HgCdTe surface can be directed to (a) remove surface impurities and insoluble products (e.g., Te from a previous Br_2-based etch) by reducing them to a soluble hydride, (b) convert the defective surface layers to the oxide, (c) sequentially reduce the oxide components, (d) remove the Hg, and (e) leave a thin protective CdTe layer, reconstructed from the lattice atoms.

III. SURFACE PROPERTIES

The electrical characteristics of an electrochemically processed interface are shown in the capacitance (C)–voltage (V) plot of Fig. 2. Circular metal–insulator–semiconductor (MIS) capacitors were fabricated by depositing a 2500 Å ZnS layer followed by a 1500 Å layer of photochemically grown SiO_2 on midwave infrared (MWIR) p-HgCdTe layers. The capacitor area was 5×10^{-4} cm^2. More than 40 samples were tested representing six layers of HgCdTe. MIS C–V data taken at 77 K from the best of these layers revealed low interface state densities (N_{ss}) of less than 10^{10} cm^{-2} eV^{-1} and flatband voltages (V_{fb}) of -1.5 V, corresponding to 1.5×10^{11} cm^{-2} fixed charges. However, the position of the V_{fb} could be shifted to zero by adding Cd^{+2} ions in the electrolyte. These samples yielded C–V curves with hysteresis ranging from zero (Fig. 2) to -0.2 V for potential excursions between -10 and $+10$ V. Control samples with conventional wet chemical treatment on the same layers were typically less consistent, with higher N_{ss} and hysteresis. The C–V characteristics of the electrochemically treated MIS device shown in Fig. 2 (solid curve) did not change significantly after 1 h 100 °C air anneal as indicated by the dashed curve.

Preliminary measurements on MWIR n/p diodes have yielded devices with R_0A values as high as 1.63×10^6 Ω cm^2 for a cutoff wavelength of 4.5 μm at 77 K. The electrochemically processed interface was notably stable to several months of air exposure. This unusual stability is attributed to the electrochemically generated CdTe interface, which may oxidize to $CdTeO_3$. Besides being chemically compatible with the substrate, CdTe is also lattice-matched to HgCdTe and is thus a desirable passivant.

IV. ELECTRODEPOSITION OF CdTe

The electrochemically reconstructed interface provides an ideal substrate for the nucleation of a thick CdTe dielectric layer. The pH 2 electrolyte used for electrochemical etching also serves as a medium for deposition of thick CdTe films. By adding $CdSO_4$ and TeO_2 to the 0.5 M K_2SO_4 electrolyte, a CdTe dielectric was deposited on an electrochemically reconstructed surface. The CdTe layer stoichiometry, crystallinity and conductivity were controlled by adjusting the relative concentrations of Cd^{2+} and Te^{4+} ions, the deposition potential and temperature.

The growth of a CdTe film on the HgCdTe surface was monitored by using photoelectrochemical techniques, as shown in Fig. 3. The photocurrent response $\Delta i_{h\nu}$ was obtained by illuminating the HgCdTe surface with a visible

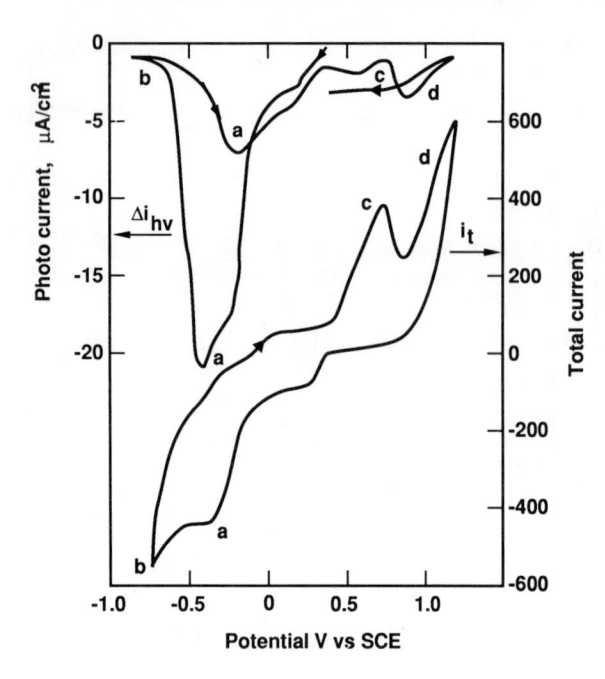

FIG. 3. Total current (i_t) and photocurrent (Δi_{hv}) as a function of HgCdTe electrode potential in 0.1 mM TeO$_2$ and 0.5 M CdSO$_4$.

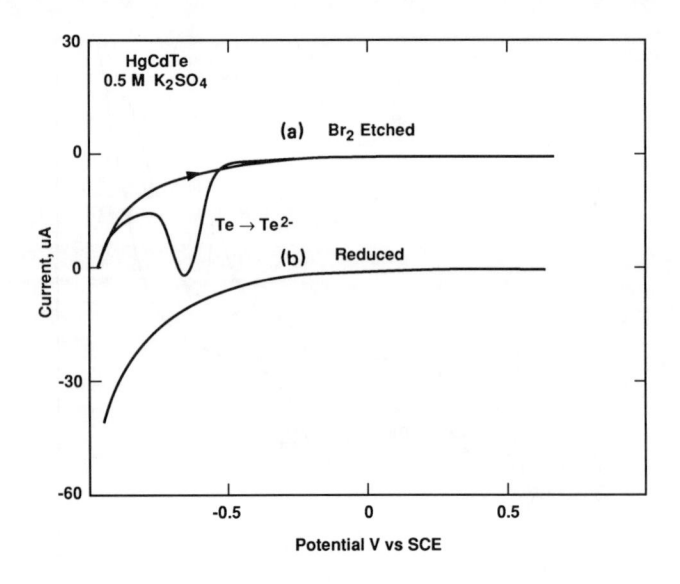

FIG. 4. Voltammetric characterization of HgCdTe surfaces in 0.2 M K$_2$SO$_4$, following Br$_2$/MeOH etch and electrochemical reduction

light (100 W tungsten halogen), chopped at a frequency of 180 Hz/s. The total current i_t is a composite of the current generated by surface processes occurring in the dark and light. The light-generated current component Δi_{hv} was extracted with a lock-in amplifier tuned to the chopping frequency.

Although HgCdTe is a p-type semiconductor, the Δi_{hv} response at an unoxidized HgCdTe surface was negligible in the potential regime of 0.25 to − 1.0 V in 0.2 M K$_2$SO$_4$. Thermalized electrons at 25 °C contribute to the relatively high conductivity of the narrow bandgap HgCdTe material. Thus the observed Δi_{hv} response was of the order of a microampere, as indicated by the level of photocurrent at the start of the Δi_{hv} curve in Fig. 3. When Cd^{2+} and Te^{4+} ions were added to the electrolyte, a large Δi_{hv} response was elicited at **a**, signaling the growth of a wide bandgap (1.5 eV) semiconductor, CdTe. The Δi_{hv} curve in Fig. 3 tracks the formation of CdTe in region **a**. The photocurrent diminishes in regime **b** due to the interference of the Cd metal deposition and H$_2$ evolution reactions. The photocurrent response is partially recovered on scan reversal, regime **a'** until the anodic decomposition of CdTe at **c** reduces it to background level. The response at **d** is attributed to the formation of the anodic oxide on HgCdTe which is also a wide band gap semiconductor.[9] At a constant potential of − 0.55 V, where the maximum photocurrent was observed, the Δi_{hv} increased by an order of magnitude while i_t remained at a steady level. A silver–grey film was deposited from this electrolyte at − 0.55 V at 70 °C on HgCdTe. The "deposited" CdTe layer of typically > 200 Å thick and could be visibly distinguished from the thin (monolayers) "reconstructed" CdTe layer. Auger electron spectroscopy indicated that the deposited film was stoichiometric CdTe.

V. VOLTAMMETRIC CHARACTERIZATION OF CHEMICALLY ETCHED SURFACES

The electrochemical method can be further extended as an *in situ* analytical tool to monitor the surface quality of HgCdTe. Br$_2$/organic solvent etches are presently used to clean HgCdTe surfaces. Considerable variation in the quality of the etched surfaces has been noted, usually only after elaborate construction and testing of devices. The reactions of HgCdTe with Br$_2$ based etchants is anisotropic, depleting the surface of Cd and to a lesser extent of Hg, leaving a Te film.

The surface oxides and other nonstoichiometric residues on chemically etched surfaces may be characterized by electrochemically oxidizing or reducing these species. The voltammogram shown in Fig. 1 can be used as a reference frame to identify the surface processes. Curve *a* in Fig. 4 shows reduction current for a Br$_2$/MeOH etched sample, as the potential is scanned negative from the open-circuit potential. A cathodic peak observed at − 0.75 V may be attributed to the reduction and dissolution of Te via

$$Te + H^+ + 2e^- \rightarrow HTe^-. \quad (1)$$

By comparison, the electrochemically reduced surface in curve *b* Fig. 4 appears clean and free of corrosion products. The difference in currents between curves *a* and *b* corresponds to about two monolayers of Te, assuming a 2e^- reduction by reaction (1).

Figure 5 shows the effects of air exposure on Br$_2$-etched surfaces. Curve *a* shows that the Te residue from the etch is converted to TeO$_2$ after 1 h of air exposure. The TeO$_2$ is reduced at − 0.25 V via the reaction

$$TeO_2 + 4H^+ + 4e \rightarrow Te + 2H_2O. \quad (2)$$

Prolonged air exposure results in the diffusion of Hg into the oxide. The reduction of HgTeO$_3$ to HgTe as

$$HgTeO_3 + 6H^+ + 6e \rightarrow HgTe + 3H_2O \quad (3)$$

and the anodic dissolution of this HgTe film as

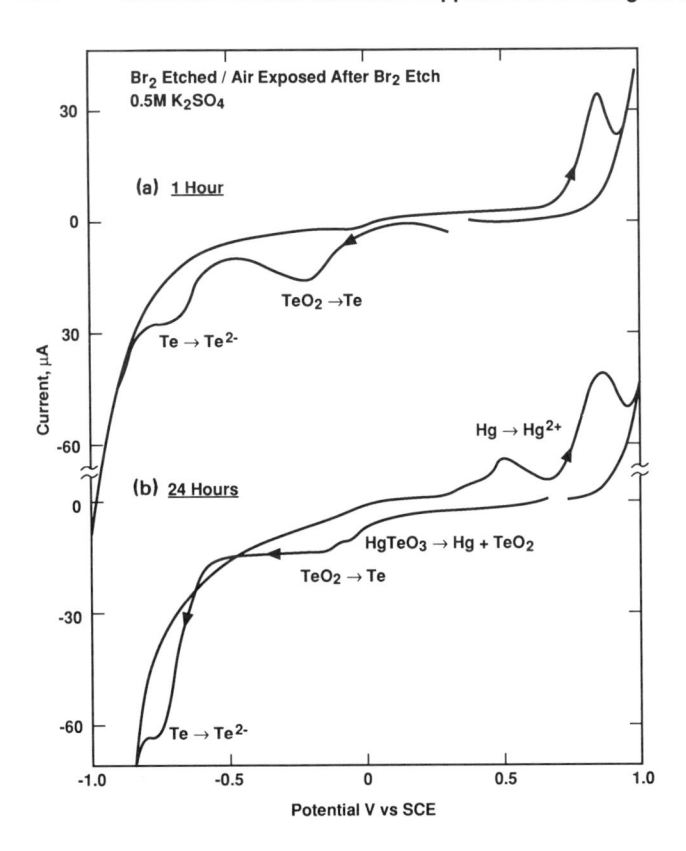

FIG. 5. Experiment of Fig. 4 after (a) 1 h air exposure of etched surface and (b) 24 h air exposure of etched surface.

$$HgTe \rightarrow Hg^{2+} + Te + 2e^- \quad (4)$$

are shown in the I–V curve b, obtained after 24-h air exposure.

Comparison of etch rates: Voltammetric methods provide a rapid quantitative estimate of the surface composition. This approach was further used to assess the effects of various Br$_2$ based etchants including post-etch air exposure on the surface composition of HgCdTe. The etchant composition and procedures are summarized in Table I. Etched samples were rinsed and N$_2$ dried before electrochemical characterization. The main difference between various etchants is evidenced in the etch rates, deduced from the coulombic charge required to reduce the excess Te on the surface. In Table I, this charge is represented in monolayers of undissolved etch products remaining on the HgCdTe surface. As anticipated the etch rates, deduced from integrating the Te and TeO$_2$ peak currents, vary with the solvents, increasing in the order of ethylene glycol (EG) < lactic acid (LA) < methanol (ME) < hydrobromic acid (HB).

The Te produced after EG or LA etch is roughly two monolayers. The presence of TeO$_2$ in the surface film implies that some residual native oxide remains on the surface after the first 30 s EG or LA etch. This oxide could be removed by repeating the treatment.

Since the etch rates are higher in ME and HB solutions, the composition of the etched surface is extremely sensitive to the etch time as well as the solution concentration. The Te produced is greater by a factor of 2 and 3 for ME and HB etches.

Roughly same thickness for the TeO$_2$ layer was found after exposing the variously etched surfaces to air for an hour but the magnitude of the reduction peak for Te dissolution varied. This indicates that some unoxidized Te remains beneath the oxide layer, particularly for samples etched in ME and HB based etches. The reduction after 20 h air exposure shows similar thickness for oxides formed on LA and EG etched and an order of magnitude higher thickness for the other two etchants. Less than a monolayer of HgTeO$_3$ is also detected in this oxide, evidenced from the reoxidation Hg^{2+} peak, as in the experiment of Fig. 5(b).

Conversion of excess Te to CdTe: An alternate approach to treat the excess Te on Br-etched surface is to use it advanta-

TABLE I. Comparison of various chemical etches. Insoluble corrosion products are represented in number of monolayers

	Air exposure	~2m Etch 1		~2m Etch 2	1 h		24 h		
Etchant/time/rinse	Surface residue	Te	TeO$_2$	Te	Te	TeO$_2$	Te	TeO$_2$	HgTeO$_3$
LA	0.5% Br$_2$/Lactic Acid 30 s H$_2$O rinse	2	<1	2	1	1.5	1	1	<0.5
EG	0.5% Br$_2$/ethylene glycol 30 s EG and IPA rinses	0.5	1	2	1	1.5	<1	1	<0.5
HB	0.5% Br$_2$/HBr 5 s H$_2$O rinse	10	4	2	1.5	1.5	1.5	3	<0.5
ME	0.5% Br$_2$/MeOH 5 s MeOH, IPA rinse	5	4	2	1.5	1.5	9	6	1

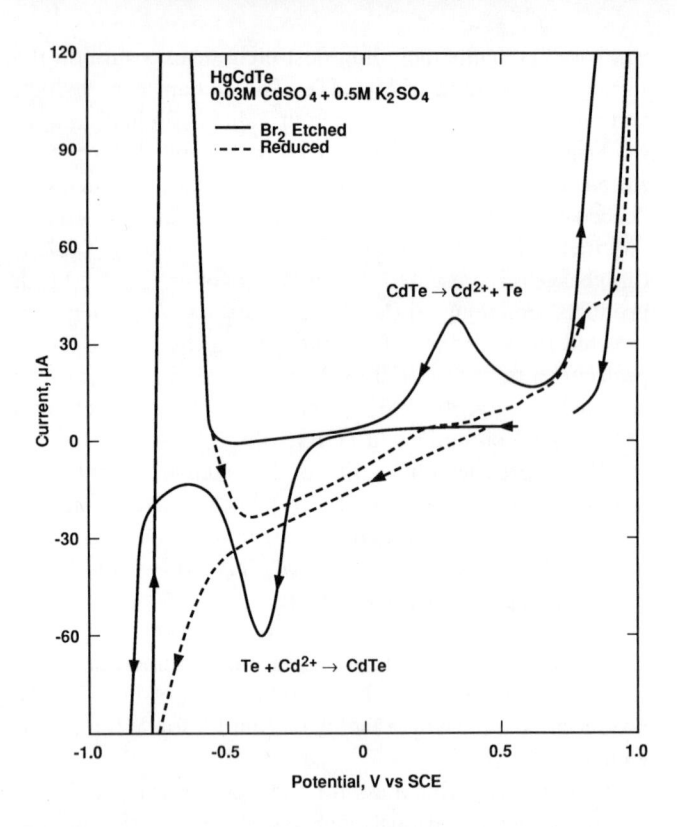

FIG. 6. Experiment of Fig. 3 in 10 mM $CdSO_4$ + 0.2 M K_2SO_4 at Br_2/MeOH etched (solid curve) and electrochemically reduced (dashed curve).

geously to grow a thin CdTe layer. If an etched specimen is reduced in a Cd^{2+}/K_2SO_4 electrolyte, a cathodic peak appears, indicating the formation of CdTe, as shown in the solid curve of Fig. 6. Note that the peak potentials are different from the oxide reduction potential and that the reverse reaction, the dissolution of CdTe, appears during the subsequent anodic scan. Neither of these reactions occurs on an electrochemically reduced, Te-free surface as shown by the dashed curve.

VI. SUMMARY

A new electrochemical method has been developed for *in situ* etching, characterization and passivation of HgCdTe surfaces. The electrochemical approach offers selectivity and flexibility to direct the path of surface reactions and thus to tailor the interface structure.

Evaluation of Br_2-etched surfaces confirms the extreme sensitivity of the Br_2-etched HgCdTe surface to the etching parameters and to the post-etch conditions. A more stable and reproducible surface may be attained by using the electrochemical cleaning procedure in Fig. 1 or by mere electrochemical reduction of excess Te (Fig. 3). The electrochemically reconstructed HgCdTe surfaces showed thermochemical stability and substantially improved electrical properties relative to the chemically etched surfaces. The electrochemical method also offers several advantages over the traditional chemical or vacuum processing techniques. Low temperature growth, precise control over surface chemistry, *in situ* cleaning and characterization capabilities, and a conformal nature are some of the unique features of electrochemical processing.

[a] Present Address: Interphases Research, 722 Rushing Creek Place, Thousand Oaks, CA 91360.
[1] P. C. Catagnus and C. T. Baker, U. S. patent No. 3,977,018, 1976.
[2] Y. Nemirovsky and L. Burstein, Appl. Phys. Lett. **44**, 443 (1984).
[3] R. L. Strong, J. D. Luttmer, D. D. Little, T. H. Teherani, and C. R. Helms, J. Vac. Sci. Technol. A **5**, 3207 (1987).
[4] T. H. Teherani and D. D. Little, U. S. Patent No. 4,726,885, 1988.
[5] E. Weiss and N. Mainzer, J. Vac. Sci. Technol. A **6**, 2765 (1988).
[6] D. R. Rhiger and R. E. Kvass, J. Vac. Sci. Technol. A **1**, 1712 (1983).
[7] C. R. Helms, J. Vac. Sci. Technol. A **8**, 1178 (1990).
[8] S. Johnson, E. R. Blazejewski, J. Bajaj, J. S. Chen, L. Bubulac and G. Williams, J. Vac. Sci. Technol. B **9**, 1661 (1991).
[9] S. Menezes, J. Electrochem Soc. (to be published).

Composition, growth mechanism, and stability of anodic fluoride films on Hg$_{1-x}$Cd$_x$Te

Eliezer Weiss[a] and C. R. Helms
Department of Electrical Engineering, Stanford University, Stanford, California 94305

(Received 2 October 1990; accepted 19 October 1990)

Fluoridic films produced by the anodization of Hg$_{1-x}$Cd$_x$Te ($x \sim 0.22$) surfaces in nonaqueous solutions consist of three distinct regions. A thick uniform region, containing the fluorides of cadmium, mercury, and tellurium, as well as HgTe, is covered by a thin CdF$_2$ layer. The third region—the film–substrate interface—poor in mercury, consists mainly of CdF$_2$ and TeF$_4$. The anodic film is grown by two mechanisms: the dominant one occurs by motion of the film–substrate interface into the semiconductor, consuming the original surface. There is, however, some growth at the film–electrolyte interface which forms the thin CdF$_2$ layer on top of the structure. The CdF$_2$ rich region acts as a diffusion barrier for the indiffusion of oxidizing species. Tellurium ions, on the other hand, diffuse to the outer surface to be oxidized there. This diffusion barrier is overcome when the fluorides are thermally stressed in the presence of traces of oxygen, and the film starts to oxidize. However, the film–substrate interface does not degrade unless the film is heavily oxidized. A model for the anodic fluoride structure is presented and discussed in light of the film stability with respect to room temperature post-growth oxidation and thermal treatments.

I. INTRODUCTION

The issue of surface passivation is important for Hg$_{1-x}$Cd$_x$Te devices, as it has been historically a major factor in limiting device performance.[1] An anodic oxide film on Hg$_{1-x}$Cd$_x$Te[2] is considered to be an effective means for passivating infrared focal plane arrays (IR FPAs) based on photoconductive or metal–insulator semiconductor (MIS) structures.[3] The two major drawbacks of the anodic oxide/Hg$_{1-x}$Cd$_x$Te structure are: (1) its poor thermal stability at temperatures above ~ 70 °C,[4,5] and (2) the large fixed positive charge it contains (typically 10^{12} cm^{-2}),[6] which render it inappropriate as passivant for IR FPAs based on photodiodes. Recently Weiss and Mainzer reported the passivation of Hg$_{1-x}$Cd$_x$Te surfaces by growing anodic layers from fluoridic solutions.[7,8] The use of anodic fluoridization or fluoro-oxidation is advantageous because the resulting interfaces are relatively thermally stable. Moreover, this method also reduces the surface recombination velocity and enables the adjustment of the amount of band bending at the surface of Hg$_{1-x}$Cd$_x$Te infrared detectors. These films can be grown anodically in either aqueous or nonaqueous solutions of KF or KF + KOH in ethylene glycol (EG). An anodic fluoride, free of oxygen, is grown from nonaqueous solutions, whereas even low hydroxyl ion concentration in an aqueous bath causes the growth of an anodic fluoro-oxide, consisting of an oxide and small amounts of fluorides.[7]

This paper presents results on the composition, growth mechanism, and the stability of anodic fluoride films on Hg$_{1-x}$Cd$_x$Te. We have concentrated only on the anodic growth in nonaqueous solutions of KF; that is, on the anodic fluoridization as opposed to the fluoro-oxidation process. A model for the anodic fluoride structure, based on our results, will be presented. It will be discussed in light of the stability of this anodic film with respect to room temperature post growth oxidation and thermal treatments.

II. EXPERIMENTAL DETAILS

The growth of the anodic films and the characterization methods were described elsewhere.[9] Briefly, single-crystalline n-type samples of random crystallographic orientation Hg$_{1-x}$Cd$_x$Te with $x \sim 0.22$ grown by solid state recrystallization were used. The anodic films were grown on bromine–methanol etched wafers in a 1 M KF in EG electrolyte using a Pt counter electrode, and a constant current density of ~ 0.5 mA/cm^2. We differentiate between two classes of growths: in one a fresh bath was used and in the second the growth was performed in a bath that had been used previously for growing an anodic film. The appropriate baths will be termed "fresh" and "used," respectively. Wafers to be heat treated were sealed in quartz ampules filled with ~ 0.5 atm of N$_2$ (99.998%). The anneal temperature ranged from 150 to 220 °C.

Auger electron spectroscopy (AES) experiments were performed in a Varian Auger spectrometer system equipped with a single pass cylindrical mirror analyzer (CMA). The electron beam was rastered on the samples with an energy of 3 keV and a current density of 0.35 mA/cm^2. Sputter profiling was performed with a rastered 1 keV, 5 μA/cm^2 Ne$^+$ ion beam, yielding a sputter rate of ~ 20 Å/min for the anodic fluoride film.[9] X-ray photoelectron spectroscopy (XPS) analyses were performed with an ESCALAB 2 (VG Scientific) using the nonmonochromatic MgKα (1253.6 eV) x-ray source and a concentric hemispherical analyzer (CHA). To obtain depth profiling, a series of XPS measurements were interdispersed with ion sputtering, during which the chamber was disconnected from the pumping system causing an increase of the base pressure to the high 10^{-9}s Torr. A 1 keV, 10 μA/cm^2 Ar$^+$ ion beam was used, yielding a sputtering rate of ~ 50 Å/min.

III. COMPOSITION OF THE ANODIC FLUORIDE

Figure 1 shows the Auger depth profiles (ADPs) of anodic fluoride films of different thicknesses: ~2000 Å [Fig. 1(a)] and ~350 Å [Fig. 1(b)] (note that in these Auger depth profiles, as well as in the rest of the ADPs shown in this work, the Auger signals were recorded for 5 min before the ion gun was turned on). The ADPs in Fig. 1 show that the structures are covered with an oxide layer resulting from interaction of the film with the ambient, as will be discussed later. Leaving this oxide film out it can be seen that the anodic film consists of three distinct regions. The near surface region of the film is rich in cadmium and fluorine and poor in both tellurium and mercury. The thick uniform intermediate region, which is the bulk of the film, contains appreciable amounts of all four elements. The third region is found at the film–substrate interface (it is not seen in the thick film of Fig. 1(a) because of the artificially broader interface formed by the longer sputtering). This interfacial region is poor in mercury and tellurium. The shift in the Cd line, which starts to appear at this stage,[7,9] is responsible to the "dip" in the Cd curve slightly deeper in the structure and screens the buildup of Cd at the interface. The low-energy proton induced nuclear reaction analysis reported in Ref. 8, being both very sensitive to fluorine and free of preferential sputtering and atomic intermixing (problems associated with ion sputtering), shows a distinct accumulation of fluorine at the interface. We deduce that the film–semiconductor interface consists of a thin CdF_2-rich layer.

The line shapes of Cd (MNN), Te (MNN), and Hg (NVV) in the fluorides were studied earlier.[7,9] While the Hg peak is not shifted at all, the Cd peak in the bulk of the film is shifted −3 eV relative to the substrate line, indicating that it is due to Cd in CdF_2.[10] The surface Cd line is shifted only by −2 eV relative to the substrate line and is due to CdTe and $CdTeO_3$ in addition to the CdF_2.[9] The only shift in the Te peak is found near the surface of the sample and is associated with TeO_2, as will be discussed later. In the XPS analysis, on the other hand, the Te $3d_{5/2}$ peak shows two distinct contributions[9] at binding energy (BE) = 572.5–572.9 eV and at BE = 575.9–576.4 eV. The first is associated with Te^{-2} (Ref. 11,12), and the second—with Te^{+4} bonded to oxygen.[11-14] Since the AES analyses have shown that the Cd is bonded exclusively to fluorine, the Te^{-2} peak comes from HgTe. Furthermore, no oxygen was found by Auger depth analyses of similar films (Fig. 1), it is an artifact due to the way the XPS analysis was performed: the high residual pressure in the vacuum chamber and the noncontinuous sputtering. Oxygen containing species, from the vacuum environment, interact with Te in the film. The HgTe in the film is not affected, since there is no Te^{+4} in the substrate line. Hence, the film contains TeF_4 (being the only solid tellurium fluoride compound at room temperature). As will be shown later, TeF_4 is very reactive toward water vapor.

The main conclusion concerning the bonding in the anodic fluoride film achieved by the AES and XPS analyses is that the film contains CdF_2, HgTe, and TeF_4. There might also be some HgF_2 in it, but due to the insensitivity of the Hg peaks to its chemical environment such a conclusion cannot be drawn from these experiments.

To better evaluate the film composition we have used an electron microprobe to analyze a thick (~2000 Å) anodic fluoride film, the ADP of which is shown in Fig. 1(a). The electron microprobe was shown to be a reliable experimental technique for obtaining the composition of the bulk anodic oxide of $Hg_{1-x}Cd_xTe$, providing thick films (≥2000 Å) were analyzed.[5] The technique gives only an average composition of the depth probed in the sample. Using this method we found the average concentrations of the various elements in the anodic fluoride film to be:[9] Hg—15%, Cd—4%, Te—19%, and F—62%. Based on these ratios we calculated the possible composition of the anodic fluoride in terms of the chemical compounds in it. The calculation was based on the AES and XPS results that the film contains CdF_2, TeF_4, and HgTe. It turns out that a fit to the results can be achieved only if we assume also the existence of HgF_2 in the film. This calculation gives: CdF_2—14%, HgF_2—21%, TeF_4—35%, and HgTe—30%, or, in other words, CdF_2:1.5HgF_2:2.5TeF_4:2.2HgTe. This yields a Te^{+4}:Te^{-2} ratio of 1.2:1, exactly what was observed in the XPS measurements.[9]

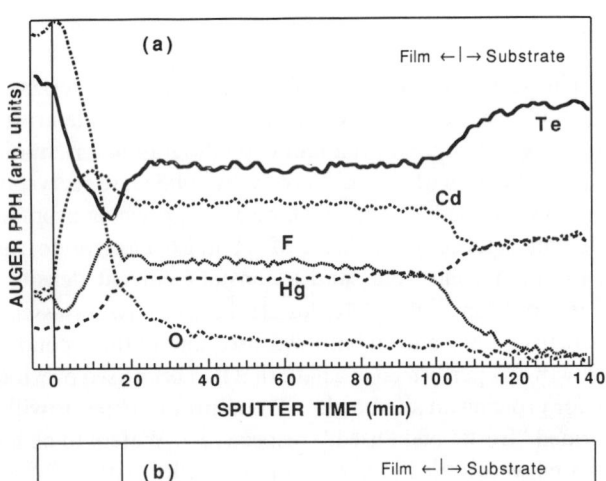

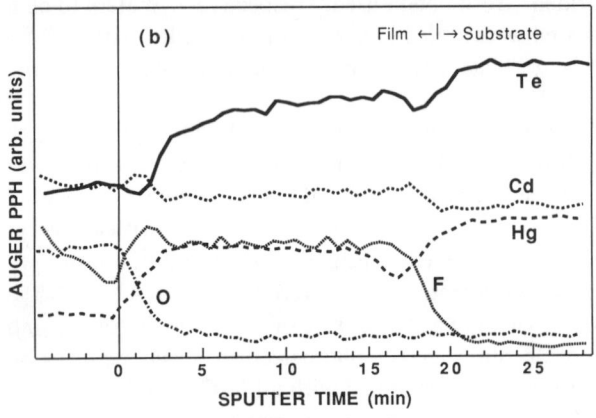

FIG. 1. ADPs (SPR ~20 Å/min) of anodic fluoride films grown on $Hg_{1-x}Cd_xTe$ (x~0.22). (a) A ~2000-Å thick film kept in the laboratory atmosphere for several hours before loading to the Auger chamber. (b) A ~350-Å thick film grown from a used bath, and kept under vacuum until it was loaded into the Auger chamber (total exposure to room ambient: ~20 min). The PPHs are for Cd MNN, Te MNN, Hg NVV, O KLL, and F KLL.

IV. ANODIC FLUORIDE GROWTH MECHANISM

The mechanism by which the film grows might be expected to have a significant effect on the electrical properties of the anodic film–substrate interface; since upon growth this interface can move into the substrate, consuming the original surface, or it occurs at the original substrate surface and the film thickens by growing at the film–electrolyte interface. It is, therefore, important to determine which of these possible mechanisms occur. To do so the substrate surface must be marked. We have used two kinds of markers: an ultrathin (~ 10 Å) Pd film and the thin CdF_2 film found on top of the anodic film (Fig. 1).

Figure 2 shows the ADP of an anodic film grown on a Pd covered $Hg_{1-x}Cd_x Te$ surface. Identical results were achieved when growing films of different thicknesses: the Pd layer is found always on top of the structure, indicating that the original surface is left outside. However, it should be born in mind that covering the surface with a thin metal film may interfere with normal growth. For instance, it can block the diffusion of the dominant reactant, thus causing an otherwise secondary mechanism to be the primary one. Indeed, higher current densities were needed in order to overcome the Pd layer and form the anodic film. Also, oxygen was liberated at the Pd surface due to electrolysis, and the voltage versus time (V–t) curves exhibited a very long plateau. However, the second marker experiment described below indicated that the ultrathin Pd film had not affected the main growth mechanism: the film grows at its substrate interface.

We recall that a thin film of CdF_2 was found on top of the anodic film (Fig. 1). How is this layer formed? The CdF_2 film was found on thick ($\geqslant 800$ Å) anodic fluoride films, and its thickness increased with the total film thickness. Thin anodic films were covered with this layer provided they had been grown from used baths. The dependence on the state of the electrolyte and the length of the anodization indicates that it is due to CdF_2 precipitation from the solution. To prove this point we grew thin films ($\leqslant 400$ Å) from fresh baths to which we had added CdF_2 in excess of its saturation level. The resulting films have profiles identical to those shown in Fig. 1(b), whereas control films grown from fresh baths without the added CdF_2 have different ADPs (see Fig. 3 below). In normal baths (without the added CdF_2) the CdF_2 formed by the anodization process is dissolved in the electrolyte during the early stages of the growth (this is indicated by the Cd depletion at the surface of the film shown in Fig. 3). Later, when its solubility limit is reached, it precipitates on top of the surface. This thin CdF_2 layer plays a crucial role in preventing the oxidation of the anodic film as will be demonstrated later. The peaks of Cd and F found on the surface in the ADPs indicate also that the main mechanism by which the film is grown is at the film-substrate interface. If this was not the case the precipitated CdF_2 would have been incorporated in the growing anodic film instead of being accumulated on its surface.

FIG. 3. ADP of a ~ 400-Å thick anodic fluoride film grown from a fresh bath.

V. STABILITY OF THE ANODIC FLUORIDE

A. Post-growth room temperature oxidation

The films discussed until now are covered by an oxide layer [Figs. 1(a), and 2]. The oxide is rich in Te and poor in Cd, with the Te (MNN) Auger peak in it being strongly shifted.[7,9] This indicates that the oxide consists mainly of TeO_2, although some oxidized Cd is also found there.[9] There might also be some HgO in the film, even though the ADP reveals only small amounts of mercury in it. It is known that AES and XPS have anomalously low Hg signals in $Hg_{1-x}Cd_x Te$ oxides.[15] The oxide is formed after exposure of the film to the room ambient. Films that were kept under vacuum and were exposed only during loading into the analysis chamber (~ 30 min) are covered by only a very thin oxide film [Fig. 1(b)]. A several hour exposure to the laboratory ambient is sufficient for the oxide to reach its final thickness: films that were kept in the room for several days (~ 10) have a top oxide layer of the same thickness as films that were exposed for only 2–3 h. The oxide thickness does change as a function of the anodic film thickness: the thicker the fluoride film the thicker the oxide on top of it. The oxide film is found on top of the surface markers, Pd (Fig. 2) or CdF_2 [Fig. 1(a)]. This implies that it is grown by out-diffusion of Te from the film to the surface. As will be discussed below, TeF_4 is the species being oxidized. The thicker the anodic fluoride film, the more TeF_4 it contains, hence the thicker the oxide layer.

The TeO_2 layer is formed on top of thick films as well as thin films that were grown from used baths. As was ex-

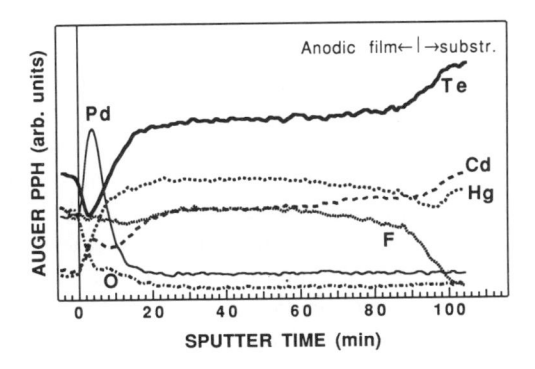

FIG. 2. ADP of a ~ 1800-Å thick anodic fluoride grown on a $Hg_{1-x}Cd_x Te$ ($x \sim 0.22$) surface covered by an ultrathin (~ 10 Å) layer of Pd.

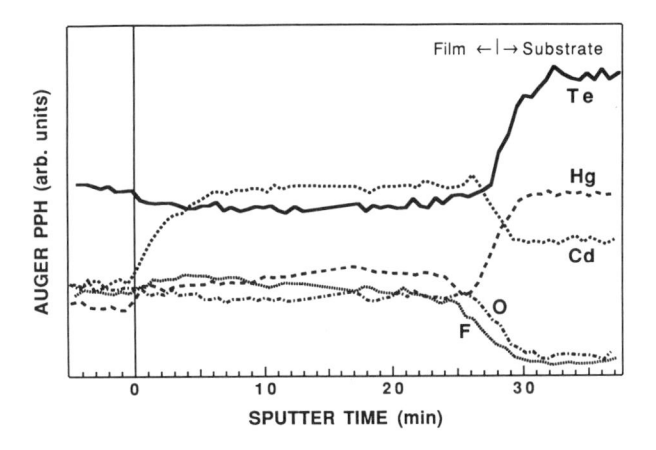

plained before, such films are characterized by a CdF_2 rich layer on their surface. We shall deal now with the oxidation of thin films grown from fresh baths. Figure 3 shows the ADP of such a thin film. The most important feature shown by it is that the film is oxidized all the way to the film–substrate interface. Also, the film does not have the CdF_2 rich layer on its surface; on the contrary, its surface is poor in Cd. This is due, as was explained earlier, to CdF_2 dissolution in the electrolyte. During such a short growth the solubility limit is not reached. Films that were kept under vacuum contain less oxygen than films that were exposed to air for longer periods. Nevertheless, they still contain an appreciable amount of oxygen. To prove that all the oxygen in these films is introduced by post-growth interaction with the room ambient, we performed the following experiment. A ~1200-Å thick film (second order gold–violet) was carefully etched until a blue color was achieved, indicating a ~400-Å thick film was left.[9] The etched film as well as a ~1200-Å thick control film were kept in air for ~1 h before being loaded into the Auger chamber. The control film had a depth profile similar to the one shown in Fig. 1(a), ensuring that the ~1200-Å thick films did not contain oxygen in their bulks. The ADP of the etched sample was identical to the one shown in Fig. 3. Clearly, after etching the top of the film oxygen penetrated it and oxidized its constituents. Films as thin as those of Fig. 3 but that are covered by the thin CdF_2 rich layer are not oxidized all the way to the film–substrate interface [Fig. 1(b)]. The CdF_2 layer serves, therefore, as a diffusion barrier for the oxidant. However, it does not prevent the out-diffusion of Te^{+4} to form the TeO_2 layer on top of it.

B. Thermal stability in the presence of oxygen

As stated above one of the major advantages of using anodic fluoride films is their better thermal stability compared to the anodic oxide.[7,8] In this subsection we present the results of our study of thermally stressed anodic fluoride-$Hg_{1-x}Cd_xTe$ interfaces.

Figure 4 shows the ADPs of anodic fluoride films which were annealed at various temperatures for 3 h in N_2. The annealed films consist of four regions that should be compared to the four original regions (including the top TeO_2 layer) in the unannealed samples (Fig. 1). It can be seen that the thickness of the top oxide and the amount of Cd in it increase as the anneal temperature increases. The Cd peak in this region (not shown) is shifted only −1 eV relative to the substrate Cd peak. Apparently the residual oxygen in the ampule reacts with the anodic fluoride, thickening the top TeO_2 layer. Furthermore, the stable $CdTeO_3$ compound is formed at the TeO_2-anodic fluoride interface, consuming the thin CdF_2 layer which is now converted to oxy-fluoride compounds. The bulk of the fluoride (the third region from the top) contains oxygen and the fluorine concentration in it decreases as the anneal temperature is increased. It should be noticed, however, that it still contains an appreciable amount of HgTe. When the anneal temperature is high enough [>195 °C, Figs. 4(b), 4(c)] the oxygen can diffuse all the way to the film–substrate interface to form the fourth region there. Judging from the Te line shape (not shown) it still contains HgTe but also Te^{+4} bonded to oxygen.

In the interface between the fourth, oxygen-rich region and the substrate the concentration of the various elements change gradually from their values in the film to those in the substrate. This is in contrast to both anodic oxides[4,5] and thin anodic fluoride films, like the one shown in Fig. 3, that are rapidly oxidized upon exposure to the room ambient. This is seen in Fig. 5 which shows an ADP of a ~400 Å anodic fluoride which was annealed at 220 °C for 3 h in N_2. A control sample grown to the same thickness and sealed in an ampule without annealing had an ADP identical to the one shown in Fig. 3. The ADP of the annealed sample (Fig. 5) is markedly different than either that of the unannealed one (Fig. 3) or those of the thick annealed fluoride films (Fig. 4). First, it contains less fluorine and more oxygen than the unannealed sample. Second, the atomic concentration of the various elements in the film–substrate interface are no longer changing gradually. On the contrary, the interface can be divided into two regions: on the film side of the interface Cd and O accumulate and Hg is somewhat depleted, wheres on the substrate side it is the mercury that accumulates and Cd is dramatically depleted. This behavior resembles very much the behavior of thermally stressed anodic oxide layers.[4,5] It is caused, most probably, by the same driving force: the formation of the very stable $CdTeO_3$ by the reaction of TeO_2, for example, with the substrate.[4,5]

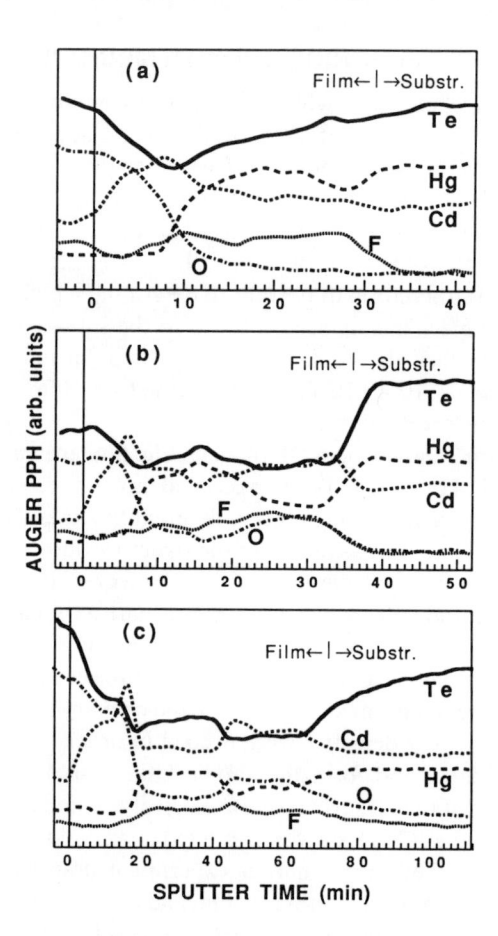

FIG. 4. ADPs of thick anodic fluoride films annealed at different temperatures for 3 h in the presence of traces of oxygen: (a) 150 °C, (b) 195 °C, and (c) 220 °C. The films were grown to the thicknesses of ~800-Å (a,b) or 1200 Å (c).

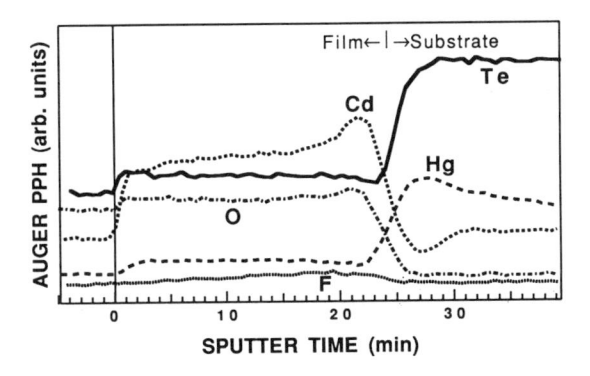

FIG. 5. ADP of a ~400-Å thick anodic fluoride grown from a fresh bath and annealed at 220 °C for 3 h in the presence of traces of oxygen.

VI. MODEL FOR THE ANODIC FLUORIDE STRUCTURE AND ITS IMPLICATIONS

Based on the results described above, we have constructed a simple model for the structure of the anodic fluoride which is presented schematically in Fig. 6(a). The bulk of the film contains CdF_2, HgF_2, TeF_4, and HgTe in ratios of 1:1.5:2.5:2.2. The amount of TeF_4 is equal to the sum of CdF_2 and HgF_2 in the film, which might imply that these compounds exist in the film as $CdTeF_6$ and $HgTeF_6$ in a ratio of 1:1.5. This seems reasonable because of the expected stability of the TeF_6^{-2} ion, the sp^3d^2 hybridization in which is expected to give rise to a stable octahedral symmetry.[16] However, the TeF_6^{-2} anion is not known, despite extensive attempts to synthesize it by a wide variety of reactions.[17] On the other hand, the TeF_5^- ion, with a structure of a deformated square pyramid, is known to exist.[17] The exact chemical compounds existing in the anodic fluoride are, therefore, not completely clear yet. We shall designate them henceforth as a mixed lattice of $MF_2 \cdot TeF_4 \cdot HgTe$ with M representing Cd and Hg in the ratio 1:1.5.

The two possible fluoride film structures [Figs. 6(a), 6(b)] oxidize upon exposure to room ambient. This is shown schematically in Figs. 6(c) and 6(d). The dramatic difference between the two structures is due to the prevention of oxygen in-diffusion by the thin CdF_2 layer on top of the structure in Figs. 6(a), 6(c). In this case the cations are diffusing to the surface to form an oxide on top of the structure. When this CdF_2 diffusion barrier is absent [Figs. 6(b), 6(d)], the oxidant diffuses into the film and oxidizes the cations there. It is not likely that the HgTe in the anodic film oxidizes upon exposure to air. The question is, therefore, which fluoride compound is responsible for the oxidation observed by the exposure to the room ambient?

It is reported[18] that TeF_4 is unstable in the presence of water:

$$TeF_4(s) + 2H_2O(g) \rightleftharpoons TeO_2(s) + 4HF(g). \quad (1)$$

Using enthalpies of formation and third law entropies from Refs. 19 and 20 it can be shown that reaction (1) reaches equilibrium (at 298 K) already at a partial pressure ratio of $P_{H_2O}/(P_{HF})^2 = 40$ atm^{-1}. At higher ratios, expected to be found in the room ambient, reaction (1) is strongly driven to the right. Repeating the calculation for the oxidation of CdF_2 and HgF_2:

$$CdF_2(s) + H_2O(g) \rightleftharpoons CdO(s) + 2HF(g), \quad (2)$$

$$HgF_2(s) + H_2O(g) \rightleftharpoons HgO(s) + 2HF(g), \quad (3)$$

it can be shown that equilibrium is reached at very high $P_{H_2O}/(P_{HF})^2$ ratios ($\sim 10^{17}$ and $\sim 10^{26}$ atm^{-1} for CdF_2 and HgF_2, respectively). These ratios are so high that under normal conditions both CdF_2 and HgF_2 remain unoxidized. The intensity of the fluorine Auger signal (relative to that of Cd) is reduced by about a third when the film is oxidized as in Fig. 3. That is, in the structure $MF_2 \cdot TeF_4$ two fluorine atoms are replaced by an oxygen atom. Since the anion $TeOF_4^{-2}$ is known to exist[17] the following reaction occurs upon the exposure:

$$MF_2 \cdot TeF_4(s) + H_2O(g) \rightleftharpoons MTeOF_4(s) + 2HF(g). \quad (4)$$

The results of the annealing experiments can be understood with the help of Fig. 7, which shows possible reactions which may occur in either a thin (and oxidized) anodic fluoride [Fig. 7(a)] or a thick (unoxidized) film [Fig. 7(b)]. The ternary phase diagram for Hg, Cd, Te, and F[21] shows that any fluoride bonded Hg or Te are unstable with respect to Te bonded Cd (the phase diagram in Ref. 21 treats TeF_6, which is a gas at room temperature, there are only minor

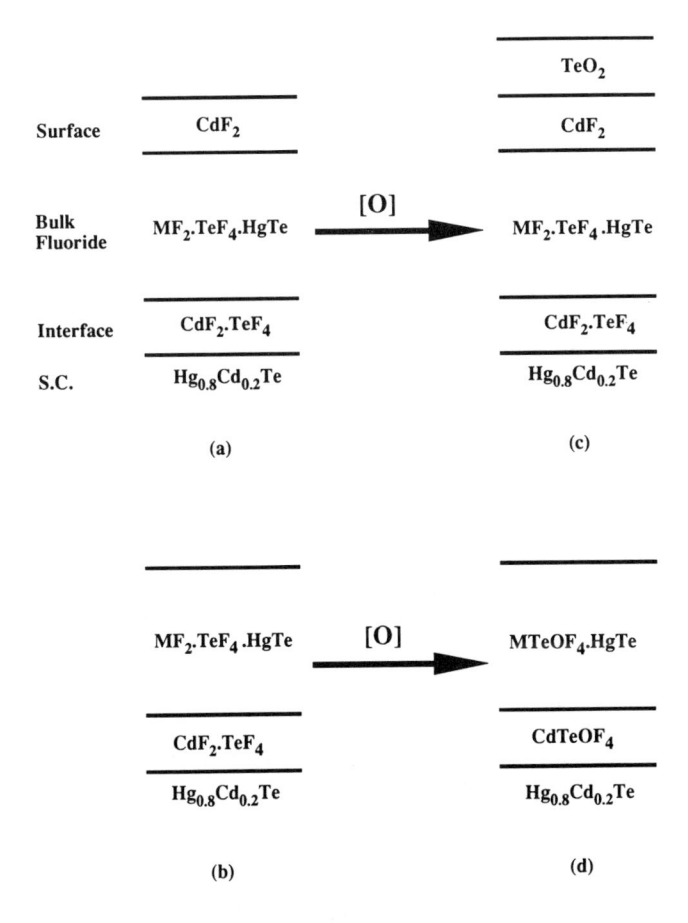

FIG. 6. Model for different film structures and their interfaces with the semiconductor. (a) "As grown" thick (≥ 800 Å) film or thin (≤ 400 Å) film grown from a used bath. (b) A film in which the top CdF_2 rich layer is absent. (c) Same as the film in (a), but after exposure to room ambient. (d) Same as the film in (b), but after exposure to the room ambient. M represents Cd and Hg in the ratio 1:1.5..

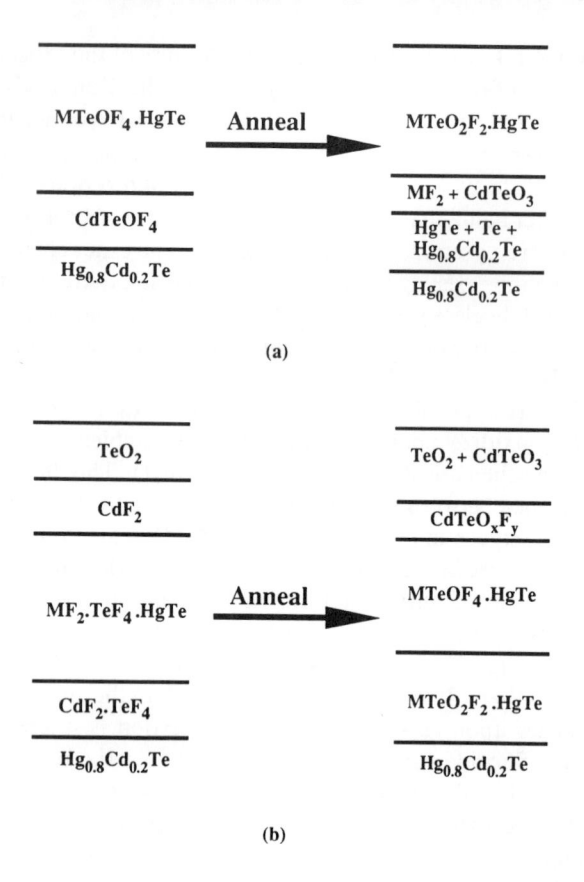

FIG. 7. Model showing the affect of annealing at 220 °C in the presence of traces of oxygen on the anodic fluoride-$Hg_{1-x}Cd_xTe$ structure for two cases: (a) a film without the top CdF_2 layer, and (b) a film with it. M represents Cd and Hg in the ratio 1:1.5..

changes if one treats TeF_4 instead). Therefore, an interaction between the anodic fluoride and the substrate is expected to take place, similar to the anodic oxide case. However, our results show that such instability is not found in the anodic fluoride samples thermally stressed to 220 °C. Nevertheless, an interaction with the residual oxygen in the annealing ambient is possible as is explained next. Figure 5 shows that when a thin oxidized sample is heated, it further reacts with oxygen and loses more fluorine. This yields another stable oxy-fluoride ion—$TeO_2F_2^{-2}$ (Ref. 17):

$$2MTeOF_4(s) + O_2(g) \rightleftharpoons 2MTeO_2F_2(s) + 2F_2(g). \quad (5)$$

From the similarity between the interface in Fig. 5 and the thermally stressed anodic oxide-$Hg_{1-x}Cd_xTe$ interface[4,5] we can deduce that $MTeO_2F_2$ reacts with the substrate according to a reaction like:

$$3MTeO_2F_2 + 10Hg_{0.8}Cd_{0.2}Te$$
$$\rightleftharpoons 3MF_2 + 2CdTeO_3 + 8HgTe + 3Te. \quad (6)$$

It is the large free energy of formation of $CdTeO_3$ that drives reaction (6) to the right.

The anodic fluoride films of Fig. 4 behave differently because oxygen has first to overcome the diffusion barrier imposed by the CdF_2 layer; second it has to react with the fluorides to form $MTeOF_4$ and then $MTeO_2F_2$. The profiles show that the oxygen-rich oxy-fluoride compound segregates at the film–substrate interface, this is probably because such a process lowers the interfacial excess energy. A reaction similar to reaction (6) might take place at higher temperatures when enough $MTeO_2F_2$ is formed at the interface.

VII. CONCLUSIONS

The lack of a compound, as stable as $CdTeO_3$, in the fluoride-$Hg_{1-x}Cd_xTe$ system makes the anodic fluoride unstable with respect to oxidation. However, a secondary growth mechanism forms a CdF_2-rich layer on top of the anodic fluoride. After it is formed this thin layer acts as a diffusion barrier and prevents the oxidation of the fluoride film when exposed to air. On the other hand, the lack of a stable compound like $CdTeO_3$ in the fluoride case is advantageous with respect to thermal stability: there is not enough driving force for possible degradation reactions to occur. Actually, the same driving force that acts on the anodic oxide is acting also in the fluoride case: the formation of $CdTeO_3$. Nevertheless, for that to happen, the film has first to react with oxygen to saturation. After enough of the oxygen-rich oxy-fluoride compound is formed, the degradation reaction can proceed in a considerable amount.

ACKNOWLEDGMENTS

This work was supported in part from gift funds from Texas Instruments Inc. who also supplied the $Hg_{1-x}Cd_xTe$ substrates. Valuable discussions with Dr. M. Seelmann-Eggebert are greatly appreciated. One of us (E.W.) would like to acknowledge partial support from SCD.

---

[a] Permanent address: SCD–Semi-Conductor Devices, D. N. Misgav 20179, Israel.
[1] Y. Nemirovsky and G. Bahir, J. Vac. Sci. Technol. A **7**, 450 (1989).
[2] P. C. Catagnus and C. T. Baker, U. S. Patent No. 3 977 018 (24 August 1976).
[3] *Semiconductors and Semimetals*, edited by R. K. Wilardson and A. C. Beer (Academic, New York, 1981), Vol. 18.
[4] C. M. Stahle, C. R. Helms, and A. Simmons, J. Vac. Sci. Technol. B **5**, 1092 (1987).
[5] C. M. Stahle, C. R. Helms, H. F. Schaake, R. L. Strong, A. Simmons, J. B. Pallix, and C. H. Becker, J. Vac. Sci. Technol. A **7**, 474 (1989); **8**, 3373 (1990).
[6] N. E. Byer, G. D. Davis, S. P. Buchner, and J. S. Ahearn, in *Insulating Films on Semiconductors*, edited by J. F. Verweij and D. R. Wolters (Elsevier, New York, 1983), p. 238.
[7] E. Weiss and N. Mainzer, J. Vac. Sci. Technol. A **6**, 2765 (1988).
[8] N. Mainzer, E. Weiss, D. Laser, and M. Shaanan, J. Vac. Sci. Technol. A **7**, 460 (1989).
[9] E. Weiss and C. R. Helms, J. Electrochem. Soc. (to be published)
[10] M. K. Bahl, R. L. Watson, and K. J. Irgolic, J. Chem. Phys. **68**, 3272 (1978).
[11] L. C. Lynn and R. L. Opila, Surf. Interface Anal. **15**, 180 (1990), and references therein.
[12] M. K. Bahl, R. L. Watson, and K. J. Irgolic, J. Chem. Phys. **66**, 5526 (1977).
[13] M. Seelmann-Eggebert, G. Brant, and H. J. Richter, J. Vac. Sci. Technol. A **2**, 11 (1984).
[14] T. S. Sun, S. P. Buchner, and N. E. Byer, J. Vac. Sci. Technol. **17**, 1067 (1980).
[15] See, for example, C. M. Stahle. D. J. Thomson, C. R. Helms, C. H. Becker, and A. Simmons, Appl. Phys. Lett. **47**, 521 (1985).
[16] R. Stendel, *Chemistry of Non-Metals*, (Walter de Guyther, Berlin, 1977), pp. 229–235.
[17] A. Engebrech and F. Sladky, Adv. Inorg. Chem. Radiochem. **24**, 189

(1981).

[18] J. H. Junkins, H. A. Bernardt, and E. J. Barber, J. Am. Chem. Soc. **74**, 5749 (1952).

[19] O. Kubaschewski and C. B. Alcak, *Metallurgical Thermochemistry*, 5th ed., (Pergamon, Oxford, 1983).

[20] K. C. Mills, *Thermodynamic Data For Inorganic Sulphides, Selenides and Tellurides* (Butterworth, London, 1974).

[21] C. R. Helms, J. Vac. Sci. Technol. A **8**, 1178 (1990).

Study of temperature dependent structural changes in molecular-beam epitaxy grown Hg$_{1-x}$Cd$_x$Te by x-ray lattice parameter measurements and extended x-ray absorption fine structure

D. Di Marzio, M. B. Lee, and J. DeCarlo
Grumman Corporate Research Center, Bethpage New York 11714

A. Gibaud[a] and S. M. Heald
Brookhaven National Laboratory, Upton, New York 11973

(Received 2 October 1990; accepted 10 January 1991)

Infrared detectors fabricated from Hg$_{1-x}$Cd$_x$Te typically operate in the 60 to 160 K range. The temperature dependence of the atomic structure of HgCdTe may influence device performance. We present the first detailed study of the x-ray diffraction lattice parameters of molecular-beam epitaxy grown Hg$_{1-x}$Cd$_x$Te epilayers between 15 and 300 K. The epilayers were grown on (100) oriented CdTe substrates, and varied in thickness (6 to 11 μm) and composition ($x = 0$–0.172). The (400) reflection was measured to determine the lattice parameter $a\perp$ normal to the film. HgTe ($x = 0$) exhibited normal lattice contraction ($\alpha = 4.7 \times 10^{-6} \pm 0.3$ K^{-1} at 300 K), with a minimum in $a\perp$ at 60 K, and an expansion of $a\perp$ below 60 K. In addition to showing a minimum in $a\perp$ at 60 K, some of the Hg$_{1-x}$Cd$_x$Te ($x \neq 0$) epilayers (10 μm thick) exhibited anomalous behavior with varying degrees of thermal hysteresis in $a\perp$. The average contraction of $a\perp$ for these epilayers from 300 to 60 K is 0.006 Å. This is compared with results we have obtained from a temperature dependent extended x-ray absorption fine structure study of these HgCdTe epilayers: whereas HgTe exhibited a normal thermal contraction of the Hg–Te bond length consistent with the lattice parameter results, in the HgCdTe epilayers this bond contracts 0.02 to 0.03 Å on cooling from 300 to 10 K. We also present lattice parameter measurements for a thin cap layer of CdTe on HgCdTe. An increase of 0.0134 Å in $a\perp$ relative to the bulk was observed for a 1000 Å layer of CdTe on HgCdTe at 300 K.

I. INTRODUCTION

High performance Hg$_{1-x}$Cd$_x$Te focal plane arrays are typically operated in the range of 60 to 160 K for reduced noise, increased mobility, and thus a high detectivity D^*. The composition of Hg$_{1-x}$Cd$_x$Te is chosen so that the band gap has the desired value at the temperature of operation. Consequently, a knowledge of the temperature dependence of the band gap is important for the application of HgCdTe devices at low temperature. HgCdTe is one of the few semiconductor materials to show a reduction in band gap with decreasing temperature. This is due to a combination of electron–phonon interactions and lattice thermal contraction.[1,2] The electronic properties (e.g., Hall effect) of bulk and thin film HgCdTe at low temperatures have been measured extensively, but few measurements have been made on the low temperature structural parameters.

Low temperature structural information is important for understanding the stresses and strains that may be present in device interfaces (HgCdTe/substrate and passivant/HgCdTe), and HgTe/CdTe superlattices. Lattice mismatch (0.33% for HgTe/CdTe) can produce strain and result in dislocations at the interface.[3] Results of hydrostatic pressure and uniaxial stress (piezoelectric) measurements have shown significant changes in the resistivity, carrier concentration, and mobility when different pressures and stresses (compressive and tensile) are applied to the material.[4,5] These stresses are not unlike the modifications experienced at the interfaces of bilayer and multilayer structures. Using the published values of the compressibility K and the hydrostatic pressure dependence of the energy gap dE_0/dP[6] an isotropic compression of a CdTe thin film which allows for a commensurate fit to a thicker HgTe substrate would require the equivalent pressure of 4.2 kbar. Differing coefficients of thermal expansion can also affect the interface. These effects in conjunction with thermal cycling may impact the ultimate device performance at operating temperatures. In addition to the lattice averaged structural information given by x-ray diffraction and thermal expansion, extended x-ray absorption fine structure (EXAFS) studies at room temperature have demonstrated that a bimodal bond length distribution exists in III–V and II–IV tetrahedrally coordinated ternary semiconductor alloys.[7] Mayanovic et al. observed little difference between the bond lengths of the parent binary compounds (HgTe and CdTe), and the same bonds in the ternary alloy HgCdTe.[8] This suggests that there is a significant local distortion of the atomic structure that is not readily apparent from x-ray diffraction.

We report here the first x-ray diffraction lattice parameter study of Hg$_{1-x}$Cd$_x$Te thin films from 300 to 15 K and in the composition range of $x = 0$ to $x = 0.172$. We compare the results to a recently completed low temperature EXAFS study of the same samples.

II. EXPERIMENTAL PROCEDURES

The (100)Hg$_{1-x}$Cd$_x$Te epilayers were grown at Grumman in a Riber 2300 molecular-beam epitaxy (MBE) system. They were deposited on 1 cm^2 (100)CdTe substrates supplied by II–VI Corporation, and varied in thickness from 1.5 to 11 μm. Compositions and area uniformity were determined by X–Y mapping of Fourier transform infrared (FTIR) cutoff wavelength measurements. The absorption coefficients were derived from the FTIR transmission data, and a 500 cm^{-1} cutoff was used in conjunction with the empirical expression derived by Hansen et al.[2] to determine compositions. Rocking curves measured using our channel cut Si crystal system (Blake) had widths from 70 to 180 arcs for the samples studied. The epilayers, which were unannealed, were n type with mobilities from 9.2×10^4 to 7.5×10^5 cm^2/V s. A complete characterization of these epilayers is described by Lee et al.[9] A Huber four circle diffractometer with a rotating anode CuKα1 beam at Brookhaven National Laboratory was used to determine the lattice parameters. The diffractometer was equipped with Si(111) monochromator and analyzer crystals, and centroid angular positions and integrated intensities with $\Delta\theta$ rocking curves at each 2θ value was measured to insure that the measured change in Bragg angle with temperature was free of geometrical effects. The sample was cooled down to 15 K with a displex equipped with Be windows. The (400) Bragg reflection was measured to determine the lattice spacing perpendicular to the film surface ($a\perp$). Bragg peak positions were determined by Gaussian fits, and the error in the lattice parameter was determined to be $\pm 2.5 \times 10^{-4}$ Å.

The Hg L_{III} edge was selected for the EXAFS measurements of the HgCdTe epilayers. This was done at beam line X-11 at the National Synchrotron Light Source (NSLS) at Brookhaven National Laboratory. A Si(111) double-crystal monochromator with a nominal energy resolution of 2.0 eV was used. The epilayers were held by a Cu sample holder with the substrate normal parallel to horizontal and 45° to the x-ray beam. The temperature dependent measurements were carried out using a displex refrigerator with a Be window. X-ray fluorescence was measured with an argon-filled ion chamber placed next to the Be window. Bragg diffraction peaks were detected with Polaroid film and subsequently masked out with lead foil. Small variations of the angle between the sample and the x-ray beam did not reveal any observable changes in fluorescence intensity from secondary effects of masked Bragg reflections. Data was taken out to $k = 14$ Å^{-1}, and the Hg–Te first shell was isolated with a typical r-space window of 1.2 Å.

III. RESULTS

Figure 1 shows the lattice parameters as a function of composition as determined by infrared (IR) cutoff at room temperature. The extinction depth of CuKα1 radiation in HgTe (or small x composition Hg$_{1-x}$Cd$_x$Te) is approximately 3 μm. The Hg$_{1-x}$Cd$_x$Te epilayers with thicknesses above 5 μm may be expected to exhibit bulk lattice values. A CdTe substrate (II–VI Corporation) served as one end point, and a 6 μm thick HgTe ($x = 0$) epilayer on CdTe served as the other end point. Significant scatter in the lattice parameter of HgCdTe of various compositions has been reported.[10] This may be attributed to variations in the independent determinations of the x value, as well as the stoichiometry and overall crystal quality (Te precipitates, etc.).[11]

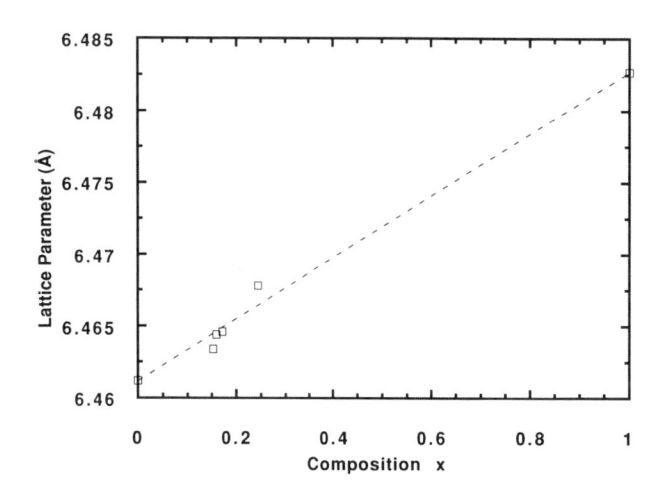

FIG. 1. Room temperature lattice parameters of Hg$_{1-x}$Cd$_x$Te epilayers as a function of composition. The compositions were determined using a 500 cm^{-1} cutoff of the IR transmission curve. The dashed line joins the two end point compounds (HgTe and CdTe).

Approximate agreement with Vegards law is observed among the various compositions in Fig. 1.

Figure 2 shows the relative lattice parameter $a\perp$ (parallel to the substrate normal) for four epilayers as a function of temperature for both cooling and heating cycles. Starting at 300 K, HgTe ($x = 0$, 6 μm thick) exhibits an approximately linear contraction of $a\perp$ with decreasing temperature, which is expected behavior of a crystal with lattice bonding anharmonicity. The thermal expansion coefficient $\alpha = (1/a\perp)\partial a\perp/\partial T$ where T is the temperature, was calculated from a polynomial fit of the lattice parameter data. The relative expansion coefficients of the epilayers is shown in Fig. 3. An average of $4.7 \times 10^{-6} \pm 0.3$ K^{-1} between cooling and heating is derived for α for HgTe at room temperature, consistent with capacitance dilatometry measurements on

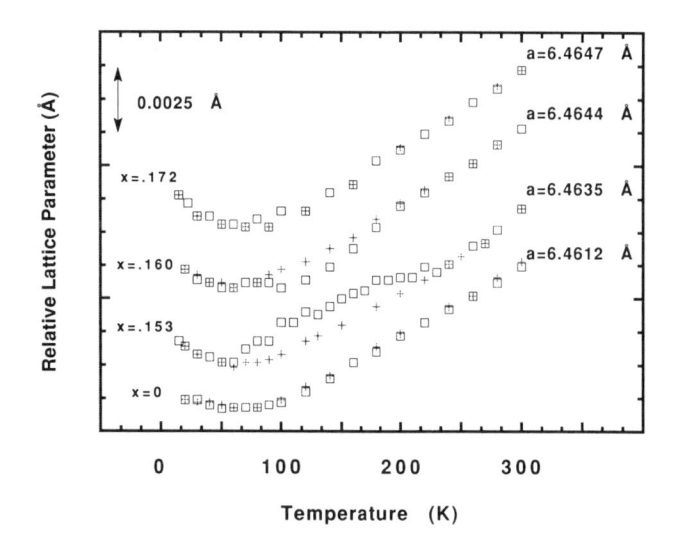

FIG. 2. Relative lattice parameters of the epilayers as a function of temperature; □: cooling cycle, +: heating cycle. For clarity the curves have been separated and placed on a relative scale. The room temperature lattice parameter a for each epilayer is indicated in the figure.

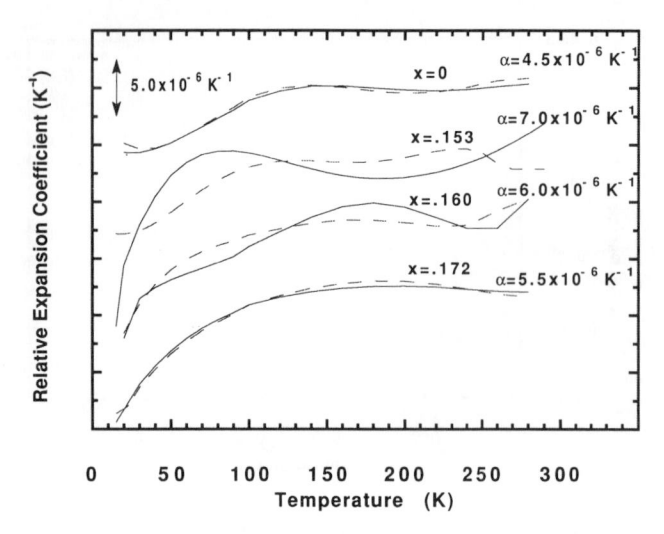

FIG. 3. Relative thermal expansion coefficient α of the epilayers as a function of temperature;—: cooling cycle; ---: heating cycle. For clarity the curves have been separated and placed on a relative scale. The room temperature thermal expansion coefficient for the cooling cycle α for each epilayer is indicated in the figure.

bulk material.[12] Near 120 K (close to the Debye temperature), α starts to decrease, and passes through zero at 62 K, at which point $a\perp$ begins to expand with decreasing temperature. Below 30 K, α decreases once again. From thermodynamic arguments, $\alpha = 0$ at 0 K. The total contraction of the lattice $\Delta a\perp$ from 300 to 60 K is 0.0054 Å. The expansion of the lattice at low temperatures has been observed in capacitance dilatometry experiments of bulk HgCdTe as well as other tetrahedrally coordinated II–IV, III–V, and column IV semiconductors. Although not well understood, this expansion is believed to be caused by a competition between temperature dependent acoustic shear and compressional phonon modes in the crystal. If the shear modes dominate the compressional modes at low temperature, the thermal expansion coefficient becomes negative.[13]

In Fig. 2, the sample $x = .153$ (11 μm thick) relative lattice parameter is shown for two complete cooling and heating cycles. Near room temperature, the thermal behavior appears somewhat anomalous with an average value of $\alpha = 5.9 \times 10^{-6} \pm 0.5$ K^{-1} (Fig. 3). This is larger than for HgTe and suggests a greater lattice anharmonicity, with Cd substitution for Hg, and thus an average bond softening in the ternary compound. When cooled below 230 K this epilayer deviates significantly from the expected linear behavior observed in HgTe down to 60 K, when the lattice begins to expand. As shown in Fig. 3, the negative coefficient of thermal expansion of $x = 0.153$ near 20 K is also an order of magnitude larger than α for HgTe. When the sample is subsequently heated, the lattice parameter follows the cooling cycle up to 60 K, upon which it expands in a normal approximately linear fashion up to 300 K. The difference in lattice parameters from 60 to 230 K between the cooling and heating cycle is greater than the experimental error. This unusual hysteretic behavior was repeated in a second cooling and heating cycle. The net contraction of the lattice $\Delta a\perp$ between 300 and 60 K is 0.0061 Å.

Similar hysteretic behavior is shown in Fig. 2 for composition $x = 0.160$ (11 μm thick). Sample $x = 0.160$ also has a 1000 Å cap layer of CdTe deposited on top of the HgCdTe epilayer. Near room temperature, the average $\alpha = 6.1 \times 10^{-6} \pm 0.3$ K^{-1} (Fig. 3) which again is larger than for HgTe. Cooling below 240 K the lattice parameter deviates from linear behavior but in contrast to $x = 0.153$, the lattice parameter drops below the linear heating portion of the cycle. Like $x = 0.153$, $x = 0.160$ lattice parameters below 60 K shows little hysteresis and exhibits a significantly more negative α in comparison with HgTe. The contraction of the epilayer $\Delta a\perp$ from 300 to 60 K is 0.0062 Å.

For sample $x = 0.172$ (10-μm thick) shown in Fig. 2, the epilayer exhibits the least amount of anomalous behavior of the 10 to 11 μm thick HgCdTe epilayers. Although not quite as smooth as HgTe, any hysteretic behavior observed for sample $x = 0.172$ in the temperature dependence of the lattice parameter is within the experimental error. Figure 3 shows the thermal expansion coefficient α near room temperature to be $5.5 \times 10^{-6} \pm 0.3$ K^{-1}, and the net lattice contraction $\Delta a\perp$ between 300 and 60 K is 0.0058 Å.

Returning to sample $x = 0.160$, the room temperature lattice parameter, $a\perp$, of the 1000 Å thick CdTe cap layer was found to be 6.4967 Å, which is 0.0140 Å larger than the room temperature value for bulk CdTe. This expansion of $a\perp$ is expected if the in-plane lattice parameter $a\|$ of the thin cap layer is assumed to contract in order to lattice match the underlying thick HgCdTe epilayer. This tetragonal distortion is in accord with the elastic behavior of cubic crystals. We can assume that the cap layer is thin enough so that the in-plane strain $\epsilon\|$ and the out-of-plane strain $\epsilon\perp$ in the CdTe cap layer can be given by

$$\epsilon\| = (a\|_c - a_c)/a_c, \quad (1)$$
$$\epsilon\perp = (a\perp_c - a_c)/a_c, \quad (2)$$

where $a\|_c$ is the in-plane lattice parameter of the CdTe cap layer, $a\perp_c$ is the out-of-plane lattice parameter of the cap layer, and a_c is the lattice parameter of the unstrained (bulk) CdTe cap layer. For a cubic crystal in the (100) direction, it can be shown that[14]

$$\epsilon\|/(\epsilon\| - \epsilon\perp) = c_{11}/(c_{11} - 2c_{12}), \quad (3)$$

where c_{11} and c_{12} are elastic constants of CdTe. Using the published values of c_{11} and c_{12} at room temperature, and assuming that $a\|_c$ strains to fit the underlying HgCdTe epilayer, we find that the predicted $a\perp_c = 6.5127$ Å. This is reasonably close (within 7%) to the measured value of 6.4967 Å if it is considered that the upper portion of the 1000 Å cap layer has partially relaxed towards its bulk dimensions.

It may be expected that dislocations due to lattice mismatch at the CdTe/HgCdTe interface would lead to a relaxation of the CdTe lattice parameter towards the bulk value. If we assume a uniform distribution of dislocations throughout the cap layer, a crude upper estimate of the dislocation density can be made.[14] The measured $a\perp$ of the cap layer would be the same as a dislocation free cap layer grown on a substrate with a lattice parameter given by $a_s(1 \pm n\langle b\rangle)$, where n is the dislocation density along a straight line, $\langle b\rangle$ is the average component of the Burgers vectors along this line, and a_s is the true substrate lattice parameter. If we assume a value of $a/\sqrt{2}$ for the Burgers vector, an upper limit of the density of misfit dislocations is calculated to be 3.6×10^4 cm^{-1}.

The critical thickness for misfit dislocations for the system CdTe/Cd$_{1-x}$Zn$_x$Te with $x \approx 0.03$–0.04 was measured by Fontaine et al. to be from 4000 to 5500 Å.[15] Since these x values for Cd$_{1-x}$Zn$_x$Te approximates the lattice parameters for the HgCdTe epilayers, this range of critical thickness would apply in our case. However, it may be expected that defect structures such as hillocks on the (100) oriented HgCdTe epilayer surface could reduce the critical thickness.

The contraction in the cap layer, between 300 and 60 K, is 0.0067 Å. For the cap layer near room temperature, $\alpha = 4.7 \times 10^{-6} \pm 0.3$ K^{-1}, which agrees with bulk measurements on CdTe. A roughly linear behavior is seen in both heating and cooling cycles down to 150 K. Below 120 K some anomalous behavior as compared to HgTe is observed.

IV. DISCUSSION

The room temperature thermal expansion coefficients of the epilayers approximately agree with previously published dilatometry measurements, but the thermally induced hysteresis in the lattice parameter for some of the epilayers has not previously been seen. The time interval between data points (temperatures) was approximately 12 min, and the rocking curve-integrated Bragg peak measurements help insure that extraneous thermal dilations of the goiniometer and refrigerator do not affect the final results. Hysteretic stress behavior with thermal cycling has been seen in other thin film systems.[16] As pointed out by Hoffman,[17] if the epilayer is elastically coupled to the substrate, a sufficient change in temperature could lead to flow stress fields in the film, and to deviations from linear and reversible thermal expansion. A nonlinear temperature dependence of the lattice parameter can be attributed to a plastic flow induced by differential thermal contraction between the epilayer and substrate. This process would begin at the thin film deposition temperature. Unlike various III–V epilayers, dislocations are highly mobile in HgCdTe ternary alloys. From the hysteresis observed in Fig. 2, some degree of reversible relaxation may occur near room temperature. This relaxation may be influenced by various defects and by thermally activated Hg vacancy creation and diffusion, which is already substantial at room temperature.

Thick layers of epitaxial HgCdTe grown on (100) oriented substrates have easily observable inverted pyramidal hillock defects which originate at the substrate/epilayer interface, and grow outwards towards the film surface. The density of these defects in the present samples is from 7.6×10^4 to 1×10^5 cm^{-2}, but it is not clear whether the volume fraction of the total epilayer this represents would be large enough to significantly affect the thermal behavior of the average lattice parameter. In addition to their effects on electrical properties, other structural defects such as Hg vacancies, Hg interstitials, threading dislocations, and twin planes can play a major role in the thermal elastic behavior of HgCdTe.

Previous studies of the temperature dependent lattice parameter of single crystal CdTe have revealed some unusual and variable behavior. Using a strain gauge technique, Greenough et al. reported an anomalous contraction in the CdTe (100) direction at 79 K with decreasing temperature.[18] Smith et al.,[13] using capacitance dilatometry, later found no anomalous contraction in the temperature range reported by Greenough et al. Lattice parameter studies of 3 μm thick metal organic chemical vapor deposited (MOCVD) CdTe(100) epilayers on GaAs by Staudenmann et al.[19] showed a large anomalous contraction of $a\perp$ (≈ 0.025 Å) in the CdTe epilayer between 120 and 60 K. Later work by the same group reported a 0.020 Å contraction of $a\parallel$ between 350 and 250 K in a similar CdTe/GaAs system, but no anomalous behavior was seen in $a\perp$.[20] All of this suggests that there may be an uncontrolled variable in the sample preparation of the bulk and epilayer materials. CdTe is known to exhibit Te microprecipitates and this may also affect the structural and electronic properties of HgCdTe. Precipitates and ternary alloy clusters in HgCdTe may have thermal expansion coefficients different from the homogeneous alloy and this may lead to nonlinear thermal behavior. The high diffusivity of Hg in HgCdTe frequently results in poorly controlled doping levels and subsequent variability in electronic properties. Excess interstitial Hg could also have an effect on the thermal expansion behavior.

We have recently completed a temperature dependent extended x-ray absorption fine structure study (EXAFS) of some of the above Hg$_{1-x}$Cd$_x$Te epilayers.[21] EXAFS displays large modulations of the absorption coefficient above the atomic core level absorption edge, and it is sensitive to the chemical state and local atomic structure.[22] Whereas conventional x-ray diffraction can give a precise measure of a macroscopically averaged lattice structure, EXAFS can provide a local probe of the number of nearest neighbors, bond lengths, and the relative local disorder of the material. In this investigation the Hg L_{III} EXAFS was measured to study the Hg–Te bond. Figure 4 shows the change in bond length between the Hg and the Te atom as a function of temperature for HgTe ($x = 0$) and $x = 0.160$. Assuming an

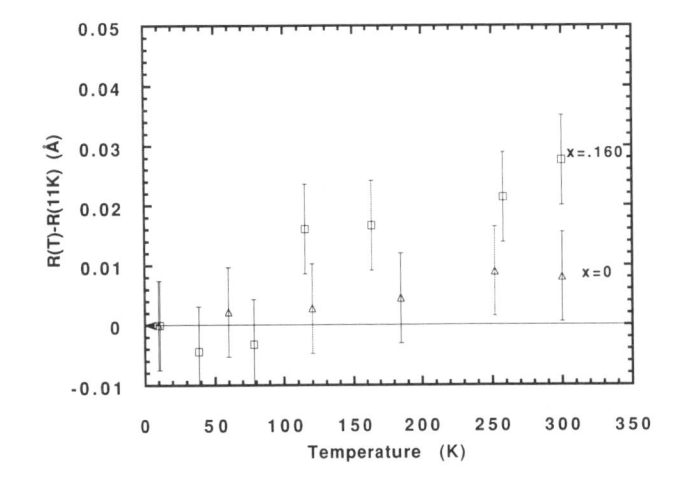

FIG. 4. The relative change of the Hg–Te bond length R as measured by EXAFS, in sample $x = 0$ and $x = 0.160$ between temperature T and 11 K.

isotropic contraction of the Hg–Te bond length proportional to the change in the measured lattice parameter $a\perp$ the observed bond length changes in Fig. 4 are significantly larger than the x-ray diffraction results would suggest. HgTe ($x = 0$) shows a relatively large but normal (approximately linear) contraction of the bond length with temperature. Since there is a significantly larger error inherent in EXAFS distance determination as compared to x-ray diffraction, the HgTe lattice parameter results fall within the experimental error of the EXAFS derived bond length contraction. This is not the case for sample $x = 0.160$ in Fig. 4. Sample $x = 0.160$ exhibits normal behavior down to 120 K, after which a pronounced contraction is observed. A slight expansion, not unlike that expected for these materials, is observed upon further cooling below 50 K, although the experimental error makes this observation tentative. Samples $x = 0.153$ and $x = 0.172$ were also analyzed by EXAFS, and the same order of contraction with temperature was observed, but a lack of a sufficient number of data points prevented any identification of a transition below 120 K, or a low temperature expansion below 60 K. At temperatures around 100 K, the predicted band gap for $x = 0.160$ is only 25 meV,[2] which would be expected to lead to a significant degree of band mixing. A change in the amount of s and p orbital character in the conduction and valence bands will affect the length and orientation of the Hg–Te bond.

A relatively large tendency for bond bending, leading to increased lattice shear, is expected for these highly ionic II–VI semiconductors. Unlike the highly directional nature of a purely covalent bond (Si–Si), ionic bonds represent an essentially non directional central force. Since the Hg–Te bond is significantly weakened in comparison to the Cd–Te bond in the $Hg_{1-x}Cd_xTe$ ternary alloy[21] a combination of internal lattice strains originating from the bimodal bond length nature of the material may result in a larger change in the Hg–Te bond length. The average lattice parameter $a\perp$ along the (100) direction, however, is a result of an average of the bimodal Hg–Te and Cd–Te structures as measured by x-ray diffraction, and thus may not show the same degree of change.

In addition to the Hg–Te bond length, we also derived the relative Debye–Waller factors as a function of temperature from the EXAFS data. From this we were able to determine the relative amount of static and thermal disorder in the epilayers. The magnitude of the Einstein temperature, which is a measure of the bond strength, was found to consistently decrease with increasing Cd substitution for Hg (increasing x).[21] This is a direct confirmation that the Hg–Te bond is destabilized with increasing Cd content. This is in contrast with the trend towards reduced hysteresis in Fig. 2 with increasing x. Initially, from $x = 0$ to $x = 0.153$ we see the onset of anomalous behavior, but as the Cd content increases, the hysteresis is reduced. Since the Cd–Te bond is significantly stronger than the Hg–Te bond, the Cd–Te bond will eventually dominate the lattice as the Cd content is increased, and x-ray diffraction will reflect this.

A destabilization of the Hg–Te bond may also be enhanced by a temperature induced closing of the band gap, especially for epilayers with small x values. The amount of gap reduction, and consequent s–p orbital mixing, with a reduction in temperature would be expected to decrease with increasing Cd content in the alloys. This would be consistent with the trends observed in Fig. 2.

The observed trend toward reduced hysteretic behavior with increasing Cd content would imply a minimum of anomalous behavior for CdTe ($x = 1$). This is in variance with some previous work which shows anomalous behavior in bulk and MOCVD deposited thin films of CdTe. The electronic changes that may occur as a result of the change in composition may be necessary but not sufficient to produce gross structural changes as revealed by the lattice parameter. The type and amount of defects, such as dislocations, hillocks, and Te precipitates, which depend on sample preparation (MBE, MOCVD, etc.), may play a role in the anomalous behavior. The epilayers in this study were grown in nearly identical conditions, so some degree of consistency was achieved. A wider range of composition (x), (111) oriented, and annealed epilayers must be studied before a clearer picture can emerge.

V. CONCLUSION

We have measured for the first time the temperature dependent lattice parameters and derived the thermal expansion coefficients for a variety of $Hg_{1-x}Cd_xTe/CdTe$ epilayer systems. These results were then compared to an EXAFS study of some of the same samples. HgTe ($x = 0$) showed normal linear contraction of the lattice parameter ($a\perp$) with an expansion of the lattice below 60 K. A 1000 Å CdTe cap layer on sample $x = 0.160$ at room temperature showed an expansion of $a\perp$ consistent with a partially commensurate fit to the underlying HgCdTe epilayer. This observation may be relevant to the use of CdTe as a passivant for HgCdTe. Some of the HgCdTe ternary alloy film lattice parameters showed a significant degree of nonlinear thermal hysteresis along with an enhanced expansion below 60 K. This unusual nonlinear and hysteretic behavior may affect the structural and electrical properties of devices fabricated from this material especially where temperature cycling is concerned.

This anomalous behavior is compared to a weakening of the Hg–Te bond with Cd substitution, and an anomalous and relatively large contraction of the Hg–Te bond length with temperature as measured by EXAFS. However, reduced hysteretic behavior is observed in the lattice parameter with increasing Cd content, and it is expected that the stronger Cd–Te bond will eventually dominate the lattice. Various static and temperature dependent defect mechanisms, such as Te inclusions, alloy clustering, and differential thermal expansion, which may vary in type and degree depending on the deposition process, are believed to be at play in these materials.

ACKNOWLEDGMENTS

The authors wish to thank A. Roberts, A. Berghmans, L. Casagrande, D. Larson, S. Sinha, and R. Silberstein for their contributions to this work. For the use of the facilities at Brookhaven National Laboratory, we acknowledge the support of the U.S. Department of Energy, Division of Materials Sciences, under Contract No. DE-AC02-76CH00016, and No. DE-AS05-80-ER10742.

[a)] Present address: Universite du Maine, Faculte des Sciences, Le Mans CEDEX, France.

[1] *Springer Series in Solid State Sciences, Electronic Structure and Optical Properties of Semiconductors*, edited by M. L. Cohen and J. R. Chelikowsky (Springer-Verlag, Berlin, Heidelberg, 1988), Vol. 75, p. 182.

[2] G. L. Hansen, J. L. Schmit, and T. N. Casselman, J. Appl. Phys. **53**, 7099 (1982).

[3] M. J. Bevan, N. J. Doyle, J. Greggi, and D. Snyder, J. Vac. Sci. Technol. A **8**, 1049 (1990).

[4] P. S. Kireev, V. V. Ptashinskii, A. M. Sokolov, and V. G. Kovalev, Sov. Phys. Semicond. **7**, 266 (1973).

[5] P. I. Baranskii, A. I. Elizarov, V. A. Kulik, and K. R. Kubanov, Sov. Phys. Semicond. **13**, 490 (1979).

[6] R. Dornhaus and G. Nimtz, *Narrow-Gap Semiconductors*, Springer Tracts in Modern Physics (Springer-Verlag Berlin, Heidelberg, 1985), Vol. 98, p. 178.

[7] J. C. Mikkelsen, Jr. and J. B. Boyce, Phys. Rev. B **28**, 7130 (1983).

[8] R. A. Mayanovic, W. F. Pong, and B. A. Bunker, Phys. Rev. B **42**, 11174 (1990).

[9] M. B. Lee, J. DeCarlo, D. Di Marzio, and M. Kesselman, in *Proceedings of the MRS Symposium on the Properties of II–VI Semiconductors: Bulk Crystals, Epitaxial Films, Quantum Well Structures, and Dilute Magnetic Systems*, Boston, 1989, edited by F. J. Bartoli, Jr., H. F. Schaake, and J. F. Schetzina (MRS, Pittsburgh, 1990), Vol. 161, p. 377.

[10] W. M. Higgins, G. N. Pultz, R. G. Roy, and R. A. Lancaster, J. Vac. Sci. Technol. A **7**, 271 (1989).

[11] J. C. Brice, *EMIS Data Review, Properties of Mercury Cadmium Telluride, EMIS Datareviews Series No. 3*, edited by J. C. Brice and P. Capper (INSPEC, London, 1987), p. 4.

[12] O. Caporaletti and G. M. Graham, Appl. Phys. Lett. **39**, 338 (1981).

[13] T. F. Smith and G. K. White, J. Phys. C **8**, 2031 (1975).

[14] J. Hornstra and W. J. Bartels, J. Cryst. Growth **44**, 513 (1978).

[15] C. Fontaine, J. P. Gaillard, S. Magli, A. Million, and J. Piaguet, Appl. Phys. Lett. **50**, 903 (1987).

[16] A. K. Sinha, H. J. Levinstein, and T. E. Smith, J. Appl. Phys. **49**, 2423 (1978).

[17] R. W. Hoffman, in *Proceedings of the MRS Symposium on Materials Characterization*, Palo Alto, 1986, edited by N. Cheung and M. Nicolet (MRS, Pittsburgh, 1986), Vol. 69, p. 95.

[18] R. D. Greenough and S. B. Palmer, J. Phys. D **6**, 587 (1973).

[19] J. L. Staudenmann, R. D. Horning, R. D. Knox, D. K. Arch, and J. L. Schmit, Appl. Phys. Lett. **48**, 994 (1986).

[20] R. D. Horning and J. L. Staudenmann, Appl. Phys. Lett. **50**, 1482 (1987).

[21] D. Di Marzio, M. B. Lee, J. DeCarlo, and S. M. Heald (in preparation).

[22] E. A. Stern and S. M. Heald, in *Handbook on Synchrotron Radiation* (North-Holland, Amsterdam, 1983), Vol. 1, p. 955.

Characterization of CdTe, (Cd,Zn)Te, and Cd(Te,Se) single crystals by transmission electron microscopy

R. S. Rai and S. Mahajan
Department of Metallurgical Engineering and Materials Science, Carnegie Mellon University, Pittsburgh, Pennsylvania 15213

S. McDevitt and C. J. Johnson
II–VI Incorporated, Saxonburg, Pennsylvania 16056

(Received 2 October 1990; accepted 1 February 1991)

CdTe, (CD,Zn)Te, and Cd(Te,Se) crystals grown by the Bridgman technique have been characterized by transmission electron microscopy. Results indicate that the Te precipitates are seen in all the crystals, but their density and size are lowest and largest in the case of Cd(Te,Se) crystals. In addition, dislocations, stacking faults, and microtwins are observed in as-grown CdTe, (Cd,Zn)Te, and Cd(Te,Se) crystals. Arguments have been developed to rationalize these observations and their ramifications on crystal perfection are discussed.

I. INTRODUCTION

CdTe and its alloys crystallize in the zinc-blende structure and belong to the space group $F\bar{4}3m$. These materials possess many attractive properties and have applications in the area of optoelectronics. Currently, there is an increased demand for the CdTe family substrates for the growth of (Hg,Cd)Te and (Hg,Zn)Te epitaxial layers, which in turn are used in the fabrication of large area infrared detector arrays. However, the major problem is the poor quality substrates because of the presence of grown-in structural defects, such as dislocations, twins, and precipitates. These structural defects which form during the crystal growth have deleterious effects on electronic properties and growth of (Hg,Cd)Te layers on CdTe family substrates by liquid phase epitaxy (LPE).[1]

At present, (Cd,Zn)Te is the leading candidate material for substrates for LPE growth of (Hg,Cd)Te epitaxial layers.[2,3] This is because of its infrared transparency which allows backside illumination of detector arrays and by varying Zn content, the lattice matching of any alloy composition of (Hg,Cd)Te can be achieved. Also, its purity is higher than the other lattice matched candidate Cd(Te,Se).

It is well established that the performance of heterostructure devices is strongly influenced by the quality of the underlying substrate.[4] Substrate defects have been observed to replicate into epitaxial layers grown by molecular-beam epitaxy,[5] organometallic vapor phase epitaxy,[8] and LPE. Therefore, in addition to growing large area single crystals, controlled low defect growth of these materials is highly desirable. Vertical Bridgman with over pressure (VBOP) and horizontal Bridgman with over pressure (HBOP) have the advantages of producing large area, high quality, and stoichiometric crystals as compared with three other growth techniques, i.e., heat exchanger method (HEM), traveling heater method (THM), and high pressure Bridgman. Further, to assess the suitability of CdTe family substrates for the growth of (Hg,Cd)Te epitaxial layers, the perfection of the substrates must be ascertained. To achieve this objective, we have carried out transmission electron microscopy studies (TEM) on CdTe, (Cd,Zn)Te, and Cd(Te,Se) crystals grown by the modified Bridgman technique. The results of this investigation constitute the present paper.

II. EXPERIMENTAL PROCEDURE

CdTe, (Cd,Zn)Te, and Cd(Te,Se) crystals were grown by the modified Bridgman technique at II–VI Incorporated. Samples for TEM studies were prepared by mechanical polishing followed by chemical thinning in a 1% bromine in methanol solution. Thin samples were examined in a Philips EM 420 electron microscope operating at 120 keV. The standard **g·b** and **g·R** criteria, where **g** is the operation reflection, were used to determine the Burgers vector (**b**) of dislocations and the displacement vectors (**R**) of the faults, respectively.

III. RESULTS

The microstructural features commonly observed in as-grown crystals are precipitates, dislocations, stacking faults, and microtwins, and are discussed in detail below.

A. Precipitate distribution and density

Shown in Fig. 1 are typical dark-field electron micrographs obtained from (a) CdTe, (b) (Cd,Zn)Te, and (c) Cd(Te,Se) crystals. The small, bright, irregularly shaped particles are precipitates which are randomly distributed within the matrix. The density and average size of the precipitates have been measured and these results are summarized in Table I.

Figure 2 shows a selected area electron diffraction pattern obtained from Fig. 1(a). In addition to the CdTe spots labeled M, precipitate spots (P) are observed. The latter have been indexed in terms of HCP Te. Similar results were obtained from the (Cd,Zn)Te and Cd(Te,Se) crystals. It is therefore inferred that the bright particles in Fig. 1 are Te precipitates. In addition, it has been ascertained using stereomicroscopy that the precipitates are contained in the bulk of the foils.

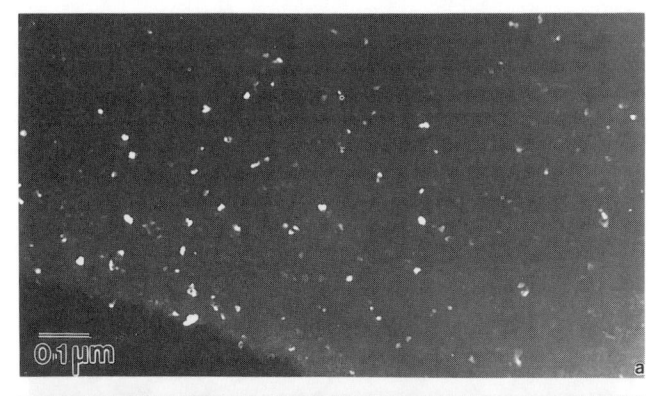

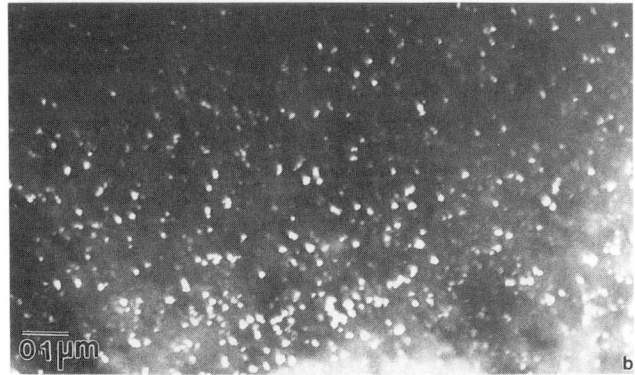

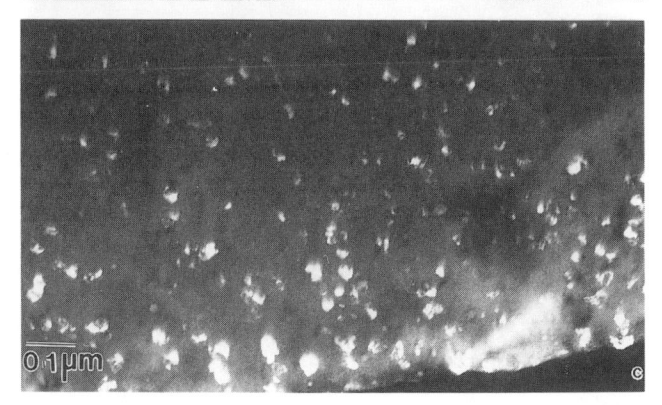

FIG. 1. A set of dark-field electron micrographs obtained using a precipitate reflection in (a) CdTe, (b) (Cd,Zn)Te, and (c) Cd(Te,Se) crystals. The small bright particles are Te precipitates within the matrix in each micrograph.

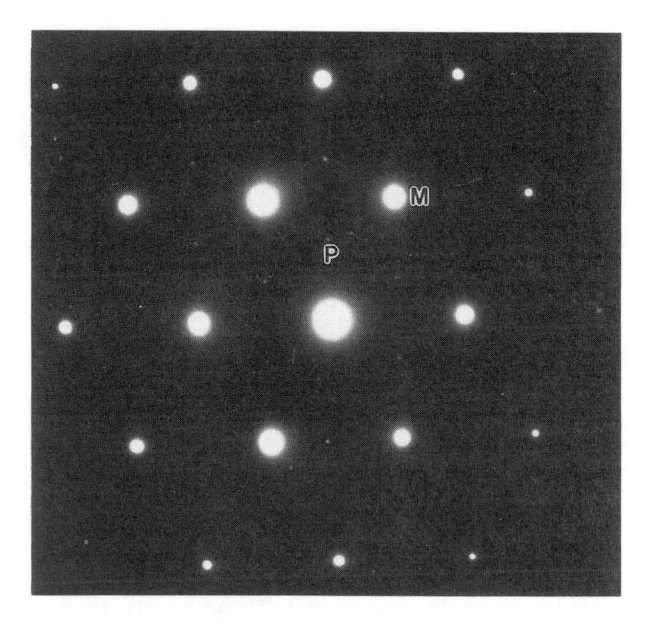

FIG. 2. A selected area electron diffraction pattern obtained from Fig. 1(a) strong spots (M) correspond to CdTe whereas weak spots (P) to Te. The orientational relationship is $[111]_{CdTe} \| [0001]_{Te}$.

B. Dislocation distributions

In addition to the precipitates, isolated dislocations and dislocation arrays are seen in the crystals. Typical results are reproduced as Fig. 3. The dislocation density in the as-grown CdTe crystals is fairly high, Fig. 3(a), whereas it is much lower in the case of (Cd,Zn)Te, Fig. 3(b), and Cd(Te,Se) crystals, Fig. 3(c). Furthermore, the density is comparable in Figs. 3(b) and 3(c).

C. Stacking faults and microtwins

In certain regions of the crystals, planar features shown in Fig. 4 are observed. Figure 4(a) obtained from an as-grown CdTe crystal shows dislocation arrays lying on parallel {111} planes and overlapping stacking faults. The $\frac{1}{2}\langle 110 \rangle$ dislocations constituting the array are dissociated into Shockley partials which are separated from each other by stacking faults. The density of the planar features is much higher in the case of (Cd,Zn)Te, Fig. 4(b), and Cd(Te,Se) crystals, Fig. 4(c).

A selected area electron diffraction pattern obtained from Fig. 4(b) is shown in Fig. 5. The observed pattern can be indexed in a consistent manner if it is assumed that the planar features in Fig. 4(b) are microtwins lying on the {111} planes. The diffraction spots emanating from the twinned regions are labeled T. It was ascertained in a similar manner that the planar features in Fig. 4(c) are also microtwins.

IV. DISCUSSION

Several interesting observations emerge from the preceding study. First, the average size of Te precipitates observed in as-grown CdTe crystals is slightly smaller than those seen in (Cd,Zn)Te crystals; Cd(Te,Se) crystals show the largest precipitates. Second, the dislocation density of (Cd,Zn)Te and Cd(Te,Se) crystals is comparable and is much lower than that of the CdTe crystals. Third, microtwins are seen in (Cd,Zn)Te and Cd(Te,Se) crystals, whereas overlapping stacking faults are observed in CdTe crystals.

TABLE I. Density and average size of precipitates observed in CdTe, (Cd,Zn)Te, and Cd(Te,Se) crystals.

Material	Precipitate density cm^{-3}	Precipitate size nm
CdTe	6×10^{15}	15
(Cd,Zn)Te	1.3×10^{16}	20
Cd(Te,Se)	3×10^{15}	30

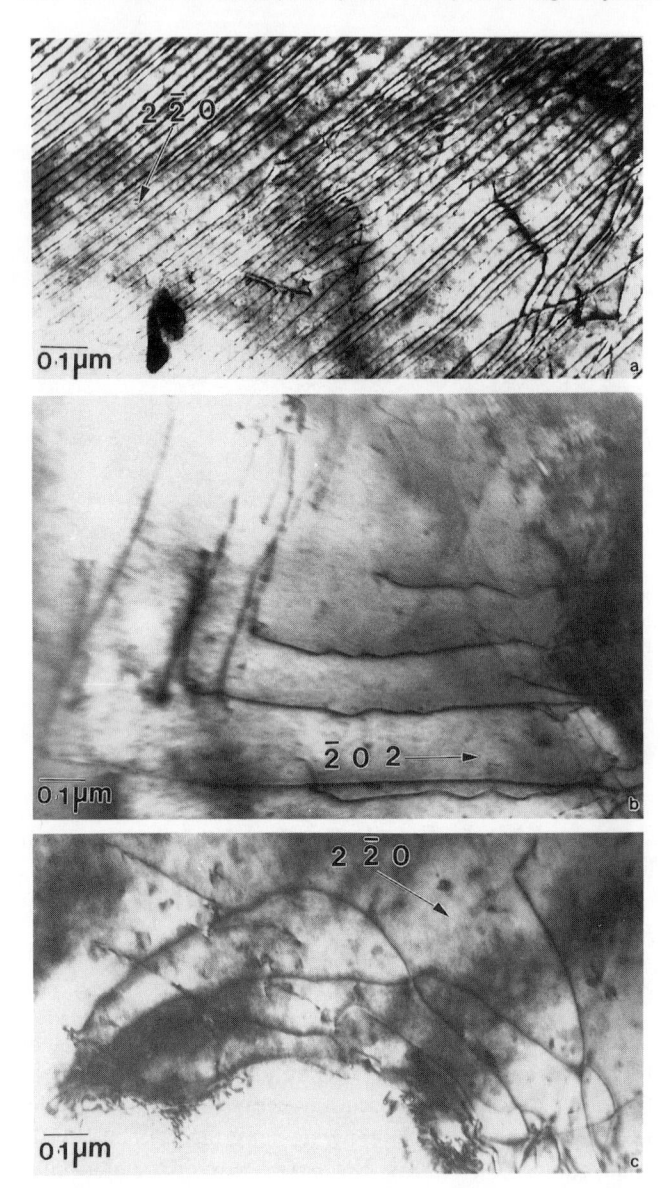

FIG. 3. (a) Electron micrograph showing dislocations in an as-grown CdTe crystal. The foil orientation is a near [111] zone and diffraction vector is [2$\bar{2}$0]. (b) Electron micrograph showing dislocations in an as-grown (Cd,Zn)Te crystal. The foil orientation is a near [111] zone and diffraction vector is [$\bar{2}$02]. (c) Electron micrograph showing dislocations in an as-grown Cd(Te,Se) crystal. The foil orientation is a near [111] zone and diffraction vector is [2$\bar{2}$0].

The presence of Te precipitates in as-grown CdTe crystals is consistent with the earlier results.[7–10] A plausible reason for their formation is that in the Cd–Te system, composition with the highest melting lies on the Te-rich side. As a result the CdTe solid becomes supersaturated with Te during cool down from the melt. This supersaturation can be eliminated by the formation of the precipitates.

Assuming that the precipitates are spherical in shape, it can be shown from the data presented in Table I that the volume of the precipitates in (Cd,Zn)Te and Cd(Te,Se) crystals is comparable, whereas it is lower in the case of CdTe crystals. It is conceivable that the addition of isoelectronic impurities like Zn and Te to CdTe shifts the composi-

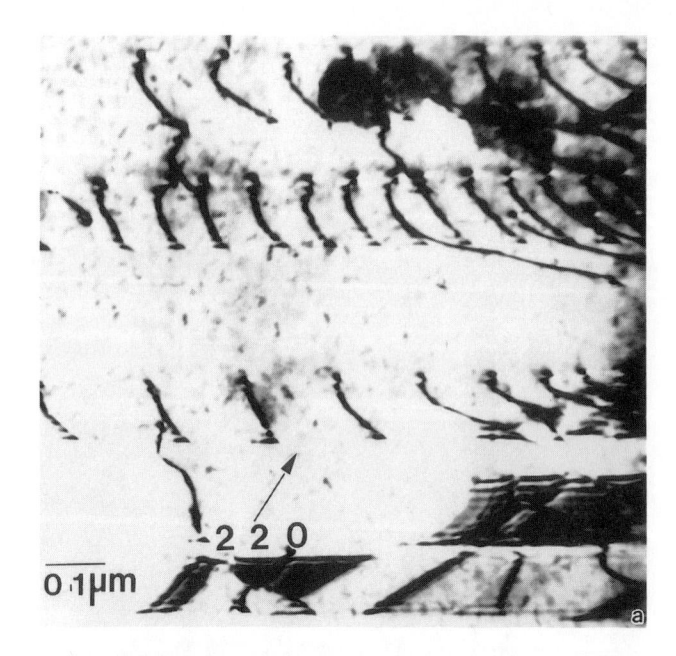

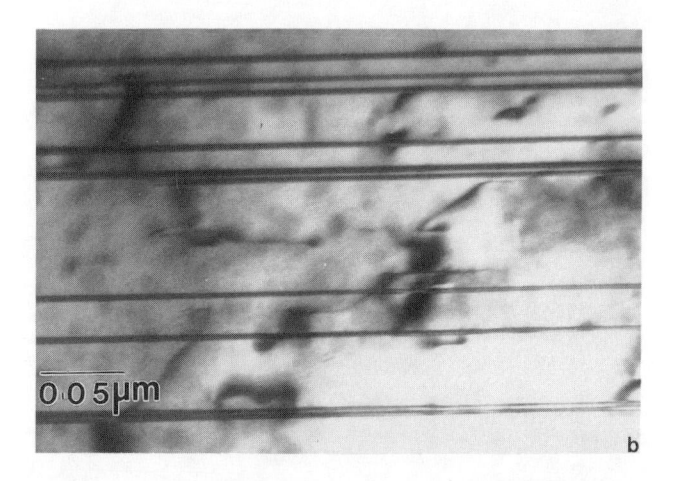

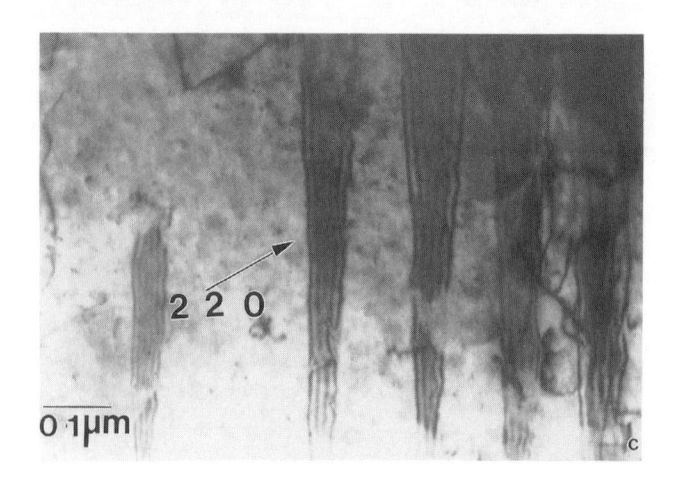

FIG. 4. (a) Electron micrograph showing dislocation arrays lying on parallel {111} planes and overlapping stacking faults in an as-grown CdTe crystal. The foil orientation is a near [$\bar{1}$11] zone and diffraction vector is [220]. (b) Electron micrograph showing microtwins in an as-grown (Cd,Zn)Te crystal. The foil orientation is a near [110] zone and diffraction vector is [2$\bar{2}$0]. (c) Electron micrograph showing microtwins imaged as fringes, and dislocations in an as-grown Cd(Te,Se) crystal. The foil orientation is a near [$\bar{1}$12] zone and diffraction vector is [220].

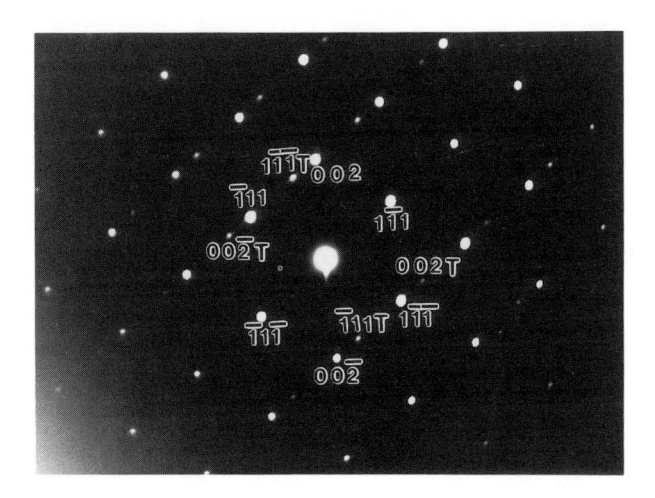

FIG. 5. A selected area electron diffraction obtained from Fig. 4(b) showing matrix as well as twin spots in [110] orientation. The matrix and twin spots (labeled T) are indexed.

tion with the highest melting point to higher Te levels. A consequence of this would be to further enhance the Te supersaturation, resulting in a larger volume of the precipitates.

An interesting question is why are the precipitates largest in size in the case of Cd(Te,Se) crystals? If it is assumed that the replacement of Te atoms with Se atoms reduces the point defect-atom interactions on the group VI sublattice and thus enhances Te diffusion, then the presence of larger precipitates is relatively easy to rationalize.

The introduction of dislocations during crystal growth could result from three different sources. First, dislocations present in seed crystals could propagate into a growing crystal. Second, dislocations generated at peripheral regions of the growing crystal could, under the influence of thermal-gradient-induced stresses, propagate into the crystal interior. Third, supersaturation of point defects could lead to clustering to form dislocation loops during cool down following growth.

The first source may have relevance in the present situation even though the growth is unseeded. Since solidification is achieved by moving the solid–liquid interface, dislocations present in the solidified portion could be replicated into the newly formed solid. Based on the recent work of Beam et al.[11] it is anticipated that all the dislocations will be incorporated into the new solid whether or not their Burgers vectors are parallel or inclined to the growth surface. However, the dislocation arrangements shown in Fig. 3 resemble those seen in semiconductor crystals deformed at low temperatures[12–14] and therefore could have formed after growth.

Since the resemblance between the observed dislocation structures and the deformed structures is striking, it is reckoned that thermal-gradient-induced stresses is the major source of dislocations in these crystals. It is envisaged that dislocation loops resulting from the condensation of point defects multiply under the influence of the above stresses, resulting in the observed dislocation structures.

Results indicate that, even though the three types of crystals undergo a similar growth sequence, the dislocation densities in these crystals are very different. This observation can be interpreted if it is assumed that the (Cd,Zn)Te and Cd(Te,Se) are stronger than the CdTe crystals. This assessment is indeed borne out by the recent work of Rai et al.[15] on the deformation characteristics of CdTe, (Cd,Zn)Te, and Cd(Te,Se) crystals. Furthermore, the strengthening could be attributed either to the tetrahedral radii differences between the Cd and Zn atoms and the Te and Se atoms as envisaged by Ehrenreich and Hirth[16] or to the stronger bonding between the Zn and Te atoms and the Cd and Se atoms as compared to the Cd–Te bonds.[17]

The coexistence of slip dislocations and microtwins in as-grown crystals, in particularly Fig. 4(a), is consistent with the model proposed by Mahajan and Chin[18] to explain the formation of deformation twins in face-centered-cubic crystals. They envisage that two coplanar $\frac{1}{2}\langle 110 \rangle$ slip dislocations of different Burgers interact with each other to form a three-layer twin according to the following reaction:

$$\tfrac{1}{2}\langle 1\bar{1}0 \rangle + \tfrac{1}{2}\langle 10\bar{1} \rangle \to 3 \times \tfrac{1}{6}\langle 2\bar{1}\bar{1} \rangle$$

Three-layer twins, located at different levels within a slip band, coalesce into each other to form a microtwin.

V. SUMMARY

The microstructure of CdTe, (Cd,Zn)Te, and Cd(Te,Se) crystals grown by the Bridgman technique have been investigated by TEM. Te precipitates are seen in all the crystals, but their density and size are lowest and largest in the case of Cd(Te,Se) crystals. Dislocation density of (Cd,Zn)Te and Cd(Te,Se) crystals is comparable and is much lower than that of CdTe crystals. In addition, microtwins are seen in (Cd,Zn)Te and Cd(Te,Se) crystals, whereas overlapping stacking faults are observed in CdTe crystals. It is inferred that thermal-gradient-induced stresses is the major source of dislocation and planar faults in these crystals.

ACKNOWLEDGMENT

The authors are grateful for the support of this work by the U.S. Army/CNVEO under Contract No. DAAB 07-87-C-F088.

[1] L. O. Bublac, W. F. Tennant, D. D. Edwall, E. R. Gertner, and J. C. Robinson, J. Vac. Sci. Technol. A **3**, 163 (1985).
[2] S. L. Bell and S. Sen, J. Vac. Sci. Technol. A **3**, 112 (1985).
[3] J. J. Kennedy, P. M. Amritraj, P. R. Boyd, and S. B. Quadri, J. Cryst. Growth **86**, 93 (1988).
[4] B. V. Dutt, S. Mahajan, R. J. Roedel, G. P. Schwartz, D. C. Miller, and L. Derick, J. Electrochem. Soc. **128**, 1573 (1981).
[5] J. P. Faurie, A. Million, and J. Piaguet, J. Cryst. Growth **59**, 10 (1982).
[6] P. D. Brown, J. E. Hails, G. J. Russel, and J. Woods, Appl. Phys. Lett. **50**, 1144 (1987).
[7] S. H. Shin, J. Bajaj, L. A. Moudy, and D. T. Cheung, Appl. Phys. Lett. **43**, 68 (1983).
[8] H. F. Schaake, J. H. Tregilas, A. J. Lewis, and P. M. Evert, J. Vac. Sci. Technol. A **1**, 1625 (1983).
[9] S. McDevitt, B. E. Dean. D. G. Ryding, F. J. Scheldens, and S. Mahajan, Mater. Lett. **4**, 451 (1986).
[10] M. A. Shahid, S. C. McDevitt, S. Mahajan, and C. J. Johnson, Inst. Phys. Conf. No. **87**, 321 (1987).

[11] E. A. Beam, S. Mahajan, and W. A. Bonner, Mater. Sci. Eng. B 7, 83 (1990).
[12] A. Alexander and P. Haasen, Solid State Phys. 22, 27 (1968).
[13] S. Mahajan, D. Barsen, and P. Haasen, Acta Metall. 27, 1165 (1979).
[14] I. Yonenaga, U. Onose, and K. Sumino, J. Mater. Res. 2, 252 (1987).
[15] R. S. Rai, S. Mahajan, D. J. Michel, H. H. Smith, S. McDevitt, and C. J. Johnson (to be published).
[16] H. Ehrenreich and J. P. Hirth, Appl. Phys. Lett. 46, 668 (1985).
[17] A. Sher, A. B. Chen. W. E. Spicer, and C. K. Shih, J. Vac. Sci. Technol. A 3, 105 (1985).
[18] S. Mahajan and G. Y. Chin, Acta Metall. 21, 1353 (1973).

Optical techniques for composition measurement of bulk and thin-film Cd$_{1-y}$Zn$_y$Te

S. M. Johnson, S. Sen, W. H. Konkel, and M. H. Kalisher
Santa Barbara Research Center, 75 Coromar Drive, Goleta, California 93117

(Received 22 October 1990; accepted 28 January 1991)

The composition of high-quality single-crystal bulk-grown Cd$_{1-y}$Zn$_y$Te ($0 \leqslant y \leqslant 0.2$) was determined from precision lattice constant measurements for a total of 22 data points. These samples were used to develop calibration curves for an accurate, contactless, nondestructive optical determination of composition using either 300 K transmission measurements or 77 K photoluminescence measurements. The 300 K transmission technique is useful for bulk CdZnTe wafers while the 77 K PL technique is applicable to both bulk and thin-film CdZnTe. Both techniques are useful in determining Cd$_{1-y}$Zn$_y$Te composition in the range ($0 \leqslant y \leqslant 0.2$) which covers the range needed for lattice-matching to Hg$_{1-x}$Cd$_x$Te and Hg$_{1-x}$Zn$_x$Te epitaxial layers. The ability to map sample composition with high precision is available with both techniques.

I. INTRODUCTION

Lattice-matched single-crystal bulk Cd$_{1-y}$Zn$_y$Te ($y < 0.05$) is the preferred substrate for growth of high-quality epitaxial HgCdTe used for the fabrication of second-generation infrared focal-plane arrays.[1-3] HgCdTe grown by liquid-phase epitaxy (LPE) is currently the best-established technology and uses CdZnTe substrates with areas as large as 30 cm^2.[4,5] Bulk Cd$_{1-y}$Zn$_y$Te ($y \approx 0.20$) is also important for lattice matching to epitaxial HgZnTe.[6,7] Additionally, thin-film alternative substrates of CdZnTe/GaAs/Si are rapidly gaining importance for improvements in size, strength, cost, and reliability of hybrid focal-plane arrays.[8,9] Since the distribution coefficient for Zn in CdZnTe exceeds unity, the Zn composition is nonuniform along the axial length of the boule;[2,3] large area wafers cut parallel or obliquely to the growth axis will be compositionally nonuniform. For thin-film CdZnTe/GaAs/Si compositional nonuniformities can result from the growth dynamics of metal-organic chemical vapor deposition (MOCVD). An accurate nondestructive measurement of the magnitude and spatial uniformity of the composition of both bulk and thin-film Cd$_{1-y}$Zn$_y$Te is needed to ensure uniform lattice matching to the HgCdTe or HgZnTe epitaxial layer.

The chemical composition of Cd$_{1-y}$Zn$_y$Te has been determined by energy-dispersive x-ray analysis (EDX),[10] x-ray fluorescence (XRF),[11] atomic absorption,[12] and direct current plasma emission spectrometry.[3] EDX measurements are inaccurate in the low composition range of $y < 0.05$, XRF has a large sampling area, and both atomic absorption and emission spectrometry require destruction of the sample. Contactless optical techniques that are sensitive to the variation in bandgap with composition are desirable.

Earlier workers have correlated both 300 K and 77 K absorption measurements with the composition of six Cd$_{1-y}$Zn$_y$Te ($0 \leqslant y \leqslant 1$) samples using composition determined by the weight of the components before they were melted together.[13] Both room temperature[14] and low temperature[15] (4.2 K and 77 K) reflection spectra of a small number of Cd$_{1-y}$Zn$_y$Te ($0 \leqslant y \leqslant 1$) samples were correlated with composition determined by x-ray powder diffraction photographs ($y \pm 0.01$). Low temperature (77 K) photoreflectance measurements of seven Cd$_{1-y}$Zn$_y$Te ($0 \leqslant y \leqslant 1$) samples were correlated with composition determined by x-ray and atomic absorption analysis.[16]

Photoluminescence (PL) measurements are more widely used, however most studies had either very few samples or the independent composition measurement technique was inaccurate or not stated. Measurements of the composition of six MBE-grown Cd$_{1-y}$Zn$_y$Te ($0 \leqslant y \leqslant 1$) layers by EDX were correlated with PL measurements done at 12 K and 300 K.[10] PL measurements of seven MBE samples were similarly correlated with composition determined by Auger electron spectroscopy in another study.[17] The composition of four MBE-grown Cd$_{1-y}$Zn$_y$Te ($0 \leqslant y \leqslant 0.065$) layers determined by XRF was correlated with PL measurements done at 2 K.[11] A relationship of PL measurements at 80 K of bulk-grown Cd$_{1-y}$Zn$_y$Te to composition was published, however, no information was reported on the number of samples or the technique used to determine composition.[18] The most extensive work in the low composition range was recently published for nine bulk Cd$_{1-y}$Zn$_y$Te ($0.02 \leqslant y \leqslant 0.07$) samples using 4.2 K PL measurements correlated with lattice constants determined from x-ray powder diffraction measurements.[19] However PL measurements in liquid He limit the utility of this technique to relatively small samples.

We have extended this photoluminescence calibration using 22 data points in the range ($0 \leqslant y \leqslant 0.2$) and also to $T = 77$ K, which allows immersion of large substrates in liquid nitrogen for rapid measurements rather than using a liquid-He dewar. We have further developed an accurate calibration curve for bulk substrates using a simple measurement of the 300 K optical transmission cut-on wavelength. The lattice constants of the high-quality single-crystal bulk CdTe, ZnTe, and Cd$_{1-y}$Zn$_y$Te calibration samples were determined using the Bond technique; the composition was calculated using Vegard's law. With these calibration curves the optical techniques are useful for mapping the composition of both bulk and thin-film substrates over composition ranges suitable for lattice matching to HgCdTe and HgZnTe epitaxial layers.

II. PRECISION LATTICE CONSTANT MEASUREMENTS

All of the $Cd_{1-y}Zn_yTe$ samples used for generating the calibration curves were grown by the vertical Bridgman technique.[1,2] Single-crystal samples were polished on both sides to remove saw damage and had a nominal thickness of 1 mm. Surface orientation was primarily {111} however a few {100} and {211} samples were also used.

Lattice constant measurements were performed on a high-resolution x-ray diffractometer having a Ge four-crystal monochromator to produce monochromatic $CuK\alpha_1$ radiation and having an optically encoded angular readout for precise angle measurements.[20] The x-ray beam size was approximately 1×4 mm^2. The lattice constants at various positions on the samples were determined using the Bond technique for measurements using symmetric x-ray reflections.[21] In this technique the Bragg angle is determined by measuring the sample rotation angle between two equivalent x-ray reflections for the x-ray beam incident from the left-hand side and the right-hand side of the sample. All samples had narrow single-peaked x-ray rocking curves in their respective measurement locations (no subgrain boundaries).

Lattice constants were used to determine the composition, y, (ZnTe mole fraction) of unknown samples and were calculated from Vegard's law using our lattice constant measurements of the binary samples ($a_{CdTe} = 6.4823$ Å, $a_{ZnTe} = 6.1004$ Å) using Eq. (1).

$$y = (6.4823 - a_{meas})/0.3819. \quad (1)$$

We believe our precision in lattice constant measurement is better than $a \pm 0.0002$ Å which translates to a composition uncertainty of $y \pm 0.0005$.

III. 300 K TRANSMISSION MEASUREMENTS

A Perkin-Elmer Lamda 9 UV/Vis/NIR spectrophotometer (double-beam, double monochromator, ratio recording) was used for absolute transmission measurements of the $Cd_{1-y}Zn_yTe$ samples at room temperature. The spot size was 3.2 mm in diameter. The spectral bandpass of the system was 0.86 nm and wavelength repeatability was $\lambda \pm 0.2$ nm. The cut-on wavelength was determined from a linear fit of the transmission (T) spectra over the range ($0.1 \leq T \leq 0.3$) and extrapolating this to $T = 0$.

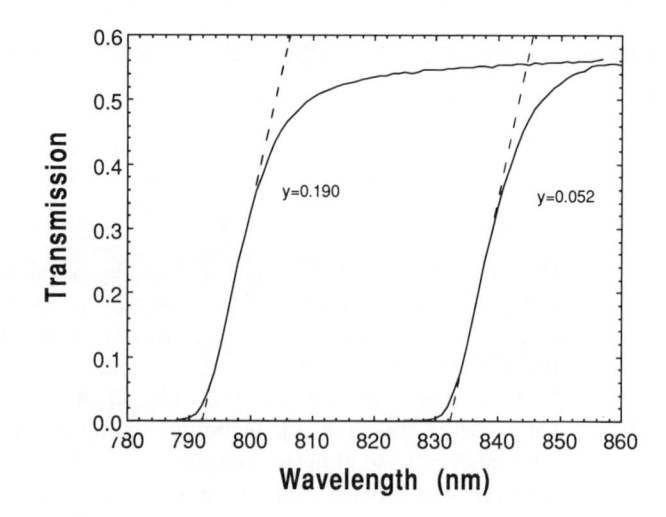

Figure 1 shows an example of two room-temperature transmission spectra for $Cd_{1-y}Zn_yTe$ samples having compositions of $y = 0.052$ and $y = 0.190$. Figure 2 shows the absorption coefficient versus wavelength for these same two samples determined from the transmission spectra in Fig. 1. The transmission spectrum is dependent on the sample thickness so a calibration using the cut-on wavelength needs to be generated using the same sample thickness. Figure 3 shows an example of transmission spectra (no reflection losses) of $Cd_{0.948}Zn_{0.052}Te$ calculated for sample thicknesses

FIG. 1. Example of two room temperature transmission spectra for $Cd_{1-y}Zn_yTe$ samples having compositions of $y = 0.052$ and $y = 0.190$.

of 0.8, 1.0, and 1.2 mm using the absorption data in Fig. 1. The cut-on wavelength for these sample thicknesses of 0.8–1.2 mm varies nonlinearly from 831.4 to 833.2 nm which would translate to a variation in composition of $y = 0.054$–0.048 using our calibration curve. By calculating a series of spectra for different sample thicknesses, as in Fig. 3, the cut-on wavelength can be plotted as a function of thickness and cut-on wavelength correction, $\Delta\lambda_{co}$, can be calculated for samples deviating from a 1 mm standard thickness. The cut-on wavelength correction to $t = 1$ mm, which is good for the range (0.8 mm $\leq t \leq$ 1.2 mm) is given by Eq. (2).

$$\Delta\lambda_{co} = -7.520 + (10.65)t - (3.143)t^2. \quad (2)$$

For a sample having a thickness, t, different than 1 mm, the corrected cut-on wavelength is determined by subtracting $\Delta\lambda_{co}$ from the measured cut-on wavelength. Equation (2) is valid over the $Cd_{1-y}Zn_yTe$ composition range ($0 \leq y \leq 0.2$).

Figure 4(a) is a plot of the 300 K cut-on energy versus composition for samples with $t = 1$ mm over the range ($0 \leq y \leq 0.2$); Fig. 4(b) is an expansion of (a) in the region of

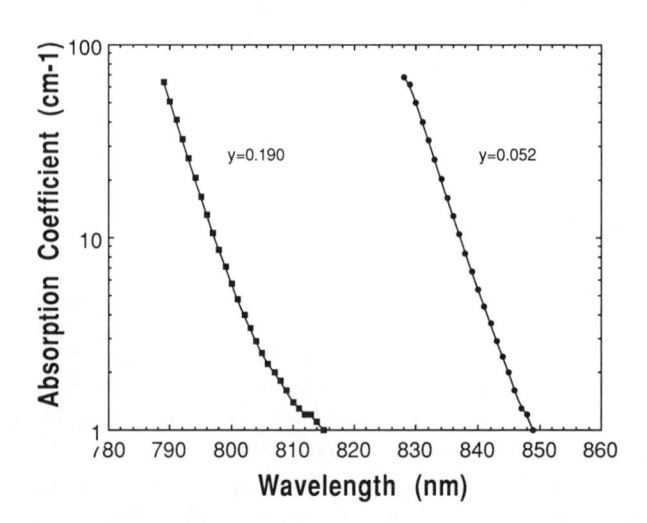

FIG. 2. Room temperature absorption coefficient vs wavelength for the same two samples in Fig. 1.

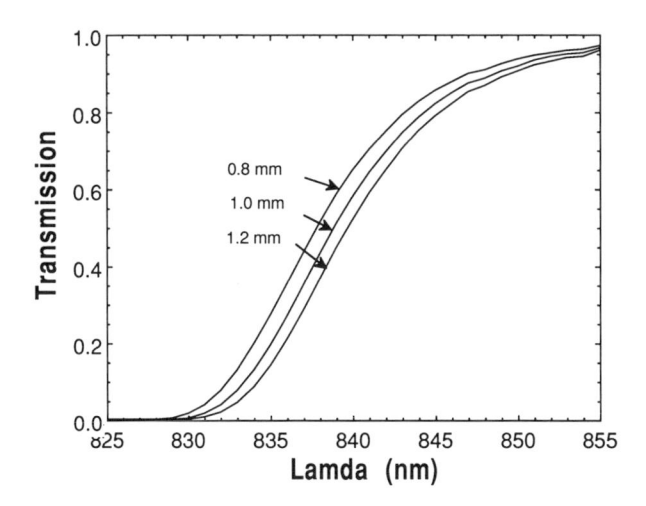

FIG. 3. Example of a transmission spectra (no reflection losses) of $Cd_{0.948}Zn_{0.052}Te$ calculated for sample thicknesses of 0.8, 1.0, and 1.2 mm using the absorption data in Fig. 1.

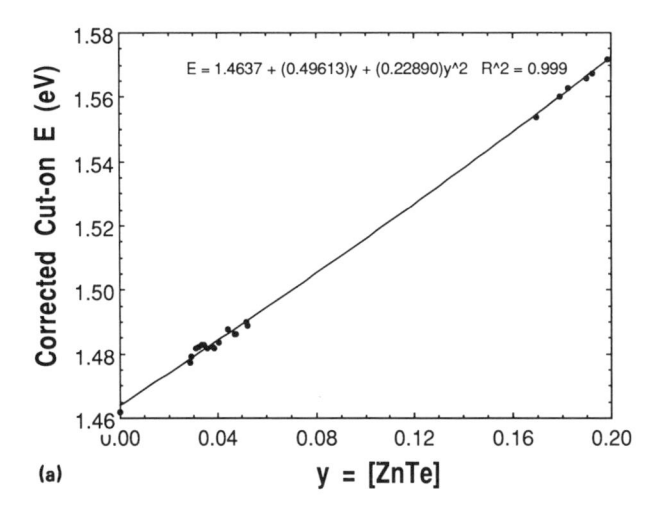

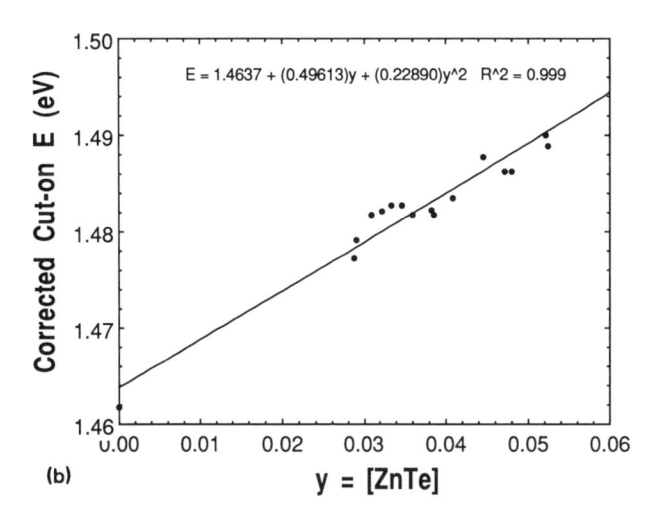

FIG. 4. (a) 300 K cut-on energy vs composition for samples with $t = 1$ mm over the range $(0 \leqslant y \leqslant 0.2)$; (b) is an expansion of (a) in the region of $(0 \leqslant y \leqslant 0.06)$

$(0 \leqslant y \leqslant 0.06)$. A quadratic least-squares fit to these data is given in Eq. (3).

$$E_{co} = 1.4637 + (0.49613)y + (0.22890)y^2. \quad (3)$$

This equation can be inverted to yield the composition in terms of the measured 300 K cut-on energy and is given by Eq. (4).

$$y = -4.1836 + (3.8140)E_{co} - (0.65223)E_{co}^2. \quad (4)$$

This expression is good for $Cd_{1-y}Zn_yTe$ over the composition range of $(0 \leqslant y \leqslant 0.2)$ which covers the range needed for lattice-matching to $Hg_{1-x}Cd_xTe$ and $Hg_{1-x}Zn_xTe$ epitaxial layers.

IV. 77 K PHOTOLUMINESCENCE MEASUREMENTS

A SPEX 1404 double spectrometer with 600 g/mm gratings was used for PL measurements of the $Cd_{1-y}Zn_yTe$ samples at 77 K. The spectral bandpass of the system was 0.04 nm and wavelength repeatability was $\lambda \pm 0.09$ nm. A CW Ar^+ ion laser (4880 Å) with a power of 50 mW was focused to a spot size of approximately 1–2 mm in diameter for excitation. Although PL spectra of bulk $Cd_{1-y}Zn_yTe$ samples can be obtained at a lower excitation power, thin-film $Cd_{1-y}Zn_yTe$ generally requires this maginitude of excitation power to maintain a good signal to noise ratio. The samples were immersed in LN_2 to maintain a constant temperature.

The energy position of the principle bound exciton (PBE) at one-half the maximum of the rapidly changing high energy side of the peak $PBE_{1/2}$ was used for correlation with the $Cd_{1-y}Zn_yTe$ composition. For room-temperature PL measurements of InGaAs this energy position on the low energy side of the PBE peak was used as a reliable measure of composition[22] (at room temperature the low energy side of the peak is sharper). Figure 5 shows examples of 77 K PL spectra for $Cd_{1-y}Zn_yTe$ samples having compositions of (a) $y = 0.052$ and (b) $y = 0.190$ (same samples used for Figs. 1 and 2).

Figure 6(a) is a plot of the 77 K $PBE_{1/2}$ energy versus composition for samples over the range $(0 \leqslant y \leqslant 0.2)$; Fig. 6(b) is an expansion of (a) in the region of $(0 \leqslant y \leqslant 0.06)$. The 77 K PL measurements shown in Fig. 6 have less scatter than the 300 K transmission measurements shown in Fig. 4 made on the same samples. A quadratic least-squares fit to these data is given in Eq. (4).

$$E_{1/2} = 1.5860 + (0.50060)y + (0.29692)y^2. \quad (4)$$

This equation can be inverted to yield the composition in terms of the measured 77 K $PBE_{1/2}$ energy and is given by Eq. (5).

$$y = -6.9101 + (6.7650)E_{1/2} - (1.5181)E_{1/2}^2 \quad (5)$$

This expression is good for $Cd_{1-y}Zn_yTe$ over the composition range of $(0 \leqslant y \leqslant 0.2)$.

To determine the composition of $Cd_{1-y}Zn_yTe$ thin-films grown by MOCVD and MBE on GaAs or GaAs/Si substrates a surface sensitive technique such as PL is needed since both GaAs and Si are strongly absorbing in the wavelength region of interest. To illustrate the utility of the PL technique, we determined the composition of two CdZnTe/

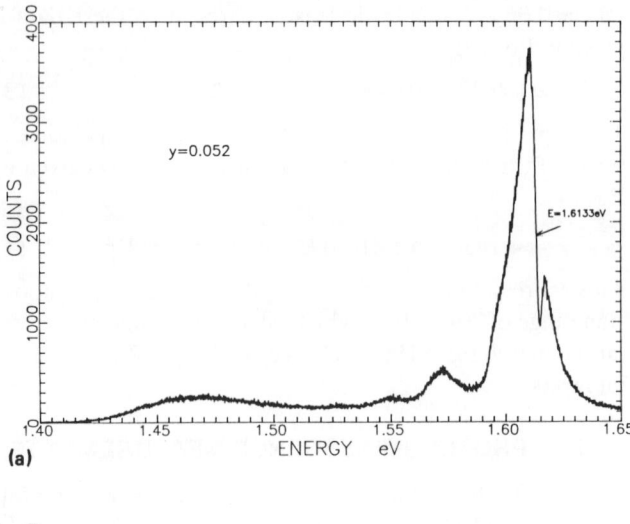

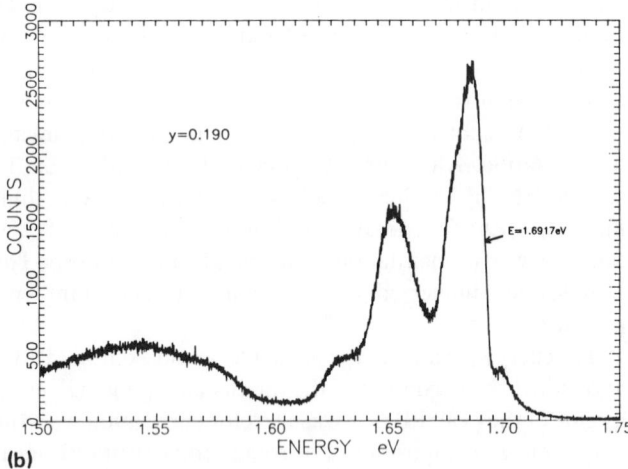

FIG. 5. 77 K photoluminescence spectra of $Cd_{1-y}Zn_yTe$ samples having compostions of (a) $y = 0.052$ and (b) $y = 0.190$.

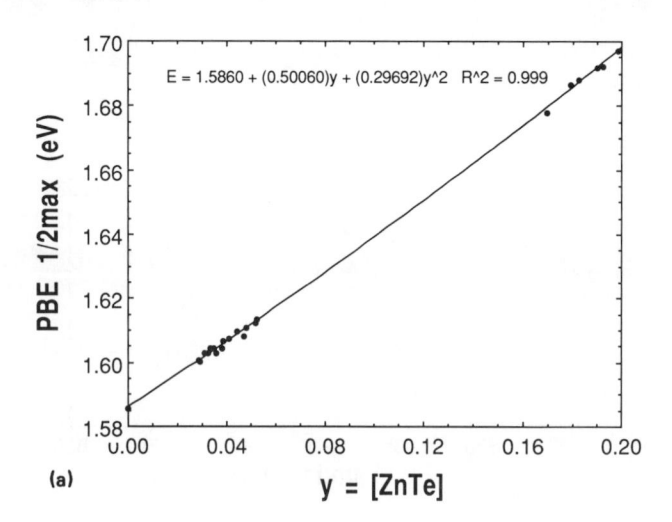

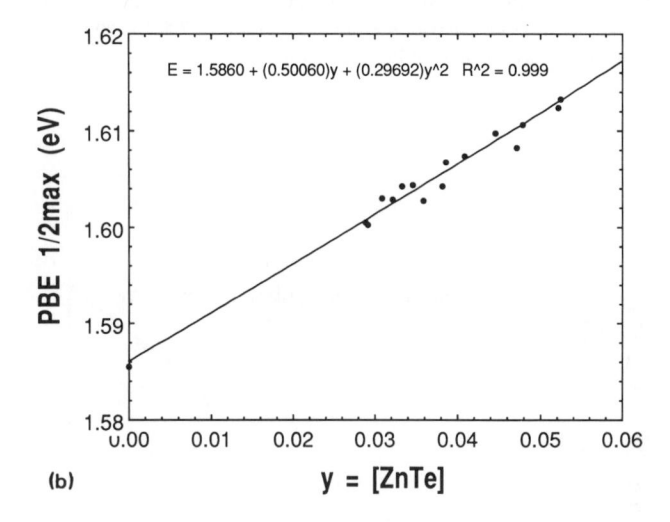

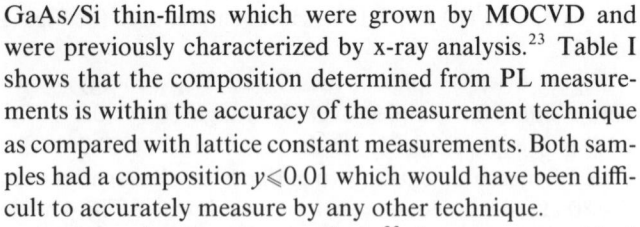

FIG. 6. (a) 77 K $PBE_{1/2}$ energy versus composition for samples over the range $(0 \leq y \leq 0.2)$; (b) is an expansion of (a) in the region of $(0 \leq y \leq 0.06)$.

GaAs/Si thin-films which were grown by MOCVD and were previously characterized by x-ray analysis.[23] Table I shows that the composition determined from PL measurements is within the accuracy of the measurement technique as compared with lattice constant measurements. Both samples had a composition $y \leq 0.01$ which would have been difficult to accurately measure by any other technique.

As determined by x-ray analysis,[23] there was no residual strain in these thin-film samples which had CdZnTe layer thicknesses of about 5 μm. However for layer thicknesses less than approximately 1 μm care must be taken in the interpretation of PL measurements since biaxial compressive strains in CdTe/GaAs were shown to cause the PBE peak to shift to lower energy.[24]

V. ACCURACY AND PRECISION OF OPTICAL TECHNIQUES

A figure of merit for the accuracy of the measurement (a measure of the true composition) is the standard deviation of the quadratic least squares fit to the calibration data. The precision (ability to distinguish composition variation) is primarily determined by the wavelength repeatability of the spectrometers used for the measurements. Table II summarizes the accuracy and precision of the optical techniques. The accuracy of the 77 K PL technique in this work compares favorably with a stated accuracy of $y \pm 0.002$ (based on the uncertainty in x-ray determined lattice constants) given in a recently published 4.2 K PL calibration study.[19] The accuracy of the 77 K PL technique is better than that of the 300 K transmission technique, however the ability to map sample composition with high precision is available with both techniques.

VI. SUMMARY AND CONCLUSIONS

The composition of high-quality single-crystal bulk-grown $Cd_{1-y}Zn_yTe$ $(0 \leq y \leq 0.2)$ was determined from preci-

TABLE I. Example of composition determined from 77 K PL measurements on thin film substrates of CdZnTe/GaAs/Si.

Sample	Lattice constant (Å)	y (Vegard)	y (PL)
{100}1° CdZnTe	6.4785	0.0100	0.0080
{111}4° CdZnTe	6.4795	0.0073	0.0072

TABLE II. Summary of the accuracy and precision of the optical techniques for determining composition of Cd_yZn_yTe.

Technique	Accuracy	Precision
300 K Transmission	$y \pm 0.0027$	$y \pm 0.000\,70$
77 K PL	$y \pm 0.0019$	$y \pm 0.000\,34$

sion lattice constant measurements for a total of 22 data points. These samples were used to develop calibration curves for an accurate, contactless, nondestructive optical determination of composition using either 300 K transmission measurements or 77 K photoluminescence measurements. The 300 K transmission technique is useful for bulk CdZnTe wafers while the 77 K PL technique is applicable to both bulk and thin-film CdZnTe. Both techniques are useful in determining $Cd_{1-y}Zn_yTe$ composition in the range ($0 \leqslant y \leqslant 0.2$) which covers the region needed for lattice-matching to $Hg_{1-x}Cd_xTe$ and $Hg_{1-x}Zn_xTe$ epitaxial layers. The ability to map sample composition with high precision is available with both techniques. Recently a scanning PL system which has the ability to rapidly map wafers as large as 75 mm with a minimum sampling grid size of 2.5 μm has been reported which can allow for high-resolution wafer composition mapping.[25] Using these simple measurement techniques to determine and map $Cd_{1-y}Zn_yTe$ composition prior to epitaxial layer growth will allow for more quantification of the effects of lattice mismatch on the properties of epitaxial HgCdTe and HgZnTe which may allow for subsequent improvements in epitaxial layer quality and device performance.

ACKNOWLEDGMENTS

The authors thank K. T. Miller for making the precision lattice constant measurements, G. A. Walter, D. C. Strickland, and K. L. Ried for their assistance with the transmission measurements and analysis, V. L. Liguori for sample preparation, and W. J. Hamilton for helpful discussions.

[1] S. L. Bell and S. Sen, J. Vac. Sci. Technol. A 3, 112 (1985).
[2] S. Sen, S. M. Johnson, J. A. Kiele, W. H. Konkel, and J. E. Stannard, in *Properties of II–VI Semiconductors: Bulk Crystals, Epitaxial Films, Quantum Well Structures, and Dilute Magnetic Systems*, edited by F. J. Bartoli, Jr., H. F. Schaake, and J. F. Schetzina (Mater. Res. Soc. Vol. 161, Pittsburgh, 1990), p. 3.
[3] S. McDevitt, D. R. John, J. L. Sepich, K. A. Bowers, J. F. Schetzina, R. S. Rai, and S. Mahajan, in *Properties of II–VI Semiconductors: Bulk Crystals, Epitaxial Films, Quantum Well Structures, and Dilute Magnetic Systems*, edited by F. J. Bartoli, Jr., H. F. Schaake, and J. F. Schetzina (Mater. Res. Soc. Vol. 161, Pittsburgh, 1990), p. 15.
[4] T. Tung, M. H. Kalisher, A. P. Stevens, and P. E. Herning, in *Materials for Infrared Detectors and Sources*, edited by R. F. C. Farrow, J. F. Schetzina, and J. T. Cheung (Mater. Res. Soc. Vol. 90, Pittsburgh, 1987), p. 321.
[5] T. Tung, J. Cryst. Growth 86, 161 (1988).
[6] E. J. Smith, T. Tung, S. Sen, W. H. Konkel, J. B. James, V. B. Harper, B. F. Zuck, and R. A. Cole, J. Vac. Sci. Technol. A 5, 3043 (1987).
[7] E. A. Patten, M. H. Kalisher, G. R. Chapman, J. M. Fulton, C. Y. Huang, P. R. Norton, M. Ray, and S. Sen, J. Vac. Sci. Technol. B 9, 1746 (1991).
[8] W. L. Ahlgren, S. M. Johnson, E. J. Smith, R. P. Ruth, B. C. Johnston, M. H. Kalisher, C. A. Cockrum, T. W. James, D. L. Arney, C. K. Ziegler, and W. Lick, J. Vac. Sci. Technol. A 7, 331 (1989).
[9] S. M. Johnson, M. H. Kalisher, W. L. Ahlgren, J. B. James, and C. A. Cockrum, Appl. Phys. Lett. 56, 946 (1990).
[10] D. J. Olego, J. P. Faurie, S. Sivananthan, and P. M. Raccah, Appl. Phys. Lett. 47, 1172 (1985).
[11] N. Magnea, F. Dal'bo, J. L. Pautrat, A. Million, L. Di Cioccio, and G. Feuillet, in *Materials for Infrared Detectors and Sources*, edited by R. F. C. Farrow, J. F. Schetzina, and J. T. Cheung (Mater. Res. Soc., Pittsburgh, 1987), Vol. 90, p. 455.
[12] M. Yoshikawa, J. Appl. Phys. 63, 1533 (1988).
[13] M. S. Brodin, M. V. Kurik, V. M. Matlak, and B. S. Oktyabr'skii, Sov. Phys. Semicond. 2, 603 (1968).
[14] K. Saito, A. Ebina, and T. Takahashi, Solid State Comm. 11, 841 (1972).
[15] E. F. Gross, G. M. Grigorovich, I. V. Pozdnyakov, V. G. Sredin, and L. G. Suslina, Sov. Phys. Solid State, 12, 2352 (1971).
[16] P. M. Amirtharaj, J. H. Dinan, J. J. Kennedy, P. R. Boyd, and O. J. Glembocki, J. Vac. Sci. Technol. A 4, 2028 (1986).
[17] C. J. Summers, A. Torabi, B. K. Wagner, J. D. Benson, S. R. Stock, and P. C. Huang, in *Materials Technologies for IR Detectors*, edited by J. Besson (Soc. Phot. Opt. Inst. Eng. 659, Bellingham, 1986), p. 153.
[18] K. Y. Lay, N. C. Giles-Taylor, J. F. Schetzina, and K. J. Bachman, J. Electrochem. Soc. 133, 1049 (1986).
[19] W. M. Duncan, R. J. Koestner, J. H. Tregilgas, H.-Y. Liu, and M. -C. Chen, in *Properties of II–VI Semiconductors: Bulk Crystals, Epitaxial Films, Quantum Well Structures, and Dilute Magnetic Systems*, edited by F. J. Bartoli, Jr., H. F. Schaake, and J. F. Schetzina (Mater. Res. Soc. Vol. 161, Pittsburgh, 1990), p. 39.
[20] K. T. Miller, Hughes Research Laboratories (unpublished).
[21] W. L. Bond, Acta Cryst. 13, 814 (1960).
[22] I. C. Bassignana, C. J. Miner, and N. Puetz, J. Appl. Phys. 65, 4299 (1989).
[23] S. M. Johnson, W. L. Ahlgren, M. H. Kalisher, J. B. James, and W. J. Hamiltion, Jr., in *Properties of II–VI Semiconductors: Bulk Crystals, Epitaxial Films, Quantum Well Structures, and Dilute Magnetic Systems*, edited by F. J. Bartoli, Jr., H. F. Schaake, and J. F. Schetzina (Mater. Res. Soc. Vol. 161, Pittsburgh, 1990), p. 351.
[24] D. J. Olego, J. Petruzzello, S. K. Ghandi, N. R. Taskar, and I. B. Bhat, Appl. Phys. Lett. 51, 127 (1987).
[25] C. J. L. Moore and C. J. Miner, J. Cryst. Growth 103, 21 (1990).

Author Index

Arias, J. M. — (3) 1646
Arnold, D. J. — (3) 1813, 1818
Astles, M. G. — (3) 1687

Baars, J. — (3) 1709
Bajaj, J. — (3) 1661
Bandara, K. M. S. V. — (3) 1789
Bartoli, F. J. — (3) 1813, 1818
Becla, P. — (3) 1777
Benz, R. G., II — (3) 1656
Berding, M. A. — (3) 1738, 1858
Bernardi, S. — (3) 1870
Bhat, I. B. — (3) 1625, 1705
Blanks, D. K. — (3) 1764
Blazejewski, E. R. — (3) 1661, 1874
Bouchut, P. — (3) 1794
Brink, D. — (3) 1709
Bubulac, L. — (3) 1661
Bubulac, L. O. — (3) 1691, 1695

Capper, P. — (3) 1667, 1682
Chamonal, J. P. — (3) 1794
Chandra, D. — (3) 1852
Chapman, G. R. — (3) 1746
Chen, J. S. — (3) 1661, 1823
Chen, J.-S. — (3) 1691
Choe, J.-W. — (3) 1789
Cinader, G. — (3) 1634
Cook, J. W., Jr. — (3) 1799, 1805, 1813
Coon, D. D. — (3) 1789

Dean, B. E. — (3) 1840
DeCarlo, J. — (3) 1886
Destefanis, G. — (3) 1794
DeWames, R. E. — (3) 1646
Di Marzio, D. — (3) 1886
Dobbyn, R. C. — (3) 1840

Easton, B. C. — (3) 1682
Edwall, D. D. — (3) 1691, 1695, 1823
Ehsani, H. — (3) 1625
Ellsworth, J. — (3) 1840

Fadley, C. S. — (3) 1870
Fastow, R. — (3) 1829
Faurie, J. P. — (3) 1651
Francombe, M. H. — (3) 1789
Freeman, Charles F. — (3) 1613

Friedman, D. J. — (3) 1870
Fulton, J. M. — (3) 1746
Furdyna, J. K. — (3) 1809

Gale, I. G. — (3) 1682
Ghandhi, S. K. — (3) 1625, 1705
Gibaud, A. — (3) 1886
Goodwin, M. W. — (3) 1731, 1852
Gough, J. S. — (3) 1687
Grainger, F. — (3) 1682
Granozzi, G. — (3) 1870
Grudzien, N. — (3) 1777

Hallock, P. — (3) 1630
Harris, K. A. — (3) 1752
Heald, S. M. — (3) 1886
Helms, C. R. — (3) 1781, 1785, 1879
Herman, G. S. — (3) 1870
Hoffman, C. A. — (3) 1813, 1818
Houlton, M. R. — (3) 1687
Huang, C. Y. — (3) 1746
Hwang, S. — (3) 1799

Irvine, S. J. C. — (3) 1687

Johnson, C. J. — (3) 1840, 1892
Johnson, S. M. — (3) 1897
Johnston, S. — (3) 1661
Johnston, S. L. — (3) 1823

Kalisher, M. H. — (3) 1746, 1897
Kennedy, J. J. — (3) 1840
Kibel, Martyn H. — (3) 1770
Kim, D. — (3) 1639
Koestner, R. J. — (3) 1731
Konkel, W. H. — (3) 1897
Korenstein, R. — (3) 1630
Krishnamurthy, Srinivasan — (3) 1858
Krueger, E. Eric — (3) 1724
Kuriyama, M. — (3) 1840

Lange, M. D. — (3) 1651
Lansari, Y. — (3) 1799, 1805, 1813
Lee, M. B. — (3) 1886
Lee, S. B. — (3) 1639
Leech, Patrick W. — (3) 1770
Lin, Y. F. — (3) 1789
Littler, C. L. — (3) 1847

Loloee, M. R. — (3) 1847
Lowney, J. R. — (3) 1847
Luo, H. — (3) 1809

MacLeod, B. — (3) 1630
Mahajan, S. — (3) 1892
Maxey, C. D. — (3) 1682
McDevitt, S. — (3) 1892
McDevitt, S. C. — (3) 1840
McLevige, W. V. — (3) 1646, 1874
Menezes, S. — (3) 1874
Mercer, D. E. — (3) 1781
Meyassed, M. — (3) 1829
Meyer, J. R. — (3) 1813, 1818
Million, A. — (3) 1794
Mohnkern, L. M. — (3) 1752
Myers, T. H. — (3) 1752

Nemirovsky, Y. — (3) 1829
Neugebauer, G. T. — (3) 1840
Norton, P. R. — (3) 1746
Norton, Peter W. — (3) 1724

O, Byungsung — (3) 1789
Osterwalder, J. — (3) 1870
Otsuka, N. — (3) 1752

Parat, K. K. — (3) 1625, 1705
Pasko, J. G. — (3) 1646
Patten, E. A. — (3) 1746
Pelliciari, B. — (3) 1794
Petruzzello, J. — (3) 1651
Piaguet, J. — (3) 1794
Piotrowski, J. — (3) 1777
Pultz, G. N. — (3) 1724

Rai, R. S. — (3) 1892
Raizman, A. — (3) 1634
Rajavel, D. — (3) 1656
Ram-Mohan, L. R. — (3) 1809, 1818
Ray, M. — (3) 1746
Reine, M. B. — (3) 1724
Richter, H. J. — (3) 1861
Rizzi, G. A. — (3) 1870
Roberts, J. A. — (3) 1682

Schaake, H. F. — (3) 1731
Schetzina, J. F. — (3) 1799, 1805, 1813

Schiebel, R. A. — (3) 1759
Scilla, G. J. — (3) 1705
Scott, G. — (3) 1781, 1785
Seelmann-Eggebert, M. — (3) 1861
Seiler, D. G. — (3) 1847
Sen, S. — (3) 1746, 1897
Sepich, J. L. — (3) 1840
Shaw, N. — (3) 1687
Sher, A. — (3) 1634, 1738, 1858
Shin, S. H. — (3) 1646
Sorar, E. — (3) 1789
Sporken, R. — (3) 1651
Stevenson, D. A. — (3) 1615, 1639
Summers, C. J. — (3) 1656

Takei, W. J. — (3) 1789
Tang, M-F. S. — (3) 1615
Taskar, N. R. — (3) 1705
Tennant, W. E. — (3) 1874
Tran, T. T. — (3) 1870
Tregilgas, J. H. — (3) 1852

Unikovsky, A. — (3) 1829

van Schilfgaarde, M. — (3) 1738
Viswanathan, C. R. — (3) 1695
Vydyanath, H. R. — (3) 1716, 1840

Wagner, B. K. — (3) 1656
Wang, C. C. — (3) 1740
Weiss, Eliezer — (3) 1879
Whiffin, P. A. C. — (3) 1682
Williams, G. — (3) 1661

Yang, G. L. — (3) 1809
Yang, Z. — (3) 1799, 1805
Yanka, R. W. — (3) 1752
Yoon, I. T. — (3) 1847
Young, M. L. — (3) 1687
Younger, C. R. — (3) 1823
Yu, Z. — (3) 1805

Zandian, M. — (3) 1646
Ziegler, J. — (3) 1709
Ziegler, J. P. — (3) 1874
Zucca, R. — (3) 1823

AIP Conference Proceedings

		L.C. Number	ISBN
No. 82	Interpretation of Climate and Photochemical Models, Ozone and Temperature Measurements (La Jolla Institute, 1981)	82-71345	0-88318-181-9
No. 83	The Galactic Center (Cal. Inst. of Tech., 1982)	82-71635	0-88318-182-7
No. 84	Physics in the Steel Industry (APS/AISI, Lehigh University, 1981)	82-72033	0-88318-183-5
No. 85	Proton-Antiproton Collider Physics –1981 (Madison, WI)	82-72141	0-88318-184-3
No. 86	Momentum Wave Functions – 1982 (Adelaide, Australia)	82-72375	0-88318-185-1
No. 87	Physics of High Energy Particle Accelerators (Fermilab Summer School, 1981)	82-72421	0-88318-186-X
No. 88	Mathematical Methods in Hydrodynamics and Integrability in Dynamical Systems (La Jolla Institute, 1981)	82-72462	0-88318-187-8
No. 89	Neutron Scattering – 1981 (Argonne National Laboratory)	82-73094	0-88318-188-6
No. 90	Laser Techniques for Extreme Ultraviolet Spectroscopy (Boulder, CO, 1982)	82-73205	0-88318-189-4
No. 91	Laser Acceleration of Particles (Los Alamos, NM, 1982)	82-73361	0-88318-190-8
No. 92	The State of Particle Accelerators and High Energy Physics (Fermilab, 1981)	82-73861	0-88318-191-6
No. 93	Novel Results in Particle Physics (Vanderbilt, 1982)	82-73954	0-88318-192-4
No. 94	X-Ray and Atomic Inner-Shell Physics – 1982 (International Conference, U. of Oregon)	82-74075	0-88318-193-2
No. 95	High Energy Spin Physics – 1982 (Brookhaven National Laboratory)	83-70154	0-88318-194-0
No. 96	Science Underground (Los Alamos, NM, 1982)	83-70377	0-88318-195-9
No. 97	The Interaction Between Medium Energy Nucleons in Nuclei – 1982 (Indiana University)	83-70649	0-88318-196-7
No. 98	Particles and Fields – 1982 (APS/DPF University of Maryland)	83-70807	0-88318-197-5
No. 99	Neutrino Mass and Gauge Structure of Weak Interactions (Telemark, 1982)	83-71072	0-88318-198-3
No. 100	Excimer Lasers – 1983 (OSA, Lake Tahoe, NV)	83-71437	0-88318-199-1
No. 101	Positron-Electron Pairs in Astrophysics (Goddard Space Flight Center, 1983)	83-71926	0-88318-200-9
No. 102	Intense Medium Energy Sources of Strangeness (UC-Santa Cruz, CA, 1983)	83-72261	0-88318-201-7
No. 103	Quantum Fluids and Solids – 1983 (Sanibel Island, FL)	83-72440	0-88318-202-5
No. 104	Physics, Technology and the Nuclear Arms Race (APS, Baltimore, MD, 1983)	83-72533	0-88318-203-3
No. 105	Physics of High Energy Particle Accelerators (SLAC Summer School, 1982)	83-72986	0-88318-304-8
No. 106	Predictability of Fluid Motions (La Jolla Institute, 1983)	83-73641	0-88318-305-6

No. 107	Physics and Chemistry of Porous Media (Schlumberger-Doll Research, 1983)	83-73640	0-88318-306-4
No. 108	The Time Projection Chamber (TRIUMF, Vancouver, 1983)	83-83445	0-88318-307-2
No. 109	Random Walks and Their Applications in the Physical and Biological Sciences (NBS/La Jolla Institute, 1982)	84-70208	0-88318-308-0
No. 110	Hadron Substructure in Nuclear Physics (Indiana University, 1983)	84-70165	0-88318-309-9
No. 111	Production and Neutralization of Negative Ions and Beams (3rd Int'l Symposium) (Brookhaven, NY, 1983)	84-70379	0-88318-310-2
No. 112	Particles and Fields – 1983 (APS/DPF, Blacksburg, VA)	84-70378	0-88318-311-0
No. 113	Experimental Meson Spectroscopy – 1983 (7th International Conference, Brookhaven, NY)	84-70910	0-88318-312-9
No. 114	Low Energy Tests of Conservation Laws in Particle Physics (Blacksburg, VA, 1983)	84-71157	0-88318-313-7
No. 115	High Energy Transients in Astrophysics (Santa Cruz, CA, 1983)	84-71205	0-88318-314-5
No. 116	Problems in Unification and Supergravity (La Jolla Institute, 1983)	84-71246	0-88318-315-3
No. 117	Polarized Proton Ion Sources (TRIUMF, Vancouver, 1983)	84-71235	0-88318-316-1
No. 118	Free Electron Generation of Extreme Ultraviolet Coherent Radiation (Brookhaven/OSA, 1983)	84-71539	0-88318-317-X
No. 119	Laser Techniques in the Extreme Ultraviolet (OSA, Boulder, CO, 1984)	84-72128	0-88318-318-8
No. 120	Optical Effects in Amorphous Semiconductors (Snowbird, UT, 1984)	84-72419	0-88318-319-6
No. 121	High Energy e^+e^- Interactions (Vanderbilt, 1984)	84-72632	0-88318-320-X
No. 122	The Physics of VLSI (Xerox, Palo Alto, CA, 1984)	84-72729	0-88318-321-8
No. 123	Intersections Between Particle and Nuclear Physics (Steamboat Springs, CO, 1984)	84-72790	0-88318-322-6
No. 124	Neutron-Nucleus Collisions: A Probe of Nuclear Structure (Burr Oak State Park, 1984)	84-73216	0-88318-323-4
No. 125	Capture Gamma-Ray Spectroscopy and Related Topics – 1984 (Int'l Symposium, Knoxville, TN)	84-73303	0-88318-324-2
No. 126	Solar Neutrinos and Neutrino Astronomy (Homestake, 1984)	84-63143	0-88318-325-0
No. 127	Physics of High Energy Particle Accelerators (BNL/SUNY Summer School, 1983)	85-70057	0-88318-326-9
No. 128	Nuclear Physics with Stored, Cooled Beams (McCormick's Creek State Park, IN, 1984)	85-71167	0-88318-327-7
No. 129	Radiofrequency Plasma Heating (Sixth Topical Conference) (Callaway Gardens, GA, 1985)	85-48027	0-88318-328-5
No. 130	Laser Acceleration of Particles (Malibu, CA, 1985)	85-48028	0-88318-329-3
No. 131	Workshop on Polarized ^{3}He Beams and Targets (Princeton, NJ, 1984)	85-48026	0-88318-330-7
No. 132	Hadron Spectroscopy–1985 (International Conference, Univ. of Maryland)	85-72537	0-88318-331-5
No. 133	Hadronic Probes and Nuclear Interactions (Arizona State University, 1985)	85-72638	0-88318-332-3

No.	Title	LCCN	ISBN
No. 134	The State of High Energy Physics (BNL/SUNY Summer School, 1983)	85-73170	0-88318-333-1
No. 135	Energy Sources: Conservation and Renewables (APS, Washington, DC, 1985)	85-73019	0-88318-334-X
No. 136	Atomic Theory Workshop on Relativistic and QED Effects in Heavy Atoms (Gaithersburg, MD, 1985)	85-73790	0-88318-335-8
No. 137	Polymer-Flow Interaction (La Jolla Institute, 1985)	85-73915	0-88318-336-6
No. 138	Frontiers in Electronic Materials and Processing (Houston, TX, 1985)	86-70108	0-88318-337-4
No. 139	High-Current, High-Brightness, and High-Duty Factor Ion Injectors (La Jolla Institute, 1985)	86-70245	0-88318-338-2
No. 140	Boron-Rich Solids (Albuquerque, NM, 1985)	86-70246	0-88318-339-0
No. 141	Gamma-Ray Bursts (Stanford, CA, 1984)	86-70761	0-88318-340-4
No. 142	Nuclear Structure at High Spin, Excitation, and Momentum Transfer (Indiana University, 1985)	86-70837	0-88318-341-2
No. 143	Mexican School of Particles and Fields (Oaxtepec, México, 1984)	86-81187	0-88318-342-0
No. 144	Magnetospheric Phenomena in Astrophysics (Los Alamos, NM, 1984)	86-71149	0-88318-343-9
No. 145	Polarized Beams at SSC & Polarized Antiprotons (Ann Arbor, MI & Bodega Bay, CA, 1985)	86-71343	0-88318-344-7
No. 146	Advances in Laser Science–I (Dallas, TX, 1985)	86-71536	0-88318-345-5
No. 147	Short Wavelength Coherent Radiation: Generation and Applications (Monterey, CA, 1986)	86-71674	0-88318-346-3
No. 148	Space Colonization: Technology and The Liberal Arts (Geneva, NY, 1985)	86-71675	0-88318-347-1
No. 149	Physics and Chemistry of Protective Coatings (Universal City, CA, 1985)	86-72019	0-88318-348-X
No. 150	Intersections Between Particle and Nuclear Physics (Lake Louise, Canada, 1986)	86-72018	0-88318-349-8
No. 151	Neural Networks for Computing (Snowbird, UT, 1986)	86-72481	0-88318-351-X
No. 152	Heavy Ion Inertial Fusion (Washington, DC, 1986)	86-73185	0-88318-352-8
No. 153	Physics of Particle Accelerators (SLAC Summer School, 1985) (Fermilab Summer School, 1984)	87-70103	0-88318-353-6
No. 154	Physics and Chemistry of Porous Media—II (Ridge Field, CT, 1986)	83-73640	0-88318-354-4
No. 155	The Galactic Center: Proceedings of the Symposium Honoring C. H. Townes (Berkeley, CA, 1986)	86-73186	0-88318-355-2
No. 156	Advanced Accelerator Concepts (Madison, WI, 1986)	87-70635	0-88318-358-0
No. 157	Stability of Amorphous Silicon Alloy Materials and Devices (Palo Alto, CA, 1987)	87-70990	0-88318-359-9
No. 158	Production and Neutralization of Negative Ions and Beams (Brookhaven, NY, 1986)	87-71695	0-88318-358-7
No. 159	Applications of Radio-Frequency Power to Plasma: Seventh Topical Conference (Kissimmee, FL, 1987)	87-71812	0-88318-359-5

No. 160	Advances in Laser Science–II (Seattle, WA, 1986)	87-71962	0-88318-360-9
No. 161	Electron Scattering in Nuclear and Particle Science: In Commemoration of the 35th Anniversary of the Lyman-Hanson-Scott Experiment (Urbana, IL, 1986)	87-72403	0-88318-361-7
No. 162	Few-Body Systems and Multiparticle Dynamics (Crystal City, VA, 1987)	87-72594	0-88318-362-5
No. 163	Pion–Nucleus Physics: Future Directions and New Facilities at LAMPF (Los Alamos, NM, 1987)	87-72961	0-88318-363-3
No. 164	Nuclei Far from Stability: Fifth International Conference (Rosseau Lake, ON, 1987)	87-73214	0-88318-364-1
No. 165	Thin Film Processing and Characterization of High-Temperature Superconductors (Anaheim, CA, 1987)	87-73420	0-88318-365-X
No. 166	Photovoltaic Safety (Denver, CO, 1988)	88-42854	0-88318-366-8
No. 167	Deposition and Growth: Limits for Microelectronics (Anaheim, CA, 1987)	88-71432	0-88318-367-6
No. 168	Atomic Processes in Plasmas (Santa Fe, NM, 1987)	88-71273	0-88318-368-4
No. 169	Modern Physics in America: A Michelson-Morley Centennial Symposium (Cleveland, OH, 1987)	88-71348	0-88318-369-2
No. 170	Nuclear Spectroscopy of Astrophysical Sources (Washington, DC, 1987)	88-71625	0-88318-370-6
No. 171	Vacuum Design of Advanced and Compact Synchrotron Light Sources (Upton, NY, 1988)	88-71824	0-88318-371-4
No. 172	Advances in Laser Science–III: Proceedings of the International Laser Science Conference (Atlantic City, NJ, 1987)	88-71879	0-88318-372-2
No. 173	Cooperative Networks in Physics Education (Oaxtepec, Mexico, 1987)	88-72091	0-88318-373-0
No. 174	Radio Wave Scattering in the Interstellar Medium (San Diego, CA, 1988)	88-72092	0-88318-374-9
No. 175	Non-neutral Plasma Physics (Washington, DC, 1988)	88-72275	0-88318-375-7
No. 176	Intersections Between Particle and Nuclear Physics (Third International Conference) (Rockport, ME, 1988)	88-62535	0-88318-376-5
No. 177	Linear Accelerator and Beam Optics Codes (La Jolla, CA, 1988)	88-46074	0-88318-377-3
No. 178	Nuclear Arms Technologies in the 1990s (Washington, DC, 1988)	88-83262	0-88318-378-1
No. 179	The Michelson Era in American Science: 1870–1930 (Cleveland, OH, 1987)	88-83369	0-88318-379-X
No. 180	Frontiers in Science: International Symposium (Urbana, IL, 1987)	88-83526	0-88318-380-3
No. 181	Muon-Catalyzed Fusion (Sanibel Island, FL, 1988)	88-83636	0-88318-381-1
No. 182	High T_c Superconducting Thin Films, Devices, and Application (Atlanta, GA, 1988)	88-03947	0-88318-382-X
No. 183	Cosmic Abundances of Matter (Minneapolis, MN, 1988)	89-80147	0-88318-383-8
No. 184	Physics of Particle Accelerators (Ithaca, NY, 1988)	89-83575	0-88318-384-6

No. 185	Glueballs, Hybrids, and Exotic Hadrons (Upton, NY, 1988)	89-83513	0-88318-385-4
No. 186	High-Energy Radiation Background in Space (Sanibel Island, FL, 1987)	89-83833	0-88318-386-2
No. 187	High-Energy Spin Physics (Minneapolis, MN, 1988)	89-83948	0-88318-387-0
No. 188	International Symposium on Electron Beam Ion Sources and their Applications (Upton, NY, 1988)	89-84343	0-88318-388-9
No. 189	Relativistic, Quantum Electrodynamic, and Weak Interaction Effects in Atoms (Santa Barbara, CA, 1988)	89-84431	0-88318-389-7
No. 190	Radio-frequency Power in Plasmas (Irvine, CA, 1989)	89-45805	0-88318-397-8
No. 191	Advances in Laser Science–IV (Atlanta, GA, 1988)	89-85595	0-88318-391-9
No. 192	Vacuum Mechatronics (First International Workshop) (Santa Barbara, CA, 1989)	89-45905	0-88318-394-3
No. 193	Advanced Accelerator Concepts (Lake Arrowhead, CA, 1989)	89-45914	0-88318-393-5
No. 194	Quantum Fluids and Solids—1989 (Gainesville, FL, 1989)	89-81079	0-88318-395-1
No. 195	Dense Z-Pinches (Laguna Beach, CA, 1989)	89-46212	0-88318-396-X
No. 196	Heavy Quark Physics (Ithaca, NY, 1989)	89-81583	0-88318-644-6
No. 197	Drops and Bubbles (Monterey, CA, 1988)	89-46360	0-88318-392-7
No. 198	Astrophysics in Antarctica (Newark, DE, 1989)	89-46421	0-88318-398-6
No. 199	Surface Conditioning of Vacuum Systems (Los Angeles, CA, 1989)	89-82542	0-88318-756-6
No. 200	High T_c Superconducting Thin Films: Processing, Characterization, and Applications (Boston, MA, 1989)	90-80006	0-88318-759-0
No. 201	QED Stucture Functions (Ann Arbor, MI, 1989)	90-80229	0-88318-671-3
No. 202	NASA Workshop on Physics From a Lunar Base (Stanford, CA, 1989)	90-55073	0-88318-646-2
No. 203	Particle Astrophysics: The NASA Cosmic Ray Program for the 1990s and Beyond (Greenbelt, MD, 1989)	90-55077	0-88318-763-9
No. 204	Aspects of Electron–Molecule Scattering and Photoionization (New Haven, CT, 1989)	90-55175	0-88318-764-7
No. 205	The Physics of Electronic and Atomic Collisions (XVI International Conference) (New York, NY, 1989)	90-53183	0-88318-390-0
No. 206	Atomic Processes in Plasmas (Gaithersburg, MD, 1989)	90-55265	0-88318-769-8
No. 207	Astrophysics from the Moon (Annapolis, MD, 1990)	90-55582	0-88318-770-1
No. 208	Current Topics in Shock Waves (Bethlehem, PA, 1989)	90-55617	0-88318-776-0
No. 209	Computing for High Luminosity and High Intensity Facilities (Santa Fe, NM, 1990)	90-55634	0-88318-786-8
No. 210	Production and Neutralization of Negative Ions and Beams (Brookhaven, NY, 1990)	90-55316	0-88318-786-8
No. 211	High-Energy Astrophysics in the 21st Century (Taos, NM, 1989)	90-55644	0-88318-803-1

No. 212	Accelerator Instrumentation (Brookhaven, NY, 1989)	90-55838	0-88318-645-4
No. 213	Frontiers in Condensed Matter Theory (New York, NY, 1989)	90-6421	0-88318-771-X 0-88318-772-8 (pbk.)
No. 214	Beam Dynamics Issues of High-Luminosity Asymmetric Collider Rings (Berkeley, CA, 1990)	90-55857	0-88318-767-1
No. 215	X-Ray and Inner-Shell Processes (Knoxville, TN, 1990)	90-84700	0-88318-790-6
No. 216	Spectral Line Shapes, Vol. 6 (Austin, TX, 1990)	90-06278	0-88318-791-4
No. 217	Space Nuclear Power Systems (Albuquerque, NM, 1991)	90-56220	0-88318-838-4
No. 218	Positron Beams for Solids and Surfaces (London, Canada, 1990)	90-56407	0-88318-842-2
No. 219	Superconductivity and Its Applications (Buffalo, NY, 1990)	91-55020	0-88318-835-X
No. 220	High Energy Gamma-Ray Astronomy (Ann Arbor, MI, 1990)	91-70876	0-88318-812-0
No. 221	Particle Production Near Threshold (Nashville, IN, 1990)	91-55134	0-88318-829-5
No. 222	After the First Three Minutes (College Park, MD, 1990)	91-55214	0-88318-828-7
No. 223	Polarized Collider Workshop (University Park, PA, 1990)	91-71303	0-88318-826-0
No. 224	LAMPF Workshop on (π, K) Physics (Los Alamos, NM, 1990)	91-71304	0-88318-825-2
No. 225	Half Collision Resonance Phenomena in Molecules (Caracus, Venezuela, 1990)	91-55210	0-88318-840-6
No. 226	The Living Cell in Four Dimensions (Gif sur Yvette, France, 1990)	91-55209	0-88318-794-9
No. 227	Advanced Processing and Characterization Technologies (Clearwater, FL, 1991)	91-55194	0-88318-910-0
No. 228	Anomalous Nuclear Effects in Deuterium/Solid Systems (Provo, UT, 1990)	91-55245	0-88318-833-3
No. 229	Accelerator Instrumentation (Batavia, IL, 1990)	91-55347	0-88318-832-1
No. 230	Nonlinear Dynamics and Particle Acceleration (Tsukuba, Japan, 1990)	91-55348	0-88318-824-4
No. 231	Boron-Rich Solids (Albuquerque, NM, 1990)	91-53024	0-88318-793-4
No. 232	Gamma-Ray Line Astrophysics (Paris–Saclay, France, 1990)	91-55492	0-88318-875-9
No. 233	Atomic Physics 12 (Ann Arbor, MI, 1990)	91-55595	088318-811-2
No. 234	Amorphous Silicon Materials and Solar Cells (Denver, CO, 1991)	91-55575	088318-831-7
No. 235	Physics and Chemistry of MCT and Novel IR Detector Materials (San Francisco, CA, 1990)	91-55493	0-88318-931-3
No. 236	Vacuum Design of Synchrotron Light Sources (Argonne, IL, 1990)	91-55527	0-88318-873-2
No. 237	Kent M. Terwilliger Memorial Symposium (Ann Arbor, MI, 1989)	91-55576	0-88318-788-4